Springer Handbook of Enzymes Volume 24

Dietmar Schomburg and
Ida Schomburg (Eds.)

Springer Handbook of Enzymes

Volume 24

Class 1 · Oxidoreductases IX

EC 1.6–1.8

coedited by Antje Chang

Second Edition

Springer

Professor Dietmar Schomburg
e-mail: d.schomburg@uni-koeln.de

Dr. Ida Schomburg
e-mail: i.schomburg@uni-koeln.de

Dr. Antje Chang
e-mail: a.chang@uni-koeln.de

University to Cologne
Institute for Biochemistry
Zülpicher Strasse 47
50674 Cologne
Germany

Library of Congress Control Number: 2005921725

ISBN-10 3-540-23855-7 2nd Edition Springer Berlin Heidelberg New York
ISBN-13 978-3-540-23855-3

The first edition was published as Volume 7 (ISBN 3-540-57246-5) of the "Enzyme Handbook".

Springer is a part of Springer Science+Business Media
springeronline.com

Printed in The Netherlands

Cover design: Erich Kirchner, Heidelberg
Typesetting: medionet AG, Berlin

Printed on acid-free paper 2/3141xv-5 4 3 2 1 0

Preface

Today, as the full information about the genome is becoming available for a rapidly increasing number of organisms and transcriptome and proteome analyses are beginning to provide us with a much wider image of protein regulation and function, it is obvious that there are limitations to our ability to access functional data for the gene products – the proteins and, in particular, for enzymes. Those data are inherently very difficult to collect, interpret and standardize as they are widely distributed among journals from different fields and are often subject to experimental conditions. Nevertheless a systematic collection is essential for our interpretation of genome information and more so for applications of this knowledge in the fields of medicine, agriculture, etc. Progress on enzyme immobilisation, enzyme production, enzyme inhibition, coenzyme regeneration and enzyme engineering has opened up fascinating new fields for the potential application of enzymes in a wide range of different areas.

The development of the enzyme data information system BRENDA was started in 1987 at the German National Research Centre for Biotechnology in Braunschweig (GBF) and is now continuing at the University at Cologne, Institute of Biochemistry. The present book "Springer Handbook of Enzymes" represents the printed version of this data bank. The information system has been developed into a full metabolic database.

The enzymes in this Handbook are arranged according to the Enzyme Commission list of enzymes. Some 3,700 "different" enzymes are covered. Frequently enzymes with very different properties are included under the same EC-number. Although we intend to give a representative overview on the characteristics and variability of each enzyme, the Handbook is not a compendium. The reader will have to go to the primary literature for more detailed information. Naturally it is not possible to cover all the numerous literature references for each enzyme (for some enzymes up to 40,000) if the data representation is to be concise as is intended.

It should be mentioned here that the data have been extracted from the literature and critically evaluated by qualified scientists. On the other hand, the original authors' nomenclature for enzyme forms and subunits is retained. In order to keep the tables concise, redundant information is avoided as far as possible (e.g. if K_m values are measured in the presence of an obvious cosubstrate, only the name of the cosubstrate is given in parentheses as a commentary without reference to its specific role).

The authors are grateful to the following biologists and chemists for invaluable help in the compilation of data: Cornelia Munaretto and Dr. Antje Chang.

Cologne
Spring 2005

Dietmar Schomburg, Ida Schomburg

List of Abbreviations

A	adenine
Ac	acetyl
ADP	adenosine 5'-diphosphate
Ala	alanine
All	allose
Alt	altrose
AMP	adenosine 5'-monophosphate
Ara	arabinose
Arg	arginine
Asn	asparagine
Asp	aspartic acid
ATP	adenosine 5'-triphosphate
Bicine	N,N'-bis(2-hydroxyethyl)glycine
C	cytosine
cal	calorie
CDP	cytidine 5'-diphosphate
CDTA	trans-1,2-diaminocyclohexane-N,N,N,N-tetraacetic acid
CMP	cytidine 5'-monophosphate
CoA	coenzyme A
CTP	cytidine 5'-triphosphate
Cys	cysteine
d	deoxy-
D-	(and L-) prefixes indicating configuration
DFP	diisopropyl fluorophosphate
DNA	deoxyribonucleic acid
DPN	diphosphopyridinium nucleotide (now NAD^+)
DTNB	5,5'-dithiobis(2-nitrobenzoate)
DTT	dithiothreitol (i.e. Cleland's reagent)
EC	number of enzyme in Enzyme Commission's system
E. coli	Escherichia coli
EDTA	ethylene diaminetetraacetate
EGTA	ethylene glycol bis(-aminoethyl ether) tetraacetate
ER	endoplasmic reticulum
Et	ethyl
EXAFS	extended X-ray absorption fine structure
FAD	flavin-adenine dinucleotide
FMN	flavin mononucleotide (riboflavin 5'-monophosphate)
Fru	fructose
Fuc	fucose
G	guanine
Gal	galactose

GDP	guanosine 5'-diphosphate
Glc	glucose
GlcN	glucosamine
GlcNAc	N-acetylglucosamine
Gln	glutamine
Glu	glutamic acid
Gly	glycine
GMP	guanosine 5'-monophosphate
GSH	glutathione
GSSG	oxidized glutathione
GTP	guanosine 5'-triphosphate
Gul	gulose
h	hour
H4	tetrahydro
HEPES	4-(2-hydroxyethyl)-1-piperazineethane sulfonic acid
His	histidine
HPLC	high performance liquid chromatography
Hyl	hydroxylysine
Hyp	hydroxyproline
IAA	iodoacetamide
Ig	immunoglobulin
Ile	isoleucine
Ido	idose
IDP	inosine 5'-diphosphate
IMP	inosine 5'-monophosphate
ITP	inosine 5'-triphosphate
K_m	Michaelis constant
L-	(and D-) prefixes indicating configuration
Leu	leucine
Lys	lysine
Lyx	lyxose
M	mol/l
mM	millimol/l
m-	*meta-*
Man	mannose
MES	2-(N-morpholino)ethane sulfonate
Met	methionine
min	minute
MOPS	3-(N-morpholino)propane sulfonate
Mur	muramic acid
MW	molecular weight
NAD^+	nicotinamide-adenine dinucleotide
NADH	reduced NAD
$NADP^+$	NAD phosphate
NADPH	reduced NADP
NAD(P)H	indicates either NADH or NADPH
NBS	N-bromosuccinimide

NDP	nucleoside 5'-diphosphate
NEM	N-ethylmaleimide
Neu	neuraminic acid
NMN	nicotinamide mononucleotide
NMP	nucleoside 5'-monophosphate
NTP	nucleoside 5'-triphosphate
o-	*ortho-*
Orn	ornithine
p-	*para-*
PBS	phosphate-buffered saline
PCMB	*p*-chloromercuribenzoate
PEP	phosphoenolpyruvate
pH	$-\log 10[H^+]$
Ph	phenyl
Phe	phenylalanine
PHMB	*p*-hydroxymercuribenzoate
PIXE	proton-induced X-ray emission
PMSF	phenylmethane-sulfonylfluoride
p-NPP	*p*-nitrophenyl phosphate
Pro	proline
Q_{10}	factor for the change in reaction rate for a 10 °C temperature increase
Rha	rhamnose
Rib	ribose
RNA	ribonucleic acid
mRNA	messenger RNA
rRNA	ribosomal RNA
tRNA	transfer RNA
Sar	N-methylglycine (sarcosine)
SDS-PAGE	sodium dodecyl sulfate polyacrylamide gel electrophoresis
Ser	serine
T	thymine
t_H	time for half-completion of reaction
Tal	talose
TDP	thymidine 5'-diphosphate
TEA	triethanolamine
Thr	threonine
TLCK	N^α-*p*-tosyl-L-lysine chloromethyl ketone
T_m	melting temperature
TMP	thymidine 5'-monophosphate
Tos-	tosyl-(*p*-toluenesulfonyl-)
TPN	triphosphopyridinium nucleotide (now $NADP^+$)
Tris	tris(hydroxymethyl)-aminomethane
Trp	tryptophan
TTP	thymidine 5'-triphosphate
Tyr	tyrosine
U	uridine
U/mg	μmol/(mg*min)

UDP	uridine 5'-diphosphate
UMP	uridine 5'-monophosphate
UTP	uridine 5'-triphosphate
Val	valine
Xaa	symbol for an amino acid of unknown constitution in peptide formula
XAS	X-ray absorption spectroscopy
Xyl	xylose

List of Deleted and Transferred Enzymes

Since its foundation in 1956 the Nomenclature Committee of the International Union of Biochemistry and Molecular Biology (NC-IUBMB) has continually revised and updated the list of enzymes. Entries for new enzymes have been added, others have been deleted completely, or transferred to another EC number in the original class or to different EC classes, catalyzing other types of chemical reactions. The old numbers have not been allotted to new enzymes; instead the place has been left vacant or cross-references given to the changes in nomenclature.

Deleted and Transferred Enzymes
For EC class 1.6–1.8 these changes are:

Recommended name	Old EC number	Alteration
$NADH_2$ cytochrome c reductase	1.6.2.1	transferred to EC 1.6.99.3
cytochrome reductase (NADPH)	1.6.2.3	deleted
cystine reductase (NADH)	1.6.4.1	transferred to EC 1.8.1.6
glutathione reductase (NADPH)	1.6.4.2	transferred to EC 1.8.1.7
dihydrolipoamide reductase (NAD^+)	1.6.4.3	transferred to EC 1.8.1.4
protein-disulfide reductase [NAD(P)H]	1.6.4.4	transferred to EC 1.8.1.8
thioredoxin reductase (NADPH)	1.6.4.5	transferred to EC 1.8.1.9
CoA-glutathione reductase (NADPH)	1.6.4.6	transferred to EC 1.8.1.10
asparagusate reductase (NADH)	1.6.4.7	transferred to EC 1.8.1.11
trypanothione reductase	1.6.4.8	transferred to EC 1.8.1.12
bis-gamma-glutamylcystine reductase (NADPH)	1.6.4.9	transferred to EC 1.8.1.13
CoA-disulfide reductase (NADH)	1.6.4.10	transferred to EC 1.8.1.14
quinone reductase	1.6.5.1	deleted
menadione reductase	1.6.5.2	transferred to EC 1.6.99.2
nitrate reductase (NADH)	1.6.6.1	transferred to EC 1.7.1.1
nitrate reductase [NAD(P)H]	1.6.6.2	transferred to EC 1.7.1.2
nitrate reductase (NADPH)	1.6.6.3	transferred to EC 1.7.1.3
nitrite reductase [NAD(P)H]	1.6.6.4	transferred to EC 1.7.1.4
– –	1.6.6.5	transferred to EC 1.7.99.3
hyponitrite reductase	1.6.6.6	transferred to EC 1.7.1.5
azobenzene reductase	1.6.6.7	transferred to EC 1.7.1.6
GMP reductase	1.6.6.8	transferred to EC 1.7.1.7
nitroquinoline-N-oxide reductase	1.6.6.10	transferred to EC 1.7.1.9
hydroxylamine reductase (NADH)	1.6.6.11	transferred to EC 1.7.1.10
4-(dimethylamino)phenylazoxybenzene reductase	1.6.6.12	transferred to EC 1.7.1.11

Recommended name	Old EC number	Alteration
N-hydroxy-2-acetamidofluorene reductase	1.6.6.13	transferred to EC 1.7.1.12
ferredoxin-NADP$^+$ reductase	1.6.7.1	transferred to EC 1.18.1.2
rubredoxin-NAD$^+$ reductase	1.6.7.2	transferred to EC 1.18.1.1
NAD(P)H dehydrogenase (FMN)	1.6.8.1	transferred to EC 1.5.1.29
NADPH dehydrogenase (flavin)	1.6.8.2	transferred to EC 1.5.1.30
nitrite reductase	1.6.99.4	transferred to EC 1.18.1.2
dihydropteridine reductase	1.6.99.7	transferred to EC 1.5.1.34
dihydropteridine reductase (NADH)	1.6.99.10	deleted, included in EC 1.6.99.7
aquacobalamin reductase (NADPH)	1.6.99.11	transferred to EC 1.6.99.11
cyanocobalamin reductase (NADPH, cyanide-eliminating)	1.6.99.12	transferred to EC 1.16.1.6
ferric-chelate reductase	1.6.99.13	transferred to EC 1.16.1.7
– –	1.7.1.8	deleted 2002
nitric-oxide reductase	1.7.99.2	deleted
nitrite reductase	1.7.99.3	deleted, now included in EC 1.7.2.1
cysteamine dehydrogenase	1.8.1.1	deleted
– –	1.8.2.3	deleted
nitrate-ester reductase	1.8.6.1	deleted, included in EC 2.5.1.18
phosphoadenosine-phosphosulfate reductase	1.8.99.4	transferred to EC 1.8.4.8

Index of Recommended Enzyme Names

EC-No.	Recommended Name	Page
1.7.3.2	acetylindoxyl oxidase	344
1.7.3.5	3-aci-nitropropanoate oxidase	368
1.8.99.2	adenylyl-sulfate reductase	694
1.8.4.9	adenylyl-sulfate reductase (glutathione)	663
1.8.4.10	adenylyl-sulfate reductase (thioredoxin)	668
1.6.99.8	*aquacobalamin reductase (transferred to EC 1.16.1.3)*	*231*
1.6.99.11	*aquacobalamin reductase (NADPH) (transferred to EC 1.6.99.11)*	*234*
1.8.1.11	asparagusate reductase	539
1.6.4.7	*asparagusate reductase (NADH) (transferred to EC 1.8.1.11)*	*100*
1.6.6.7	*azobenzene reductase (transferred to EC 1.7.1.6)*	*154*
1.7.1.6	azobenzene reductase	288
1.6.5.6	*p*-benzoquinone reductase (NADPH)	142
1.8.1.13	bis-γ-glutamylcystine reductase	558
1.6.4.9	*bis-γ-glutamylcystine reductase (NADPH) (transferred to EC 1.8.1.13)*	*102*
1.8.1.14	CoA-disulfide reductase	561
1.6.4.10	*CoA-disulfide reductase (NADH) (transferred to EC 1.8.1.14)*	*103*
1.8.1.10	CoA-glutathione reductase.	535
1.6.4.6	*CoA-glutathione reductase (NADPH) (transferred to EC 1.8.1.10)*	*99*
1.6.99.9	*cob(II)alamin reductase (transferred to EC 1.16.1.4)*	*232*
1.6.99.12	*cyanocobalamin reductase (NADPH, cyanide-eliminating) (transferred to EC 1.16.1.6)*	*235*
1.8.1.1	*cysteamine dehydrogenase (deleted)*	*451*
1.8.1.6	cystine reductase	486
1.6.4.1	*cystine reductase (NADH) (transferred to EC 1.8.1.6)*	*94*
1.6.2.3	*cytochrome reductase (NADPH) (deleted)*	*57*
1.6.2.2	cytochrome-b_5 reductase	35
1.8.1.4	dihydrolipoamide dehydrogenase	463
1.6.4.3	*dihydrolipoamide reductase (NAD^+) (transferred to EC 1.8.1.4)*	*96*
1.6.99.7	*dihydropteridine reductase (transferred to EC 1.5.1.34)*	*230*
1.6.99.10	*dihydropteridine reductase (NADH) (deleted, included in EC 1.6.99.7)*	*233*
1.7.1.11	4-(dimethylamino)phenylazoxybenzene reductase	319
1.6.6.12	*4-(dimethylamino)phenylazoxybenzene reductase (transferred to EC 1.7.1.11)*	*173*
1.8.4.7	enzyme-thiol transhydrogenase (glutathione-disulfide)	656
1.6.7.1	*ferredoxin-$NADP^+$ reductase (transferred to EC 1.18.1.2)*	*175*
1.7.7.2	ferredoxin-nitrate reductase	381
1.7.7.1	ferredoxin-nitrite reductase	370
1.6.99.13	*ferric-chelate reductase (transferred to EC 1.16.1.7)*	*236*
1.8.5.1	glutathione dehydrogenase (ascorbate)	670
1.8.3.3	glutathione oxidase	604
1.6.4.2	*glutathione reductase (NADPH) (transferred to EC 1.8.1.7)*	*95*
1.8.4.3	glutathione-CoA-glutathione transhydrogenase	632
1.8.4.4	glutathione-cystine transhydrogenase	635
1.8.1.7	glutathione-disulfide reductase	488
1.8.4.1	glutathione-homocystine transhydrogenase	615

1.6.6.8	*GMP reductase (transferred to EC 1.7.1.7)*	*155*
1.7.1.7	GMP reductase	299
1.8.99.3	hydrogensulfite reductase	708
1.6.5.7	2-hydroxy-1,4-benzoquinone reductase	146
1.6.6.13	*N-hydroxy-2-acetamidofluorene reductase (transferred to EC 1.7.1.12)*	*174*
1.7.1.12	N-hydroxy-2-acetamidofluorene reductase	322
1.7.3.4	hydroxylamine oxidase	360
1.7.99.1	hydroxylamine reductase.	389
1.6.6.11	*hydroxylamine reductase (NADH) (transferred to EC 1.7.1.10)*	*172*
1.7.1.10	hydroxylamine reductase (NADH).	310
1.7.1.5	hyponitrite reductase	286
1.6.6.6	*hyponitrite reductase (transferred to EC 1.7.1.5)*	*153*
1.8.1.3	hypotaurine dehydrogenase	461
1.6.2.6	leghemoglobin reductase.	87
1.6.5.2	*menadione reductase (transferred to EC 1.6.99.2)*	*105*
1.8.3.4	methanethiol oxidase	609
1.8.4.5	methionine-S-oxide reductase	639
1.7.99.5	5,10-methylenetetrahydrofolate reductase ($FADH_2$)	421
1.6.5.4	monodehydroascorbate reductase (NADH)	126
1.8.1.15	mycothione reductase	563
1.6.1.2	$NAD(P)^+$ transhydrogenase (AB-specific)	10
1.6.1.1	$NAD(P)^+$ transhydrogenase (B-specific)	1
1.6.8.1	*NAD(P)H dehydrogenase (FMN) (transferred to EC 1.5.1.29)*	*177*
1.6.99.2	NAD(P)H dehydrogenase (quinone)	187
1.6.99.3	NADH dehydrogenase	207
1.6.99.5	NADH dehydrogenase (quinone)	219
1.6.5.3	NADH dehydrogenase (ubiquinone)	106
1.6.2.1	*$NADH_2$ cytochrome c reductase (transferred to EC 1.6.99.3)*	*34*
1.6.99.1	NADPH dehydrogenase	179
1.6.8.2	*NADPH dehydrogenase (flavin) (transferred to EC 1.5.1.30)*	*178*
1.6.99.6	NADPH dehydrogenase (quinone).	225
1.6.3.1	NADPH oxidase	92
1.6.5.5	NADPH:quinone reductase	135
1.6.2.5	NADPH-cytochrome-c_2 reductase	84
1.6.2.4	NADPH-ferrihemoprotein reductase	58
1.7.99.4	nitrate reductase	396
1.7.1.1	nitrate reductase (NADH)	237
1.6.6.1	*nitrate reductase (NADH) (transferred to EC 1.7.1.1)*	*148*
1.6.6.3	*nitrate reductase (NADPH) (transferred to EC 1.7.1.3)*	*150*
1.7.1.3	nitrate reductase (NADPH).	267
1.6.6.2	*nitrate reductase [NAD(P)H] (transferred to EC 1.7.1.2)*	*149*
1.7.1.2	Nitrate reductase [NAD(P)H]	260
1.8.6.1	*nitrate-ester reductase (deleted, included in EC 2.5.1.18)*	*678*
1.7.99.7	nitric-oxide reductase	441
1.7.99.2	*nitric-oxide reductase (deleted)*	*394*
1.6.99.4	*nitrite reductase (transferred to EC 1.18.1.2)*	*218*
1.7.99.3	*nitrite reductase (deleted, now included in EC 1.7.2.1)*	*395*
1.7.2.2	nitrite reductase (cytochrome; ammonia-forming)	331
1.7.2.1	nitrite reductase (NO-forming)	325
1.7.1.4	nitrite reductase [NAD(P)H]	277
1.6.6.4	*nitrite reductase [NAD(P)H] (transferred to EC 1.7.1.4)*	*151*
1.7.3.1	nitroethane oxidase	341
1.7.1.9	nitroquinoline-N-oxide reductase	307

1.6.6.10	*nitroquinoline-N-oxide reductase (transferred to EC 1.7.1.9)*	*171*
1.7.99.6	nitrous-oxide reductase	432
1.8.1.5	2-oxopropyl-CoM reductase (carboxylating)	483
1.8.99.4	*phosphoadenosine-phosphosulfate reductase (transferred to EC 1.8.4.8)*	*717*
1.8.4.8	phosphoadenylyl-sulfate reductase (thioredoxin)	659
1.8.3.5	prenylcysteine oxidase	612
1.8.1.8	protein-disulfide reductase	514
1.8.4.2	protein-disulfide reductase (glutathione)	617
1.6.4.4	*protein-disulfide reductase [NAD(P)H] (transferred to EC 1.8.1.8)*	*97*
1.8.4.6	protein-methionine-S-oxide reductase	643
1.6.5.1	*quinone reductase (deleted)*	*104*
1.8.2.3	*recommended Name never specified (deleted)*	*583*
1.6.6.5	*recommended Name never specified (transferred to EC 1.7.99.3)*	*152*
1.7.1.8	*recommended Name never specified (deleted 2002)*	*306*
1.6.7.2	*rubredoxin-NAD^+ reductase (transferred to EC 1.18.1.1)*	*176*
1.8.2.1	sulfite dehydrogenase	566
1.8.3.1	sulfite oxidase	584
1.8.99.1	sulfite reductase	685
1.8.7.1	sulfite reductase (ferredoxin)	679
1.8.1.2	sulfite reductase (NADPH)	452
1.8.3.2	thiol oxidase	594
1.6.4.5	*thioredoxin reductase (NADPH) (transferred to EC 1.8.1.9)*	*98*
1.8.1.9	thioredoxin-disulfide reductase	517
1.8.2.2	thiosulfate dehydrogenase	574
1.6.6.9	trimethylamine-N-oxide reductase	156
1.7.2.3	trimethylamine-N-oxide reductase (cytochrome c)	336
1.6.4.8	*trypanothione reductase (transferred to EC 1.8.1.12)*	*101*
1.8.1.12	trypanothione-disulfide reductase	543
1.7.3.3	urate oxidase	346

Description of Data Fields

All information except the nomenclature of the enzymes (which is based on the recommendations of the Nomenclature Committee of IUBMB (International Union of Biochemistry and Molecular Biology) and IUPAC (International Union of Pure and Applied Chemistry) is extracted from original literature (or reviews for very well characterized enzymes). The quality and reliability of the data depends on the method of determination, and for older literature on the techniques available at that time. This is especially true for the fields Molecular Weight and Subunits.
The general structure of the fields is: **Information – Organism – Commentary – Literature**
The information can be found in the form of numerical values (temperature, pH, K_m etc.) or as text (cofactors, inhibitors etc.).
Sometimes data are classified as Additional Information. Here you may find data that cannot be recalculated to the units required for a field or also general information being valid for all values. For example, for Inhibitors, Additional Information may contain a list of compounds that are not inhibitory.
The detailed structure and contents of each field is described below. If one of these fields is missing for a particular enzyme, this means that for this field, no data are available.

1 Nomenclature

EC number

The number is as given by the IUBMB, classes of enzymes and subclasses defined according to the reaction catalyzed.

Systematic name

This is the name as given by the IUBMB/IUPAC Nomenclature Committee

Recommended name

This is the name as given by the IUBMB/IUPAC Nomenclature Committee

Synonyms

Synonyms which are found in other databases or in the literature, abbreviations, names of commercially available products. If identical names are frequently used for different enzymes, these will be mentioned here, cross references are given. If another EC number has been included in this entry, it is mentioned here.

CAS registry number

The majority of enzymes have a single chemical abstract (CAS) number. Some have no number at all, some have two or more numbers. Sometimes two enzymes share a common number. When this occurs, it is mentioned in the commentary.

2 Source Organism

For listing organisms their systematic name is preferred. If these are not mentioned in the literature, the names from the respective literature are used. For example if an enzyme from yeast is described without being specified further, yeast will be the entry. This field defines the code numbers for the organisms in which the enzyme with the respective EC number is found. These code numbers (form <_>) are displayed together with each entry in all fields of Brenda where organism-specific information is given.

3 Reaction and Specificity

Catalyzed reaction

The reaction as defined by the IUBMB. The commentary gives information on the mechanism, the stereochemistry, or on thermodynamic data of the reaction.

Reaction type

According to the enzyme class a type can be attributed. These can be oxidation, reduction, elimination, addition, or a name (e.g. Knorr reaction)

Natural substrates and products

These are substrates and products which are metabolized in vivo. A natural substrate is only given if it is mentioned in the literature. The commentary gives information on the pathways for which this enzyme is important. If the enzyme is induced by a specific compound or growth conditions, this will be included in the commentary. In *Additional information* you will find comments on the metabolic role, sometimes only assumptions can be found in the references or the natural substrates are unknown.

In the listings, each natural substrate (indicated by a bold **S**) is followed by its respective product (indicated by a bold **P**). Products are given with organisms and references included only if the respective authors were able to demonstrate the formation of the specific product. If only the disappearance of the substrate was observed, the product is included without organisms of references. In cases with unclear product formation only a ? as a dummy is given.

Substrates and products

All natural or synthetic substrates are listed (not in stoichiometric quantities). The commentary gives information on the reversibility of the reaction,

on isomers accepted as substrates and it compares the efficiency of substrates. If a specific substrate is accepted by only one of several isozymes, this will be stated here.
The field *Additional Information* summarizes compounds that are not accepted as substrates or general comments which are valid for all substrates.
In the listings, each substrate (indicated by a bold **S**) is followed by its respective product (indicated by a bold **P**). Products are given with organisms and references included if the respective authors demonstrated the formation of the specific product. If only the disappearance of the substrate was observed, the product will be included without organisms or references. In cases with unclear product formation only a ? as a dummy is given.

Inhibitors

Compounds found to be inhibitory are listed. The commentary may explain experimental conditions, the concentration yielding a specific degree of inhibition or the inhibition constant. If a substance is activating at a specific concentration but inhibiting at a higher or lower value, the commentary will explain this.

Cofactors, prosthetic groups

This field contains cofactors which participate in the reaction but are not bound to the enzyme, and prosthetic groups being tightly bound. The commentary explains the function or, if known, the stereochemistry, or whether the cofactor can be replaced by a similar compound with higher or lower efficiency.

Activating Compounds

This field lists compounds with a positive effect on the activity. The enzyme may be inactive in the absence of certain compounds or may require activating molecules like sulfhydryl compounds, chelating agents, or lipids. If a substance is activating at a specific concentration but inhibiting at a higher or lower value, the commentary will explain this.

Metals, ions

This field lists all metals or ions that have activating effects. The commentary explains the role each of the cited metal has, being either bound e.g. as Fe-S centers or being required in solution. If an ion plays a dual role, activating at a certain concentration but inhibiting at a higher or lower concentration, this will be given in the commentary.

Turnover number (min^{-1})

The k_{cat} is given in the unit min^{-1}. The commentary lists the names of the substrates, sometimes with information on the reaction conditions or the type of reaction if the enzyme is capable of catalyzing different reactions with a single substrate. For cases where it is impossible to give the turnover number in the defined unit (e.g., substrates without a defined molecular weight, or an undefined amount of protein) this is summarized in Additional Information.

Specific activity (U/mg)

The unit is micromol/minute/milligram of protein. The commentary may contain information on specific assay conditions or if another than the natural substrate was used in the assay. Entries in Additional Information are included if the units of the activity are missing in the literature or are not calculable to the obligatory unit. Information on literature with a detailed description of the assay method may also be found.

K_m-Value (mM)

The unit is mM. Each value is connected to a substrate name. The commentary gives, if available, information on specific reaction condition, isozymes or presence of activators. The references for values which cannot be expressed in mM (e.g. for macromolecular, not precisely defined substrates) are given in Additional Information. In this field we also cite literature with detailed kinetic analyses.

K_i-Value (mM)

The unit of the inhibition constant is mM. Each value is connected to an inhibitor name. The commentary gives, if available, the type of inhibition (e.g. competitive, non-competitive) and the reaction conditions (pH-value and the temperature). Values which cannot be expressed in the requested unit and references for detailed inhibition studies are summerized under Additional information.

pH-Optimum

The value is given to one decimal place. The commentary may contain information on specific assay conditions, such as temperature, presence of activators or if this optimum is valid for only one of several isozymes. If the enzyme has a second optimum, this will be mentioned here.

pH-Range

Mostly given as a range e.g. 4.0–7.0 with an added commentary explaining the activity in this range. Sometimes, not a range but a single value indicating the upper or lower limit of enzyme activity is given. In this case, the commentary is obligatory.

Temperature optimum (°C)

Sometimes, if no temperature optimum is found in the literature, the temperature of the assay is given instead. This is always mentioned in the commentary.

Temperature range (°C)

This is the range over which the enzyme is active. The commentary may give the percentage of activity at the outer limits. Also commentaries on specific assay conditions, additives etc.

4 Enzyme Structure

Molecular weight

This field gives the molecular weight of the holoenzyme. For monomeric enzymes it is identical to the value given for subunits. As the accuracy depends on the method of determination this is given in the commentary if provided in the literature. Some enzymes are only active as multienzyme complexes for which the names and/or EC numbers of all participating enzymes are given in the commentary.

Subunits

The tertiary structure of the active species is described. The enzyme can be active as a monomer a dimer, trimer and so on. The stoichiometry of subunit composition is given. Some enzymes can be active in more than one state of complexation with differing effectivities. The analytical method is included.

Posttranslational modifications

The main entries in this field may be proteolytic modification, or side-chain modification, or no modification. The commentary will give details of the modifications e.g.:

- proteolytic modification <1> (<1>, propeptide Name) [1];
- side-chain modification <2> (<2>, N-glycosylated, 12% mannose) [2];
- no modification [3]

5 Isolation / Preparation / Mutation / Application

Source / tissue

For multicellular organisms, the tissue used for isolation of the enzyme or the tissue in which the enzyme is present is given. Cell-lines may also be a source of enzymes.

Localization

The subcellular localization is described. Typical entries are: cytoplasm, nucleus, extracellular, membrane.

Purification

The field consists of an organism and a reference. Only references with a detailed description of the purification procedure are cited.

Renaturation

Commentary on denaturant or renaturation procedure.

Crystallization

The literature is cited which describes the procedure of crystallization, or the X-ray structure.

Cloning

Lists of organisms and references, sometimes a commentary about expression or gene structure.

Engineering

The properties of modified proteins are described.

Application

Actual or possible applications in the fields of pharmacology, medicine, synthesis, analysis, agriculture, nutrition are described.

6 Stability

pH-Stability

This field can either give a range in which the enzyme is stable or a single value. In the latter case the commentary is obligatory and explains the conditions and stability at this value.

Temperature stability

This field can either give a range in which the enzyme is stable or a single value. In the latter case the commentary is obligatory and explains the conditions and stability at this value.

Oxidation stability

Stability in the presence of oxidizing agents, e.g. O_2, H_2O_2, especially important for enzymes which are only active under anaerobic conditions.

Organic solvent stability

The stability in the presence of organic solvents is described.

General stability information

This field summarizes general information on stability, e.g., increased stability of immobilized enzymes, stabilization by SH-reagents, detergents, glycerol or albumins etc.

Storage stability

Storage conditions and reported stability or loss of activity during storage.

References

Authors, Title, Journal, Volume, Pages, Year.

NAD(P)$^+$ Transhydrogenase (B-specific) 1.6.1.1

1 Nomenclature

EC number
1.6.1.1

Systematic name
NADPH:NAD$^+$ oxidoreductase (B-specific)

Recommended name
NAD(P)$^+$ transhydrogenase (B-specific)

Synonyms
H$^+$-thase
NAD transhydrogenase
NAD(P) transhydrogenase
NAD(P)(+) transhydrogenase [B-specific]
NADH transhydrogenase
NADH-NADP-transhydrogenase
NADPH-NAD oxidoreductase
NADPH-NAD transhydrogenase
NADPH:NAD$^+$ transhydrogenase
STH
nicotinamide adenine dinucleotide (phosphate) transhydrogenase
nicotinamide nucleotide transhydrogenase
non-energy-linked transhydrogenase
pyridine nucleotide transferase
pyridine nucleotide transhydrogenase
transhydrogenase
transhydrogenase, nicotinamide adenine dinucleotide (phosphate)

CAS registry number
9014-18-0 (not distinguished from EC 1.6.1.2)
9072-60-0 (not distinguished from EC 1.6.1.2)

2 Source Organism

<1> *Pseudomonas fluorescens* [1, 2, 20]
<2> *Azotobacter vinelandii* [1-6, 13-16, 19]
<3> *Pseudomonas aeruginosa* [1, 2, 7, 9-12, 17, 18]
<4> *Azotobacter chroococcum* [1]
<5> *Azotobacter agilis* [1]

<6> *Beneckea natriegens* (formerly Pseudomonas natriegens [8]) [8]
<7> *Chromatium sp.* [1]
<8> *Solanum tuberosum* (potato tubers [21]) [21]
<9> *Pisum sativum* (pea [21]) [21]

3 Reaction and Specificity

Catalyzed reaction

NADPH + NAD$^+$ = NADP$^+$ + NADH (<1, 4, 5> mechanism [1])

Reaction type

oxidation
redox reaction
reduction

Natural substrates and products

S NADH + NADP$^+$ <3> (Reversibility: ? <3> [12]) [12]
P NADPH + NAD$^+$ <3> [12]

Substrates and products

S NAD(P)H + 2,6-dichlorophenolindophenol <2> (Reversibility: ? <2> [14]) [14]
P NAD(P)$^+$ + reduced 2,6-dichlorophenolindophenol <2> [14]
S NAD(P)H + $K_4Fe(CN)_6$ <2> (Reversibility: ? <2> [14]) [14]
P NAD(P)$^+$ + $K_3Fe(CN)_6$ <2> [14]
S NADH + 2'-NADP$^+$ <3> (Reversibility: ? <3> [11]) [11]
P NAD$^+$ + 2'-NADPH <3> [11]
S NADH + 3'-NADP$^+$ <3> (Reversibility: ? <3> [11]) [11]
P NAD$^+$ + 3'-NADPH <3> [11]
S NADH + thio-NAD$^+$ <2> (Reversibility: ? <2> [14]) [14]
P NAD$^+$ + thio-NADH <2> [14]
S NADH + thio-NADP$^+$ <2> (Reversibility: ? <2> [14]) [14]
P NAD$^+$ + thio-NADPH <2> [14]
S NADP$^+$ + NADH <1-9> (<1, 4, 5> degree of reversibility depends on source of enzyme [1]; <3, 6> reduction of NADP$^+$ is preferred [8, 12]; <7> poorly reversible [1]; <3, 8, 9> 4B-specific for NAD(P)H [11, 20, 21]; <2, 3> diaphorase-type reactions with NAD(P)H, $K_3Fe(CN)_6$ and 2,6-dichlorophenol indophenol [14, 18]) (Reversibility: r <1-7> [1, 8, 12] ? <8, 9> [20, 21]) [1, 8, 11, 12, 14, 18, 19, 20, 21]
P NADPH + NAD$^+$ <1-6> [1, 8, 11, 12, 14, 18, 19]
S NADPH + 3-acetylpyridine-NAD$^+$ <1, 3> (Reversibility: ? <1, 3> [1, 18]) [1, 18]
P NADP$^+$ + 3-acetylpyridine-NADH <1, 3> [1, 18]
S NADPH + deamino-NAD$^+$ <1-3> (Reversibility: ? <1-3> [1, 18, 19]) [1, 18, 19]
P NADP$^+$ + deamino-NADH <1-3> [1, 18, 19]

S NADPH + pyridine aldehyde-NAD^+ <1, 3> (Reversibility: ? <1, 3> [1, 18]) [1, 18]
P $NADP^+$ + pyridine aldehyde-NADH <1, 3> [1, 18]
S NADPH + thio-NAD^+ <2> (Reversibility: ? <2> [14]) [14]
P $NADP^+$ + thio-NADH <2> [14]
S NADPH + thio-$NADP^+$ <2, 3> (Reversibility: ? <2, 3> [14, 18, 19]) [14, 18, 19]
P $NADP^+$ + thio-NADPH <2, 3> [14, 18, 19]

Inhibitors

2'-AMP <6, 7> [1, 8]
5'-AMP <6> (<6> 5 mM, 92% inhibition of $NADP^+$ reduction, complete inhibition of NAD^+ reduction [8]) [8]
ADP <6> (<6> 5 mM, 94% inhibition of $NADP^+$ reduction, complete inhibition of NAD^+ reduction [8]) [8]
ATP <6> (<6> 5 mM, 81% inhibition of $NADP^+$ reduction, complete inhibition of NAD^+ reduction [8]) [8]
CTP <6> (<6> 5 mM, 86% inhibition of $NADP^+$ reduction [8]) [8]
GTP <6> (<6> 5 mM, 89% inhibition of $NADP^+$ reduction [8]) [8]
NAD^+ <2> (<2> competitive to thio-NAD^+, uncompetitive with respect to NADPH [19]) [19]
$NADP^+$ <2, 3> (<3> inhibition of 2'-AMP activated reaction [9]; <3> 0.01 mM, 28% inhibition of NAD^+ reduction with NADPH, 12.5% inhibition of $NADP^+$ reduction with NADPH [9]; <2> uncompetitive to thio-NAD^+, [19]; <3> inhibition in absence of Ca^{2+} [12]; <3> strong inhibition in the absence of Ca^{2+}, saturation with Ca^{2+} completely abolishes inhibition [12]) [9, 12, 19]
TTP <6> (<6> 2 mM, 71% inhibition of $NADP^+$ reduction [8]) [8]
arsenate <6> (<6> complete inhibition of activity in either direction [8]) [8]
deoxycholate <2> [14]
diphosphate <6> (<6> 5 mM, 91% inhibition of $NADP^+$ reduction [8]) [8]
p-aminophenylarsenoxide <2> (<2> 0.1 mM, 40-60% inhibition in the absence of either phosphate or magnesium ions, reduction of NAD^+ by NADPH in cell-free extracts is rapidly and completely inhibited in the presence of 20 mM phosphate [5]) [5]
p-chloromercuribenzoate <2> (<2> 0.044 mM, 40-50% inhibition after 30 min, activity can be restored by adding 2-mercaptoethanol [5]) [5]
p-hydroxymercuribenzoate <1-3> (<1> dependent on presence of oxidized or reduced substrate [1]) [1, 5, 18]
phosphate <2, 6> (<2> 10-25 mM, 60-70% inhibition of purified enzyme, complete inhibition of enzyme in cell-free extracts by 5-10 mM phosphate, $NADP^+$ reduction by NADH is inhibited, reduction of NAD^+ by NADPH is hardly affected [5]; <6> 5 mM, complete inhibition of activity in either direction [8]) [5, 8]
phosphoenolpyruvate <6> (<6> 87% inhibition of $NADP^+$ reduction [8]) [8]
pyridoxal 5'-phosphate <6> (<6> 5 mM, 91% inhibition of $NADP^+$ reduction [8]) [8]

Additional information <3> (<3> not inhibited by palmitoyl-CoA, not affected by treatment with 0.2 mg trypsin/mg protein [11]) [11]

Cofactors/prosthetic groups

FAD <1-5> (<1-5> inactivation by heat treatment can be reversed by addition of FAD [1]; <2> reduction of FAD leads to its dissociation at enzyme concentrations of 10-100 nM and 30°C-40°C, the apoenzyme can be reactivated to 10-15% by addition of FAD [3]; <2> FAD dissociates from the enzyme by heat treatment at 100°C or by treatment with 5% trichloroacetic acid at 0°C [14]) [1, 3, 13, 14]

Activating compounds

2'-AMP <1-5> (<3> partially replaceable by 0.3 mM 2',3'-cyclic AMP or coenzyme A [12]; <3> 0.5 mM, 26% activation of NAD$^+$ reduction by NADPH, 12.5% activation of NADP$^+$ reduction by NADPH, 1100% activation of NAD$^+$ reduction by NADH, 21% activation of NADP$^+$ reduction by NADH [9]; <3> activation of NADP$^+$ reduction by NADH is strongly Ca^{2+} dependent at nonsaturating concentrations of 2'-AMP [12]; <2> almost no effect [14]; <2> activation of reduction of NADP$^+$ or thio-NAD$^+$ by NADH, no effect on reduction of NAD$^+$ or thio-NAD$^+$ by NADPH [19]) [1, 9, 10, 11, 12, 14, 19]

p-hydroxymercuribenzoate <1> (<1> activation, depending on presence of oxidized or reduced substrates [1]) [1]

Metals, ions

Ca^{2+} <1-5> (<2> 10-20fold activation of NAD$^+$ reduction by NADPH at alkaline pH in cell-free extracts [5]; <3> Ca^{2+}-dependent allosteric conformational change, competitive inhibition of activation by Mg^{2+} [7]) [1, 5, 7, 10, 11, 12]

EDTA <2> (<2> slight activating at low buffer concentrations [14]) [14]

K^+ <3> (<3> no full activation compared to Ca^{2+} [12]) [12]

Mg^{2+} <2, 3> (<2> 10-20fold activation of NAD$^+$ reduction by NADPH at alkaline pH in cell-free extracts [5]; <3> no full activation even at saturation concentrations [12]) [5, 12]

Mn^{2+} <3> (<3> compared to Ca^{2+} 10 times higher concentrations are required to obtain the same degree of activation [12]) [12]

Turnover number (min^{-1})

14000-20000 <3> (NADPH) [18]

Specific activity (U/mg)

0.0094 <9> (<9> activity in inside-out mitochondrial particles from leaf [21]) [21]

0.0131 <8> (<8> activity in inside-out mitochondrial particles from tuber [21]) [21]

11.5 <1> [1]

24 <7> [1]

217 <3> [1]

220 <3> [10]

220-260 <2> [14]

220-362 <2> [1]
251 <2> [19]
270 <2> [15]
362 <2> [13]
575 <2> [4]

K_m-Value (mM)

0.0003 <2> (FAD, <2> at 25°C [14]) [14]
0.0025 <2> (FAD, <2> at 0°C [14]) [14]
0.015 <2> (NADPH, <2> + NAD$^+$ [16]) [16]
0.025 <2> (NADH, <2> + NADP$^+$ [16]) [16]
0.03 <2> (thio-NADP$^+$, <2> + NADH [16]) [16]
0.04 <2> (thio-NAD$^+$, <2> + NADH [19]; <2> NADPH + thio-NAD$^+$ [16]) [16, 19]
0.05 <2> (thio-NAD$^+$, <2> + NADH [16]) [16]
0.06 <2> (NADH, <2> + thio-NAD$^+$ [16]) [16]
0.075 <2> (thio-NAD$^+$, <2> + NADPH [16]) [16]
0.077 <2> (NADH, <2> + thio-NAD$^+$ [19]) [19]
0.085 <2> (NADH, <2> + thio-NADP$^+$ [16]) [16]
0.11 <2> (NAD$^+$, <2> + NADPH [16]) [16]
0.25 <2> (thio-NAD$^+$, <2> + NADPH [19]) [19]
0.38 <2> (NAD$^+$, <2> + NADPH [19]) [19]
0.4 <2> (deamino-NAD$^+$, <2> + NADPH [19]) [19]
Additional information <2, 3> (<2> dependency on Mg^{2+} concentration [16]; <3> kinetic studies [9,17]) [9, 16, 17]

pH-Optimum

7 <6> (<6> NADH formation [8]) [8]
7-8 <1, 2, 4, 5> [1, 5]
9.6 <6> (<6> NADPH formation [8]) [8]

pH-Range

8.4-8.7 <2> (<2> half maximal activity in this pH range in the absence of Ca^{2+} or Mg^{2+}, no activity above [5]) [5]
9.1-9.3 <2> (<2> half maximal activity in this pH range in the presence of 5 mM Ca^{2+} or Mg^{2+}, no activity above [5]) [5]

Temperature optimum (°C)

30-35 <2> [19]

4 Enzyme Structure

Molecular weight

421000 <2> (<2> octameric form at pH 8.5-9.0, sedimentation equilibrium [6]; <2> purified enzyme: exceptional stability of polymers at neutral pH [5]) [5, 6]
1600000 <3> (<3> sedimentation equilibrium in presence of 2'-AMP [10]) [10]

6400000 <3> (<3> sedimentation equilibrium in presence of $NADP^+$ [10]) [10]
27000000 <2> (<2> rod-like polymers, electron micrography [15]) [15]
30000000-50000000 <2> (<2> rod-like structure, light scattering [14]) [14]

Subunits

octamer <1-3> (<1> x * 52000, SDS-PAGE, immunoblot [2]; <1> x * 54000, SDS-PAGE [20]; <3> x * 52000, SDS-PAGE, immunoblot [2]; <3> x * 54000, SDS-PAGE [10]; <2> x * 52000, SDS-PAGE, immunoblot [2, 4, 13]; <2> 8 * 54000, four subunits are placed in a rhomb, a second tetramer is located staggered on top of the first one, deduced from amino acid analysis and electron micrography of purified enzyme [4]; <2> x * 52000, model of quarternary structure, SDS-PAGE, immunoblot [3]; <2> 8 * 54000, SDS-PAGE [6]; <2> octameric in cell-free extract, polymeric in purified form [5]; <2> x * 58000, SDS-PAGE [13]) [2-6, 10, 13]

5 Isolation/Preparation/Mutation/Application

Source/tissue

leaf <8> [21]
tuber <8> [21]

Localization

mitochondrion <8, 9> (<8, 9> inside-out submitochondrial particles [21]) [21]

Purification

<1> (overview [1]; affinity chromatography, native and recombinant enzyme [20]) [1, 20]
<2> (affinity chromatography [4, 6]) [4, 6, 13, 14]
<3> (affinity chromatography on immobilized 2'-AMP [10]) [18]
<6> (partial [8]) [8]

Renaturation

<2> (inactivation by 4 M guanidinium HCl, 4-6% reactivation after transfer in 100 mM Tris buffer, pH 7.5 containing 100 mM FAD [3]) [3]
<3> (urea inactivation: 3% reactivation after dialysis against buffer containing 0.1 mM FAD, 35% after dialysis against buffer containing 1% mercaptoethanol and 95% reactivation after dialysis against buffer containing both 0.1 mM FAD and 1% mercaptoethanol [10]) [10]

Crystallization

<2> [4, 15]

Cloning

<1> (expression in Escherichia coli [20]) [20]

6 Stability

pH-Stability

7 <2> (<2> polymeric enzyme stable [5]) [5]

Temperature stability

50 <2> (<2> 1 h stable, inactivation is dramatically accelerated by NADH and NADPH, partial protection by $NADP^+$ and FMN, almost full protection by FAD [14]) [14]
51 <3> (<3> 25 min, 50% inactivation, accelerated by addition of NADPH, reactivation by FAD [18]) [18]
55 <2> (<2> approx. 50% activity lost after about 2 min, almost complete loss of activity after 20 min, biphasic inactivation: 70% activity lost with a first-order inactivation constant, 30% is lost much more rapidly, rate of thermal inactivation depends on concentration of NAD^+, $NADP^+$, NADH, NADPH, free FAD, Mg^{2+} and phosphate, independent of pH between pH 5 and pH 9, significant acceleration outside this range, addition of 1 mM FAD lowers inactivation rate about 20fold [3]) [3]
65 <2> (<2> 15 min, complete inactivation, protection by FAD [14]) [14]

General stability information

<2>, 50% inactivation after 10 min in 1 M guanidinium HCl [3]
<2>, bovine serum albumin, 0.2%, stabilization of diluted solutions [14]
<2>, urea, 5 min, 50% inactivation [14]
<3>, urea, 8 M, no dissociation, complete inactivation of enzyme activity in 5 M urea [10]

Storage stability

<1>, -20°C, 50 mM Tris-HCl, pH 7.0, 2 mM dithiothreitol, several weeks, no loss of activity, great loss of activity after several months [20]
<2>, 4°C, 0.1 M phosphate buffer, pH 7.5, 1 mM EDTA, several months, no loss of activity, storage at -20°C yields a partly insoluble enzyme [14]
<6>, -15°C, 50 mM Tris-HCl buffer, pH 7.5, no loss of activity in 12 weeks [8]

References

[1] Rydström, J.; Hoek, J.B.; Ernster, L.: Nicotinamide nucleotide transhydrogenases. The Enzymes, 3rd Ed. (Boyer, P.D., ed.), **13**, 51-88 (1976)
[2] Voordouw, G.; van der Vies, S.M.; Themmen, A.P.N.: Why are two different types of pyridine nucleotide transhydrogenase found in living organisms?. Eur. J. Biochem., **131**, 527-533 (1983)
[3] Voordouw, G.; de Haard, H.; Timmermans, J.A.M.; Veeger, C.; Zabel, P.: Dissociation and assembly of pyridine nucleotide transhydrogenase from Azotobacter vinelandii. Eur. J. Biochem., **127**, 267-274 (1982)
[4] Voordouw, G.; van der Vies, S.M.; Eweg, J.K.; Veeger, C.; van Breemen, J.F.L.; van Bruggen, E.F.J.: Pyridine nucleotide transhydrogenase from Azotobac-

ter vinelandii. Improved purification, physical properties and subunit arrangement in purified polymers. Eur. J. Biochem., **111**, 347-355 (1980)
[5] Voordouw, G.; van der Vies, S.; Scholten, J.W.; Veeger, C.: Pyridine nucleotide transhydrogenase from Azotobacter vinelandii. Differences in properties between the purified and the cell-free extract enzyme. Eur. J. Biochem., **107**, 337-344 (1980)
[6] Voordouw, G.; Veeger, C.; van Breemen, J.F.L.; van Bruggen, E.F.J.: Structure of pyridine nucleotide transhydrogenase from Azotobacter vinelandii. Eur. J. Biochem., **98**, 447-454 (1979)
[7] Höjeberg, B.; Rydström, J.: Ca^{2+}-dependent allosteric regulation of nicotinamide nucleotide transhydrogenase from Pseudomonas aeruginosa. Eur. J. Biochem., **77**, 235-241 (1977)
[8] Collins, P.A.; Knowles, C.J.: Transhydrogenase activity in the marine bacterium Beneckea natriegens. Biochim. Biophys. Acta, **480**, 77-82 (1977)
[9] Widmer, F.; Kaplan, N.O.: Regulatory properties of the pyridine nucleotide transhydrogenase from Pseudomonas aeruginosa. Kinetic studies and fluorescence titration. Biochemistry, **15**, 4693-4699 (1976)
[10] Wermuth, B.; Kaplan, N.O.: Pyridine nucleotide transhydrogenase from Pseudomonas aeruginosa: purification by affinity chromatography and physicochemical properties. Arch. Biochem. Biophys., **176**, 136-143 (1976)
[11] Hoek, J.B.; Rydström, J.; Höjeberg, B.: Comparative studies on nicotinamide nucleotide transhydrogenase from different sources. Biochim. Biophys. Acta, **333**, 237-245 (1974)
[12] Rydström, J.; Hoek, J.B.; Höjeberg, B.: Ca^{2+}-dependent allosteric regulation of nicotinamide nucleotide transhydrogenase from Pseudomonas aeruginosa. Biochem. Biophys. Res. Commun., **52**, 421-429 (1973)
[13] Middleditch, L.E.; Atchison, R.W.; Chung, A.E.: Pyridine nucleotide transhydrogenase from Azotobacter vinelandii. Some aspects of its structure. J. Biol. Chem., **247**, 6802-6809 (1972)
[14] Van den Broek, H.W.J.; Santema, J.S.; Wassink, J.H.; Veeger, C.: Pyridine-nucleotide transhydrogenase. 1. Isolation, purification and characterisation of the transhydrogenase from Azotobacter vinelandii. Eur. J. Biochem., **24**, 31-45 (1971)
[15] Van den Broek, H.W.J.; van Breemen, J.F.L.; van Bruggen, E.F.J.; Veeger, C.: Pyridine-nucleotide transhydrogenase. 2. Electron-microscopic studies on the transhydrogenase from Azotobacter vinelandii. Eur. J. Biochem., **24**, 46-54 (1971)
[16] Van den Broek, H.W.J.; Veeger, C.: Pyridine-nucleotide transhydrogenase. 5. Kinetic studies on transhydrogenase from Azotobacter vinelandii. Eur. J. Biochem., **24**, 72-82 (1971)
[17] Cohen, P.T.; Kaplan, N.O.: Kinetic characteristics of the pyridine nucleotide transhydrogenase from Pseudomonas aeruginosa. J. Biol. Chem., **245**, 4666-4672 (1970)
[18] Cohen, P.T.; Kaplan, N.O.: Purification and properties of the pyridine nucleotide transhydrogenase from Pseudomonas aeruginosa. J. Biol. Chem., **245**, 2825-2836 (1970)

[19] Chung, A.E.: Pyridine nucleotide transhydrogenase from Azotobacter vinelandii. J. Bacteriol., **102**, 438-447 (1970)
[20] French, C.E.; Boonstra, B.; Bufton, K.A.; Bruce, N.C.: Cloning, sequence, and properties of the soluble pyridine nucleotide transhydrogenase of Pseudomonas fluorescens. J. Bacteriol., **179**, 2761-2765 (1997)
[21] Bykova, N.V.; Rasmusson, A.G.; Igamberdiev, A.U.; Gardestrom, P.; Moller, I.M.: Two separate transhydrogenase activities are present in plant mitochondria. Biochem. Biophys. Res. Commun., **265**, 106-111 (1999)

NAD(P)$^+$ Transhydrogenase (AB-specific) 1.6.1.2

1 Nomenclature

EC number
1.6.1.2

Systematic name
NADPH:NAD$^+$ oxidoreductase (AB-specific)

Recommended name
NAD(P)$^+$ transhydrogenase (AB-specific)

Synonyms
H$^+$-thase
NAD transhydrogenase
NAD(P) transhydrogenase
NADH transhydrogenase
NADH-NADP-transhydrogenase
NADPH-NAD oxidoreductase
NADPH-NAD transhydrogenase
NADPH:NAD$^+$ transhydrogenase
dII
dIII
energy-linked transhydrogenase
nicotinamide adenine dinucleotide (phosphate) transhydrogenase
nicotinamide nucleotide transhydrogenase
pyridine nucleotide transferase
pyridine nucleotide transhydrogenase
transhydrogenase
transhydrogenase, nicotinamide adenine dinucleotide (phosphate)

CAS registry number
9014-18-0 (not distinguished from EC 1.6.1.1)
9072-60-0 (not distinguished from EC 1.6.1.1)

2 Source Organism

<1> *mammalia* [1, 2, 9, 43, 59, 60]
<2> *Bos taurus* [2, 4, 7, 8, 10, 11, 13-15, 18, 21-24, 27-30, 32, 33, 35-44, 47, 49-51, 53, 57, 59-62, 63, 68, 75]
<3> *Escherichia coli* [1, 3, 17, 20, 25, 26, 43, 46, 50, 52, 64, 65, 69, 70, 71, 72, 74, 75, 78, 79]

<4> *Rhodobacter capsulatus* (formerly Rhodopseudomonas capsulata [5, 6]) [5, 6]
<5> *Rhodospirillum rubrum* [19, 34, 45, 48, 50, 54, 66, 67, 68, 70, 72, 73, 74, 75, 76, 77, 81, 82]
<6> *Rhodopseudomonas sphaeroides* [56]
<7> *Rattus norvegicus* [12, 16, , 23, 31, 50]
<8> *Salmonella typhimurium* [55]
<9> *Micrococcus denitrificans* [58]
<10> *Azotobacter sp.* [43]
<11> *Klebsiella pneumonia* [64]
<12> *Homo sapiens* [73]
<13> *Entamoeba histolytica* [80]

3 Reaction and Specificity

Catalyzed reaction

NADPH + NAD^+ = $NADP^+$ + NADH (<2> proposed proton-pump mechanism [33]; <2> stereochemistry [35]; <2,3,5> enzyme is stereospecific for the 4A hydrogen of NADH and the 4B hydrogen of NADPH [50]; <1,2,5,7> mechanism [2,50]; <2> mitochondrial proton-motive force accelerates the rate of NADPH formation 10-12fold and shifts the equilibrium in the direction of product formation [63]; <2,3,5> hypothetical mechanism [75])

Reaction type

oxidation
redox reaction
reduction

Natural substrates and products

S NADH + $NADP^+$ <1, 2> (<1> forward reaction [9]; <1> physiological role [9]; <2> coupled to transmembrane transport of protons from cytosol to mitochondria [59, 60]) (Reversibility: r <1, 2> [2, 9, 59, 60]) [2, 9, 59, 60]

P NADPH + NAD^+ <1, 2> [2, 9, 59, 60]

Substrates and products

S NADH + $NADP^+$ <1-3, 5, 7, 11, 12, 13> (<1-3, 5, 7> specific for 4A site of NADH, i.e. pro-R hydrogen and 4B site of NADPH, i.e. pro-S hydrogen [2, 3, 50]; <1, 2> the reaction is coupled to a transmembrane proton translocation from cytosol to mitochondria [1, 9, 33]; <2> stereospecificity of $NADP^+$ reduction, [35]; <1, 3> inactive against 3'-analogs of $NADP^+$ [2, 3]; <5> synthesis of diphosphate from phosphate in chromatophores by reverse reaction [19]; <4> solubilized and purified enzyme does not catalyze reduction of acetyl pyridine adenine dinucleotide by NADH in absence of $NADP^+$ [5]; <3, 5> reaction is catalyzed by a mixture of recombinant domains dI and dIII of either species or by a hybrid mixture of domains I and III from both species [72]; <5, 12> reaction is catalyzed by a mixture of recombinant human domain III and recombinant Rhodospirillum rubrum domain I [73]) (Reversibility: r <1, 2, 5, 7, 11, 12, 13> [9, 50, 57, 64, 73, 80]) [1, 2, 3, 5, 9, 15, 19, 33, 35, 50, 57, 64, 72, 73, 80]

P NAD$^+$ + NADPH <1-3, 5, 7, 11, 12, 13> [1, 2, 3, 5, 9, 15, 19, 33, 35, 50, 57, 64, 73, 80]

S NADH + thio-NADP$^+$ <1, 2, 4, 5, 7, 10> (Reversibility: r <1, 2, 4, 5, 7, 10> [2, 6, 15, , 43, 50]) [2, 6, 15, 43, 50]

P NAD$^+$ + thio-NADPH <1, 2, 4, 5, 7, 10> [2, 6, 15, 43, 50]

S NADPH + oxidized 3-acetylpyridine adenine dinucleotide <2, 3, 5, 10, 11> (<3> the uncoupler carbonylcyanide-*m*-chlorophylhydrazone stimulates approx. 2fold [65]; <5> catalyzed by a mixture of purified recombinant domains I and III [66]; <2, 3, 5> enzyme also catalyzes a rapid, so called cyclic reaction, i.e. the reduction of acetylpyridine adenine dinucleotide in the presence of either NADP$^+$ or NADPH: the NADPH/NADP$^+$ remain permanently bound to domain III and are alternately oxidized by acetylpyridine adenine dinucleotide and then reduced by NADH in domain I [68, 75]) (Reversibility: r <2, 3, 5, 10, 11> [15, 43, 46, 64, 66]) [15, 43, 46, 64, 65, 66, 68, 75]

P NADP$^+$ + reduced 3-acetylpyridine adenine dinucleotide <2, 3, 5, 10, 11> [15, 43, 46, 64, 66, 68, 75]

S NMNH + thio-NADP$^+$ <2> (Reversibility: ? <2> [22]) [22]

P NMN$^+$ + thio-NADPH <2> [22]

Inhibitors

2',5'-ADP <3> (<3> bacteriorhodopsin co-reconstituted enzyme, 36.8% inhibition of thio-NADP$^+$ reduction by NADH in the dark, 34.4% in the light [65]) [65]

2'-AMP <1, 2, 3> (<2> noncompetitive vs. NAD$^+$, competitive vs. NADPH [33]; <3> bacteriorhodopsin co-reconstituted enzyme, 36.8% inhibition of thio-NADP$^+$ reduction by NADH in the dark, 31.2% in the light [65]) [2, 65]

2,2'-dithiodipyridine <2> (<2> 0.2 mM, 45% inhibition [41]) [41]

2,2'-thiodiethanethiole <2> (<2> 0.5 mM, 37% inhibition [41]) [41]

2,4-dinitrophenyl-3'-dephospho-CoA <2> (<2> competitive vs. NAD$^+$, noncompetitive vs. NADPH [27]) [27]

2-(4'-maleimideylanilino)-naphthalene-6-sulfonic acid <3> (<3> 0.004 mM, 2 h incubation, 75% inhibition of reverse reaction catalyzed by A348C mutant enzyme, 95% inhibition of A390C mutant enzyme after 1 h, 90% inhibition of K424C mutant enzyme after 1 h, 55% inhibition of R425C mutant enzyme after 1h [70]) [70]

3',5'-AMP <1> [2]

3'-AMP <1> [2]

3-aminopyridine adenine dinucleotide phosphate <3> (<3> competitive vs. NADP(H) [70]) [70]

5'-AMP <2-4, 7> (<2> competitive vs. NAD$^+$, noncompetitive vs. NADPH [33]; <3> competitive vs acetylpyridine adenine dinucleotide, noncompetitive vs. NADPH [46]; <3> bacteriorhodopsin co-reconstituted enzyme,

54.6% inhibition of thio-NADP$^+$ reduction by NADH in the dark, 24.3% in the light [65]) [5, 33, 46, 50]
5'-[*p*-(fluorosulfonyl)benzoyl]-adenosine <2> (<2> structural analog of adenosine, 2 mM, almost complete inactivation after 25 min, acetylpyridine adenine dinucleotide, NADP$^+$, 5'-AMP, 5'-ADP or a mixture of 2'-AMP and 3'-AMP protect from inactivation, NADPH accelerates the inhibition rate, inhibition rate constant increases 50fold by increasing the pH from 6.0 to 8.5 [24]) [24]
5'-adenosine diphosphate ribose <3> (<3> bacteriorhodopsin co-reconstituted enzyme, 29.6% inhibition of thio-NADP$^+$ reduction by NADH in the dark, 13.2% in the light [65]) [65]
5,5'-dithiobis(2-nitrobenzoate) <2> (<2> 0.02 mM, 51% inhibition, 92% inhibition in the presence of 2 mM Mg^{2+} [41]; <2> NADPH, NADP$^+$, or NADP site-specific inhibitors protect from inactivation [51]) [41, 51]
ADP <1> [2]
Ca^{2+} <1, 2, 7> (<2> 20 mM, 50% inhibition at pH 7.0, maximal inhibition at pH 9.0 [41]) [2, 41, 50]
CoA <1, 2> (<2> competitive vs. NAD(P)H [27]) [2, 27]
D_2O <1> [2]
K^+ <2> (<2> 200 mM, 50% inhibition at pH 7.9, 300 mM, 40% inhibition at pH 5.5, 95% at pH 8.5 [41]) [41]
La^{3+} <2> (<2> 0.1 mM, 50% inhibition at pH 7.0, maximal inhibition at pH 8.0 [41]) [41]
Mg^{2+} <1, 2, 5, 7> (<2> 20 mM, 50% inhibition at pH 7.0, competitive vs. NADPH and thio-NADP$^+$, noncompetitive vs. acetylpyridine adenine dinucleotide and NADH [41]; <2> 5 mM, aprrox. 20% inhibition, 20 mM, approx. 60% inhibition [47]; <5> 1 mM, half-maximal inhibition, 10 mM, 80% inhibition [54]) [2, 41, 47, 50, 54]
Mn^{2+} <1, 2, 7> (<2> 10 mM, 50% inhibition at pH 7.0, maximal inhibition at pH 9.0 [41]) [2, 41, 50]
N,N'-dicylclohexylcarbodiimide <2, 3, 7> (<7> NAD$^+$ and acetylpyridine adenine dinucleotide protect, NADP$^+$ accelerates inhibition [16]; <2> NADH, acetylpyridine adenine dinucleotide, 5'-AMP and 5'-ADP offer nearly complete protection [22, 28]; <2> NADPH enhances the rate of inhibition by 3-4fold or more [28]; <3> NADH protects from inhibition, NADP$^+$ and to a lesser extent NADPH increase the rate of inhibition [25]; <2> maximal inhibition at pH 6.5, NAD(H) and oxidized or reduced acetylpyridine adenine dinucleotide completely protect from inactivation, NADP$^+$ increases the inhibition rate [37]) [16, 22, 25, 28, 29, 31, 37, 63]
N-(4-azido-2-nitrophenyl)-2-aminoethylsulfonate <2> (<2> trivial name NAP-taurine, time-dependent inactivation of reconstituted enzyme after photolysis in NAP-taurine loaded vesicles, acetylpyridine adenine dinucleotide stimulates inactivation [30]) [30]
N-(ethoxycarbonyl)-2-ethoxy-1,2-dihydroquinoline <2> (<2> 1.8 mM, complete inactivation after 55 min, approx. 25% inactivation after 55 min in the presence of 4 mM NMNH [63]) [22, 63]

N-ethylmaleimide <2, 3, 5> (<2> at neutral pH NADP$^+$ and 2'-AMP partially protect while NADPH accelerates the inactivation rate, little inactivation below pH 7.5, rapid inactivation above, modification of Cys-893 is responsible for inactivation [7]; <5> inactivation of integral membrane bound component and soluble protein factor of enzyme from chromatophore, NADPH potentiates inactivation of the enzyme complex at low concentrations, NADP$^+$ partially protects the intact complex and fully protects both components [45]; <3> cysteine-free mutant enzyme is not inhibited [69]; <3> 0.2 mM, 57% inhibition of reverse reaction in A348C mutant enzyme, 60% inhibition of K424C mutant enzyme, more than 95% inhibition of A390C mutant enzyme after 1h, R425C mutant enzyme is not inhibited [70]) [7, 45, 69, 70]
NAD$^+$ <1, 3> (<1> product inhibition of forward reaction i.e. reduction of NADP$^+$ by NADH [9]; <3> bacteriorhodopsin co-reconstituted enzyme, 61.5% inhibition of thio-NADP$^+$ reduction by NADH in the dark, 18.4% in the light [65]) [9, 65]
NADH <1, 2> (<1> product inhibition of reverse reaction i.e. reduction of NAD$^+$ by NADPH [9]; <2> competitive vs. NAD$^+$, noncompetitive vs. NADPH [33]) [9, 33]
NADP$^+$ <1, 2, 4> (<4> mixed product inhibition vs. acetylpyridine adenine dinucleotide, competitive vs. NADPH [5]; <1> product inhibition of reverse reaction i.e. reduction of NAD$^+$ by NADPH [9]; <2> noncompetitive vs. NAD$^+$, competitive vs. NADPH [33]) [5, 9, 33]
NADPH <1, 3> (<1> product inhibition of forward reaction i.e. reduction of NADP$^+$ by NADH [9]; <3> bacteriorhodopsin co-reconstituted enzyme, 76.9% inhibition of thio-NADP$^+$ reduction by NADH in the dark, 57.4% in the light [65]) [9, 65]
Na$^+$ <2> (<2> 200 mM, 50% inhibition at pH 7.9, 300 mM, 40% inhibition at pH 5.5, 95% at pH 8.5 [41]) [41]
S-7-nitrobenzofuran-4-yl-3'-dephospho-CoA <2> (<2> strong inhibitor, competitive vs. NAD$^+$ and NADPH [27]) [27]
S-7-nitrobenzofuran-4-yl-CoA <2> (<2> strong inhibitor, competitive vs. NADPH, non-competitive vs. NAD$^+$ [27]) [27]
Sr^{2+} <2> (<2> 25 mM, 50% inhibition at pH 7.0 [41]) [41]
Tl$^+$ <2> (<2> 20 mM, 50% inhibition at pH 7.9 [41]) [41]
acetyl-CoA <1> [2]
acetyl-dephospho-CoA <1> (<1> competitive vs. NAD(H) [2]) [2]
adenosine <1, 3> (<3> bacteriorhodopsin co-reconstituted enzyme, 54.2% inhibition of thio-NADP$^+$ reduction by NADH in the dark, 16.0% in the light [65]) [2, 65]
butane-2,3-dione <5, 7> (<5> 40 mM, almost complete inactivation of enzyme activity in chromatophores after 12 min, NAD$^+$ and NADP$^+$ partially protect, complete protection by a combination of NAD$^+$ and NADP$^+$ [34]) [31, 34]
cardiolipin <2> (<2> noncompetitive vs. NAD$^+$ and NADPH [33]) [33]
dansyl chloride <2> (<2> 0.25 mM, almost complete inactivation after 8 min, NADP$^+$ or NADPH accelerate inhibition rate [21]) [21]
dephospho-CoA <1> (<1> competitive vs. NAD(H) [2]) [2]

diethyldicarbonate <2> (<2> inhibition is approx. 50% accelerated in the presence of NAD(H) [37]) [37]
ethoxyformic anhydride <2> (<2> 2 mM, almost complete inactivation after 6 min, $NADP^+$ or NADPH accelerate inhibition rate [21]) [21]
formamide disulfide dihydrochloride <2> (<2> 0.2 mM, 43% inhibition [41]) [41]
glutathione <2, 3> (<2> inhibition of forward and reverse reaction in the presence of NADPH, no inhibition of forward reaction in the presence of NADH, 40% inhibition of reverse reaction, little or no inhibition in the absence of substrates [15]; <3> protection by $NADP^+$ or NAD^+, presence of NADPH accelerates inhibition [52]) [15, 52]
glutathione disulfide <2> (<2> strong, time dependent inhibition of thio-$NADP^+$ reduction by NADH and acetylpyridine adenine dinucleotide reduction by NADPH, 50% inhibition after 40 min incubation in 26.7 mM glutathione disulfide, presence of NADPH accelerates inhibition 20fold [15]) [15]
methylmethane thiosulfonate <2> (<2> modification of Cys-893 [7]; <3> 10 mM, approx. 80% inactivation after 320 min, approx. 40% in the presence of $NADP^+$ or NAD^+, approx. 90% in the presence of NADPH [52]) [7, 52]
p-chloromercuribenzoate <2, 6> (<2> 0.003 mM, 25% inhibition [41]) [41, 56]
p-chlororomercuriphenyl sulfonate <6> [56]
palmitoyl-CoA <1, 2, 5, 11> (<2> competitive inhibition vs. NADPH [33,47]; <11> competitive vs. NADPH [64]) [2, 19, 33, 47, 54, 64]
pentane-2,4-dione <5> (<5> inactivation of chromatophore complex, 156 mM, approx. 85% inactivation after 30 min, NADPH and $NADP^+$ partially protect, half-maximal protection with 0.015 mM NADPH and 0.030 mM $NADP^+$ respetively [45]) [45]
phenylarsine oxide <2, 11> (<2> 0.448 mM, 50% inhibition of acetylpyridine adenine dinucleotide reduction by NADPH after 1 min, 97% inhibition after 60 min, addition of glutathione restores about 50% of activity [15]; <11> 0.48 mM, 50% inhibition of acetylpyridine adenine dinucleotide reduction in inside-out membrane vesicles after 5 min [64]) [15, 64]
phospholipase A <2> (<2> 74% inhibition of activity in submitochondrial particles [53]) [53]
phospholipase C <2> (<2> 10-20% inhibition of activity in submitochondrial particles [53]) [53]
pyridoxal 5'-phosphate <2> (<2> 0.8 mM, almost complete inactivation after 5 min, 0.4 mM $NADP^+$ or NADPH protect from inactivation, inhibition can be reversed to a considerable extent by L-lysine [21]) [21]
reduced acetylpyridine adenine dinucleotide <3> (<3> competitive vs. oxidized acetylpyridine dinucleotide, noncompetitive vs. NADPH [46]) [46]
reduced acetylpyridine adenine dinucleotide <4> (<4> competitive inhibition vs. oxidized acetylpyridine adenine dinucleotide, mixed inhibition vs. NADPH [5]) [5]
triiodothyronine <1, 2, 7> [2, 50]

Cofactors/prosthetic groups

NAD^{+} <1-13> [1-82]
NADH <1-13> [1-82]
NADP^{+} <1-13> [1-82]
NADPH <1-13> [1-82]
Additional information <2, 4> (<2, 4> no flavin cofactor, differentiation from EC 1.6.1.1. [5, 24]) [5, 24]

Activating compounds

asolectin <3> (<3> 10fold stimulation of partially purified enzyme [52]) [52]
carbonyl cyanide *m*-chlorophenylhydrazone <2> (<2> 20fold activation of thio-NADP^{+} reduction by NADH in enzyme-bacteriorhodopsin vesicles in the light [13]) [13, 14]
cardiolipin <3> (<3> 10fold stimulation of partially purified enzyme with E. coli cardiolipin [52]) [52]
lecithin <3> (<3> 5fold stimulation of partially purified enzyme [52]) [52]
light <4> (<4> forward reaction in chromatophores is accelerated more than 20fold during illumination with photosynthetically active light [6]) [6]
lipids <2> [50]
lysolecithin <5> (<5> aprrox. 0.3%, 150% activation of activity in chromatophore extracts [48]) [48]
lysophosphatidylcholine <4> [5]
phospholipids <2, 3> (<2> approx. 2.5fold activation 120 min after sonication with mitochondrial phospholipids [47]; <3> approx. 15fold stimulation of partially purified enzyme with crude E. coli lipid fraction [52]) [47, 52, 53]

Metals, ions

Ca^{2+} <3, 4> (<4> required for catalytic activity [6]; <4> strong activation at concentrations of approx. 7-8 mM [5]; <4> strong activation of forward reaction in the light at concentrations above 0.1 mM [6]; <3> no activation [3]) [3, 5, 6]
K^{+} <4> (<4> strong activation at concentrations of approx. 100 mM [5]; <4> strong activation of forward reaction, i.e. the reduction of NADP^{+} by NADH, in the light at concentrations of approx. 50-80 mM, half-maximal activation at 10 mM [6]) [5, 6]
Li^{+} <11> (<11> not clear whether this reflects a general salt effect or a Li^{+} specific effect [64]) [64]
Mg^{2+} <4> (<4> required for catalytic activity [6]; <4> strong activation at concentrations of approx. 3-7 mM [5]; <4> strong activation of forward reaction, i.e. the reduction of NADP^{+} by NADH, in the light at concentrations above 0.1 mM [6]) [5, 6]
Na^{+} <4, 11> (<4> strong activation of forward reaction, i.e. the reduction of NADP^{+} by NADH, in the light at concentrations of approx. 50-80 mM, half-maximal activation at 10 mM [6]; <11> not clear whether this reflects a general salt effect or a Na^{+} specific effect [64]) [6, 64]

Turnover number (min^{-1})

1.2 <5> (oxidized acetylpyridine adenine dinucleotide, <5> k_{cat} for recombinant domain III in the presence of saturating amounts of domain I, reduction of acetylpyridine adenine dinucleotide by NADPH [66]) [66]
4 <3> (reduced acetylpyridine adenine dinucleotide, <3> reverse reaction of wild-type domain III/R. rubrum domain I mixture [74]) [74]
6 <3> (reduced acetylpyridine adenine dinucleotide, <3> reverse reaction of A432C mutant domain III/R. rubrum domain I mixture [74]) [74]
7 <3> (reduced acetylpyridine adenine dinucleotide, <3> reverse reaction of D393C mutant domain III/R. rubrum domain I mixture [74]) [74]
10 <3> (reduced acetylpyridine adenine dinucleotide, <3> reverse reaction of H345C mutant domain III/R. rubrum domain I mixture [74]) [74]
10 <3> (reduced acetylpyridine adenine dinucleotide, <3> reverse reaction of R350C mutant domain III/R. rubrum domain I mixture [74]) [74]
17 <3> (reduced acetylpyridine adenine dinucleotide, <3> reverse reaction of K424C mutant domain III/R. rubrum domain I mixture [74]) [74]
17 <3> (reduced acetylpyridine adenine dinucleotide, <3> reverse reaction of R425C mutant domain III/R. rubrum domain I mixture [74]) [74]
33 <3> (reduced acetylpyridine adenine dinucleotide, <3> reverse reaction of A348C mutant domain III/R. rubrum domain I mixture [74]) [74]
34 <3> (reduced acetylpyridine adenine dinucleotide, <3> reverse reaction of G430C mutant domain III/R. rubrum domain I mixture [74]) [74]
55 <3> (reduced acetylpyridine adenine dinucleotide, <3> reverse reaction of D392C mutant domain III/R. rubrum domain I mixture [74]) [74]
86.4 <5> (oxidized acetylpyridine adenine dinucleotide, <5> k_{cat} for recombinant domain I in the presence of saturating amounts of domain III, reduction of acetylpyridine adenine dinucleotide by NADPH [66]) [66]
600 <3> (reduced acetylpyridine adenine dinucleotide, <3> cyclic reaction of R425C mutant domain III/R. rubrum domain I mixture [74]) [74]
700 <3> (reduced acetylpyridine adenine dinucleotide, <3> cyclic reaction of G430C mutant domain III/R. rubrum domain I mixture [74]) [74]
900 <3> (reduced acetylpyridine adenine dinucleotide, <3> cyclic reaction of D392C mutant domain III/R. rubrum domain I mixture [74]) [74]
1250 <3> (reduced acetylpyridine adenine dinucleotide, <3> cyclic reaction of H345C mutant domain III/R. rubrum domain I mixture [74]) [74]
2000 <3> (reduced acetylpyridine adenine dinucleotide, <3> cyclic reaction of R350C mutant domain III/R. rubrum domain I mixture [74]) [74]
2200 <3> (reduced acetylpyridine adenine dinucleotide, <3> cyclic reaction of K424C mutant domain III/R. rubrum domain I mixture [74]) [74]
2400 <3> (reduced acetylpyridine adenine dinucleotide, <3> cyclic reaction of D393C mutant domain III/R. rubrum domain I mixture [74]) [74]
3100 <3> (reduced acetylpyridine adenine dinucleotide, <3> cyclic reaction of A432C mutant domain III/R. rubrum domain I mixture [74]) [74]
3300 <3> (reduced acetylpyridine adenine dinucleotide, <3> cyclic reaction of A348C mutant domain III/R. rubrum domain I mixture [74]) [74]
4900 <3> (reduced acetylpyridine adenine dinucleotide, <3> cyclic reaction of wild-type domain III/R. rubrum domain I mixture [74]) [74]

Specific activity (U/mg)

0.044 <3> (<3> activity in membranes of strain AB1450 [20]) [20]
0.1 <3> (<3> membrane bound mutant enzyme with a direct linker between α and β subunits [79]) [79]
0.1 <3> (<3> purified mutant enzyme with a direct linker between α and β subunits, forward reaction [79]) [79]
0.19 <2> (<2> reduction of acetylpyridine adenine dinucleotide by NADPH [28]) [28]
0.2 <3> (<3> reduction of $NADP^+$ by NADH driven by electron transport, cysteine-free enzyme reconstituted in membrane vesicles [69]) [69]
0.26 <11> (<11> activity in inside-out membrane vesicles at pH 7.4, presence of an uncoupler i.e. carbonylcyanide-*m*-chlorophylhydrazone results in 2fold stimulation [64]) [64]
0.3 <3> (<3> membrane bound mutant enzyme with a 18 residues long linker between α and β subunits [79]) [79]
0.4 <3> (<3> purified mutant enzyme with a 18 residues long linker between α and β subunits, forward reaction [79]) [79]
0.42 <3> (<3> reduction of $NADP^+$ by NADH driven by electron transport, wild-type enzyme reconstituted in membrane vesicles [69]) [69]
0.6 <3> (<3> membrane bound mutant enzyme with a 32 residues long linker between α and β subunits [79]) [79]
0.7 <3> (<3> purified mutant enzyme with a 32 residues long linker between α and β subunits, forward reaction [79]) [79]
0.8 <3> (<3> purified mutant enzyme with a direct linker between α and β subunits, reverse reaction [79]) [79]
0.9 <3> (<3> purified enzyme, forward reaction [79]) [79]
1.4 <3> (<3> membrane bound enzyme, reverse reaction [79]) [79]
1.9 <3> (<3> reduction of acetylpyridine adenine dinucleotide by NADPH, cysteine-free enzyme reconstituted in membrane vesicles [69]) [69]
2.6 <7> (<7> partially purified enzyme [23]) [23]
3 <3> (<3> reduction of acetylpyridine adenine dinucleotide by NADPH, wild-type enzyme reconstituted in membrane vesicles [69]) [69]
3.78 <4> [5]
3.9 <3> (<3> purified mutant enzyme with a 18 residues long linker between α and β subunits, reverse reaction [79]) [79]
4.4 <2> [36, 44, 49]
5.5 <3> (<3> purified mutant enzyme with a 32 residues long linker between α and β subunits, reverse reaction [79]) [79]
7 <2> [36, 47]
7 <3> (<3> purified mutant enzyme with a direct linker between α and β subunits, cyclic reaction [79]) [79]
9.6 <3> (<3> purified enzyme, reverse reaction [79]) [79]
10 <3> [46]
12-15 <3> [65]
13.6 <3> (<3> partially purified enzyme, assay in the presence of Escherichia coli phospholipids [52]) [52]

14.2 <2> (<2> reduction of acetylpyridine adenine dinucleotide [39]) [36, 39]
16.2 <2> [29]
22 <3> (<3> partially purified enzyme from strain W6 [25]) [25]
24.6 <2> (<2> reduction of 3-acetylpyridine adenine dinucleotide [4]) [4]
29.9 <3> (<3> purified enzyme from strain JM83 [25]) [25]
35.6 <2> (<2> reduction of 3-acetylpyridine adenine dinucleotide [18]) [18]
42 <3> (<3> purified mutant enzyme with a 32 residues long linker between α and β subunits, cyclic reaction [79]) [79]
46 <3> (<3> purified mutant enzyme with a 18 residues long linker between α and β subunits, cyclic reaction [79]) [79]
62.3 <2> [36]
63 <3> (<3> purified enzyme, cyclic reaction [79]) [79]

K_m-Value (mM)

0.0017 <3> (NADH, <3> wild-type enzyme [69]) [69]
0.0017 <2> (NADP$^+$, <2> + NADH [33]) [33]
0.002 <3> (NADH, <3> cysteine-free enzyme [69]) [69]
0.003 <3> (NADH, <3> enzyme co-reconstituted with bacteriorhodopsin, in the light [65]) [65]
0.0051 <3> (NADPH, <3> cysteine-free enzyme [69]) [69]
0.006 <4> (NADPH) [5]
0.0065 <2> (NADP$^+$, <2> energy-linked reaction [50]) [50]
0.008 <3> (thio-NADP$^+$, <3> wild-type enzyme [69]) [69]
0.009 <2> (NADH) [2]
0.009 <2> (NADH, <2> nonenergy-linked reaction [50]) [50]
0.009 <4> (thio-NADP$^+$, <4> pH 6.3, in the light [6]) [6]
0.01 <2> (NADH, <2> + NADP$^+$ [33]) [33]
0.012 <3> (thio-NADP$^+$, <3> enzyme co-reconstituted with bacteriorhodopsin, in the dark [65]) [65]
0.0126 <2> (NADH, <2> energy-linked reaction [50]) [50]
0.013 <3> (thio-NADP$^+$, <3> enzyme co-reconstituted with bacteriorhodopsin, in the light [65]) [65]
0.013 <4> (thio-NADP$^+$, <4> pH 7.6, in the light [6]) [6]
0.014 <3> (NADPH) [46]
0.014 <3> (NADPH, <3> cysteine-free mutant enzyme [70]) [70]
0.015 <3> (NADPH, <3> wild-type enzyme [69]) [69]
0.015 <5> (NADPH, <5> Y235N and Y235F mutants of domain I [67]) [67]
0.018 <3> (thio-NADP$^+$, <3> cysteine-free enzyme [69]) [69]
0.019 <3> (NADH, <3> enzyme co-reconstituted with bacteriorhodopsin, in the dark [65]) [65]
0.019 <4> (thio-NADP$^+$, <4> pH 7.6, in the dark [6]) [6]
0.02 <4> (NADH, <4> pH 6.3, in the light [6]) [6]
0.02 <2> (NADPH, <2> + NAD$^+$ [33]) [2, 33]
0.02 <2> (NADPH, <2> energy-linked reaction [50]) [50]
0.02 <2> (NADPH, <2> nonenergy-linked reaction [50]) [50]

0.02 <3> (oxidized acetylpyridine adenine dinucleotide, <3> cysteine-free enzyme [69]) [69]
0.022 <4> (NADH, <4> pH 6.3, in the dark [6]) [6]
0.023 <3> (NADPH) [25]
0.023 <3> (NADPH, <3> A348C mutation introduced into cysteine-free mutant enzyme [70]) [70]
0.025 <4> (NADH, <4> pH 7.6, in the light [6]) [6]
0.026 <3> (oxidized acetylpyridine adenine dinucleotide, <3> wild-type enzyme [69]) [69]
0.028 <2> (NAD^+) [2]
0.028 <2> (NAD^+, <2> nonenergy-linked reaction [50]) [50]
0.028 <3> (acetylpyridine adenine dinucleotide) [46]
0.029 <2> (NADPH, <2> + acetylpyridine adenine dinucleotide [33]) [33]
0.029 <3> (NADPH, <3> membrane-bound enzyme [25]) [25]
0.029 <4> (thio-$NADP^+$, <4> pH 9.0, in the light [6]) [6]
0.03 <5> (NADPH, <5> wild-type domain I [67]) [67]
0.032 <4> (thio-$NADP^+$, <4> pH 6.3, in the dark [6]) [6]
0.033 <4> (NADH, <4> pH 9.0, in the light [6]) [6]
0.033 <3> (NADPH, <3> A390C mutation introduced into cysteine-free mutant enzyme [70]) [70]
0.033 <3> (acetylpyridine adenine dinucleotide) [25]
0.04 <2> ($NADP^+$) [2]
0.04 <2> ($NADP^+$, <2> nonenergy-linked reaction [50]) [50]
0.04 <3> (NADPH, <3> K424C mutation introduced into cysteine-free mutant enzyme [70]) [70]
0.041 <3> (acetylpyridine adenine dinucleotide, <3> membrane-bound enzyme [25]) [25]
0.0435 <2> (NAD^+, <2> energy-linked reaction [50]) [50]
0.045 <4> (thio-$NADP^+$, <4> pH 9.0, in the dark [6]) [6]
0.055 <4> (thio-$NADP^+$, <4> pH 10.0, in the light [6]) [6]
0.06 <5> (oxidized acetylpyridine adenine dinucleotide, <5> wild-type domain I [67]) [67]
0.063 <3> (NADH, <3> enzyme co-reconstituted with bacteriorhodopsin, in the presence of the uncoupler carbonyl cyanide *m*-chloro-phenylhydrazone [65]) [65]
0.063 <3> (thio-$NADP^+$, <3> enzyme co-reconstituted with bacteriorhodopsin, in the presence of the uncoupler carbonyl cyanide *m*-chloro-phenylhydrazone [65]) [65]
0.064 <4> (NADH, <4> pH 7.6, in the dark [6]) [6]
0.065 <4> (acetylpyridine adenine dinucleotide) [5]
0.066 <4> (NADH, <4> pH 9.0, in the dark [6]) [6]
0.12 <3> (NADPH, <3> R425C mutation introduced into cysteine-free mutant enzyme [70]) [70]
0.125 <2> (NAD^+, <2> + NADPH [33]) [33]
0.166 <2> (acetylpyridine adenine dinucleotide, <2> + NADPH [33]) [33]
0.6 <5> (oxidized acetylpyridine adenine dinucleotide, <5> Y235F mutant of domain I [67]) [67]

0.8 <5> (oxidized acetylpyridine adenine dinucleotide, <5> Y235N mutant of domain I [67]) [67]
2.6 <2> (NMNH, <2> reduction of thio-NADP$^+$ in energized submitochondrial particles [22]) [22]
3.4 <2> (NMNH, <2> reduction of thio-NADP$^+$ in non energized submitochondrial particles [22]) [22]

K_i-Value (mM)

0.00015 <1> (palmitoyl-CoA) [2,47]
0.0003 <2> (S-7-nitrobenzofuran-4-yl-3'-dephospho-CoA) [27]
0.00125 <11> (palmitoyl-CoA) [64]
0.0026 <2> (S-7-nitrobenzofuran-4-yl-CoA) [27]
0.003 <2> (2,4-dinitrophenyl-3'-dephospho-CoA) [27]
0.0035 <3> (NADPH) [46]
0.007 <3> (acetylpyridine adenine dinucleotide) [46]
0.007 <3> (palmitoyl-CoA) [64]
0.009 <1> (dephospho-CoA) [2,27]
0.01 <2> (palmitoyl-CoA) [33]
0.011 <4> (NADPH) [5]
0.011 <1> (acetyl-dephospho-CoA) [2]
0.093 <3> (reduced acetylpyridine adenine dinucleotide, <3> vs. NADPH [46]) [46]
0.1 <3> (aminoadenine dinucleotide phosphate, <3> approximate value [70]) [70]
0.116 <4> (acetylpyridiene adenine dinucleotide) [5]
0.12 <3> (reduced acetylpyridine adenine dinucleotide, <3> vs. oxidized acetylpyridine adenine dinucleotide [46]) [46]
0.17 <3> (NADP$^+$, <3> vs. NADPH [46]) [46]
0.2 <1,2> (CoA) [2,27]
0.2 <1> (acetyl-CoA) [2]
0.26 <3> (NADPH, <3> vs. oxidized acetylpyridine adenine dinucleotide [46]) [46]
0.3 <1> (5'-AMP) [2]
0.3 <1> (ADP) [2]
0.4 <1> (3',5'-AMP) [2]
0.5 <1> (adenosine) [2]
0.51 <4> (5'-AMP, <4> uncompetitive inhibition vs. NADPH [5]) [5]
0.59 <4> (2'-AMP, <4> competitive inhibition vs. NADPH [5]) [5]
0.68 <4> (5'-AMP, <4> competitive inhibition vs. acetylpyridine adenine dinucleotide [5]) [5]
0.7 <1> (2'-AMP) [2]
0.7 <1> (3'-AMP) [2]
0.75 <4> (2'-AMP, <4> competitive inhibition vs. acetylpyridine adenine dinucleotide [5]) [5]
2.03 <4> (2'-AMP, <4> uncompetitive inhibition vs. acetylpyridine adenine dinucleotide [5]) [5]
2.3 <3> (2'-AMP, <3> vs. NADPH [46]) [46]

2.5 <2> (Mg^{2+}, <2> at pH 7.0 [41]) [41]
2.63 <4> (5'-AMP, <4> competitive inhibition vs. NADPH [5]) [5]
3.4 <3> (5'-AMP, <3> vs. oxidized acetylpyridine adenine dinucleotide [46]) [46]
3.7 <3> (5'-AMP, <3> vs. NADPH [46]) [46]
8 <3> (2'-AMP, <3> vs. oxidized acetylpyridine adenine dinucleotide [46]) [46]

pH-Optimum

5.5 <1> (<1> reduction of $NADP^+$ [2]) [2]
5.5 <13> (<13> rapid decline above, reverse reaction [80]) [80]
5.5 <3> (<3> reduction of oxidized acetylpyridine adenine dinucleotide by NADH i.e. forward reaction [65]; <3> H91E mutant enzyme, optima for forward reaction below pH 5.5, 7% of wild-type enzyme activity at pH 6.0 [71]) [65, 71]
5.5-6.5 <13> (<13> rapid decline above, cyclic reaction [80]) [80]
6 <3> (<3> reverse reaction catalyzed by H91E mutant enzyme, 20% of wild-type enzyme activity [71]) [71]
6.2-6.3 <2> [47]
7 <1, 2> (<1> reduction of NAD^+ [2]; <2> ATP-driven proton translocation in co-reconstituted enzyme-ATPase vesicles [14]) [2, 14]
7-8 <11> (<11> reduction of oxidized acetylpyridine adenine dinucleotide by NADPH in inside-out membrane vesicles [64]) [64]

pH-Range

4.7-6.7 <2> (<2> less than 50% of maximal activity above and below [47]) [47]

4 Enzyme Structure

Molecular weight

155000 <13> (<13> gel filtration [80]) [80]
196000 <1> [1]
206000-249000 <2> (<2> cross-linking of purified enzyme with 10.9 mM dimethyl suberimidate dihydrochloride [32]) [32]
210000-230000 <2> (<2> SDS-PAGE after cross-linking with bifunctional reagents dimethyl adipimidate, dimethyl pimelimidate, dimethyl suberimidate and dithiobis(succinimidyl propionate) [38]) [11, 38]
250000 <2> [50]
278000 <2> (<2> radiation inactivation, hydrodynamic properties [11]) [11]

Subunits

? <4, 5, 6, 11> (<4> x * 53000 + x * 48000, SDS-PAGE [5]; <6> x * 6784, soluble component, amino acid analysis [56]; <11> x * 54000, immunoblot with antibodies against beef heart enzyme [64]; <5> domain I exists as a separate polypeptide that can be removed from everted membrane vesicle

i.e. chromatophores [68]; <5> domain I contains the binding site for NAD$^+$ and NADH, domain III for NADP$^+$ and NADPH [68]) [1, 5, 56, 64, 68]
dimer <1, 2, 7, 13> (<2> 2 * 109212, calculated from cDNA sequence [10]; <2,7> 2 * 110000, SDS-PAGE [1,10,12]; <2> 2 * 120000, SDS-PAGE [47,49]; <2> signal peptide MW 4816, sequence of mRNA [8]; <2> 2 * 115000, SDS-PAGE [11]; <7> 2 * 111500, SDS-PAGE [16]; <2> x * 97000, SDS-PAGE [49]; <2> 2 * 109065, monomer is composed of three domains: a 430 residue long N-terminal hydrophilic domain called dI, a 400 residue long central hydrophobic domain that intercalates into the membrane called dII, and a 200 residue long C-terminal hydrophilic domain called dIII [63]; <13> 2 * 66000, SDS-PAGE [80]) [1, 4, 8, 10, 12, 16, 32, 47, 49, 63, 75, 80]
tetramer <3> (<3> α_2,β_2, 2 * 53906 + 2 * 48667, calculation from nucleotide sequence [17]; <3> α_2,β_2, 2 * 50000 + 2 * 47000 [1]; <3> domain I, i.e. α_1 to α_{404}, and III i.e. β_{260} to β_{462}, are exposed to the cytosol and contain the binding sites for NAD(H) and NAD(P)H, respectively; domain II, i.e. α_{405} to α_{510}, spans the membrane [70]) [1, 17, 70, 75]
trimer <5> (<5> dimer of domain I combined with a domain III monomer, deduced from crystal structure [76]) [76]

Posttranslational modification

phospholipoprotein <2> (<2> 5 mol loosely bound phospholipids, 9 mol tightly bound phospholipids [61]) [61]

5 Isolation/Preparation/Mutation/Application

Source/tissue

artery <1, 2> [2]
heart <1, 2, 7> [2, 4, 8, 13-15, 18, 21-24, 27-33, 35-44, 47, 49, 51, 53, 62]
kidney <2> [2]
liver <1, 2, 7> [2, 12, 16]
Additional information <1> (<1> no or low activity in brain, prostate, seminal vesicle, spleen, testis [2]) [2]

Localization

chromatophore <1, 4-6> (<6> soluble and insoluble component [56]; <5> enzyme is compromised of a membrane-bound component and an easily dissociable soluble factor [48]) [1, 5, 19, 34, 48, 56]
cytoplasmic membrane <3> [25]
mitochondrial membrane <1, 2, 7, 12> (<2> membrane topography [4]; <7> orientation in the inner mitochondrial membrane [12]) [1, 2, 4, 7, 9-12, 15, 16, 18, 21-24, 27-33, 35-41, 47, 49, 62, 73]

Purification

<2> (affinity chromatography [18, 23, 36]; NaCl wash, Triton X-100 extraction, affinity chromatography on immobilized NAD$^+$ [36]; FPLC in the presence of 0.05 or 0.1% Triton X-100, comparison of methods [29]; immunoexclusion chromatography [39]; overview of early procedures [50]; fractiona-

tion of submitochondrial particles, DEAE-Sepharose, hydroxyapatite [44]) [18, 23, 29, 36, 39, 44, 47, 49, 50]
<3> (recombinant enzyme [17]; partially purified [52]; His-tagged H91E mutant enzyme [71]; recombinant domain I [72]; recombinant wild-type, T393C, R425C, G430C and A432C mutant domain III [74]) [25, 46, 52, 65, 71, 72, 74]
<4> [5]
<5> (recombinant domains I and III [66, 72]; recombinant domain I [73]; recombinant domain I [74]; recombinant domain I and recombinant E155W and Y171W mutant domain III [81]) [66, 68, 72, 74, 81]
<6> (soluble component [56]) [56]
<7> (affinity chromatography [16]; affinity chromatography on NADP$^+$/ NAD$^+$-agarose gels, partial purification [23]) [16, 23]
<12> (recombinant domain III [73]; recombinant domain III [81]) [73, 81]
<13> (recombinant domain III-domain I protein [80]) [80]

Renaturation

<2, 5, 7> (reconstitution into liposomes [13, 14, 18, 30, 33, 39, 42, 45, 50]; co-reconstitution with bacteriorhodopsin [65]) [13, 14, 18, 30, 33, 39, 42, 45, 50, 65]

Crystallization

<4, 5> (crystal structure of domain I/domain III complex [76]; solution structure solved by NMR [77]; crystal structure of domain I i.e. α_1 subunit, with and without bound NADH, 1.8 A resolution without NADH, 1.9 A with NADH bound [82]) [5, 76, 77, 82]

Cloning

<2> (expression in Escherichia coli [10]) [10]
<3> (expression in Escherichia coli on multicopy plasmid [17]; expression of cysteine mutants A348C, A390C, K424C, and R425C in Escherichia coli [70]; His-tagged H91E mutant enzyme expressed in Escherichia coli [71]; expression of domain I in Escherichia coli [72]; expression of wild-type domain III, T393C, R425C, G430C and A432C mutant domain III in Escherichia coli [74]) [17, 20, 26, 71, 72, 74]
<5> (expression of domains I and III in Escherichia coli [66, 72]; expression of domain I Y235N and Y235F mutants in Escherichia coli [67, 68]; wild-type and cysteine-free enzyme [69]; expression of domain I in Escherichia coli [73, 74]; expression of domain I and of E155W and Y171W mutant domain III [81]) [66, 67, 68, 69, 70, 73, 74, 81]
<12> (<12> epresssion of domain III in Escherichia coli [73]) [73]
<13> (<13> expression of domain III-domain I protein in Escherichia coli [80]) [80]

Engineering

A348C <3> (<3> mutation introduced into a cysteine-free mutant enzyme, mutant shows markedly reduced activity [70]) [70]
A390C <3> (<3> mutation introduced into a cysteine-free mutant enzyme [70]) [70]

A432C <3> (<3> mutation in domain III, reverse reaction in the presence of domain I from R. rubrum, 150% higher reaction rate than wild-type domain III/R. rubrum domain I mixture [74]) [74]
C292T,C339T,C395S,C397T,C435S <3> (<3> cysteine of the α subunits replaced, similar activity as wild-type [69]) [69]
C292T/C339T/C395S/C397T/C435S/C147S/C260S <3> (<3> all 7 cysteines of the enzyme, 5 localized in the α subunit and 2 in the β subunit, are replaced, the cysteine-free mutant shows about 5fold more activity in the reduction of acetylpyridine adenine dinucleotide by NADH than wild-type, the cyclic reduction of acetylpyridine adenine dinucleotide by NADH via NADPH is 2-2.5fold more activ [69]) [69]
D213K <3> (<3> mutation in domain II [78]) [78]
D213R <3> (<3> mutation in domain II [78]) [78]
D401E <3> (<3> mutation in β subunit [71]) [71]
D401G <3> (<3> mutation in β subunit [71]) [71]
D401V <3> (<3> mutation in β subunit [71]) [71]
E155W <5> (<5> mutation in domain III, similar catalytic activities as wild-type, used for tryptophan fluorescence measurements [81]) [81]
E413D <3> (<3> mutation in β subunit [71]) [71]
E413G <3> (<3> mutation in β subunit [71]) [71]
E413V <3> (<3> mutation in β subunit [71]) [71]
G430C <3> (<3> mutation in domain III, reverse reaction in the presence of domain I from Rhodospirillum rubrum, 850% higher reaction rate than wild-type domain III/R. rubrum domain I mixture [74]) [74]
H91E <3> (<3> mutation in β subunit [71]) [71]
H91K <3> (<3> mutation in β subunit [71]) [71]
H91K <3> (<3> mutation in domain II, leads to occlusion of NADP(H) at the NADP(H)-binding site of domain III [78]) [78]
H91R <3> (<3> mutation in domain II, leads to occlusion of NADP(H) at the NADP(H)-binding site of domain III [78]) [78]
K416G <3> (<3> mutation in β subunit [71]) [71]
K424C <3> (<3> mutation introduced into a cysteine-free mutant enzyme, mutant shows markedly reduced activity [70]) [70]
K424G <3> (<3> mutation in β subunit [71]) [71]
K424R <3> (<3> mutation in β subunit [71]) [71]
K452D <3> (<3> mutation in β subunit [71]) [71]
K452G <3> (<3> mutation in β subunit [71]) [71]
N222K <3> (<3> mutation in domain II, leads to occlusion of NADP(H) at the NADP(H)-binding site of domain III [78]) [78]
N222R <3> (<3> mutation in domain II, leads to occlusion of NADP(H) at the NADP(H)-binding site of domain III [78]) [78]
R425C <3> (<3> mutation in domain III, reverse reaction in the presence of domain I from R. rubrum, 425% higher reaction rate than wild-type domain III/R. rubrum domain I mixture [74]) [74]
R425C <3> (<3> mutation introduced into a cysteine-free mutant enzyme, mutant shows markedly reduced activity [70]) [70]
R425E <3> (<3> mutation in β subunit [71]) [71]

R425G <3> (<3> mutation in β subunit [71]) [71]
R425K <3> (<3> mutation in β subunit [71]) [71]
T393C <3> (<3> mutation in domain III, reverse reaction in the presence of domain I from Rhodospirillum rubrum, 175% higher reaction rate than wild-type domain III/R. rubrum domain I mixture [74]) [74]
Y171W <5> (<5> mutation in domain III, similar catalytic activities as wild-type, used for tryptophan fluorescence measurements [81]) [81]
Y235F <5> (<5> mutation resides in the soluble NAD(H)-binding peripheral membrane subunit, i.e. domain I, reconstitution of depleted membranes with mutant domain I gives 44% of activity that is obtained with wild-type domain I reconstituted membranes [67]; <5> mutant domain I/wild-type domain III mixtures catalyse acetylpyridine adenine dinucleotide reduction with similar rates as wild-type domain I/wild-type domain III mixtures [68]) [67, 68]
Y235N <5> (<5> mutation resides in the soluble NAD(H)-binding peripheral membrane subunit, i.e. domain I, reconstitution of depleted membranes with mutant domain I gives 18% of activity that is obtained with wild-type domain I reconstituted membranes [67]; <5> mutant domain I/wild-type domain III mixtures catalyse acetylpyridine adenine dinucleotide reduction with similar rates as wild-type domain I/wild-type domain III mixtures [68]) [67, 68]
Additional information <3> (<3> properties of a variety of mutant enzymes containing modified conserved and semiconserved basic and acidic residues in the β subunit [71]) [71]

6 Stability

Temperature stability

40 <6> (<6> 5 min, soluble component, complete inactivation [56]) [56]
44 <5> (<5> 2 min, solubilized membrane component, 50% inactivation [48]) [48]
45 <5> (<5> incubation of depleted membranes, inactivation of membrane transhydrogenase component [54]) [54]
48 <5> (<5> 2 min, membrane particles, 50% inactivation [48]) [48]
50 <5> (<5> complete protection in the presence of NADPH, half-maximal protection with 0.01 mM NADPH [54]) [54]
53 <5> (<5> incubation for 3 min, 10 mM Mg^{2+} almost completely protect from inactivation [54]) [54]
Additional information <2> (<2> proteolytic inactivation as a function of temperature [40]; <2> protection against thermal inactivation by cations, half-maximal protection is afforded at: 1-2 mM Mg^{2+}, 2 mM Ca^{2+}, 0.5 mM Mn^{2+} and 0.025 mM La^{3+} [41]; <2> NADPH protects native enzyme from thermal inactivation, $NADP^+$ accelerates thermal inactivation, both native and S-thionitrobenzoate derivatized enzyme are protected by Mg^{2+} against thermoinactivation [51]) [40, 41, 51]

Organic solvent stability

1,1-dimethylbutanol <2> (<2> inactivation [62]) [62]
acetone <2> (<2> used for conversion of mitochondria to an acetone powder causes inactivation [62]) [62]
ethanol <2> (<2> 10%, 40°C, inactivation [62]) [62]
n-butanol <2> (<2> inactivation [62]) [62]

General stability information

<2>, cations prevent from tryptic inactivation, 95% protection with Mn^{2+}, 89% with Ca^{2+} and 79% with Mg^{2+} [41]
<2>, inactivation during prolonged column chromatography [44]
<2>, unstable, loses 30-50% activity within 48 h at 4°C, remainder of the enzyme becomes inactive after 2 weeks, complete inactivation after freezing at -20°C or -70°C overnight, enzyme is very susceptible to trypsinolysis, NADH protects from tryptic inactivation, NADPH promotes tryptic inactivation [47]
<2>, urea, 6 M, inactivation [38]
<4>, purified enzyme is inactivated at 4°C even in presence of dithiothreitol [5]
<5>, solubilized membrane component is unstable losing all activity when stored overnight at 4°C or -70°C, stable at -70°C in the presence of 0.025 mM $NADP^+$ for several months, inactivation by refreezing after thawing [48]
<5>, trypsinolysis is stimulated several fold by NADPH and $NADP^+$, half-maximal stimulation with 0.001-0.002 mM NADPH and 0.002-0.003 mM $NADP^+$, Mg^{2+} protects from $NADP^+$ stimulated inactivation [48, 54]

Storage stability

<2>, 4°C, 0.1 M sodium phosphate buffer, pH 7.5, 1 mM dithiothreitol, 0.05% sodium cholate [39]
<2>, 4°C, reconstituted with phospholipids, stable for at least 2 months [44]
<3>, 4°C, 50 mM Tris-HCl, pH 7.8, 1 mM, EDTA, 1 mM, dithiothreitol, 1 week, 10% loss of activity [25]
<4>, -15°C, 20 mM sodium tricine buffer, pH 7.6, 1 mM dithiothreitol, 0.2% Triton X-100, 30% glycerol, 4 weeks, no loss of activity [5]
<5>, -70°C or 4°C, Rhodospirillum rubrum membrane component, inactivation in 1 day, stabilization by $NADP^+$ [48]
<6>, 4°C, unsoluble component: 0.1 M glycyl-glycine buffer, pH 8.0, 10% sucrose, unstable, soluble factor: 120 h with 0.015 mM dithiothreitol stable [56]

References

[1] Jackson, J.B.; Lever, T.M.; Rydström, J.; Persson, B.; Carlenor, E.: Proton-translocating transhydrogenase from photosynthetic bacteria. Biochem. Soc. Trans., **19**, 573-575 (1991)
[2] Rydström, J.; Hoek, J.B.; Ernster, L.: Nicotinamide nucleotide transhydrogenases. The Enzymes, 3rd Ed. (Boyer, P.D., ed.), **13**, 51-88 (1976)

[3] Hoek, J.B.; Rydström, J.; Höjeberg, B.: Comparative studies on nicotinamide nucleotide transhydrogenase from different sources. Biochim. Biophys. Acta, **333**, 237-245 (1974)
[4] Yamaguchi, M.; Hatefi, Y.: Mitochondrial energy-linked nicotinamide nucleotide transhydrogenase. Membrane topography of the bovine enzyme. J. Biol. Chem., **266**, 5728-5735 (1991)
[5] Lever, R.M.; Palmer, T.; Cunningham, I.J.; Cotton, N. P.J.; Jackson, J.B.: Purification and properties of the H(+)-nicotinamide nucleotide transhydrogenase from Rhodobacter capsulatus. Eur. J. Biochem., **197**, 247-255 (1991)
[6] Cotton, N.P.J.; Lever, T.M.; Nore, B.F.; Jones, M.R.; Jackson, J.B.: The coupling between protonmotive force and the $NAD(P)^+$ transhydrogenase in chromatophores from photosynthetic bacteria [published erratum appears in Eur J Biochem 1989 Oct 1;184(3):729]. Eur. J. Biochem., **182**, 593-603 (1989)
[7] Yamaguchi, M.; Hatefi, Y.: Mitochondrial nicotinamide nucleotide transhydrogenase: NADPH binding increases and NADP binding decreases the acidity and susceptibility to modification of cysteine-893. Biochemistry, **28**, 6050-6056 (1989)
[8] Yamaguchi, M.; Hatefi, Y.; Trach, K.; Hoch, J.A.: Amino acid sequence of the signal peptide of mitochondrial nicotinamide nucleotide transhydrogenase as determined from the sequence of its messenger RNA. Biochem. Biophys. Res. Commun., **157**, 24-29 (1988)
[9] Hoek, J.B.; Rydström, J.: Physiological roles of nicotinamide nucleotide transhydrogenase. Biochem. J., **254**, 1-10 (1988)
[10] Yamaguchi, M.; Hatefi, Y.; Trach, K.; Hoch, J.A.: The primary structure of the mitochondrial energy-linked nicotinamide nucleotide transhydrogenase deduced from the sequence of cDNA clones. J. Biol. Chem., **263**, 2761-2767 (1988)
[11] Persson, B.; Ahnström, G.; Rydström, J.: Energy-linked nicotinamide nucleotide transhydrogenase: hydrodynamic properties and active form of purified and membrane-bound mitochondrial transhydrogenase from beef heart. Arch. Biochem. Biophys., **259**, 341-349 (1987)
[12] Weis, J.K.; Wu, L.N.Y.; Fisher, R.R.: The orientation of transhydrogenase in the inner mitochondrial membrane of rat liver. Arch. Biochem. Biophys., **257**, 424-429 (1987)
[13] Eytan, G.D.; Eytan, E.; Rydström, J.: Energy-linked nicotinamide-nucleotide transhydrogenase. Light-driven transhydrogenase catalyzed by transhydrogenase from beef heart mitochondria reconstituted with bacteriorhodopsin. J. Biol. Chem., **262**, 5015-5019 (1987)
[14] Eytan, G.D.; Persson, B.; Ekebacke, A.; Rydström, J.: Energy-linked nicotinamide-nucleotide transhydrogenase. Characterization of reconstituted ATP-driven transhydrogenase from beef heart mitochondria. J. Biol. Chem., **262**, 5008-5014 (1987)
[15] Persson, B.; Rydström, J.: Evidence for a role of a vicinal dithiol in catalysis and proton pumping in mitochondrial nicotinamide nucleotide transhydrogenase. Biochem. Biophys. Res. Commun., **142**, 573-578 (1987)

[16] Moody, A.J.: Identification of the nicotinamide nucleotide transhydrogenase of rat liver mitochondria. Biochem. Soc. Trans., **14**, 1210-1212 (1986)
[17] Clarke, D.M.; Loo, T.W.; Gillam, S.; Bragg, P.D.: Nucleotide sequence of the pntA and pntB genes encoding the pyridine nucleotide transhydrogenase of Escherichia coli. Eur. J. Biochem., **158**, 647-653 (1986)
[18] Wu, L.N.Y.; Alberta, J.A.; Fisher, R.R.: Purification and reconstitution of bovine heart mitochondrial transhydrogenase. Methods Enzymol., **126**, 353-360 (1986)
[19] Nore, B.F.; Husain, I.; Nyren, P.; Baltscheffsky, M.: Synthesis of pyrophosphate coupled to the reverse energy-linked transhydrogenase reaction in Rhodospirillum rubrum chromatophores. FEBS Lett., **200**, 133-138 (1986)
[20] Clarke, D.M.; Bragg, P.D.: Expression of the cloned subunits of Escherichia coli transhydrogenase from separate replicons. FEBS Lett., **200**, 23-26 (1986)
[21] Yamaguchi, M.; Hatefi, Y.: Mitochondrial nicotinamide nucleotide transhydrogenase: inhibition by ethoxyformic anhydride, dansyl chloride, and pyridoxal phosphate. Arch. Biochem. Biophys., **243**, 20-27 (1985)
[22] Phelps, D.C.; Hatefi, Y.: Mitochondrial nicotinamide nucleotide transhydrogenase: nonidentical modification by N,N-dicyclohexylcarbodiimide and N-(ethoxycarbonyl)-2-ethoxy-1,2-dihydroquinoline at the NAD(H) binding site. Arch. Biochem. Biophys., **243**, 298-304 (1985)
[23] Carlenor, E.; Tang, H.L.; Rydström, J.: Affinity chromatography of mitochondrial nicotinamide nucleotide transhydrogenase. Anal. Biochem., **148**, 518-523 (1985)
[24] Phelps, D.C.; Hatefi, Y.: Mitochondrial nicotinamide nucleotide transhydrogenase: active site modification by 5-[*p*-(fluorosulfonyl)benzoyl]adenosine. Biochemistry, **24**, 3503-3507 (1985)
[25] Clarke, D.M.; Bragg, P.D.: Purification and properties of reconstitutively active nicotinamide nucleotide transhydrogenase of Escherichia coli. Eur. J. Biochem., **149**, 517-523 (1985)
[26] Clarke, D.M.; Bragg, P.D.: Cloning and expression of the transhydrogenase gene of Escherichia coli. J. Bacteriol., **162**, 367-373 (1985)
[27] Kozlov, I.A.; Milgrom, Y.M.; Saburova, L.A.; Sobolev, A.Y.: The interaction of mitochondrial transhydrogenase with derivatives of coenzyme A. Eur. J. Biochem., **145**, 413-416 (1984)
[28] Phelps, D.C.; Hatefi, Y.: Interaction of purified nicotinamidenucleotide transhydrogenase with dicyclohexylcarbodiimide. Biochemistry, **23**, 4475-4480 (1984)
[29] Persson, B.; Enander, K.; Tang, H.L.; Rydström, J.: Energy-linked nicotinamide nucleotide transhydrogenase. Properties of proton-translocating mitochondrial transhydrogenase from beef heart purified by fast protein liquid chromatography. J. Biol. Chem., **259**, 8626-8632 (1984)
[30] Pennington, R.M.; Fisher, R.R.: Reconstituted mitochondrial transhydrogenase is a transmembrane protein. FEBS Lett., **164**, 345-349 (1983)
[31] Moody, A.J.; Reid, R.A.: Inhibition of nicotinamide nucleotide transhydrogenase in rat liver submitochondrial particles by dicyclohexylcarbodiimide and butanedione. Biochem. J., **209**, 889-892 (1983)

[32] Wu, L.N.W.; Fisher, R.R.: Subunit structure of submitochondrial particle membrane transhydrogenase. J. Biol. Chem., **258**, 7847-7851 (1983)
[33] Enander, K.; Rydström, J.: Energy-linked nicotinamide nucleotide transhydrogenase. Kinetics and regulation of purified and reconstituted transhydrogenase from beef heart mitochondria. J. Biol. Chem., **257**, 14760-14766 (1982)
[34] McFadden, B.J.; Fisher, R.R.: Resolution and reconstitution of Rhodospirillum rubrum pyridine dinucleotide transhydrogenase: localization of substrate binding sites. Arch. Biochem. Biophys., **190**, 820-828 (1978)
[35] Wu, L.N.Y.; Fisher, R.R.: Stereochemistry of NADPH leads to $NADP^+$ transhydrogenation catalyzed by bovine heart mitochondrial pyridine dinucleotide transhydrogenase. J. Biol. Chem., **257**, 11680-11683 (1982)
[36] Wu, L.N.Y.; Pennington, R.M.; Everett, T.D.; Fisher, R.R.: An improved method for the purification of bovine heart mitochondrial transhydrogenase. J. Biol. Chem., **257**, 4052-4055 (1982)
[37] Phelps, D.C.; Hatefi, Y.: Inhibition of the mitochondrial nicotinamide nucleotide transhydrogenase by dicyclohexylcarbodiimide and diethylpyrocarbonate. J. Biol. Chem., **256**, 8217-8221 (1981)
[38] Anderson, W.M.; Fisher, R.R.: The subunit structure of bovine heart mitochondrial transhydrogenase. Biochim. Biophys. Acta, **635**, 194-199 (1981)
[39] Anderson, W.M.; Fowler, W.T.; Pennington, R.M.; Fisher, R.R.: Immunochemical characterization and purification of bovine heart mitochondrial pyridine dinucleotide transhydrogenase. J. Biol. Chem., **256**, 1888-1895 (1981)
[40] Blazyk, J.F.; Blazyk, J.M.; Kline, C.M.: Temperature-dependent changes in the activity and tryptic susceptibility of membrane-bound transhydrogenase. J. Biol. Chem., **256**, 691-694 (1981)
[41] O'Neal, S.G.; Earle, S.R.; Fisher, R.R.: The effect of metal ions on mitochondrial pyridine dinucleotide transhydrogenase. Biochim. Biophys. Acta, **589**, 217-230 (1980)
[42] Rydström, J.; Fleischer, S.: Reconstitution of mitochondrial nicotinamide nucleotide transhydrogenase from beef heart with synthetic phospholipids. Methods Enzymol., **55**, 811-816 (1979)
[43] Rydström, J.: Assay of nicotinamide nucleotide transhydrogenases in mammalian, bacterial, and reconstituted systems. Methods Enzymol., **55**, 261-273 (1979)
[44] Höjeberg, B.; Rydström, J.: Purification of mitochondrial nicotinamide nucleotide transhydrogenase from beef heart. Methods Enzymol., **55**, 275-283 (1979)
[45] Jacobs, E.; Fisher, R.R.: Resolution and reconstitution of Rhodospirillum rubrum pyridine dinucleotide transhydrogenase: chemical modification with N-ethylmaleimide and 2,4-pentanedione. Biochemistry, **18**, 4315-4322 (1979)
[46] Hanson, R.L.: The kinetic mechanism of pyridine nucleotide transhydrogenase from Escherichia coli. J. Biol. Chem., **254**, 888-893 (1979)

[47] Anderson, W.M.; Fisher, R.R.: Purification and partial characterization of bovine heart mitochondrial pyridine dinucleotide transhydrogenase. Arch. Biochem. Biophys., **187**, 180-190 (1978)
[48] Jacobs, E.; Heriot, K.; Fisher, R.R.: Resolution and reconstitution of Rhodospirillum rubrum pyridine dinucleotide transhydrogenase. II. Solubilization of the membrane-bound component. Arch. Microbiol., **115**, 151-156 (1977)
[49] Höjeberg, B.; Rydström, J.: Purification and molecular properties of reconstitutively active nicotinamide nucleotide transhydrogenase from beef heart mitochondria. Biochem. Biophys. Res. Commun., **78**, 1183-1190 (1977)
[50] Rydström, J.: Energy-linked nicotinamide nucleotide transhydrogenases. Biochim. Biophys. Acta, **463**, 155-184 (1977)
[51] O'Neal, S.G.; Fisher, R.R.: Studies on sulfhydryl group modification of mitochondrial pyridine dinucleotide transhydrogenase. J. Biol. Chem., **252**, 4552-4556 (1977)
[52] Houghton, R.L.; Fisher, R.R.; Sanadi, D.R.: Dependence of Escherichia coli pyridine nucleotide transhydrogenase on phospholipids and its sensitivity. Biochem. Biophys. Res. Commun., **73**, 751-757 (1976)
[53] Rydström, J.; Heok, J.B.; Ericson, B.G.; Hundal, T.: Evidence for a lipid dependence of mitochondrial nicotinamide nucleotide transhydrogenase. Biochim. Biophys. Acta, **430**, 419-425 (1976)
[54] Fisher, R.R.; Rampey, S.A.; Sadighi, A.; Fisher, K.: Resolution and reconstitution of Rhodospirillum rubrum pyridine dinucleotide transhydrogenase. Proteolytic and thermal inactivation of the membrane component. J. Biol. Chem., **250**, 819-825 (1975)
[55] Singh, A.P.; Bragg, P.D.: The pyridine-nucleotide transhydrogenase of Salmonella typhimurium. J. Gen. Microbiol., **82**, 237-246 (1974)
[56] Berger, T.J.; Orlando, J.A.: Purification and some properties of a protein factor required for light-dependent transhydrogenase in Rhodopseudomonas spheroides. Arch. Biochem. Biophys., **159**, 25-31 (1973)
[57] Fisher, R.R.; Kaplan, N.O.: Studies on the mitochondrial energy-linked pyridine nucleotide transhydrogenase. Biochemistry, **12**, 1182-1188 (1973)
[58] Asano, A.; Imai, K.; Sato, K.: Oxidative phosphorylation in Micrococcus dentrificans. II. The properties of pyridine nucleotide transhydrogenase. Biochim. Biophys. Acta, **143**, 477-486 (1967)
[59] Fisher, R.R.; Earle, S.R.: Membrane-bound pyridine dinucleotide transhydrogenases. The Pyridine Nucleotide Coenzymes (Everse, J., Andersson, B., You, K.S., eds.) Academic Press, New York, 279-324 (1982)
[60] Rydström, J.; Persson, B.; Carlenor, E.: Transhydrogenases linked to pyridine nucleotides. Pyridine Nucleotide Coenzymes, Chem. Biochem. Med. Aspects (Dolphin, D., Poulson, R., Avramovic, O., eds.) Wiley and Sons, New York, **2B**, 433-461 (1987)
[61] Rydström, J.: Mitochondrial nicotinamide nucleotide transhydrogenase. Mitochondria and Microsomes (Lee, C.P., Schatz, G., Dallner, G. eds.) Addison-Wesley, Reading MA, 317-335 (1981)

[62] Kaufman, B.; Kaplan, N.O.: Pyridine nucleotide transhydrogenase. VIII. Properties of the transhydrogenase reactions of an enzyme complex isolated from beef heart mitochondria. J. Biol. Chem., **236**, 2133-2139 (1961)
[63] Yamaguchi, M.; Hatefi, Y.: Energy-transducing nicotinamide nucleotide transhydrogenase. Nucleotide binding properties of the purified enzyme and proteolytic fragments. J. Biol. Chem., **268**, 17871-17877 (1993)
[64] Fristedt, U.; Rydström, J.; Persson, B.: Evidence for a nicotineamide nucleotide transhydrogenase in Klebsiella pneumoniae. Biochem. Biophys. Res. Commun., **198**, 928-932 (1994)
[65] Hu, X.; Zhang, J.W.; Persson, A.; Rydstrom, J.: Characterization of the interaction of NADH with proton pumping E. coli transhydrogenase reconstituted in the absence and in the presence of bacteriorhodopsin. Biochim. Biophys. Acta, **1229**, 64-72 (1995)
[66] Diggle, C.; Bizouarn, T.; Cotton, N.P.; Jackson, J.B.: Properties of the purified, recombinant, NADP(H)-binding domain III of the proton-translocating nicotinamide nucleotide transhydrogenase from Rhodospirillum rubrum. Eur. J. Biochem., **241**, 162-170 (1996)
[67] Diggle, C.; Quirk, P.G.; Bizouarnt, T.; Grimley, R.L.; Cottont, N.P.J.; Thomas, C.M.; Jackson, J.B.: Mutation of Tyr235 in the NAD(H)-binding subunit of the proton-translocating nicotinamide nucleotide transhydrogenase of Rhodospirillum rubrum affects the conformational dynamics of a mobile loop and lowers the catalytic activity of the enzyme. J. Biol. Chem., **271**, 10109-10115 (1996)
[68] Bizouarn, T.; Grimley, R.; Diggle, C.; Thomas, C.M.; Jackson, J.B.: Mutations at tyrosine-235 in the mobile loop region of domain I protein of transhydrogenase from Rhodospirillum rubrum strongly inhibit hydride transfer. Biochim. Biophys. Acta, **1320**, 265-274 (1997)
[69] Meuller, J.; Zhang, J.; Hou, C.; Bragg, P.D.; Rydstrom, J.: Properties of a cysteine-free proton-pumping nicotinamide nucleotide transhydrogenase. Biochem. J., **324**, 681-687 (1997)
[70] Fjellström, O.; Axelsson, M.; Bizouarn, T.; Hu, X.; Johansson, C.; Meuller, J.; Rydström, J.: Mapping of residues in the NADP(H)-binding site of proton-translocating nicotinamide nucleotide transhydrogenase from Escherichia coli. J. Biol.Chem., **274**, 6350-6359 (1999)
[71] Hu, X.; Zhang, J.; Fjellstrom, O.; Bizouarn, T.; Rydstrom, J.: Site-directed mutagenesis of charged and potentially proton-carrying residues in the β subunit of the proton-translocating nicotinamide nucleotide transhydrogenase from Escherichia coli. Characterization of the β H91, β D392, and β K424 mutants. Biochemistry, **38**, 1652-1658 (1999)
[72] Fjellstrom, O.; Bizouarn, T.; Zhang, J.W.; Rydstrom, J.; Venning, J.D.; Jackson, J.B.: Catalytic properties of hybrid complexes of the NAD(H)-binding and NADP(H)-binding domains of the proton-translocating transhydrogenases from Escherichia coli and Rhodospirillum rubrum. Biochemistry, **38**, 415-422 (1999)
[73] Peake, S.J.; Venning, J.D.; Jackson, J.B.: A catalytically active complex formed from the recombinant dI protein of Rhodospirillum rubrum trans-

hydrogenase, and the recombinant dIII protein of the human enzyme. Biochim. Biophys. Acta, **1411**, 159-169 (1999)
[74] Bergkvist, A.; Johansson, C.; Johansson, T.; Rydström, J.; Karlsson, G.: Interactions of the NADP(H)-binding domain III of proton-translocating transhydrogenase from Escherichia coli with NADP(H) and the NAD(H)-binding domain I studied by NMR and site-directed mutagenesis. Biochemistry, **39**, 12595-12605 (2000)
[75] Bizouarn, T.; Fjellstrom, O.; Meuller, J.; Axelsson, M.; Bergkvist, A.; Johansson, C.; Goran Karlsson, B.; Rydstrom, J.: Proton translocating nicotinamide nucleotide transhydrogenase from E. coli. Mechanism of action deduced from its structural and catalytic properties. Biochim. Biophys. Acta, **1457**, 211-228 (2000)
[76] Cotton, N.P.; White, S.A.; Peake, S.J.; McSweeny, s.; Jackson, J.B.: The crystal structure of an asymmetric complex of the two nucleotide binding components of proton-translocating transhydrogenase. Structure, **7**, 165-176 (2001)
[77] Jeeves, M.; Smith, K.J.; Quirk, P.G.; Cotton, N.P.; Jackson, J.B.: Solution structure of the NADP(H)-binding component (dIII) of proton-translocating transhydrogenase from Rhodospirillum rubrum. Biochim. Biophys. Acta, **15**, 248-257 (2000)
[78] Bragg, P.D.; Hou, C.: The presence of an aqueous cavity in the proton-pumping pathway of the pyridine nucleotide transhydrogenase of Escherichia coli is suggested by the reaction of the enzyme with sulfhydryl inhibitors. Arch. Biochem. Biophys., **380**, 141-150 (2000)
[79] Meuller, J.; Mjörn, K.; Karlsson, J.; Tigerström, a.; Rydström, J.; Hou, C.; Bragg, P.D.: Properties of a proton-translocating nicotinamide nucleotide transhydrogenase from Escherichia coli with α and β subunits linked through fused transmembrane helices. Biochim. Biophys. Acta, **1506**, 163-171 (2001)
[80] Weston, C.J.; White, S.A.; Jackson, B.: The unusual transhydrogenase of Entamoeba histolytica. FEBS Lett., **488**, 51-54 (2001)
[81] Rodrigues, D.J.; Venning, J.D.; Quirk, P.G.; Jackson, J.B.: A change in ionization of the NADP(H)-binding component (dIII) if proton-translocating transhydrogenase regulates both hydride transfer and nucleotide release. Eur. J. Biochem., **268**, 1430-1438 (2001)
[82] Prasad, G.S.; Wahlberg, M.; Sridhar, V.; Sundaresan, V.; Yamaguchi, M.; Hatefi, Y.; Stout, C.D.: Crystal structures of transhydrogenase domain I with and without bound NADH. Biochemistry, **41**, 12745-12754 (2002)

$NADH_2$ Cytochrome c reductase 1.6.2.1

1 Nomenclature

EC number

1.6.2.1 (transferred to EC 1.6.99.3)

Recommended name

$NADH_2$ cytochrome c reductase

Cytochrome-b_5 reductase 1.6.2.2

1 Nomenclature

EC number
1.6.2.2

Systematic name
NADH:ferricytochrome-b_5 oxidoreductase

Recommended name
cytochrome-b_5 reductase

Synonyms
B5R
NADH-cytochrome b_5 reductase
NADH-cytochrome-b_5 reductase
NADH-ferricytochrome b_5 oxidoreductase
P34/P32
P35
cytochrome b_5 reductase
dihydronicotinamide adenine dinucleotide-cytochrome b_5 reductase
reduced nicotinamide adeninedinucleotide-cytochrome b_5 reductase
reductase, cytochrome b_5

CAS registry number
9032-25-1

2 Source Organism

<1> *Homo sapiens* (deficiency leads to 2 different types of recessive congenital methemoglobinemia, in type 1 the soluble enzyme in erythrocytes is affected, in type 2 both soluble and membrane-bound isoforms are affected [46]; different cancer cell lines [52]) [2, 3, 6, 7, 10, 17, 20, 23, 27, 29, 32, 41, 42, 44, 45, 46, 52, 58]
<2> *Sus scrofa* [4, 11, 19, 39, 40, 49, 55, 61, 66]
<3> *Rattus norvegicus* (antibodies against the liver enzyme activate NADH 5-α reductase [18]; enzyme is expressed through the use of at least 4 different promotors [46]) [5, 14, 18, 24, 37, 39, 40, 46, 47, 48, 50, 51, 54, 57, 65, 67]
<4> *Phascolopsis gouldii* (sipuncilid worm [1]) [1, 9]
<5> *Pisum sativum* [12]
<6> *Solanum tuberosum* (potato [13, 43]) [13, 43]

<7> *Zea mays* [13]
<8> *Bos taurus* (calf [38, 39]) [8, 15, 16, 21, 22, 28, 38, 39, 59]
<9> *Tetrahymena pyriformis* [26]
<10> *Oryctolagus cuniculus* [25, 30, 31, 33, 34, 36, 39, 10, 47]
<11> *Saccharomyces cerevisiae* (grown anaerobically [35]) [35, 60]
<12> *Mus musculus* [53]
<13> *Musca domestica* (learn-pyrethroid-resistance strain, house fly [56]) [56]
<14> *Brassica napus* (turnip [59]) [59]
<15> *Ovis aries* (akkaraman sheep [62]) [62]
<16> *Zea mays* [63]
<17> *Zea mays* [64]

3 Reaction and Specificity

Catalyzed reaction

NADH + H^+ + 2 ferricytochrome b_5 = NAD^+ + 2 ferrocytochrome b_5 (<2> electron-transport chain from NADH to a terminal oxidase desaturase, proposed electron transfer mechanism [19]; <10> ordered bi bi mechanism [25]; <2, 3, 8, 10> hypothetical mechanism [39])

Reaction type

oxidation
redox reaction
reduction

Natural substrates and products

S 5α-dihydrotestosterone + acceptor <3> (Reversibility: r <3> [18]) [18, 18]
P ?
S NADH + ferricytochrome b_5 <1, 3, 6, 9, 11, 17> (<9> membrane bound form of somatic cells: essential for lipid metabolism [26]; <3, 6> involved in desaturation of fatty acids [40, 43]; <1> involved in metabolism of endogenous compounds such as steroids, drugs, carcinogens, environmental pollutants [41]; <1> soluble form of erythrocytes: reduction of methemoglobin [6, 42]; <1> enzyme participates in methemoglobin reduction in erythrocytes, in other tissues it plays a role in elongation and desaturation of fatty acids, P-450 mediated drug metabolism and cholesterol biosynthesis as part of the microsomal electron transfer system [45]; <11> enzyme complex drives the entire sterol 14-demethylation reaction [60]; <17> enzyme is assumed to be part of an endoplasmic reticulum associated redox chain that oxidizes NADH to provide electrons via cytochrome b_5 to endoplasmic reticulum associated fatty acyl desaturase and related hydroxylases as in mammals [64]) (Reversibility: ? <1, 3, 6, 9, 11, 17> [6, 26, 40-43, 60, 64]) [6, 26, 40-43, 45, 60, 64]
P NAD^+ + ferrocytochrome b_5 <1, 3, 6, 9, 11, 17> [6, 26, 40-43, 60, 64]
S deoxyhemerythrin + O_2 <4> (Reversibility: ? <4> [9]) [9, 9]
P ?

Substrates and products

S 1,2-dihydro-8-(4-methylpiperazin-1-yl)-4-phenylimidazol[3,2-e]pyrazine 5-oxide + NADH <12> (<12> potential bioreductive drug, trivial name RB90740 [53]) (Reversibility: ? <12> [53]) [53]

P 1,2-dihydro-8-(4-methylpiperazin-1-yl)-4-phenylimidazol[3,2-e]pyrazine + NAD^+ <12> [53]

S Cu^{2+}-citrate + NADH <16, 17> (<16> reduces also ferricyanide [63]; <16, 17> isoforms I and II, 59 and 47% of Fe^{3+}-citrate reduction respectively [63,64]) (Reversibility: ? <16, 17> [63, 64]) [63, 64]

P Cu^+-citrate + NAD^+ <16, 17> [63, 64]

S Fe^{3+}-ATP + NADH <3> (<3> ferric-EDTA is not reduced [47]; <3> reconstituted system containing NADH, cytochrome b_5 reductase, cytochrome b_5 and microsomal lipids catalyzes lipid peroxidation in the presence of ferric-ATP, ferric-histidine and ferric-ammonium sulfate [57]) (Reversibility: ? <3> [47]) [47, 57]

P Fe^{2+}-ATP + NAD^+ <3> [47, 57]

S Fe^{3+}-EDTA + NADH <16, 17> (<16> reduces also ferricyanide [63]; <16, 17> isoforms I and II [63, 64]) (Reversibility: ? <16, 17> [63, 64]) [63, 64]

P Fe^{2+}-EDTA + NAD^+ <16, 17> [63, 64]

S Fe^{3+}-ammonium sulfate + NADH <3> (<3> strongly elevated by the addition of cytochrome b_5 [47]) (Reversibility: ? <3> [47]) [47]

P Fe^{2+}-ammonium sulfate + NAD^+ <3> [47]

S Fe^{3+}-citrate + NADH <16, 17> (<16> reduces also ferricyanide [63]; <16, 17> isoforms I and II [63, 64]) (Reversibility: ? <16, 17> [63, 64]) [63, 64]

P Fe^{2+}-citrate + NAD^+ <16, 17> [63, 64]

S Fe^{3+}-histidine + NADH <3> (<3> strongly elevated by the addition of cytochrome b_5 [47]) (Reversibility: ? <3> [47]) [47]

P Fe^{2+}-histidine + NAD^+ <3> [47]

S Fe^{3+}-nitrilotriacetate + NADH <3> (<3> in the presence of cytochrome b_5, iron chelate reduction in descending order: Fe^{3+}-nitrolotriacetate, Fe^{3+}-ADP, Fe^{3+}-diphosphate, Fe^{3+}-citrate [50]) (Reversibility: ? <3> [50]) [50]

P Fe^{2+}-nitrilotriacetate + NAD^+ <3> [50]

S NADH + ferricytochrome b_5 <1, 3, 4, 8, 10, 11, 13, 14, 15> (<1, 3, 8, 10, 11> specific for NADH as electron donor, artificial acceptors: ferricyanide, 2,6-dichlorphenolindophenol [7, 21, 23, 25, 27, 32-37]; <4> additional acceptor: methemerythrin [9]; <1, 10> additional acceptor: methemoglobin-ferrocyanide complex [27, 34]; <10> additional acceptor: hemin [25]; <1> artificial acceptors: *p*-benzoquinone, 5-hydroxy-1,4-naphthoquinone, nitroblue-tetrazolium [23]; <1> additional electron donors: deamino-NADH, 3-acetylpyridine-NADH [23]; <8,10> poor donor: NADPH [21, 25]; <1> poor electron acceptors: methylene blue, ferricytochrome c, O_2, oxidized glutathione, methemoglobin [32]; <8> no acceptors: ubiquinone-30, menadione, dihydrofolate, lipoamide [21]; <10> low activity with trypsin-solubilized cytochrome b_5 [33]; <1> artificial electron acceptor in the presence of menadione:cytochrome c [52]; <3> recombinant enzyme, very low activity with NADPH [54]; <13> uses both NADH and NADPH as electron donors, artificial acceptors: ferricyanide, 2,6-dichlophenolindophenol [56]; <14> artificial acceptor: ferricyanide, high reactivity with NADPH as electron donor [59]; <15> artificial acceptor: ferricyanide [62]) (Reversibility: ? <1, 3, 4, 8, 10, 11, 13, 14, 15> [7, 9, 18, 21, 25, 27, 32, 34-37, 56, 59, 62]) [7, 9, 18, 21, 23, 25, 27, 32-37, 52, 54, 56, 59, 62]

P NAD^+ + H^+ + ferrocytochrome b_5 <1, 3, 4, 8, 10, 11> [7, 9, 18, 21, 25, 27, 32-37, 52]

S aquacobalamin + NADH <3> (<3> in the presence of outer membrane cytochrome b, no activity with cyanocobalamin [48]) (Reversibility: ? <3> [48]) [48]

P reduced aquacobalamin + NAD^+ <3> [48]

S cytochrome c + NADH <3, 16, 17> (<16> reduces also ferricyanide [63]; <16,17> isoforms I and II, 16% and 27% of Fe^{3+}-citrate reduction respectively [63, 64]; <3> cytochrome b_5/cytochrome b_5 reductase FAD-domain-fusion protein, NADPH is preferred [67]) (Reversibility: ? <16, 17> [63, 64]) [63, 64, 67]

P reduced cytochrome c + NAD^+ <3, 16, 17> [63, 64, 67]

S cytochrome c + NADH <3, 15> (<3> in the presence of outer membrane cytochrome b [48]; <15> increases in the presence of cytochrome b_5 [62]) (Reversibility: ? <3, 15> [48, 62]) [48, 62]

P reduced cytochrome c + NAD^+ <3> [48]

S methemerythrin + NADH <4> (Reversibility: ? <4> [1]) [1, 9]

P deoxymethemerythrin + NAD^+ <4> [1, 9]

S methemoglobin-ferrocyanide complex + NADH <10> (Reversibility: ? <10> [34]) [34]

P reduced methemoglobin-ferrocyanide complex + NAD^+ <10> [34]

Inhibitors

1,10-phenanthroline <11> [35]

2,2'-dipyridyl <11> [35]

5'-(*p*-fluorosulfonylbenzoyl)-adenosine <3> [18]

5-propyl-2-thiouracil <12> (<12> 25 mM, almost complete inhibition [53]) [53]

ADP <1> (<1> 5 mM, 66% inhibition [23]) [23, 27]

Br^- <8> (<8> competitive vs. cytochrome b_5, reversible by dilution [16]) [16]

$CaCl_2$ <8> (<8> 8 mM, 50% inhibition, competitive vs. cytochrome b_5 [8]) [8]

Cl^- <8> (<8> competitive vs. cytochrome b_5, reversible by dilution [16]) [8, 16, 21]

F^- <8> (<8> competitive vs. cytochrome b_5, reversible by dilution [16]) [16]

Hg^{2+} <11> [35]

I^- <8> (<8> competitive vs cytochrome b_5, reversible by dilution [16]) [16]

K^+ high ionic strength <1, 11> (<1> reduction of cytochrome b_5 or dichlorophenolindophenol [32]) [32, 35]

$MgCl_2$ <8> (<8> 78.1 mM, 50% inhibition, competitive vs. cytochrome b_5 [8]) [8]

N-ethylmaleimide <1, 8, 10> (<8> 10 mM, 89% inhibition [21]; <1> 8 mM, 70% inhibition [23]; <10> 1 mM, 90% inhibition [25]) [21, 23, 25, 27, 38]

NAD^+ <11> (<11> competitive [35]) [35]
$NADP^+$ <3> [18]
Tris <1> (<1> reduction of cytochrome b_5 [32]) [32]
acetate <8> [16]
acrynol <1, 8> (<8> 0.1 mM, 75% inhibition [21]; <1> 0.1 mM, 88% inhibition [29]) [21, 27, 29]
adenine nucleotides <1, 8> [21, 27]
amytal <3> [18]
atebrin <1, 8, 10> (<8> 1 mM, complete inhibition [21]; <10> 0.5 mM, complete inhibition [25]; <1> 0.1 mM, 44% inhibition [29]) [21, 25, 27, 29]
benzyl alcohol <1> (<1> 100 mM, 52% inhibition, reversible, may be due to changes in membrane fluidity [3]) [3]
citrate <8> [16]
dicoumarol <1, 3> (<1> 0.3 mM, 57% inhibition [23]) [18, 23]
hemin <1> [3]
inositol hexaphosphate <1> [3]
iodoacetamide <10> (<10> 1 mM, complete inhibition [25]) [25]
iodoacetic acid <8> (<8> 5 mM, complete inhibition [21]) [21]
mepacrin <3> [18]
p-chloromercuribenzoate <1, 8, 10, 11> (<8> 0.001 mM, complete inhibition [21]; <1> 0.005 mM, complete inhibition [23]; <10> 0.001 mM, complete inhibition [25]) [21, 23, 25, 27, 35]
p-hydroxymercuribenzoate <10, 12> (<10> 0.1 mM, complete inhibition [34]; <12> 0.1 mM, almost complete inhibition [53]) [34, 53]
pentachlorophenol <3> [18]
phosphate <8> (<8> competitive inhibition [21]) [21]
phytohemagglutinin <1> [3]
proflavin <1, 8, 10> (<8> 0.1 mM, 86% inhibition [21]; <10> 0.1 mM, 74% inhibition [25]; <1> 0.1 mM, 98% inhibition [29]) [21, 25, 29]
succinate <8> [16]
taurodeoxycholate <1, 8> (<1> 20 mM, 84% inhibition [23]) [21, 23]
thenoyltrifluoroacetone <3> [18]
wheat germ agglutinin <1> [3]
Additional information <4> (<4> not inhibited by NAD^+ and ferrocyanide [1]) [1]

Cofactors/prosthetic groups

FAD <1, 4, 10> (<1> 1 mol FAD per mol of recombinant enzyme [6]; <1, 10> 1 mol FAD per mol enzyme [10, 25]) [1, 6, 10, 25, 27, 29, 32]
NADH <1-11> [1-43]
NADPH <1, 3, 13, 14> (<1> 7-9% of NADH activity [17]; <3> 1/300 of NADH linked activity [37]) [17, 37, 56, 59]

Activating compounds

9-amino-1,2,3,4-tetrahydroacridine <1> (<1> activation [3]) [3]
detergents <1, 2> (<1, 2> e.g. Triton X-100, activation [4, 23]) [4, 23]
spermine <1> (<1> activation [3]) [3]

Metals, ions

Ca^{2+} <2> (<2> 2 mM, 23fold activation [4]) [4]
Mg^{2+} <2> (<2> half maximal activation at 1 mM [4]) [4]
Additional information <2, 5, 8, 10> (<2, 5, 10> rate of reduction depends on ionic strength [12, 19, 25]; <8> enzyme does not contain appreciable amounts of iron, copper, manganes, silver, mercury, lead, zinc, cobalt or nickel [39]) [12, 19, 25, 39]

Turnover number (min^{-1})

60 <3> (cytochrome b_5, <3> K110E mutant enzyme [65]) [65]
240 <3> (cytochrome b_5, <3> K110Q mutant enzyme [65]) [65]
498 <3> (cytochrome c, <3> cytochrome b_5/cytochrome b_5 reductase FAD-domain fusion protein [67]) [67]
1280 <1> (cytochrome b_5) [32]
1302 <3> (ferricyanide, <3> cytochrome b_5/cytochrome b_5 reductase FAD-domain fusion protein [67]) [67]
2400 <3> (ferricyanide, <3> K110Q mutant enzyme [65]) [65]
3010 <13> (2,6-dichlorophenolindophenol) [56]
3180 <2> (NADH, <2> G273 mutant enzyme [61]) [61]
3420 <2> (cytochrome b_5, <2> G273 mutant enzyme [61]) [61]
4560 <2> (ferricyanide, <2> G273 mutant enzyme [61]) [61]
4620 <2> (NADH, <2> H49E mutant enzyme [61]) [61]
5400 <3> (cytochrome b_5, <3> K110A mutant enzyme [65]) [65]
5900 <1> (ferricyanide, <1> enzyme from liver microsomes, protease solubilized [20]) [20]
6300 <2> (cytochrome b_5, <2> R63Q mutant enzyme [66]) [66]
6420 <2> (cytochrome b_5, <2> Y65A mutant enzyme [66]) [66]
6600 <3> (cytochrome b_5, <3> K110H mutant enzyme [65]) [65]
6600 <2> (ferricyanide, <2> H49E mutant enzyme [61]) [61]
7200 <3> (ferricyanide, <3> K110E mutant enzyme [65]) [65]
7260 <2> (cytochrome b_5, <2> R63A mutant enzyme [66]) [66]
7500 <1> (cytochrome b_5, <1> L41A mutant enzyme [58]) [58]
9600 <2> (cytochrome b_5, <2> H49E mutant enzyme [61]) [61]
9900 <2> (cytochrome b_5, <2> S99V mutant enzyme [66]) [66]
10800 <2> (NADH, <2> Y65A mutant enzyme [66]) [66]
11460 <2> (cytochrome b_5, <2> K97A mutant enzyme [66]) [66]
12000 <3> (cytochrome b_5, <3> K110R mutant enzyme [65]) [65]
16200 <3> (cytochrome b_5, <3> recombinant wild-type enzyme [65]) [65]
16200 <3> (ferricyanide, <3> K110A mutant enzyme [65]) [65]
17700 <2> (cytochrome b_5, <2> S99A mutant enzyme [66]) [66]
17800 <13> (NADPH) [56]
18300 <2> (cytochrome b_5, <2> Y65F mutant enzyme [66]) [66]
18400 <13> (ferricyanide) [56]
18600 <2> (NADH, <2> ΔF272 mutant enzyme [61]) [61]
19900 <13> (NADH) [56]
20400 <3> (ferricyanide, <3> K110H mutant enzyme [65]) [65]
20820 <2> (cytochrome b_5, <2> R63K mutant enzyme [66]) [66]

22000 <5> (ferricyanide) [12]
22500 <2> (NADH, <2> S99V mutant enzyme [66]) [66]
22800 <2> (cytochrome b_5, <2> H49A mutant enzyme [61]) [61]
23400 <2> (NADH, <2> native enzyme [61]) [61]
24000 <2> (ferricyanide, <2> ΔF272 mutant enzyme [61]) [61]
24180 <2> (NADH, <2> R63A mutant enzyme [66]) [66]
24720 <2> (NADH, <2> R63Q mutant enzyme [66]) [66]
24900 <2> (cytochrome b_5, <2> K97R mutant enzyme [66]) [66]
25200 <2> (cytochrome b_5, <2> H49K mutant enzyme [61]) [61]
26100 <2> (NADH, <2> S99A mutant enzyme [66]) [66]
26400 <2> (NADH, <2> H49A mutant enzyme [61]) [61]
27000 <2> (cytochrome b_5, <2> ΔF272 mutant enzyme [61]) [61]
28200 <2> (ferricyanide, <2> H49A mutant enzyme [61]) [61]
28200 <3> (ferricyanide, <3> K110R mutant enzyme [65]) [65]
28320 <1> (cytochrome b_5, <1> L125A mutant enzyme [58]) [58]
29400 <2> (ferricyanide, <2> H49K mutant enzyme [61]) [61]
30000 <8> (NADH) [38]
30000 <1> (ferricyanide) [17]
30060 <2> (NADH, <2> Y65F mutant enzyme [66]) [66]
30600 <2> (cytochrome b_5, <2> native enzyme [61]) [61]
30900 <2> (NADH, <2> R63K mutant enzyme [66]) [66]
31200 <2> (cytochrome b_5, <2> S99T mutant enzyme [66]) [66]
33600 <2> (cytochrome b_5, <2> recombinant wild-type enzyme [61]) [61]
33780 <1> (cytochrome b_5, <1> recombinant K110M mutant enzyme [45]) [45]
34200 <2> (NADH, <2> recombinant wild-type enzyme [61]) [61]
34800 <2> (cytochrome b_5, <2> H49Y mutant enzyme [61]) [61]
37800 <2> (ferricyanide, <2> native enzyme [61]) [61]
39660 <2> (cytochrome b_5, <2> recombinant wild-type enzyme [66]) [66]
40000 <10> (ferricyanide) [33]
40600 <1> (ferricyanide, <1> enzyme from erythrocyte membrane [20]) [20]
42000 <1> (ferricyanide, <1> enzyme from erythrocyte cytosol [20]) [20]
42600 <2> (NADH, <2> H49K mutant enzyme [61]) [61]
43600 <2> (ferricyanide, <2> H49Y mutant enzyme [61]) [61]
44520 <1> (cytochrome b_5, <1> recombinant K110A mutant enzyme [45]) [45]
45000 <2> (NADH, <2> H49Y mutant enzyme [61]) [61]
48000 <3> (ferricyanide, <3> recombinant wild-type enzyme [65]) [65]
48600 <2> (ferricyanide, <2> recombinant wild-type enzyme [61]) [61]
49600 <1> (ferricyanide, <1> enzyme from liver microsomes, detergent solubilized [20]) [20]
51060 <2> (NADH, <2> S99T mutant enzyme [66]) [66]
52320 <1> (cytochrome b_5, <1> recombinant wild-type enzyme [45]) [45]
52620 <1> (cytochrome b_5, <1> recombinant K110R mutant enzyme [45]) [45]
54540 <2> (NADH, <2> K97R mutant enzyme [66]) [66]
63600 <2> (NADH, <2> K97A mutant enzyme [66]) [66]
66000 <2> (NADH, <2> recombinant wild-type enzyme [66]) [66]

Specific activity (U/mg)

0.00072 <12> (<12> liver microsomes, reduction of 1,2-dihydro-8-(4-methylpiperazin-1-yl)-4-phenylimidazol[3,2-e]pyrazine 5-oxide [53]) [53]
0.0039 <2> (<2> activity in liver microsomes [4]) [4]
0.036-0.109 <1> (<1> activity in lysates of different cancer cell lines [52]) [52]
0.0648 <2> (<2> activity in liver microsomes in the presence of 2 mM $MgCl_2$ [4]) [4]
0.0888 <2> (<2> activity in liver microsomes in the presence of 2 mM $CaCl_2$ [4]) [4]
0.109 <3> (<3> reduction of aquacobalamin [48]) [48]
0.18 <5> [12]
0.372 <2> (<2> in the presence of 0.1% Triton X-100 [4]) [4]
1.69 <10> (<10> methemoglobin-ferrocyanide reductases assay [34]) [34]
2.042 <13> (<13> reduction of cytochrome b_5 [56]) [56]
3.5 <4> (<4> low MW aggregate, nitro blue tetrazolium as electron acceptor [1]) [1]
10.6 <4> (<4> low MW aggregate, cytochrome b as electron acceptor [1]) [1]
69.8 <11> [35]
96.3 <1> (<1> recombinant enzyme [6]) [6]
137 <8> [21]
138 <4> (<4> low MW aggregate, dichlorophenolindophenol as electron acceptor [1]) [1]
230.5 <4> (<4> high MW aggregate [1]) [1]
497 <10> [25]
611.1 <4> (<4> low MW aggregate, ferricyanide as electron acceptor [1]) [1]
622.5 <10> [33]
628 <6, 7> (<6,7> purified on Mono Q, ferricyanide reduction [13]) [13]
670 <15> [62]
785 <1> [17]
790 <1> [27]
1007 <13> (<13> reduction of ferricyanide [56]) [56]
1200-1300 <10> (<10> ferricyanide reduction [36]) [36]
1400 <3> (<3> recombinant enzyme [54]) [54]
1470 <3> (<3> enzyme from liver plasma membrane [51]) [51]
1510 <1> (<1> liver microsomal enzyme, ferricyanide as electron acceptor [7]) [7]
1575 <10> [47]
1679 <3> [47]
2480 <3> (<3> reduction of ferricyanide [37]) [37]
Additional information <1, 8, 14> (<1> 0.00017 mmol ferricyanide reduced/min/10000000 cells, activity in neutrophils [23]; <8> 102.0 units/mg, 1 unit is defined as the change of 1 absorbance unit at 600 nm/min [28]; <14> 41.25 units, 1 unit is defined as the amount of enzyme changing 0.001 of Δabsorbance/min [59]; <8> 28.6 units, 1 unit is defined as the amount of enzyme changing 0.001 of Δabsorbance/min [59]) [23, 59]

K_m-Value (mM)

0.000007 <5> (cytochrome b_5) [12]
0.000088 <15> (cytochrome b_5) [62]
0.00016 <1> (NADH, <1> enzyme from erythrocyte [32]) [32]
0.0003 <2> (NADH, <2> H49E mutant enzyme [61]) [61]
0.0004-0.0005 <2> (NADH, <2> lysosome- and detergent-solubilized enzyme [19]) [19]
0.0006 <1> (NADH, <1> enzyme from erythrocyte membrane [27]) [27]
0.0006 <1> (NADH, <1> recombinant wild-type enzyme [45]) [45]
0.0006 <2> (ferricyanide, <2> H49K mutant enzyme [61]) [61]
0.0006-0.0007 <1> (NADH, <1> enzyme from erythrocyte and liver [20]) [20]
0.00064 <2> (NADH, <2> G273 mutant enzyme [61]) [61]
0.0008 <1> (NADH) [6]
0.0008 <11> (cytochrome b_5, <11> yeast cytochrome b_5 [35]) [35]
0.0008 <11> (ferricytochrome b_5) [35]
0.00084 <2> (NADH, <2> ΔF272 mutant enzyme [61]) [61]
0.00091 <8> (NADH) [16]
0.001 <3> (NADPH, <3> cytochrome b_5/cytochrome b_5 reductase FAD-domain fusion protein [67]) [67]
0.0011 <8> (NADH) [21]
0.00114 <1> (cytochrome b_5) [6]
0.00148 <2> (cytochrome b_5, <2> S99A mutant enzyme [66]) [66]
0.0015 <2> (NADH, <2> recombinant wild-type enzyme [61]) [61]
0.0015 <11> (cytochrome b_5, <11> calf cytochrome b_5 [35]) [35]
0.0016 <2> (NADH, <2> native enzyme [61]) [61]
0.0019 <2> (NADH, <2> H49A mutant enzyme [61]) [61]
0.0019 <2> (NADH, <2> S99T mutant enzyme [66]) [66]
0.0021 <2> (NADH, <2> R63K mutant enzyme [66]) [66]
0.0022 <2> (ferricyanide, <2> H49E mutant enzyme [61]) [61]
0.0023 <1> (NADH, <1> recombinant K110R mutant enzyme [45]) [45]
0.0025 <10> (NADH) [25]
0.0025 <2> (cytochrome b_5, <2> G273 mutant enzyme [61]) [61]
0.0025 <2> (ferricyanide, <2> native enzyme [61]) [61]
0.0026 <2> (ferricyanide, <2> H49Y mutant enzyme [61]) [61]
0.0027 <2> (NADH, <2> H49Y mutant enzyme [61]) [61]
0.0028 <4> (cytochrome b_5) [1]
0.0028 <2> (ferricyanide, <2> H49A mutant enzyme [61]) [61]
0.0028 <2> (ferricyanide, <2> recombinant wild-type enzyme [61]) [61]
0.0029 <2> (NADH, <2> K97R mutant enzyme [66]) [66]
0.003 <2> (NADH, <2> Y65F mutant enzyme [66]) [66]
0.003 <3> (cytochrome b_5, <3> K110Q mutant enzyme [65]) [65]
0.0031 <2> (NADH, <2> recombinant wild-type enzyme [66]) [66]
0.0031 <2> (cytochrome b_5, <2> H49K mutant enzyme [61]) [61]
0.004 <3> (cytochrome b_5, <3> K110E mutant enzyme [65]) [65]
0.0042 <2> (NADH, <2> Y65A mutant enzyme [66]) [66]
0.0042 <8> (ferricyanide) [21]

0.0044 <2> (NADH, <2> K97A mutant enzyme [66]) [66]
0.005 <2> (cytochrome b_5, <2> Y65A mutant enzyme [66]) [66]
0.005 <3> (ferricyanide, <3> K110R mutant enzyme [65]) [65]
0.0052 <2> (ferricyanide, <2> ΔF272 mutant enzyme [61]) [61]
0.0052 <2> (ferricyanide, <2> G273 mutant enzyme [61]) [61]
0.0053 <2> (cytochrome b_5, <2> in the presence of 2 mM Ca^{2+} [4]) [4]
0.006 <3> (NADH, <3> recombinant enzyme [54]) [54]
0.006 <3> (ferricyanide, <3> recombinant wild-type enzyme [65]) [65]
0.0062 <2> (cytochrome b_5, <2> Y65F mutant enzyme [66]) [66]
0.0063 <11> (NADH) [35]
0.0066 <1> (cytochrome b_5, <1> recombinant wild-type enzyme [45]) [45]
0.0069 <2> (cytochrome b_5, <2> K97R mutant enzyme [66]) [66]
0.007 <3> (cytochrome c, <3> cytochrome b_5/cytochrome b_5 reductase FAD-domain fusion protein [67]) [67]
0.007 <3> (ferricyanide, <3> recombinant enzyme [54]) [54]
0.0073 <2> (cytochrome b_5, <2> R63K mutant enzyme [66]) [66]
0.008 <3> (ferricyanide, <3> cytochrome b_5/cytochrome b_5 reductase FAD-domain fusion protein [67]) [67]
0.008 <10> (methemoglobin-ferrocyanide complex) [34]
0.0088 <2> (cytochrome b_5, <2> native enzyme [61]) [61]
0.0089 <2> (cytochrome b_5, <2> H49A mutant enzyme [61]) [61]
0.009 <10> (cytochrome b_5) [25]
0.009 <3> (cytochrome b_5, <3> K110A mutant enzyme [65]) [65]
0.009 <3> (cytochrome b_5, <3> K110R mutant enzyme [65]) [65]
0.0091 <2> (cytochrome b_5, <2> recombinant wild-type enzyme [61]) [61]
0.0094 <2> (NADH, <2> S99V mutant enzyme [66]) [66]
0.0099 <2> (cytochrome b_5, <2> S99V mutant enzyme [66]) [66]
0.01 <2> (NADH, <2> H49K mutant enzyme [61]) [61]
0.01 <3> (cytochrome b_5, <3> recombinant wild-type enzyme [65]) [65]
0.01 <10> (ferricyanide) [34]
0.01 <10> (ferricyanide) [34]
0.0104 <2> (cytochrome b_5, <2> recombinant wild-type enzyme [66]) [66]
0.011 <2> (cytochrome b_5, <2> H49Y mutant enzyme [61]) [61]
0.011 <3> (cytochrome b_5, <3> recombinant enzyme [54]) [54]
0.0112 <2> (NADH, <2> in the presence of 2 mM Ca^{2+} [4]) [4]
0.0118 <2> (cytochrome b_5, <2> S99T mutant enzyme [66]) [66]
0.012 <3> (cytochrome b_5, <3> K110H mutant enzyme [65]) [65]
0.013 <13> (cytochrome b_5) [56]
0.0136 <2> (ferricyanide, <2> in the presence of 2 mM Ca^{2+} [4]) [4]
0.0142 <1> (cytochrome b_5, <1> recombinant K110A mutant enzyme [45]) [45]
0.0143 <8> (cytochrome b_5) [21]
0.0144 <3> (NADH) [48]
0.015 <10> (cytochrome b_5) [34]
0.0161 <2> (cytochrome b_5, <2> K97A mutant enzyme [66]) [66]
0.0167 <1> (cytochrome b_5, <1> recombinant K110R mutant enzyme [45]) [45]

0.017 <2> (cytochrome b_5, <2> ΔF272 mutant enzyme [61]) [61]
0.017 <13> (ferricyanide) [56]
0.0185 <2> (cytochrome b_5, <2> R63Q mutant enzyme [66]) [66]
0.0195 <2> (NADH) [4]
0.02 <1> (cytochrome b_5, <1> enzyme from erythrocyte membrane [27]) [27]
0.02 <1> (cytochrome b_5, <1> enzyme from eythrocyte membrane [20]) [20]
0.021 <2> (NADH, <2> R63Q mutant enzyme [66]) [66]
0.0231 <2> (ferricyanide) [4]
0.024 <15> (ferricyanide) [62]
0.025-0.089 <3> (testosterone, <3> depending on phosphate concentration [18]) [18]
0.026 <2> (NADH, <2> S99A mutant enzyme [66]) [66]
0.0285 <1> (cytochrome b_5, <1> recombinant K110M mutant enzyme [45]) [45]
0.03 <2> (cytochrome b_5, <2> H49E mutant enzyme [61]) [61]
0.03-0.04 <2> (cytochrome b_5, <2> lysosome- and detergent-solubilized enzyme [19]) [19]
0.0349 <2> (cytochrome b_5) [4]
0.035 <1> (cytochrome b_5, <1> enzyme from liver microsome membrane, solubilized with Triton X-100 [20]) [20]
0.0362 <1> (cytochrome b_5, <1> L125A mutant enzyme [58]) [58]
0.0384 <1> (ferricyanide) [23]
0.04 <1> (cytochrome b_5, <1> enzyme from eythrocyte cytosol [20]) [20]
0.0419 <3> (aquacobalamin) [48]
0.042 <1> (cytochrome b_5, <1> L41A mutant enzyme [58]) [58]
0.0421 <2> (cytochrome b_5, <2> R63A mutant enzyme [66]) [66]
0.044 <13> (NADH) [56]
0.045 <1> (cytochrome b_5, <1> enzyme from liver microsome membrane, cathepsin D solubilized [20]) [20]
0.0478 <1> (NADH) [23]
0.05 <3> (ferricyanide, <3> K110H mutant enzyme [65]) [65]
0.06 <4> (NADH) [1]
0.09 <4> (ferricyanide) [1]
0.104 <2> (NADH, <2> R63A mutant enzyme [66]) [66]
0.121 <1> (NADH, <1> recombinant K110A mutant enzyme [45]) [45]
0.37 <3> (ferricyanide, <3> K110E mutant enzyme [65]) [65]
0.38 <3> (ferricyanide, <3> K110A mutant enzyme [65]) [65]
0.672 <1> (NADH, <1> recombinant K110M mutant enzyme [45]) [45]
0.76 <3> (ferricyanide, <3> K110Q mutant enzyme [65]) [65]
0.83 <13> (2,6-dichlorophenolindophenol) [56]
2.17 <8> (NADH) [59]
2.34 <8> (NADPH) [59]
3.32 <14> (NADH) [59]
3.85 <8> (ferricyanide) [59]
4.35 <14> (ferricyanide) [59]
7.69 <14> (NADPH) [59]
14 <13> (NADPH) [56]

K_i-Value (mM)

0.08 <1> (ADP) [27]
0.23 <11> (NAD^+) [35]

pH-Optimum

5-5.7 <5> [12]
5.2 <1> (<1> NADH-methemoglobin-ferrocyanide [27]) [27]
5.5 <4> (<4> acceptor ferricyanide [1]) [1]
5.6 <11> (<11> steady decrease above [35]) [35]
5.7-8 <1> (<1> rapid decrease above [23]) [23]
6-8 <4> (<4> acceptor Phascolopsis gouldii cytochrome b_5 [1]) [1]
6.4 <10> (<10> citrate-phosphate buffer [25]) [25]
6.5 <10> (<10> Tris-phosphate buffer [34]) [34]
6.5-8.5 <1> (<1> NADH-ferricyanide [27]) [27]
6.6 <10> (<10> Tris-maleate buffer [25]) [25]
6.8 <1> (<1> reduction of cytochrome c [52]) [52]
6.8-7.5 <15> [62]
7 <10> (<10> triethanolamine buffer [34]) [34]
7-7.5 <8> [21]
8 <4> (<4> acceptor dichlorphenolindophenol [1]) [1]
Additional information <10> (<10> no distinct optimum with phosphate or Tris-HCl buffer [25]) [25]

pH-Range

5-8 <1> (<1> depending on acceptor [27]) [27]

4 Enzyme Structure

Molecular weight

25000 <10> (<10> liver, cytosol, gel filtration, sucrose density gradient centrifugation [34]) [34]
27000 <3, 11> (<11> gel filtration [35]; <3> sedimentation equilibrium [37]) [35, 37]
28000 <3> (<3> gel filtration [37]) [37]
30000 <3, 10> (<3> erythrocytes, cytosol, gel filtration [24]; <10> enzyme from liver cytosol, gel filtration [34]) [24, 34]
33000 <1, 10> (<1> recombinant enzyme, gel filtration [6]; <10> gel filtration [25]) [6, 25]
34700 <10> (<10> erythrocytes, cytosol, gel filtration, calculation from FAD content [25]) [25]
44000 <6, 7> (<6,7> gel filtration [13]) [13]
135000 <5> (<5> gel filtration [12]) [12]
200000 <4> (<4> oligomeric aggregate of detergent-solubilized enzyme in aqueous media, gel filtration [1]) [1]
365000 <10> (<10> oligomeric aggregates of detergent solubilized enzyme, sedimentation equilibrium [36]) [30, 36]

Subunits

? <1, 2, 4, 8, 9, 10, 13, 14, 15, 16> (<4> x * 34000, SDS-PAGE [1]; <2> x * 30000, lysosome-solubilized enzyme, SDS-PAGE [19]; <2> x * 35000, detergent-solubilized enzyme, SDS-PAGE [19]; <1> x * 36000, enzyme from erythrocyte membrane, SDS-PAGE [20]; <1> x * 32000, enzyme from liver microsomes solubilized with Triton X-100, SDS-PAGE [20]; <1> x * 32000, enzyme from erythrocyte cytosol and liver microsomal enzyme solubilized with cathepsin D, SDS-PAGE [20]; <8> x * 31000, SDS-PAGE [21]; <9> x * 22000, SDS-PAGE [26]; <10> x * 33000, SDS-PAGE, gel filtration in the presence of 6 M guanidine hydrochloride, gel filtration in the presence of 0.5% sodium deoxycholate [33]; <8> x * 38400, liver microsomes, sedimentation equilibrium [39]; <13> x * 33000, SDS-PAGE [56]; <13> x * 31000, may occur from protease cleavage of the 33000 Da protein during purification [56]; <8,14> x * 35000, SDS-PAGE [59]; <15> x * 34000, SDS-PAGE [62]; <16> x * 31170, deduced from nucleotide sequence [63]) [1, 19, 20, 21, 26, 33, 36, 39, 56, 59, 62, 63]

monomer <3, 10, 11> (<10> 1 * 33000, soluble enzyme from erythrocyte, SDS-PAGE [25]; <11> 1 * 28000, SDS-PAGE [35]; <3,10> 1 * 33100, SDS-PAGE [47]; <3> 1 * 32000, SDS-PAGE [51]; <3> 1 * 30000, recombinant enzyme, SDS-PAGE [54]; <3> 1 * 30588, recombinant enzyme, MALDI-TOF mass spectrometry [54]) [25, 35, 47, 51, 54]

tetramer <5> (<5> 4 * 34500, SDS-PAGE [12]) [12]

Additional information <8, 10> (<8> liver: membrane binding domain located at NH_2-terminal site of protein, 6400-6500 Da [22]; <10> liver: hydrophobic domain of approx. 30 amino acids located at COOH-terminal site of protein [31]) [22, 31]

Posttranslational modification

lipoprotein <1, 3, 8> (<1> myristic acid at NH_2-terminus, amide bond, no ester bond [6, 7]; <8> amphipathic membrane protein containing a hydrophilic, catalytic domain and a smaller hydrohobic membrane-binding domain [22]) [5, 6, 7, 22]

5 Isolation/Preparation/Mutation/Application

Source/tissue

brain <3, 8> (<3> transcripts of L, X and Y mRNAs are detected [46]) [21, 46]

erythrocyte <1, 3, 4, 10> (<1, 3, 4, 10> soluble form [1, 3, 6, 9, 10, 20, 24, 25, 27, 29, 32]; <3> transcript of R mRNA is detected [46]) [1, 3, 6, 9, 10, 20, 24, 25, 27, 29, 32, 46]

kidney <3> (<3> transcripts of L, X and Y mRNAs are detected [46]) [46]

liver <1-3, 8, 10, 12, 15> (<3> transcripts of L, X and Y mRNAs are detected [46]) [4-6, 11, 14-16, 18-20, 22, 28, 30, 31, 33, 34, 36-39, 46, 48, 50, 51, 53, 54, 57, 62]

lung <3> (<3> transcripts of L, X and Y mRNAs are detected [46]) [46]

neutrophil <1> [17, 23]
reticulocyte <3> (<3> transcripts of L, X, R and Y mRNAs are detected [46]) [46]
root <16, 17> [63, 64]
seedling <5, 7> [12, 13]
tuber <6> [13]

Localization

cytosol <1, 3, 10> (<1> may be a proteolytic product of the membrane enzyme [20]; <3> 2 forms in erythrocytes: a soluble and a membrane-bound enzyme which represents the majority [24]) [20, 24, 27, 32, 34]
endoplasmic reticulum <1-3, 5-11, 12, 13, 15> [8, 11-13, 14, 18-21, 26, 28, 30, 32, 33, 35-39, 50, 53, 55, 56, 57, 60, 62]
membrane <1, 3, 4, 8, 10, 14> (<1> cytoplasmic side of erythrocyte membrane [3]; <3> membrane of endoplasmic reticulum, outer mitochondrial membrane, tightly bound to cytoplasmic face of membrane [5]; <3> subcellular distribution, high concentration in outer mitochondrial membrane [14]; <8> membrane-binding domain resides in the NH_2-terminal region [21]; <10> microsomal enzyme, COOH-terminal domain constitutes the hydrophobic moiety which is responsible for membrane-binding [31]; <10> enzyme binds in vitro to a variety of membrane preparations, protease treatment removes a hydrophobic segment that is responsible for membrane binding [33]; <3> plasma membrane [51]) [1, 3, 5, 6, 9, 14, 17, 21, 27, 31, 33, 36, 51, 59]
mitochondrion <3, 14> (<3> outer membrane enzyme complex [48]) [48, 59]
tonoplast <16, 17> [63, 64]
Additional information <8> (<8> possible conformation of membrane binding domain [15]) [15]

Purification

<1> (recombinant enzyme [6]; recombinant wild-type, K110A, K110M and K110R mutant enzyme [45]; recombinant wild-type, K41A, K125A and K163A mutant enzyme [58]) [6, 7, 17, 27, 29, 32, 45, 58]
<2> (lysosome- and detergent-solubilization [19]; recombinant wild-type, H49A, H49E, H49K, H49Y, ΔF272 and ΔG273 mutant enzyme [61]; recombinant wild-type, K97A, K97R, S99A, S99T, S99V, R63A, R63Q, R63K, Y65A and Y65F mutant enzymes [66]) [19, 61, 66]
<3> (detergent-solubilization [33]; lysosome-solubilization [37]; polyethylene glycol precipitation, DEAE-cellulose, hydroxylapatite, ADP-agarose [47]; Triton X-100, anion exchange chromatography, gel filtration [51]; recombinant enzyme, affinity chromatography on 5'-ADP-agarose [54]; wild-type, K110R, K110H, K110A, K110E and K110Q mutant enzymes [65]) [33, 37, 47, 51, 54, 65]
<4> [1, 9]
<5> [12]
<8> (lysosome-solubilization [21]; separation of membrane binding and catalytic domain [22]; affinity chromatography on ADP-agarose [28]) [21, 22, 28]

<9> [26]
<10> (polyethylene glycol precipitation, ADP-agarose [47]) [25, 34, 36, 47]
<11> (recombinant enzyme [60]) [35]
<13> (Triton X-100, DEAE-cellulose, CM-cellulose, 5'-ADP-agarose [56]) [56]
<14> (DEAE-cellulose, DEAE-Sephadex, hydroxyapatite [59]) [59]
<15> (DEAE-cellulose, 5'-ADP-agarose affinity chromatography [62]) [62]
<6, 7> (DEAE-trisacryl, Blue ultrogel, Mono Q [13]) [13]

Renaturation
<5> (reconstitution [12]) [12]

Crystallization
<1> (vapor equilibrium method, 3.6% protein solution, 30% polyethyleneglycol 4000, preliminary X-ray data [10]) [10]
<2> (vapor diffusion method, 5 mg/ml protein, 12.5% polyethylenglycol 4000, 50 mM potassium phosphate pH 6.0-8.0, 0.1 mM EDTA, preliminary X-ray data [11]; X-ray structure, 2.4 A resolution [49]; x-ray structure, 2.1 A resolution, enzyme is composed of a FAD-binding domain and a NADH domain, electrostatic properties of NADH binding site [54]) [11, 49, 54]
<3> (sitting drop method, reservoir: 8% polyethylene glycol 6000, 5% 2-methyl-2,4-pentanediol in 100 mM 4-(2-hydroxyethyl)piperazine-1-ethanesulfonic acid, pH 7.5, X-ray structure, resolution: enzyme 2.0 A, enzyme-NAD^+ complex, 2.3 A [65]) [65]

Cloning
<1> (expression in Escherichia coli [2,6]; cloning of cDNA [44]; expression of wild-type, K110A, K110M and K110R mutant enzymes in Escherichia coli [45]; expression of wild-type, K41A and K125A mutant enzyme in Escherichia coli [58]) [2, 6, 44, 45, 58]
<2> (expression of wild-type, H49A, H49E, H49K, H49Y, ΔF272 and ΔG273 mutant enzymes in Escherichia coli [61]; expression of wild-type, K97A, K97R, S99A, S99T, S99V, R63A, R63Q, R63K, Y65A and Y65F mutant enzymes in Escherichia coli [66]) [61, 66]
<3> (cloning of cDNA [2]; expression in Escherichia coli [54]; expression of wild-type, K110R, K110H, K110A, K110E and K110Q mutant enzymes in Escherichia coli [65]; expression of a soluble fusion protein comprising a b-type cytochrome containing domain and a FAD-containing domain [67]) [2, 54, 65, 67]
<11> (heterologous expression in Saccharomyces cerevisiae [60]) [60]
<16> (cloning of cDNA, expression in Saccharomyces cerevisiae [63]) [63]
<17> (expression in Saccharomaces cerevisiae [64]) [64]

Engineering
H49A <2> (<2> reduced k_{cat} [61]) [61]
H49E <2> (<2> elevated K_m value for cytochrome b_5, strongly reduced k_{cat} [61]) [61]
H49K <2> (<2> reduced K_m value for cytochrome b_5 [61]) [61]
H49Y <2> (<2> similar to wild-type [61]) [61]
K110A <1> (<1> 200fold elevated K_m value for NADH, 85% of k_{cat} [45]) [45]

K110A <3> (<3> strongly reduced k_{cat} for ferricyanide and cytochrome b_5 [65]) [65]
K110E <3> (<3> strongly reduced k_{cat} for ferricyanide and cytochrome b_5 [65]) [65]
K110H <3> (<3> strongly reduced k_{cat} for ferricyanide and cytochrome b_5 [65]) [65]
K110M <1> (<1> 1120fold elevated K_m value for NADH [45]) [45]
K110Q <3> (<3> very low k_{cat} for ferricyanide and cytochrome b_5 [65]) [65]
K110R <1> (<1> similar kinetic properties as wild-type [45]) [45]
K110R <3> (<3> reduced k_{cat} for ferricyanid and cytochrome b_5 [65]) [65]
K125A <1> (<1> 5.3fold elevated K_m value for cytochrome b_5 [58]) [58]
K163A <1> (<1> 5.7fold elevated K_m value for cytochrome b_5 [58]) [58]
K41A <1> (<1> 6.3fold elevated K_m value for cytochrome b_5 [58]) [58]
K97A <2> (<2> mutation in flavin-binding motif [66]) [66]
K97R <2> (<2> mutation in flavin-binding motif [66]) [66]
R63A <2> (<2> mutation in flavin-binding motif [66]) [66]
R63K <2> (<2> mutation in flavin-binding motif [66]) [66]
R63Q <2> (<2> mutation in flavin-binding motif [66]) [66]
S99A <2> (<2> mutation in flavin-binding motif [66]) [66]
S99T <2> (<2> mutation in flavin-binding motif [66]) [66]
S99V <2> (<2> mutation in flavin-binding motif [66]) [66]
Y65A <2> (<2> mutation in flavin-binding motif [66]) [66]
Y65F <2> (<2> mutation in flavin-binding motif [66]) [66]

Application

medicine <1, 12> (<1> potentially an important enzyme required for the reductive activation of bioreductive drugs that can be used in the treatment of solid tumours [52]) [52, 53]

6 Stability

pH-Stability

7-9 <8> [38]
7.5-8.1 <1> [32]

Temperature stability

-12 <8> (<8> stable [38]) [38]
0-5 <8> (<8> several days [38]) [38]
21 <8> (<8> several h [38]) [38]

Oxidation stability

<4>, O_2, solubilized enzyme very sensitive to atmospheric oxygen [1]

General stability information

<1>, enzyme from erythrocyte loses its activity in the absence of EDTA after 5 d at 4°C [32]
<3>, inactivation caused by solubilization with detergents, activity restored by phosphatidylcholine [18]

<3>, phosphate stabilizes [18]
<4>, EDTA is essential for stability [1]
<4>, dithiothreitol is essential for stability [1]
<8>, inactivation during dialysis is prevented by NADH [21]
<10>, 0.2% phosphatidycholine liposomes or Triton X-100 stabilize [33]
<10>, Triton X-100 stabilizes [33]
<10>, freezing/thawing: 50% inactivation [33]
<10>, hydrophilic domain is unfolded at approx. 1 M guanidinium hydrochloride whereas much higher concentrations are required for denaturation of the hydrophobic domain [30]
<11>, unstable, 40% loss of activity at 0°C overnight, 1 mg/ml gelatin stabilizes, FAD protects against inactivation [35]
<6, 7>, application of HPLC during purification causes inactivation [13]

Storage stability

<1>, -20°C, pH 7.5-8.1, 0.5 mM EDTA, protein concentration above 0.02 mg/ml [32]
<3>, frozen, 50 mM potassium phosphate buffer, pH 7.5, 1 mM EDTA, several weeks [37]
<4>, 4°C, anaerobic conditions, EDTA, dithiothreitol, several months [9]
<4>, liquid N_2 [1]
<8>, 0°C, at least 1 month [16]
<8>, 0°C, at least 1 week [21]
<9>, -90°C, oxidized state, 1 month [26]
<10>, -20°C, 25 mM phosphate buffer, pH 7.6, 1 mM EDTA, 0.1 mM dithiothreitol, at least a few months [25]
<10>, -20°C, several months, no loss of activity [34]
<10>, -70°C, concentrated solution, several months, no loss of activity [33]
<10>, -70°C, several months, no loss of activity [33, 36]

References

[1] Bonomi, F.; Long, R.C.; Kurtz, D.M.: Purification and properties of a membrane-bound NADH-cytochrome-b_5 reductase from erythrocytes of the sipunculid worm, Phascolopsis gouldii. Biochim. Biophys. Acta, **999**, 147-156 (1989)

[2] Rigby, J.S.; Bull, P.C.; Ashworth, A.; Shephard, E.A.; Santisteban, I.; Phillips, I.R.: Isolation and characterization of genes coding for cytochrome b_5 and cytochrome-b_5 reductase. Biochem. Soc. Trans., **17**, 194-195 (1989)

[3] Palmieri, D.A.; Rangachari, A.; Butterfield, D.A.: Effects of domain-specific erythrocyte membrane modulators on acetylcholinesterase and NADH:cytochrome b_5 reductase activities. Arch. Biochem. Biophys., **280**, 224-228 (1990)

[4] Tamura, M.; Yoshida, S.; Tamura, T.; Saitoh, T.; Takeshita, M.: Effect of divalent cations on NADH-dependent and NADPH-dependent cytochrome b_5

reduction by hepatic microsomes. Arch. Biochem. Biophys., **280**, 313-319 (1990)
[5] Borgese, N.; Longhi, R.: Both the outer mitochondrial membrane and the microsomal forms of cytochrome b_5 reductase contain covalently bound myristic acid. Quantitative analysis on the polyvinylidene difluoride-immobilized proteins. Biochem. J., **266**, 341-347 (1990)
[6] Shirabe, K.; Yubisui, T.; Takeshita, M.: Expression of human erythrocyte NADH-cytochrome b_5 reductase as an α-thrombin-cleavable fused protein in Escherichia coli. Biochim. Biophys. Acta, **1008**, 189-192 (1989)
[7] Murakami, K.; Yubisui, T.; Takeshita, M.; Miyata, T.: The NH_2-terminal structures of human and rat liver microsomal NADH-cytochrome b_5 reductases. J. Biochem., **105**, 312-317 (1989)
[8] Tamura, M.; Yubisui, T.; Takeshita, M.: The opposite effect of bivalent cations on cytochrome b_5 reduction by NADH:cytochrome b_5 reductase and NADPH:cytochrome c reductase. Biochem. J., **251**, 711-715 (1988)
[9] Utecht, R.E.; Kurtz, D.M.: Cytochrome b_5 and NADH-cytochrome-b_5 reductase from sipunculan erythrocytes; a methemerythrin reduction system from Phascolopsis gouldii. Biochim. Biophys. Acta, **953**, 164-178 (1988)
[10] Takano, T.; Ogawa, K.; Sato, M.; Bando, S.; Yubisui, T.: Preliminary X-ray data of NADH-cytochrome b_5 reductase from human erythrocytes. J. Mol. Biol., **195**, 749-750 (1987)
[11] Miki, K.; Kaida, S.; Kasai, N.; Iyanagi, T.; Kobayashi, K.; Hayashi, K.: Crystallization and preliminary X-ray crystallographic study of NADH-cytochrome b_5 reductase from pig liver microsomes. J. Biol. Chem., **262**, 11801-11802 (1987)
[12] Jollie, D.R.; Sligar, S.G.; Schuler, M.: Purification and characterization of microsomal cytochrome b_5 and NADH cytochrome b_5 reductase from Pisum sativum. Plant Physiol., **85**, 457-462 (1987)
[13] Galle, A.M.; Kader, J.C.: High performance liquid chromatography of plant membrane proteins. NADH-cytochrome b_5 reductase as a model. J. Chromatogr., **366**, 422-426 (1986)
[14] Borgese, N.; Pietrini, G.: Distribution of the integral membrane protein NADH-cytochrome b_5 reductase in rat liver cells, studied with a quantitative radioimmunoblotting assay. Biochem. J., **239**, 393-403 (1986)
[15] Kensil, C.R.; Strittmatter, P.: Binding and fluorescence properties of the membrane domain of NADH-cytochrome-b_5 reductase. Determination of the depth of Trp-16 in the bilayer. J. Biol. Chem., **261**, 7316-7321 (1986)
[16] Tamura, M.; Yubisui, T.; Takeshita, M.: The inhibitory effect of halides and carboxylates on hepatic NADH: cytochrome b_5 oxidoreductase. Biochem. J., **230**, 273-276 (1985)
[17] Tauber, A.I.; Wright, J.; Higson, F.K.; Edelman, S.A.; Waxman, D.J.: Purification and characterization of the human neutrophil NADH-cytochrome b_5 reductase. Blood, **66**, 673-678 (1985)
[18] Golf, S.W.; Graf, V.; Rempeters, G.; Mersdorf, S.: Properties and biochemical characterization of NADH 5 α-reductase from rat liver microsomes. Biol. Chem. Hoppe-Seyler, **366**, 647-653 (1985)

[19] Iyanagi, T.; Watanabe, S.; Anan, K.F.: One-electron oxidation-reduction properties of hepatic NADH-cytochrome b_5 reductase. Biochemistry, **23**, 1418-1425 (1984)

[20] Kitajima, S.; Minakami, S.: Human NADH-cytochrome b_5 reductases: comparison among those of erythrocyte membrane, erythrocyte cytosol, and liver microsomes. J. Biochem., **93**, 615-620 (1983)

[21] Tamura, M.; Yubisui, T.; Takeshita, M.: Microsomal NADH-cytochrome b_5 reductase of bovine brain: purification and properties. J. Biochem., **94**, 1547-1555 (1983)

[22] Kensil, C.R.; Hediger, M.A.; Ozols, J.; Strittmatter, P.: Isolation and partial characterization of the NH_2-terminal membrane-binding domain of NADH-cytochrome b_5 reductase. J. Biol. Chem., **258**, 14656-14663 (1983)

[23] Badwey, J.A.; Tauber, A.I.; Karnovsky, M.L.: Properties of NADH-cytochrome-b_5 reductase from human neutrophils. Blood, **62**, 152-157 (1983)

[24] Borgese, N.; Macconi, D.; Parola, L.; Pietrini, G.: Rat erythrocyte NADH-cytochrome b_5 reductase. Quantitation and comparison between the membrane-bound and soluble forms using an antibody against the rat liver enzyme. J. Biol. Chem., **257**, 13854-13861 (1982)

[25] Yubisui, T.; Takeshita, M.: Purification and properties of soluble NADH-cytochrome b_5 reductase of rabbit erythrocytes. J. Biochem., **91**, 1467-1477 (1982)

[26] Fukushima, H.; Umeki, S.; Watanabe, T.; Nozawa, Y.: Purification and partial characterization of cytochrome b_5 from Tetrahymena pyriformis. Biochem. Biophys. Res. Commun., **105**, 502-508 (1982)

[27] Kitajima, S.; Yasukochi, Y.; Minakami, S.: Purification and properties of human erythrocyte membrane NADH-cytochrome b_5 reductase. Arch. Biochem. Biophys., **210**, 330-339 (1981)

[28] Schafer, D.A.; Hultquist, D.E.: Purification of bovine liver microsomal NADH-cytochrome b_5 reductase using affinity chromatography. Biochem. Biophys. Res. Commun., **95**, 381-387 (1980)

[29] Yubisui, T.; Takeshita, M.: Characterization of the purified NADH-cytochrome b_5 reductase of human erythrocytes as a FAD-containing enzyme. J. Biol. Chem., **255**, 2454-2456 (1980)

[30] Tajima, S.; Mihara, K.; Sato, R.: Two-domain structure of microsomal reduced nicotinamide adenine dinucleotide-cytochrome b_5 reductase. Arch. Biochem. Biophys., **198**, 137-144 (1979)

[31] Mihara, K.; Sato, R.; Sakakibara, R.; Wada, H.: Reduced nicotinamide adenine dinucleotide-cytochrome b_5 reductase: location of the hydrophobic, membrane-binding region at the carboxyl-terminal end and the masked amino terminus. Biochemistry, **17**, 2829-2834 (1978)

[32] Hultquist, D.E.: Methemoglobin reduction system of erythrocytes. Methods Enzymol., **52**, 463-473 (1978)

[33] Mihara, K.; Sato, R.: Detergent-solubilized NADH-cytochrome b_5 reductase. Methods Enzymol., **52**, 102-108 (1978)

[34] Lostanlen, D.; De Barro, A.V.; Leroux, A.; Kaplan, J. C.: Soluble NADH-cytochrome b_5 reductase from rabbit liver cytosol: partial purification and characterization. Biochim. Biophys. Acta, **526**, 42-51 (1978)

[35] Kubota, S.; Yoshida, Y.; Kumaoka, H.: Studies on the microsomal electron-transport system of anaerobically grown yeast. IV. Purification and characterization of NADH-cytochrome b_5 reductase. J. Biochem., **81**, 187-195 (1977)
[36] Mihara, K.; Sato, R.: Purification and properties of the intact form of NADH-cytochrome b_5 reductase from rabbit liver microsomes. J. Biochem., **78**, 1057-1073 (1975)
[37] Takesue, S.; Omura, T.: Purification and properties of NADH-cytochrome b_5 reductase solubilized by lysosomes from rat liver microsomes. J. Biochem., **67**, 267-276 (1970)
[38] Strittmatter, P.: NADH-cytochrome b_5 reductase. Methods Enzymol., **10**, 561-565 (1967)
[39] Strittmatter, P.: Microsomal cytochrome b_5 and cytochrome b_5 reductase. The Enzymes, 2nd Ed.(Boyer, P.D., Lardy, H., Myrbäck, K., eds.), **8**, 113-145 (1963)
[40] Oshino, N.; Sato, R.: Stimulation by phenols of the reoxidation microsomal bound cytochrome b_5 and its implication to fatty acid desaturation. J. Biochem., **69**, 169-180 (1971)
[41] Ortiz de Montellano, P.R.: Cytochrome P_{450}-structure, mechanism and biochemistry. Cytochrome P_{450}. Structure, Mechanism Biochem. (Ortiz de Montellano, P.R. ed.) Plenum Press, New York, (1986)
[42] Hultquist, D.E.; Passon, P.G.: Catalysis of methaemoglobin reduction by erythrocyte cytochrome b_5 and cytochrome b_5 reductase. Nature, **229**, 252-254 (1971)
[43] Kader, J.C.: Cyanide sensitivity and induction of the microsomal oleoyl-CoA desaturase of potato tuber. Biochim. Biophys. Acta, **486**, 429-436 (1977)
[44] Yubisui, T.; Naitoh, Y.; Zenno, S.; Tamura, M.; Takeshita, M.; Sakaki, Y.: Molecular cloning of cDNAs of human liver and placenta NADH-cytochrome b_5 reductase. Proc. Natl. Acad. Sci. USA, **84**, 3609-3613 (1987)
[45] Fujimoto, Y.; Shirabe, K.; Nagai, T.; Yubisui, T.; Takeshita, M.: Role of Lys-110 of human NADH-cytochrome b_5 reductase in NADH binding as probed by site-directed mutagenesis. FEBS Lett., **322**, 30-32. (1993)
[46] Mota Vieira, L.; Kaplan, J.C.; Kahn, A.; Leroux, A.: Heterogeneity of the rat NADH-cytochrome-b_5-reductase transcripts resulting from multiple alternative first exons. Eur. J. Biochem., **220**, 729-737 (1994)
[47] Yang, M.X.; Cederbaum, A.I.: Fractionation of liver microsomes with polyethylene glycol and purification of NADH-cytochrome b_5 oxidoreductase and cytochrome b_5. Arch. Biochem. Biophys., **315**, 438-444 (1994)
[48] Saido, H.; Watanabe, F.; Tamura, Y.; Miyatake, K.; Ito, A.; Yubisui, T.; Nakano, Y.: Cytochrome b_5-like hemoprotein/cytochrome b_5 reductase complex in rat liver mitochondria has NADH-linked aquacobalamin reductase activity. J. Nutr., **124**, 1037-1040 (1994)
[49] Nishida, H.; Inaka, K.; Yamanaka, M.; Kaida, S.; Kobayashi, K.; Miki, K.: Crystal structure of NADH-cytochrome b_5 reductase from pig liver at 2.4 A resolution. Biochemistry, **34**, 2763-2767 (1995)

[50] Miura, A.; Tampo, Y.; Yonaha, M.: The reducing ability of iron chelates by NADH-cytochrome b_5 reductase or cytochrome b_5 responsible for NADH-supported lipid peroxidation. Biochem. Mol. Biol. Int., **37**, 141-150 (1995)
[51] Kim, C.; Crane, F.L.; Becker, G.W.; Morre, D.J.: Purification of NADH-cytochrome b_5 reductase from rat liver plasma membranes. Protoplasma, **184**, 111-117 (1995)
[52] Barham, H.M.; Inglis, R.; Chinje, E.C.; Stratford, I.J.: Development and validation of a spectrophotometric assay for measuring the activity of NADH: cytochrome b_5 reductase in human tumour cells. Br. J. Cancer, **74**, 1188-1193 (1996)
[53] Barham, H.M.; Stratford, I.J.: Enzymology of the reduction of the novel fused pyrazine mono-N-oxide bioreductive drug, RB90740 roles for P450 reductase and cytochrome b_5 reductase. Biochem. Pharmacol., **51**, 829-837 (1996)
[54] Barber, M.J.; Quinn, G.B.: High-level expression in Escherichia coli of the soluble, catalytic domain of rat hepatic cytochrome b_5 reductase. Protein Expr. Purif., **8**, 41-47 (1996)
[55] Nishida, H.; Miki, K.: Electrostatic properties deduced from refined structures of NADH-cytochrome b_5 reductase and the other flavin-dependent reductases: pyridine nucleotide-binding and interaction with an electron-transfer partner. Proteins Struct. Funct. Genet., **26**, 32-41 (1996)
[56] Zhang, M.; Scott, J.G.: Purification and characterization of cytochrome b_5 reductase from the house fly, Musca domestica. Comp. Biochem. Physiol. B, **113**, 175-183 (1996)
[57] Yang, M.X.; Cederbaum, A.I.: Interaction of ferric complexes with NADH-cytochrome b_5 reductase and cytochrome b_5: lipid peroxidation, H_2O_2 generation, and ferric reduction. Arch. Biochem. Biophys., **331**, 69-78 (1996)
[58] Kawano, M.; Shirabe, K.; Nagai, T.; Takeshita, M.: Role of carboxyl residues surrounding heme of human cytochrome b_5 in the electrostatic interaction with NADH-cytochrome b_5 reductase. Biochem. Biophys. Res. Commun., **245**, 666-669 (1998)
[59] Lee, J.Y.; Kim, Y.H.; Lee, S.J.: Purification and comparison of NADH-cytochrome b_5 reductase from mitochondrial outer membrane of bovine heart and turnip. Bull. Chem. Soc. Kor., **19**, 160-164 (1998)
[60] Lamb, D.C.; Kelly, D.E.; Manning, N.J.; Kaderbhai, M.A.; Kelly, S.L.: Biodiversity of the P450 catalytic cycle: yeast cytochrome b_5/NADH cytochrome b_5 reductase complex efficiently drives the entire sterol 14-demethylation (CYP51) reaction. FEBS Lett., **462**, 283-288 (1999)
[61] Kimura, S.; Emi, Y.; Ikushiro, S.; Iyanagi, T.: Systematic mutations of highly conserved His49 and carboxyl-terminal of recombinant porcine liver NADH-cytochrome b_5 reductase solubilized domain. Biochim. Biophys. Acta, **1430**, 290-301 (1999)
[62] Arinc, E.; Cakir, D.: Simultaneous purification and characterization of cytochrome b_5 reductase and cytochrome b_5 from sheep liver. Int. J. Biochem. Cell Biol., **31**, 345-362 (1999)

[63] Bagnaresi, P.; Thoiron, S.; Mansion, M.; Rossignol, M.; Pupillo, P.; Briat, J.F.: Cloning and characterization of a maize cytochrome-b_5 reductase with Fe^{3+}-chelate reduction capability. Biochem. J., **338**, 499-505 (1999)
[64] Bagnaresi, P.; Mazars-Marty, D.; Pupilld, P.; Marty, F.; Briat, J.F.: Tonoplast subcellular localization of maize cytochrome b_5 reductases. Plant J., **24**, 645-654 (2000)
[65] Bewley, M.C.; Marohnic, C.C.; Barber, M.J.: The structure and biochemistry of NADH-dependent cytochrome b_5 reductase are now consistent. Biochemistry, **40**, 13574-13582 (2001)
[66] Kimura, S.; Nishida, H.; Iyanagi, T.: Effects of flavin-binding motif amino acid mutations in the NADH-cytochrome b_5 reductase catalytic domain on protein stability and catalysis. J. Biochem., **130**, 481-490 (2001)
[67] Davis, C.A.; Dhawan, I.K.; Johnson, M.K.; Barber, M.J.: Heterologous expression of an endogenous rat cytochrome b_5/cytochrome b_5 reductase fusion protein: Identification of histidines 62 and 85 as the heme axial ligands. Arch. Biochem. Biophys., **400**, 63-75 (2002)

Cytochrome reductase (NADPH) 1.6.2.3

1 Nomenclature

EC number
1.6.2.3 (deleted)

Recommended name
cytochrome reductase (NADPH)

1 Nomenclature

EC number
1.6.2.4

Systematic name
NADPH:ferrihemoprotein oxidoreductase

Recommended name
NADPH-ferrihemoprotein reductase

Synonyms
CPR
EC 1.6.99.2 (formerly)
FAD-cytochrome c reductase
NADP-cytochrome c reductase
NADP-cytochrome reductase
NADPH-cytochrome P_{450} (CYP) oxidoreductase
NADPH-cytochrome c oxidoreductase
NADPH-cytochrome c reductase
NADPH-cytochrome *p*-450 reductase
NADPH-dependent cytochrome c reductase
NADPH-ferricytochrome c oxidoreductase
P450R
TPNH-cytochrome c reductase
$TPNH_2$ cytochrome c reductase
aldehyde reductase (NADPH-dependent)
cytochrome c reductase (reduced nicotinamide adenine dinucleotide phosphate, NADPH, NADPH-dependent)
dihydroxynicotinamide adenine dinucleotide phosphate-cytochrome c reductase
ferrihemprotein P_{450} reductase
reduced nicotinamide adenine dinucleotide phosphate-cytochrome c reductase
reductase, cytochrome c (reduced nicotinamide adenine dinucleotide phosphate)

CAS registry number
9023-03-4

2 Source Organism

<1> *Sus scrofa* [1, 6, 11, 15, 29, 31, 32, 35, 38, 47, 48, 59, 65]
<2> *Rattus norvegicus* (Sprague-Dawley strain [2, 5, 16, 21, 33, 34, 41, 62, 63, 72]; Wistar strain [10, 36, 61, 68]; albino animals [27]) [2, 5, 6, 8, 10, 16, 21, 27, 29, 31, 33, 34, 36, 41, 43, 46, 51, 58, 61, 62-64, 68, 72, 75, 76]
<3> *Oryctolagus cuniculus* [3, 4, 30, 39]
<4> *Homo sapiens* [45, 70, 73, 74]
<5> *Helianthus tuberosus* (L. var. Blanc commun., Jerusalem artichoke [6, 9]) [6, 9]
<6> *Nitrobacter winogradskyi* [7]
<7> *Mesocricetus auratus* [8]
<8> *Trypanosoma cruzi* [12]
<9> *Trichosporon cutaneum* [13]
<10> *Lodderomyces elongisporus* [17]
<11> *Aspergillus ochraceus* [18]
<12> *Saccharomyces cerevisiae* (grown anaerobically [28]; grown anaerobically, wild type strain [37, 56]; JL20 strain [74]) [6, 19, 23, 28, 37, 42, 56, 74]
<13> *Candida tropicalis* (grown on alkanes or glycerol [20, 22, 26]) [20, 22, 26]
<14> *Spodoptera eridania* (southern Armyworm [21]) [21]
<15> *Catharanthus roseus* [24]
<16> *Tetrahymena pyriformis* [40]
<17> *plants* (e.g. maize, potato, avocado, bramble, tulip, leek, Vicia faba, sunflower, bramble, Gingko biloba [6]) [6]
<18> *Musca domestica* [44]
<19> *Equus caballus* [5]
<20> *Gallus gallus* [25]
<21> *Locusta migratoria migratorioides* [6, 57]
<22> *Cavia porcellus* [14, 60]
<23> *Brassica rapa* [50]
<24> *Phaseolus aureus* [52]
<25> *Dictyostelium discoideum* (axenic mutant, strain Ax-2 [53, 54]) [53, 54]
<26> *Mus musculus* (NMRI/Kisslegg strain [49, 55]) [49, 55]
<27> *Fusarium oxysporum* [66]
<28> *Sparus aurata* [67]
<29> *Delphinapterus leucas* [69]
<30> *Spodoptera frugiperda* [70]
<31> *Trichoplusia ni* [70]
<32> *Bacillus megaterium* [71]
<33> *Petunia hybrida* [77]
<34> *Tulipa fosteriana* [78]

3 Reaction and Specificity

Catalyzed reaction

NADPH + H^+ + 2 ferricytochrome = $NADP^+$ + 2 ferrocytochrome

Reaction type

oxidation
redox reaction
reduction

Natural substrates and products

S NADPH + cytochrome P_{450} <1, 5, 13, 17, 21, 29, 32> (<5, 13>, part of the microsomal electron-transport chains and responsible for the transfer of reducing equivalents between NADPH or NADH and cytochrome P-450 [9, 20]; <13>, component of the hydroxylation of lauric acid [22]; <13>, during hydroxylation of alkanes [26]; <21>, during ecdysone hydroxylation [57]; <32>, during hydroxylation of fatty acids through the bacterial fusion protein P450BM3 [71]) (Reversibility: ? <1, 5, 13, 17, 21, 29, 32> [6, 9, 20, 22, 26, 57, 69, 71]) [6, 9, 20, 22, 26, 57, 69, 71]

P reduced cytochrome P_{450} + $NADP^+$

S Additional information <1, 2, 3, 10, 12, 13, 21, 26, 27-29, 32> (<2, 13>, monooxygenase system composed of cytochrome P_{450}, NADPH-cytochrome c reductase, phospholipids [16, 26]; <13>, monooxygenase system only present in cells grown on alkanes [26]; <2>, detoxification of drugs, inactivation of procarcinogens [16]; <3>, mixed function oxidase system composed of cytochrome P_{450}, NADPH-cytochrome c reductase and phospholipids, biotransformation of airborne compounds [30]; <10>, acts as an electron transfer component of the alkane monoxygenase system [17]; <12> microsomal electron transport system composed of cytochrome b_5, cytochrome P_{450}, NADH-cytochrome b_5 reductase and NADPH-cytochrome c reductase [28, 37, 56]; <26>, component of the microsomal electron transport system [49]; <21>, monooxygenase system composed of ecdysone 20-monooxygenase, cytochrome P_{450} and NADPH-cytochrome c reductase [57]; <1, 2>, microsomal heme oxygenase system composed of heme oxygenase and NADPH-cytochrome c reductase catalyzes the oxidative degradation of heme to biliverdin, essential role in the physiological heme catabolism [59, 64]; <27>, electron transport system composed of cytochrome P_{450}, cytochrome b_5-like hemeproteins, NADPH cytochrome c reductase, and NADH-ferricyanide reductase [66]; <28>, mixed-function oxidase system composed of cytochrome P_{450}, NADPH cytochrome P_{450} reductase and lipid [67]; <29>, monooxygenase system composed of cytochrome P_{450}, cytochrome b_5 and NADPH cytochrome c P450 reductase [69]; <32>, part of the bacterial fusion protein P450BM3 composed of cytochrome P_{450} fatty acid hydroxylase and NADPH cytochrome P_{450} reductase [71]; <12>, involved in the sterol biosynthesis pathways [74]) [16, 17, 26, 28, 30, 37, 49, 56, 57, 59, 64, 66, 67, 69, 71, 74]

P ?

Substrates and products

S NADPH + O_2 <22, 12> (<22, 12>, slow reaction, presence of menadione, or duroquinone, or vitamin K_3 essential [14, 28]) (Reversibility: ? <22, 12> [14, 28]) [14, 28]

P $NADP^+$ + O_2^- <12, 22> (<22, 12>, superoxide anion [14, 28]; <22>, replacement of NADPH by NADH lowers the O_2^- generation [14]) [14, 28]

S NADPH + ferricytochrome c <1-8, 10, 12-15, 19, 20-26, 28, 30-34> (<1>, NADH less than 5% of NADPH activity [1]; <3>, additional electron donor: azidonitrophenyl-γ-aminobutyryl-NADPH [4]; <1, 2, 5, 6, 8, 10, 12-15, 22, 32>, additional electron acceptor: 2,6-dichlorophenolindophenol [6, 7, 12, 14, 17, 20, 21, 23, 24, 26, 28, 36, 37, 41, 65, 71]; <1, 2, 3, 5, 7, 12, 13, 15, 21, 32>, additional electron acceptor: cytochrome P-450 [1, 8, 9, 20, 22, 24, 31, 34, 36-39, 56-58, 68, 71, 72]; <2, 5, 8, 10, 12-15, 32>, additional electron acceptor: ferricyanide [6, 9, 12, 17, 20, 21, 24, 26, 28, 37, 41, 61, 71, 75]; <6, 8, 12, 13, 22>, additional electron acceptor: menadione [7, 12, 14, 20, 26, 28]; <6>, additional electron acceptor: duroquinone [7]; <1, 2, 13, 14, 33>, additional electron acceptor: neotetrazolium chloride [20, 21, 26, 32, 36, 38, 77]; <1, 22>, additional electron acceptor: nitroblue tetrazolium salt [14, 65]; <2>, additional electron acceptor: vitamin K_3 [36]; <2>, additional electron acceptor: benzoquinone [36]; <3, 6>, additional electron acceptor: benzylviologen [3, 7]; <2>, additional electron acceptor: cytochrome b_5 [41]; <2>, additional electron acceptor: formylated cytochrome c [51]) (Reversibility: ? <1-8, 10, 12-15, 19, 20-26, 28, 30-34> [1-12, 14, 15, 17, 19, 20, 21, 22-28, 30-34, 36-39, 41, 48-58, 60-62, 64, 65, 67, 68, 70-73, 75-78]) [1-12, 14, 15, 17, 19, 20, 21, 22, 24-28, 30-34, 36-39, 41, 48-58, 60-62, 64, 65, 67, 68, 70-73, 75-78]

P $NADP^+$ + ferrocytochrome c <1-8, 10, 12-15, 19, 20-26, 28, 30-34> [1-12, 14, 15, 17, 19, 20, 21, 22-28, 30-34, 36-39, 41, 48-58, 60-62, 64, 65, 67, 68, 70-73, 75-78]

S NADPH + hexadecanal <2> (<2>, brain, hexadecanal replaceable by *p*-nitroacetophenone, or *p*-pyridinecarboxaldehyde, benzalacetone or *p*-nitrobenzaldehyde [10]) (Reversibility: ? <2> [10]) [10]

P $NADP^+$ + hexadecanol <2> [10]

S NADPH + octanal <2> (Reversibility: ? <2> [10]) [10]

P $NADP^+$ + octanol <2> [10]

S Additional information <1-5, 21, 22, 32, 34> (<5>, hydroxylation of coumarate [9]; <3>, O-deethylation of 7-ethoxycoumarin, [30]; <1, 2, 3, 22>, N-demethylation of benzphetamine [29, 30, 33, 60]; <2>, N-demethylation of aminopyrine [33]; <2>, aniline hydroxylase [33]; <2>, as part of the microsomal ethanol-oxidizing system composed of NADPH-cytochrome c reductase, cytochrome P-450, phospholipids [33]; <1>, ω-hydroxylation of fatty acids together with cytochrome P-450 [38]; <2>, component of the electron transport chain required for activity of the 17,20-lyase in testis microsomes, removal of 2-carbon side chain from 17-position of 21-carbon steroids [34]; <2>, involved in the reduction of testosterone [36]; <21>, hydroxylation of ecdysone [57]; <2>, hydroxylation of benzo-a-pyrene, testosterone and progesterone [58]; <2>, O-dealkylation of pentoxyresorufin [68]; <4>, coexpressed with CYP3A4 in insect cell lines, testosterone 6-β-hydroxylase activity [70]; <32>, simultaneous catalysis of reduction of cytochrome c and hydroxylation of laurate [71]; <2>, denitration of glyceryl trinitrate [72]; <4>, coexpressed with two differently modified CYP2D6 cDNAs in Escherichia coli, functional recombinant monooxygenase, bufuralol and metoprolol metabolism [73]; <34>, oxidation of N-nitrosodimethylamine, N-nitroso-N-methylaniline, aminopyrine and 1-phenylazo-2-naphtol [78]) (Reversibility: ? <1-5, 21, 22, 32, 34> [9, 29, 30, 33, 34, 36, 38, 57, 58, 60, 68, 70-73, 78]) [9, 29, 30, 33, 34, 36, 38, 57, 58, 60, 68, 70-73, 78]

P ?

Inhibitors

2'-AMP <14> (<14>, competitive inhibition [21]) [21]
2,6-dichlorophenolindophenol <22> (<22>, formation of superoxide anion, in the presence of menadione [14]) [14]
3-((3-cholamidopropyl)dimethylammonio)propanesulfonic acid <7> [8]
3-aminonicotinamide adenine dinucleotide phosphate <5> [6]
5,5'-dithiobis(2-nitrobenzoate) <12> (<12>, in absence of FAD or NADPH [23]) [23]
EDTA <22> (<22>, reduction of cytochrome c and formation of superoxide anion [14]) [14]
$HgCl_2$ <12, 13> (<13>, cells grown on alkane or glycerol, presence of NADPH provides protection against inhibition [26]; <12>, incubation with NADPH provides protection against inhibition [28]) [20, 26, 28]
Lubrol WX <1> [48]
N-ethylmaleimide <22> (<22>, reduction of cytochrome c and formation of superoxide anion [14]) [14]
NAD^+ <6> (<6>, cosubstrate NADH [7]) [7]
$NADP^+$ <1, 5, 6, 13, 14, 15, 21> (<6>, cosubstrate NADPH [7]; <5>, noncompetitive inhibition [9]; <1, 14, 15>, competitive inhibition [21, 24, 32]) [6, 7, 9, 21, 24, 26, 32, 57]
Triton N-101 <7> [8]
alizarin <14> [21]
cetyltrimethyl ammonium bromide <15> [24]
diphenyleneiodonium sulfate <2> (<2>, time-dependent and concentration-dependent inhibition of cDNA-expressed liver enzyme as well as aortic and hepatic microsomal enzyme activity [72]) [72]
mersalyl <3, 14> [3, 21]
p-chloromercuribenzoate <12-13, 15> (<13>, cells grown on alkane or glycerol, presence of NADPH provides protection against inhibition [26]; <12>, incubation with NADPH provides protection against inhibition [28]) [20, 24, 26, 28]
p-hydroxymercuribenzoate <14> (<14>, presence of mercaptoethanol or NADPH provides slight protection against inhibition [21]) [21]
quercetin <22> (<22>, reduction of cytochrome c and formation of superoxide anion [14]) [14]

steapsin <2> (<2>, brain [10]) [10]
Additional information <5, 12, 13> (<5>, high ionic strength [9]; <13>, increasing ionic strength, grown on alkane or glycerol [26]) [9, 26, 28]

Cofactors/prosthetic groups

FAD <1-3, 5-8, 10, 12-15, 20, 32, 33> (<1, 2, 3, 5, 7, 10, 12-14>, ratio FAD : FMN 1 : 1 [1, 8, 9, 11, 17, 20, 21, 25, 26, 28, 29, 31, 32, 35, 39]; <6, 8, 12>, 1 mol per subunit [7, 12, 23]; <12, 13>, tightly bound [20, 28]; <12>, loosely bound, [42]) [1-3, 7-9, 11, 12, 14, 17, 20, 21, 23-26, 28, 31-33, 35, 39, 41, 42, 61, 71, 72, 75-77]
FMN <1-3, 5-7, 10, 12-15, 20, 32, 33> (<1, 2, 3, 5, 7, 10, 12-14>, ratio FAD : FMN 1 : 1 [1, 8, 9, 11, 17, 20, 21, 25, 26, 28, 29, 31, 32, 35, 39]; <13>, tightly bound [20]; <12>, loosely bound [28, 42]; <10,13>, 1 mol per mol of enzyme [17, 20]; <6>, 0.87 nmol per mg protein [7]; <5>, not implicated in plant NADPH-ferricyanide reductase [9]; <2>, riboflavin-deficient microsomes: 40-66% of that of the control microsomes [61]) [1-3, 7-9, 11, 17, 20, 21, 23-26, 28, 31, 32, 35, 39, 42, 61, 71, 72, 75-77]
NADH <2, 5, 6> (<2, 6>, less effective than NADPH [7, 41]; <5>, low affinity [9]; <33>, activity not detected [77]) [7, 9, 41]
NADPH <1-8, 10, 12-15, 17, 19, 20, 21, 24, 32, 33, 34> (<1>, NADH less than 5% of NADPH activity [1]; <2>, not replaceable by NADH [26, 28, 33]; <3>, higher activity towards NADPH than towards NADH [3]) [1, 3-12, 14, 15, 17, 20, 21, 24, 25, 26, 28, 33, 36, 38, 41, 52, 57, 64, 71, 72, 74, 76-78]
azido nitrophenyl-γ-aminobutyryl-NADPH <3> (<3>, in the absence of photoirradiation equally efficient as NADPH [4]) [4]

Activating compounds

CO <32> (<32>, stimulation [71]) [71]
hexadecanol <32> (<32>, stimulation [71]) [71]
lanosterol <12> (<12>, activation [56]) [56]
laurate <32> (<32>, plus carbon monoxide, greater stimulation compared to laurate alone or carbon monoxide alone, decrease of stimulation in the presence of NADPH [71]) [71]
laurate <32> (<32>, stimulation [71]) [71]
menadione <14> (<14>, slight stimulation [21]) [21]
nagarse <12> (<12>, activation [37]) [37]
nonionic detergent <10> (<10>, activation [17]) [17]
palmitate <32> (<32>, stimulation [71]) [71]
phenobarbital <8> (<8>, stimulation) [12]
potassium phosphate <33> (<33>, activity slightly higher than with sodium chloride or sodium sulfate [77]) [77]
sodium cholate/Lubrol PX <7> (<7>, increases activity [8]) [8]
spinach ferredoxin <6> (<6>, stimulation of the NADPH-reduction of cytochrome c-550 and horse cytochrome c [7]) [7]
tetradecanol <32> (<32>, stimulation [71]) [71]
Additional information <2, 5, 14, 15, 33> (<5>, low ionic strength, slight stimulation [9]; <2, 14, 15>, increasing ionic strength, increase in activity [21, 24]; <33>, storage of the enzyme at -75°C for 4 months results in a slight activation of the enzyme activity [77]) [9, 21, 24, 77]

Metals, ions

Additional information <5, 15> (<5, 15>, stimulation by increasing ionic strength [9, 24]) [9, 24]

Turnover number (min^{-1})

87 <8> (menadione) [12]
458 <8> (2,6-dichlorophenolindophenol) [12]
897 <8> (cytochrome c) [12]
1230 <1> (cytochrome c) [32]
1250-1350 <1, 2> (cytochrome c) [31]
1280 <2> (cytochrome c, <2>, aortic microsomal enzyme [72]) [72]
1316 <2> (cytochrome c, <2>, purified enzyme [72]) [72]
3200 <32> (cytochrome c) [71]
3870 <8> (ferricyanide) [12]
6100 <13> (cytochrome c, <13>, grown on alkanes [26]) [26]
25000 <32> (cytochrome c, <32>, in the presence of laurate and carbon monoxide [71]) [71]
Additional information <1, 2, 6> [7, 29, 33]

Specific activity (U/mg)

0.002 <23> (<23>, outer membrane of mitochondria [50]) [50]
0.00271 <34> (<34>, microsomes [78]) [78]
0.0041 <12> (<12>, nuclei [19]) [19]
0.0085 <16> (<16>, whole homogenate, prepared from 6 day-organisms, centrifugation for 165 min [40]) [40]
0.0095 <16> (<16>, post-mitochondrial supernatant, prepared from 6 day-organisms, centrifugation for 165 min [40]) [40]
0.0097 <16> (<16>, post-mitochondrial supernatant, prepared from 1 day-organisms, centrifugation for 6 h [40]) [40]
0.0104 <2> (<2>, aortic microsomal preparation [72]) [72]
0.01087 <25> (<25>, after cell breakage with glass beads and for 30 min [53]) [53]
0.0116 <16> (<16>, whole homogenate, prepared from 1 day-organisms, centrifugation for 165 min [40]) [40]
0.0128 <30> (<30>, endogenous enzyme [70]) [70]
0.0133 <26> (<26>, fraction precipitated with 40-50% ammonium sulfate, 3-methylcholantrene treated-animals [49]) [49]
0.0135 <16> (<16>, whole homogenate, prepared from 1 day-organisms, centrifugation for 6 h [40]) [40]
0.0144 <16> (<16>, post-mitochondrial supernatant, prepared from 1 day-organisms, centrifugation for 165 min [40]) [40]
0.0173 <31> (<31>, endogenous enzmye [70]) [70]
0.0196 <1> (<1>, mitochondria [48]) [48]
0.02 <2, 21, 23> (<2>, brain [10]; <23> microsomes [50]; <21>, midgut and hindgut [57]) [10, 50, 57]
0.02-0.12 <33> (<33>, flower buds, open flowers exhibits a similar value [77]) [77]
0.0204 <1> (<1>, inner membrane plus matrix [48]) [48]

0.021 <2> (<2>, riboflavin-deficient animals, feeding period 7 weeks, electron acceptor: cytochrome c [61]) [61]
0.0223 <12> (<12>, spheroplasts [19]) [19]
0.0224 <1> (<1>, inner membrane [48]) [48]
0.0233 <28> [67]
0.025 <2> (<2>, electron donor: NADH, electron acceptor: dichlorophenolindophenol [41]) [41]
0.029 <21> (<21>, malpighian tubules [57]) [57]
0.03 <2> (<2>, electron donor: NADPH, electron acceptor: cytochrome b_5 [41]) [41]
0.033 <25> (<25>, endoplasmic reticulum [54]) [54]
0.034 <2> (<2>, β-naphthoflavone-treated animals [63]) [63]
0.036 <2> (<2>, control microsomes [63]) [63]
0.036 <25> (<25>, mitochondrial fraction [54]) [54]
0.04 <26> (<26>, microsomes, 3-methylcholantrene treated-animals [55]) [55]
0.0438 <25> (<25>, after cell breakage through filter screen and centrifugation for 30 min [53]) [53]
0.048 <26> (<26>, microsomes, control animals [55]) [55]
0.05 <8> (<8>, cholate and Lubrol Px [12]) [12]
0.051 <21> (<21>, fat body [57]) [57]
0.055 <2> (<2>, electron donor: NADH, electron acceptor: cytochrome c [41]) [41]
0.059 <8> (<8>, no detergent [12]) [12]
0.06572 <2> (<2>, purified microsomal ethanol-oxidizing system fraction [33]) [33]
0.066 <2> (<2>, riboflavin-deficient animals, feeding period 6 weeks, electron acceptor: cytochrome c [61]) [61]
0.075 <2> (<2>, riboflavin deficient animals, feeding period 7 weeks, electron acceptor: ferricyanide [61]) [61]
0.076 <29> (<29>, males [69]) [69]
0.0775 <34> (<34>, purified enzyme [78]) [78]
0.079 <29> (<29>, females [69]) [69]
0.081 <12> [37]
0.0857 <1> (<1>, outer membrane [48]) [48]
0.08572 <2> (<2>, microsomes [33]) [33]
0.087 <1> (<1>, microsomes [48]) [48]
0.097 <2> (<2>, riboflavin-deficient-animals, feeding period 6 weeks, electron acceptor: ferricyanide [61]) [61]
0.1-0.3 <13> (<13>, grown on glycerol, microsomal fraction [26]) [26]
0.105 <26> (<26>, microsomes, phenobarbital-treated animals [55]) [55]
0.119 <11> (<11>, 105000 x g pellet, microsomes [18]) [18]
0.124-0.138 <26> (<26>, fraction precipitated with 40-50% ammonium sulfate, 3-methylcholantrene-treated animals [55]) [55]
0.144 <26> (<26>, fraction precipitated with 40-50% ammonium sulfate, control animals [55]) [55]

0.157 <2> (<2>, control animals, feeding period 6 weeks, electron acceptor: ferricyanide [61]) [61]

0.165 <2> (<2>, control animals, feeding period 6 weeks, electron acceptor: cytochrome c [61]) [61]

0.17 <2> (<2>, R454E mutant, electron acceptor: cytochrome c [76]) [76]

0.171 <2> (<2>, nuclear envelope [27]) [27]

0.189 <2> (<2>, electron acceptor: formylated cytochrome c [51]) [51]

0.211 <2> (<2>, brain, electron acceptor: hexadecanal [10]) [10]

0.225-0.38 <4> (<4>, endogenous enzyme [70]) [70]

0.23 <2> (<2>, Y456S mutant, electron acceptor: cytochrome c [76]) [76]

0.245 <15> (<15>, purification by ion exchange chromatography and gel filtration, electron acceptor: dichlorophenolindophenol [24]) [24]

0.258-0.276 <26> (<26>, fraction precipitated with 40-50% ammonium sulfate, phenobarbital-treated animals [55]) [55]

0.28 <2> (<2>, Y140D/178D double mutant, electron acceptor: cytochrome c [75]) [75]

0.3-0.7 <4> [73]

0.333 <2> (<2>, microsomes [27]) [27]

0.361 <4> (<4>, coexpression of 17α-2D6 + pJR7, supplementation of cultures with δ-aminolevulinic acid [73]) [73]

0.414 <4> (<4>, coexpression of 17α-2D6 + pJR7, without supplementation of cultures with δ-aminolevulinic acid [73]) [73]

0.44 <2> (<2>, Y178D mutant, electron acceptor: cytochrome c [75]) [75]

0.517 <4> (<4>, coexpression of ompA-2D6 + pJR7, supplementation of cultures with δ-aminolevulinic acid [73]) [73]

0.58 <2> (<2>, peroxidation of microsomes for 18 min [62]) [62]

0.601 <4> (<4>, coexpression of ompA-2D6 + pJR7, without supplementation of cultures with δ-aminolevulinic acid [73]) [73]

0.676 <2> (<2>, purified enzyme, liver [34]) [34]

0.68 <22> [14]

0.7 <9> (<9>, under carbon limitation [13]) [13]

0.72 <2> (<2>, control microsomes [62]) [62]

0.75-1.1 <13> (<13>, grown on alkane, microsomal fraction [26]) [26]

0.788 <20> (<20>, mitochondria [25]) [25]

0.928 <2> (<2>, electron acceptor: cytochrome c [51]) [51]

1 <9> (<9>, under oxygen limitation [13]) [13]

1.12 <15> (<15>, purification by ion exchange chromatography and gel filtration, electron acceptor: ferricyanide [24]) [24]

1.2 <27> (<27>, soybean oil or malt extract as carbon source, high aeration [66]) [66]

1.35 <2> (<2>, G488L mutant, electron acceptor: cytochrome c [76]) [76]

1.429 <15> (<15>, purification by ion exchange chromatography and gel filtration [24]) [24]

1.696 <4> (<4>, recombinant enzyme coexpressed with CYP3A4 in Spodoptera frugiperda cell lines [70]) [70]

1.7 <2> (<2>, W677X mutant, electron acceptor: cytochrome c [76]) [76]

2.6 <27> (<27>, glucose as carbon source, low aeration [66]) [66]

3.904 <4> (<4>, recombinant enzyme coexpressed with CYP3A4 in Trichoplusia ni cell lines [70]) [70]
4-8 <3> [4]
5.5 <1> (<1>, preparative polyacrylamide disc gel electrophoresis [38]) [38]
5.7 <27> (<27>, soybean oil as carbon source, low aeration [66]) [66]
5.9 <2> (<2>, brain, electron acceptor: *p*-nitrobenzaldehyde [10]) [10]
8.64 <6> (<6>, electron acceptor: dichlorophenolindophenol [7]) [7]
10.5 <27> (<27>, glycerol as carbon source, low aeration [66]) [66]
11 <2> (<2>, Y140D mutant, electron acceptor: cytochrome c [75]) [75]
12.5 <13> (<13>, grown on alkane, electron acceptor: neotetrazolium chloride, purified enzyme [26]) [26]
13.1 <24> (<24>, endoplasmic reticulum [52]) [52]
15.1 <2> [36]
15.2 <1> (<1>, polymorphonuclear leukocytes [1]) [1]
16.5 <2> (<2>, cDNA-expressed liver enzyme [72]) [72]
17.1 <1> (<1>, liver [1]) [1]
17.14 <15> (<15>, purification by affinity chromatography on 2',5'-ADP-Sepharose 4B [24]) [24]
17.9 <2> (<2>, electron donor: NADPH, electron acceptor: dichlorophenolindophenol [41]) [41]
18.1 <24> (<24>, mitochondria [52]) [52]
18.4 <10> (<10>, electron acceptor: 2,6-dichlorophenolindophenol [17]) [17]
18.5 <13> (<13>, grown on alkane, electron acceptor: ferricyanide, purified enzyme [26]) [26]
19 <13> (electron acceptor: neotetrazolium chloride [20]) [20]
22 <15> (<15>, in the presence of 0.005 mM FMN [24]) [24]
24.3 <12> (<12>, purified truncated protein [74]) [74]
25 <2> [5]
26 <7> [8]
26 <13> (electron acceptor: ferricyanide [20]) [20]
26.1 <1> [11]
28 <13> (electron acceptor: menadione [20]) [20]
29 <13> (<13>, grown on alkane, electron acceptor: menadione, purified enzyme [26]) [26]
30-50 <2> [64]
32.7 <2> (<2>, T491V mutant, electron acceptor: cytochrome c [76]) [76]
33 <5> [9]
34.2 <22> [60]
35 <2> [68]
35 <2> [8]
36 <2> (<2>, W677Y mutant, electron acceptor: cytochrome c [76]) [76]
36 <4> (<4>, purified truncated protein [74]) [74]
37.6 <2> (<2>, purified enzyme [41]) [41]
38-41 <3> [39]
40 <19> [5]
40.5 <2, 14> (<2>, electron donor: NADPH, electron acceptor: ferricyanide [41]) [21, 41]

40.7 <24> (<24>, Golgi apparatus [52]) [52]
41.1 <24> (<24>, plasma membrane [52]) [52]
41.2 <2> (<2>, electron donor: NADPH, electron acceptor: cytochrome c [41]) [41]
42 <13> (electron acceptor: dichlorophenolindophenol [20]) [20]
46 <2> (<2>, Y140F/178F double mutant, electron acceptor: cytochrome c [75]) [75]
47.4 <2> (<2>, Y178F mutant, electron acceptor: cytochrome c [75]) [75]
48.1 <2> (<2>, Y178D mutant, electron acceptor: ferricyanide [75]) [75]
50 <13> (<13>, grown on alkane, electron acceptor: dichlorophenolindophenol, purified enzyme [26]) [26]
51 <2> (<2>, Y140D/178D double mutant, electron acceptor: ferricyanide [75]) [75]
51.5 <2> (<2>, bacterially expressed reductase protein, electron acceptor: cytochrome c [75]) [75]
52.9 <2> (<2>, electron acceptor: ferricyanide [75]) [75]
53.3 <2> (<2>, electron acceptor: cytochrome c [75]) [75]
55.2 <2> (<2>, Y140F mutant, electron acceptor: cytochrome c [75]) [75]
56 <13> [20]
56.1 <2> [21]
57.4 <2> (<2>, wild type, electron acceptor: cytochrome c [76]) [76]
60 <2> (<2>, purified enzyme, microsomes [27]) [27]
60-120 <33> (<33>, purified enzyme [77]) [77]
62.5 <2> (<2>, S678X mutant, electron acceptor: cytochrome c [76]) [76]
62.5 <2> (<2>, purification by affinity chromatography on agarose-hexane-adenosine 2',5'-diphosphate [16]) [16]
63.8 <2> (<2>, purification by n-octylamino-Sepharose 4B and 2',5'-ADP column chromatography [16]) [16]
65.4 <2> (<2>, C472T mutant, electron acceptor: cytochrome c [76]) [76]
68.9 <10> [17]
70 <13> (<13>, grown on alkane, electron acceptor: cytochrome c, purified enzyme [26]) [26]
94.2 <2> (<2>, Y140F/178F double mutant, electron acceptor: ferricyanide [75]) [75]
94.8 <2> (<2>, Y140F mutant, electron acceptor: ferricyanide [75]) [75]
98.3 <10> (<10>, electron acceptor: ferricyanide [17]) [17]
102 <2> (<2>, Y140D mutant, electron acceptor: ferricyanide [75]) [75]
102 <2> (<2>, bacterially expressed reductase protein, electron acceptor: ferricyanide [75]) [75]
113 <2> (<2>, Y178F mutant, electron acceptor: ferricyanide [75]) [75]
150 <12> [23]
150-180 <12> [28]
Additional information <1, 2, 3, 20> [3, 25, 27, 29, 30]

K_m-Value (mM)

0.000001 <2> (NADPH, <2>, wild type, cosubstrate FAD, K_m below [76]) [76]
0.000001 <2> (cytochrome c, <2>, wild type, cosubstrate FAD, K_m below [76]) [76]
0.00012 <2> (NADPH, <2>, T491V mutant, cosubstrate FAD [76]) [76]
0.00012 <2> (cytochrome c, <2>, T491V mutant, cosubstrate FAD [76]) [76]
0.0008 <2> (NADPH, <2>, G488L mutant, cosubstrate FAD [76]) [76]
0.0008 <2> (cytochrome c, <2>, G488L mutant, cosubstrate FAD [76]) [76]
0.001 <3> (NADPH) [4]
0.001-0.003 <2> (NADPH) [72]
0.0015 <2> (cytochrome c, <2>, W677X mutant, cosubstrate NADPH [76]) [76]
0.0017 <20> (NADPH, <20>, mitochondrial reductase [25]) [25]
0.00189 <6> (NADPH) [7]
0.0019 <3> (azidonitrophenyl-γ-aminobutyryl-NADPH) [4]
0.0028 <2> (cytochrome c, <2>, W677Y mutant, cosubstrate NADPH [76]) [76]
0.00303 <8> (NADPH) [12]
0.0033 <1> (NADPH, <1>, kidney [35]) [35]
0.0034 <12> (cytochrome c) [28]
0.0034 <20> (cytochrome c, <20>, mitochondrial reductase [25]) [25]
0.0036 <1> (NADPH) [1]
0.0041 <2> (cytochrome c, <2>, R454E mutant, cosubstrate NADPH [76]) [76]
0.0043 <13> (cytochrome c, <13>, grown on glycerol [26]) [26]
0.00455 <13> (cytochrome c, <13>, grown on alkanes [26]) [26]
0.0046 <1> (NADPH) [32]
0.0047 <1> (NADPH, <1>, spleen [35]) [35]
0.0053 <22> (menadione) [14]
0.0053 <22> (menadione) [14]
0.0055 <20> (NADPH, <20>, microsomal reductase [25]) [25]
0.0056 <2> (cytochrome c, <2>, S678X mutant, cosubstrate NADPH [76]) [76]
0.0057 <15> (NADPH) [24]
0.0058 <2> (cytochrome c, <2>, Y456S mutant, cosubstrate NADPH [76]) [76]
0.0059 <2> (cytochrome c, <2>, T491V mutant, cosubstrate NADPH [76]) [76]
0.006 <1> (cytochrome c, <1>, spleen [35]) [35]
0.0062 <2> (cytochrome c, <2>, wild type, cosubstrate NADPH [76]) [76]
0.0063 <2> (NADPH) [21]
0.0063 <2> (cytochrome c, <2>, C472T mutant, cosubstrate NADPH [76]) [76]
0.0064 <2> (NADPH, <2>, wild type, cosubstrate cytochrome c [75]) [75]
0.0066 <2> (NADPH, <2>, Y140D mutant, cosubstrate cytochrome c [75]) [75]

0.0066 <2> (NADPH, <2>, native enzyme, liver, cosubstrate cytochrome c [75]) [75]
0.0071 <2> (NADPH, <2>, Y178F mutant, cosubstrate cytochrome c [75]) [75]
0.0077 <2> (cytochrome c, <2>, G488L mutant, cosubstrate NADPH [76]) [76]
0.0078 <2> (NADPH, <2>, Y140F mutant, cosubstrate cytochrome c [75]) [75]
0.0078 <2> (NADPH, <2>, Y456S mutant, cosubstrate FAD [76]) [76]
0.0078 <15> (cytochrome c) [24]
0.0078 <2> (cytochrome c, <2>, Y456S mutant, cosubstrate FAD [76]) [76]
0.0082 <19> (cytochrome c) [5]
0.0085 <2> (NADPH, <2>, W677X mutant, cosubstrate cytochrome c [76]) [76]
0.0095 <2> (NADPH, <2>, Y178D mutant, cosubstrate cytochrome c [75]) [75]
0.0095 <1> (cytochrome c, <1>, kidney [35]) [35]
0.01 <1> (cytochrome c) [32]
0.0105 <20> (cytochrome c, <20>, microsomal reductase [25]) [25]
0.0114 <2> (NADPH, <2>, Y140F/178F double mutant, cosubstrate cytochrome c [75]) [75]
0.013 <21> (NADPH) [57]
0.013 <2, 14> (cytochrome c) [5, 21]
0.0131 <2> (cytochrome c) [21]
0.0131 <2> (cytochrome c, <2>, Y140F/178F double mutant [75]) [75]
0.0133 <20> (NADH, <20>, microsomal reductase [25]) [25]
0.0134 <2> (NADPH, <2>, G488L mutant, cosubstrate cytochrome c [76]) [76]
0.014 <2> (cytochrome c, <2>, Y178F mutant [75]) [75]
0.0143 <2> (NADPH, <2>, W677Y mutant, cosubstrate cytochrome c [76]) [76]
0.0163 <2> (NADPH, <2>, wild type, cosubstrate cytochrome c [76]) [76]
0.0166 <2> (NADPH, <2>, R454E mutant, cosubstrate cytochrome c [76]) [76]
0.0178 <2> (cytochrome c, <2>, native enzyme, liver [75]) [75]
0.0182 <5> (cytochrome c, <5>, determined on the basis of a sequential mechanism [9]) [9]
0.0185 <2> (cytochrome c, <2>, Y178D mutant [75]) [75]
0.0188 <2> (cytochrome c, <2>, Y140F mutant [75]) [75]
0.0191 <2> (cytochrome c, <2>, Y140D mutant [75]) [75]
0.0199 <2> (NADPH, <2>, S678X mutant, cosubstrate cytochrome c [76]) [76]
0.02 <5> (NADPH) [9]
0.0201 <2> (NADPH, <2>, T491V mutant, cosubstrate cytochrome c [76]) [76]
0.0211 <2> (cytochrome c, <2>, wild type [75]) [75]

0.0217 <2> (NADPH, <2>, C472T mutant, cosubstrate cytochrome c [76]) [76]
0.022 <1> (NADPH) [11]
0.025 <22> (NADPH) [14]
0.0251 <2> (NADPH, <2>, R454E mutant, cosubstrate FAD [76]) [76]
0.0251 <2> (cytochrome c, <2>, R454E mutant, cosubstrate FAD [76]) [76]
0.029 <13> (NADPH, <13>, grown on alkanes [26]) [26]
0.03 <2> (hexadecanal) [10]
0.0308 <5> (NADPH, <5>, determined on the basis of a sequential mechanism [9]) [9]
0.0324 <12> (NADPH) [28]
0.033 <13> (NADPH, <13>, grown on glycerol [26]) [26]
0.0341 <14> (NADPH, <14>, cosubstrate cytochrome c [21]) [21]
0.0466 <14> (cytochrome c, <14>, cosubstrate NADPH [21]) [21]
0.0548 <2> (NADPH, <2>, Y456S mutant, cosubstrate cytochrome c [76]) [76]
0.055 <13> (2,6-dichlorophenolindophenol, <13>, grown on glycerol [26]) [26]
0.057 <1> (cytochrome c) [11]
0.06289 <8> (cytochrome c) [12]
0.077 <13> (2,6-dichlorophenolindophenol, <13>, grown on alkanes [26]) [26]
0.21 <2> (octanal) [10]
0.23 <22> (NADH) [14]
0.23 <22> (NADH) [14]
0.31 <2> (*p*-nitrobenzaldehyde) [10]
1.4 <2, 6> (*p*-nitroacetophenone, <6> electron donor NADH, electron acceptor dichlorophenolindophenol [7]) [7, 10]
2.5 <2> (benzalacetone) [10]
7.2 <2> (ethanol, <2>, microsomal ethanol oxidizing system [33]) [33]
20 <20> (NADH, <20>, mitochondrial reductase [25]) [25]
24 <1> (NADH) [32]
40 <1> (NADH, <1>, kidney [35]) [35]
50 <1> (NADH, <1>, spleen [35]) [35]
Additional information <22, 5, 8> (<22>, O_2^--generation [14]; <5>, cosubstrate NADH, biphasic kinetic [9]) [9, 11, 12, 14]

pH-Optimum

6.9-7.5 <2> (<2>, microsomal ethanol-oxidizing system [33]) [33]
7-7.4 <22> (<22>, O_2^--generation [14]) [14]
7.4 <33> [77]
7.5-9 <14> [21]
7.7 <11> [18]
7.8 <13> [20]
7.8-8 <12, 20> [25, 28]
7.8-8 <13> (<13>, grown on alkane or glycerol [26]) [26]
8-9 <1> [35]

pH-Range

6-8 <33> [77]
6.5-9 <14> [21]
7-8.5 <13> (<13>, less than 50% of maximal activity above and below, grown on alkane or glycerol [26]) [26]

4 Enzyme Structure

Molecular weight

65000-68000 <1, 3, 12, 13> (<12>, sedimentation equilibrium centrifugation, values depending on pH [23]; <13>, gel filtration, sedimentation equilibrium centrifugation [26]; <1>, gel filtration, kidney [32, 35]; <3>, gel filtration, liver [39]) [23, 26, 32, 35, 39]
70000 <1, 6, 12> (<6>, gel filtration [7]; <1>, testis, gel filtration [11]; <12>, gel filtration [28]) [7, 11, 28]
75000 <12> (<12>, truncated protein [74]) [74]
78000 <12> [74]
78000-79000 <2, 3> (<3>, sedimentation equilibrium centrifugation [39]; <2>, liver [46]) [39, 46]
82000-85000 <4, 5, 12, 18> (<5>, SDS-PAGE followed by Western blotting [6]; <12>, calculation from FAD content [28]) [6, 28, 44, 45]
84000 <33> (<33>, western blot analysis of fresh microsomal extracts [77]) [77]
100000 <8> (<8>, gel filtration [12]) [12]
400000 <1> (<1>, polymorphonuclear leukocytes, gel filtration [1]) [1]
Additional information <5, 12, 13, 15> (<5,13>, differences in MW partially due to method of solubilization [9, 20]; <12>, a weak band at the level of the Jerusalem artichoke reductase [6]; <1>, liver microsomes, gel filtration [38]) [6, 9, 20, 24, 38]

Subunits

? <1-3, 5, 7, 10, 13, 14, 19, 26> (<1>, x * 72000-87000, SDS-PAGE [1, 11, 14, 29]; <3>, x * 77000, SDS-PAGE [3,4]; <2,19>, x * 85000, SDS-PAGE [5]; <2>, x * 75000-76000, SDS-PAGE [10, 16, 27, 29]; <7,13>, x * 76000, SDS-PAGE [8, 20]; <5>, x * 82000, SDS-PAGE [9]; <10>, x * 79000, SDS-PAGE [17]; <14>, x * 70000, SDS-PAGE [21]; <26>, x * 62000 [49]; <2>, wild type, x * 80000, SDS-PAGE [75]) [1, 3, 4, 5, 8-11, 14, 16, 17, 20, 21, 27, 29, 49, 75]
dimer <6, 8, 12> (<6>, 2 * 36000, SDS-PAGE [7]; <8>, 2 * 52000, SDS-PAGE [12]; <12>, 2 * 34300-40000, SDS-PAGE, sedimentation equilibrium centrifugation after treatment with guanidine-HCl [23]) [7, 12, 23]
monomer <3, 12, 13> (<12>, 1 * 72000, SDS-PAGE [28]; <3>, 1 * 75000, Triton-solubilized, SDS-PAGE [39]; <3>, 1 * 68000, trypsin-solubilized, SDS-PAGE [39]; <13>, 1 * 67000, SDS-PAGE, trypsin- or lipase-solubilized [26]) [26, 28, 39]
Additional information <5, 6, 13> (<5, 13>, differences in subunit weight partially due to method of solubilization [9, 20]; <6>, N-terminal amino acid

sequence is 27% homologous to that of the Spirulina ferredoxin-NADP reductase [7]) [7, 9, 20]

5 Isolation/Preparation/Mutation/Application

Source/tissue

aorta <2> [72]
brain <2> [10]
bulb <17, 34> [6, 78]
epimastigote <8> [12]
fat body <21> (<21>, of larvae of both sexes [57]) [57]
flower <33> (<33>, buds and nascent, open flowers [77]) [77]
heart <1> [48]
hypocotyl <24> [52]
kidney <1, 20> [25, 32, 35]
leukocyte <1> (<1>, polymorphonuclear [1, 14]) [1, 14]
liver <1-4, 7, 22, 26, 28, 29> (<2, 7, 26>, phenobarbital-treated animals [2, 5, 8, 10, 16, 21, 29, 34, 55, 62, 68]; <26>, 3-methylcholantrene-treated animals [55]; <2>, riboflavin-deficient animals [61]) [1, 2, 5, 6, 8, 10, 16, 21, 27, 29, 31, 33, 34, 36, 38, 39, 41, 43, 46, 47, 49, 55, 58-62, 64, 67-70, 72, 75, 76]
lung <3> [30]
malpighian tubule <21> (<21>, of larvae of both sexes [57]) [57]
mesocarp <17> [6]
midgut <14, 21> (<14>, of larvae, induction of larvae with pentamethylbenzene [21]; <21>, of larvae of both sexes [57]) [6, 21, 57]
neutrophil <3> (peritoneal) [3, 4]
placenta <4, 19> [5, 45]
seedling <15, 17> (<17>, Vicia faba, leek, sunflower [6]) [6, 24]
small intestine <2> (<2>, β-naphtoflavone-treated animals [63]) [63]
spleen <1> [35]
testis <1, 2> [11, 34]
thyroid gland <1> [65]
tuber <5> [6, 9]
uterine endometrium <1> [15]

Localization

Golgi apparatus <24> [52]
cytoplasm <1, 2> [1-5]
cytosol <4, 6, 8, 12> (<4, 12>, heterologously expressed native and truncated protein [74]) [7, 12, 74]
endoplasmic reticulum <2, 24, 25, 33> [43, 52, 54, 77]
microsome <7> [8]
microsome <1, 2, 3, 4, 5, 7, 9-12, 14, 16, 17, 19, 20-23, 26, 28, 29, 30, 31, 33, 34> (<2>, 0.03 to 0.1% of aortic microsomal protein [72]; <4, 12>, heterologously expressed native and truncated protein [74]) [2, 5, 6, 8-11, 13, 15-18, 21, 25, 27, 29, 30-41, 49, 50, 55, 57-63, 67-70, 72, 74, 77, 78]

mitochondrial outer membrane <20> (<20>, outer membrane of vitamin D_3-deficient mitochondrion [25]) [25]
mitochondrial outer membrane <23> (<23>, outer membrane of mitochondria [50]) [50]
mitochondrion <21, 24, 25> [52-54, 57]
nuclear envelope <2> [27]
nucleus <12> (<12>, low activity [19]) [19]
plasma membrane <1, 22, 24> [14, 52, 65]
spheroplast <4, 12> [19, 73]
Additional information <15> [24]

Purification

<1> (solubilization with lipase, column chromatography on Sephadex G-100 superfine, DEAE-cellulose and Sephadex G-200 superfine [32]; chromatography in DEAE-cellulose, DEAE-Sephadex A-50 and hydroxylapatite [38]) [1, 11, 29, 32, 35, 38, 47]
<2> (FAD-depleted enzyme [2]; chromatography on 2',5'-ADP agarose [5]; single-step procedure involving affinity chromatography on agarose-hexane-adenosine 2',5'-diphosphate or two-column procedure involving chromatography on n-octylamino-Sepharose 4B followed by chromatography on 2',5'-ADP [16]; column chromatography on Sephadex G-100 and DEAE-cellulose [21]; DEAE-cellulose followed by sequential chromatography on Sephadex G-150, DEAE-cellulose and hydroxylapatite [27]; liver microsomes, column chromatography on Sephadex G-150 and hydroxylapatatite [34]; column chromatography on DEAE-cellulose and affinity chromatography on cytochrome-c-Sepharose 4B [36]; trypsin digestion followed by Sephadex filtration and column chromatography on DEAE-cellulose [41]; column chromatography on DEAE-cellulose and 2',5'-ADP-Sepharose 4B [61]; column chromatography on 2',5'-ADP-Sepharose, phospholipid composition relatively enriched in L-α-1,2-diacyl-sn-glycero-3-phosphoserine and L-α-1,2-diacyl-sn-glycero-3-phosphoinositol [68]; expressed in Escherichia coli [75]) [2, 5, 10, 16, 21, 27, 29, 33, 34, 36, 41, 58, 61, 68, 75]
<3> (<3>, sequential chromatography on DE 52 cellulose and 2',5'-ADP Sepharose [3,4]) [3, 4, 30, 39]
<5> (anion exchange chromatography and affinity chromatography [6, 9]) [6, 9]
<6> [7]
<7> [8]
<8> (DEAE-cellulose and 2',5'-ADP-agarose column chromatography [12]) [12]
<10> (DEAE-Sephacel and hydroxylapatite column chromatography [17]) [17]
<12> (<12>, column chromatography on Sephadex G-150 and DE-52 [28]) [23, 28]
<13> (column chromatography on phenyl-sepharose CL-4B, DEAE cellulose, hydroxylapatite and Sephadex G-150 [20]; solubilization with sodium cholate followed by column chromatography on phenyl-Sepharose CL-4B and DEAE-

cellulose [22]; DEAE-cellulose and Sephadex G-100 column chromatography [26]) [20, 22, 26]
<14> (column chromatography on Sephadex G-100 and DEAE-cellulose [21]) [21]
<15> (purified 120fold by a combination of ion exchange chromatography and gel filtration and 745fold by affinity chromatography on 2',5'-ADP-Sepharose 4B [24]) [24]
<19> (chromatography on 2',5'-ADP agarose [5]) [5]
<20> (mitochondria [25]) [25]
<22> (chromatofocusing in combination with affinity chromatography [11]; solubilization with 0.2% Triton X-100, gel filtration on Sephacryl S-300 and affinity chromatography on 2',5'-ADP agarose [14]) [14, 60]
<32> (expressed in Escherichia coli, column chromatography on 2',5'-ADP-agarose [71]) [71]
<33> (using column chromatography on adenosine 2',5'-biphosphate-Sepharose, followed by high-performance anion-exchange chromatography [77]) [77]
<34> (using column chromatography on DEAE-Sepharose CL 6B and affinity chromatography on a 2',5'-ADP Sepharose 4B column [78]) [78]

Renaturation

<1, 2, 3, 5, 12, 22, 32, 34> (<3>, reconstitution of O_2^--generating system [3]; <5>, reconstitution of monooxygenase system [9]; <3>, reconstitution of O-deethylation and N-demethylation activity [30]; <1>, reconstitution of laurate ω-hydroxylation activity [38]; <2>, reconstitution of hydroxylation of benzo-α-pyrene [58]; <22>, reconstitution of N-demethylation of benzphetamine [60]; <2>, riboflavin-deficient microsomes, reconstitution of activity with FAD, FMN and riboflavin [61]; <32>, oxidation of the reduced P450BM3 by cytochrome c, ferricyanide or 2,6-dichlorophenolindophenol rapidly restores electron transfer and hydroxylase activity [71]; <12>, native or truncated protein reconstituts ergosterol biosynthesis when added to a cell-free system, as well as reconstitution of activity with purified Saccharomyces cerevisiae CYP61, sterol 22-desaturase, Candida albicans and human CYP51 [74]; <4>, only native protein reconstituted CYP61, Candida albicans CYP51, and human CYP51 activity [74]; <34>, reconstitution of metabolism of xenobiotics by plant CYPs, reconstituted electron transport chain consisting of partially purified CYP fraction, phospholipid and isolated tulip NADPH: CYP reductase [78]) [3, 9, 30, 38, 58, 60, 61, 71, 74, 78]

Crystallization

<12> [23]

Cloning

<4> (coexpression with CYP3A4 in Spodoptera frugiperda and Trichoplusia ni cell lines [70]) [70]
<4> (coexpression with two differently modified CYP2D6 cDNAs in Escherichia coli, 17α-2D6 + pJR7 and ompA-2D6 + pJR7 [73]) [73]

<4> (expression in transformed strains of Saccharomyces cerevisiae [74]) [74]
<12> (expression in transformed strains [74]) [74]
<2, 32> (expression in Escherichia coli [71, 75, 76]) [71, 75, 76]

Engineering

C472T <2> (<2>, substitution does not affect FAD or FMN incorporation, substitution has no effect on activity, K_m for NADPH or K_m for cytochrome c [76]) [76]
G488L <2> (<2>, substitution decreases FAD binding by approximately 80% but does not affect FMN incorporation, 42fold decrease in catalytic activity compared to wild type, substitution does not affect either K_m for NADPH or K_m for cytochrome c, addition of FAD to the mutant results in partial restoration of catalytic activity [76]) [76]
R454E <2> (<2>, substitution decreases both FAD binding and FMN incorporation, suggesting interaction between the two flavin domains and/or the interconnecting region, FAD content ranges from undetectable to approximately 0.1 mol of FAD/mol of enzyme, 338fold decrease in catalytic activity compared to wild type, substitution does not affect either K_m for NADPH or K_m for cytochrome c, addition of FAD to the mutant resulted in partial restoration of catalytic activity [76]) [76]
S678X <2> (<2>, substitution does not affect FAD or FMN incorporation, substitution has no effect on the catalytic activity or kinetic properties [76]) [76]
T491V <2> (<2>, substitution decreases FAD binding by approximately 50% but does not affect FMN incorporation, 2fold decrease in catalytic activity compared to wild type, substitution does not affect either K_m for NADPH or K_m for cytochrome c, addition of FAD to the mutant results in full restoration of catalytic activity [76]) [76]
W677X <2> (<2>, substitution does not affect FAD or FMN incorporation, 34fold decrease in catalytic activity compared to wild type, substitution does not alter significantly K_m for cytochrome c but decreases K_m for NADPH [76]) [76]
W677Y <2> (<2>, substitution does not affect FAD or FMN incorporation, 2fold decrease in catalytic activity compared to wild type, substitution does not alter significantly K_m for cytochrome c but decreases K_m for NADPH [76]) [76]
Y140D <2> (<2>, substitution does not eliminate FMN binding but reduces cytochrome c reductase activity, K_m-value for cytochrome c or NADPH similar to wild type [75]) [75]
Y140D/178D <2> (<2>, substitution abolishes FMN binding and cytochrome c reductase activity [75]) [75]
Y140F <2> (<2>, substitution has no effect on FMN content or catalytic activity, K_m-value for cytochrome c or NADPH similar to wild type [75]) [75]
Y140F/178F <2> (<2>, substitution has no effect on FMN content or catalytic activity, slightly decreases K_m for cytochrome c, NADPH K_m-value slightly higher than wild type [75]) [75]

Y178D <2> (<2>, substitution abolishes FMN binding and cytochrome c reductase activity, K_m-value for cytochrome c similar to wild type, NADPH K_m-value slightly higher than wild type [75]) [75]
Y178F <2> (<2>, substitution has no effect on FMN content or catalytic activity, slightly decreases K_m for cytochrome c, NADPH K_m-value similar to wild type [75]) [75]
Y456S <2> (<2>, substitution decreases FAD binding but did not affect FMN incorporation, 250fold decrease in catalytic activity compared to wild type, substitution increases K_m for cytochrome c, addition of FAD to the mutant results in full restoration of catalytic activity [76]) [76]

6 Stability

Temperature stability

-90 <8> (<8>, inactivation [12]) [12]
2-4 <8> (<8>, 1 day, inactivation [12]) [12]
20 <8> (<8>, 1 day, from 5% to 10% loss of activity [12]) [12]
25-30 <2> (<2>, diluted solutions: gradual loss of activity [41]) [41]
36 <14> (<14>, inactivation above [21]) [21]
40 <14> (<14>, 50% activity [21]) [21]
50 <2> (<2>, 40% activity remaining after 3 min [34]) [34]
60 <2> (<2>, brain, inactivation [10]) [10]
100 <22> (<22>, 10 min, inhibition of O_2^--formation [14]) [14]
Additional information <1, 12> (<12>, FAD and NADPH: protection against thermal inactivation [23]; <1>, stable in a frozen state at least for one month [38]) [23, 38]

General stability information

<1>, unstable during purification [32]
<2>, stable to multiple freeze-thaw cycles, FAD-depleted enzyme [2]
<3>, ability to support benzphetamine N-demethylation activity decreased by about 30% after 1 week of storage at -20°C [30]
<12, 13>, FAD, FMN necessary for stabilization during purification [20, 28]

Storage stability

<2>, -15°C [21]
<2>, -15°C or-20°C, 10 mM phosphate buffer, pH 7.5, several months [41]
<2>, -20°C, several weeks [5]
<2>, -78°C, 30 mM potassium phosphate buffer, pH 7.7, 0.1 mM EDTA, 20% glycerol, 0.4 mM PMSF [16]
<2>, -80°C [10]
<2>, 0°C, some days [41]
<2>, 4°C or room temperature, at both low and high ionic strength, in the presence or absence of NADP, FAD-depleted enzyme [2]
<3>, -20°C, N_2-atmosphere, several weeks [30]
<3>, -20°C, under nitrogen, several weeks [30]
<7>, -80°C, 4 weeks [8]

<7>, -80°C, 50 mM phosphate buffer, pH 7.4, 0.1 mM EDTA, 20% glycerol [8]
<8>, -90°C or -20°C, 5-10% loss of activity, reactivation by FAD, 24 h [12]
<10>, -20°C [17]
<13>, -70°C, 0.15 mM potassium phosphate buffer, pH 7, 1 mM mercaptoethanol, 1 mM EDTA, 1 micromol FMN, 1 micromol FAD, 0.3% Mulgofen BC-720, 30% glycerol, several months [20]
<14>, -15°C, more than 1 year [21]
<14>, 0-4°C, several months [21]
<19>, -20°C, several weeks [5]
<32>, expressed in E. coli, -80°C [71]
<33>, -75°C, months, 5 mM 2'-AMP [77]
<34>, -80°C, 10 mM potassium phosphate, pH 7.4, 0.02% Emulgen 911, 0.002 mM flavinmononucleotide [78]

References

[1] Kojima, H.; Takahashi, K.; Sakane, F.; Koyama, J.: Purification and characterization of NADPH-cytochrome c reductase from porcine polymorphonuclear leukocytes. J. Biochem., **102**, 1083-1088 (1987)
[2] Kurzban, G.P.; Howarth, J.; Palmer, G.; Strobel, H.W.: NADPH-cytochrome P-450 reductase. Physical properties and redox behavior in the absence of the FAD moiety. J. Biol. Chem., **265**, 12272-12279 (1990)
[3] Laporte, F.; Doussiere, J.; Vignais, P.V.: Properties of the NADPH dehydrogenase component of the oxidase complex from rabbit peritoneal neutrophils: reconstitution of an oxidase activity with the dehydrogenase component and a membrane extract. Biochem. Biophys. Res. Commun., **167**, 790-797 (1990)
[4] Laporte, F.; Doussiere, J.; Vignais, P.V.: Characterization of multiple active forms of the NADPH dehydrogenase component of the oxidase complex from rabbit peritoneal neutrophils by photolabeling with an arylazido derivative of $NADP^+$. Biochem. Biophys. Res. Commun., **168**, 78-84 (1990)
[5] Vibet, A.; Dintinger, T.; Maboundou, J.C.; Gaillard, J.L.; Divoux, D.; Silberzahn, P.: Estrogen synthetase in the horse. Comparison of equine placental and rat liver NADPH-cytrochrome c (P-450) reductase activities. FEBS Lett., **261**, 31-34 (1990)
[6] Benveniste, I.; Lesot, A.; Hasenfratz, M.; Durst, F.: Immunochemical characterization of NADPH-cytochrome P-450 reductase from Jerusalem artichoke and other higher plants. Biochem. J., **259**, 847-853 (1989)
[7] Kurokawa, T.; Fukumori, Y.; Yamanaka, T.: Purification of a flavoprotein having NADPH-cytochrome c reductase and transhydrogenase activities from Nitrobacter winogradskyi and its molecular and enzymatic properties. Arch. Microbiol., **148**, 95-99 (1987)
[8] Ardies, C.M.; Lasker, J.M.; Bloswick, B.P.; Lieber, C. S.: Purification of NADPH:cytochrome c (cytochrome P-450) reductase from hamster liver microsomes by detergent extraction and affinity chromatography. Anal. Biochem., **162**, 39-46 (1987)

[9] Benveniste, I.; Gabriac, B.; Durst, F.: Purification and characterization of the NADPH-cytochrome P-450 (cytochrome c) reductase from higher-plant microsomal fraction. Biochem. J., **235**, 365-373 (1986)
[10] Takahashi, N.; Saito, T.; Goda, Y.; Tomita, K.: Characterization of microsomal NADPH-dependent aldehyde reductase from rat brain. J. Biochem., **99**, 513-519 (1986)
[11] Kuwada, M.; Ohsawa, Y.; Horie, S.: Purification of NADPH-cytochrome c reductase from swine testis microsomes by chromatofocusing and characterization of the purified reductase. Biochim. Biophys. Acta, **830**, 45-51 (1985)
[12] Kuwahara, T.; White, R.A.; Agosin, M.: A cytosolic FAD-containing enzyme catalyzing cytochrome c reduction in Trypanosoma cruzi. I. Purification and some properties. Arch. Biochem. Biophys., **239**, 18-28 (1985)
[13] Laurila, H.; Käppeli, O.; Fiechter, A.: The cytochrome P450-containing monooxygenase of Trichosporon cutaneum: occurrence and properties. Arch. Microbiol., **140**, 257-259 (1984)
[14] Sakane, F.; Takahashi, K.; Koyama, J.: Purification and characterization of a membrane-bound NADPH-cytochrome c reductase capable of catalyzing menadione-dependent O_2^- formation in guinea pig polymorphonuclear leukocytes. J. Biochem., **96**, 671-678 (1984)
[15] Sierralta, W.D.; Szendro, P.I.: Characterization of microsomal subfractions from porcine endometrium cells. Hoppe-Seyler's Z. Physiol. Chem., **364**, 1329-1335 (1983)
[16] Shephard, E.A.; Pike, S.F.; Rabin, B.R.; Phillips, I. R.: A rapid one-step purification of NADPH-cytochrome c (P-450) reductase from rat liver microsomes. Anal. Biochem., **129**, 430-433 (1983)
[17] Honeck, H.; Schunck, W.H.; Riege, P.; Muller, H.G.: The cytochrome P-450 alkane monooxygenase system of the yeast Lodderomyces elongisporus: purification and some properties of the NADPH-cytochrome P-450 reductase. Biochem. Biophys. Res. Commun., **106**, 1318-1324 (1982)
[18] Jayanthi, C.R.; Madyastha, P.; Madyastha, K.M.: Microsomal 11α-hydroxylation of progesterone in Aspergillus ochraceus: Part I: Characterization of the hydroxylase system. Biochem. Biophys. Res. Commun., **106**, 1262-1268 (1982)
[19] Kamimura, K.; Tsuchiya, E.; Miyakawa, T.; Fuki, S.; Hirata, A.: Isolation and characterization of the nuclei from Saccharomyces cerevisiae. Curr. Microbiol., **6**, 175-180 (1981)
[20] Bertrand, J.C.; Gilewicz, M.; Bazin, H.; Azoulay, E.: Purified detergent-solubilized NADPH-cytochrome c (P-450) reductase from Candida tropicalis grown on alkanes. Biochem. Biophys. Res. Commun., **94**, 889-893 (1980)
[21] Crankshaw, D.L.; Hetnarski, K.; Wilkinson, C.F.: Purification and characterization of NADPH–cytochrome c reductase from the midgut of the southern armyworm (Spodoptera eridania). Biochem. J., **181**, 593-605 (1979)
[22] Bertrand, J.C.; Gilewicz, M.; Bazin, H.; Zacek, M.; Azoulay, E.: Partial purification of cytochrome P-450 of Candida tropicalis and reconstitution of hydroxylase activity. FEBS Lett., **105**, 143-146 (1979)

[23] Tryon, E.; Cress, M.C.; Hamada, M.; Kuby, S.A.: Studies on NADPH-cytochrome c reductase I: Isolation and several properties of the crystalline enzyme from ale yeast. Arch. Biochem. Biophys., **197**, 104-118 (1979)
[24] Madyastha, K.M.; Coscia, C.J.: Detergent-solubilized NADPH-cytochrome c(P-450) reductase from the higher plant, Catharanthus roseus. Purification and characterization. J. Biol. Chem., **254**, 2419-2427 (1979)
[25] Kulkoski, J.A.; Weber, J.L.; Ghazarian, J.G.: NADPH-cytochrome c reductase in outer membrane of kidney mitochondria. Purification and properties. Arch. Biochem. Biophys., **192**, 539-547 (1979)
[26] Bertrand, J.C.; Bazin, H.; Zacek, M.; Gilewicz, M.; Azoulay, E.: NADPH-cytochrome c reductase of Candida tropicalis grown on alkane. Eur. J. Biochem., **93**, 237-243 (1979)
[27] Zimmermann, J.J.; Kasper, C.B.: Immunological and biochemical characterization of nuclear envelope reduced nicotinamide adenine dinucleotide phosphate-cytochrome c oxidoreductase. Arch. Biochem. Biophys., **190**, 726-735 (1978)
[28] Kubota, S.; Yoshida, Y.; Kumaoka, H.; Furumichi, A.: Studies on the microsomal electron-transport system of anaerobically grown yeast. V. Purification and characterization of NADPH-cytochrome c reductase. J. Biochem., **81**, 197-205 (1977)
[29] Yasukochi, Y.; Masters, B.S.S.: Some properties of a detergent-solubilized NADPH-cytochrome c(cytochrome P-450) reductase purified by biospecific affinity chromatography. J. Biol. Chem., **251**, 5337-5344 (1976)
[30] Arinc, E.; Philpot, R.M.: Preparation and properties of partially purified pulmonary cytochrome P-450 from rabbits. J. Biol. Chem., **251**, 3213-3220 (1976)
[31] Masters, B.S.S.; Prough, R.A.; Kamin, H.: Properties of the stable aerobic and anaerobic half-reduced states of NADPH-cytochrome c reductase. Biochemistry, **14**, 607-613 (1975)
[32] Fan, L.L.; Masters, B.S.S.: Properties of purified kidney microsomal NADPH-cytochrome c reductase. Arch. Biochem. Biophys., **165**, 665-671 (1974)
[33] Teschke, R.; Hasumura, Y.; Lieber, C.S.: Hepatic microsomal ethanol-oxidizing system: solubilization, isolation, and characterization. Arch. Biochem. Biophys., **163**, 404-415 (1974)
[34] Betz, G.; Roper, M.; Tsai, P.: Steroid 17,20-lyase from testis microsomes: participation of NADPH cytochrome c reductase. Arch. Biochem. Biophys., **163**, 318-323 (1974)
[35] Iyanagi, T.: Some properties of kidney cortex and splenic microsomal NADPH-cytochrome c reductase. FEBS Lett., **46**, 51-54 (1974)
[36] Golf, S.W.; Graef, V.; Staudinger, H.: Solubilization and purification of the NADPH-cytochrome reductase from rat liver microsomes. Hoppe-Seyler's Z. Physiol. Chem., **355**, 1063-1069 (1974)
[37] Yoshida, Y.; Kumaoka, H.; Sato, R.: Studies on the microsomal electron-transport system of anaerobically grown yeast. I. Intracellular localization and characterization. J. Biochem., **75**, 1201-1210 (1974)

[38] Ichihara, K.; Kusunose, E.; Kusunose, M.: Some properties of NADPH-cytochrome c reductase reconstitutively active in fatty-acid ω-hydroxylation. Eur. J. Biochem., **38**, 463-472 (1973)
[39] Iyanagi, T.; Mason, H.S.: Some properties of hepatic reduced nicotinamide adenine dinucleotide phosphate-cytochrome c reductase. Biochemistry, **12**, 2297-2308 (1973)
[40] Poole, R.K.; Nicholl, W.G.; Howells, L.; Lloyd, D.: The microsomal fraction from Tetrahymena pyriformis strain ST: characterization and subfractionation. J. Gen. Microbiol., **68**, 283-294 (1971)
[41] Omura, T.; Takesue, S.: A new method for simultaneous purification of cytochrome b_5 and NADPH-cytochrome c reductase from rat liver microsomes. J. Biochem., **67**, 249-257 (1970)
[42] Aoyama, Y.; Yoshida, Y.; Kubota, S.; Kamaoka, H.; Furumichi, A.: NADPH-cytochrome P-450 reductase of yeast microsomes. Arch. Biochem. Biophys., **185**, 362-369 (1978)
[43] Cooper, M.B.; Craft, J.A.; Estall, M.R.; Rabin, B.R.: Asymmetric distribution of cytochrome P-450 and NADPH-cytochrome P-450 (cytochrome c) reductase in vesicles from smooth endoplasmic reticulum of rat liver. Biochem. J., **190**, 737-746 (1980)
[44] Mayer, R.T.; Durrant, J.L.: Purified NADPH cytochrome P-450 reductase. Interaction with hepatic microsomes and phospholipid vesicles. J. Biol. Chem., **254**, 4177-4185 (1979)
[45] Osawa, Y.; Higashiyama, T.; Nakamura, T.: Species specificity of estrogen biosynthesis in pregnancy. Immunochemical difference of placental NADPH-cytochrome c (P-450) reductase in human, baboon and horse. J. Steroid Biochem., **15**, 449-452 (1981)
[46] Yasukochi, Y.; Masters, B.S.S.: Tetramer-dimer dissociation in homoglobin and the Bohr effect. J. Biol. Chem., **251**, 5537-5544 (1976)
[47] Masters, B.S.S.; Williams, C.H.; Kamin, H.: Preparation and properties of microsomal TPNH-cytochrome c reductase from pig liver. Methods Enzymol., **10**, 565-573 (1967)
[48] Maisterrena, B.; Comte, J.; Gautheron, D.C.: Purification of pig heart mitochondrial membranes, enzymatic and morphological characterization as compared to microsomes. Biochim. Biophys. Acta, **367**, 115-126 (1974)
[49] Mull, R.H.; Voigt, T.; Flemming, K.: Characterization of microsomal electron transport components from control, phenobarbital, and 3-methylcholantrene treated mice: I. Distribution of electron transport components in ammonium sulfate fractions from mouse liver microsomes. Biochem. Biophys. Res. Commun., **64**, 1098-1106 (1975)
[50] Day, D.A.; Wiskich, J.T.: Isolation and properties of the outer membrane of plant mitochondria. Arch. Biochem. Biophys., **171**, 117-123 (1975)
[51] Erecinska, M.: Cytochrome c interaction with membranes, formylated cytochrome c1. Arch. Biochem. Biophys., **169**, 199-208 (1975)
[52] Bowles, D.J.; Kauss, H.: Characterization, enzymatic and lectin properties of isolated membranes from Phaseolus aureus. Biochim. Biophys. Acta, **443**, 360-374 (1976)

[53] Gilkes, N.R.; Weeks, G.: The purification and characterization of Dictyostelium discoideum plasma membranes. Biochim. Biophys. Acta, **464**, 142-156 (1977)
[54] Gilkes, N.R. and Weeks, G.: An improved procedure for the purification of plasma membranes from Dictyostelium discoideum. Can. J. Biochem., **55**, 1233-1236 (1977)
[55] Mull, R.H.; Schgaguler, M.. Mönig, H.; Voigt, T.; Flemming, K.: Characterization of microsomal electron transport components from control, phenobarbital- and 3-methylcholantrene-treated mice, III. Improved resolution and quantitation of major components in ammonium sulfate fractions from total liver microsomes. Biochim. Biophys. Acta, **462**, 671-688 (1977)
[56] Yoshida, Y.; Aoyama, Y.; Kumaoka, H.; Kubota, S.: A highly purified preparation of cytochrome P-450 from microsomes of anaerobically grown yeast. Biochem. Biophys. Res. Commun., **78**, 1005-1010 (1977)
[57] Feyereisen, R.; Durst, F.: Ecdysterone Biosynthesis: A microsomal cytochrome-P-450-linked ecdysone 20-monooxygenase from tissues of the african migratory locust. Eur. J. Biochem., **88**, 37-47 (1978)
[58] Dialameh, G.H.: Isolation and partial purification of cytochrome-P-450 from induced rat liver. Biochem. Biophys. Res. Commun., **81**, 19-23 (1978)
[59] Yoshida, T.; Kikuchi, G.: Purification and properties of heme oxygenase from rat liver microsomes. J. Biol. Chem., **254**, 4487-4491 (1979)
[60] Kitada, M.; Yamazaki, C.; Hirota, K.; Kitagawa, H.: Purification of cytochrome P-450 from liver microsomes of phenobarbital-treated guinea pigs. Biochem. Biophys. Res. Commun., **93**, 1020-1026 (1980)
[61] Hara, T.; Taniguchi, M.: Relationship between changes in properties and contents of riboflavin derivatives of NADPH-cytochrome P-450 reductase in the liver microsomes of riboflavin-deficient rats. J. Biochem., **97**, 473-482 (1985)
[62] Gut, J.; Kawato, S.; Cherry, R.J.; Winterhalter, K.H.; Richter, C.: Lipid peroxidation decreases the rotational mobility of cytochrome P-450 in rat liver microsomes. Biochim. Biophys. Acta, **817**, 217-228 (1985)
[63] Lindeskog, P.; Haaparanta, T.; Norgard, M.; Glaumann, H.; Hansson, T.; Gustafsson J.: Isolation of rat intestinal microsomes: Partial characterisation of mucosal cytochrome P-450. Arch. Biochem. Biophys., **244**, 492-501 (1986)
[64] Kutty, R.K.; Maines, M.D.: Characterization of an NADH-dependent haem-degrading system in ox heart mitochondria. Biochem. J., **246**, 467-474 (1987)
[65] Dupuy, C.; Kaniewski, J.; Deme, D.; Pommier, J.; Virion, A.: NADPH-dependent H_2O_2 generation catalyzed by thyroid plasma membranes. Eur. J. Biochem., **185**, 597-603 (1989)
[66] Shoun, H.; Sudo, Y.; Sato, K.; Beppu, T.: Components of the cytochrome P-450 monooxygenase system of the fungus, Fusarium oxysporun: Conditions for induction and isolation of cytochrome b_5. Agric. Biol. Chem., **53**, 2153-2161 (1989)

[67] Arinc, E.; Sen, A.: Characterization of cytochrome P_{450} dependent mixed-fuction oxidase system of gilthead seabream (Sparus aurata; sparidae) liver. Comp. Biochem. Physiol. B, **104**, 133-139 (1993)
[68] Balvers, W.G.; Boersma, M.G.; Vervoort, J.; Ouwehand, A.; Rietjens, I.M.C.M.: A specific interaction between NADPH-cytochrome reductase and phosphatidylserine and phosphatidylinositol. Eur. J. Biochem., **218**, 1021-1029 (1993)
[69] White, R.D.; Hahn, M.E.; Lockhart, W.L.; Stegeman, J.J.: Catalytic and Immunochemical Characterization of Hepatic Microsomal cytochromes P_{450} in Beluga Whale (Delphinapterus leucas). Toxicol. Appl. Pharmacol., **126**, 45-57 (1994)
[70] Lee, C.A.; Kadwell, S.H.; Kost, T.A.; Serabjit-Singh, C.J.: CYP3A4 expressed by insect cells infected with a recombinant baculovirus containing both CYP3A4 and human NADPH-cytochrome P_{450} reductase is catalytically similar to human liver microsomal CYP3A4. Arch. Biochem. Biophys., **319**, 157-167 (1995)
[71] Murataliev, M.B.; Feyereisen, R.: Functional interactions in cytochrome P_{450}BM3. Fatty acid substrate binding alters electron-transfer properties of the flavoprotein domain. Biochemistry, **35**, 15029-15037 (1996)
[72] McGuire, J.J.; Anderson, D.J.; McDonald, B.J.; Narayanasami, R.; Bennett, B.M.: Inhibition of NADPH-cytochrome P_{450} reductase and glyceryl trinitrate biotransformation by diphenyleneiodonium sulfate. Biochem. Pharmacol., **56**, 881-893 (1998)
[73] Pritchard, M.P.; Glancey, M.J.; Blake, J.A.R.; Gilham, D.E.; Burchell, B.; Wolf, C.R.; Friedberg, T.: Functional co-expression of CYP2D6 and human NADPH-cytochrome P_{450} reductase in Escherichia coli. Pharmacogenetics, **8**, 33-42 (1998)
[74] Venkateswarlu, K.; Lamb, D.C.; Kelly, D.E.; Manning, N.J.; Kelly, S.L.: The N-terminal membrane domain of yeast NADPH-cytochrome P_{450} (CYP) oxidoreductase is not required for catalytic activity in steril biosynthesis or in reconstitution of CYP activity. J. Biol. Chem., **273**, 4492-4496 (1998)
[75] Shen, A.L.; Porter, T.D.; Wilson, T.E.; Kasper, C.B.: Structural analysis of the FMN binding domain of NADPH-cytochrome P-450 oxidoreductase by site-directed mutagenesis. J. Biol. Chem., **264**, 7584-7589 (1989)
[76] Shen, A.L.; Kasper, C.B.: Differential contributions of NADPH-Cytochrome P_{450} oxidoreductase FAD binding site residues to flavin binding and catalysis. J. Biol. Chem., **275**, 41087-41091 (2000)
[77] Menting, J.G.T.; Cornish, E.; Scopes, R.K.: Purification and partial characterization of NADPH-cytochrome c reductase from Petunia hybrida flowers. Plant Physiol., **106**, 643-650 (1994)
[78] Stiborova, M.; Schmeiser, H.H.; Frei, E.: Oxidation of xenobiotics by plant microsomes, a reconstituted cytochrome P450 system and peroxidase: a comparative study. Phytochemistry, **54**, 353-362 (2000)

NADPH-Cytochrome-c_2 reductase 1.6.2.5

1 Nomenclature

EC number
1.6.2.5

Systematic name
NADPH:ferricytochrome-c_2 oxidoreductase

Recommended name
NADPH-cytochrome-c_2 reductase

Synonyms
NADPH-cytochrome f reductase <5> [1]
cytochrome c_2 reductase (reduced nicotinamide adenine dinucleotide phosphate, NADPH)
reductase, cytochrome c_2 (reduced nicotinamide adenine dinucleotide phosphate)

CAS registry number
37256-32-9

2 Source Organism

<1> *Homo sapiens* (human, Morris hepatoma 7800 C_1 cells [3]) [3]
<2> *Mus musculus* (mouse, strain C57BL76J [2]) [2]
<3> *Rattus norvegicus* (rat, Wistar [3]) [3]
<4> *Rhodopseudomonas palustris* [1]
<5> *Rhodopseudomonas sphaeroides* [1]

3 Reaction and Specificity

Catalyzed reaction
NADPH + H^+ + 2 ferricytochrome c_2 = $NADP^+$ + 2 ferrocytochrome c_2 (a flavoprotein (FAD))

Reaction type
oxidation
redox reaction
reduction

Natural substrates and products

S ferricytochrome c_2 + NADPH <1-5> (<5> native cytochrome is the natural physiological electron acceptor [1]) (Reversibility: ? <1-5> [1-3]) [1-3]

P ferrocytochrome c_2 + $NADP^+$

Substrates and products

S 2,6-dichloroindophenol + NADPH <5> (Reversibility: ? <5> [1]) [1]

P ? + $NADP^+$

S $Fe(CN)_6^{3-}$ + NADPH <5> (Reversibility: ? <5> [1]) [1]

P $Fe(CN)_6^{4-}$ + $NADP^+$

S Rhodospirillum rubrum cytochrome c_2 + NADPH <5> (Reversibility: ? <5> [1]) [1]

P ? + $NADP^+$

S ferricytochrome c_2 + NADPH <1-5> (<4, 5> contains additional $NADP^+$ diaphorase activity [1]) (Reversibility: ? <1-5> [1-3]) [1-3]

P ferrocytochrome c_2 + $NADP^+$

S horse heart cytochrome c + NADPH <5> (Reversibility: ? <5> [1]) [1]

P ? + $NADP^+$

S mammalian cytochrome c + NADPH <5> (Reversibility: ? <5> [1]) [1]

P ? + $NADP^+$

S Additional information <5> (<5> cytochrome c-553 is ineffective as an electron acceptor [1]) [1]

P ?

Inhibitors

N-ethylmaleimide <5> [1]
NADH <5> (<5> competitive inhibitor [1]) [1]
iodoacetate <5> [1]
p-chloromercuribenzoate <5> [1]
thyroxine <5> [1]

Cofactors/prosthetic groups

FAD <5> (<5> nonmetalloflavoprotein with FAD as prosthetic group [1]) [1]

Specific activity (U/mg)

164 <5> [1]

K_m-Value (mM)

0.0125 <5> (2,6-dichloroindophenol) [1]
0.037 <5> (ferricytochrome c_2) [1]
0.125 <5> ($K_3Fe(CN)_6$) [1]

K_i-Value (mM)

0.055 <5> (NADH) [1]

pH-Optimum

7.5 <5> [1]

pH-Range

5.6-8 <5> [1]

4 Enzyme Structure

Molecular weight

43000 <5> (<5> gel filtration [1]) [1]

5 Isolation/Preparation/Mutation/Application

Source/tissue

hepatocyte <3> [3]
liver <1, 2> (<1> hepatoma cell line Morris 7800 C1 [3]) [2, 3]

Purification

<5> [1]

6 Stability

General stability information

<5>, activity is lost overnight if the enzyme is diluted [1]

Storage stability

<5>, -15°C, can be stored as lyophilized powder for at least 3 months [1]
<5>, 4°C, 0.4 M Tris-HCl, pH 8.0 purified concentrated enzyme is stable for approximately 2 weeks, gradual loss in activity can be restored by addition of glutathione to a final concentration of 0.0001 M, partially purified enzyme is stable for approximately 1 month [1]

References

[1] Sabo, D.J.; Orlando, J.A.: Isolation, purification, and some properties of reduced nicotinamide adenine dinucleotide phosphate-cytochrome c_2 reductase from Rhodopseudomonas spheroides. J. Biol. Chem., **243**, 3742-3749 (1968)

[2] Feuers, R.J.; Delongchamp, R.R.; Casciano, D.A.; Burkhart, J.G.; Mohrenweiser, H.W.: Assay for mouse tissue enzymes: levels of activity and statistical variation for 29 enzymes of liver or brain. Anal. Biochem., **101**, 123-130 (1979)

[3] Norrheim, L.; Sorensen, H.; Gautvik, K.; Bremer, J.; Spydevold O.: Synergistic actions of tetradecylthioacetic acid (TTA) and dexamethasone on induction of the peroxisomal β-oxidation and on growth inhibition of Morris hepatoma cells. Both effects are counteracted by insulin. Biochim. Biophys. Acta, **1051**, 319-323 (1990)

Leghemoglobin reductase 1.6.2.6

1 Nomenclature

EC number
1.6.2.6

Systematic name
NAD(P)H:ferrileghemoglobin oxidoreductase

Recommended name
leghemoglobin reductase

Synonyms
ferric leghemoglobin reductase

CAS registry number
60440-35-9

2 Source Organism

<1> *Glycine max* (soybean [1-3]) [1-7]
<2> *Phaseolus vulgaris* (bean [2]) [2]
<3> *Vigna unguiculata* (cowpea [2]) [2, 8]

3 Reaction and Specificity

Catalyzed reaction
NAD(P)H + H^+ + 2 ferrileghemoglobin = $NAD(P)^+$ + 2 ferroleghemoglobin

Reaction type
oxidation
redox reaction
reduction

Natural substrates and products
S NAD(P)H + ferrileghemoglobin <1> (<1> proposed to be involved in maintaining leghemoglobin in an active form in legume nodules [6]) (Reversibility: ? <1> [1, 2, 3, 4]) [1, 2, 3, 4, 6]
P $NAD(P)^+$ + ferroleghemoglobin <1> [1, 2, 3, 4, 6]

Substrates and products
S 2,6-dichloroindophenol + NADH <1, 3> (<1> no activity in the absence of [4]) (Reversibility: ? <1, 3> [4, 8]) [4, 8]

P reduced 2,6-dichloroindophenol + NAD^+ <1, 3> [4, 8]
S NAD(P)H + ferricytochrome c <1> (<1> reduced at 53% of leghemoglobin rate [1]) (Reversibility: ? <1> [1]) [1]
P $NAD(P)^+$ + ferrocytochrome c <1> [1]
S NAD(P)H + ferrileghemoglobin <1, 3> (<1> 31% activity with NADPH [1]; <1> no activity in the absence of O_2, formation of H_2O_2 or peroxide intermediates during NADH oxidation [4]; <3> reduces both cowpea ferrihemoglobin and soybean ferrileghemoglobin [8]) (Reversibility: ? <1, 3> [1, 2, 3, 4, 8]) [1, 2, 3, 4, 8]
P $NAD(P)^+$ + ferroleghemoglobin <1, 3> [1, 2, 3, 4, 8]
S NAD(P)H + ferrimyoglobin <1> (Reversibility: ? <1> [1]) [1]
P $NAD(P)^+$ + ferromyoglobin <1> [1]
S NADH + lipoamide <1> (<1> recombinant enzyme, lipoamide-dependent NADH oxidation [6]) (Reversibility: ? <1> [6]) [6]
P NAD^+ + lipoamide <1> [6]

Inhibitors

5,5'-dithiobis(2-nitrobenzoic acid) <1> [7]
$CdSO_4$ <1> [7]
$Cu(NO_3)_2$ <1> [7]
acetate <1> (<1> slight [1]) [1]
amobarbital <1> (<1> 1 mM, 35% after 0.5-1 h, 80% after 4 h [2]) [2]
arsenic acid <1> [7]
arsenic trioxide <1> [7]
catalase <1> (<1> possibly H_2O_2 as reaction intermediate [2]) [2]
iodoacetamide <1> (<1> 1 mM, 95-100% inhibition [2]) [1, 2]
nicotinate <1> (<1> ferrileghemoglobin-nitrite complex [2]) [1, 2]
p-chloromercuribenzoate <1> [7]
p-hydroxymercuribenzoate <1> (<1> 0.5 mM, slight inhibition [2]) [2]
phenylarsine oxide <1> [7]
quinacrine <1> (<1> 0.5 mM, 30-40% inhibition [2]) [2]

Cofactors/prosthetic groups

FAD <1> [3]
NADH <1-3> [1, 2]
NADPH <1-3> (<1> 31% of NADH activity [1]; <1-3> 54-80% of NADH activity [2]) [1, 2]

Turnover number (min^{-1})

150 <3> (ferrileghemoglobin, <3> soybean ferrileghemoglobin [8]) [8]
186 <3> (ferrileghemoglobin, <3> cowpea ferrileghemoglobin [8]) [8]
372 <1> (ferrileghemoglobin, <1> native enzyme [6]) [6]
372 <1> (ferrileghemoglobin, <1> recombinant enzyme [6]) [6]
1860 <1> (lipoamide, <1> recombinant enzyme [6]) [6]

Specific activity (U/mg)

0.215 <1> (<1> reduction of ferrileghemoglobin [2]) [2]
0.216 <3> (<3> reduction of ferrileghemoglobin [8]) [8]
0.218 <1> (<1> reduction of ferrileghemoglobin [1]) [1]

0.388 <1> (<1> reduction of ferrileghemoglobin in the presence of O_2 [4]) [4]
0.486 <1> (<1> recombinant enzyme, reduction of ferrileghemoglobin [6]) [6]
0.5 <1> (<1> reduction of ferrileghemoglobin [3]) [3]
1.85 <1> (<1> oxidation of NADH in the presence of O_2, 30fold higher than in the absence of O_2 [4]) [4]
2 <1> (<1> reduction of 2,6-dichlorophenolindophenol [2]) [2]
2.565 <3> (<3> reduction of 2,6-dichloroindophenol [8]) [8]
4.47 <1> (<1> recombinant enzyme, reduction of 2,6-dichloroindophenol [6]) [6]
4.5 <1> (<1> reduction of 2,6-dichloroindophenol in the presence of O_2 [4]) [4]
4.89 <1> (<1> reduction of 2,6-dichlorophenolindophenol [3]) [3]

K_m-Value (mM)
0.0063 <1> (ferrileghemoglobin, <1> recombinant enzyme [6]) [6]
0.0092 <1> (ferrileghemoglobin, <1> native enzyme [6]) [6]
0.0095 <1> (ferrileghemoglobin) [1]
0.0104 <3> (ferrileghemoglobin, <3> cowpea ferrileghemoglobin [8]) [8]
0.0124 <3> (ferrileghemoglobin, <3> soybean ferrileghemoglobin [8]) [8]
0.0133 <2, 3> (ferrileghemoglobin) [2]
0.0188 <1> (NADH) [1]
0.046 <1> (NADH, <1> lipoamide-dependent NADH oxidation [6]) [6]
0.051 <1> (NADH) [3]
0.057 <3> (NADH) [8]
0.088 <1> (ferrileghemoglobin) [2]
0.716 <1> (lipoamide, <1> lipoamide-dependent NADH oxidation [6]) [6]

pH-Optimum
5.2 <1> (<1> decrease of activity from 100% at pH 5.2, to 30% at pH 6.7, no measurements below 5.2 [1]) [1]
6.5 <3> (<3> reduction of ferrileghemoglobin [8]) [8]

pH-Range
4.8-8.5 <3> (<3> no activity below and above [8]) [8]

4 Enzyme Structure

Molecular weight
83000 <1> (<1> equilibrium ultracentrifugation [1]) [1]
100000 <1> (<1> gel filtration [1]) [1]
110000 <1, 3> (<1, 3> gel filtration [3, 8]) [3, 8]

Subunits
dimer <1> (<1> 2 * 54000, SDS-PAGE [1, 3, 7]; <1,3> 2 * 55000, SDS-PAGE [2, 8]; <3> 2 * 56000, deduced from amino acid sequence [8]) [1-3, 7, 8]

5 Isolation/Preparation/Mutation/Application

Source/tissue

leaf <1> (<1> high mRNA level [5]) [5]
nodule <1> (<1> high mRNA level [5]) [5]
root nodule <1-3> [1-5, 8]
stem <1> (<1> low mRNA level [5]) [5]

Localization

cytosol <1> [1, 3]

Purification

<1> (recombinant enzyme, Sepharose 6B, Probond [6]) [1, 3, 6, 7]
<3> (ammonium sulfate, hydroxylapatite, ion exchange, gel filtration [8]) [8]

Cloning

<1> (cloning of cDNA [5]; expression in Escherichia coli [6, 7]) [5, 6, 7]
<3> (cloning of cDNA [8]) [8]

6 Stability

Temperature stability

Additional information <1> (<1> heat labile [1]) [1]

Storage stability

<1>, -70°C [3]
<1>, 0°C, 37 days, 96% activity retains [1]

References

[1] Saari, L.L.; Klucas, R.V.: Ferric leghemoglobin reductase from soybean root nodules. Arch. Biochem. Biophys., **231**, 102-113 (1984)

[2] Becana, M.; Klucas, R.V.: Enzymatic and nonenzymatic mechanisms for ferric leghemoglobin reduction in legume root nodules. Proc. Natl. Acad. Sci. USA, **87**, 7295-7299 (1990)

[3] Ji, L.; Wood, S.; Becana, M.; Klucas, R.V.: Purification and characterization of soybean root nodule ferric leghemoglobin reductase. Plant Physiol., **96**, 32-37 (1991)

[4] Ji, L.; Becana, M.; Klucas, R.V.: Involvement of molecular oxygen in the enzyme-catalyzed NADH oxidation and ferric leghemoglobin reduction. Plant Physiol., **100**, 33-39 (1992)

[5] Ji, L.; Becana, M.; Sarath, G.; Klucas, R.V.: Cloning and sequence analysis of a cDNA encoding ferric leghemoglobin reductase from soybean nodules. Plant Physiol., **104**, 453-459 (1994)

[6] Ji, L.; Becana, M.; Sarath, G.; Shearman, L.; Klucas, R.V.: Overproduction in Escherichia coli and characterization of a soybean ferric leghemoglobin reductase. Plant Physiol., **106**, 203-209 (1994)
[7] Kim, H.M.: Redox potential of a soybean ferric leghemoglobin reductase. J. Biochem. Mol. Biol., **31**, 444-452 (1998)
[8] Luan, P.; Arechaga-Ocampo, E.; Sarath, G.; Arredondo-Peter, R.; Klucas, R.V.: Analysis of a ferric leghemoglobin reductase from cowpea (Vigna unguiculata) root nodules. Plant Sci., **154**, 161-170 (2000)

NADPH Oxidase 1.6.3.1

1 Nomenclature

EC number
1.6.3.1

Systematic name
NAD(P)H:oxygen oxidoreductase

Recommended name
NADPH oxidase

Synonyms
HOX2
ThOX
dual oxidase
p138tox
thyroid NADPH oxidase
thyroid oxidase
thyroid oxidase 2

CAS registry number
9032-22-8

2 Source Organism

<1> *Homo sapiens* [1-6]
<2> *Sus scrofa* [4]

3 Reaction and Specificity

Catalyzed reaction
NAD(P)H + H^+ + O_2 = $NAD(P)^+$ + H_2O_2 (Requires FAD, heme and calcium. When calcium is present, this transmembrane glycoprotein generates H_2O_2 by transfering electrons from intracellular NAD(P)H to extracellular molecular oxygen. The electron bridge within the enzyme contains one molecule of FAD and probably two heme groups. This flavoprotein is expressed at the apical membrane of thyrocytes, and provides H_2O_2 for the thyroid peroxidase-catalysed biosynthesis of thyroid hormones)

Reaction type
oxidation
redox reaction
reduction

Natural substrates and products
S NAD(P)H + H^+ + O_2 <1, 2> (Reversibility: ? <1, 2> [1-6]) [1-6]
P $NAD(P)^+$ + H_2O_2

Substrates and products
S NAD(P)H + H^+ + O_2 <1, 2> (Reversibility: ? <1, 2> [1-6]) [1-6]
P $NAD(P)^+$ + H_2O_2

K_i-Value (mM)
0.0002 <5> (*p*-chloromercuribenzoate) [8]
0.003 <5> (phenylarsine oxide, <5> 3 nmol/mg protein [11]) [11]
0.026 <2> (propylthiouracil) [13]
0.031 <2> (methimazol) [13]

References

[1] Moreno, J.C.; Bikker, H.; Kempers, M.J.; van Trotsenburg, A.S.; Baas, F.; de Vijlder, J.J.; Vulsma, T.; Ris-Stalpers, C.: Inactivating mutations in the gene for thyroid oxidase 2 (THOX2) and congenital hypothyroidism. N. Engl. J. Med., **347**, 95-102 (2002)

[2] De Deken, X.; Wang, D.; Dumont, J.E.; Miot, F.: Characterization of ThOX proteins as components of the thyroid H_2O_2-generating system. Exp. Cell Res., **273**, 187-196 (2002)

[3] De Deken, X.; Wang, D.; Many, M.C.; Costagliola, S.; Libert, F.; Vassart, G.; Dumont, J.E.; Miot, F.: Cloning of two human thyroid cDNAs encoding new members of the NADPH oxidase family. J. Biol. Chem., **275**, 23227-23233 (2000)

[4] Dupuy, C.; Ohayon, R.; Valent, A.; Noel-Hudson, M.S.; Deme, D.; Virion, A.: Purification of a novel flavoprotein involved in the thyroid NADPH oxidase. Cloning of the porcine and human cDNAs. J. Biol. Chem., **274**, 37265-37269 (1999)

[5] Leseney, A.M.; Deme, D.; Legue, O.; Ohayon, R.; Chanson, P.; Sales, J.P.; Pires de Carvalho, D.; Dupuy, C.; Virion, A.: Biochemical characterization of a Ca^{2+}/NAD(P)H-dependent H_2O_2 generator in human thyroid tissue. Biochimie, **81**, 373-380 (1999)

[6] Dupuy, C.; Virion, A.; Ohayon, R.; Kaniewski, J.; Deme, D.; Pommier, J.: Mechanism of hydrogen peroxide formation catalyzed by NADPH oxidase in thyroid plasma membrane. J. Biol. Chem., **266**, 3739-3743 (1991)

Cystine reductase (NADH) 1.6.4.1

1 Nomenclature

EC number
1.6.4.1 (transferred to EC 1.8.1.6)

Recommended name
cystine reductase (NADH)

Glutathione reductase (NADPH) 1.6.4.2

1 Nomenclature

EC number
1.6.4.2 (transferred to EC 1.8.1.7)

Recommended name
glutathione reductase (NADPH)

Dihydrolipoamide reductase (NAD^+) 1.6.4.3

1 Nomenclature

EC number
1.6.4.3 (transferred to EC 1.8.1.4)

Recommended name
dihydrolipoamide reductase (NAD^+)

Protein-disulfide reductase [NAD(P)H] 1.6.4.4

1 Nomenclature

EC number

1.6.4.4 (transferred to EC 1.8.1.8)

Recommended name

protein-disulfide reductase [NAD(P)H]

Thioredoxin reductase (NADPH) 1.6.4.5

1 Nomenclature

EC number
1.6.4.5 (transferred to EC 1.8.1.9)

Recommended name
thioredoxin reductase (NADPH)

CoA-Glutathione reductase (NADPH) 1.6.4.6

1 Nomenclature

EC number
1.6.4.6 (transferred to EC 1.8.1.10)

Recommended name
CoA-glutathione reductase (NADPH)

Asparagusate reductase (NADH) 1.6.4.7

1 Nomenclature

EC number
1.6.4.7 (transferred to EC 1.8.1.11)

Recommended name
asparagusate reductase (NADH)

Trypanothione reductase 1.6.4.8

1 Nomenclature

EC number
1.6.4.8 (transferred to EC 1.8.1.12)

Recommended name
trypanothione reductase

Bis-γ-glutamylcystine reductase (NADPH) 1.6.4.9

1 Nomenclature

EC number
1.6.4.9 (transferred to EC 1.8.1.13)

Recommended name
bis-γ-glutamylcystine reductase (NADPH)

CoA-Disulfide reductase (NADH) 1.6.4.10

1 Nomenclature

EC number

1.6.4.10 (transferred to EC 1.8.1.14)

Recommended name

CoA-disulfide reductase (NADH)

Quinone reductase 1.6.5.1

1 Nomenclature

EC number
1.6.5.1 (deleted)

Recommended name
quinone reductase

Menadione reductase 1.6.5.2

1 Nomenclature

EC number
1.6.5.2 (transferred to EC 1.6.99.2)

Recommended name
menadione reductase

NADH Dehydrogenase (ubiquinone) 1.6.5.3

1 Nomenclature

EC number
1.6.5.3

Systematic name
NADH:ubiquinone oxidoreductase

Recommended name
NADH dehydrogenase (ubiquinone)

Synonyms
13 kDa differentiation-associated protein
AMAPOR <25> [49]
CDA016
CGI-39
CI-11KD
CI-12KD
CI-14.8KD
CI-14KD
CI-15 kDa
CI-16KD
CI-17.3KD
CI-17.8KD
CI-18 kDa
CI-18Kd
CI-19.3KD
CI-19KD
CI-20KD
CI-21KD
CI-22.5Kd
CI-23KD
CI-27KD
CI-28.5KD
CI-29.9KD
CI-29KD
CI-30KD
CI-31KD
CI-38.5KD
CI-39KD
CI-40KD
CI-42.5KD

CI-42KD
CI-49KD
CI-51KD
CI-75KD
CI-78KD
CI-9.5
CI-9KD
CI-AGGG
CI-AQDQ
CI-ASHI
CI-B12
CI-B14
CI-B14.5a
CI-B14.5b
CI-B15
CI-B16.6
CI-B17
CI-B17.2
CI-B18
CI-B22
CI-B8
CI-B9
CI-KFYI
CI-MLRQ
CI-MNLL
CI-MWFE
CI-PDSW
CI-PGIV
CI-SGDH
CIB17.2
cell adhesion protein SQM1
cell death-regulatory protein GRIM-19
complex I-11KD
complex I-12KD
complex I-14.8KD
complex I-14KD
complex I-15 kDa
complex I-16KD
complex I-17.3KD
complex I-17.8KD
complex I-18 kDa
complex I-18Kd
complex I-19.3KD
complex I-19KD
complex I-20KD
complex I-21KD
complex I-22.5Kd

complex I-23KD
complex I-27KD
complex I-28.5KD
complex I-29.9KD
complex I-29KD
complex I-30KD
complex I-38.5KD
complex I-39KD
complex I-40KD
complex I-42.5KD
complex I-42KD
complex I-49KD
complex I-51KD
complex I-75KD
complex I-78KD
complex I-9.5KD
complex I-9KD
complex I-AGGG
complex I-AQDQ
complex I-ASHI
complex I-B12
complex I-B14
complex I-B14.5a
complex I-B14.5b
complex I-B15
complex I-B16.6
complex I-B17
complex I-B17.2
complex I-B18
complex I-B22
complex I-B8
complex I-B9
complex I-KFYI
complex I-MLRQ
complex I-MNLL
complex I-MWFE
complex I-PDSW
complex I-PGIV
complex I-SGDH
DPNH-coenzyme Q reductase
DPNH-ubiquinone reductase
GGHPW
GRIM-19
gene associated with retinoic-interferon-induced mortality 19 protein
hypothetical protein Walter
internal NADH dehydrogenase
NADH coenzyme Q1 reductase

NADH-CoQ oxidoreductase
NADH-CoQ reductase
NADH-Q_6 oxidoreductase
NADH-coenzyme Q oxidoreductase
NADH-coenzyme Q reductase
NADH-ferricyanide reductase
NADH-quinone oxidoreductase <4, 17> [30, 31, 32]
NADH-quinone reductase
NADH-ubiquinone oxidoreductase
NADH-ubiquinone reductase
NADH-ubiquinone-1 reductase
NADH:Q oxidoreductase
NADH:Caldariella quinone oxidoreductase <24> [48]
NADH:ubiquinone oxidoreductase complex
NDH-1 <2, 4, 17, 18> [7, 31, 32, 37]
NDH-1 <1> (<1>, proton-pumping NADH-ubiquinone oxidoreductase [17]) [17]
NDH-2 <2, 10> (<2, 10>, non-proton-pumping NADH-ubiquinone oxidoreductase [17]) [17]
NDH-II <24> [48]
Na^+-NQR <19, 20> [38, 40]
protein P1
ubiquinone-binding protein
artificial mediator accepting pyridine nuclöeotide oxidoreductase <25> [49]
coenzyme Q reductase
complex I (NADH:Q_1 oxidoreductase)
complex I (electron transport chain)
complex I (mitochondrial electron transport)
complex I dehydrogenase
dihydronicotinamide adenine dinucleotide-coenzyme Q reductase
electron transfer complex I
mitochondrial electron transport complex 1
mitochondrial electron transport complex I
proton-translocating NADH-quinone oxidoreductase <17> [32]
reduced nicotinamide adenine dinucleotide-coenzyme Q reductase
reductase, ubiquinone
sodium motive NADH:quinone oxidoreductase <19> [38]
type I dehydrogenase
ubiquinone reductase
Additional information (the complex, present in mitochondria, can be degraded to form EC 1.6.99.3)

CAS registry number

9028-04-0

2 Source Organism

<1> *Bos taurus* (rotenone-sensitive enzyme complex [16]; NDH-1 [17]) [1, 2, 9, 14, 16, 17, 24, 28, 42, 43, 46]
<2> *Escherichia coli* (GR19N [3]; wild-type and cells defective for NADH dehydrogenase I [6]; enzyme form NDH-1 [7]; NDH-1 and NDH-2 [17]) [3, 6, 7, 17, 37, 43]
<3> *Saccharomyces cerevisiae* (rotenone-insensitive enzyme [4, 25, 33, 34]) [4, 25, 33, 34, 43]
<4> *Paracoccus denitrificans* [5, 10, 30, 31, 35, 36, 42, 43]
<5> *Beta vulgaris* [8, 26]
<6> *Phaseolus aureus* [10]
<7> *Arum maculatum* [11]
<8> *Neurospora crassa* (rotenone-sensitive enzyme [13]) [12, 13]
<9> *Yarrowia lipolytica* [15]
<10> *Solanum tuberosum* (NDH-1 and NDH-2 [17]) [17, 23]
<11> *Rattus norvegicus* [18, 19]
<12> *Vibrio alginolyticus* [20, 37, 41]
<13> *Trypanosoma brucei* (brucei [27]) [21, 27]
<14> *Homo sapiens* [22]
<15> *Chlamydomonas reinhardtii* [29]
<16> *Corynebacterium glutamicum* [50]
<17> *Thermus thermophilus* (HB-8 [32, 35, 39, 44]) [32, 35, 39, 43, 44]
<18> *Klebsiella pneumonia* [37]
<19> *Vibrio harveyi* [38]
<20> *Vibrio cholerae* [40]
<21> *Bacillus subtilis* [43]
<22> *Sulfolobus acidocaldarius* [45]
<23> *Candida utilis* [47]
<24> *Acidianus ambivalens* [48]
<25> *Clostridium thermoaceticum* [49]

3 Reaction and Specificity

Catalyzed reaction

NADH + H^+ + ubiquinone = NAD^+ + ubiquinol (<3, 13>, ping-pong mechanism [21, 25])

Reaction type

oxidation
redox reaction
reduction

Natural substrates and products

S NADH + ubiquinone <1, 2, 4, 5, 9, 12, 16, 17, 24> (<2>, mutations in NADH:ubiquinone oxidoreductase NADH of Escherichia coli affect growth on mixed amino acids, because the large NADH/NAD^+ ratio inhibits enzymes, e.g. citrate synthase and malate dehydrogenase, shared by the tricarboxylic acid cycle and the glyoxylate shunt [6]; <9>, proton-translocating NADH-ubiquinone oxidoreductase is the largest multiprotein complex of the respiratory chain [15]; <12>, the enzyme functions as a primary Na^+ pump [20]; <3>, the enzyme catalyzes the transfer of electrons without translocation of protons across the membrane [25]; <5>, NADH-ubiquinone reductase complex I is involved in the respiratory chain [26]; <1, 4, 17>, proton-translocating enzyme [17, 30, 32, 36, 39]; <2, 12, 18, 19, 20>, Na^+-translocation [37, 38, 40]; <12>, the enzyme couples the exergonic oxidation of NADH with ubiquinone to the transport of Na^+ ions from the inside of a bacterial cell to the periplasmic space [37]; <12>, the Na^+ pump is coupled to the respiratory chain at the step of NADH:quinone oxidoreductase [41]; <1>, vectorial electron translocation is coupled to electron transfer [42]; <16, 24>, enzyme is involved in respiratory chain [48, 50]) (Reversibility: ? <1, 2, 4, 5, 9, 12, 16, 17, 24> [6, 15, 17, 20, 25, 26, 30, 32, 36, 37, 38, 39, 40, 48, 50]) [6, 15, 17, 20, 25, 26, 30, 32, 36, 37, 38, 39, 40, 41, 42, 48, 50]

P ?

Substrates and products

S NADH + 1,1'-carbamoylmethylviologen <25> (Reversibility: r <25> [49]) [49]

P NAD^+ + reduced 1,1'-carbamoylmethylviologen

S NADH + 2,6-dichlorophenolindophenol <1, 3, 7, 13, 22, 23, 25> (<7>, 41.7% of the activity with ubiquinone-1 [11]; <3>, 2,6-dichlorophenol indophenol [25]) (Reversibility: ? <1, 3, 7, 13, 22, 23, 25> [11, 21, 25, 46, 47, 49]) [11, 21, 25, 45, 46, 47, 49]

P NAD^+ + reduced 2,6-dichlorophenol indophenol

S NADH + 2-methylnaphthoquinone <1, 12> (<1>, 7.6% of the activity with ubiquinone-1 [14]) (Reversibility: ? <1, 12> [14, 20, 46]) [14, 20, 46]

P NAD^+ + ?

S NADH + FAD <25> (Reversibility: ? <25> [49]) [49]

P NAD^+ + $FADH_2$

S NADH + O_2 <13> (Reversibility: ? <13> [21]) [21]

P NAD^+ + superoxide radical <13> [21]

S NADH + anthraquinone-2,6-disulfonate <25> (Reversibility: ? <25> [49]) [49]

P NAD^+ + ?

S NADH + benzoquinone <22> (Reversibility: ? <22> [45]) [45]

P NAD^+ + benzoquinol

S NADH + caldariellaquinone <22, 24> (Reversibility: ? <22, 24> [45, 48]) [45, 48]

P NAD^+ + caldariellaquinol

S NADH + cytochrome c <1, 7, 23> (<7>, 8.3% of the activity with ubiquinone-1 [11]; <1>, 12%-16% of the activity with ubiquinone-1 [14]) (Reversibility: ? <1, 7, 23> [11, 14, 46, 47]) [11, 14, 46, 47]

P NAD^+ + reduced cytochrome c

S NADH + decylubiquinone <1, 2, 9> (<1>, no activity with the enzymatically active subcomplexes Ilambda, IS, and IlamdaS at similar rates to complex I [2]; <9>, n-decylubiquinone [15]) (Reversibility: ? <1, 2, 9> [2, 7, 9, 15]) [2, 7, 9, 15]
P NAD^+ + decylubiquinol
S NADH + duroquinone <7> (<7>, 64.5% of the activity with ubiquinone-1 [11]) (Reversibility: ? <7> [11]) [11]
P NAD^+ + duroquinol
S NADH + ferricyanide <1, 5, 7, 23, 24> (<1>, activity is 27.4fold of the activity with ubiquinone-1 [14]; <7>, 232.7% of the activity with ubiquinone-1 [11]) (Reversibility: ? <1, 5, 7, 23, 24> [2, 8, 11, 14, 46, 47]) [2, 8, 11, 14, 46, 47, 48]
P NAD^+ + ferrocyanide
S NADH + hexamineruthenium(III)-chloride <9> (Reversibility: ? <9> [15]) [15]
P NAD^+ + ?
S NADH + naphthoquinone <22> (Reversibility: ? <22> [45]) [45]
P NAD^+ + naphthoquinol
S NADH + ubiquinone <1-25> (<1, 4, 17>, proton-translocating enzyme [17, 30, 32, 36, 39]; <2, 12, 18, 19, 20>, Na^+-translocation [37, 38, 40]) (Reversibility: ? <1-25> [1-50]) [1-50]
P NAD^+ + ubiquinol
S NADH + ubiquinone-0 <11, 13, 25> (Reversibility: ? <11, 13, 25> [18, 21, 49]) [18, 21, 49]
P NAD^+ + ubiquinol-0
S NADH + ubiquinone-1 <1, 7, 12, 13, 16, 23, 24> (<1>, reaction is catalyzed by the enzymatically active subcomplexes Ilambda, IS, and IlamdaS at similar rates to complex I [2]; <12>, electron transfer to ubiquinone-1 involves the semiquinone radical as an intermediate [20]) (Reversibility: ? <1, 7, 12, 13, 16, 23, 24> [2, 7, 11, 14, 16, 20, 21, 42, 47, 48, 50]) [2, 7, 11, 14, 16, 20, 21, 42, 47, 48, 50]
P NAD^+ + ubiquinol-1
S NADH + ubiquinone-10 <11> (Reversibility: ? <11> [19]) [19]
P NAD^+ + ubiquinol-10
S NADH + ubiquinone-2 <2, 3, 10, 13, 24> (Reversibility: ? <2, 3, 10, 13, 24> [4, 7, 21, 23, 48]) [4, 7, 21, 23, 48]
P NAD^+ + reduced ubiquinol-2
S NADH + ubiquinone-6 <1, 3> (<1>, 86% of the activity with ubiquinone-1 [14]) (Reversibility: ? <1, 3> [4, 14, 25]) [4, 14, 25]
P NAD^+ + ubiquinol-6
S NADH + ubiquinone-9 <11> [19]
P NAD^+ + ubiquinol-9
S NADPH + 2-methylnaphthoquinone <1> (<1>, about 1% of the activity with NADH [46]) (Reversibility: ? <1> [46]) [46]
P $NADP^+$ + ?
S NADPH + ferricyanide <1, 22, 25> (<1>, 0.13% of the activity with NADH and ferricyanide [14]; <1>, ping pong bi bi mechanism [42];

<1>, about 1% of the activity with NADH [46]) (Reversibility: ? <1, 22, 25> [14, 42, 45, 46, 48, 49]) [14, 42, 45, 46, 48, 49]

P $NADP^+$ + ferrocyanide

S NADPH + ubiquinone-1 <1, 5, 13, 16> (<5>, at a slow rate [8]) (Reversibility: ? <1, 5, 13, 16> [8, 21, 46, 50]) [8, 21, 46, 50]

P $NADP^+$ + ubiquinol-1

S Additional information <1, 5> (<5>, no transhydrogenase activity with NADPH [8]; <1>, NADH-NADP transhydrogenation at a very slow rate [46]) [8, 46]

P ?

Inhibitors

2-heptyl-4-hydroxyquinoline <12> [41]
4-(4-benzyl-phenoxy)-N-(3,4-dimethoxy-benzyl)-benzamide <2> [17]
4-(4-tert-butyl-phenoxy)-N-(3,4-dimethoxy-benzyl)-benzamide <1, 10> [17]
AMP <3> (<3>, dead-end inhibitor, linear competitive inhibitor of NADH, linear uncompetitive inhibtor of oxidized 2,6-dichlorophenol indophenol [25]) [25]
KCN <3> [33]
Mg^{2+} <1> (<1>, inhibition of the enzyme in its inactive form, in its active form the enzyme complex is not sensitive [24]) [24]
N-(3,4-dimethoxy-benzyl)-4-[4-(1,1-dimethyl-propyl)-phenoxy]-benzamide <2> [17]
NAD^+ <5> (<5>, 4 mM, 19% inhibition of NADH-ubiquinone-1 reductase activity [8]) [8]
NADH <1, 12> (<12>, incubation of the aerobic enzyme with NADH in the absence of an electron acceptor, the enzyme is destroyed with a half-inactivation time of about 2 min [20]; <1>, inhibits the interaction of the reduced enzyme with ferricyanide [42]) [20, 42]
NEM <1> (<1>, inhibition of the enzyme in its inactive form, in its active form the enzyme complex is not sensitive [24]) [24]
Triton X-100 <1, 22> (<1>, specific inhibitor of ubiquinone reduction by complex I [16]) [16, 45]
barbiturates <1> [14]
bullatacin <4, 17> [35]
capsaicin <1, 2, 4, 10, 13, 17> (<1, 2, 10>, and synthetic capsaicin analogue. Several synthetic capsaicins discriminate between NDH-1 and NDH-2 much better than natural capsaicin [17]; <1, 2, 4, 17>, inhibition of enzymes involved in energy coupling site [43]; <3, 21>, no inhibition of enzymes where the energy couling site is absent [43]) [17, 27, 43]
demerol <1> [14]
dicyclohexylcarbodiimide <5> [26]
diphenyl iodonium <13> (<13>, maximal inhibition after preincubation with NADH, more complete inhibition with the more hydrophobic electron acceptors such as ubiquinone-1 or ubiquinone-2 as electron acceptor compared to the more hydrophilic ones, such as ubiquinone-0 or dichloroindophenol [21]) [21]

doxorubicin <11> [18]
ferricyanide <1> (<1>, inhibits the interaction of the oxidized enzyme with NADH [42]) [42]
flavone <3, 7> (<7>, 0.165 mM, 50% reduction of activity with duroquinone [11]; <13>, insensitive to [21]; <3>, dead-end inhibitor [25]; <3>, partial inhibitor displaying a hyperbolic uncompetitive inhibition with respect to oxidized 2,6-dichlorophenol indophenol [25]) [11, 21, 25]
mersalyl <5> (<5>, 0.3 mM, 77% inhibition of NADH-ubiquinone-1 reductase activity, 96% inhibition of NADH-potassium ferricyanide reductase activity [8]) [8]
molvizarin <1> [28]
otivarin <1> [28]
p-chloromercuriphenyl sulfonic acid <23> (<23>, 0.1 mM, 50% inhibition [47]) [47]
piercidin A <1, 4, 13, 17> [14, 27, 18, 35]
pyridaben <4, 17> [35]
reduced 2,6-dichlorophenolindophenol <3> [25]
rolliniastatin-1 <1> [28]
rolliniastatin-2 <1> [28]
rotenone <1, 4, 5, 6, 8, 10, 11, 13, 17> (<10>, strong [23]; <8>, mitochondria contain two different NADH:ubiquinone reductases. One enzyme oxidizes endogenous NADH, couples electron transfer to proton translocation and is inhibited by rotenone, the other enzyme oxidizes exogenous NADH without proton tranlocation and is insensitive to rotenone. The activity of the rotenone-insensitive enzyme highly exceeds the activity of the rotenone-sensitive enzyme [13]; <3, 13>, rotenone-insensitive enzyme [21, 25, 33, 34]) [10, 13, 14, 19, 23, 26, 27, 35, 42]
squamocin <1> [28]

Cofactors/prosthetic groups

FAD <3, 12, 19, 20, 23> (<3, 12>, enzyme contains FAD [20, 25, 33]; <19>, enzyme contains noncovalently bound FAD [38]; <3>, contains a single molecule of FAD per subunit [4]; <12>, the NQRF subunit carries FAD [37]; <20>, enzyme contains one molecule of noncovalently bound FAD [40]; <1>, enzyme contains 13.47 nM of FMN per mg of enzyme [46]) [4, 20, 25, 33, 37, 38, 40, 44, 46, 47]
FMN <1, 4, 10, 13, 17, 19, 20> (<1, 19>, enzyme contains FMN [1, 2, 38, 42]; <13, 17>, enzyme contains noncovalently bound FMN [21, 44]; <4>, enzyme contains one noncovalently bound FMN [30]; <1>, contains 1 FMN per 650000 Da enzyme [14]; <10>, contains 0.5 mol of FMN per mol of isolated enzyme [23]; <12>, does not contain FMN [20]; <4>, the enzyme complex contains one molecule of FMN [36]; <19>, enzyme contains covalently bound FMN [38]; <20>, enzyme contains 2 molecules of covalently bound FMN [40]) [1, 2, 14, 21, 23, 30, 36, 38, 40, 42, 43]
NADH <1-25> [1-50]
NADPH <1, 16, 22> (<16>, high activity [50]; <23>, no activity [47]) [14, 42, 45, 46, 50]

flavin <22> (<22>, enzyme contains 1.4 mol flavin per mol of enzyme [45]) [45]
ubiquinone <20> (<20>, enzyme contains approximately one bound ubiquinone [40]) [40]
ubiquinone-10 <1> (<1>, the enzyme contains 4.2-4.5 mol of ubiquinone-10 per 650000 Da enzyme [14]) [14]

Activating compounds

FAD <22> (<22>, 0.1 mM, about 2fold stimulation [45]) [45]
Additional information <1> (<1>, brief exposure of the thermally deactivated mitochondria with malate/glutamate, NAD^+ and cytochrome c induced reactivation of the inactivated enzyme, exposure to 37°C induces deactivation of complex I [24]) [24]

Metals, ions

Na^+ <12, 19> (<12>, activates [20]; <19>, the Na^+-dependent step of the Na^+-translocating NADH:ubiquinone oxidoreductase is located between the noncovalently bound FAD and the covalently bound FMN [38]) [20, 38]
iron <1, 2, 3, 4, 12, 17, 23, 25> (<1,4>, enzyme contains Fe-S clusters [1, 2, 3, 4, 5, 46]; <1>, contains 4 major iron centers, contains 16-18 gatoms of iron per 650000 Da enzyme [14]; <2>, at least 4 different species of low-potential iron-sulfur clusters, characterization of the Fe-S clusters [3]; <1>, the NADH:ubiquinone oxidoreductase complex is composed of three distinct fragments: 1. IP: a water soluble Fe-S-protein that is composed of 5-6 polypeptides and contains four Fe-S clusters, 2. FP: a water-soluble Fe-S-flavoprotein that is composed of three polypeptides and contains FMN and two Fe-S clusters, 3. a water-insoluble fraction containing phospholipids and hydrophobic polypeptides [1]; <3>, contains 0.08 mol Fe per mol of flavin after dialysis [4]; <4>, 2 binuclear and 3 tetranuclear EPR detectable iron-sulfur clusters. Cluster N-1a is the most labile component among the five iron sulfur clusters [5]; <1>, one binuclear and three tetranuclear NADH-reducible iron-sulfur clusters [9]; <12>, an Fe-S-cluster of the [2Fe-2S] type is involved in electron translocation [20]; <4>, the enzyme contains at least 5 EPR-visible iron-sulfur clusters. The NQO3 subunits contains at least two distinct iron-sulfur clusters: 1 [2Fe-2S] cluster with axial EPR signals and a [4Fe-4S] cluster with rhombic symmetry. The midpoint redox potentials of [2Fe-2S] and [4Fe-4S] clusters at pH 8.6 is -472 mV and -391 mV, respectively [30]; <17>, the enzyme contains at most nine putative iron-sulfur cluster binding sites, the NQO2 subunit bears a single [2Fe-2S] cluster, the NQO3 subunit contains multiple iron-sulfur clusters: one [2Fe-2S] cluster, one [4Fe-4S] cluster and possibly another [4Fe-4S] cluster [32]; <4>, enzyme complex contains up to 8 iron-sulfur clusters, 2 * [2Fe-2S] and 6 * [4Fe-4S] [36]; <17>, the CXXCXXXCX27C motif in the NQO3 subunit most likely ligates the [4Fe-4S] cluster [39]; <1>, 5 binuclear and three tetranuclear iron-sulfur clusters [42]; <17>, the enzyme contains at least three iron-sulfur clusters [44]; <23>, contains approximately 0.06 mol of iron per mol of enzyme [47]; <24>, contains no Fe-S clusters [48]; <25>, enzyme contains 10.6 atoms of iron and 12.7 atoms of sulfur per dodecamer [49]) [1, 2, 3, 4, 5, 9, 20, 30, 32, 36, 37, 39, 42, 44, 46, 47, 49]

Turnover number (min^{-1})

4800-6000 <3> (NADH, <3>, with ubiquinone-6 as acceptor, Triton assay medium [4]) [4]
30000-33000 <3> (NADH, <3>, with ubiquinone-6 as acceptor, mitochondria [4]) [4]

Specific activity (U/mg)

0.186 <11> (<11>, NADH-ubiquinone-0 oxidoreductase activity [18]) [18]
3.4 <5> [8]
4.8 <13> [21]
10.3 <11> (<11>, NADH-ferricyanide oxidoreductase activity [18]) [18]
11.5 <10> (<10>, activity with ubiquinone-2 [23]) [23]
46.5 <16> [50]
58.2 <2> [7]
61.9 <3> (<3>, reaction with ubiquinone-6 [4]) [4]
64 <9> (<9>, NADH:hexamineruthenium(III)-chloride oxidoreductase activity [15]) [15]
1671 <3> (<3>, reaction with ubiquinone-2 [4]) [4]
Additional information <12> [20]

K_m-Value (mM)

0.001 <5> (NADH) [26]
0.006 <24> (NADH) [48]
0.0062 <3> (2,6-dichlorophenolindophenol) [25]
0.007 <1> (NADH, <1>, reaction with ferricyanide [14]) [14]
0.0094 <3> (NADH) [25]
0.012 <1> (cytochrome c) [14]
0.014 <1> (NADH, <1>, reaction with cytochrome c [14]) [14, 42]
0.016 <23> (2,6-dichlorophenolindophenol) [47]
0.019 <5> (NADH, <5>, reaction with ubiquinone-1 [8]) [8]
0.031 <3> (NADH, <3>, with ubiquinone-6 as acceptor [4]) [4]
0.04 <1> (ubiquinone-1) [42]
0.05 <22> (naphthoquinone) [45]
0.055 <5> (ferricyanide) [8]
0.066 <2> (decylubiquinone) [7]
0.083 <2> (ubiquinone-2) [7]
0.1 <22> (benzoquinone) [45]
0.143 <5> (ubiquinone-1) [8]
0.25 <25> (NAD^+) [49]
0.35 <25> (ubiquinone-0) [49]
0.4 <25> (ferricyanide) [49]
0.4 <2> (ubiquinone-1) [7]
0.55 <25> (1,1'-carbamoylmethylviologen) [49]
0.55 <25> (2,6-dichlorophenolindophenol) [49]
1 <25> (1,1'-carbamoylmethylviologen) [49]
1.1 <25> (NADH) [49]
4 <1> (ferricyanide) [14]

K_i-Value (mM)

0.0003 <1> (rolliniastatin-1) [28]
0.0004 <1> (squamocin) [28]
0.0006 <1> (rolliniastatin-2) [28]
0.0008 <1> (otivarin) [28]
0.001 <1> (molvizarin) [28]
0.001 <1> (piercidin) [28]
0.004 <1> (rotenone) [28]
0.0083 <3> (NADH) [33]
0.01 <1> (Triton X-100, <1>, pH 8.0, 25°C [16]) [16]
0.015 <3> (2,6-dichlorophenolindophenol) [25]
5.5 <3> (AMP, <3>, versus NADH [25]) [25]
11.5 <3> (AMP, <3>, versus oxidized 2,6-dichlorophenolindophenol [25]) [25]

pH-Optimum

4.5 <16, 22> (<22>, reaction with ferricyanide [45]; <16>, reaction with NADPH and ubiquinone-1 [50]) [45, 50]
4.5-9.5 <3> (<3>, reaction with ubiquinone-6 is almost constant between pH 4.5 and pH 9.5 [4]) [4]
5 <25> (<25>, reduction of NAD^+ with reduced methyl viologen [49]) [49]
6-6.5 <13> [21]
6.2 <3> (<3>, reaction with ubiquinone-2 [4]) [4]
6.5 <16> (<16>, reaction with NADH and ubiquinone-1 [50]) [50]
7.1 <7> (<7>, reaction with ubiquinone-1 [11]) [11]
7.5 <22, 23> (<22>, reaction with 2,6-dichlorophenolindophenol [45]) [45, 47]
10 <25> (<25>, reaction with ferricyanide or 1,1'-carbamoylmethylviologen [49]) [49]

pH-Range

3.5-7 <22> (<22>, 50% of maximal activity at pH 3.5 and 7.0, reaction with ferricyanide [45]) [45]
6.7-7.9 <7> (<7>, about 90% of maximal activity at pH 6.7 and at pH 7.9, reaction with ubiquinone-1 [11]) [11]

Temperature range (°C)

35-80 <24> (<24>, activity increases 4fold from 35°C to 80°C, the maximal possible operating temperature [48]) [48]

4 Enzyme Structure

Molecular weight

60000 <24> (<24>, gel filtration [48]) [48]
65000 <13> (<13>, gel filtration [21]) [21]
69000 <1> (<1>, gel filtration [46]) [46]
87000 <23> (<23>, disc gel electrophoresis [47]) [47]

95000 <22> (<22>, gel filtration [45]) [45]
200000 <25> (<25>, non-denaturing PAGE [49]) [49]
210000 <25> (<25>, gel filtration [49]) [49]
360000 <20> (<20>, it is likely that the protein is a monomer, calculated molecular mass of 215000 Da with bound detergent, gel filtration [40]) [40]
610000 <8> (<8>, calculation from sedimentation data [12]) [12]
900000 <10> [23]
Additional information <5> (<5>, several molecular mass forms of the purified complex exist [26]) [26]

Subunits

? <3, 11, 16, 23> (<23>, x * 37000, SDS-PAGE [47]; <3>, x * 53000, SDS-PAGE [4, 25]; <16>, x * 55000, SDS-PAGE [50]; <11>, x * 57000, SDS-PAGE [18]) [4, 18, 25, 47, 50]
dimer <13, 22> (<13>, 2 * 33000, SDS-PAGE [21]; <22>, 2 * 50000, SDS-PAGE [45]) [21, 45]
dodecamer <25> (<25>, 12 * 17000, SDS-PAGE [49]) [49]
monomer <24> (<24>, 1 * 47000, SDS-PAGE [48]) [48]
Additional information <1, 2, 4, 5, 6, 8, 10, 12, 13, 14, 15, 17> (<2>, the enzyme contains 13 different subunits [7]; <1>, the NADH:ubiquinone oxidoreductase complex is composed of three distinct fragments: 1. IP: a water soluble Fe-S-protein that is composed of 5-6 polypeptides and contains four Fe-S clusters, 2. FP: a water-soluble FeS-flavoprotein that is composed of three polypeptides and contains FMN and two FeS clusters, 3. a water-insoluble fraction containing phospholipids and hydrophobic polypeptides [1]; <5>, isolated complex contains 14 major polypeptides [8]; <6>, at least five major polypeptides with MW of 76000 Da, 46000 Da, 39000 Da, 33000 Da and 27000 Da [10]; <8>, consists of about 25 different subunits [12]; <1>, the enzyme complex is composed of at least 10 polypeptides ranging in MW from 10000 Da to 70000 Da [14]; <12>, the enzyme consists of three different hydrophilic and probably three different hydrophobic subunits [20]; <14>, the enzyme complex consists of at least 43 proteins, seven are encoded by the mitochondrial genome, while the remainder are encoded by the nuclear genome [22]; <10>, composed of at least 32 individual subunits [23]; <5>, about 15 polypeptides including that at 80000 Da, 54000 Da, 53000 Da, 51000 Da, 27000 Da, 25000 Da and 22000 Da crossreact with polyclonal antibodies raised against complex I from Neurospora crassa [26]; <13>, four polypeptides, of 15000 Da, 20000 Da, 38000 Da and 51000 Da are identified as subunits of complex I by SDS-PAGE [27]; <15>, the enzyme complex comprises more than 35 subunits, the majority of which are encoded by the nucleus. Only five components ND1, ND2, ND4, ND5, and ND6 are coded for by the mitochondrial genome. Dum 5 mutant has a 1T deletion in the 3ÚTR of nd5 whereas dum 17 is a 1T deletion in the coding sequence of nd6. Absence of intact ND1 or ND6 subunits prevents the assembly of the 850000 Da whole complex. Loss of ND4 or ND4/ND5 leads to the formation of a subcomplex of 650000 Da [29]; <4>, proton-translocating NADH-quinone oxidoreductase is composed of at least 14 dissimilar subunits, designated NQO1-14. NQO1, NQO2, NQO3,

NQO9, and probably NQO6 subunits are cofactor binding subunits [30]; <4, 17>, the NQO6 subunits plays a key role in electron transfer by functionally coupling iron-sulfur cluster N2 to quinone [35]; <4>, enzyme is composed of at least 14 subunits [36]; <1>, enzyme complex contains approximately 25 unlike polypeptides in three distinct fragments: 1. HP: contains the bulk of phospholipids of the complex I. 2. IP: contains polypeptides of 75000 Da, 49000 Da, 30000 Da, 18000 Da, 15000 Da and 13000 Da, 3. FP: contains polypeptides of 9000 Da, 24000 Da, and 51000 Da. FP and IP are surrounded by HP polypeptides [42]; <17>, enzyme complex is composed of approximately 10 unlike polypeptides, the NADH-binding subunit has a MW of 47000 Da determined by SDS-PAGE [43]) [1, 7, 8, 10, 12, 14, 20, 22, 23, 26, 27, 29, 30, 35, 36, 42, 43]

Posttranslational modification

lipoprotein <1> (<1>, the enzyme complex contains 0.22 mg of lipid per 650000 Da enzyme [14]; <1>, the NADH:ubiquinone oxidoreductase complex is composed of three distinct fragments: 1. IP: a water soluble Fe-S-protein that is composed of 5-6 polypeptides and contains four Fe-S clusters, 2. FP: a water-soluble FeS-flavoprotein that is composed of three polypeptides and contains FMN and two FeS clusters, 3. a water-insoluble fraction containing phospholipids and hydrophobic polypeptides [1]; <1>, the enzyme complex contains phospholipids [42]) [1, 14, 42]

5 Isolation/Preparation/Mutation/Application

Source/tissue

chorionic villus <14> [22]
heart <1> [1, 9, 14, 16, 17, 24, 28, 42, 43, 46]
liver <11> [18, 19]
muscle <14> (<14>, skeletal [22]) [22]
skin fibroblast <14> (cell culture) [22]
spadix <7> [11]
tuber <10> [17]
Additional information <13> (<13>, procyclic form of the organism [21,27]) [21, 27]

Localization

cytoplasmic membrane <4> [36]
membrane <2, 3, 4, 12, 16, 17, 20, 21> (<20>, associated to [40]) [3, 6, 20, 31, 40, 41, 42, 43, 47, 50]
microsome <10, 11> [19, 23]
mitochondrial membrane <3, 5, 6, 7, 9, 13> (<5, 13>, inner membrane [8, 21]) [8, 10, 11, 15, 21, 27, 33]
mitochondrion <1, 3, 5, 6, 7, 8, 9, 10, 13, 17> (<5, 13>, inner membrane [8, 21]; <3, 6, 7, 9, 13>, membrane [10, 11, 15, 27, 33]; <8>, the mitochondrial multienzyme-complex is of dual origin. 6 of the at least 22 subunits with MW of 70000 Da, 48000 Da, 37000 Da, 25000 Da, 22000 Da and 18000 Da are

synthesized in mitochondria. 11 subunits are synthesized in the cytoplasm [13]; <9>, constitutively high content [15]; <13>, loosly bound to inner mitochondrial membrane [21]) [1, 2, 4, 8, 9, 10, 11, 12, 13, 14, 15, 17, 21, 24, 25, 26, 27, 28, 32, 33, 34, 42, 43, 46]
plasma membrane <2, 11> [17, 18]
Additional information <1> (<1>, submitochondrial particles [16]) [16]

Purification
<1> (enzymically active subcomplexes Ilambda, IS, and IlambdaS [2]) [1, 2, 46]
<2> [7]
<3> [4]
<4> (recombinant NQO3 subunit [30]) [30]
<5> [8, 26]
<6> [10]
<9> (efficient large scale purification of His-tagged proton translocating complex [15]) [15]
<10> [23]
<11> [18]
<12> [20]
<13> (partial [27]) [21, 27]
<16> [50]
<19> [38]
<20> [40]
<22> [45]
<23> [47]
<25> [49]

Crystallization
<8> (membrane crystals of the enzyme complex are prepared by adding mixed phospholipid-Triton X-100 micelles and then removing the Triton by dialysis [12]) [12]

Cloning
<3> (the enzyme overexpressed in Escherichia coli acts as a member of the respiratory chain in the host cell [33]; the NDI1 gene encoding the rotenone-insensitive internal NADH-quinone oxidoreductase is cotransfected into complex-I-deficient Chinese hamster CCL16-B2 cells. The enzyme is expressed functionally and catalyzes electron transfer from NADH in the matrix to ubiquinone 10 in the inner mitochondrial membranes [34]) [33, 34]
<4> (NQO3 subunit overexpressed in NQO3 [30]; expression of the subunits NQO4, NQO5 and NQO6 in Escherichia coli [31]) [30, 31]
<17> (NQO2 subunit expressed in Escherichia coli [32]) [32]
<20> (cloned under the regulation of the PBAD promoter and expressed in Vibrio cholera [40]) [40]

Application
medicine <3, 14> (<14>, a deficient activity of the NADH:ubiquinone oxidoreductase enzyme complex is frequently observed in clinical heterogeneous

group of mitochondrial disorders with Leigh (-like) disease as the main contributor. Enzyme complex activity measurement in skeletal muscle is the mainstay of the diagnostic process. Fibroblast studies are a prerequisite whenever prenatal enzyme diagnostic is considered [22]; <3>, the NDI1 gene provides a potentially useful tool for gene therapy of mitochondrial diseases caused by complex I deficiency [34]) [22, 34]

6 Stability

pH-Stability

4 <22> (<22>, 25°C, stable for at least 1 h [45]) [45]
4.5 <3> (<3>, activity is irreversibly destroyed below [4]) [4]
4.5-9.5 <13> (<13>, stable [21]) [21]
7 <22> (<22>, 100°C, 15 min, 30% loss of activity [45]) [45]
9.5 <3> (<3>, activity is irreversibly destroyed above [4]) [4]

Temperature stability

25 <22> (<22>, pH 4, stable for at least 1 h [45]) [45]
65 <17> (<17>, 3 h, Fe-S cluster remains intact [32]) [32]
100 <22> (<22>, pH 7, 15 min, 30% loss of activity [45]) [45]
Additional information <1> (<1>, Triton X-100 partially protects complex I against thermally induced deactivation and partially activates the thermally deactivated enzyme [16]) [16]

Oxidation stability

<4>, the tetranuclear cluster in the isolated NQO3 subunit is sensitive towards oxidants and converts into [3Fe-4S] form [30]
<25>, storage of 0.3 mg purified enzyme in 2 ml 0.1 M Tris/HCl, pH 8.0, for 24 h at 4°C, results in 15% inactivation under anaerobic conditions, 55% inactivation in presence of air [49]

Storage stability

<2>, -80°C, stable for 6 months or longer [7]
<25>, -15°C, 0.07 mg/ml enzyme, 0.1 M Tris/HCl, pH 8.0, 72 h, under anaerobic conditions, 70% loss of activity [49]
<25>, 4°C, 0.07 mg/ml enzyme, 0.1 M Tris/HCl, pH 8.0, 72 h, under anaerobic conditions, 75% loss of activity [49]
<25>, in liquid nitrogen, 5% loss of activity after 72 h [49]

References

[1] Ragman, C.I.; Hatefi, Y.: Isolation of the iron-sulfur-containing polypeptides of NADH: oxidoreductase ubiquinone. Methods Enzymol., **126**, 360-369 (1986)
[2] Finel, M.; Majander, A.S.; Tyynelae, J.; De Jong, A.M.P.; Albracht, S.P.J.; Wikstroem, M.: Isolation and characterization of subcomplexes of the mi-

tochondrial NADH:ubiquinone oxidoreductase (complex I). Eur. J. Biochem., **226**, 237-242 (1994)
[3] Meinhardt, S.W.; Matsushita, K.; Kaback, H.R.; Ohnishi, T.: EPR characterization of the iron-sulfur-containing NADH-ubiquinone oxidoreductase of the Escherichia coli aerobic respiratory chain. Biochemistry, **28**, 2153-2160 (1989)
[4] De Vries, S.; Grivell, L.A.: Purification and characterization of a rotenone-insensitive NADH:Q_6 oxidoreductase from mitochondria of Saccharomyces cerevisiae. Eur. J. Biochem., **176**, 377-384 (1988)
[5] Meinhardt, S.W.; Kula, T.; Yagi, T.; Lillich, T.; Ohnishi, T.: EPR characterization of the iron-sulfur clusters in the NADH: ubiquinone oxidoreductase segment of the respiratory chain in Paracoccus denitrificans. J. Biol. Chem., **262**, 9147-9153 (1987)
[6] Pruess, B.M.; Nelms, J.M.; Park, C.; Wolfe, A.J.: Mutations in NADH:ubiquinone oxidoreductase of Escherichia coli affect growth on mixed amino acids. J. Bacteriol., **176**, 2143-2150 (1994)
[7] David, P.; Baumann, M.; Wikstrom, M.; Finel, M.: Interaction of purified NDH-1 from Escherichia coli with ubiquinone analogues. Biochim. Biophys. Acta, **1553**, 268-278 (2002)
[8] Soole, K.L.; Dry, I.B.; Wiskich, J.T.: Partial purification and characterization of complex I, NADH:ubiquinone reductase, from the inner membrane of beetroot mitochondria. Plant Physiol., **98**, 588-594 (1991)
[9] Kowal, A.T.; Morningstar, J.E.; Johnson, M.K.; Ramsay, R.R.; Singer, T.P.: Spectroscopic characterization of the number and type of iron-sulfur clusters in NADH:ubiquinone oxidoreductase. J. Biol. Chem., **261**, 9239-9245 (1986)
[10] Cottingham, I.R.; Moore, A.L.: Analysis of NADH dehydrogenases from plant [mung bean (Phaseolus aureus)] mitochondrial membranes on non-denaturing polyacrylamide gels and purification of complex I by band excision. Biochem. J., **254**, 303-305 (1988)
[11] Cook, N.D.; Cammack, R.: Properties of a soluble rotenone-insensitive NADH dehydrogenase released from Arum maculatum. Biochim. Biophys. Acta, **827**, 30-35 (1985)
[12] Leonard, K.; Haiker, H.; Weiss, H.: Three-dimensional structure of NADH: ubiquinone reductase (complex I) from Neurospora mitochondria determined by electron microscopy of membrane crystals. J. Mol. Biol., **194**, 277-286 (1987)
[13] Ise, W.; Haiker, H.; Weiss, H.: Mitochondrial translation of subunits of the rotenone-sensitive NADH:ubiquinone reductase in Neurospora crassa. EMBO J., **4**, 2075-2080 (1985)
[14] Hatefi, Y.: The enzymes and the enzyme complexes of the mitochondrial oxidative phosphorylation system. The Enzymes of Biological Membranes (Martonosi, A.V., ed.) Plenum, New York, **4**, 3-41 (1976)
[15] Kashani-Poor, N.; Kerscher, S.; Zickermann, V.; Brandt, U.: Efficient large scale purification of His-tagged proton translocating NADH:ubiquinone oxidoreductase (complex I) from the strictly aerobic yeast Yarrowia lipolytica. Biochim. Biophys. Acta, **1504**, 363-370 (2001)

[16] Ushakova, A.V.; Grivennikova, V.G.; Ohnishi, T.; Vinogradov, A.D.: Triton X-100 as a specific inhibitor of the mammalian NADH-ubiquinone oxidoreductase (complex I). Biochim. Biophys. Acta, **1409**, 143-153 (1999)
[17] Satoh, T.; Miyoshi, H.; Sakamoto, K.; Iwamura, H.: Comparison of the inhibitory action of synthetic capsaicin analogs with various NADH-ubiquinone oxidoreductases. Biochim. Biophys. Acta, **1273**, 21-30 (1996)
[18] Kim, C.; Crane, F.L.; Faulk, W.P.; Morre, D.J.: Purification and characterization of a doxorubicin-inhibited NADH-quinone (NADH-ferricyanide) reductase from rat liver plasma membranes. J. Biol. Chem., **277**, 16441-16447 (2002)
[19] Shigemura, T.; Kang, D.; Nagata-Kuno, K.; Takeshige, K.; Hamasaki, N.: Characterization of NAD(P)H-dependent ubiquinone reductase activities in rat liver microsomes. Biochim. Biophys. Acta, **1141**, 213-220 (1993)
[20] Pfenninger-Li, X.D.; Albracht, S.P.J.; van Belzen, R.; Dimroth, P.: NADH:ubiquinone oxidoreductase of Vibrio alginolyticus: Purification, properties, and reconstitution of the Na^+ pump. Biochemistry, **35**, 6233-6242 (1996)
[21] Fang, J.; Beattie, D.S.: Novel FMN-containing rotenone-insensitive NADH dehydrogenase from Trypanosoma brucei mitochondria: isolation and characterization. Biochemistry, **41**, 3065-3072 (2002)
[22] Smeitink, J.; Sengers, R.; Trijbels, F.; Van den Heuvel, L.: Human NADH:ubiquinone oxidoreductase. J. Bioenerg. Biomembr., **33**, 259-266 (2001)
[23] Herz, U.; Schroeder, W.; Liddell, A.; Leaver, C.J.; Brennicke, A.; Grohmann, L.: Purification of the NADH:ubiquinone oxidoreductase (complex I) of the respiratory chain from the inner mitochondrial membrane of Solanum tuberosum. J. Biol. Chem., **269**, 2263-2269 (1994)
[24] Grivennikova, V.G.; Kapustin, A.N.; Vinogradov, A.D.: Catalytic activity of NADH-ubiquinone oxidoreductase (complex I) in intact mitochondria: evidence for the slow active/inactive transition. J. Biol. Chem., **276**, 9038-9044 (2001)
[25] Velazquez, I.; Pardo, J.P.: Kinetic characterization of the rotenone-insensitive internal NADH:ubiquinone oxidoreductase of mitochondria from Saccharomyces cerevisiae. Arch. Biochem. Biophys., **389**, 7-14 (2001)
[26] Rasmusson, A.G.; Mendel-Hartvig, J.; Meller, I.M.; Wiskich, J.T.: Isolation of the rotenone-sensitive NADH-ubiquinone reductase (complex I) from red beet mitochondria. Physiol. Plant., **90**, 607-615 (1994)
[27] Fang, J.; Wang, Y.; Beattie, D.S.: Isolation and characterization of complex I, rotenone-sensitive NADH:ubiquinone oxidoreductase, from the procyclic forms of Trypanosoma brucei. Eur. J. Biochem., **268**, 3075-3082 (2001)
[28] Degli Esposti, M.; Ghelli, A.; Ratta, M.; Cortes, D.; Estornell, E.: Natural substances (acetogenins) from the family Annonaceae are powerful inhibitors of mitochondrial NADH dehydrogenase (complex I). Biochem. J., **301**, 161-167 (1994)
[29] Cardol, P.; Matagne, R.F.; Remacle, C.: Impact of mutations affecting ND mitochondria-encoded subunits on the activity and assembly of complex I in Chlamydomonas. Implication for the structural organization of the enzyme. J. Mol. Biol., **319**, 1211-1221 (2002)

[30] Yano, T.; Yagi, T.; Sledæ, V.D.; Ohnishi, T.: Expression and characterization of the 66-kilodalton (NQO3) iron-sulfur subunit of the proton-translocating NADH-quinone oxidoreducrase of Paracoccus denitrificans. J. Biol. Chem., **270**, 18264-18270 (1995)
[31] Takano, S.; Yano, T.; Yagi, T.: Structural studies of the proton-tramslocating NADH-quinone oxidoreductase (NDH-1) of Paracoccus denitrificans: identity, property, and stoichiometry of the peripheral subunits. Biochemistry, **35**, 9120-9127 (1996)
[32] Yano, T.; Chu, S.S.; Sled, V.D.; Ohnishi, T.; Yagi, T.: The proton-translocating NADH-quinone oxidoreductase (NDH-1) of the thermophilic bacterium Thermus thermophilus HB-8. J. Biol. Chem., **272**, 4201-4211 (1997)
[33] Kitajima-Ihara, T.; Yagi, T.: Rotenone-insensitive internal NADH-quinone oxidoreductase of Saccharomyces cerevisiae mitochondria: the enzyme expressed in Escherichia coli acts as a member of the respiratory chain in the host cells. FEBS Lett., **421**, 37-40 (1998)
[34] Seo, B.B.; Kitajima-Ihara, T.; Chan, E.K.L.; Scheffler, I.E.; Matsuno-Yagi, A.; Yagi, T.: Molecular remedy of complex I defects: rotenone-insensitive internal NADH-quinone oxidoreductase of Saccharomyces cerevisiae mitochondria restores the NADH oxidase activity of complex I-deficient mammalian cells. Proc. Natl. Acad. Sci. USA, **95**, 9167-9171 (1998)
[35] Schuler, F.; Yano, T.; di Bernardo, S.; Yagi, T.; Yankovskaya, V.: NADH-quinone oxidoreductase: PSST subunit couples electron transfer from iron-sulfur cluster N2 to quinone. Proc. Natl. Acad. Sci. USA, **96**, 4149-4153 (1999)
[36] Yano, T.; Yagi, T.: H^+-Translocating NADH-quinone oxidoreductase (NDH-1) of Paracoccus denitrificans. Studies on topology and stoichiometry of the peripheral subunits. J. Biol. Chem., **274**, 28606-18611 (1999)
[37] Steuber, J.: Na^+ translocation by bacterial NADH:quinone oxidoreductases: an extension to the complex-I family of primary redox pumps. Biochim. Biophys. Acta, **1505**, 45-56 (2001)
[38] Bogachev, A.V.; Bertsova, Y.V.; Barquera, B.; Verkhovsky, M.I.: : Sodium-dependent steps in the redox reaction of the Na^+ motive NADH:quinone oxidoreductase from Vibrio harveyi. Biochemistry, **40**, 7318-7323 (2001)
[39] Nakamaru-Ogiso, E.; Yano, T.; Ohnishi, T.; Yagi, T.: Characterization of the iron-sulfur cluster coordinated by a cysteine cluster motif (CXXCXXXCX27C) in the Nqo3 subunit in the proton-translocating NADH-quinone oxidoreductase (NDH-1) of Thermus thermophilus HB-8. J. Biol. Chem., **277**, 1680-1688 (2002)
[40] Barquera, B.; Hellwig, P.; Zhou, W.; Morgan, J.E.; Häse, C.C.; Gosink, K.K.; Nilges, M.; Bruesehoff, P.J.; Roth, A.; Lancaster, C.R.D.; Gennis, R.B.: Purification and characterization of the recombinant Na^+-translocating NADH:quinone oxidoreductase from Vibrio cholera. Biochemistry, **41**, 3781-3789 (2002)
[41] Tokuda, H.; Unemoto, T.: Na^+ is translocated at NADH:quinone oxidoreductase segment in the respiratory chain of Vibrio alginolyticus. J. Biol. Chem., **259**, 7785-7790 (1984)
[42] Hatefi, Y.: The mitochondrial electron transport and oxidative phosphorylation system. Annu. Rev. Biochem., **54**, 1015-1069 (1985)

[43] Yagi, T.: Inhibition by capsaicin of NADH-quinone oxidoreductase is correlated with the presence of energy-coupling site 1 in various organisms. Arch. Biochem. Biophys., **281**, 305-311 (1990)
[44] Xu, X.M.; Yagi, T.: Identification of the NADH-binding subunit of energy-transducing NADH-quinone oxidoreductase (NDH-1) of Thermus thermophilus. Biochem. Biophys. Res. Commun., **174**, 667-672 (1991)
[45] Wakao, H.; Wakagi, T.; Oshima, T.: Purification and properties of NADH dehydrogenase from a thermoacidophilic archaebacterium, Sulfolobus acidocaldarius. J. Biochem., **102**, 255-262 (1987)
[46] Galante, Y.M.; Hatefi, Y.: Purification and molecular and enzymic properties of mitochondrial NADH dehydrogenase. Arch. Biochem. Biophys., **192**, 559-568 (1979)
[47] Mackler, B.; Bevan, C.; Person, R.; Davis, K.A.: Purification and properties of the respiratory chain-linked internal NADH dehydrogenase of Candida utilis. Biochem. Int., **3**, 9-17 (1981)
[48] Gomes, C.M.; Bandeiras, T.M.; Teixeira, M.: A new type-II NADH dehydrogenase from the archaeon Acidianus ambivalens: characterization and in vitro reconstitution of the respiratory chain. J. Bioenerg. Biomembr., **33**, 1-8 (2001)
[49] Bayer, M.; Walter, K.; Simon, H.: Purification and partial characterization of a reversible artificial mediator accepting NADH oxidoreductase from Clostridium thermoaceticum. Eur. J. Biochem., **239**, 686-691 (1996)
[50] Matsushita, K.; Otofuji, A.; Iwahashi, M.; Toyama, H.; Adachi, O.: NADH dehydrogenase of Corynebacterium glutamicum. Purification of an NADH dehydrogenase II homolog able to oxidize NADPH. FEMS Microbiol. Lett., **204**, 271-276 (2001)

Monodehydroascorbate reductase (NADH) 1.6.5.4

1 Nomenclature

EC number
1.6.5.4

Systematic name
NADH:monodehydroascorbate oxidoreductase

Recommended name
monodehydroascorbate reductase (NADH)

Synonyms
AFR-reductase
MDA reductase
MDAsA reductase (NADPH)
MDHA
NADH-semidehydroascorbate oxidoreductase
NADH:AFR oxidoreductase
NADH:ascorbate radical oxidoreductase
NADH:semidehydroascorbic acid oxidoreductase
SDA reductase
SOR
ascorbate free radical reductase
ascorbate free-radical reductase
ascorbic free radical reductase
semidehydroascorbate reductase

CAS registry number
9029-26-9

2 Source Organism

<1> *Neurospora crassa* [1]
<2> *Glycine max* [2]
<3> *Nicotiana tabacum* [3]
<4> *Ricinus communis* [4]
<5> *Cuscuta reflexa* (in the parasitic plant the activity is ten times lower than in etiolated and green plants [5]) [5, 29]
<6> *Solanum tuberosum* [6, 10, 19, 22]
<7> *Rattus norvegicus* [7, 9, 11, 13, 15, 16, 17, 18, 20, 27, 28]
<8> *Cucumis sativus* [8, 19, 23, 24]

<9> *Homo sapiens* [12]
<10> *Sus scrofa* [14, 27]
<11> *Mus musculus* [19]
<12> *Helianthus tuberosus* [19]
<13> *Allium cepa* [19]
<14> *Vicia faba* [19]
<15> *Brassica oleracea* [19]
<16> *Pisum sativum* [19, 25]
<17> *Pterocladia sp.* [19]
<18> *Gigartina sp.* [19]
<19> *Hypnea sp.* [19]
<20> *Gracilaria sp.* [19]
<21> *Arabidopsis sp.* (var. Columbia [21]) [21]
<22> *Pleurotus ostreatus* [26]
<23> *Rana catesbeiana* [27]
<24> *Cavia porcellus* [27]
<25> *Oryctolagus cuniculus* [27]
<26> *Bos taurus* [27]

3 Reaction and Specificity

Catalyzed reaction

NADH + H^+ + 2 monodehydroascorbate = NAD^+ + 2 ascorbate (<8, 22>, ping pong mechanism [8, 26]; <8>, mechanism [23])

Reaction type

oxidation
redox reaction
reduction

Natural substrates and products

S NADH + monodehydroascorbate <2, 4, 5, 6, 7, 8, 9, 11, 12, 13, 14, 15, 16, 17, 18, 19, 20> (<2>, enzyme of the ascorbate-glutathione cycle [2]; <4>, the enzyme sustains glyoxysomal NAD^+ during β-oxidation, the glyoxylate cycle and gluconeogenesis in the endosperm during germination [4]; <5>, key enzyme for maintaining the ascorbic acid system in the reduced state [5]; <9>, enzyme may constitute part of cellular mechanisms which protect against and repair oxidant injury to biomolecules [12]; <7>, involvement of semidehydroascorbate reductase in the oxidation of NADH by lipid peroxide in mitochondria and microsomes [16]; <7>, the only physiologically important system for the regeneration of ascorbate [17]; <6, 11, 12, 13, 14, 15, 16, 17, 18, 19, 20>, the enzyme is involved in maintaining the ascorbic acid system in the reduced state [19]; <7>, the enzyme supports cholesterol side chain cleavage, the enzyme functions to provide cytoplasmic reducing equivalents to intramitochondrial cytochrome P-450SCC [20]; <6>, enzyme participates in the regeneration of ascorbate for scavenging toxic oxygen species in potato tuber mitochondria [22]; <6>, the enzyme is an effective scavenger of the monodehydroascorbate radical [22]; <16>, key enzyme in maintaining reduced pools of ascorbate [25]; <7>, the enzyme is involved in part of the transplasmamembrane redox system [28]) (Reversibility: ? <2, 4, 5, 6, 7, 8, 9, 11, 12, 13, 14, 15, 16, 17, 18, 19, 20> [2, 4, 5, 12, 16, 17, 19, 20, 22, 25, 28]) [2, 4, 5, 12, 16, 17, 19, 20, 22, 25, 28]

P NAD^+ + ascorbate

Substrates and products

S NADH + 1,4-benzoquinone <22> (Reversibility: ? <22> [26]) [26]

P NAD^+ + 1,4-benzoquinol

S NADH + 2,6-dichlorophenol indophenol <1, 8, 22> (<8>, about 10% of the activity with monodehydroascorbate radical [24]) (Reversibility: ? <1, 8, 22> [1, 8, 24, 25, 26]) [1, 8, 24, 25, 26]

P NAD^+ + ?

S NADH + cytochrome b_5 <1> (Reversibility: ? <1> [1]) [1]

P NAD^+ + ?

S NADH + cytochrome c <1> (Reversibility: ? <1> [1]) [1]

P NAD^+ + ?

S NADH + ferricyanide <8, 22> (Reversibility: ? <8, 22> [8, 24, 26]) [8, 24, 26]

P NAD + ferrocyanide

S NADH + monodehydroascorbate <1-26> (<8>, the monodehydroascorbate radical reacts with the fully reduced form of the enzyme with a diffusion-controlled rate [23]) (Reversibility: ? <1-26> [1-29]) [1-29]

P NAD^+ + ascorbate

S NADH + monodehydroisoascorbate <8> (Reversibility: ? <8> [24]) [24]

P NAD^+ + isoascorbate

S NADPH + monodehydroascorbate <6, 8> (Reversibility: ? <6, 8> [8, 10]) [8, 10]

P $NADP^+$ + ascorbate

S Additional information <1, 5> (<1>, no activity with NADPH [1]; <5>, enzyme is also capable of reducing phenoxyl radicals which are generated by horseradish peroxidase with H_2O_2 [29]) [1, 29]

P ?

Inhibitors

1,10-phenanthroline <22> (<22>, 10 mM, 16% inhibition [26]) [26]

1-(2-thenoyl)-3,3,3-trifluoraceton <1> (<1>, 0.3 mM, about 10% inhibition [1]) [1]

2,2'-dipyridyl <1, 22> (<1>, 1 mM, about 10% inhibition [1]; <22>, 10 mM, 21% inhibition [26]) [1, 26]

2,4,6-trinitrobenzenesulfonic acid <22> (<22>, 1 mM, 54% inhibition [26]) [26]

2-[[2-hydroxy-1,1-bis(hydroxymethyl)ethyl]'amino]ethanesulfonic acid <6> [10]

2-iodoacetamide <1, 8> (<8>, 25 mM, 21% inhibition [8]) [1, 8]

5,5'-dithiobis(2-nitrobenzoic acid) <8, 22> (<22>, 1 mM, 23% inhibition [26]) [8, 26]
8-hydroxyquinoline <22> (<22>, 10 mM, 15% inhibition [26]) [26]
Cu^{2+} <22> (<22>, 0.5 mM, complete inhibition [26]) [26]
EDTA <22> (<22>, 10 mM, 11% inhibition [26]) [26]
GSH <7> [28]
Hg^{2+} <22> (<22>, 0.5 mM, complete inhibition [26]) [26]
NEM <1, 6, 7, 8, 22> (<8>, 25 mM, 34% inhibition [8]; <22>, 1 mM, 82% inhibition [26]) [1, 6, 8, 10, 26, 28]
NaN_3 <22> (<22>, 16 mM, 16% inhibition [26]) [26]
PCMB <1, 6, 8> (<8>, 0.05 mM, 82% inhibition [8]) [1, 8, 10, 24]
Tris <6> [10]
citrate <1> (<1>, 1 mM, about 10% inhibition [1]) [1]
diacetyl <22> (<22>, 1 mM, 55% inhibition [26]) [26]
dicoumarol <8> [24]
diethyl dicarbonate <22> (<22>, 0.1 mM, 62% inhibition [26]) [26]
imidazole buffer <6> [10]
insulin <7> (<7>, inhibition of the enzyme in plasma membrane [11]) [11]
iodoacetamide <7, 22> (<22>, 5 mM, 32% inhibition [26]) [26, 28]
iodoacetate <7> [28]
lectin <7> (<7>, ConA, WGA and LFA [7]) [7]
mersalyl <6> (<6>, partially reversed by thiol-containing reagents, pyridine mucleotides protect, when added prior to the inhibitor [6]) [6, 10]
p-chloromercuribenzoate <22> (<22>, 0.1 mM, complete inhibition [26]) [26]
phenyl mercuric acetate <22> (<22>, 0.1 mM, 85% inhibition [26]) [26]
phenylglyoxal <22> (<22>, 100 mM, 46% inhibition [26]) [26]
phosphate <6> [10]
pyridoxal 5'-phosphate <22> (<22>, 1 mM, 46% inhibition [26]) [26]
sodium diethyldithiocarbaminate <1> (<1>, 1 mM, about 10% inhibition [1]) [1]
thenoyltrifluoroacetone <7> (<7>, 0.15 mM, complete inhibition [9]) [9]
trinitromethane <22> (<22>, 10 mM, 11% inhibition [26]) [26]

Cofactors/prosthetic groups

FAD <16> (<16>, flavin enzyme [25]) [25]
FMN <22> (<22>, contains 1 mol of FMN per subunit, flavin is non-covalently bound [26]) [26]
NADH <1-26> [1-29]
NADPH <6, 8> (<22>, no activity [26]) [8, 10]

Activating compounds

1,10-phenanthroline <1> (<1>, 1 mM, stimulates [1]) [1]
diphosphate <1> (<1>, 1 mM, stimulates [1]) [1]

Turnover number (min^{-1})

9000 <8> (NADPH) [8]
12000 <8> (NADH) [8]

Specific activity (U/mg)

11.09 <22> [26]
61 <6> [10]
256 <8> [8]
Additional information <1, 6, 8> [1, 22, 24]

K_m-Value (mM)

0.0012 <1> (semidehydroascorbic acid) [1]
0.0021 <22> (monodehydroascorbate) [26]
0.0025 <6> (monodehydroascorbate) [22]
0.0044 <8> (NADH) [24]
0.0046 <8> (NADH) [8]
0.005 <7> (semidehydroascorbic acid) [18]
0.0053 <16> (NADH, <16>, maltose-binding protein-monodehydroascorbate reductase fusion protein [25]) [25]
0.0057 <16> (monodehydroascorbate, <16>, maltose-binding protein-monodehydroascorbate reductase fusion protein [25]) [25]
0.0077 <6> (NADH) [10]
0.012 <1> (NADH) [1]
0.012 <8> (ferricyanide) [8]
0.0123 <6> (NADH) [22]
0.0215 <16> (NADPH, <16>, maltose-binding protein-monodehydroascorbate reductase fusion protein [25]) [25]
0.023 <8> (NADPH) [8]
0.03 <8> (2,6-dichloroindophenol) [8]
0.03 <6> (NADPH) [10]
0.035 <22> (NADH) [26]
0.05 <7> (NADH, <7>, reaction with semidehydro-D(-)-ascorbic acid and semidehydro-L(+)-ascorbic acid [18]) [18]
0.052 <16> (2,6-dichlorophenol indophenol, <16>, maltose-binding protein-monodehydroascorbate reductase fusion protein [25]) [25]
0.0573 <6> (NADPH) [22]
0.21 <8> (NADPH) [24]

pH-Optimum

7-7.2 <1> [1]
7-9 <22> [26]
7.4 <7> [18]
7.9 <8> [8]
8 <6> [10, 22]

pH-Range

5.5-9.5 <8> (<8>, pH 5.5: about 60% of maximal activity, pH 9.5: about 75% of maximal activity [8]) [8]
6.7-7.9 <1> (<1>, about 50% of maximal activity at pH 6.7 and at pH 7.9 [1]) [1]

Temperature optimum (°C)

39 <1> [1]

4 Enzyme Structure

Molecular weight

42000 <6> (<6>, gel filtration [10]) [10]
47000 <8> (<8>, gel filtration [8]) [8]
66000 <1> (<1>, gel filtration [1]) [1]
127000 <22> (<22>, gel filtration [26]) [26]

Subunits

? <6, 16> (<16>, x * 47000, SDS-PAGE [25]; <6>, x * 54000, mitochondrial enzyme, SDS-PAGE [22]) [22, 25]
dimer <22> (<22>, 2 * 62000, SDS-PAGE [26]) [26]
monomer <6, 8> (<6>, x * 41500, SDS-PAGE [10]; <8>, 1 * 47000, SDS-PAGE [8]) [8, 10]

5 Isolation/Preparation/Mutation/Application

Source/tissue

adrenal gland <7> [9, 20]
brain <7> [9]
bud <15> [19]
cell culture <3> [3]
endosperm <4> [4]
fibroblast <11> (<11>, normal and virus-transformed [19]) [19]
fruit <8> [8, 24]
heart <7> [9]
kidney <7> [9]
leaf <3> [3]
lens <7, 10, 23, 24, 26> (<23>, high activity [27]; <7,10>, weak activity [27]) [27]
leukocyte <9> [12]
liver <7, 10> [7, 9, 11, 13, 14, 15, 16, 17, 18, 28]
lung <7> [9]
mycelium <22> [26]
nodule <2> [2]
root <2, 13, 14> [2, 19]
seedling <5> [5]
spleen <7> [9]
tendril <16> (<16>, etiolated hook [19]) [19]
tuber <6, 12> [10, 19, 22]
Additional information <21> (<21>, leaf anti-oxidative system, antisense suppression of 2-cysteine peroxiredoxin does not enhances activity [21]) [21]

Localization

cytoplasm <1> [1]
cytosol <22> [26]

endoplasmic reticulum <7> (<7>, activity of the enzyme depends on the interaction of proteins and lipids and is therefore of necessity dependent upon the integrity of a certain micellar structure in the endoplasmic reticulum [15]) [15]
membrane <4, 7, 10, , 23, 24, 25, 26> (<4>, glyoxysomal membrane [4]; <7>, plasma membrane [7,11]; <7>, integrity of glucan moieties of the cell surface glycoconjugates are necessary for the optimal function of the enzyme activity [7]; <7>, outer mitochondrial membrane, highest activity in the cis-Golgi membrane fraction [9, 20]; <7>, plasma membrane [28]) [4, 7, 9, 11, 13, 20, 27, 28]
microsome <7, 10> [14, 15, 16, 17, 18]
mitochondrion <6, 7> (<7>, outer mitochondrial membrane [9, 20]) [9, 16, 20, 22]
Additional information <7> (<7>, post-nuclear particulate fraction [9]; <7>, vesicle fraction [13]) [9, 13]

Purification
<1> [1]
<6> [10, 22]
<8> [8, 24]
<22> [26]

Crystallization
<8> [24]

Cloning
<8> (high level expression in Escherichia coli, using the T7 RNA polymerase expression system [24]) [24]
<16> (expression in Escherichia coli fused to maltose-binding protein [25]) [25]

6 Stability

pH-Stability
7-11 <22> (<22>, stable [26]) [26]

Temperature stability
60 <1, 6> (<1>, 14 min, complete inactivation [1]; <6>, 1 min, 20% loss of activity [10]) [1, 10]

General stability information
<4>, in whole glyoxysomes the enzyme is insensitive to trypsin [4]
<16>, 4°C, maltose-binding protein-monodehydroascorbate reductase fusion protein, in presence of 2-mercaptoethanol, 1 month, stable [25]
<22>, freezing without glycerol, complete loss of activity [26]

Storage stability
<6>, 4°C, pH 8, 30 d, almost complete loss of activity, residual activity up to 38% when stored in presence of 1-10 mM $MgSO_4$ [10]

<22>, -20°C, 10 mM Tris/HCl buffer, pH 7.4, 2 mM dithiothreitol, 25% glycerol, 1 month, stable [26]
<22>, 4°C, 4°C, in absence of dithiothreitol, complete loss of activity within 1 day [26]

References

[1] Schulze, H.U.; Schott, H.H.; Staudinger, H.: The isolation and characterization of a NADH:semidehydroascorbic acid oxidoreductase from Neurospora crassa. Hoppe-Seyler's Z. Physiol. Chem., **353**, 1931-1942 (1972)
[2] Dalton, D.A.; Post, C.J.; Langeberg, L.: Effects of ambient oxygen and of fixed nitrogen on concentrations of glutathione, ascorbate, and associated enzymes in soybean root nodules. Plant Physiol., **96**, 812-818 (1991)
[3] Tanaka, K.; Masuda, R.; Sugimoto, T.; Kawamura, Y.; Kuboi, T.: An H_2O_2-decomposing system in cultured tobaco cells. Agric. Biol. Chem., **54**, 2003-2008 (1990)
[4] Bowditch, M.I.; Donaldson, R.P.: Ascorbate free-radical reduction by glyoxysomal membranes. Plant Physiol., **94**, 531-537 (1990)
[5] Tommasi, F.; De Gara, L.; Liso, R.; Arrigoni, O.: The ascorbic acid system in Cuscuta reflexa Roxb.. J. Plant Physiol., **135**, 766-768 (1990)
[6] Borraccino, G.; Dipierro, S.; Arrigoni, O.: Interaction of ascorbate free radical reductase with sulphhydryl reagents. Phytochemistry, **28**, 715-717 (1989)
[7] Navas, P.; Estevez, A.; Villalba, J.M.; Buron, M.I.; Crane, F.L.: Cell surface glycoconjugates control the activity of the NADH-ascorbate free radical reductase of rat liver plasma membrane. Biochem. Biophys. Res. Commun., **154**, 1029-1033 (1988)
[8] Hossain, M.A.; Asada, K.: Monodehydroascorbate reductase from cucumber is a flavin adenine dinucleotide enzyme. J. Biol. Chem., **260**, 12920-12926 (1985)
[9] Nishino, H.; Ito, A.: Subcellular distribution of OM cytochrome b-mediated NADH-semidehydroascorbate reductase activity in rat liver. J. Biochem., **100**, 1523-1531 (1986)
[10] Borraccino, G.; Dipierro, S.; Arrigoni, O.: Purification and properties of ascorbate free-radical reductase from potato tubers. Planta, **167**, 521-526 (1986)
[11] Goldenberg, H.: Insulin inhibits NADH-semidehydroascorbate reductase in rat liver plasma membrane. Biochem. Biophys. Res. Commun., **94**, 721-726 (1980)
[12] Stankova, L.; Bigley, R.; Wyss, S.R.; Aebi, H.: Catalase and dehydroascorbate reductase in human polymorphonuclear leukocytes (PMN): possible functional relationship. Experientia, **35**, 852-853 (1979)
[13] Gei-, D.; Schulze, H.U.: Isolation and chemical composition of the NADH: semidehydroascorbate oxidoreductase rich membranes from rat liver. FEBS Lett., **60**, 374-379 (1975)
[14] Weber, H.; Weis, W.; Wolf, B.: Monodehydro-L(plus)-ascorbate reducing systems in differently prepared pig liver microsomes. Hoppe-Seyler's Z. Physiol. Chem., **355**, 595-599 (1974)

[15] Schulze, H.U.; Gallenkamp, H.; Staudinger, H.: Effect of membrane lipids on the activity of NADH:semidehydroascorbic acid oxidoreductase (EC 1.6.5.4) of rat liver microsomes. Hoppe-Seyler's Z. Physiol. Chem., **354**, 391-406 (1973)
[16] Green, R.C.; O'Brien, P.J.: The involvement of semidehydroascorbate reductase in the oxidation of NADH by lipid peroxide in mitochondria and microsomes. Biochim. Biophys. Acta, **293**, 334-342 (1973)
[17] Schulze, H.U.; Gallenkamp, H.; Staudinger, H.: Microsomal NADH-dependent electron transport. Hoppe-Seyler's Z. Physiol. Chem., **351**, 809-817 (1970)
[18] Oehler, G.; Weis, W.; Staudinger, H.: Semidehydro-D(-)-ascorbic acid as a substrate of microsomal NADH: semidehydroascorbate oxidoreductase (EC 1.6.5.4). Hoppe-Seyler's Z. Physiol. Chem., **353**, 495-496 (1972)
[19] Arrigoni, O.; Dipierro, S.; Borraccino, G.: Ascorbate free radical reductase, a key enzyme of the ascorbic acid system. FEBS Lett., **125**, 242-244 (1981)
[20] Natarajan, R.D.; Harding, B.W.: Cholesterol side chain cleavage in rat adrenal supported by outer mitochondrial membrane NADH-semidehydroascorbate reductase. J. Biol. Chem., **260**, 3902-3905 (1985)
[21] Baier, M.; Noctor, G.; Foyer, C.H.; Dietz, K.J.: Antisense suppression of 2-cysteine peroxiredoxin in Arabidopsis specifically enhances the activities and expression of enzymes associated with ascorbate metabolism but not glutathione metabolism. Plant Physiol., **124**, 823-832 (2000)
[22] De Leonardis, S.; De Lorenzo, G.; Borraccino, G.; Dipierro, S.: A specific ascorbate free radical reductase isoenzyme participates in the regeneration of ascorbate for scavenging toxic oxygen species in potato tuber mitochondria. Plant Physiol., **109**, 847-851 (1995)
[23] Kobayashi, K.; Tagawa, S.; Sano, S.; Asada, K.: A direct demonstration of the catalytic action of monodehydroascorbate reductase by pulse radiolysis. J. Biol. Chem., **270**, 27551-27554 (1995)
[24] Sano, S.; Miyake, C.; Mikami, B.; Asada, K.: Molecular characterization of monodehydroascorbate radical reductase from cucumber highly expressed in Escherichia coli. J. Biol. Chem., **270**, 21354-21361 (1995)
[25] Murthy, S.S.; Zilinskas, B.A.: Molecular cloning and characterization of a cDNA encoding pea monodehydroascorbate reductase. J. Biol. Chem., **269**, 31129-31133 (1994)
[26] Yu, S.W.; Kim, Y.R.; Kang, S.O.: Spectral characterization and chemical modification of FMN-containing ascorbyl free-radical reductase from Pleurotus ostreatus. Biochem. J., **341**, 755-763 (1999)
[27] Matsukura, S.; Bando, M.; Obazawa, H.; Oka, M.; Takehana, M.: Ascorbate free radical reductase activity in vertebrate lenses of certain species. Jpn. J. Ophthalmol., **45**, 233-239 (2001)
[28] Villalba, J.M.; Canalejo, A.; Buron, M.I.; Cordoba, F.; Navas, P.: Thiol groups are involved in NADH-ascorbate free radical reductase activity of rat liver plasma membrane. Biochem. Biophys. Res. Commun., **192**, 707-713 (1993)
[29] Sakihama, Y.; Mano, J.; Sano, S.; Asada, K.; Yamasaki, H.: Reduction of phenoxyl radicals mediated by monodehydroascorbate reductase. Biochem. Biophys. Res. Commun., **279**, 949-954 (2000)

NADPH:Quinone reductase 1.6.5.5

1 Nomenclature

EC number
1.6.5.5

Systematic name
NADPH:quinone oxidoreductase

Recommended name
NADPH:quinone reductase

Synonyms
DT-diaphorase
dehydrogenase, reduced nicotinamide adenine dinucleotide (phosphate) (quinone)
EC 1.6.99.2 (formerly)
flavoprotein NAD(P)H:quinone reductase
menadione oxidoreductase
menadione reductase
NAD(P)H dehydrogenase
NAD(P)H dehydrogenase (quinone)
NAD(P)H menadione reductase
NAD(P)H quinone reductase
NAD(P)H-quinone dehydrogenase
NAD(P)H-quinone oxidoreductase
NAD(P)H-quinone reductase
NAD(P)H:(quinone-acceptor) oxidoreductase
NAD(P)H:menadione oxidoreductase
NAD(P)H:paraquat diaphorase
NADH-menadione reductase
NADH-menaquinone reductase
NADH:quinone reductase
NADPH DT-diaphorase
NADPH:quinone oxidoreductase
NADPH:quinone reductase
naphthoquinone reductase
P36
phylloquinone reductase
quinone reductase
reduced NAD(P)H dehydrogenase
viologen accepting pyridine nucleotide oxidoreductase
vitamin K reductase

ζ-crystallin homolog protein
p-benzoquinone reductase
ζ-Crystallin
ζ-Crystallin/NADPH:quinone oxidoreductase
ζ-Crystallin/quinone reductase

CAS registry number
9032-20-6

2 Source Organism

<1> *Cavia porcellus* [1, 9, 13]
<2> *Rattus norvegicus* [2, 4]
<3> *Camelus dromedarius* (camel) [3, 5-8, 10-12]
<4> *Homo sapiens* [9]
<5> *Bos taurus* (ζ crystallin shows minimal NADPH:quinone reductase activity) [13]

3 Reaction and Specificity

Catalyzed reaction
NADPH + H^+ + quinone = $NADP^+$ + semiquinone (<1>, enzyme acts through a one-electron transfer process [1]; <3>, reaction with 2,6-dichlorophenolindophenol and NADPH proceeds through a ping-pong mechanism [7]; <3>, reaction with 9,10-phenanthrenequinone and NADPH proceeds through a ping-pong mechanism [12])

Reaction type
oxidation
reduction

Natural substrates and products
S Additional information <1, 4> (<1, 4>, although in the lens the enzyme is considered to be a crystallin, or lens structural protein, because of its high abundance its enzymatic activity and expression at catalytic levels in other tissues of various species suggest that it has a fundamental physiological role outside the lens, perhaps in the detoxification of xenobiotics [9]) [9]

Substrates and products
S 1,2-naphthoquinone + NADH <1> (<1>, 10% of the activity with NADPH) [9]
P 1,2-naphthoquinol + NAD^+
S 1,2-naphthoquinone + NADPH <1, 3> [1, 5, 9, 12]
P 1,2-naphthoquinol + $NADP^+$
S 1,4-benzoquinone + NADPH <1, 3> (<1>, 15.3% of the activity with 1,2-naphthoquinone [1]) [1, 5, 12]

P 1,4-benzoquinol + $NADP^+$
S 1,4-naphthoquinone + NADPH <1, 3> (<1>, 2.6% of the activity with 1,2-naphthoquinone [1]) [1, 5, 12]
P 1,4-naphthoquinol + $NADP^+$
S 2,6-dichlorophenolindophenol + NADH <1> (<1>, 10% of the activity with NADPH) [9]
P ?
S 2,6-dichlorophenolindophenol + NADPH <1, 3> (<1>, 3.8% of the activity with 1,2-naphthoquinone [1]) [1, 5, 7, 9, 12]
P ?
S 5-hydroxy-1,4-naphthoquinone + NADH <1> (<1>, 10% of the activity with NADPH) [9]
P 5-hydroxy-1,4-naphthoquinol + NAD^+
S 5-hydroxy-1,4-naphthoquinone + NADPH <1> (<1>, i.e. juglone, 12.5% of the activity with 1,2-naphthoquinone [1]) [1, 9]
P 5-hydroxy-1,4-naphthoquinol + $NADP^+$
S 5-hydroxy-2-methyl-1,4-naphthoquinone + NADH <1> (<1>, i.e. plumbagin, 0.9% of the activity with 1,2-naphthoquinone [1]) [1]
P 5-hydroxy-2-methyl-1,4-naphthoquinol + NAD^+
S 9,10-phenanthrenequinone + NADH <1> (<1>, 10% of the activity with NADPH) [9]
P 9,10-phenanthrenequinol + NAD^+
S 9,10-phenanthrenequinone + NADPH <1, 3> (<1>, 50% of the activity with 1,2-naphthoquinone [1]; <3>, best substrate [12]) [1, 5, 6, 8-10, 12]
P 9,10-phenanthrenequinol + $NADP^+$
S methyl-1,4-benzoquinone + NADPH <1> (<1>, 20.6% of the activity with 1,2-naphthoquinone [1]) [1]
P methyl-1,4-benzoquinol + $NADP^+$
S Additional information <1> (<1>, inactive with: menadione, ubiquinone, 9,10-anthraquinone, vitamin K_1, vitamin K_2 [1]; <3>, no activity with menadione and 9,10-anthraquinone [5]) [1, 5]
P ?

Inhibitors

2,3-dimercaptopropanol <3> [12]

2,5-dichloro-3,6-dihydroxy-1,4-benzoquinone <3> (<3>, i.e. chloranilic acid, noncompetitive with respect to both NADPH and 9,10-phenanthrenequinone [8]) [8]

4-hydroxycoumarin <3> (<3>, reversible time-independent inhibition. Only dicoumarol, 4-hydroxycoumarin and warfarin inhibit in micromolar ranges. 7-Hydroxy-4-methylcoumarin is ineffective. Competitive inhibition with respect to 2,6-dichlorophenolindophenol, uncompetitive with respect to NADPH. Phenolic hydroxyl group at the C-4 position in the coumarin skeleton is important for the maximal inhibition. Sequence of potency for the inhibitors in descending order: dicoumarol, 4-hydroxycoumarin, warfarin, coumarin [11]) [11]

5,5'-dithiobis(2-nitrobenzoate) <1, 3> (<3>, inactivation is caused by a modification of one Cys per subunit, reactivation by dithiothreitol or KCN. NADPH partially protects from inactivation, 9,10-phenanthrenequinone enhances the modification [3]) [1, 3, 12]
ADP <3> (<3>, 10% inhibition at 0.2 mM) [12]
Cu^{2+} <3> (<3>, 25% inhibition at 1 mM) [12]
FAD <3> (<3>, 60% inhibition at 0.2 mM) [12]
FMN <3> (<3>, 40% inhibition at 0.2 mM) [12]
H_2O_2 <1> [1]
NAD^+ <3> (<3>, 5% inhibition at 0.2 mM) [12]
$NADP^+$ <1, 3> (<3>, mixed-type inhibition with respect to NADPH, competitive with respect to 9,10-phenanthrenequinone [12]) [1, 12]
NEM <1, 3> [1, 12]
Zn^{2+} <3> (<3>, 25% inhibition at 1 mM) [12]
Cibacron blue 3GA <1, 3> (<3>, inhibits the reaction with NADPH and 9,10-phenanthrenequinone. Linear mixed type inhibition with respect to NADPH and noncompetitive with respect to 9,10-phenanthrenequinone [6]) [1, 6]
coumarin <3> (<3>, reversible time-independent inhibition. Only dicoumarol, 4-hydroxycoumarin and warfarin inhibit in micromolar ranges. 7-Hydroxy-4-methylcoumarin is ineffective. Competitive inhibition with respect to 2,6-dichlorophenolindophenol, uncompetitive with respect to NADPH. Phenolic hydroxyl group at the C-4 position in the coumarin skeleton is important for the maximal inhibition. Sequence of potency for the inhibitors in descending order: dicoumarol, 4-hydroxycoumarin, warfarin, coumarin [11]) [11]
dicoumarol <1, 3> (<3>, reversible time-independent inhibition. Only dicoumarol, 4-hydroxycoumarin and warfarin inhibit in micromolar ranges. 7-Hydroxy-4-methylcoumarin is ineffective. Competitive inhibition with respect to 2,6-dichlorophenolindophenol, uncompetitive with respect to NADPH. Phenolic hydroxyl group at the C-4 position in the coumarin skeleton is important for the maximal inhibition. Sequence of potency for the inhibitors in descending order: dicoumarol, 4-hydroxycoumarin, warfarin, coumarin [11]; <3>, competitive with respect to 2,6-dichlorophenol-indophenol, uncompetitive with respect to NADPH [7]) [1, 7, 11, 12]
dithiothreitol <3> (<3>, inhibition is completely prevented by preincubation with 9,10-phenanthrenequinone but not by NADPH [3]; <3>, strong competitive inhibition with respect to 9,10-phenanthrenequinone [12]) [3, 12]
nitrofurantoin <1> [1]
o-phthalaldehyde <3> [10]
p-chloromercuriphenylsulfonate <1> [1]
pyridoxal-5'-phosphate <3> (<3>, inactivation follows pseudo-first-order kinetics. NADPH protects against inactivation, 9,10-phenanthrenequinone does not protect. Inhibition is uncompetitive with NADPH and non-competitive with respect to 9,10-phenanthrenequinone [10]) [10]
warfarin <3> (<3>, reversible time-independent inhibition. Only dicoumarol, 4-hydroxycoumarin and warfarin inhibit in micromolar ranges. 7-Hydroxy-4-methylcoumarin is ineffective. Competitive inhibition with respect to

2,6-dichlorophenolindophenol, uncompetitive with respect to NADPH. Phenolic hydroxyl group at the C-4 position in the coumarin skeleton is important for the maximal inhibition. Sequence of potency for the inhibitors in descending order: dicoumarol, 4-hydroxycoumarin, warfarin, coumarin [11]) [11, 12]
Additional information <1, 3> (<1>, no inactivation by 0.5 mM iodoacetate [1]; <3>, the results of the inhibition studies suggest that an essential Lys is located in the vicinity of the NADPH binding site [10]; <3>, inhibition studies suggest that an essential disulfide-bridge is present at the binding site of zeta-crystallin [3]) [1, 3, 10]

Cofactors/prosthetic groups
NADPH <1, 3> (<1>, specific for [1,5]) [1, 5]

Turnover number (min^{-1})
156 <1> (5-hydroxy-1,4-naphthoquinone, <1>) [1]
204 <1> (1,4-benzoquinone, <1>) [1]
600 <1> (9,10-phenanthrenequinone, <1>) [1]

Specific activity (U/mg)
6 <1> [9]
25.1 <3> [5]
830 <2> [4]

K_m-Value (mM)
0.0024 <3> (1,2-naphthoquinone, <3>) [12]
0.005 <1> (NADPH, <1>) [1]
0.0069 <3> (NADPH, <3>) [5, 12]
0.007 <3> (NADPH, <3>) [7]
0.01 <1> (9,10-phenanthrenequinone, <1>) [1]
0.013 <3> (2,6-dichlorophenolindophenol, <3>) [12]
0.0153 <3> (2,6-dichlorophenolindophenol, <3>) [7]
0.017 <3> (9,10-phenanthrenequinone, <3>) [5, 12]
0.027 <1> (5-hydroxy-1,4-naphthoquinone, <1>) [1]
0.065 <3> (1,4-benzoquinone, <3>) [12]
0.143 <1> (1,4-benzoquinone, <1>) [1]
0.26 <3> (1,4-naphthoquinone, <3>) [12]

pH-Optimum
7 <3> [5]

pH-Range
4.5-10 <3> (<3>, pH 4.5: about 60% of maximal activity, pH 10: about 50% of maximal activity [5]) [5]

Temperature optimum (°C)
35 <3> [5]

Temperature range (°C)
20-60 <3> (<3>, about 60% of maximal activity at 20°C and at 60°C) [5]

4 Enzyme Structure

Molecular weight

61000 <2> (<2>, cytosolic and mitochondrial enzyme, gel filtration [4]) [4]
140000 <1, 3> (<1, 3>, gel filtration [5,9]) [5, 9]

Subunits

dimer <2> (<2>, 2 * 31000, cytosolic and mitochondrial enzyme, gel filtration [4]) [4]
tetramer <1, 3> (<1, 3>, 4 * 35000, SDS-PAGE [5,9]) [5, 9]

5 Isolation/Preparation/Mutation/Application

Source/tissue

Walker carcinoma 256 <2> [2]
lens <1, 3, 5> (<3>, cortex [5]) [1, 3, 5-8, 10-13]
liver <1, 2, 4> [4, 9]

Localization

cytosol <1, 2> [4, 13]
microsome <2> [4]

Purification

<1> [9]
<2> [4]

Crystallization

<2> [2]
<3> [5]

6 Stability

Temperature stability

45 <1> (<1>, 10 min, 65% loss of activity) [1]
51 <1> (<1>, 10 min, complete loss of activity) [1]

General stability information

<1>, NADPH protects against inactivation caused by heat, NEM or H_2O_2 [1]
<1>, very stable to freezing [1]

Storage stability

<1>, -20°C, less than 10% loss of activity after 4-6 weeks [1]
<3>, -20°C, 5 months, stable [5]
<3>, 4°C, stable for at least 1 month [5]

References

[1] Rao, P.V.; Krishna, C.M.; Zigler, J.S.: Identification and characterization of the enzymatic activity of ζ-crystallin from guinea pig lens. J. Biol. Chem., **267**, 96-102 (1992)

[2] Skelly, J.V.; Suter, D.A.; Knox, R.J.; Garman, E.; Stuart, D.I.; Sanderson, M.R.; Roberts, J.J.; Neidle, S.: Preliminary crystallographic data for NAD(P)H quinone reductase isolated from the Walker 256 rat carcinoma cell line. J. Mol. Biol., **205**, 623-624 (1989)

[3] Duhaiman, A.S.; Rabbani, N.: Involvement of a disulfide bridge in catalytic activity of camel lens ζ-crystallin. Biochem. Biophys. Res. Commun., **221**, 229-233 (1996)

[4] Sharkis, D.H.; Swenson, R.P.: Purification by cibacron blue F3GA dye affinity chromatography and comparison of NAD(P)H:quinone reductase (E.C.1.6.99.2) from rat liver cytosol and microsomes. Biochem. Biophys. Res. Commun., **161**, 434-441 (1989)

[5] Duhaiman, A.S.; Rabbani, N.; AlJafari, A.A.; Alhomida, A.S.: Purification and characterization of ζ-crystallin from the camel lens. Biochem. Biophys. Res. Commun., **215**, 632-640 (1995)

[6] Duhaiman, A.S.: Inhibition of camel lens ζ-crystallin/NADPH:quinone oxidoreductase activity by cibacron blue. J. Enzyme Inhib., **10**, 263-269 (1996)

[7] Duhaiman, A.S.: Camel lens ζ-crystallin kinetics and its inhibition by dicoumarol. Biochem. Mol. Biol. Int., **38**, 251-258 (1996)

[8] Abdulaziz, A.A A.; Riskuwa, A.S.; Duhaiman, A.S.: Inhibition of camel lens ζ-crystallin/NADPH:quinone oxidoreductase activity by chloranilic acid. Biochem. Mol. Biol. Int., **41**, 415-421 (1997)

[9] Rao, P.V.; Zigler, J.S.: Purification and characterization of ζ-crystallin/quinone reductase from guinea pig liver. Biochim. Biophys. Acta, **1117**, 315-320 (1992)

[10] Rabbani, N.; Duhaiman, A.S.: Inhibition of camel lens ζ-crystallin/NADPH:quinone oxidoreductase by pyridoxal-5'-phosphate. Biochim. Biophys. Acta, **1388**, 175-180 (1998)

[11] Duhaiman, A.S.: Inhibition of ζ-crystallin by coumarins: a structure-activity study. J. Protein Chem., **15**, 261-264 (1996)

[12] Duhaiman, A.S.: Kinetic properties of camel lens ζ-crystallin. Int. J. Biochem. Cell Biol., **28**, 1163-1168 (1996)

[13] Rao, P.V.; Gonzalez, P.; Persson, B.; Jornvall, H.; Garland, D.; Zigler, J.S.: Guinea pig and bovine ζ-crystallins have distinct functional characteristics highlighting replacements in otherwise similar structures. Biochemistry, **36**, 5353-5362 (1997)

p-Benzoquinone reductase (NADPH) 1.6.5.6

1 Nomenclature

EC number
1.6.5.6

Systematic name
NADPH:*p*-benzoquinone oxidoreductase

Recommended name
p-benzoquinone reductase (NADPH)

Synonyms
carbonyl reductase

CAS registry number
9032-20-6

2 Source Organism

<1> *Pisum sativum* [1]
<2> *Mucor ambiguus* (strain AKU 3006 [2]) [2]

3 Reaction and Specificity

Catalyzed reaction
NADPH + H^+ + *p*-benzoquinone = $NADP^+$ + hydroquinone

Reaction type
oxidation
redox reaction
reduction

Substrates and products
S 1-methylisatin + NADPH <2> (Reversibility: ir <2> [2]) [2]
P $NADP^+$ + ?
S 5-bromoisatin + NADPH <2> (Reversibility: ir <2> [2]) [2]
P $NADP^+$ + ?
S 5-methylisatin + NADPH <2> (Reversibility: ir <2> [2]) [2]
P $NADP^+$ + ?
S acenaphthenequinone + NADPH <2> (<2> reduced velocity [2]) (Reversibility: ir <2> [2]) [2]

P $NADP^+$ + ?

S alloxan + NADPH <2> (Reversibility: ir <2> [2]) [2]

P $NADP^+$ + ?

S α-naphthoquinone + NADPH <2> (Reversibility: ir <2> [2]) [2]

P $NADP^+$ + ?

S β-naphthoquinone + NADPH <2> (Reversibility: ir <2> [2]) [2]

P $NADP^+$ + ?

S dihydro-4-methyl-4-propyl-2,3-furanedione + NADPH <2> (Reversibility: ir <2> [2]) [2]

P $NADP^+$ + ?

S dihydro-5-(2-butyl)-4-methyl-4-ethyl-2,3-furanedione + NADPH <2> (<2> reduced velocity [2]) (Reversibility: ir <2> [2]) [2]

P $NADP^+$ + ?

S dihydro-5-(2-pentyl)-4,4-dimethyl-2,3-furanedione + NADPH <2> (Reversibility: ir <2> [2]) [2]

P $NADP^+$ + ?

S dihydro-5-(2-pentyl)-4-methyl-4-propyl-2,3-furanedione + NADPH <2> (<2> reduced velocity [2]) (Reversibility: ir <2> [2]) [2]

P $NADP^+$ + ?

S dihydro-5-(3-pentyl)-4,4-dimethyl-2,3-furanedione + NADPH <2> (Reversibility: ir <2> [2]) [2]

P $NADP^+$ + ?

S dihydro-5-isopropyl-4,4-dimethyl-2,3-furanedione + NADPH <2> (Reversibility: ir <2> [2]) [2]

P $NADP^+$ + ?

S isatin + NADPH <2> (Reversibility: ir <2> [2]) [2]

P $NADP^+$ + ?

S ketopantoyl lactone + NADPH <2> (Reversibility: ir <2> [2]) [2]

P $NADP^+$ + ?

S ninhydrin + NADPH <2> (Reversibility: ir <2> [2]) [2]

P $NADP^+$ + ?

S *p*-benzoquinone + NADPH <2> (Reversibility: ir <2> [2]) [2]

P $NADP^+$ + hydroquinone

S *p*-toruquinone + NADPH <2> (Reversibility: ir <2> [2]) [2]

P $NADP^+$ + ?

S phenanthrenequinone + NADPH <2> (Reversibility: ir <2> [2]) [2]

P $NADP^+$ + ?

Inhibitors

1,3-cyclopentadione <2> (<2> 1.0 mM [2]) [2]

3-methyl-1,2-cyclohexadione <2> (<2> 1.0 mM [2]) [2]

5,5'-dithiobis(2-nitrobenzoic acid) <2> (<2> 0.1 mM [2]) [2]

8-hydroxyquinoline <2> (<2> 0.1 mM [2]) [2]

$CdCl_2$ <2> (<2> 1.0 mM [2]) [2]

$FeCl_2$ <2> (<2> 1.0 mM [2]) [2]

$NADP^+$ <2> (<2> 1.0 mM [2]) [2]

cyclohexadione <2> (<2> 1.0 mM [2]) [2]

dicoumarol <2> (<2> 0.1 mM [2]) [2]
o-phenanthroline <2> (<2> 1.0 mM [2]) [2]
quercetin <2> (<2> 0.01 mM [2]) [2]
Additional information <2> (<2> overview [2]) [2]

Cofactors/prosthetic groups

NADPH <2> [2]

K_m-Value (mM)

0.0123 <2> (1-methylisatin) [2]
0.05 <2> (isatin) [2]
0.053 <2> (β-naphthoquinone) [2]
0.135 <2> (NADPH) [2]
0.21 <2> (acenaphthenequinone) [2]
0.24 <2> (dihydro-5-(2-pentyl)-4-methyl-4-propyl-2,3-furanedione) [2]
0.38 <2> (dihydro-5-(2-pentyl)-4,4-dimethyl-2,3-furanedione) [2]
0.41 <2> (*p*-toruquinone) [2]
0.58 <2> (dihydro-5-(3-pentyl)-4,4-dimethyl-2,3-furanedione) [2]
0.71 <2> (ketopantoyl lactone) [2]
1.61 <2> (dihydro-5-(2-butyl)-4-methyl-4-ethyl-2,3-furanedione) [2]
2.5 <2> (alloxan) [2]
3.3 <2> (dihydro-4-methyl-4-propyl-2,3-furanedione) [2]
4.17 <2> (dihydro-5-isopropyl-4,4-dimethyl-2,3-furanedione) [2]

pH-Optimum

6 <2> (<2> for ketopantoyl lactone reduction [2]) [2]
6-7 <2> (<2> for isatin reduction [2]) [2]

pH-Range

5-9 <2> (<2> with different buffers [2]) [2]

Temperature optimum (°C)

40 <2> (<2> reduction of ketopantoyl lactone [2]) [2]
45 <2> (<2> reduction of isatin [2]) [2]

Temperature range (°C)

15-50 <2> (<2> for isatin and ketopantoyl lactone as substrates [2]) [2]

4 Enzyme Structure

Molecular weight

44000 <2> (<2> high-performance gel-permeation liquid chromatography, TSK G-3000 SW [2]) [2]
54000 <2> (<2> gel filtration, Sephadex G-100 [2]) [2]

Subunits

dimer <2> (<2> 2 * 27500, SDS-PAGE [2]) [2]

5 Isolation/Preparation/Mutation/Application

Source/tissue

leaf <1> [1]

Localization

cytosol <1> (<1> 80% of total activity [1]) [1]
mitochondrion <1> (<1> highest specific activity [1]) [1]

Purification

<2> (purification to apparent homogenity [2]) [2]

6 Stability

pH-Stability

6-7 <2> (<2> 91% of control activity at pH 9, 79% of control activity at pH 12, no activity at pH 4.0 [2]) [2]

Temperature stability

0-40 <2> (<2> 15% of control activity at 50°C [2]) [2]

References

[1] Kolesnikov, P.A.; Zore, S.V.; Pshenova, K.V.; Petrochenko, E.I.; Pletnikova, N.K.; Makovkina, L.E.; Mutuskin, A.A.: Localization of $NADH_2$($NADPH_2$): *p*-benzoquinone oxidoreductases in cell fractions of pea leaves. Fiziol. Rast. (Moscow), **20**, 170-174 (1973)
[2] Shimizu, S.; Hattori, S.; Hata, H.; Yamada, H.: A novel fungal enzyme, NADPH-dependent carbonyl reductase, showing high specificity to conjugated polyketones. Eur. J. Biochem., **174**, 37-44 (1988)

2-Hydroxy-1,4-benzoquinone reductase 1.6.5.7

1 Nomenclature

EC number
1.6.5.7

Systematic name
1,2,4-trihydroxybenzene:NAD^+ oxidoreductase

Recommended name
2-hydroxy-1,4-benzoquinone reductase

Synonyms
hydroxybenzoquinone reductase <1> [1]

CAS registry number
214466-94-1

2 Source Organism

<1> *Burkholderia cepacia* (strain AC1100 [1]) [1]

3 Reaction and Specificity

Catalyzed reaction
1,2,4-trihydroxybenzene + NAD^+ = hydroxybenzoquinone + NADH + H^+

Reaction type
oxidation
redox reaction
reduction

Substrates and products
- **S** 5-chloro-hydroxybenzoquinone + NADH <1> (Reversibility: r <1> [1]) [1]
- **P** 5-chloro-benzoquinol + NAD^+
- **S** hydroxybenzoquinone + NADH <1> (Reversibility: r <1> [1]) [1]
- **P** benzoquinol + NAD^+
- **S** methylbenzoquinone + NADH <1> (Reversibility: r <1> [1]) [1]
- **P** methylbenzoquinol + NAD^+

Cofactors/prosthetic groups
FMN <1> [1]

Specific activity (U/mg)
3 <1> (<1> with succinate as carbon source [1]) [1]
68 <1> (<1> after induction with 2,4,5-trichlorophenoxyacetic acid [1]) [1]

4 Enzyme Structure

Molecular weight
50000 <1> (<1> gel filtration, Sephacryl S100 [1]) [1]

Subunits
dimer <1> (<1> 2 * 22000, SDS-PAGE [1]) [1]

5 Isolation/Preparation/Mutation/Application

Purification
<1> (3-step purification, 27fold enrichment [1]) [1]

References

[1] Zaborina, O.; Daubaras, D.L.; Zago, A.; Xun, L.; Saido, K; Klem, T.; Nikolic, D.; Chakrabarty, A.M.: Novel pathway for conversion of chlorohydroxyquinol to maleylacetate in Burkholderia cepacia AC1100. J. Bacteriol., **180**, 4667-4675 (1998)

Nitrate reductase (NADH) 1.6.6.1

1 Nomenclature

EC number
1.6.6.1 (transferred to EC 1.7.1.1)

Recommended name
nitrate reductase (NADH)

Nitrate reductase [NAD(P)H] 1.6.6.2

1 Nomenclature

EC number
1.6.6.2 (transferred to EC 1.7.1.2)

Recommended name
nitrate reductase [NAD(P)H]

Nitrate reductase (NADPH) 1.6.6.3

1 Nomenclature

EC number

1.6.6.3 (transferred to EC 1.7.1.3)

Recommended name

nitrate reductase (NADPH)

Nitrite reductase [NAD(P)H] 1.6.6.4

1 Nomenclature

EC number

1.6.6.4 (transferred to EC 1.7.1.4)

Recommended name

nitrite reductase [NAD(P)H]

1 Nomenclature

EC number

1.6.6.5 (transferred to EC 1.7.99.3)

Recommended name

Recommended Name never specified

Hyponitrite reductase 1.6.6.6

1 Nomenclature

EC number
1.6.6.6 (transferred to EC 1.7.1.5)

Recommended name
hyponitrite reductase

Azobenzene reductase 1.6.6.7

1 Nomenclature

EC number
1.6.6.7 (transferred to EC 1.7.1.6)

Recommended name
azobenzene reductase

GMP Reductase 1.6.6.8

1 Nomenclature

EC number
1.6.6.8 (transferred to EC 1.7.1.7)

Recommended name
GMP reductase

Trimethylamine-N-oxide reductase 1.6.6.9

1 Nomenclature

EC number
1.6.6.9

Systematic name
NADH:trimethylamine-N-oxide oxidoreductase

Recommended name
trimethylamine-N-oxide reductase

Synonyms
TMAO reductase
TOR
reductase, trimethylamine N-oxide
trimethylamine N-oxide reductase
trimethylamine oxide reductase

CAS registry number
37256-34-1

2 Source Organism

<1> *Erythrobacter sp.* (aerobic photosynthetic bacterium [10]) [10]
<2> *Vibrio parahaemolyticus* [6]
<3> *Escherichia coli* (a constitutive enzyme and an inducible enzyme with different properties [2, 8]; TMAO reductase I, the major enzyme among inducible TMAO reductases [3]; reduction of trimethylamine N-oxide is catalyzed by at least 2 enzymes in Escherichia coli: trimethylamine-N-oxide reductase, which is anaerobically induced by trimethylamine N-oxide and the constitutive enzyme dimethylsulfoxide reductase [7]; inducible enzyme [11]; inducible enzyme [19]; wild-type and mutant lacking a protein, TorD, which is a chaperone specific for the enzyme [20]) [2, 3, 5, 7, 8, 11, 15, 19, 20, 22-25]
<4> *Shewanella sp.* (marine bacterium, 2 enzyme forms [4]) [4]
<5> *Rhodobacter capsulatus* (a single enzyme responsible for both trimethylamine-N-oxide and dimethylsulfoxide reductase [12]) [9, 12]
<6> *Proteus vulgaris* (a single enzyme responsible for both trimethylamine-N-oxide and dimethylsulfoxide reductase [13]) [13]
<7> *Alteromonas sp.* (inducible [14]) [14]

<8> *Salmonella typhimurium* (inducible enzyme [1]; wild-type and mutants defective in trimethylamine oxide reduction [16]) [1, 16]
<9> *Shewanella massilia* [17]
<10> *Rhodospirillum rubrum* [18]
<11> *Shewanella massilia* [21]
<12> *Shewanella oneidensis* [26]

3 Reaction and Specificity

Catalyzed reaction

NADH + H^+ + trimethylamine-N-oxide = NAD^+ + trimethylamine + H_2O

Reaction type

oxidation
redox reaction
reduction

Natural substrates and products

S trimethylamine-N-oxide + electron donor <1, 3, 5, 7, 9, 11> (<3> cytochrome $c_{554,557}$ may be the physiological electron donor [15]; <3, 7> trimethylamine N-oxide acts as a terminal electron acceptor for an anaerobic respiratory chain which requires, in addition to a primary dehydrogenase, cytochromes and quinones [5, 14, 15]; <1, 3, 5> anaerobic respiration [8, 9, 10]; <5> cytochrome c_{556} may be the physiological electron donor [9, 12]) (Reversibility: ? <1, 3, 5, 7, 9, 11> [5, 8-10, 12, 14, 15, 17, 21]) [5, 8-10, 12, 14, 15, 17, 21]
P trimethylamine + oxidized electron donor + H_2O <1, 3, 5, 7, 9, 11> [5, 8-10, 12, 14, 15, 17, 21]

Substrates and products

S 2-chloropyridine N-oxide + electron donor <3> (Reversibility: ? <3> [19]) [19]
P 2-chloropyridine + oxidized electron donor + H_2O <3> [19]
S 3-hydroxypyridine N-oxide + electron donor <3> (Reversibility: ? <3> [19]) [19]
P 3-hydroxypyridine + oxidized electron donor + H_2O <3> [19]
S 3-picoline N-oxide + electron donor <3> (Reversibility: ? <3> [19]) [19]
P 3-picoline + oxidized electron donor + H_2O <3> [19]
S 4-chloropyridine N-oxide + electron donor <3> (Reversibility: ? <3> [19]) [19]
P 4-chloropyridine + oxidized electron donor + H_2O <3> [19]
S 4-methylmorpholine N-oxide + electron donor <3, 11> (<3> good substrate [19]; <3> wild-type enzyme, molybdenum-restored enzyme and tungsten-restored enzyme [22]) (Reversibility: ? <3, 11> [19, 21, 22]) [19, 21, 22]
P 4-methylmorpholine + oxidized electron donor + H_2O <3, 11> [19, 21, 22]

S L-methionine sulfoxide + electron donor <10, 11> (<10> 26% relative activity to trimethylamine N-oxide [18]) (Reversibility: ? <10, 11> [18, 21]) [18, 21]

P L-methionine + oxidized electron donor + H_2O <10, 11> [18, 21]

S N,N-dimethyldodecylamine N-oxide + electron donor <3, 4, 11> (<4> 63% relative activity to trimethylamine N-oxide with the periplasmic enzyme and 41% relative activity to trimethylamine N-oxide with the purified enzyme [4]; <11> N,N-dimethyldodecylamine N-oxide is not an efficient substrate for the enzyme [21]) (Reversibility: ? <3, 4, 11> [4, 19, 21]) [4, 19, 21]

P N,N-dimethyldodecylamine + H_2O + oxidized electron donor <3, 4, 11> [4, 19, 21]

S N,N-dimethylhexylamine N-oxide + electron donor <3> (Reversibility: ? <3> [19]) [19]

P N,N-dimethylhexylamine + oxidized electron donor + H_2O <3> [19]

S N-cyclohexyldimethylamine N-oxide + electron donor <3> (Reversibility: ? <3> [19]) [19]

P N-cyclohexyldimethylamine + oxidized electron donor + H_2O <3> [19]

S adenosine N-oxide + electron donor <3, 4> (<4> 40% relative activity to trimethylamine N-oxide with the periplasmic enzyme and 35% relative activity to trimethylamine N-oxide with the purified enzyme [4]) (Reversibility: ? <3, 4> [4, 11]) [4, 11]

P adenosine + H_2O + oxidized electron donor <3, 4> [4, 11]

S α-picoline N-oxide + electron donor <3, 11> (<11> α-picoline N-oxide is not an efficient substrate for the enzyme [21]; <3> wild-type enzyme, molybdenum-restored enzyme and tungsten-restored enzyme [22]) (Reversibility: ? <3, 11> [3, 19, 21, 22]) [3, 19, 21, 22]

P α-picoline + H_2O + oxidized electron donor <3, 11> [3, 19, 21, 22]

S chlorate + electron donor <3, 5, 8> (Reversibility: ? <3, 5, 8> [1, 3, 11, 12]) [1, 3, 11, 12]

P ?

S dimethylsulfoxide + electron donor <3, 6, 10, 11> (<10> 9% relative activity to trimethylamine N-oxide [18]; <3> tungsten-substituted enzyme [22]) (Reversibility: ? <3, 6, 10, 11> [13, 18, 21, 22]) [13, 18, 21, 22]

P ? + oxidized electron donor + H_2O <3, 6, 10, 11> [13, 18, 21, 22]

S diphenylsulfoxide + electron donor <3> (<3> tungsten-substituted enzyme [22]) (Reversibility: ? <3> [22]) [22]

P diphenylsulfide + oxidized electron donor + H_2O <3> [22]

S γ-picoline N-oxide + electron donor <3, 4> (<4> 11% relative activity to trimethylamine N-oxide with the periplasmic enzyme and 12% relative activity to trimethylamine N-oxide with the purified enzyme [4]; <3> wild-type enzyme, molybdenum-restored enzyme and tungsten-restored enzyme [22]) (Reversibility: ? <3, 4> [4, 19, 22]) [4, 19, 22]

P γ-picoline + H_2O + oxidized electron donor <3, 4> [4, 19, 22]

S hydroxylamine N-oxide + electron donor <3, 4> (<4> 19% relative activity to trimethylamine N-oxide with the periplasmic enzyme and 3% relative activity to trimethylamine N-oxide with the purified enzyme [4]; <3>

wild-type enzyme, molybdenum-restored enzyme and tungsten-restored enzyme [22]) (Reversibility: ? <3, 4> [3, 4, 11, 19, 22]) [3, 4, 11, 19, 22]

P hydroxylamine + H_2O + oxidized electron donor <3, 4> [3, 4, 11, 19, 22]

S pyridine N-oxide + electron donor <3> (<3> wild-type enzyme, molybdenum-restored enzyme and tungsten-restored enzyme [22]) (Reversibility: ? <3> [19, 22]) [19, 22]

P pyridine + oxidized electron donor + H_2O <3> [19, 22]

S tetramethylene sulfoxide + electron donor <3> (<3> tungsten-substituted enzyme [22]) (Reversibility: ? <3> [22]) [22]

P tetrahydrothiophene + oxidized electron donor + H_2O <3> [22]

S trimethylamine N-oxide + electron donor <2, 3, 7, 8, 10, 11> (<2, 3, 8> NADH as electron donor [1-3, 5-7, 15]; <3, 4, 8, 10> benzyl viologen as electron donor [1-4, 7, 16, 18-20, 22]; <3, 4, 6-8> methyl viologen as electron donor [4, 8, 13, 14, 16]; <4> reduced methylene blue, reduced N-methylphenazonium methosulfate as electron donors [4]; <3, 7> formate as electron donor [14, 15]; <8> the activity with 3.4 mM FMN is 17% of that obtained with 0.33 mM benzyl viologen [1]; <8> vitamin K_5 or FMN as electron donors [16]; <3> quinone as electron donor [20]; <3> wild-type enzyme, molybdenum-restored enzyme and tungsten-restored enzyme [22]) (Reversibility: ? <2, 3, 7, 8, 10, 11> [1-3, 5-7, 14, 16, 18-22]) [1-3, 5-7, 14-16, 18-22]

P trimethylamine + oxidized electron donor + H_2O <2, 3, 7, 8, 10, 11> [1-3, 5-7, 14-16, 18-22]

S Additional information <4-8, 10> (<5, 6> a single enzyme responsible for both trimethylamine N-oxide and dimethylsulfoxide reductase [12, 13]; <8> the enzyme has very slight activity with nitrate, but no activity with nitrite, sulfite or thiosulfate [1]; <8> when FMN, FAD, cytochrome c, NADPH, NADH or ascorbate are substituted at equimolar concentration for benzyl viologen, none facilitates the oxidation of dithionite by the trimethylamine N-oxide [1]; <4> no activity with $FADH_2$, NADH, NADPH and the reduced forms of methylene blue and N-methylphenazonium methosulphate [4]; <7> other electron acceptors: nitrate and fumarate [14]; <10> nitrate does not act as electron acceptor [18]) [1, 4, 12-14, 18]

P ?

Inhibitors

1,10-phenanthroline <2-4> (<3, 4> slight [3, 4]) [3, 4, 6]

2,2-dipyridyl <2, 4> (<4> slight [4]; <2> 97% inhibition at 1 mM [6]) [4, 6]

5,5'-dithiobis(2-nitrobenzoate) <4> [4]

8-hydroxyquinoline <2> (<2> slight [6]) [6]

Co^{2+} <2> (<8> not inhibitory at 1 mM [1]; <2> 30% inhibition at 1 mM [6]) [6]

Cu^{2+} <3, 8> (<8> 50% inhibition at 0.1 mM [1]; <3> 52% inhibition at 0.1 mM [11]) [1, 11]

$CuSO_4$ <2> (<2> 6% inhibition at 0.01 mM with 0.1 mM sodium diethyldithiocarbamate [6]) [6]

EDTA <2, 3> (<3> 39% inhibition at 10 mM [3]; <2> slight [6]; <8> not inhibitory at 1 mM [1]) [3, 6]
EGTA <3> (<3> 38% inhibition at 10 mM [11]) [11]
FAD <8> (<8> inhibition, when added to the standard assay mixture containing benzyl viologen as electron acceptor [1]) [1]
FMN <8> (<8> inhibition, when added to the standard assay mixture containing benzyl viologen as electron acceptor [1]) [1]
Fe^{2+} <2> (<2> 30% inhibition at 0.1 mM [6]) [6]
Hg^{2+} <8> (<8> 80% inhibition at 1 mM [1]) [1]
KCN <3, 4> (<2> not inhibitory [6]; <4> partial [4]; <3> 10-15% inhibition of the inducible and of the constitutive enzyme at 10 mM [2]; <3> 71% inhibition at 10 mM [3]; <3> 20% inhibition at 1 mM [11]; <8> not inhibitory at 5 mM [1]) [2-4, 11]
Mn^{2+} <2> (<8> not inhibitory at 1 mM [1]; <2> 24% inhibition at 1 mM [6]) [6]
Mo^{6+} <2> (<2> 18% inhibition at 1 mM [6]) [6]
N-ethylmaleimide <3> (<3> both inducible and constitutive enzyme are affected equally at 5 mM [2]) [2]
NADH <2> (<2> slight inhibition at 14 mM [6]) [6]
Na_2MoO_4 <4> [4]
Na_2WO_4 <4> [4]
NaN_3 <3> (<3> 5% inhibition of inducible enzyme and 50-60% inhibition of constitutive enzyme at 10 mM [2]; <3> 39% inhibition at 10 mM [3]; <2> not inhibitory [6]; <8> not inhibitory at 10 mM [1]; <3> slight [11]) [2, 3, 11]
Ni^{2+} <2> (<2> 27% inhibition at 1 mM [6]) [6]
cupferron <2> (<2> 82% inhibition at 1 mM [6]) [6]
hydroxylamine <2> (<2> 17% inhibition at 1 mM [6]) [6]
iodoacetate <2, 4> (<2,4> partial [4,6]) [4, 6]
p-chloromercuribenzoate <2-4> (<3> 2% inhibition of inducible enzyme and 50-60% inhibition of constitutive enzyme at 5 mM [2]; <2> 100% inhibition at 0.3 mM, 70% inhibition at 0.01 mM, 13% inhibition at 0.01 mM with 0.2 mM 2-mercaptoethanol [6]; <3> slight [11]) [2, 4, 6, 11]
semicarbazide <2> (<2> 30% inhibition at 1 mM [6]) [6]
sodium chlorate <4> (<4> partial [4]) [4]
urea <3> (<3> 50% inhibition at 6 mM [11]) [11]

Cofactors/prosthetic groups

FAD <2, 4> (<2, 4> stimulation [4, 6]) [4, 6]
FMN <2, 4> (<2, 4> stimulation [4, 6]) [4, 6]
NADH <2, 3, 8> [1-3, 5-7, 15]
Additional information <1, 3, 5> (<3> the enzyme contains more than 0.4 atoms of acid-labile sulfur per molecular weight of 200000 [3]; <5> the enzyme and cytochrome c-556 may form a structural and functional association in the periplasm [9]; <1> soluble c-type cytochrome, cytochrome c-551 and cytochrome b-c1 complex may be involved in the trimethylamine N-oxide respiration system [10]; <3> a pentaheme c-type cytochrome, TorC, is involved in electron transfer to the periplasmic terminal reductase, TorC ne-

gatively autoregulates the enzyme operon [23]; <3> a pentaheme c-type cytochrome, TorC, is involved in electron transfer to the periplasmic terminal reductase, TorC belongs to the NirT/NapC c-type cytochrome family based on homologies of its N-terminal tetrahemic domain, TorCN to the cytochromes of this family, but TorC contains a C-terminal extension, TorCC with an additional heme-binding site, both domains are required for the anaerobic bacterial growth with trimethylamine N-oxide, TorCC directly transfers electrons to the periplasmic terminal reductase, whereas TorCN, which probably receives electrons from the menaquinone pool, is involved in both the electron transfer to TorCC and the binding to the periplasmic terminal reductase [25]) [3, 9, 10, 23, 25]

Metals, ions

iron <2, 3> (<3> enzyme contains 0.96 atoms of iron [3]; <2> slight stimulation by Fe^{3+} [6]; <3> 50% stimulation by 0.1 mM Fe^{3+} [11]) [3, 6, 11]
molybdenum <3, 4, 9-11> (<3> enzyme contains 1.96 atoms of molybdenum [3]; <3> contains molybdopterin [3]; <4> enzyme contains 1.32 mol molybdenum per mol of enzyme [4]; <10> deduced from inhibition of enzyme synthesis by tungstate and reduction of enzyme levels in charcoal-treated media [18]; <3> in vivo tungsten substitution of the molybdoenzyme, 0.89 atoms of tungsten per mol of purified enzyme [22]) [3-5, 17, 18, 20-22]
zinc <3> (<3> enzyme contains 1.52 atoms of zinc [3]) [3]

Turnover number (min^{-1})

240 <11> (DL-methioninesulfoxide, <11> value below [21]) [21]
240 <11> (dibutylsulfoxide, <11> value below [21]) [21]
240 <11> (dimethylsulfoxide, <11> value below [21]) [21]
240 <11> (diphenylsulfoxide, <11> value below [21]) [21]
826.2 <3> (N,N-dimethylhexylamine N-oxide) [19]
1606 <3> (diphenylsulfoxide, <3> tungsten-restored enzyme [22]) [22]
2410 <3> (dimethylsulfoxide, <3> tungsten-restored enzyme [22]) [22]
2533 <3> (pyridine N-oxide, <3> tungsten-restored enzyme [22]) [22]
2617 <3> (hydroxylamine N-oxide, <3> tungsten-restored enzyme [22]) [22]
2684 <3> (pyridine N-oxide, <3> molybdenum-restored enzyme [22]) [22]
2737 <3> (N-cyclohexyldimethylamine N-oxide) [19]
2954 <3> (3-hydroxypyridine N-oxide) [19]
2973 <3> (γ-picoline N-oxide) [19]
3132 <3> (4-methylmorpholine N-oxide, <3> tungsten-restored enzyme [22]) [22]
3207 <3> (pyridine N-oxide) [19]
3282 <3> (α-picoline N-oxide, <3> tungsten-restored enzyme [22]) [22]
3341 <3> (γ-picoline N-oxide, <3> molybdenum-restored enzyme [22]) [22]
3357 <3> (tetramethylene sulfoxide, <3> tungsten-restored enzyme [22]) [22]
3430 <3> (γ-picoline N-oxide, <3> tungsten-restored enzyme [22]) [22]
3482 <3> (trimethylamine N-oxide, <3> tungsten-restored enzyme [22]) [22]
4001 <3> (4-chloropyridine N-oxide) [19]
4197 <3> (hydroxylamine N-oxide) [19]

4237 <3> (hydroxylamine N-oxide, <3> molybdenum-restored enzyme [22]) [22]
4282 <3> (4-methylmorpholine N-oxide, <3> molybdenum-restored enzyme [22]) [22]
4680 <11> (α-picoline N-oxide) [21]
5053 <3> (α-picoline N-oxide, <3> molybdenum-restored enzyme [22]) [22]
5790 <3> (4-methylmorpholine N-oxide) [19]
6057 <3> (2-chloropyridine N-oxide) [19]
6138 <3> (α-picoline N-oxide) [19]
6564 <11> (4-methylmorpholine N-oxide) [21]
6854 <3> (trimethylamine N-oxide, <3> molybdenum-restored enzyme [22]) [22]
7188 <11> (trimethylamine N-oxide) [21]
8209 <3> (3-picoline N-oxide) [19]
9032 <3> (trimethylamine N-oxide) [19]
9060 <11> (N,N-dimethyldodecylamine N-oxide) [21]
9277 <3> (N,N-dimethyldodecylamine N-oxide) [19]
15460 <3> (trimethylamine N-oxide, <3> wild-type enzyme [22]) [22]

Specific activity (U/mg)

0.0402 <3> (<3> cells grown in the presence of L-methionine sulfoxide, fraction S100: supernatant after centrifugation at 100000 g [7]) [7]
0.0751 <3> (<3> cells grown in the absence of oxide compounds, fraction S100: supernatant after centrifugation at 100000 g [7]) [7]
0.0856 <3> (<3> cells grown in the presence of L-methionine sulfoxide, fraction S10: supernatant after centrifugation at 10000 g [7]) [7]
0.096 <7> (<7> anaerobic growth conditions, grown without trimethylamine N-oxide [14]) [14]
0.099 <3> (<3> cells grown in the presence of γ-picoline N-oxide, fraction S100: supernatant after centrifugation at 100000 g [7]) [7]
0.115 <3> (<3> cells grown in the presence of γ-picoline N-oxide, fraction S10: supernatant after centrifugation at 10000 g [7]) [7]
0.123 <3> (<3> cells grown in the absence of oxide compounds, fraction S10: supernatant after centrifugation at 10000 g [7]) [7]
0.13 <10> (<10> periplasmic fraction [18]) [18]
0.163 <7> (<7> microaerobic growth conditions, grown without trimethylamine N-oxide [14]) [14]
0.25 <7> (<7> aerobic growth conditions, grown with trimethylamine N-oxide [14]) [14]
0.268 <3> (<3> cells grown in the presence of L-methionine sulfoxide, fraction P100: solubilized portion of the precipitates [7]) [7]
0.374 <7> (<7> anaerobic growth conditions, grown with trimethylamine N-oxide [14]) [14]
0.41 <8> (<8> mutants, which reduce little or no trimethylamine N-oxide in vivo, but their extracts retain full capacity to reduce it with methyl viologen, formate as electron donor [16]) [16]

0.428 <3> (<3> cells grown in the presence of γ-picoline N-oxide, fraction P100: solubilized portion of the precipitates [7]) [7]
0.473 <3> (<3> cells grown in the the absence of oxide compounds, fraction P100: solubilized portion of the precipitates [7]) [7]
0.56 <7> (<7> microaerobic growth conditions, grown with trimethylamine N-oxide [14]) [14]
0.644 <3> (<3> cells grown in the presence of dimethylsulfoxide, fraction S100: supernatant after centrifugation at 100000 g [7]) [7]
0.66 <8> (<8> mutants partially defective in trimethylamine N-oxide reductase activity, which lack the inducible enzyme but retain the constitutive activity, FMN or vitamin K5 as electron donor [16]) [16]
0.687 <3> (<3> cells grown in the presence of dimethylsulfoxide, fraction S10: supernatant after centrifugation at 10000 g [7]) [7]
0.748 <3> (<3> cells grown in the presence of trimethylamine N-oxide, fraction S100: supernatant after centrifugation at 100000 g [7]) [7]
0.797 <3> (<3> cells grown in the presence of trimethylamine N-oxide, fraction S10: supernatant after centrifugation at 10000 g [7]) [7]
1.29 <8> (<8> mutants partially defective in trimethylamine N-oxide reductase activity, which lack the inducible enzyme but retain the constitutive activity, formate as electron donor [16]) [16]
1.56 <8> (<8> wild-type strain, formate as electron donor [16]) [16]
2.46 <3> (<3> cells grown in the presence of trimethylamine N-oxide, fraction P100: solubilized portion of the precipitates [7]) [7]
2.63 <8> (<8> mutants, which reduce little or no trimethylamine N-oxide in vivo, but their extracts retain full capacity to reduce it with methyl viologen, vitamin K_5 as electron donor [16]) [16]
2.78 <3> (<3> cells grown in the presence of dimethylsulfoxide, fraction P100: solubilized portion of the precipitates [7]) [7]
3.73 <8> (<8> wild-type strain, vitamin K_5 as electron donor [16]) [16]
3.77 <8> (<8> mutants, which reduce little or no trimethylamine N-oxide in vivo, but their extracts retain full capacity to reduce it with methyl viologen, FMN as electron donor [16]) [16]
4.31 <8> (<8> mutants partially defective in trimethylamine N-oxide reductase activity, which lack the inducible enzyme but retain the constitutive activity, methyl viologen as electron donor [16]) [16]
4.7 <8> (<8> wild-type strain, FMN as electron donor [16]) [16]
9.92 <8> (<8> wild-type strain, methyl viologen as electron donor [16]) [16]
10.2 <10> (<10> purified enzyme [18]) [18]
11 <3> (<3> periplasmic fraction of wild-type cells, initial activity [20]) [20]
11.2 <8> (<8> mutants, which reduce little or no trimethylamine N-oxide in vivo, but their extracts retain full capacity to reduce it with methyl viologen, methyl viologen as electron donor [16]) [16]
33 <3> (<3> purified enzyme, assay performed in 100 mM phosphate buffer, pH 6.8 with a trimethylamine N-oxide concentration 0.5 mM and a concentration 20 mM of 2-hydroxypyridine N-oxide [19]) [19]
34 <3> (<3> purified enzyme, assay performed in 100 mM phosphate buffer, pH 6.8 with a trimethylamine N-oxide concentration 0.5 mM [19]) [19]

35 <3> (<3> purified enzyme, assay performed in 100 mM phosphate buffer, pH 6.8 with a trimethylamine N-oxide concentration 0.5 mM and a concentration 20 mM of tetramethylenesulfoxide [19]) [19]
36 <3> (<3> purified enzyme, assay performed in 100 mM phosphate buffer, pH 6.8 with a trimethylamine N-oxide concentration 0.5 mM and a concentration 20 mM of picolinic acid N-oxide or a concentration 20 mM of dimethylsulfoxide [19]) [19]
41 <3> (<3> purified tungsten-substituted enzyme [22]) [22]
105 <3> (<3> purified enzyme, with trimethylamine N-oxide as substrate [19]) [19]
211.5 <4> [4]
237 <3> [11]
250 <3> (<3> purified enzyme [20]) [20]
1350 <3> (<3> crude extract of mutant strain lacking the TorD protein [20]) [20]
1620 <3> [3]
1720 <3> (<3> crude extract of wild-type strain [20]) [20]
1810 <3> (<3> crude extract of mutant strain lacking the TorD protein but complemented by pTorD [20]) [20]
Additional information <1, 3, 7, 8, 11> [1, 10, 11, 14, 16, 21]

K_m-Value (mM)

0.00754 <3> (4-methylmorpholine N-oxide, <3> tungsten-restored enzyme [22]) [22]
0.00765 <3> (trimethylamine N-oxide, <3> tungsten-restored enzyme [22]) [22]
0.00838 <3> (γ-picoline N-oxide, <3> tungsten-restored enzyme [22]) [22]
0.00991 <3> (α-picoline N-oxide, <3> tungsten-restored enzyme [22]) [22]
0.0108 <3> (pyridine N-oxide, <3> tungsten-restored enzyme [22]) [22]
0.017 <11> (trimethylamine N-oxide) [21]
0.02 <3> (4-methylmorpholine N-oxide) [19]
0.02 <4> (trimethylamine N-oxide) [4]
0.022 <11> (4-methylmorpholine N-oxide) [21]
0.0252 <3> (tetramethylene sulfoxide, <3> tungsten-restored enzyme [22]) [22]
0.0278 <3> (diphenylsulfoxide, <3> tungsten-restored enzyme [22]) [22]
0.0278 <3> (trimethylamine N-oxide, <3> wild-type enzyme [22]) [22]
0.0316 <3> (trimethylamine N-oxide, <3> molybdenum-restored enzyme [22]) [22]
0.0421 <3> (4-methylmorpholine N-oxide, <3> molybdenum-restored enzyme [22]) [22]
0.07 <3> (trimethylamine N-oxide) [19]
0.15 <3> (methyl viologen, <3> with trimethylamine N-oxide, pH 6.9 [3]) [3]
0.26 <3> (α-picoline N-oxide) [19]
0.3 <6> (trimethylamine-N-oxide) [13]
0.32 <3> (γ-picoline N-oxide) [19]
0.33 <3> (benzyl viologen, <3> with trimethylamine N-oxide [3]) [3]

0.61 <11> (α-picoline N-oxide) [21]
0.631 <3> (γ-picoline N-oxide, <3> molybdenum-restored enzyme [22]) [22]
0.653 <3> (α-picoline N-oxide, <3> molybdenum-restored enzyme [22]) [22]
0.67 <10> (dimethylsulfoxide) [18]
0.67 <3> (trimethylamine-N-oxide, <3> with methyl viologen, pH 6.9 [3]) [3]
0.75 <3> (3-hydroxypyridine N-oxide) [19]
0.89 <8> (trimethylamine N-oxide) [1]
0.95 <3> (trimethylamine-N-oxide, <3> with benzyl viologen, pH 5.5 [3]) [3]
0.974 <3> (dimethylsulfoxide, <3> tungsten-restored enzyme [22]) [22]
1 <11> (N,N-dimethyldodecylamine N-oxide) [21]
1.14 <3> (N,N-dimethylhexylamine N-oxide) [19]
1.25 <3> (N,N-dimethyldodecylamine N-oxide) [19]
1.7 <2> (trimethylamine N-oxide) [6]
1.86 <3> (pyridine N-oxide) [19]
2.01 <3> (2-chloropyridine N-oxide) [19]
2.15 <3> (4-chloropyridine N-oxide) [19]
2.2 <8> (chlorate) [1]
2.25 <3> (pyridine N-oxide, <3> molybdenum-restored enzyme [22]) [22]
2.41 <4> (N,N-dimethyldodecylamine N-oxide) [4]
4.74 <3> (N-cyclohexyldimethylamine N-oxide) [19]
5.43 <3> (3-picoline N-oxide) [19]
6 <6> (dimethylsulfoxide) [13]
6.95 <4> (γ-picoline N-oxide) [4]
7.39 <3> (hydroxylamine N-oxide) [19]
10.9 <3> (hydroxylamine N-oxide, <3> tungsten-restored enzyme [22]) [22]
11.8 <3> (hydroxylamine N-oxide, <3> molybdenum-restored enzyme [22]) [22]
24 <10> (trimethylamine-N-oxide) [18]

pH-Optimum

5 <3> (<3> for tungsten-substituted enzyme, a significant decrease in activity is found at pH 5.5 [22]) [22]
5-5.5 <3> (<3> for molybdoenzyme [22]) [22]
5.5 <3> (<3> for inducible and constitutive enzyme [2]; <3> with benzyl viologen [3]) [2, 3]
5.65 <8> [1]
6.9 <3> (<3> with methyl viologen [3]) [3]
7-7.5 <2> [6]

pH-Range

6-8 <2> (<2> pH 6.0: about 35% of activity maximum, pH 8.0: about 80% of activity maximum [6]) [6]

Temperature optimum (°C)

45 <8> [1]
60 <3> (<3> for molybdoenzyme, maximal activity remains the same at 80°C [22]) [22]
80 <3> (<3> for tungsten-substituted enzyme, maximal level above [22]) [22]

4 Enzyme Structure

Molecular weight

47000 <4> (<4> 2 MW forms, MW: 47000, SDS-PAGE in absence of reducing agents and MW: 84000, gel filtration, SDS-PAGE under reducing conditions [4]) [4]
80000 <5> (<5> SDS-PAGE [12]) [12]
83000 <10> (<10> gel electrophoresis [18]) [18]
84000 <4> (<4> 2 MW forms, MW: 47000, SDS-PAGE in absence of reducing conditions and MW: 84000, gel filtration, SDS-PAGE under reducing conditions [4]) [4]
87000 <10> (<10> gel filtration [18]) [18]
95000 <6> (<6> gradient gel electrophoresis, gel isoelectric focusing [13]) [13]
110000 <3> (<3> a single inducible trimethylamine-N-oxide reductase which can exist as a dimer, MW: 230000, or a monomer, MW: 110000, gel filtration [11]) [11]
200000 <3> (<3> gel filtration [3]) [3]
230000 <3> (<3> a single inducible trimethylamine-N-oxide reductase which can exist as a dimer, MW: 230000 or a monomer, MW: 110000, gel filtration [11]) [11]
332000 <8> (<8> gel filtration [1]) [1]

Subunits

dimer <3, 10> (<3> 2 * 95000, SDS-PAGE [3]; <3> a single inducible trimethylamine-N-oxide reductase which can exist as a dimer, MW 230000 or a monomer, MW 110000 [11]; <10> 1 * 40000 + 1 * 36000, SDS-PAGE [18]) [3, 11, 18]
monomer <3, 4, 11> (<4> 1 * 84000, SDS-PAGE under reducing conditions [4]; <3> a single inducible trimethylamine-N-oxide reductase which can exist as a dimer, MW 230000 or a monomer, MW 110000 [11]; <11> 1 * 90000, SDS-PAGE [21]) [4, 11, 21]
tetramer <8> (<8> 4 * 84000, SDS-PAGE [1]) [1]

5 Isolation/Preparation/Mutation/Application

Localization

cytoplasm <1> (<1> low activity [10]) [10]
membrane <2, 3> (<3> constitutive enzyme [2, 8]; <2> enzyme associated with the membrane portion of lysed spheroplasts [6]) [2, 6, 8]
periplasm <1, 3-5, 7, 9-11> (<3> inducible enzyme [8]; <3> the enzyme is translocated across the cytoplasmic membrane into the periplasm in a folded conformation via the Tat pathway, the translocation is dependent on the anionic phospholipids and on lipid polymorphism [24]) [4, 8, 10, 12, 14, 15, 17, 18, 20, 21, 23, 24]
soluble <3> (<3> inducible enzyme [2]) [2]

Purification

<3> (using heat treatment, ammonium sulfate precipitation and chromatography on Bio-Gel A-1.5m, DEAE-cellulose and reactive blue-agarose [3]; using heat treatment, column chromatography on DEAE-cellulose, Sephadex G-200, DEAE-Sepharose CL-6B and hydroxyapatite [11]; using ion exchange chromatography on DE52, chromatography on a Mono Q HR 16/10 column and preparative electrophoresis [20]; using ammonium sulfate treatment and column chromatography on Mono Q HR 5/5 and Superdex G200 HR 10/30 [25]) [3, 11, 20, 25]
<4> (of 2 enzyme forms, using ammonium sulfate treatment, ion exchange buffer treatment, column chromatography on DEAE-Sepharose CL6B and hydroxyapatite and gel filtration on Sephacryl S-300 [4]) [4]
<8> (using polyethyleneimine to remove nucleic acids, ammonium sulfate treatment, column chromatography on DEAE-Sephacel and Bio-Gel A-5m, chromatofocusing and chromatography on Sephacryl-S-300 column [1]) [1]
<10> (using ammonium sulfate treatment and chromatography on Sephacryl S-200 column and Mono-Q column [18]) [18]
<11> (using ion exchange chromatography on DE52 column, chromatography on a Mono Q HR 16/10 column and preparative electrophoresis [21]) [21]

Crystallization

<9> (crystal structure solved by molecular replacement with the coordinates of the dimethylsulfoxide reductase from Rhodobacter sphaeroides, the overall fold of the protein structure is essentially the same as the dimethylsulfoxide reductase structures, organized into four domains [17]) [17]

Cloning

<3> (identification of TorC as a possible negative regulator of the tor operon by using a genomic library cloned into a multicopy plasmid [23]; cloning of the torAD genes alone or together with torC [25]) [23, 25]
<11> (amplification of the structural torA gene encoding the enzyme and its flanking regions using PCR techniques, the torA gene is the third gene of a TMAO-inducible operon, torECAD, encoding the TMAO respiratory components, the torC gene encodes a pentahemic c-type cytochrome, likely to be involved in electron transfer to the TorA terminal reductase, the role of the torE gene is unknown, torD gene encodes the TorD protein, a cytoplasmic protein involved in the folding process of the TorA precursor protein [21]) [21]
<12> (the trimethylamine N-oxide reductase respiratory system is encoded by the torECAD operon, a three-gene cluster, torSTR, encoding a complex two-component regulatory system is present downstream of the torECAD operon, introduction of the torSTR gene cluster into Escherichia coli shows that this regulatory gene cluster is involved in trimethylamine N-oxide induction of the torE promoter, the torR operator site is required for induction of the tor structural promoter, TorS of Escherichia coli , the trimethylamine N-oxide sensor of Escherichia coli is able to transphosphorylate TorR of Shewanella oneidensis, the trimethylamine N-oxide response regulator of Shewanella oneidensis [26]) [26]

6 Stability

pH-Stability

4.3 <8> (<8> not stable [1]) [1]
6-8.5 <8> (<8> stability peak at pH 7.3 [1]) [1]

Temperature stability

60 <3> (<3> 20 min, 10% loss of activity [2]) [2]
65 <3> (<3> 5 min, 90% loss of constitutive activity, 10% loss of inducible activity [8]) [8]
70 <3> (<3> 30 min, enzyme activity does not change [3]) [3]
75 <3, 8> (<8> 10 min, pH 7.3, quite stable at temperatures below [1]; <3> 60% loss of activity [3]) [1, 3]
80 <3> (<3> molybdoenzyme, after 4 min: 50% loss of activity, after 90 min: 97% loss of activity [22]) [22]
80 <3> (<3> tungsten-substituted enzyme, after 90 min: more than 50% activity remains [22]) [22]
80 <8> (<8> 10 min, pH 7.3, rapid loss of activity [1]) [1]
90 <8> (<8> 10 min, pH 7.3, complete loss of activity [1]) [1]

Oxidation stability

<3>, aeration for 10 min causes 10% inactivation [2]

General stability information

<2>, 2-mercaptoethanol protects against inactivation during preparation of cell extract, ineffective as stabilizer during storage [6]
<3>, about 50% of the activity of the molybdoenzyme is lost in the presence of 0.1 M NaCl, only about 4% of its initial activity is left in the presence of 2 M NaCl, 15% of the initial activity of the tungsten-substituted enzyme is left at 2 M NaCl [22]

Storage stability

<2>, -20°C, 30% loss of activity after 24 hours, 50% loss of activity after 48 hours [6]
<2>, 2°C, 30% loss of activity after 24 hours, 50% loss of activity after 48 hours [6]
<3>, -80°C, 45% loss of activity after 30 days [11]
<3>, -80°C, inducible and constitutive enzyme, 10% loss of activity after 15 days [2]
<8>, -80°C, 10% loss of activity after 15 days [1]
<8>, 0°C, 50 mM sodium phosphate buffer, pH 7.3, 30% loss of activity after 17 days [1]

References

[1] Kwan, H.S.; Barrett, E.L.: Purification and properties of trimethylamine oxide reductase from Salmonella typhimurium. J. Bacteriol., **155**, 1455-1458 (1983)

[2] Violet, M.; Medani, C.L.; Giordano, G.: Trimethylamine N-oxide (TMAO) reductases from Escherichia coli K-12. FEMS Microbiol. Lett., **27**, 85-91 (1985)

[3] Yamamoto, I.; Okubo, N.; Ishimoto, M.: Further characterization of trimethylamine N-oxide reductase from Escherichia coli, a molybdoprotein. J. Biochem., **99**, 1773-1779 (1986)

[4] Clarke, G.J.; Ward, F.B.: Purification and properties of trimethylamine N-oxide reductase from Shewanella sp. NCMB 400. J. Gen. Microbiol., **13**, 379-386 (1988)

[5] Cox, J.C.; Knight, R.: Trimethylamine N-oxide (TMAO) reductase activity in chlorate-resistant or respiration-deficient mutants of Escherichia coli. FEMS Microbiol. Lett., **12**, 249-252 (1981)

[6] Unemoto, T.; Hayashi, M.; Miyaki, K.; Hayashi, M.: Intracellular localization and properties of trimethylamine-N-oxide reductase in Vibrio parahaemolyticus. Biochim. Biophys. Acta, **110**, 319-328 (1965)

[7] Yamamoto, I.; Hinakura, M.; Seki, S.; Seki, Y.; Kondo, H.: Anaerobic induction of trimethylamine N-oxide reductase and cytochromes by dimethyl sulfoxide in Escherichia coli. Curr. Microbiol., **20**, 245-249 (1990)

[8] Silvestro, A.; Pommier, J.; Pascal, M.C.; Giordano, G.: The inducible trimethylamine N-oxide reductase of Escherichia coli K12: its localization and inducers. Biochim. Biophys. Acta, **999**, 208-216 (1989)

[9] McEwan, A.G.; Richardson, D.J.; Hudig, H.; Ferguson, S.J.; Jackson, J.B.: Identification of cytochromes involved in electron transport to trimethylamine N-oxide/dimethyl sulfoxide reductase in Rhodobacter capsulatus. Biochim. Biophys. Acta, **973**, 308-314 (1989)

[10] Arata, H.; Serikawa, Y.; Takamiya, K.: Trimethylamine N-oxide respiration by aerobic photosynthetic bacterium, Erythrobacter sp. OCh 114. J. Biochem., **103**, 1011-1015 (1988)

[11] Silvestro, A.; Pommier, J.; Giordano, G.: The inducible trimethylamine-N-oxide reductase of Escherichia coli K12: biochemical and immunological studies. Biochim. Biophys. Acta, **954**, 1-13 (1988)

[12] McEwan, A.G.; Richardson, D.J.; Jackson, J.B.; Ferguson, S.J.: The periplasmic nitrate reductase and trimethylamine N-oxide reductase of the photosynthetic bacterium Rhodobacter capsulatus. Biochem. Soc. Trans., **16**, 182-183 (1988)

[13] Styrvold, O.B.; Strom, A.R.: Dimethylsulfoxide and trimethylamine oxide respiration of Proteus vulgaris. Evidence for a common terminal reductase system. Arch. Microbiol., **140**, 74-78 (1984)

[14] Easter, M.C.; Gibson, D.M.; Ward, F.B.: The induction and location of trimethylamine N-oxide reductase in Alteromonas sp. NCMB 400. J. Gen. Microbiol., **129**, 3689-3696 (1983)

[15] Bragg, P.D.; Hackett, N.R.: Cytochromes of the trimethylamine N-oxide anaerobic respiratory pathway of Escherichia coli. Biochim. Biophys. Acta, **725**, 168-177 (1983)
[16] Kwan, H.S.; Barrett, E.L.: Roles for menaquinone and the two trimethylamine oxide (TMAO) reductases in TMAO respiration in Salmonella typhimurium: Mu d(Apr lac) insertion mutations in men and tor. J. Bacteriol., **155**, 1147-1155 (1983)
[17] Czjzek, M.; Dos Santos, J.P.; Pommier, J.; Giordano, G.; Mejean, V.; Haser, R.: Crystal structure of oxidized trimethylamine N-oxide reductase from Shewanella massilia at 2.5 A resolution. J. Mol. Biol., **284**, 435-447 (1998)
[18] Sajitz, P.; Klemme, J.; Koch, H.; Molitor, M.: Isolation and properties of trimethylamine N-oxide/dimethylsulfoxide reductase from the purple bacterium Rhodospirillum rubrum. Z. Naturforsch. C, **48**, 812-814 (1993)
[19] Iobbi-Nivol, C.; Pommier, J.; Simala-Grant, J.; Mejean, V.; Giordano, G.: High substrate specificity and induction characteristics of trimethylamine N-oxide reductase of Escherichia coli. Biochim. Biophys. Acta, **1294**, 77-82 (1996)
[20] Pommier, J.; Mejean, V.; Giordano, G.; Iobbi-Nivol, C.: TorD, a cytoplasmic chaperone that interacts with the unfolded trimethylamine N-oxide reductase enzyme (TorA) in Escherichia coli. J. Biol. Chem., **273**, 16615-16620 (1998)
[21] Dos Santos, J.; Iobbi-Nivol, C.; Couillault, C; Giordano, G.; Mejean, V.: Molecular analysis of the trimethylamine N-oxide (TMAO) reductase respiratory system from a Shewanella species. J. Mol. Biol., **284**, 421-433 (1998)
[22] Buc, J.; Santini, C.; Giordani, R.; Czjzek, M.; Wu, L.; Giordano, G.: Enzymatic and physiological properties of the tungsten-substituted molybdenum TMAO reductase from Escherichia coli. Mol. Microbiol., **32**, 159-168 (1999)
[23] Ansaldi, M.; Bordi, C.; Lepelletier, M.; Mejean, V.: TorC apocytochrome negatively autoregulates the trimethylamine N-oxide (TMAO) reductase operon in Escherichia coli. Mol. Microbiol., **33**, 284-295 (1999)
[24] Mikhaleva, N.I.; Santini, C.; Giordano, G.; Nesmeyanova, M.A.; Wu, L.: Requirement for phospholipids of the translocation of the trimethylamine N-oxide reductase through the Tat pathway in Escherichia coli. FEBS Lett., **463**, 331-335 (1999)
[25] Gon, S.; Giudici-Orticoni, M.T.; Mejean, V.; Iobbi-Nivol, C.: Electron transfer and binding of the c-type cytochrome TorC to the trimethylamine N-oxide reductase in Escherichia coli. J. Biol. Chem., **276**, 11545-11551 (2001)
[26] Gon, S.; Patte, J.; Dos Santos, J.; Mejean, V.: Reconstitution of the trimethylamine oxide reductase regulatory elements of Shewanella oneidensis in Escherichia coli. J. Bacteriol., **184**, 1262-1269 (2002)

Nitroquinoline-N-oxide reductase 1.6.6.10

1 Nomenclature

EC number
1.6.6.10 (transferred to EC 1.7.1.9)

Recommended name
nitroquinoline-N-oxide reductase

Hydroxylamine reductase (NADH) 1.6.6.11

1 Nomenclature

EC number
1.6.6.11 (transferred to EC 1.7.1.10)

Recommended name
hydroxylamine reductase (NADH)

4-(Dimethylamino)phenylazoxybenzene reductase 1.6.6.12

1 Nomenclature

EC number
1.6.6.12 (transferred to EC 1.7.1.11)

Recommended name
4-(dimethylamino)phenylazoxybenzene reductase

N-Hydroxy-2-acetamidofluorene reductase 1.6.6.13

1 Nomenclature

EC number
1.6.6.13 (transferred to EC 1.7.1.12)

Recommended name
N-hydroxy-2-acetamidofluorene reductase

Ferredoxin-NADP$^+$ reductase 1.6.7.1

1 Nomenclature

EC number

1.6.7.1 (transferred to EC 1.18.1.2)

Recommended name

ferredoxin-NADP$^+$ reductase

Rubredoxin-NAD$^+$ reductase 1.6.7.2

1 Nomenclature

EC number
1.6.7.2 (transferred to EC 1.18.1.1)

Recommended name
rubredoxin-NAD$^+$ reductase

NAD(P)H Dehydrogenase (FMN) 1.6.8.1

1 Nomenclature

EC number
1.6.8.1 (transferred to EC 1.5.1.29)

Recommended name
NAD(P)H dehydrogenase (FMN)

NADPH Dehydrogenase (flavin) 1.6.8.2

1 Nomenclature

EC number
1.6.8.2 (transferred to EC 1.5.1.30)

Recommended name
NADPH dehydrogenase (flavin)

NADPH Dehydrogenase 1.6.99.1

1 Nomenclature

EC number
1.6.99.1

Systematic name
NADPH:(acceptor) oxidoreductase

Recommended name
NADPH dehydrogenase

Synonyms
EBP
estrogen-binding protein
NADPH diaphorase
NADPH-dehydrogenase
NADPH-dependent diaphorase
NADPH-diaphorase
NADPH-methemoglobin reductase
$NADPH_2$-dehydrogenase
OYE
old yellow enzyme homolog
TPNH dehydrogenase
TPNH-diaphorase
dehydrogenase, reduced nicotinamide adenine dinucleotide phosphate
diaphorase
dihydronicotinamide adenine dinucleotide phosphate dehydrogenase
nitric oxide synthase <4, 5, 7> (<5>, identical to [10]) [10, 12, 15]
old yellow enzyme
reduced nicotinamide adenine dinucleotide phosphate dehydrogenase
triphosphopyridine diaphorase
triphosphopyridine nucleotide diaphorase

CAS registry number
9001-68-7

2 Source Organism

<1> *Saccharomyces carlsbergensis* [1, 3, 4, 5, 6, 7, 19, 20, 21, 23, 24]
<2> *Spinacia oleracea* [2]
<3> *Euglena gracilis* (var. bacillaris [8]) [8]

<4> *Rattus norvegicus* (diaphorase activity of nitric oxide synthase [12]) [9, 12, 13, 14, 16]
<5> *Gallus gallus* (diaphorase activity of nitric oxide synthase [10]) [10]
<6> *Xenopus laevis* [11]
<7> *Manduca sexta* (diaphorase activity of nitric oxide synthase [15]) [15]
<8> *Saccharomyces cerevisiae* [17]
<9> *Sus scrofa* [18]
<10> *Bos taurus* (diaphorase activity of neutrophil NADPH oxidase [22]) [22]

3 Reaction and Specificity

Catalyzed reaction

NADPH + H^+ + acceptor = $NADP^+$ + reduced acceptor (<1>, mechanism [1, 6])

Reaction type

oxidation
redox reaction
reduction

Natural substrates and products

S NADPH + acceptor <1, 4, 7> (<1>, physiological role as electron transfer component [5]; <7>, the enzyme is transiently activated by pheromone [15]; <4>, intracerebral hemorrhage in the internal capsule may induce NADPH-diaphorase activity [16]) (Reversibility: ? <1, 4, 7> [15, 16]) [5, 15, 16]
P $NADP^+$ + reduced acceptor

Substrates and products

S NADH + acceptor <1> (<1>, at 10% of the activity with NADPH [1]; <2>, no activity [2]; <4>, no activity with β-NADH [12]) [1]
P NADH + reduced acceptor
S NADPH + acceptor <1-10> (<4, 5, 6>, acceptor: nitroblue tetrazolium hydrochloride [9, 10, 11, 12]; <6>, nitroblue tetrazolium is reduced to formazan [11]; <2>, 2,3',6-trichlorophenolindophenol is reduced to leuco-2,3',6-trichlorophenolindophenol [2]; <2>, acceptor: ferricyanide [2]; <2>, acceptor: methylene blue [2]; <2,8>, acceptor: menadione [2,17]; <2>, acceptor: FMN [2]; <2>, acceptor: benzoquinone [2]; <2>, acceptor: ferric chloride [2]; <2>, aceptor: 1,10-phenanthroline [2]; <1,8>, acceptor: O_2 [1, 17, 20, 24]; <1>, acceptor: ferricytochrome [1]; <1>, most efficient electron acceptor: $Fe(CN)_6^{3-}$ [1]; <1>, both α-NADPH and β-NADPH are capable of reducing the enzyme. α-NADPH is more effective [7]; <1>, completely inactive towards cytochrome c [1]; <1,8>, acceptor: 2-cyclohexenone [17, 20, 21, 24]; <8>, acceptor: duroquinone [17]; <4>, diaphorase activity of neuron somata shows no activity with α-NADPH [12]) (Reversibility: ? <1-10> [1-24]) [1-24]
P $NADP^+$ + reduced acceptor

Inhibitors

2'-AMP <2> [2]
Co^{2+} <2> (<2>, 0.5 mM [2]) [2]
EDTA <5> [10]
EGTA <5> [10]
Fe^{3+} <1> [1]
$HgCl_2$ <2> (<2>, 0.02 mM [2]) [2]
L-N^5-(1-iminoethyl)Orn <5> (<5>, 1 mM 63% inhibition [10]) [10]
NEM <3> (<3>, 3 mM, 40% loss of activity [8]) [8]
N^γ-nitro-L-Arg <5> (<5>, 3 mM, 53% inhibition [10]) [10]
N^γ-monomethyl-L-Arg <5> (<5>, 1 mM, 14% inhibition [10]) [10]
PCMB <2> (<2>, 0.1 mM [2]) [2]
Zn^{2+} <2> (<2>, 0.5 mM [2]) [2]
α-NADPH <4> (<4>, diaphorase activity of neuron somata [12]) [12]
iodoacetate <2> (<2>, 3 mM [2]) [2]
monomethyl-L-Arg acetate <5> (<5>, 1 mM, 14% inhibition [10]) [10]
p-hydroxymercuribenzoate <3> (<3>, 3 mM, 55% loss of activity [8]) [8]
Additional information <4> (<4>, the traditional chinese medicine Nao Yi An, a complex derived from materials of animal and plant origin, can selectively inhibit the induction and thus, potentially play a neuroprotective role [16]) [16]

Cofactors/prosthetic groups

FAD <2> (<2>, one FAD per enzyme molecule of 35000 Da [2]) [2]
FMN <1> (<1>, prosthetic group is FMN or riboflavin 5'-phosphate [1]; <1>, each of the two subunits contains one molecule of FMN [4]; <1>, study of the flavin environment [19]; <1>, binds to the enzyme at the carboxy-terminal end of the barrel in anextended conformation [23]) [1, 4, 19, 23]
NADH <1> (<1>, at 10% of the activity with NADPH [1]; <2>, no activity [2]; <4>, no activity with β-NADH [12]) [1]
NADPH <1-10> [1-24]
riboflavin 5'-monophosphate <1> (<1>, prosthetic group is FMN or riboflavin 5'-phosphate [1]) [1]

Metals, ions

Ca^{2+} <5> (<5>, 2 mM $CaCl_2$, 48% stimulates. The effect is reversed by 1 mM EGTA or EDTA [10]) [10]

Turnover number (min^{-1})

84 <1> (NADPH, <1>, with 2-cyclohexenone as cosubstrate, mutant enzyme H191N [24]) [24]
165 <1> (NADPH, <1>, with O_2 as cosubstrate, wild-type enzyme OYE1 [24]) [24]
186 <1> (NADPH, <1>, with O_2 as cosubstrate, wild-type enzyme [20]) [20]
252 <1> (NADPH, <1>, with 2-cyclohexenone as cosubstrate, wild-type enzyme [20, 24]) [20, 24]

294 <1> (NADPH, <1>, with 2-cyclohexenone as cosubstrate, mutant enzyme Q114N [20]) [20]
336 <1> (NADPH, <1>, with O_2 as cosubstrate, mutant enzyme Q114N [20]) [20]

Specific activity (U/mg)

28.1 <2> [2]

K_m-Value (mM)

0.0011 <4> (NADPH) [9]
0.0032 <1> (NADPH, <1>, wild-type enzyme, with O_2 as cosubstrate [20]) [20]
0.006 <2> (NADPH) [2]
0.0073 <4> (nitroblue tetrazolium hydrochloride) [9]
0.01 <1> (NADPH, <1>, mutant enzyme Q114N, with O_2 as cosubstrate [20]) [20]
0.018 <1> (NADPH, <1>, with O_2 as cosubstrate, wild-type enzyme OYE1 [24]) [24]
0.033 <1> (NADPH, <1>, with O_2 as cosubstrate, mutant enzyme H191N [24]) [24]
0.05 <1> (NADPH, <1>, with 2-cyclohexenone as cosubstrate, mutant enzyme H191N [24]) [24]
0.13 <1> (O_2, <1>, wild-type enzyme OYE1 [24]) [24]
0.2 <3> (NADPH) [8]
0.26 <1> (O_2, <1>, mutant enzyme H191N [24]) [24]
0.72 <1> (O_2, <1>, wild-type enzyme [20]) [20]
6.7 <1> (O_2, <1>, mutant enzyme Q114N [20]) [20]
Additional information <1> (<1>, kinetic studies [6]) [1, 5, 6]

pH-Optimum

7 <3> [8]

4 Enzyme Structure

Molecular weight

30000-40000 <2> (<2>, ultracentrifugal analysis [2]) [2]
76000-80000 <1> (<1>, calculation from sedimentation and diffusion data [1]) [1]
102000 <1> (<1>, low speed sedimentation without reaching equilibrium [1]) [1]
235000 <6> (<6>, gel filtration [11]) [11]

Subunits

? <1> (<1>, x * 44889, calculation from nucleotide sequence [3]) [3]
dimer <1> (<1>, 2 * 50000 [4]) [4]

5 Isolation/Preparation/Mutation/Application

Source/tissue

antenna <7> (<7>, in a subpopulation of olfactory receptor neurones and in mechano-, thermo- and hygroreceptors on the antenna and in non-neuronal cells [15]) [15]
brain <4> [9]
cerebral cortex <4> [16]
embryo <5> (<5>, retina [10]) [10]
ganglion <4, 9> (<4>, sympathetic autonomic ganglia, no activity in paravertebral superior cervical ganglia or in perikarya of any sympathetic ganglion. The superior mesenteric ganglia of aged and diabetic rats, in which dysplasia is prominent, fail to show involvement of diaphorase containing nerve terminals [14]; <9>, ganglion cells of Auerbach's plexus [18]) [14, 18]
intestinal mucosa <9> (<9>, of piglet, crypt cells, glandular epithelium in the deeper submucosa has a very strong activity, villous epithelial cells show the most intense activity in the mucosa [18]) [18]
kidney <6> [11]
leaf <2> [2]
nervous system <9> (<9>, parallel to the circular and to the longitudinal muscle [18]) [18]
neutrophil <10> [22]
olfactory bulb <4> (<4>, localized in discrete populations of neuronal somata and in olfactory glomeruli. NADPH-diaphorase activity is due to nitric oxide synthase, the glomerular NADPH-diaphorase activity is not due to nitric oxide synthase [12]) [12]
retina <5> (<5>, from embryo [10]) [10]
spinal cord <4> [13]

Localization

chloroplast <2> [2]

Purification

<1> [2, 23]
<8> [17]

Crystallization

<1> (Q114N mutant and its complex with *p*-hydroxybenzaldehyde [20]; mutant enzymes H191N and H191N/N194H [24]) [1, 4, 20, 23, 24]

Cloning

<1> (expression of mutant enzyme OYE-Q114N, mutant plasmid is transformed into competent Escherichia coli BL21(DE3) cells [20]) [3, 20]
<8> (OYE2 gene [17]; expression of mutant enzymes H191N and H191N/N194H in Escherichia coli [24]) [17, 24]

Engineering

H191N <1> (<1>, reduction of the mutant enzyme by NADPH is similar to that of the wild-type enzyme, reduction rate constant for NADH is greatly decreased, reactivity with 2-cyclohexenone is greatly decreased [24]) [24]
H191N/N194H <1> (<1>, increased rate constant for reduction of NADPH, reduction rate constant with NADH is lower by a factor of 15 compared to the wild-type enzyme, reactivity with cyclohexenone is greatly decreased [24]) [24]
Q114N <1> (<1>, the mutation results in little change in the protein structure, moving the amide group of residue 114 out of H-bonding distance, allowing repositioning of the FMN prosthetic group to form new interactions that replace the lost H-bonds. The mutation decreases the ability to bind ligands, as all dissociation constants for the substituted phenols are larger than for the wild type enzyme [20]) [20]
Y196F <1> (<1>, slowing by nearly 6 orders of magnitude of its oxidative half-reaction with 2-cyclohexenone [21]) [21]

6 Stability

Temperature stability

-5 <3> (<3>, 96 h, 14% loss of activity [8]) [8]
Additional information <1> (<1>, the apoenzyme is quite stable in the cold, but it is much more sensitive to heat denaturation than the holoenzyme [1]) [1]

Storage stability

<3>, -5°C, 96 h, 14% loss of activity [8]

References

[1] Akeson, A.; Ehrenberg, A.; Theorell, H.: Old yellow enzyme. The Enzymes, 2nd Ed (Boyer, P.D., Lardy, H., Myrbäck, K., eds.), **7**, 477-494 (1963)
[2] Jagendorf, A.T.: Chloroplast TPNH diaphorase. Methods Enzymol., **6**, 430-434 (1963)
[3] Saito, K.; Davio, M.; Lockridge, O.; Massey, V.: The cloning of old yellow enzyme. Flavins and Flavoproteins (Proc. Int. Symp., 10th, Meeting Date 1990, Curti, B., Ronchi S., Zanetti, G., eds.) de Gruyter, Berlin, New York, 349-352 (1990)
[4] Fox, K.M.; Jacques, S.M.; Karplus, P.A.: Crystallization and characterization of old yellow enzyme. Flavins and Flavoproteins (Proc. Int. Symp., 10th, Meeting Date 1990, Curti, B., Ronchi S., Zanetti, G., eds.) de Gruyter, Berlin, New York, 353-356 (1990)
[5] Miura, R.; Yamaichi, K.; Tagawa, K.; Miyake, Y.: On the structure of old yellow enzyme studied by specific limited proteolysis. J. Biochem., **102**, 1311-1320 (1987)

[6] Honma, T.; Ogura, Y.: Kinetic studies of the old yellow enzyme. I. The reaction mechanism of the enzyme with reduced nicotinamide adenine dinucleotide. Biochim. Biophys. Acta, **484**, 9-23 (1977)
[7] Massey, V.; Schopfer, L.M.: Reactivity of old yellow enzyme with α-NADPH and other pyridine nucleotide derivatives. J. Biol. Chem., **261**, 1215-1222 (1986)
[8] Mohanty, M.K.; Hunter, F.R.; Myers, J.B.: Reduced pyridine nucleotide oxidases of Eugena gracilis var. bacillaris. J. Protozool., **24**, 335-340 (1977)
[9] Kemp, M.C.; Kuonene, D.R.; Roberts, P.J.: Kinetics of rat brain NADPH-diaphorase. Biochem. Soc. Trans., **15**, 501 (1987)
[10] De-Faria, M.H.; do-Nascimento, J.L.M.; Paes-de-Carvalho, R.: Biochemical characterization of NADPH-diaphorase in the chick embryo retina: Stimulation by calcium ions and inhibition by arginine analogs. Braz. J. Med. Biol. Res., **28**, 252-255 (1995)
[11] Kerschbaum, H.H.; Huang, S.; Xie, M.; Hermann, A.: NADPH-diaphorase activity and nitric oxide synthase activity in the kidney of the clawed frog, Xenopus laevis. Cell Tissue Res., **301**, 405-411 (2000)
[12] Spessert, R.; Layes, E.: Fixation conditions affect the intensity but not the pattern of NADPH-diaphorase staining as a marker for neuronal nitric oxide synthase in rat olfactory bulb. J. Histochem. Cytochem., **42**, 1309-1315 (1994)
[13] Nakamura, H.: NADPH-diaphorase and cytosolic urea cycle enzymes in the rat spinal cord. J. Comp. Neurol., **385**, 616-626 (1997)
[14] Schmidt, R.E.; Dorsey, D.A.; McDaniel, M.L.; Corbett, J.A.: Characterization of NADPH diaphorase activity in rat sympathetic autonomic ganglia–effect of diabetes and aging. Brain Res., **617**, 343-348 (1993)
[15] Stengl, M.; Zintl, R.: NADPH diaphorase activity in the antennae of the hawk moth Manduca sexta. J. Exp. Biol., **199**, 1063-1072 (1996)
[16] Peng, Z.C.; Li, X.Q.; Liang, Q.H.; Zhu, C.X.; Luo, T.L.; Yan, X.X.; Hu, S.Y.: Induction of NADPH-diaphorase activity in the forebrain in a model of intracerebral hemorrhage and its inhibition by the traditional Chinese medicine complex Nao Yi An. Brain Res. Bull., **42**, 119-128 (1997)
[17] Stott, K.; Saito, K.; Thiele, D.J.; Massey, V.: Old yellow enzyme. The discovery of multiple isozymes and a family of related proteins. J. Biol. Chem., **268**, 6097-6106 (1993)
[18] Rabinowitz, S.S.; Lucas, J.J.; Gootman, P.M.: NADPH-diaphorase activity in piglet intestinal mucosa. Cell. Mol. Biol. Res., **41**, 73-80 (1995)
[19] Fox, K.M.; Karplus, P.A.: The flavin environment in old yellow enzyme. An evaluation of insights from spectroscopic and artificial flavin studies. J. Biol. Chem., **274**, 9357-9362 (1999)
[20] Brown, B.J.; Hyun, J.W.; Duvvuri, S.; Karplus, P.A.; Massey, V.: The role of glutamine 114 in old yellow enzyme. J. Biol. Chem., **277**, 2138-2145 (2002)
[21] Kohli, R.M.; Massey, V.: The oxidative half-reaction of old yellow enzyme. The role of tyrosine 196. J. Biol. Chem., **273**, 32763-32770 (1998)
[22] Poinas, A.; Gaillard, J.; Vignais, P.; Doussiere, J.: Exploration of the diaphorase activity of neutrophil NADPH oxidase. Critical assessment of the inter-

action of iodonitrotetrazolium with the oxidase redox components. Eur. J. Biochem., **269**, 1243-1252 (2002)
[23] Fox, K.M.; Karplus, P.A.: Old yellow enzyme at 2 ANG resolution: Overall structure, ligand binding, and comparison with related flavoproteins. Structure, **2**, 1089-1105 (1994)
[24] Brown, B.J.; Deng, Z.; Karplus, P.A.; Massey, V.: On the active site of old yellow enzyme. Role of histidine 191 and asparagine 194. J. Biol. Chem., **273**, 32753-32762 (1998)

NAD(P)H Dehydrogenase (quinone) 1.6.99.2

1 Nomenclature

EC number
1.6.99.2

Systematic name
NAD(P)H:(quinone-acceptor) oxidoreductase

Recommended name
NAD(P)H dehydrogenase (quinone)

Synonyms
azoreductase
DT-diaphorase
DTD
EC 1.6.5.2 (formerly)
NAD(P)H dehydrogenase
NAD(P)H menadione reductase
NAD(P)H-quinone dehydrogenase
NAD(P)H-quinone oxidoreductase
NAD(P)H: (quinone-acceptor)oxidoreductase (EC 1.6.99.2)
NAD(P)H: menadione oxidoreductase
NADH-menadione reductase
QR1
QR2
dehydrogenase, reduced nicotinamide adenine dinucleotide (phosphate, quinone)
diaphorase
flavoprotein NAD(P)H-quinone reductase
menadione oxidoreductase
menadione reductase
naphthoquinone reductase
p-benzoquinone reductase
phylloquinone reductase
quinone reductase
reduced NAD(P)H dehydrogenase
reduced nicotinamide-adenine dinucleotide (phosphate) dehydrogenase
viologen accepting pyridine nucleotide oxidoreductase
vitamin K reductase
vitamin-K reductase

CAS registry number

9032-20-6

2 Source Organism

<1> *Phanerochaete chrysosporium* [22]
<2> *Halobacterium cutirubrum* [21]
<3> *Bos taurus* [1, 4, 18]
<4> *Canis familiaris* [2]
<5> *Octopus vulgaris* [3]
<6> *Clostridium tyrobutyricum* [5]
<7> *Rattus norvegicus* (isoforms II, III, IV, V, VI and VII in liver [8]; 3 to 4fold induction by 3-methylcholanthrene [15]) [4, 6, 7, 8, 9, 13-15, 18, 23, 24, 26, 27, 28, 34]
<8> *Microcystis aeruginosa* [10, 19]
<9> *Escherichia coli* [11]
<10> *Ricinus communis* (castor bean, enzymes DI and DII [20]) [20]
<11> *Sporotrichum pulverulentum* [12]
<12> *Mus musculus* (hydrophilic and hydrophobic isoform [16]; 2 forms have the same primary sequence [24]) [16, 17, 24, 27, 32]
<13> *Sus scrofa* [4]
<14> *Oryctolagus cuniculus* [4]
<15> *Gallus gallus* [4]
<16> *Xenopus laevis* [4]
<17> *Tinca vulgaris* [4]
<18> *Homo sapiens* (NAD(P)H:quinone acceptor oxidoreductase 2 [29]; breast cancer cell lines [33]) [24, 27, 29, 31, 32, 33, 36, 37]
<19> *Glycine max* [25]
<20> *Triphysaria versicolor* [30]
<21> *Giardia lamblia* (amitochondriate eukaryote [35]) [35]

3 Reaction and Specificity

Catalyzed reaction

NAD(P)H + H^+ + acceptor = $NAD(P)^+$ + reduced acceptor (<8> ordered bi-bi mechanism, [10]; <6,9> ping-pong mechanism [5,11])

Reaction type

oxidation
redox reaction
reduction

Natural substrates and products

S quinone + NAD(P)H <7, 10> (<7> protection of cells against damage by reactive oxygen species generated during oxidative cycling of quinones

and semiquinone radicals [15]; <10> possible role in seed germination [20]) (Reversibility: ? <7, 10> [15, 20]) [15, 20]
P hydroquinone + $NAD(P)^+$ <7, 10> [15, 20]
S vitamin K + NAD(P)H <3> (Reversibility: ? <3> [4]) [4]
P reduced vitamin K + $NAD(P)^+$ <3> [4]

Substrates and products
S 1,4-dimethoxymethylbenzoquinone + NAD(P)H <3> (Reversibility: ? <3> [4]) [4]
P 1,4-dimethoxymethylhydrobenzoquinone + $NAD(P)^+$ <3> [4]
S 1,4-naphthoquinone + NAD(P)H <3, 6, 8, 12, 21> (<8> specific for NADPH [19]) (Reversibility: ? <3, 6, 8, 12, 21> [4, 5, 16, 19, 35]) [4, 5, 16, 19, 35]
P 1,4-naphthohydroquinone + $NAD(P)^+$ <3, 6, 8, 12, 21> [4, 5, 16, 19, 35]
S 1,4-tetramethylbenzoquinone + NAD(P)H <3> (Reversibility: ? <3> [4]) [4]
P 1,4-tetramethylhydrobenzoquinone + $NAD(P)^+$ <3> [4]
S 2,3-dimethyl-1,4-naphthoquinone + NAD(P)H <3, 9> (Reversibility: ? <3, 9> [4, 11]) [4, 11]
P 2,3-dimethyl-1,4-hydronaphthoquinone $NAD(P)^+$ <3, 9> [4, 11]
S 2,6-dichlorophenolindophenol + NAD(P)H <3, 5, 6, 7, 8, 9, 12, 19> (<5, 6, 9> low activity [3, 5, 11]; <8> very low activity with NADH [19]; <7> microsomal enzyme [34]) (Reversibility: ? <3, 5, 6, 7, 8, 9, 12, 19> [1, 3, 4, 5, 11, 15, 16, 18, 19, 25]) [1, 3, 4, 5, 11, 15, 16, 18, 19, 25, 34]
P reduced 2,6-dichlorophenolindophenol + $NAD(P)^+$ <3, 5, 6, 7, 8, 9, 12, 19> [1, 3, 4, 5, 11, 15, 16, 18, 19, 25, 34]
S 2,6-dimethoxybenzoquinone + NAD(P)H <20> (Reversibility: ? <20> [30]) [30]
P 2,6-dimethoxyhydrobenzoquinone + $NAD(P)^+$ <20> [30]
S 2-ethyl-1,4-naphthoquinone + NAD(P)H <3> (Reversibility: ? <3> [4]) [4]
P 2-ethyl-1,4-hydronaphthoquinone $NAD(P)^+$ <3> [4]
S 2-hydroxymethyl-5-methoxy-2,5-cyclohexadiene-1,4-dione + NAD(P)H <1> (Reversibility: ? <1> [22]) [22]
P 2-hydroxymethyl-5-methoxy-2,5-cyclohexadiene-1,4-diol + $NAD(P)^+$ <1> [22]
S 2-methoxy-2,5-cyclohexadiene-1,4-dione + NAD(P)H <1> (Reversibility: ? <1> [22]) [22]
P 2-methoxy-2,5-cyclohexadiene-1,4-diol + $NAD(P)^+$ <1> [22]
S 2-methoxy-5-(methoxymethyl)-2,5-cyclohexadiene-1,4-dione + NAD(P)H <1> (Reversibility: ? <1> [22]) [22]
P 2-methoxy-5-methoxymethyl-2,5-cyclohexadiene-1,4-diol + $NAD(P)^+$ <1> [22]
S 2-methoxyquinone + NAD(P)H <11> (Reversibility: ? <11> [12]) [12]
P 2-methoxyhydroquinone + $NAD(P)^+$ <11> [12]
S 2-methyl-1,4-naphthoquinone + NAD(P)H <3> (Reversibility: ? <3> [4]) [4]

P 2-methyl-1,4-hydronaphthoquinone NAD(P)$^+$ <3> [4]

S 2-methyl-1,4-naphthoquinone + NAD(P)H <3, 4, 7-9, 12, 18, 20, 21> (<3, 4, 7-9, 12> artificial acceptor, trivial name menadione [1, 2, 4, 8, 11, 15, 16, 18, 19]; <9> specific for NADH [11]; <8> specific for NADPH [19]; <18> involved in the process of reductive activation of several cytotoxic antitumor quinones, such as mitomycins, anthracyclines and azaridinylbenzoquinones [27]; <7> microsomal enzyme [34]) (Reversibility: ? <3, 4, 7-9, 12, 18, 20, 21> [1, 2, 4, 8, 11, 15, 16, 18, 19, 27, 30, 35]) [1, 2, 4, 8, 11, 15, 16, 18, 19, 27, 30, 34, 35]

P 2-methyl-1,4-naphthoquinol + NAD(P)$^+$ <3, 4, 7-9, 12, 18, 20, 21> [1, 2, 4, 8, 11, 15, 16, 18, 19, 27, 30, 34, 35]

S 2-methyl-3-phythyl-1,4-naphthoquinone + NAD(P)H <4, 7> (<4, 7> trivial name vitamin K_1 [2, 8, 15]; <4> no activity with vitamin K_1 diphosphate, menadione diphosphate, vitamin K_1 oxide, d-α-tocopherol and d-α-tocoquinone [2]) (Reversibility: ? <4, 7> [2, 8, 15]) [2, 8, 15]

P 2-methyl-3-phythyl-1,4-naphthohydroquinone <4, 7> [2, 8, 15]

S 2-methyl-5-hydroxy-1,4-naphthoquinone + NADPH <8> (Reversibility: ? <8> [19]) [19]

P 2-methyl-5-hydroxy-1,4-naphthohydroquinone + NADP$^+$ <8> [19]

S 3,4-dihydro-2,2-dimethyl-2H-naphthol[1,2-b]pyran-5,6-dione + NAD(P)H <18> (<18> trivial name β-lapachone, potent cytotoxic agent against various cancer cell lines [33]) (Reversibility: ? <18> [33]) [33]

P 3,4-dihydro-2,2-dimethyl-2H-naphthol[1,2-b]pyran-5,6-diol + NAD(P)$^+$ <18> [33]

S 3,6-quinone + NAD(P)H <7> (Reversibility: ? <7> [8]) [8]

P 3,6-hydroquinone + NAD(P)$^+$ <7> [8]

S 3-(4,5-dimethylthiazol-2-yl)-2,5-diphenyltetrazolium bromide + NAD(P)H <21> (Reversibility: ? <21> [35]) [35]

P reduced 3-(4,5-dimethylthiazol-2-yl)-2,5-diphenyltetrazolium bromide + NAD(P)$^+$ <21> [35]

S 4,5-dimethoxy-3,5-cyclohexadiene-1,2-dione <1> (Reversibility: ? <1> [22]) [22]

P 4,5-dimethoxy-3,5-cyclohexadiene-1,2-diol + NAD(P)$^+$ <1> [22]

S 5,8-dihydroxynaphthoquinone + NAD(P)H <19> (Reversibility: ? <19> [25]) [25]

P 5,8-dihydroxynaphthohydroquinone + NAD(P)$^+$ <19> [25]

S 5-(1-aziridinyl)-3-(hydroxymethyl)-2-(3-hydroxy-1-propenyl)-1-methyl-1H-indole-4,7-dione + NAD(P)H <7, 12, 18> (Reversibility: ? <7, 12, 18> [27]) [27]

P 5-(1-aziridinyl)-3-(hydroxymethyl)-2-(3-hydroxy-1-propenyl)-1-methyl-1H-indole-4,7-diol + NAD(P)$^+$ <7, 12, 18> [27]

S 5-(aziridin-1-yl)-2,4-dinitrobenzamide + NAD(P)H <7, 12, 18> (Reversibility: ? <7, 12, 18> [27]) [27]

P 5-(aziridin-1-yl)-4-hydroxylamino-2-benzamide + NAD(P)$^+$ <7, 12, 18> [27]

S 5-(aziridin-1-yl)-2,4-dinitrobenzamide + reduced dihydronicotineamide riboside <7, 12, 18> (Reversibility: ? <7, 12, 18> [27]) [27]

P 5-(aziridin-1-yl)-4-hydroxylamino-2-benzamide + oxidized reduced dihydronicotineamide riboside <7, 12, 18> [27]
S 5-hydroxy-1,4-naphthoquinone + NADPH <8> (Reversibility: ? <8> [19]) [19]
P 5-hydroxy-1,4-naphthohydroquinone + $NADP^+$ <8> [19]
S 6-hydroxydopaminequinone + NAD(P)H <19> (Reversibility: ? <19> [25]) [25]
P 6-hydroxydopaminehydroquinone + $NAD(P)^+$ <19> [25]
S 6-hydroxydopaquinone + NAD(P)H <19> (Reversibility: ? <19> [25]) [25]
P 6-hydroxydopahydroquinone + $NAD(P)^+$ <19> [25]
S $K_3[Fe(CN)_6]$ + NAD(P)H <3, 5, 8, 9, 19, 21> (<3, 5> no activity with lipoic acid, cytochrome c and vitamin K_1 [1, 3, 4, 5]; <5> no activity with O_2, vitamin K_3, ubiquinone-30 and 2-*p*-iodophenyl-3-*p*-nitrophenyl-5-phenyltetrazolium chloride [3]; <3, 9> no activity with coenzyme Q_{10} [1, 11]; <8> specific for NADPH [19]) (Reversibility: ? <3, 5, 8, 9, 19, 21> [1, 3, 4, 11, 19, 25, 35]) [1, 3, 4, 5, 11, 19, 25, 35]
P $K_4[Fe(CN)_6]$ + $NAD(P)^+$ <3, 5, 8, 9, 19, 21> [1, 3, 4, 5, 11, 19, 25, 35]
S benzo(a)pyrene + NAD(P)H <7> (Reversibility: ? <7> [8]) [8]
P reduced benzo(a)pyrene <7> [8]
S benzoquinone + $NAD(P)^+$ <3, 4, 5, 6, 8, 9, 12> (<5> no reaction with *p*-benzoquinone [3]; <6> no activity with coenzyme Q_6 and vitamin K_1 [5]; <8> specific for NADPH [19]) (Reversibility: ? <3, 4, 5, 6, 8, 9, 12> [2, 3, 4, 5, 11, 16, 19]) [2, 3, 4, 5, 11, 16, 19]
P benzohydroquinone + $NAD(P)^+$ <3, 4, 6, 8, 9, 12> [2, 3, 4, 5, 11, 16, 19]
S coenzyme Q_1 + NAD(P)H <9> (Reversibility: ? <9> [11]) [11]
P reduced coenzyme Q_1 + $NAD(P)^+$ <9> [11]
S coenzyme Q_{10} + NAD(P)H <4> (Reversibility: ? <4> [2]) [2]
P reduced coenzyme Q_{10} + $NAD(P)^+$ <4> [2]
S cyclized-dopamine *ortho*-quinone + NAD(P)H <7> (Reversibility: ? <7> [8]) [8]
P cyclized-dopamine *ortho*-hydroquinone + $NADP^+$ <7> [8]
S indolequinones + NAD(P)H <18> (<18> compounds with electron-withdrawing groups at the indole 3-position are among the best substrates, groups larger than methyl at N-1 are tolerated, compounds with a leaving group at the 3-indolyl methyl position inactivate the enzyme [36]) (Reversibility: ? <18> [36]) [36]
P indolehydroquinones + $NAD(P)^+$ <18> [36]
S juglone + NAD(P)H <19, 20> (Reversibility: ? <19, 20> [25, 30]) [25, 30]
P dihydrojuglone + $NAD(P)^+$ <19, 20> [25, 30]
S menadione + reduced nicotinamide 2-azidoadenine dinucleotide <7> (Reversibility: ? <7> [9]) [9]
P reduced menadione + oxidized nicotinamide 2-azidoadenine dinucleotide <7> [9]
S menadione + reduced nicotinamide 8-azidoadenine dinucleotide <7> (Reversibility: ? <7> [9]) [9]
P reduced menadione + oxidized nicotinamide 8-azidoadenine dinucleotide <7> [9]

S methoxyquinone + NAD(P)H <11> (Reversibility: ? <11> [12]) [12]
P methoxyhydroquinone + $NAD(P)^+$ <11> [12]
S methyl red + NAD(P)H <21> (Reversibility: ? <21> [35]) [35]
P reduced methyl red + $NAD(P)^+$ <21> [35]
S methylene blue + NAD(P)H <3> (Reversibility: ? <3> [1]) [1]
P reduced methylene blue + $NAD(P)^+$ <3> [1]
S toluquinone + NAD(P)H <11> (Reversibility: ? <11> [12]) [12]
P toluhydroquinone + $NAD(P)^+$ <11> [12]
S vitamin K_2 + NAD(P)H <3> (Reversibility: ? <3> [4]) [4]
P reduced vitamin K_2 + $NAD(P)^+$ <3> [4]

Inhibitors

(A)-2-azido-NAD^+ <7> (<7> photodependent inhibition [9]) [9]
(A)-8-azido-NAD^+ <7> (<7> photodependent inhibition [9]) [9]
1,10-phenanthroline <3, 4, 8, 21> (<4> 1 mM, 50% inhibition [2]; <8> 1 mM, 72% inhibition [19]) [1, 2, 19, 35]
1,4-diaminoanthraquinone-2-sulfonic acid <7> [7]
1,4-naphthohydroquinone <8> (<8> competitive vs. 1,4-naphthohydroquinone, noncompetitive vs. NADPH [10]) [10]
1-(2-thenoyl)-3,3,3-trifluoracetone <8> (<8> 1 mM, 89% inhibition [19]) [19]
1-amino-4-[(3-carboxyphenyl)amino]-anthraquinone-2-sulfonic acid <7> [7]
1-amino-4-[(4-amino-3-sulfophenyl)amino]-anthraquinone-2-sulfonic acid <7> [7]
1-aminoanthraquinone <7> [7]
1-aminoanthraquinone-2-carboxylic acid <7> [7]
2,4-dinitroaniline <4> (<4> 0.1 mM, 33% inhibition [2]) [2]
2,4-dinitronaphthol <4> (<4> 0.1 mM, 90% inhibition [2]) [2]
2,4-dinitrophenol <3, 4, 8, 10> (<3> 0.067 mM, 56% inhibition [1]; <4> 0.1 mM, 39% inhibition [2]; <8> 1 mM, 86% inhibition [19]) [1, 2, 19, 20]
2-(4'-chlorophenyl)-indan-1,3-dione <3> (<3> 0.0002 mM, 50% inhibition [4]) [4]
2-heptyl-4-hydroxyquinoline-N-oxide <2> [21]
2-pivaloyl-1,3-indanedione <7> [6]
3,3'-methylene-bis(4-hydroxycoumarin) <1, 3-7, 9, 12, 18, 19, 20, 21> (<1> trivial name dicoumarol, slight inhibition [22]; <3> 0.000033 mM, 56% inhibition [1]; <4> 0.1 mM, 97% inhibition [2]; <5> only in the presence of glutathione [3]; <7> competitive vs. NADH, uncompetitive vs. vitamin K_1 [15]; <12> competitive vs. NADH [24]; <19> 0.04 mM, complete inhibition [25]; <18> 0.01 mM, 25% inhibition, recombinant NAD(P)H:quinone acceptor oxireductase 2 [29]; <20> competitive vs. NADH [30]) [1-6, 11, 14-16, 22, 24, 25, 27, 29, 30, 35]
3-(α-phenylpropyl)-4-hydroxycoumarin <3> (<3> 0.01 mM, 50% inhibition [4]) [4]
3-[α-(4'-nitrophenyl)-β-acetylethyl]-4-hydroxycoumarin <3> (<3> 0.02 mM, 50% inhibition [4]) [4]
4',5,7-trihydroxyflavone <18> (<18> 0.01 mM, 61% inhibition, recombinant NAD(P)H:quinone acceptor oxidoreductase 2 [29]) [29]

4,4'-dichlorodicoumarol <3> (<3> 0.001 mM, 50% inhibition [4]) [4]
4-hydroxycoumarin <3> [4]
4-hydroxymercuribenzoate <3> (<3> 0.067 mM, 50% inhibition [1]) [1]
4-methoxycoumarin <3> [4]
5,5'-dithiobis(2-nitrobenzoic acid) <7, 10> (<7> 1 mM, 67% inhibition, 10 mM, complete inhibition [23]; <10> [20]) [20, 23]
5,6-dimethylxanthenone-4-acetic acid <18> (<18> tumour blood flow inhibitor, competitive vs. NADH [31]) [31]
5,7-dihydroxyflavone <7, 12, 18> (<12> 0.009 mM, approx. 20% inhibition [24]; <7> 0.009 mM, approx. 80% inhibition [24]; <18> 0.009 mM, approx. 90% inhibition [24]; <7> 0.00018 mM, 50% inhibition [28]) [24, 28]
5-hydroxy-7-bromoacetylflavone <7> (<7> 0.00018 mM, 50% inhibition, irreversible, useful as affinity label [28]) [28]
5-methoxy-1,2-dimethyl-3-[(4-nitrophenoxy)methyl]indole-4,7-dione <18> (<18> mechanism-based inhibitor, time- and concentration dependent inhibition that requires the presence of NAD(P)H, 0.0015 mM, complete inhibition after 4 min [37]) [37]
7,8-dibromoacetylflavone <7> (<7> 0.00003 mM, 50% inhibition [28]) [28]
7-bromoacetylflavone <7> (<7> time-dependent inactivation within 30 s, 0.00074 mM, 50% inhibition [28]) [28]
7-hydroxylflavone <7> (<7> competitive inhibition, 0.00074 mM, 50% inhibition [28]) [28]
8-hydroxyquinoline <4, 8> (<4> 1 mM, 50% inhibition [2]; <8> 0.3 mM, 80% inhibition [19]) [2, 19]
ATP <8> (<8> 0.2 mM, 75% inhibition [19]) [19]
Alizarin Red S <19> (<19> 0.04 mM, 38% inhibition [25]) [25]
Cibacron Blue <7> (<7> and related anthraquinone dyes [7]) [7, 14]
Cibacron Blue F3G-A <19> (<19> 0.04 mM, 75% inhibition [25]) [25]
DTT <10> [20]
EDTA <8, 10> (<8> 0.15 mM, 80% inhibition [19]) [19, 20]
$HgCl_2$ <7> (<7> 0.2 mM, 94% inhibition [23]) [23]
N-ethylmaleimide <3, 21> (<3> 0.33 mM, complete inhibition [1]) [1, 35]
$NADP^+$ <8> (<8> noncompetitive vs. NADPH [10]) [10]
Procion Blue MX-R <19> (<19> 0.04 mM, 75% inhibition [25]) [25]
Procion Blue M-3GS <7> [7]
acid blue 25 <7> [7]
α,α'-dipyridyl <4> (<4> 1 mM, 505 inhibition [2]) [2]
α-naphthoflavone <18> (<18> 0.01 mM, 75% inhibition, recombinant NAD(P)H:quinone acceptor oxidoreductase 2 [29]) [29]
amytal <3> (<3> 6.6 mM, 55% inhibition [1]) [1]
anthraqinone <7> [7]
anthraquinone-2-carboxylic acid <7> [7]
anthraquinone-2-sulfonic acid <7> [7]
β-naphthoflavone <18> (<18> 0.01 mM, 76% inhibition, recombinant NAD(P)H:quinone acceptor oxidoreductase 2 [29]) [29]

bis-3,3'-(4-hydroxycoumarinyl)acetoacetate <3, 4> (<4> trivial tromexan, 0.1 mM, 79% inhibition [2]) [2, 4]
bromaminic acid <7> [7]
chrysin <18> (<18> 0.01 mM, 100% inhibition, recombinant NAD(P)H:quinone acceptor oxidoreductase 2 [29]) [29]
dicoumarol dimethylether <3> (<3> 0.0005, 50% inhibition [4]) [4]
flavone-8-acetic acid <18> (<18> tumour blood flow inhibitor, competitive vs. NADH [31]) [31]
galangin <18> (<18> 0.01 mM, 100% inhibition, recombinant NAD(P)H:quinone acceptor oxidoreductase 2 [29]) [29]
iodoacetic acid <10> [20]
methylene-5,5'-bis(4,6-dioxo-2-methyldihydropyran) <3> (<3> 0.003 mM, 50% inhibition [4]) [4]
morin <18> (<18> 0.01 mM, 95% inhibition, recombinant NAD(P)H:quinone acceptor oxidoreductase 2 [29]) [29]
p-chloromercuribenzoate <6> [5]
phenindione <4> (<4> 0.1 mM, 92% inhibition [2]) [2]
phenylmercuric acetate <2> [21]
quercetin <18, 21> (<18> 0.01 mM, 100% inhibition, recombinant NAD(P)H:quinone acceptor oxidoreductase 2 [29]) [29, 35]
quinacrine dihydrochloride <21> (<21> 0.05 mM, 65-95% inhibition depending on substrate [35]) [35]
salicylate <4> (<4> 1 mM, 62% inhibition [2]) [2]
thenoyltrifluoroacetone <10> [20]
warfarin <4> (<4> 1 mM, 70% inhibition [2]) [2]
Additional information <3, 6, 7> (<6> not inhibited by *o*-phenanthroline, KCN, quinacrine, EDTA and 2,3-dimercapto-1-propanol [5]; <3> not inhibited by iodoacetate, 4-chloromercuribenzoate, Cu^{2+}, $FeSO_4$, $FeCl_3$, $CdSO_4$, $ZnSO_4$, $CuSO_4$, EDTA, α,α'-dipyridyl, *o*-phenanthroline, dimethylglyoxime, dithizon, NaCN, amytal, veronal, antimycin, 2,4-dinitrophenol and thyroxin [4]; <7> microsomal enzyme: not inhibited by dicoumarol, lapachol and *p*-chloromercuribenzoate [34]) [4, 5, 34]

Cofactors/prosthetic groups

FAD <3, 4, 7, 9, 12, 18> (<3, 7> enzyme contains FAD [1, 4, 6, 14]; <3, 7> contains 1 mol of FAD per mol of enzyme [4, 6]; <3, 4> stimulates [1, 2]; <9> requires FMN or FAD as cofactor, FMN is more effective [11]; <7> required [15]; <12> purified recombinant enzyme contains FAD whereas purified native enzyme does not contain FAD [24]; <18> NAD(P)H:quinone acceptor oxidoreductase 2, contains 1 FAD per subunit [29]) [1, 2, 4, 6, 11, 14, 15, 24, 29]
FMN <3-6, 9> (<9> enzyme requires FMN or FAD as cofactor, FMN is 30times more effective [11]; <3, 4> stimulation [1, 2]; <5, 6> contains FNN as prosthetic group [3, 5]) [1-3, 5, 11]
NAD^+ <3> [1]
NADH <3, 4, 6, 7, 9-11> [2, 4, 5, 11, 12, 15, 20]
$NADP^+$ <3> [1]

NADPH <3, 4, 6, 7, 10, 11> (<9> no cofactor [11]) [2, 4, 5, 12, 15, 20]
deamino-NADH <4> [2]
nicotinamide riboside <18> (<18> recombinant NAD(P)H:quinone acceptor oxidoreductase 2 [29]) [29]

Activating compounds

2-mercaptoethanol <5> (<5> stimulates [3]) [3]
cysteine <5> (<5> stimulates [3]) [3]
GSH <5> (<5> maximal stimulation with 0.5 mM, approx. 3fold activity [3]) [3]

Turnover number (min^{-1})

0.6 <18> (5-(aziridin-1-yl)-2,4-dinitrobenzamide, <18> recombinant NAD(P)H:quinone acceptor oxidoreductase 2 [29]) [29]
20 <18> (methyl red, <18> recombinant NAD(P)H:quinone acceptor oxidoreductase 1 [29]) [29]
230 <18> (methyl red, <18> recombinant NAD(P)H:quinone acceptor oxidoreductase 2 [29]) [29]
360 <18> (5-(aziridin-1-yl)-2,4-dinitrobenzamide, <18> recombinant NAD(P)H:quinone acceptor oxidoreductase 2 [29]) [29]
660 <18> (3-(4,5-dimethylthiazol-2-yl)-2,5-diphenyltetrazolium, <18> recombinant NAD(P)H:quinone acceptor oxidoreductase 2 [29]) [29]
1384 <21> (menadione, <21> cofactor NADH [35]) [35]
1796 <21> (naphthoquinone, <21> cofactor NADH [35]) [35]
1903 <21> (NADH, <21> reduction of menadione [35]) [35]
2090 <21> (menadione, <21> cofactor NADPH [35]) [35]
2300 <18> (menadione, <18> recombinant NAD(P)H:quinone acceptor oxidoreductase 2 [29]) [29]
2357 <21> (naphthoquinone, <21> cofactor NADPH [35]) [35]
2600 <18> (dihydronicotinamide riboside, <18> recombinant NAD(P)H:quinone acceptor oxidoreductase 2 [29]) [29]
2800 <18> (2,6-dichlorophenolindophenol, <18> recombinant NAD(P)H:quinone acceptor oxidoreductase 2 [29]) [29]
2851 <21> (NADPH, <21> reduction of naphthoquinone [35]) [35]
3161 <21> (NADPH, <21> reduction of menadione [35]) [35]
3760 <21> (NADH, <21> reduction of naphthoquinone [35]) [35]
5460 <7> (reduced nicotinamide 8-azidoadenine dinucleotide) [9]
10020 <7> (reduced nicotinamide 2-azidoadenine dinucleotide) [9]
11520 <7> (NADH) [9]
31000 <18> (dihydronicotinamide riboside, <18> recombinant NAD(P)H:quinone acceptor oxidoreductase 1 [29]) [29]
33000 <18> (menadione, <18> recombinant NAD(P)H:quinone acceptor oxidoreductase 1 [29]) [29]
36000 <18> (2,6-dichlorophenolindophenol, <18> recombinant NAD(P)H:quinone acceptor oxidoreductase 1 [29]) [29]
700000 <3> (menadione) [4]

Specific activity (U/mg)

0.039 <7> (<7> activity in liver microsomes, reduction of menadione, cofactor NADPH [34]) [34]
0.041 <7> (<7> activity in liver cytosol, reduction of menadione, cofactor NADH [34]) [34]
0.051 <7> (<7> activity in liver cytosol, reduction of menadione, cofactor NADPH [34]) [34]
0.121 <7> (<7> activity in liver cytosol, reduction of 2,6-dichlorophenolindophenol, cofactor NADPH [34]) [34]
0.128 <7> (<7> activity in liver cytosol, reduction of 2,6-dichlorophenolindophenol, cofactor NADH [34]) [34]
0.153 <7> (<7> activity in liver microsomes, reduction of menadione, cofactor NADH [34]) [34]
0.157 <7> (<7> activity in liver microsomes, reduction of 2,6-dichlorophenolindophenol, cofactor NADPH [34]) [34]
0.63 <7> (<7> activity in liver microsomes, reduction of 2,6-dichlorophenolindophenol, cofactor NADH [34]) [34]
2.8 <5> [3]
2.9 <4> [2]
46.2 <6> [5]
371 <7> [6]
435.3 <18> (<18> recombinant enzyme [31]) [31]
505 <7> (<7> isoform III, towards menadione [8]) [8]
634 <7> (<7> isoform IV, towards menadione [8]) [8]
710 <7> (<7> cytosol [13]) [13]
830 <7> (<7> microsomes [13]) [13]
1129 <7> (<7> isoform VII, towards menadione [8]) [8]
1186 <7> (<7> isoform V, towards menadione [8]) [8]
1212 <7> (<7> dichloroindophenol [18]) [18]
1720 <12> (<12> hydrophilic isoform [16]) [16]
1770 <12> (<12> hydrophobic isoform [16]) [16]
2220 <7> [14]
2228 <7> (<7> isoform II, towards menadione [8]) [8]
2452 <7> (<7> 2-methyl-1,4-naphthoquinone [18]) [18]
2459 <7> (<7> isoform VIb, towards menadione [8]) [8]
2469 <19> [25]
3205 <9> [11]
5860 <7> (<7> isoform VIa, towards menadione [8]) [8]
Additional information <3, 8> (<3> 230000.0 units/mg, 1 unit is defined as the amount of enzyme causing a decrease in absorbance at 340 nm of 0.001/min [1]; <3> 1400000.0, 1 unit is defined as the amount of enzyme causing a decrease in absorbance at 340 nm of 0.001/min [4]; <8> 90300 units/mg, 1 unit is defined as that amount of enzyme causing an absorption change of 0.001 units/min at the specific wavelength [19]) [1, 4, 19]

K_m-Value (mM)

0.0000017 <12> (FAD, <12> hydrophobic isoform [16]) [16]
0.0000027 <12> (FAD, <12> hydrophilic isoform [16]) [15]
0.000188 <21> (menadione, <21> cofactor NADH [35]) [35]
0.00024 <21> (menadione, <21> cofactor NADPH [35]) [35]
0.000437 <21> (naphthoquinone, <21> cofactor NADH [35]) [35]
0.00059 <21> (naphthoquinone, <21> cofactor NADPH [35]) [35]
0.00196 <12> (menadione, <12> hydrophilic isoform [16]) [16]
0.002 <7, 9> (menadione) [11, 15]
0.002 <18> (menadione) [24]
0.0023 <18> (menadione, <18> recombinant NAD(P)H:quinone acceptor oxidoreductase 2 [29]) [29]
0.00247 <12> (menadione, <12> hydrophobic isoform [15]) [16]
0.0025 <6> (FMN) [5]
0.0025 <3> (menadione, <3> cofactor NADH [1]) [1]
0.0026 <7> (menadione) [24]
0.0027 <18> (menadione, <18> recombinant NAD(P)H:quinone acceptor oxidoreductase 1 [29]) [29]
0.0041 <12> (menadione) [24]
0.0047 <18> (methyl red, <18> recombinant NAD(P)H:quinone acceptor oxidoreductase 2 [29]) [29]
0.005 <1> (NADH) [22]
0.0062 <7> (NADH) [9]
0.007 <18> (2,6-dichlorophenolindophenol, <18> recombinant NAD(P)H:-quinone acceptor oxidoreductase 2 [29]) [29]
0.012 <18> (3-(4,5-dimethylthiazol-2-yl)-2,5-diphenyltetrazolium, <18> recombinant NAD(P)H:quinone acceptor oxidoreductase 2 [29]) [29]
0.0143 <7> (reduced nicotinamide 2-azidoadenine dinucleotide) [9]
0.0175 <7> (NADH) [15]
0.018 <18> (2,6-dichlorophenolindophenol, <18> recombinant NAD(P)H:-quinone acceptor oxidoreductase 1 [29]) [29]
0.0185 <7> (reduced nicotinamide 8-azidoadenine dinucleotide) [9]
0.02-0.05 <2> (NADH) [21]
0.0233 <7> (2,6-dichloroindophenol) [15]
0.024 <3> (methylene blue, <3> cofactor NADPH [1]) [1]
0.025 <18> (5-(1-aziridinyl)-3-(hydroxymethyl)-2-(3-hydroxy-1-propenyl)-1-methyl-1H-indole-4,7-dione) [27]
0.027 <7> (5-(1-aziridinyl)-3-(hydroxymethyl)-2-(3-hydroxy-1-propenyl)-1-methyl-1H-indole-4,7-dione) [27]
0.028 <18> (dihydronicotinamide riboside, <18> recombinant NAD(P)H:quinone acceptor oxidoreductase 2 [29]) [29]
0.028 <18> (methyl red, <18> recombinant NAD(P)H:quinone acceptor oxidoreductase 1 [29]) [29]
0.03 <21> (NADPH, <21> reduction of menadione [35]) [35]
0.0305 <7> (NADH, <7> in the presence of 0.026 mM dicoumarol [15]) [15]
0.033 <21> (NADPH, <21> reduction of naphthoquinone [35]) [35]
0.034 <5> (NADH) [3]

0.035 <6> (menadione, <6> cofactor NADH [5]) [5]
0.04 <12> (5-(1-aziridinyl)-3-(hydroxymethyl)-2-(3-hydroxy-1-propenyl)-1-methyl-1H-indole-4,7-dione) [27]
0.04 <3> (menadione, <3> cofactor NADPH [1]) [1]
0.04-0.05 <2> (menadione, <2> in untreated vesicles, in low salt disrupted membranes and in French-press treated preparations [21]) [21]
0.045 <7> (NADH) [18]
0.045 <21> (NADH, <21> reduction of naphthoquinone [35]) [35]
0.0581 <12> (NADPH, <12> hydrophobic isoform [16]) [16]
0.0641 <7> (vitamin K_1) [15]
0.0643 <12> (NADH, <12> hydrophilic isoform [16]) [16]
0.066 <12> (NADPH, <12> hydrophilic isoform [16]) [16]
0.07 <18> (dihydronicotinamide riboside, <18> recombinant NAD(P)H:quinone acceptor oxidoreductase 1 [29]) [29]
0.074 <7> (NADH) [23]
0.085 <7> (NADPH) [18]
0.0881 <12> (NADH, <12> hydrophobic isoform [16]) [16]
0.1 <18> (NADH) [27]
0.11 <7> (NADH) [27]
0.115 <1> (2-methoxy-5-(methoxymethyl)-2,5-cyclohexadiene) [22]
0.115 <1> (2-methoxy-5-(methoxymethyl)-2,5-cyclohexadiene-1,4-dione) [22]
0.12 <12> (NADH) [27]
0.12 <7> (NADPH) [24]
0.125 <2> (menadione, <2> in low-salt and detergent-solubilized fractions [21]) [21]
0.132 <9> (NADH) [11]
0.17 <12> (NADPH) [24]
0.19 <7> (NADH) [24]
0.19 <19> (NADPH) [25]
0.22 <5> (NADPH) [3]
0.223 <6> (NADH, <6> cofactor menadione [5]) [5]
0.24 <18> (NADPH) [24]
0.26 <18> (5-(aziridin-1-yl)-2,4-dinitrobenzamide, <18> recombinant NAD(P)H:quinone acceptor oxidoreductase 2 [29]) [29]
0.29 <18> (NADH) [24]
0.3 <3> (vitamin K_1, <3> cofactor NADPH [1]) [1]
0.32 <12> (NADH) [24]
0.33 <3> (vitamin K_1, <3> cofactor NADH [1]) [1]
0.34 <6> (NADPH, <6> acceptor menadione [5]) [5]
0.341 <19> (NADH) [25]
0.549 <21> (NADH, <21> reduction of menadione [35]) [35]
0.84 <7> (5-(aziridin-1-yl)-4-hydroxylamino-2-benzamide) [27]
1.28 <12> (5-(aziridin-1-yl)-4-hydroxylamino-2-benzamide) [27]
1.37 <18> (5-(aziridin-1-yl)-2,4-dinitrobenzamide, <18> recombinant NAD(P)H:quinone acceptor oxidoreductase 2 [29]) [29]
1.37 <18> (5-(aziridin-1-yl)-4-hydroxylamino-2-benzamide) [27]

1.4 <3> (potassium ferricyanide, <3> cofactor NADPH [1]) [1]
1.7 <3> (potassium ferricyanide, <3> cofactor NADH [1]) [1]

K_i-Value (mM)

0.0000062 <7> (Cibacron Blue) [7]
0.000022 <7> (acid blue 25) [7]
0.000028 <7> (Procion Blue M-3GS) [7]
0.00005 <7> (1,amino-4-[(4-amino-3-sulfophenyl)amino]-anthraquinone-2-sulfonic acid) [7]
0.000077 <7> (1-amino-4-[(3-carboxyphenyl)amino]-anthraquinone-2-sulfonic acid) [7]
0.0001-0.001 <2> (2-heptyl-4-hydroxyquinoline-N-oxide, <2> depending on enzyme solubilization [21]) [21]
0.00011 <7> (1-aminoanthraquinone-2-carboxylic acid) [7]
0.00012 <7> (bromoaminic acid) [7]
0.000131 <12> (dicoumarol, <12> hydrophilic isoform [16]) [16]
0.00017 <7> (Cibacron Blue, <7> competitive vs. NADH [14]) [14]
0.00022 <9> (dicoumarol) [11]
0.000288 <12> (dicoumarol, <12> hydrophobic isoform [16]) [16]
0.00045 <18> (5-methoxy-1,2-dimethyl-3-[(4-nitrophenoxy)methyl]indole-4,7-dione) [37]
0.00046 <7> (1,4-diaminoanthraquinone-2-sulfonic acid) [7]
0.00046 <7> (1-aminoanthraquinone-2-carboxylic acid) [7]
0.00054 <7> (Cibacron Blue, <7> noncompetitive vs. menadione [14]) [14]
0.00095 <21> (dicoumarol, <21> cofactor NADH [35]) [35]
0.001-0.002 <2> (phenylmercuric acetate, <2> depending on enzyme solubilization [21]) [21]
0.00109 <21> (dicoumarol, <21> cofactor NADH [35]) [35]
0.0013 <7> (anthraquinone-2-carboxylic acid) [7]
0.0015 <7> (1-aminoanthraquinone) [7]
0.0036 <7> (anthraquinone-2-sulfonic acid) [7]
0.00412 <21> (dicoumarol, <21> cofactor NADPH [35]) [35]
0.00511 <21> (dicoumarol, <21> cofactor NADPH [35]) [35]
0.0075 <8> (1,4-naphthohydroquinone, <8> vs. 1,4-naphthoquinone, value derived from from intercept [10]) [10]
0.008 <8> (1,4-naphthohydroquinone, <8> vs. NADPH, value derived from from slope [10]) [10]
0.0106 <20> (dicoumarol) [30]
0.02 <18> (5,6-dimethylxanthenone-4-acetic acid) [31]
0.039 <8> ($NADP^+$, <8> vs. NADPH, value derived from from intercept [10]) [10]
0.05 <7> (anthraquinone) [7]
0.058 <8> ($NADP^+$, <8> vs. NADPH, value derived from from slope [10]) [10]
0.058 <5> (dicoumarol, <5> cofactor NADH [3]) [3]
0.062 <5> (dicoumarol, <5> cofactor NADPH [3]) [3]
0.07 <18> (5,7-dihydroxyflavone) [27]
0.075 <18> (flavone-8-acetic acid) [31]

0.077 <8> ($NADP^+$, <8> vs. 1,4-naphthoquinone, value derived from from slope [10]) [10]
0.08 <7> (5,7-dihydroxyflavone) [27]
0.14 <8> ($NADP^+$, <8> vs. 1,4-naphthoquinone, value derived from intercept [10]) [10]
0.68 <12> (5,7-dihydroxyflavone) [27]

pH-Optimum
4.4 <6> (<6> plateau region from pH 6.0-7.5 with 70% activity [5]) [5]
5.7 <3> (<3> menadione [4]) [4]
7-8 <9> [11]
7-8.5 <10> [20]
7.5 <19> (<19> 60% activity at pH 9.0 [25]) [25]
8 <1> (<1> 2-methoxy-5-(methoxymethyl)-2,5-cyclohexadiene [22]) [22]
8-9 <7> [23]
8.5 <1> (<1> 2-hydroxymethyl-5-methoxy-2,5-cyclohexadiene-1,4-dione [22]) [22]
9-9.5 <8> [19]
9.7 <4> (<4> decrease above [2]) [2]

pH-Range
4.8-9.8 <3> [4]
5-9.5 <10> [20]
5.5-9 <20> (<20> reduction of 2,6-dimethoxybenzoquinone [30]) [30]
6-10.2 <7> (<7> inactive above or below [23]) [23]

Temperature optimum (°C)
25 <7, 8> (<7,8> assay at [13,19]) [13, 19]
30-43 <19> (<19> decrease above [25]) [25]

4 Enzyme Structure

Molecular weight
36000 <2> (<2> gel filtration [21]) [21]
39700-41700 <8> (<8> gel filtration, amino acid composition [19]) [19]
47000-69000 <1> (<1> gel filtration, SDS-PAGE [22]) [22]
50000 <9, 18> (<9> gel filtration [11]; <18> NAD(P)H:quinone acceptor oxidoreductase 2, gel filtration [29]) [11, 29]
52000 <3, 12> (<3> sedimentation equilibrium, sedimentation velocity [4]; <12> hydrophilic and hydrophobic isoforms, gel filtration [16]) [4, 16]
55000 <7> (<7> gel filtration [6]) [6]
60000 <6> (<6> gel filtration [5]) [5]
60260 <10> (<10> enzyme DII, gel filtration, SDS-PAGE, amino acid analysis [20]) [20]
61000 <7> (<7> cytosolic and microsomal enzyme, gel filtration [13]) [13]
61600 <19> (<19> gel filtration [25]) [25]
66070 <10> (<10> enzyme DI, gel filtration, SDS-PAGE, amino acid analysis [20]) [20]

Subunits

? <20> (<20> x * 25000, recombinant enzyme, SDS-PAGE [30]) [30]
dimer <7, 9, 12, 18, 19> (<9> 2 * 24000, SDS-PAGE [11]; <7> 2 * 27000, SDS-PAGE [6]; <12> 2 * 30000, hydrophilic and hydrophobic isoform, SDS-PAGE [16]; <7> 2 * 32000, cytosolic and microsomal enzyme, SDS-PAGE [13]; 2 * 30835, electrospray mass spectrometry [24]; <19> 2 * 21900, SDS-PAGE [25]; <18> 2 * 26000, NAD(P)H:quinone acceptor oxidoreductase 2, SDS-PAGE [29]) [6, 11, 13, 14, 16, 23, 24, 25, 29]
monomer <6, 10> (<6> 1 * 60000, SDS-PAGE [5]; <10> 1 * 63000, enzyme DII, SDS-PAGE [20]; <10> 1 * 67000, enzyme DI, SDS-PAGE [20]) [5, 20]
octamer <8> (<8> 8 * 5100, SDS-PAGE [19]) [19]
tetramer <7> (<7> 2 * 18000 + 2 * 8000-9000, SDS-PAGE [23]) [23]

Posttranslational modification

glycoprotein <7> (<7> all 6 isoforms are glycoproteins [8]) [8]
no glycoprotein <7, 12> (<12> very unstable posttranslational modification is suggested that is not detectable by mass spectrometry [24]) [23, 24]

5 Isolation/Preparation/Mutation/Application

Source/tissue

brain <3, 7> [1, 18]
heart <3, 7, 14> [4, 18]
hepatopancreas <5> [3]
kidney <3, 7, 13, 14, 7, 15> [4, 18]
liver <3, 4, 7, 12, 13-17> [2, 4, 6, 8, 9, 13-16, 18, 23, 34]
lung <3, 7, 14> [4, 18]
mammary gland <3, 7> [18]
mycelium <1, 11> [12, 22]
root <20> [30]
seed <10, 19> [20, 25]
skeletal muscle <3, 7> [18]
small intestine <3, 7> [18]

Localization

cytosol <3, 7, 12> [8, 13, 16, 18, 23]
intracellular <1, 11> [12, 22]
microsome <3, 7> [13, 18, 34]
mitochondrion <3, 4, 7> [2, 18]
soluble <8> [19]
vesicular fraction <2> (<2> interior surface [21]) [21]

Purification

<1> [22]
<3> [1, 4, 18]
<4> [2]
<5> [3]

<6> [5]
<7> (partial, isoforms [8]; affinity chromatography on dicoumarol-Sepharose 4B [6]; affinity chromatography on Cibacron blue F3GA [13, 14, 26]; recombinant enzyme [27]; microsomal enzyme, partial [34]) [6, 8, 13, 14, 18, 23, 26, 27, 28, 34]
<8> [19]
<10> (2 isoenzymes: D-I, D-II [20]) [20]
<12> (affinity chromatography on Cibacron blue-Agarose [16]; recombinant enzyme [24, 27]) [16, 17, 24, 27]
<18> (recombinant enzyme [27]; recombinant NAD(P)H:quinone acceptor oxidoreductase 2, affinity chromatography on Affi-gel blue [29]) [27, 27, 31, 37]
<19> (Agarose cibacron blue, AH-Sepharose, CM-cellulose, Sephacryl S-200 HR [25]) [25]
<20> (recombinant enzyme [30]) [30]
<21> (recombinant enzyme [35]) [35]

Crystallization

<7> (1.6 M ammonium sulfate, 150 mM imidazole, pH 7.0, small needles [14]; hanging drop method, crystals from a solution of 10 mg/ml enzyme, 1.45 M ammonium sulfate, 0.5% polyethylene glycol 8000, 30 mM Cibacron blue and 150 mM imidazole, pH 7.0, complex with $NADP^+$ or duroquinone, 2.1 A resolution, structure and substrate binding mechanism [26]) [14, 26]
<12> (hydrophilic isoform in the presence of FAD, hanging drop diffusion, 10 mg/ml enzyme, in 200 mM potassium phosphate, 0.5 mM EDTA, 0.5 mM, FAD, pH 8.0, 45-50% saturation of ammonium sulfate [16]; hanging drop vapor diffusion from a solution containing 10-15 mg/ml enzyme in 25 mM Tris-HCl, pH 8.0, 0.005 mM FAD, mixed with equal volumes of reservoir solution consisting of 30% polyethylene glycol 3350, 200 mM sodium acetate and 100 mM sodium tricine, pH 8.5, X-ray structure, 2.8 A resolution [32]) [16, 32]
<18> (hanging drop vapor diffusion from a solution containing 10-15 mg/ml enzyme in 25 mM Tris-HCl, pH 8.0, 0.005 mM FAD, mixed with equal volumes of reservoir solution consisting of 30% polyethylene glycol 3350, 200 mM sodium acetate and 100 mM sodium tricine, pH 8.5, X-ray structure, 1.7 A resolution [32]; crystal structure of enzyme complexed with 5-methoxy-1,2-dimethyl-3-[(4-nitrophenoxy)methyl]indole-4,7-dione, 1.8 A resolution [37]) [32, 37]

Cloning

<7> (expression in Escherichia coli [24]) [24, 27, 28]
<12> (expression in Escherichia coli [24]) [24, 27]
<18> (expression in Escherichia coli [24]; NAD(P)H:quinone acceptor oxidoreductase 2 [29]) [24, 27, 29, 31, 37]
<20> (heterologous expression in Pichia pastoris [30]) [30]
<21> (expression in Escherichia coli [35]) [35]

Application

medicine <12> (<12> activation of antitumor prodrugs [27]) [27]

6 Stability

Temperature stability

0 <4> (<4> some days [2]) [2]
20 <4> (<4> several h [2]) [2]
57 <4> (<4> 5 min, 90% loss of activity [2]) [2]

General stability information

<4>, stable to dialysis against 0.01 M potassium phosphate buffer, pH 7.5 and 0.01 M Tris buffer, pH 8.2 and 8.9, 16 h or more at 3°C [2]
<7>, sucrose, 0.25 M, essential for stabilization during storage at -20°C and purification [18]

Storage stability

<3>, -20°C, some weeks [4]
<7>, -20°C, 0.25 M sucrose, several months [18]
<10>, 5°C, 0.05 M sodium phosphate, pH 8, enzyme D-II is stable for 1 month, enzyme D-I loses 30% of its activity after 1 week [20]

References

[1] Giuditta, A.; Strecker, H.J.: Purification and some properties of a brain diaphorase. Biochim. Biophys. Acta, **48**, 10-19 (1961)
[2] Wosilait, W.D.: The reduction of vitamin K_1 by an enzyme from dog liver. J. Biol. Chem., **235**, 1196-1201 (1960)
[3] Di Prisco, G.; Casola, L.; Giuditta, A.: Purification and properties of a soluble reduced nicotinamide-adenine dinucleotide (phosphate) dehydrogenase from the hepatopancreas of Octopus vulgaris. Biochem. J., **105**, 455-460 (1967)
[4] Märki, F.; Martius, C.: Vitamin K-Reduktase, Darstellung und Eigenschaften. Biochem. Z., **333**, 111-135 (1960)
[5] Petitdemange, H.; Marczak, R.; Raval, G.; Gay, R.: Menadione reductase from Clostridium tyrobutyricum. Can. J. Microbiol., **26**, 324-329 (1980)
[6] Rase, B.; Bartfai, T.; Ernster, L.: Purification of DT-diaphorase by affinity chromatography. Occurrence of two subunits and nonlinear Dixon and Scatchard plots of the inhibition by anticoagulants. Arch. Biochem. Biophys., **172**, 380-386 (1976)
[7] Prestera, T.; Prochaska, H.J.; Talalay, P.: Inhibition of NAD(P)H:(quinone-acceptor) oxidoreductase by cibacron blue and related anthraquinone dyes: a structure-activity study. Biochemistry, **31**, 824-833 (1992)
[8] Segura-Aguilar, J.; Kaiser, R.; Lind, C.: Separation and characterization of isoforms of DT-diaphorase from rat liver cytosol. Biochim. Biophys. Acta, **1120**, 33-42 (1992)

[9] Deng, P.S.K.; Zhao, S.H.; Iyanagi, T.; Chen, S.: Photodependent inhibition of rat liver NAD(P)H:quinone acceptor oxidoreductase by (A)-2-azido-NAD^+ and (A)-8-azido-NAD. Biochemistry, **30**, 6942-6948 (1991)

[10] Cloete, F.; Viljoen, C.C.; Scott, W.E.; Oosthuizen, M.M.J.: A steady-state kinetic investigation of the mechanism of action of NAD(P)H dehydrogenase of the cyanobacterium, Microcystis aeruginosa. Biochim. Biophys. Acta, **870**, 279-291 (1986)

[11] Hayashi, M.; Hasegawa, K.; Oguni, Y.; Unemoto, T.: Characterization of FMN-dependent NADH-quinone reductase induced by menadione in Escherichia coli. Biochim. Biophys. Acta, **1035**, 230-236 (1990)

[12] Buswell, J.A.; Eriksson, K.E.: NAD(P)H dehydrogenase (quinone) from Sporotrichum pulverulentum. Methods Enzymol., **161**, 271-274 (1988)

[13] Sharkis, D.H.; Swenson, R.P.: Purification by cibacron blue F3GA dye affinity chromatography and comparison of NAD(P)H:quinone reductase (E.C.1.6.99.2) from rat liver cytosol and microsomes. Biochem. Biophys. Res. Commun., **161**, 434-441 (1989)

[14] Prochaska, H.J.: Purification and crystallization of rat liver NAD(P)H:(quinone-acceptor) oxidoreductase by cibacron blue affinity chromatography: identification of a new and potent inhibitor. Arch. Biochem. Biophys., **267**, 529-538 (1988)

[15] Fasco, M.J.; Principe, L.M.: Vitamin K_1 hydroquinone formation catalyzed by DT-diaphorase. Biochem. Biophys. Res. Commun., **104**, 187-192 (1982)

[16] Prochaska, H.J.; Talalay, P.: Purification and characterization of two isofunctional forms of NAD(P)H: quinone reductase from mouse liver. J. Biol. Chem., **261**, 1372-1378 (1986)

[17] Amzel, L.M.; Bryant, S.H.; Prochaska, H.J.; Talalay, P.: Preliminary crystallographic X-ray data for an NAD(P)H:quinone reductase from mouse liver. J. Biol. Chem., **261**, 1379 (1986)

[18] Lind, C.; Cadenas, E.; Hochstein, P.; Ernster, L.: DT-diaphorase: purification, properties, and function. Methods Enzymol., **186**, 287-301 (1990)

[19] Viljoen, C.C.; Cloete, F.; Scott, W.E.: Isolation and characterization of an NAD(P)H dehydrogenase from the cyanobacterium, Microcystis aeruginosa. Biochim. Biophys. Acta, **827**, 247-259 (1985)

[20] Viljoen, C.C.; Cloete, F.; Botes, D.P.; Kruger, H.: Isolation and characterization of NAD(P)H-dehydrogenases from seeds of the castor bean. Phytochemistry, **22**, 365-370 (1983)

[21] Lanyi, J.K.: Studies of the electron transport chain of extremely halophilic bacteria. VII. Solubilization properties of menadione reductase. J. Biol. Chem., **247**, 3001-3007 (1972)

[22] Constam, D.; Muheim, A.; Zimmermann, W.; Fiechter, A.: Purification and partial characterization of an intracellular NADH:quinone oxidoreductase from Phanerochaete chrysosporium. J. Gen. Microbiol., **137**, 2209-2214 (1991)

[23] Wallin, R.: Some molecular properties of NAD(P)H dehydrogenase from rat liver. Biochem. J., **181**, 127-135 (1979)

[24] Chen, S.; Clarke, P.E.; Martino, P.A.; Deng, P.S.K.; Yeh, C.H.; Lee, T.D.; Prochaska, H.J.; Talalay, P.: Mouse liver NAD(P)H:quinone acceptor oxidore-

ductase: protein sequence analysis by tandem mass spectrometry, cDNA cloning, expression in Escherichia coli, and enzyme activity analysis. Protein Sci., **3**, 1296-1304 (1994)

[25] Rescigno, A.; Sollai, F.; Masala, S.; Porcu, M.C.; Sanjust, E.; Rinaldi, A.C.; Curreli, N.; Grifi, D.; Rinaldi, A.: Purification and characterization of an NAD(P)H:quinone oxidoreductase from Glycine max seedlings. Prep. Biochem., **25**, 57-67 (1995)

[26] Li, R.; Bianchet, M.A.; Talalay, P.; Amzel, L.M.: The three-dimensional structure of NAD(P)H:quinone reductase, a flavoprotein involved in cancer chemoprotection and chemotherapy: mechanism of the two-electron reduction. Proc. Natl. Acad. Sci. USA, **92**, 8846-8850 (1995)

[27] Chen, S.; Knox, R.; Lewis, A.D.; Friedlos, F.; Workman, P.; Deng, P.S.K.; Fung, M.; Ebenstein, D.; Wu, K.; Tsai, T.M.: Catalytic properties of NAD(P)H:quinone acceptor oxidoreductase: study involving mouse, rat, human, and mouse-rat chimeric enzymes. Mol. Pharmacol., **47**, 934-939 (1995)

[28] Chen, S.; Deng, P.S.K.; Swiderek, K.; Li, M.; Chan, S.I.: Interaction of flavones and their bromoacetyl derivatives with NAD(P)H:quinone acceptor oxidoreductase. Mol. Pharmacol., **47**, 419-424 (1995)

[29] Wu, K.; Knox, R.; Sun, X.Z.; Joseph, P.; Jaiswal, A.K.; Zhang, D.; Deng, P.S.K.; Chen, S.: Catalytic properties of NAD(P)H:quinone oxidoreductase-2 (NQO2), a dihydronicotinamide riboside dependent oxidoreductase. Arch. Biochem. Biophys., **347**, 221-228 (1997)

[30] Wrobel, R.L.; Matvienko, M.; Yoder, J.I.: Heterologous expression and biochemical characterization of an NAD(P)H:quinone oxidoreductase from the hemiparasitic plant Triphysaria versicolor. Plant Physiol. Biochem., **40**, 265-272 (2002)

[31] Phillips, R.M.: Inhibition of DT-diaphorase (NAD(P)H:quinone oxidoreductase, EC 1.6.99.2) by 5,6-dimethylxanthenone-4-acetic acid (DMXAA) and flavone-8-acetic acid (FAA): implications for bioreductive drug development. Biochem. Pharmacol., **58**, 303-310 (1999)

[32] Faig, M.; Bianchet, M.A.; Talalay, P.; Chen, S.; Winski, S.; Ross, D.; Amzel, L.M.: Structures of recombinant human and mouse NAD(P)H:quinone oxidoreductases: species comparison and structural changes with substrate binding and release. Proc. Natl. Acad. Sci. USA, **97**, 3177-3182 (2000)

[33] Pink, J.J.; Planchon, S.M.; Tagliarino, C.; Varnes, M.E.; Siegel, D.; Boothman, D.A.: NAD(P)H:quinone oxidoreductase activity is the principal determinant of β-lapachone cytotoxicity. J. Biol. Chem., **275**, 5416-5424 (2000)

[34] Jaiswal, A.K.: Characterization and partial purification of microsomal NAD(P)H:quinone oxidoreductases. Arch. Biochem. Biophys., **375**, 62-68 (2000)

[35] Sanchez, L.B.; Elmendorf, H.; Nash, T.E.; Muller, M.: NAD(P)H:menadione oxidoreductase of the amitochondriate eukaryote Giardia lamblia: a simpler homologue of the vertebrate enzyme. Microbiology, **147**, 561-570 (2001)

[36] Swann, E.; Barraja, P.; Oberlander, A.M.; Gardipee, W.T.; Hudnott, A.R.; Beall, H.D.; Moody, C.J.: Indolequinone antitumor agents: correlation between quinone structure and rate of metabolism by recombinant human

NAD(P)H:quinone oxidoreductase. Part 2. J. Med. Chem., **44**, 3311-3319 (2001)

[37] Winski, S.L.; Faig, M.; Bianchet, M.A.; Siegel, D.; Swann, E.; Fung, K.; Duncan, M.W.; Moody, C.J.; Amzel, L.M.; Ross, D.: Characterization of a mechanism-based inhibitor of NAD(P)H:quinone oxidoreductase 1 by biochemical, X-ray crystallographic, and mass spectrometric approaches. Biochemistry, **40**, 15135-15142 (2001)

NADH Dehydrogenase 1.6.99.3

1 Nomenclature

EC number
1.6.99.3

Systematic name
NADH:(acceptor) oxidoreductase

Recommended name
NADH dehydrogenase

Synonyms
alkyl hydroperoxide reductase
DPNH dehydrogenase
DPNH diaphorase
EC 1.6.2.1 (formerly)
H_2O_2 forming NADH oxidase
NADH diaphorase
NADH hydrogenase
NADH oxidoreductase
NADH-menadione oxidoreductase
NADH:oxygen oxidoreductase
NOXase
β-NADH dehydrogenase
cytochrome c reductase
dehydrogenase, reduced nicotinamide adenine dinucleotide
diaphorase
dihydrocodehydrogenase I dehydrogenase
dihydronicotinamide adenine dinucleotide dehydrogenase
diphosphopyridine diaphorase
proteins, specific or class, gene MURF3
reduced diphosphopyridine nucleotide diaphorase
type I dehydrogenase
Additional information (enzyme is present in a mitochondrial complex as EC 1.6.5.3)

CAS registry number
9079-67-8

2 Source Organism

<1> *Bos taurus* [1, 15]
<2> *halophilic bacterium* (strain AR-I [2]) [2, 13]
<3> *Sus scrofa* [3]
<4> *Rhodopseudomonas capsulata* (strain 37b4 [4]) [4, 5]
<5> *Homo sapiens* [6, 7, 20]
<6> *Acinetobacter calcoaceticus* [8]
<7> *Bacillus sp.* (YN-1 [9]) [9]
<8> *Photobacterium phosphoreum* [10]
<9> *Drosophila hydei* [11]
<10> *Escherichia coli* [12, 18, 19]
<11> *Propionibacterium shermanii* [14]
<12> *Bacillus subtilis* (W23 [16]) [16]
<13> *Agrobacterium tumefaciens* [17]
<14> *Bacillus megaterium* [21]

3 Reaction and Specificity

Catalyzed reaction

NADH + H^+ + acceptor = NAD^+ + reduced acceptor (<8>, ping-pong mechanism [10])

Reaction type

oxidation
redox reaction
reduction

Natural substrates and products

S Additional information <5, 10, 14> (<10>, enzyme is involved in the respiratory chain [18, 19]; <14>, the enzyme is localized at the main point of electron entry into the respiratory chain. The enzyme catalyzes the transfer of electrons from NADH to the respiratory electron transport components and hence links the major catabolic and energy-producing pathways [21]; <5>, the enzyme transfers electrons from NADH to external oxidants like ferricyanide through pathways which are linked to metabolic processess in the cell [20]) [18, 19, 20, 21]
P ?

Substrates and products

S 3-acetylpyridine-NADH + 2,6-dichlorophenolindophenol <2> (Reversibility: ? <2> [2]) [2]
P 3-acetylpyridine-NAD^+ + reduced 2,6-dichlorophenolindophenol
S 3-acetylpyridine-deamino-NADH + 2,6-dichlorophenol indophenol <2> (Reversibility: ? <2> [2]) [2]
P 3-acetylpyridine-deamino-NAD^+ + reduced 2,6-dichlorophenolindophenol

S NADH + 2,6-dichlorophenolindophenol <1, 2, 3, 4, 5, 7, 8, 10, 11, 13, 14> (<4, 6>, specific for NADH [4, 5, 8]) (Reversibility: ? <1, 2, 3, 4, 5, 7, 8, 10, 11, 13, 14> [1, 2, 3, 4, 5, 7, 9, 10, 12, 13, 14, 15, 17, 18, 19, 21]) [1, 2, 3, 4, 5, 7, 9, 10, 12, 13, 14, 15, 17, 18, 19, 21]

P NAD^+ + reduced 2,6-dichlorophenolindophenol

S NADH + 2-methyl-1,4-naphthoquinone <2, 4, 7, 8, 10, 11> (<2>, i.e. menadione, about 60% of the activity with 2,6-dichlorophenolindophenol [2]) (Reversibility: ? <2, 4, 7, 8, 10, 11> [2, 5, 9, 10, 12, 14, 18, 19]) [2, 5, 9, 10, 12, 14, 18, 19]

P NAD^+ + 2-methyl-1,4-naphthoquinol

S NADH + 3-(4',5'-dimethyl-thiazol-2-yl)2,4-diphenyltetrazolium bromide <12, 14> (Reversibility: ? <12, 14> [16, 21]) [16, 21]

P NAD^+ + ?

S NADH + 5-hydroxy-1,4-naphthoquinone <2> (<2>, i.e. juglone, about 55% of the activity with 2,6-dichlorophenolindophenol [2]) (Reversibility: ? <2> [2]) [2]

P NAD^+ + 5-hydroxy-1,4-naphthoquinol

S NADH + ferricyanide <1, 3, 4, 5, 6, 7, 8, 9, 10, 13, 14> (Reversibility: ? <1, 3, 4, 5, 6, 7, 8, 9, 10, 13, 14> [3, 5, 6, 7, 8, 9, 10, 11, 13, 14, 15, 17, 18, 19, 20, 21]) [3, 5, 6, 7, 8, 9, 10, 11, 13, 14, 15, 17, 18, 19, 20, 21]

P NAD^+ + ferrocyanide

S NADH + ferricytochrome b_5 <1, 5> (Reversibility: ? <1, 5> [7, 15]) [7, 15]

P NAD^+ + ferrocytochrome b_5

S NADH + ferricytochrome c <1, 3, 4, 5, 7, 8, 11, 13> (Reversibility: ? <1, 3, 4, 5, 7, 8, 11, 13> [3, 5, 7, 9, 10, 14, 15, 17]) [3, 5, 7, 9, 10, 14, 15, 17]

P NAD^+ + ferrocytochrome c

S NADH + methemoglobin-ferrocyanide complex <5> (Reversibility: ? <5> [7]) [7]

P NAD^+ + ?

S NADPH + 2,6-dichlorophenolindophenol <1, 5, 11> (<1>, activity with NADPH is about 35% of the activity with NADH [1]; <5>, less active as NADH [7]; <4>, less than 0.01% of the activity with NADH [4]; <1>, no activity [15]) (Reversibility: ? <1, 5, 11> [1, 7, 14]) [1, 7, 14]

P $NADP^+$ + reduced 2,6-dichlorophenolindophenol

S NADPH + 3-(4',5'-dimethyl-thiazol-2-yl)2,4-diphenyltetrazolium bromide <12, 14> (<14>, rate of oxidation is less than 5% of the reaction with NADH [21]; <12>, less than 6% of the activity with NADH [16]) (Reversibility: ? <12, 14> [16, 21]) [16, 21]

P $NADP^+$ + ?

S thionicotinamide-NADH + 2,6-dichlorophenolindophenol <2> (Reversibility: ? <2> [2]) [2]

P thionicotinamide-NAD^+ + reduced 2,6-dichlorophenolindophenol

Inhibitors

2,6-dichlorophenol indophenol <2> (<2>, above 0.07 mM [2]; <2>, product inhibitor [13]) [2, 13]
3-pyridine aldehyde NADH <2> (<2>, competitive with respect to NADH [2]) [2]
ADP <5, 10> (<5>, strong competitive [7]; <10>, 10 mM, 90% inhibition in the reaction with 0.04 mM 2,6-dichlorophenolindophenol and 0.032 mM NADH, competitive [19]) [7, 19]
AMP <5, 10, 14> (<10>, 10 mM, 98% inhibition in the reaction with 0.04 mM 2,6-dichlorophenolindophenol and 0.032 mM NADH, competitive [19]) [7, 18, 21]
AMP <2, 14> (<2>, competitive inhibition of NADH oxidation [2]; <14>, 10 mM, 41% inhibition [21]) [2, 21]
ATP <5, 10, 14> (<10>, 10 mM, 50% inhibition in the reaction with 0.04 mM 2,6-dichlorophenolindophenol and 0.032 mM NADH, competitive [19]; <14>, 10 mM, 17% inhibition [21]) [7, 19, 21]
Cl^- <8> [10]
Cs^+ <8> (<8>, activates [10]) [10]
EDTA <14> (<14>, 10 mM, 21% inhibition [21]) [21]
FMN <4> (<4>, inactivates [5]) [5]
$HgCl_2$ <14> (<14>, 10 mM, complete inhibition [21]) [21]
KCN <14> (<14>, 10 mM, 36% inhibition [21]) [21]
Li^+ <8> (<8>, activates [10]) [10]
NAD^+ <1, 2, 5, 10, 14> (<2>, product inhibitor, competitive with NADH, non-competitive with 2,6-dichlorophenolindophenol [13]; <2>, competitive inhibition of NADH oxidation [2]; <1>, weakly competitive [15]; <10>, effectively competitive [19]; <14>, competitive [21]) [2, 6, 13, 14, 15, 19, 21]
NADH <4, 11> (<4>, inactivates [5]) [5, 14]
NEM <1, 5> (<5>, above 1 mM [7]; <1>, 0.5 mM, 12% inhibition [15]) [7, 15]
NO_3^- <8> [10]
PCMB <1, 3, 5, 14> (<3>, 0.1 mM, complete inhibition [3]; <5>, complete inhibition at 0.001 mM [7]; <1>, 0.005 mM, 50% inhibition [15]; <14>, 10 mM, 90% inhibition [21]) [3, 7, 15, 21]
Rb^+ <8> (<8>, activates [10]) [10]
SCN^+ <8> [10]
acrinol <5> (<5>, 1 mM, complete inhibition [7]) [7]
adenine <10> (<10>, 3 mM, 60% inhibition in the reaction with 0.04 mM 2,6-dichlorophenolindophenol and 0.032 mM NADH [19]) [19]
adenosine <12> (<12>, slight inhibition [16]) [16]
atebrin <5> (<5>, 1 mM, complete inhibition [7]) [7]
cAMP <10> (<10>, 10 mM, 76% inhibition in the reaction with 0.04 mM 2,6-dichlorophenolindophenol and 0.032 mM NADH [19]) [19]
deoxyAMP <10> (<10>, 10 mM, 41% inhibition in the reaction with 0.04 mM 2,6-dichlorophenolindophenol and 0.032 mM NADH [19]) [19]
deoxyadenosine <10, 12> (<10>, 10 mM, 32% inhibition in the reaction with 0.04 mM 2,6-dichlorophenolindophenol and 0.032 mM NADH [19]; <12>, slight inhibition [16]) [16, 19]

ferricyanide <5, 11> (<5>, competitive substrate inhibition due to its direct binding to the enzyme and forming an inactive binary complex [6]) [6, 14]
ferrocyanide <5> [6]
nicotinic acid amide <12> (<12>, slight inhibition [16]) [16]
p-hydroxymercuribenzoate <13> [17]
p-hydroxymercuriphenyl sulfonate <11> (<11>, 2 mM, 20% inhibition of activity with ferricyanide [14]) [14]
phosphate <3> (<3>, 0.033 M, pH 8.5, 70% inhibition [3]; <8>, no deactivation [10]) [3]
riboflavin <4> (<4>, inactivates [5]) [5]
sodium arsenite <14> (<14>, 10 mM, 44% inhibition [21]) [21]
sodium azide <14> (<14>, 10 mM, 24% inhibition [21]) [21]
Additional information <8, 14> (<8>, activation by monovalent cations is due to increase of affinity of the enzyme for substrates [10]; <14>, no inhibition by rotenone [21]) [10, 21]

Cofactors/prosthetic groups

3-acetylpyridine-NADH <2> [2]
3-acetylpyridine-deamino-NADH <2> [2]
FAD <2, 5, 7, 8, 10, 12, 13, 14> (<2>, FAD represents 85% of the total flavin [2]; <5>, contains one mol FAD per 36000 Da subunit [7]; <7>, contains 1 mol FAD per subunit [9]; <12>, contains 1 molecule FAD per molecule [16]; <13>, natural prosthetic group of the enzyme, apoenzyme catalyzed reaction with ferricyanide, FAD is required for reaction with 2,6-dichlorophenol or with cytochrome c, K_m: 0.00056 mM [17]; <10>, purified enzyme has an absolute requirement for FAD. K_m-value for the reaction with 2,6-dichlorophenol indophenol is 0.004 mM, the K_m-value for the reaction with ferricyanide is 0.006 mM, the K_M-value for the reaction with menadione is 0.003 mM [19]) [2, 7, 9, 10, 16, 17, 19, 21]
FMN <13> (<13>, is about 10% as effective as FAD [17]) [17]
NADH <1-14> (<4, 6, 7, 8, 10, 14>, specific for NADH [4, 5, 8, 9, 10, 18, 21]) [1-21]
NADPH <1, 5, 11, 12, 13> (<1>, activity with NADPH is about 35% of the activity with NADH [1]; <5>, less active than NADH [7]; <11>, activity is below 2% of the NADH-oxidizing activity [14]; <12>, less than 6% of the activity with NADH in reaction with 3-(4',5'-dimethyl-thiazol-2-yl)2,4-diphenyltetrazolium bromide [16]; <13>, oxidized at about 5% the rate of NADH [17]; <10>, no activity [18, 19]) [1-21]
thionicotinamide-NADH <2> [2]
Additional information <1, 10> (<1>, no flavin detectable [15]; <10>, purified enzyme contains no flavin [19]) [15, 19]

Activating compounds

Cl^- <2> (<2>, stimulates [13]) [13]
FMN <4, 6> (<4>, activation of soluble enzyme [5]; <6>, required [8]) [5, 8]
NO_3^- <2> (<2>, stimulates [13]) [13]
SO_4^{2-} <2> (<2>, stimulates [13]) [13]

Triton X-100 <10> (<10>, enzyme is inactive in absence of detergent or phospholipid. Triton X-100 gives optimal activity [12]) [12]
diphosphatidylglycerol <10> (<10>, most effective lipid activator. Maximal activation at pH 7 and at 0°C [12]) [12]
phosphatidylglycerol <10> (<10>, most effective lipid activator. Maximal activation at pH 7 and 0°C [12]) [12]
phospholipid <7> (<7>, slightly increases activity [9]) [9]
spermidine <10> (<10>, enzyme optimally activated by phospholipids is additionally activated 2fold [12]) [12]
Additional information <5> (<5>, activation of the enzyme in erythrocytes by β-adrenergic agonists [20]) [20]

Metals, ions

Ca^{2+} <10> (<10>, Triton X-100-activated enzyme form is stimulated by several divalent cations without specificity [12]) [12]
K^+ <7, 8, 10> (<7, 8>, stimulates activity with electron acceptors other than cytochrome c [9, 10]; <10>, KCN stimulates [19]) [9, 10, 19]
Mg^{2+} <10> (<10>, Triton X-100-activated enzyme form is stimulated by several divalent cations without specificity [12]) [12]
Mn^{2+} <10> (<10>, Triton X-100-activated enzyme form is stimulated by several divalent cations without specificity [12]) [12]
Na^+ <2, 7> (<2>, NaCl stimulates, 43% of maximal activity in presence of 2.5 M NaCl [2]; <7, 8>, stimulates activity with electron acceptors other than cytochrome c [9, 10]) [2, 9, 10]
iron <11> (<11>, contains 9.3 nM per mg protein [14]) [14]

Turnover number (min^{-1})

40600 <5> (ferricyanide) [7]

Specific activity (U/mg)

0.305 <13> [17]
0.43 <6> [8]
0.725 <14> [21]
24.2 <8> [10]
24.67 <12> [16]
376 <2> [2]
790 <5> (<5>, reaction with ferricyanide [7]) [7]
Additional information <1, 4, 9> [1, 4, 5, 11, 15]

K_m-Value (mM)

0.0006 <5> (NADH) [7]
0.002 <8> (cytochrome c) [10]
0.0029 <4> (2,6-dichlorophenolindophenol) [4]
0.004 <4> (NADH) [5]
0.0079 <4> (FMN) [4]
0.0095 <1> (ferricyanide) [15]
0.014 <5> (NADH) [6]
0.015 <4> (cytochrome c) [4]

0.0157 <14> (NADH, <14>, reaction with 2,6-dichlorophenolindophenol [21]) [21]
0.017 <2> (NADH, <2>, reaction with 5-hydroxy-1,4-naphthoquinone [2]) [2]
0.018 <1> (NADH) [15]
0.02 <5> (cytochrome b_5) [7]
0.02 <13> (NADH) [17]
0.021 <2> (2-methyl-1,4-naphthoquinone) [2]
0.022 <2> (NADH, <2>, reaction with 2,6-dichlorophenolindophenol [2]) [2]
0.03 <2> (2,6-dichlorophenol indophenol) [2]
0.034 <2> (NADH, <2>, reaction with 2-methyl-1,4-naphthoquinone [2]) [2]
0.0348 <14> (NADH, <14>, reaction with ferricyanide [21]) [21]
0.035 <8> (2-methyl-1,4-naphthoquinone) [10]
0.045 <4> (NADH) [4]
0.049 <5> (ferricyanide) [6]
0.063 <4> (2,6-dichlorophenolindophenol) [5]
0.0692 <14> (NADH, <14>, reaction with 3-(4',5'-dimethylthiazol-2-yl)-diphenyltetrazolium bromide [21]) [21]
0.087 <11> (NADH) [14]
0.11 <8> (2,6-dichlorophenolindophenol) [10]
0.133 <2> (5-hydroxy-1,4-naphthoquinone) [2]
0.94 <11> (ferricyanide) [14]
1.035 <14> (NADPH, <14>, reaction with 3-(4',5'-dimethylthiazol-2-yl)-diphenyltetrazolium bromide [21]) [21]
2.1 <8> (ferricyanide) [10]

K_i-Value (mM)

0.02 <10> (NAD^+, <10>, inhibition of-2,6-dichlorophenolindophenol reductase activity, purified enzyme [19]) [19]
0.037 <14> (NAD^+) [21]
0.07 <10> (NAD^+, <10>, inhibition of NADH-2,6-dichlorophenolindophenol reductase activity, membrane bound enzyme [19]) [19]
0.08 <5> (ADP) [7]
0.118 <1> (NAD^+) [15]
0.18 <11> (NAD^+) [14]
0.4 <2> (3-pyridine aldehyde NADH) [2]
0.5 <10> (AMP, <10>, inhibition of NADH-2,6-dichlorophenolindophenol reductase activity, membrane bound enzyme [19]) [19]
0.6 <10> (AMP, <10>, inhibition of NADH-2,6-dichlorophenolindophenol reductase activity, purified enzyme [19]) [19]
0.7 <10> (ADP, <10>, inhibition of NADH-2,6-dichlorophenolindophenol reductase activity, membrane bound enzyme [19]) [19]
1 <10> (ADP, <10>, inhibition of NADH-2,6-dichlorophenolindophenol reductase activity, purified enzyme [19]) [19]
1.5 <2> (AMP) [2]
1.7 <2> (NAD^+) [2]

9 <10> (ATP, <10>, inhibition of NADH-2,6-dichlorophenolindophenol reductase activity, membrane bound enzyme [19]) [19]

pH-Optimum

5.2 <5> (<5>, NADH-methemoglobin-ferrocyanide reductase activity [7]) [7]
5.6 <9> [11]
6.5-7 <10> (<10>, reaction with 2,6-dichlorophenolindophenol [19]) [19]
6.5-8.5 <5> (<5>, NADH-ferricyanide reductase activity [7]) [7]
7.4 <13> [17]
7.5 <4> [4, 5]
7.5-7.8 <14> [21]
7.5-8 <12> (<12>, activity with 3-(4',5'-dimethyl-thiazol-2-yl)2,4-diphenyltetrazolium bromide [16]) [16]
8.5-9 <7> (<7>, reaction with 2,6-dichlorophenolindophenol [9]) [9]
Additional information <3> [3]

Temperature optimum (°C)

37 <12> (<12>, activity with 3-(4',5'-dimethyl-thiazol-2-yl)2,4-diphenyltetrazolium bromide [16]) [16]

4 Enzyme Structure

Molecular weight

29000 <1> (<1>, equilibrium sedimentation [1]) [1]
37000 <4> (<4>, soluble enzyme, gel filtration [4]) [4]
64000 <2> (<2>, gel filtration [2]) [2]
79000 <9> (<9>, gel filtration [11]) [11]
97000 <4> (<4>, membrane-bound enzyme, gel filtration [5]) [5]
138000 <7> (<7>, gel filtration [9]) [9]
144000 <5> (<5>, gel filtration [7]) [7]
215000 <11> (<11>, gel filtration [14]) [14]
480000 <1> (<1>, highly aggregated enzyme form, non-denaturing PAGE [15]) [15]

Subunits

? <1, 5, 9, 10, 12, 14> (<1>, x * 17500, SDS-PAGE [15]; <5>, x * 36000, SDS-PAGE [7]; <10>, x * 38000, SDS-PAGE [18]; <14>, x * 42000, SDS-PAGE [21]; <9>, x * 57000, SDS-PAGE [11]; <12>, x * 64000, SDS-PAGE [16]) [7, 11, 15, 16, 18, 21]
dimer <7> (<7>, 2 * 65000, SDS-PAGE [9]) [9]
hexamer <4> (<4>, 6 * 15500, membrane-bound enzyme, SDS-PAGE [5]) [5]
monomer <4> (<4>, 1 * 38000, soluble enzyme, SDS-PAGE [4]) [4]

Posttranslational modification

lipoprotein <6> (<6>, 0.0004 mg phospholipid per mg of protein, most of which is cardiolipin with traces of phosphatidylethanolamine and phosphatidylglycerol [8]) [8]

5 Isolation/Preparation/Mutation/Application

Source/tissue

cardiac muscle <3> (<3>, digestion of muscle preparation at 37°C with sake-venom phospholipase [3]) [3]
embryo <9> [11]
erythrocyte <1, 5> (<5>, membrane [7]) [1, 6, 7, 20]
larva <9> [11]
neutrophil <1> [15]

Localization

membrane <1, 4, 5, 6, 7, 8, 10, 12, 14> (<14>, can be solubilized by sonication [21]; <1>, associated [15]; <4, 6, 8, 10>, bound to [5, 8, 10, 18]) [5, 7, 8, 9, 10, 15, 16, 18, 21]
mitochondrion <9> [11]
plasma membrane <5> [20]
soluble <4, 11, 13> [5, 14, 17]
vesicular fraction <12> [16]

Purification

<1> [1, 15]
<2> [2]
<4> [4, 5]
<5> [7]
<6> [8]
<7> [9]
<8> [10]
<9> [11]
<10> [18]
<11> (partial [14]) [14]
<13> [17]
<14> [21]

Crystallization

<1> (crystallization with ammonium sulfate [1]) [1]

6 Stability

Temperature stability

25 <4> (<4>, 5 min, in presence of FMN, riboflavin, and NADH, stable [4]) [4]
40 <11> (<11>, 40 min, 90% loss of the activity with ferricyanide [14]) [14]

General stability information

<2>, NADH enhances stability [2]
<2>, in presence of 100 mM NaCl, immediate and exponential loss of activity. In presence of 250 mM NaCl stable for about 30 min before the loss of activity

becomes exponential. In presence of 500 mM NaCl the enzyme is stable during 150 min [2]
<10>, activated complex prepared with diphosphatidylglycerol is stable, activated complex prepared with phosphatidylglycerol loses activity rapidly [12]
<11>, stable to repeated freezing and thawing [14]

Storage stability
<1>, -20°C, stable for several months [15]
<2>, 4°C, 1 month, less than 10% loss of activity [2]
<11>, -25°C, stable for several months [14]

References

[1] Adachi, K.; Okuyama, T.: Study on the reduced pyridine nucleotide dehydrogenase of bovine erythrocytes. I. Crystallization and properties of the reduced pyridine nucleotide dehydrogenase of bovine erythrocytes. Biochim. Biophys. Acta, **268**, 629-637 (1972)
[2] Hochstein, L.I.; Dalton, B.P.: Studies of a halophilic NADH dehydrogenase. I. Purification and properties of the enzyme. Biochim. Biophys. Acta, **302**, 216-228 (1973)
[3] Kaniuga, Z.: The transformation of mitochondrial NADH dehydrogenase into NADH:cytochrome c oxidoreductase. Biochim. Biophys. Acta, **73**, 550-564 (1963)
[4] Ohshima, T.; Ohshima, M.; Drews, G.: Purification and partial characterization of the soluble NADH dehydrogenase from a phototrophic bacterium Rhodopseudomonas capsulata. Z. Naturforsch. C, **39**, 68-72 (1984)
[5] Ohshima, T.; Drews, G.: Isolation and partial characterization of the membrane-bound NADH dehydrogenase from the phototrophic bacterium Rhodopseudomonas capsulata. Z. Naturforsch. C, **36**, 400-406 (1981)
[6] Wang, C.S.: Human erythrocyte NADH: (acceptor) oxidoreductase. Kinetic properties and competitive substrate inhibition by ferricyanide. Biochim. Biophys. Acta, **616**, 22-29 (1980)
[7] Kitajima, S.; Yasukochi, Y.; Minakami, S.: Purification and properties of human erythrocyte membrane NADH-cytochrome b_5 reductase. Arch. Biochem. Biophys., **210**, 330-339 (1981)
[8] Borneleit, P.; Kleber, H.P.: Purification and properties of the membrane-bound NADH dehydrogenase from hydrocarbon-grown Acinetobacter calcoaceticus. Biochim. Biophys. Acta, **722**, 94-101 (1983)
[9] Hisae, N.; Aizawa, K.; Koyama, N.; Sekiguchi, T.; Nosoh, Y.: Purification and properties of NADH dehydrogenase from an alkalophilic Bacillus. Biochim. Biophys. Acta, **743**, 232-238 (1983)
[10] Imagawa, T.; Nakamura, T.: Properties and kinetics of salt activation of a membrane-bound NADH dehydrogenase from a marine bacterium Photobacterium phosphoreum. J. Biochem., **84**, 547-557 (1978)

[11] Hermans, L.: Purification of mitochondrial NADH dehydrogenase from Drosophila hydei and comparison with the heat-shock polypeptides. Biochim. Biophys. Acta, **567**, 125-134 (1979)

[12] Dancey, G.F.; Shapiro, B.M.: Specific phospholipid requirement for activity of the purified respiratory chain NADH dehydrogenase of Escherichia coli. Biochim. Biophys. Acta, **487**, 368-377 (1977)

[13] Hochstein, L.I.: Studies of a halophilic NADH dehydrogenase. II. Kinetic properties of the enzyme in relation to salt activation. Biochim. Biophys. Acta, **403**, 58-66 (1975)

[14] Schwartz, A.C.; Krause, A.E.: Partial purification and properties of NADH dehydrogenase from Propionibacterium shermanii. Z. Allg. Mikrobiol., **15**, 99-110 (1975)

[15] Nisimoto, Y.; Wilson, E.; Heyl, B.L.; Lambeth, J.D.: NADH dehydrogenase from bovine neutrophil membranes. Purification and properties. J. Biol. Chem., **261**, 285-290 (1986)

[16] Bergsma, J.; van Dongen, M.B.M.; Konings, W.N.: Purification and characterization of NADH dehydrogenase from Bacillus subtilis. Eur. J. Biochem., **128**, 151-157 (1982)

[17] Ramanarayanan, M.; Rao, N.A.; Vaidyanathan, C.S.: Purification & properties of a soluble reduced nicotinamide adenine dinucleotide (NADH) dehydrogenase from Agrobacterium tumefaciens. Indian J. Biochem. Biophys., **8**, 214-218 (1971)

[18] Dancey, G.F.; Levine, A.E.; Shapiro, B.M.: The NADH dehydrogenase of the respiratory chain of Escherichia coli. I. Properties of the membrane-bound enzyme, its solubilization, and purification to near homogeneity. J. Biol. Chem., **251**, 5911-5920 (1976)

[19] Dancey, G.F.; Shapiro, B.M.: The NADH dehydrogenase of the respiratory chain of Escherichia coli. II. Kinetics of the purified enzyme and the effects of antibodies elicited against it on membrane-bound and free enzyme. J. Biol. Chem., **251**, 5921-5928 (1976)

[20] Marques, F.; Bicho, M.P.: Activation of a NADH dehydrogenase in the human erythrocyte by β-adrenergic agonists: possible involvement of a G protein in enzyme activation. Biol. Signals, **6**, 52-61 (1997)

[21] Thiagalingam, S.; Yang, T.: Purification and characterization of NADH dehydrogenase from Bacillus megaterium. Can. J. Microbiol., **39**, 826-833 (1993)

Nitrite reductase 1.6.99.4

1 Nomenclature

EC number
1.6.99.4 (transferred to EC 1.18.1.2)

Recommended name
nitrite reductase

NADH Dehydrogenase (quinone) 1.6.99.5

1 Nomenclature

EC number
1.6.99.5

Systematic name
NADH:(quinone-acceptor) oxidoreductase

Recommended name
NADH dehydrogenase (quinone)

Synonyms
D-diaphorase
DPNH-menadione reductase
NADH-quinone oxidoreductase
NUO1
NUO10
NUO11
NUO12
NUO13
NUO14
NUO2
NUO3/NUO4
NUO5
NUO6
NUO7
NUO8
NUO9
dehydrogenase, reduced nicotinamide adenine dinucleotide (quinone)

CAS registry number
37256-36-3

2 Source Organism

<1> *Sus scrofa* (hog [1]) [1]
<2> *Zea mays* [2]
<3> *Cucurbita pepo* (var. melopepo, zucchini [2]) [2]
<4> *Aphanocapsa* [3]
<5> *Rattus norvegicus* [4, 7]
<6> *Pediococcus halophilus* [5]
<7> *Escherichia coli* [6]
<8> *Gloeophyllum trabeum* [8]

3 Reaction and Specificity

Catalyzed reaction

NADH + H^+ + acceptor = NAD^+ + reduced acceptor

Reaction type

oxidation
redox reaction
reduction

Natural substrates and products

S NADH + 2,5-dimethoxy-1,4-benzoquinone <8> (<8> 2,5-dimethoxy-1,4-benzoquinone produced by Gleophyllum trabeum [8]) (Reversibility: <8> ? [8]) [8]
P NAD^+ + 2,5-dimethoxy-1,4-benzoquinol <8> [8]
S NADH + 4,5-dimethoxy-1,2-benzoquinone <8> (<8> 4,5-dimethoxy-1,2-benzoquinone produced by Gleophyllum trabeum [8]) (Reversibility: <8> ? [8]) [8]
P NAD^+ + 4,5-dimethoxy-1,2-benzoquinol <8> [8]
S Additional information <8> (<8> the enzyme may function in quinone detoxification, it is also hypothesized that the fungus has recruited it to drive extracellular oxyradical production [8]) [8]
P ?

Substrates and products

S NADH + 1,2-naphthoquinone <1> (Reversibility: ? <1> [1]) [1]
P NAD^+ + 1,2-naphthoquinol <1> [1]
S NADH + 1,4-benzoquinone <8> (Reversibility: ? <8> [8]) [8]
P NAD^+ + 1,4-benzoquinol <8> [8]
S NADH + 1,4-naphthoquinone <1, 6> (<6> 1,4-naphthoquinone is one of the best acceptors [5]) (Reversibility: ? <1, 6> [1, 5]) [1, 5]
P NAD^+ + 1,4-naphthoquinol <1, 6> [1, 5]
S NADH + 2,5-dimethoxy-1,4-benzoquinone <8> (<8> 2,5-dimethoxy-1,4-benzoquinone produced by Gleophyllum trabeum [8]) (Reversibility: ? <8> [8]) [8]
P NAD^+ + 2,5-dimethoxy-1,4-benzoquinol <8> [8]
S NADH + 2,5-toluquinone <6> (Reversibility: ? <6> [5]) [5]
P NAD^+ + 2,5-toluquinol <6> [5]
S NADH + 2,6-dichlorophenolindophenol <6, 7> (<7> 2,6-dichlorophenol-indophenol is a poor electron acceptor [6]) (Reversibility: ? <6, 7> [5, 6]) [5, 6]
P NAD^+ + ? <6, 7> [5, 6]
S NADH + 2,6-dimethoxy-1,4-benzoquinone <8> (Reversibility: ? <8> [8]) [8]
P NAD^+ + 2,6-dimethoxy-1,4-benzoquinol <8> [8]
S NADH + 2-methoxy-1,4-benzoquinone <8> (Reversibility: ? <8> [8]) [8]
P NAD^+ + 2-methoxy-1,4-benzoquinol <8> [8]

S NADH + 4,5-dimethoxy-1,2-benzoquinone <8> (<8> 4,5-dimethoxy-1,2-benzoquinone produced by Gleophyllum trabeum [8]) (Reversibility: ? <8> [8]) [8]
P NAD^+ + 4,5-dimethoxy-1,2-benzoquinol <8> [8]
S NADH + benzoquinone <1, 4, 6> (<1> benzoquinone is the best acceptor [1]) (Reversibility: ? <1, 4, 6> [1, 3, 5]) [1, 3, 5]
P NAD^+ + benzoquinol <1, 4, 6> [1, 3, 5]
S NADH + duroquinone <3> (Reversibility: ? <3> [2]) [2]
P NAD^+ + duroquinol <3> [2]
S NADH + ferricyanide <6, 7> (Reversibility: ? <6, 7> [5, 6]) [5, 6]
P NAD^+ + ferrocyanide <6, 7> [5, 6]
S NADH + menadione <1, 5-7> (<1,7> menadione is a poor electron acceptor [1,6]; <6> menadione is one of the best acceptors [5]) (Reversibility: ? <1, 5-7> [1, 4-7]) [1, 4-7]
P NAD^+ + menadiol <1, 5-7> [1, 4-7]
S NADH + methyl-*p*-benzoquinone <1> (Reversibility: ? <1> [1]) [1]
P NAD^+ + methyl-*p*-benzoquinol <1> [1]
S NADH + paraquat <5> (<5> the enzyme is a novel oxidation-reduction system participating in paraquat toxicity [7]) (Reversibility: ? <5> [7]) [7]
P NAD^+ + ? <5> [7]
S NADH + ubiquinone-1 <3, 7> (Reversibility: ? <3, 7> [2, 6]) [2, 6]
P NAD^+ + ubiquinol-1 <3, 7> [2, 6]
S Additional information <6> (<6> the enzyme can not utilize free flavins, cytochrome c or molecular oxygen as an electron acceptor [5]) [5]
P ?

Inhibitors

2,4-dinitrophenol <1> (<1> 52% inhibition at 0.16 mM and 100% inhibition at 0.33 mM [1]; <5> not inhibitory [4]) [1]
AMP <1> (<1> at high concentrations [1]) [1]
$HgCl_2$ <6> (<6> completely inhibits at 0.1 mM [5]) [5]
Liponox DCH <7> (<7> inhibitory at concentrations higher than 0.02% [6]) [6]
dicoumarol <5> (<5> at high concentrations [7]) [7]
o-phenanthroline <4> [3]
p-chloromercuribenzoate <6> (<6> completely inhibits at 0.1 mM [5]) [5]
p-hydroxymercuribenzoate <5> (<5> less than 40% inhibition at 0.001 mM, 100% inhibition at 0.5 mM [7]) [7]
rotenone <4> (<5> not inhibitory [7]) [3]
thenoyltrifluoroacetone <4> [3]

Cofactors/prosthetic groups

FAD <7> (<7> 2.97 nmol per mg of enzyme [6]) [6]
FMN <8> (<8> 0.9 mol of FMN per mol of subunit [8]) [8]
NADH <1-8> (<1,4,6,8> specific for [1,3,5,8]) [1-8]

Activating compounds

Liponox DCH <7> (<7> required at 0.02% for optimum activity [6]) [6]
Additional information <1> (<1> no stimulation with FAD or FMN [1]) [1]

Metals, ions

Mg^{2+} <7> (<7> required for activity [6]) [6]

Turnover number (min^{-1})

66000-96000 <8> (2,5-dimethoxy-1,4-benzoquinone) [8]
66000-96000 <8> (4,5-dimethoxy-1,2-benzoquinone) [8]

Specific activity (U/mg)

0.035 <4> (<4> soluble fraction with 0.1 mM *o*-phenanthroline [3]) [3]
0.085 <4> (<4> soluble fraction with 0.1 mM rotenone [3]) [3]
0.123 <4> (<4> soluble fraction with 0.1 mM thenoyltrifluoroacetone [3]) [3]
0.682 <4> (<4> soluble fraction control [3]) [3]
7.4 <6> (<6> partially purified enzyme [5]) [5]
Additional information <1, 7, 8> [1, 6, 8]

K_m-Value (mM)

0.005-0.007 <8> (2,5-dimethoxy-1,4-benzoquinone) [8]
0.005-0.007 <8> (4,5-dimethoxy-1,2-benzoquinone) [8]
0.012 <1> (NADH) [1]
0.0438 <7> (NADH, <7> enzyme activated before assay, electron acceptor: ferricyanide [6]) [6]
0.0452 <7> (NADH, <7> electron acceptor: ferricyanide [6]) [6]
0.0483 <7> (NADH, <7> enzyme activated before assay, electron acceptor: ubiquinone-1 [6]) [6]
0.085-0.09 <8> (NADH) [8]
0.0965 <7> (NADH, <7> electron acceptor: ubiquinone-1 [6]) [6]
0.1 <1> (menadione) [1]

pH-Optimum

5.5 <1> (<1> 2 optima: pH 5.5 and pH 7.5 [1]) [1]
6 <6> [5]
7.5 <1> (<1> 2 optima: pH 5.5 and pH 7.5 [1]) [1]

4 Enzyme Structure

Molecular weight

50000 <8> (<8> gel filtration [8]) [8]

Subunits

dimer <8> (<8> 2 * 22000, SDS-PAGE [8]) [8]

5 Isolation/Preparation/Mutation/Application

Source/tissue

hypocotyl <3> [2]
liver <1, 5> [1, 4, 7]
whole plant <2, 3> (etiolated cell) [2]

Localization

mitochondrion <5> (<5> inner membrane [4]; <5> outer membrane [7]) [4, 7]
soluble <1, 4, 8> (<8> 100000 g supernatant of a day 7 mycelial extract [8]) [1, 3, 8]
Additional information <3> (<3> extramitochondrial membrane [2]) [2]

Purification

<1> (using ammonium sulfate treatment, pH adjustment to 5, centrifugation, adjustement of the supernatant solution to pH 6, protamine-$CHCl_3$ treatment, column chromatography on DEAE-Sephadex and second ammonium sulfate treatment [1]) [1]
<4> (using 1% Triton X-100 treatment, ammonium sulfate precipitation, sucrose density gradient and column chromatography on DEAE-cellulose [3]) [3]
<6> (partial, using dialysis and column chromatography on DEAE-Sepharose and Phenyl-Sepharose [5]) [5]
<7> (using column chromatography on DEAE-Sephacel and DEAE-5PW [6]) [6]
<8> (using column chromatography on Phenyl-Sepharose, hydroxyapatite, DEAE-Sepharose and Superdex 200 [8]) [8]

Cloning

<8> (characterization of the gene that encodes the enzyme, the gene has substancial similarity to NAD(P)H:quinone reductase genes from other fungi [8]) [8]

6 Stability

Temperature stability

62 <2> (<2> irreversibly inactictivated after 1 minute [2]) [2]

Storage stability

<8>, -20°C, phosphate buffer, pH 6, loss of activity within several days [8]
<8>, -20°C, sodium citrate buffer, pH 6, 20% v/v ethylene glycol, stable [8]
<8>, -20°C, sodium citrate buffer, pH 6, loss of activity within several days [8]

References

[1] Koli, A.K.; Yearby, C.; Scott, W.; Donaldson, K.O.: Purification and properties of three separate menadione reductases from hog liver. J. Biol. Chem., **244**, 621-629 (1969)

[2] Pupillo, P.; de Luca, L.: Pyridine nucleotide-linked dehydrogenases (quinone dependent) in plasma membrane and endoplasmatic reticulum of plant cells. Dev. Plant Biol., **7**, 321-328 (1982)

[3] Sandmann, G.; Malkin, R.: NADH and NADPH dehydrogenases from the blue-green alga, Aphanocapsa. Arch. Microbiol., **136**, 49-53 (1983)

[4] Frei, B.; Winterhalter, K.H.; Richter, C.: Menadione-(2-methyl-1,4-naphtoquinone-) dependent enzymatic redox cycling and calcium release by mitochondria. Biochemistry, **25**, 4438-4443 (1986)

[5] Kanbe, C.; Uchida, K.: NADH dehydrogenase activity of Pediococcus halophilus as a factor determining its reducing force. Agric. Biol. Chem., **51**, 507-514 (1987)

[6] Hayashi, M.; Miyoshi, T.; Takashina, S.; Unemoto, T.: Purification of NADH-ferricyanide dehydrogenase and NADH-quinone reductase from Escherichia coli membranes and their roles in the respiratory chain. Biochim. Biophys. Acta, **977**, 62-69 (1989)

[7] Shimada, H.; Hirai, K.; Simamura, E.; Pan, J.: Mitochondrial NADH-quinone oxidoreductase of the outer membrane is responsible for paraquat cytotoxicity in rat livers. Arch. Biochem. Biophys., **351**, 75-81 (1998)

[8] Jensen, K.A.; Ryan, Z.C.; Vanden Wymelenberg, A.; Cullen, D.; Hammel, K.E.: An NADH:quinone oxidoreductase active during biodegradation by the brown-rot basidiomycete Gloeophyllum trabeum. Appl. Environ. Microbiol., **68**, 2699-2703 (2002)

NADPH Dehydrogenase (quinone) 1.6.99.6

1 Nomenclature

EC number
1.6.99.6

Systematic name
NADPH:(quinone-acceptor) oxidoreductase

Recommended name
NADPH dehydrogenase (quinone)

Synonyms
NADPH oxidase
dehydrogenase, reduced nicotinamide adenine dinucleotide phosphate (quinone)

CAS registry number
37256-37-4

2 Source Organism

<1> *Sus scrofa* (hog [1]) [1, 5]
<2> *Zea mays* [2]
<3> *Cucurbita pepo* [2]
<4> *Bos taurus* (multienzyme complex [3]) [3, 5]
<5> *Homo sapiens* (multienzyme complex [4]; multicomponent NADPH-oxidase consisting of a heterodimeric flavocytochrome b_{558}, localized in the membrane and at least two cytosolic components p47-phox and p67-phox, enzyme plays a role in chronic granulomatous diseases, enzyme system is activated upon exposure to the appropriate stimuli leading to assembly in the plasma membrane of membrane-bound and cytosolic components [5]) [4, 5, 6, 7, 8]
<6> *Cavia porcellus* [5, 7]

3 Reaction and Specificity

Catalyzed reaction
NADPH + H^+ + acceptor = $NADP^+$ + reduced acceptor (<5> model of electron transport from NADPH to the terminal electron acceptor O_2 [5])

Reaction type

oxidation
redox reaction
reduction

Natural substrates and products

S NADPH + O_2 <1, 4, 5, 6> (<5> enzyme is stimulated by phagocytizable particles [4]; <5> part of defense mechanism against a wide variety of bacteria [5]) (Reversibility: ? <1, 4, 5, 6> [3, 4, 5]) [3, 4, 5]
P $NADP^+$ + O_2^- <1, 4, 5, 6> [3, 4, 5]

Substrates and products

S NADPH + 1,4-naphthoquinone <1> (Reversibility: ? <1> [1]) [1]
P $NADP^+$ + 1,4-naphthohydroquinone <1> [1]
S NADPH + O_2 <1, 4, 5, 6> (<5> enzyme is stimulated by phagocytizable particles [4]) (Reversibility: ? <1, 4, 5, 6> [3, 4, 5]) [3, 4, 5]
P $NADP^+$ + O_2^- <1, 4, 5, 6> [3, 4, 5]
S NADPH + duroquinone <3> (Reversibility: ? <3> [2]) [2]
P $NADP^+$ + durohydroquinone <3> [2]
S NADPH + menadione <1> (Reversibility: ? <1> [1]) [1]
P $NADP^+$ + menadiole <1> [1]
S NADPH + methyl-*p*-benzoquinone <1> (Reversibility: ? <1> [1]) [1]
P $NADP^+$ + methyl-*p*-benzohydroquinone <1> [1]
S NADPH + *p*-benzoquinone <1> (Reversibility: ? <1> [1]) [1]
P $NADP^+$ + *p*-benzohydroquinone <1> [1]
S NADPH + ubiquinone-1 <3> (Reversibility: ? <3> [2]) [2]
P $NADP^+$ + ubihydroquinone-1 <3> [2]

Inhibitors

15-cis-(4-propyl-cyclohexyl)-16,17,18,19,20-pentanor-9-deoxy-9α,6-nitrilo-prostaglandin F1 methyl ester <5> (<5> 0.021 mM, 50% inhibition of the enzyme in neutrophils possible due to scavenging of O_2^-, inhibition of SDS-induced activation in cell free extracts, 0.22 mM, 50% inhibition [4]) [4]
4-(2-aminoethyl)benzenesulfonylfluoride <5, 6> (<5> 0.206 mM, 50% inhibition, noncompetitive vs. NADPH, competitive vs. p47-phox in the concentration range of 0.5-2.0 mM [7]) [7]
4-(amidino)benzenesulfonylfluoride <5> (<5> 1.49 mM, 50% inhibition [7]) [7]
5-(E)-6,9-deoxa-6,9α-methylene-15-cyclopentyl-16,17,18,19,20-pentanor-prostaglanin I_2 <5> (<5> inhibition of sodiumdodecylsulfate-induced activation in cell free extracts, 0.17 mM, 50% inhibition [4]) [4]
dicoumarol <1> [1]
folic acid derivatives <1> [1]
lucensomycin <4> (<4> 0.02 mM, 50% inhibition [3]) [3]
prostaglandin E_1 <5> (<5> inhibition of sodiumdodecylsulfate-induced activation in cell free extracts, 0.044 mM, 50% inhibition [4]) [4]
Additional information <1> (<1> not inhibited by 2,4-dinitrophenol [1]) [1]

Cofactors/prosthetic groups

FAD <1, 4, 5, 6> (<1> activity depends on added FAD [1]; <5> cytochrome b_{558} of the multienzyme complex may contain FAD and heme [5]) [1, 5]

Activating compounds

GTP-γS <4> (<4> optimal concentration approx. 0.015 mM [3]) [3]
arachidonic acid <4> (<4> maximal enzyme activity in the presence of 0.25-0.35 mM, inhibition above [3]) [3]
lithium dodecylsulfate <5> [7]
protein kinase C <5> [5]
sodium dodecylsulfate <5> [4]

Turnover number (min^{-1})

840 <5> (O_2, <5> turnover in membranes of Epstein-Barr virus immortalized B lymphocytes [8]) [8]
6240 <5> (O_2, <5> turnover in neutrophil membranes [8]) [8]
10620 <5> (O_2, <5> turnover of purified cytochrome b_{558} from neutrophils reconstituted with 0.3 mM neutrophil cytosol, 0.04 mM GTPγS, 5 mM $MgCl_2$ and 0.01 mM FAD [8]) [8]
11220 <5> (O_2, <5> turnover of purified cytochrome b_{558} from Epstein-Barr virus immotalized B lymphcytes reconstituted with 0.3 mM neutrophil cytosol, 0.04 mM GTPγS, 5 mM $MgCl_2$ and 0.01 mM FAD [8]) [8]

Specific activity (U/mg)

0.015 <4> (<4> activity in membranes after 1 h at 25°C [3]) [3]
0.033 <4> (<4> activity in freshly prepared membranes at 25°C [3]) [3]
1.097 <5> (<5> activity of purified cytochrome b_{558} reconstituted with neutrophil membrane phospholipids, activation with arachidonic acid [6]) [6]
1.258 <5> (<5> activity of purified cytochrome b_{558} reconstituted with neutrophil membrane phospholipids, activation with sodiumdodecylsulfate [6]) [6]

K_m-Value (mM)

0.011 <1> (menadione) [1]
0.04 <5> (NADPH) [4]
0.046 <1> (NADPH) [1]

pH-Optimum

6.5-7.5 <1> [1]

4 Enzyme Structure

Subunits

? <5> (<5> x * 47000-48000, p47-phox, cytosolic component of the multicomponent complex, SDS-PAGE [5]) [5]
? <5> (<5> x * 65000-67000, p67-phox, cytosolic component of the multicomponent complex, SDS-PAGE [5]) [5]

dimer <5> (<5> 1 * 22000-23000 + 1 * 76000-91000, subunits of the membrane component of the enzyme complex, cytochrome b_{558}, SDS-PAGE [5]) [5]

Posttranslational modification

glycoprotein <5> (<5> cytochrome b_{558} subunit gp91-phox of the multicomponent complex is heavily glycosylated [5]; <5> 3 putative sites for N-linked glycosylation at amino-acid positions 131, 148 and 239 [8]; <5> cytochrome b_{558} isolated from Epstein-Barr virus immortalized B lymphocytes is higher glycosylated than that isolated from neutrophils [8]) [5, 8]
side-chain modification <5> (<5> phosphorylation of cytosolic component p47-phox [5]) [5]

5 Isolation/Preparation/Mutation/Application

Source/tissue

blood <5> (<5> Epstein-Barr virus immortalized B lymphocytes [8]) [8]
liver <1> [1]
neutrophil <4> (<4> resting neutrophils [3]) [3, 5, 6, 7]
plant <2, 3> (<2, 3> etiolated [2]) [2]

Localization

membrane <1, 2, 4, 5, 6> (<5> subunit gp91-phox of cytochrome b_{558} of the multicomponent complex is an integral membrane protein with several membrane-spanning regions [5]; <2> mitochondrial outer membrane [2]) [2, 3, 5]

Purification

<1> [1]
<5> (cytochrome b_{558} and cytosolic components p47-phox and p67-phox [5]; cytochrome b_{558} from neutrophils [7, 8]; cytochrome b_{558} from Epstein-Barr virus immortalized B lymphocytes [8]) [5, 6, 7, 8]

Cloning

<5> (cytosolic components p67-phox and p47-phox of the multicomponent NADPH oxidase [6]; expression of p47-phox and p67-phox in Sf9 cells [7]) [6, 7]

6 Stability

Temperature stability

25 <4> (<4> activity in membranes, approx. 50% activity are lost after 1 h, freeze/thawing accelerates inactivation, loss of 90% activity after 1 h [3]) [3]

References

[1] Koli, A.K.; Yearby, C.; Scott, W.; Donaldson, K.O.: Purification and properties of three separate menadione reductases from hog liver. J. Biol. Chem., **244**, 621-629 (1969)

[2] Pupillo, P.; de Luca, L.: Pyridine nucleotide-linked dehydrogenases (quinone dependent) in plasma membrane and endoplasmatic reticulum of plant cells. Dev. Plant Biol., **7**, 321-328 (1982)

[3] Capuozzo, E.; Crifo, C.; Salerno, C.: Polyene antibiotics inhibit superoxide-producing NADPH oxidase in a polymorphonuclear cell-free system. Biochem. Pharmacol., **45**, 766-768 (1993)

[4] Umeki, S.: Prostaglandin E1 and analogs of prostacyclin influencing superoxide production by the human neutrophil NADPH oxidase system. Int. J. Biochem., **26**, 1003-1008 (1994)

[5] Umeki, S.: Mechanisms for the activation/electron transfer of neutrophil NADPH-oxidase complex and molecular pathology of chronic granulomatous disease. Annu. Hematol., **68**, 267-277 (1994)

[6] Batot, G.; Martel, C.; Capdeville, N.; Wientjes, F.; Morel, F.: Characterization of neutrophil NADPH oxidase activity reconstituted in a cell-free assay using specific monoclonal antibodies raised against cytochrome b_{558}. Eur. J. Biochem., **234**, 208-215 (1995)

[7] Diatchuk, V.; Lotan, O.; Koshkin, V.; Wikstroem, P.; Pick, E.: Inhibition of NADPH oxidase activation by 4-(2-aminoethyl)-benzenesulfonyl fluoride and related compounds. J. Biol. Chem., **272**, 13292-13301 (1997)

[8] Paclet, M.H.; Coleman, A.W.; Burritt, J.; Morel, F.: NADPH oxidase of Epstein-Barr-virus immortalized B lymphocytes. Effect of cytochrome b_{558} glycosylation. Eur. J. Biochem., **268**, 5197-5208 (2001)

Dihydropteridine reductase 1.6.99.7

1 Nomenclature

EC number

1.6.99.7 (transferred to EC 1.5.1.34)

Recommended name

dihydropteridine reductase

Aquacobalamin reductase 1.6.99.8

1 Nomenclature

EC number
1.6.99.8 (transferred to EC 1.16.1.3)

Recommended name
aquacobalamin reductase

Cob(II)alamin reductase 1.6.99.9

1 Nomenclature

EC number
1.6.99.9 (transferred to EC 1.16.1.4)

Recommended name
cob(II)alamin reductase

Dihydropteridine reductase (NADH) 1.6.99.10

1 Nomenclature

EC number
1.6.99.10 (deleted, included in EC 1.6.99.7)

Recommended name
dihydropteridine reductase (NADH)

Aquacobalamin reductase (NADPH) 1.6.99.11

1 Nomenclature

EC number
1.6.99.11 (transferred to EC 1.6.99.11)

Recommended name
aquacobalamin reductase (NADPH)

Cyanocobalamin reductase (NADPH, cyanide-eliminating) 1.6.99.12

1 Nomenclature

EC number

1.6.99.12 (transferred to EC 1.16.1.6)

Recommended name

cyanocobalamin reductase (NADPH, cyanide-eliminating)

Ferric-chelate reductase 1.6.99.13

1 Nomenclature

EC number
1.6.99.13 (transferred to EC 1.16.1.7)

Recommended name
ferric-chelate reductase

Nitrate reductase (NADH) 1.7.1.1

1 Nomenclature

EC number
1.7.1.1

Systematic name
nitrite:NAD^+ oxidoreductase

Recommended name
nitrate reductase (NADH)

Synonyms
EC 1.6.6.1 (formerly)
NADH-dependent nitrate reductase
NADH-nitrate reductase
NADH:nitrate oxidoreductase
NR1
NR2
assimilatory NADH:nitrate reductase
assimilatory nitrate reductase
reductase, nitrate

CAS registry number
9013-03-0

2 Source Organism

<1> *Hordeum vulgare* (L. cv Steptoe [55]) [1, 5, 6, 40, 52, 55, 68, 72]
<2> *Chlorella vulgaris* [2, 3, 7, 13, 28, 41, 44, 45, 65]
<3> *Spinacia oleracea* (recombinant flavin domain [82]) [4, 10, 19, 20, 26, 29, 37, 38, 39, 47, 49, 51, 53, 58, 64, 66, 69, 70, 82, 83, 84]
<4> *Cucurbita pepo* [8, 12, 58]
<5> *Triticum aestivum* (L. cv. Olympic [9]; L. cv. Bindawarra [48]) [9, 22, 30, 33, 34, 48, 58]
<6> *Gracilaria tenuistipitata* [87]
<7> *Zea mays* (cv W64A * W182E [36, 45, 67]) [12, 22, 25, 32, 36, 43, 45, 50, 53, 58, 60, 61, 67, 74]
<8> *Oryza sativa* [14, 46, 56]
<9> *Agrostemma githago* [15]
<10> *Chlorella pyrenoidosa* [16, 57]
<11> *Chlorella sp.* (Berlin strain [17]) [17, 82]

<12> *Cucurbita maxima* [18, 21, 61, 71, 81]
<13> *Lycopersicon esculentum* [23, 58]
<14> *Nicotiana tabacum* [24, 35]
<15> *Ankistrodesmus braunii* [27]
<16> *Thalassiosira pseudonana* [31]
<17> *Cucurbita sp.* [42]
<18> *Porphyra yezoensis* [86]
<19> *Pisum arvense* [50]
<20> *Glycine max* (wild-type and nr1-mutant plants [78]) [11, 53, 76, 78, 85]
<21> *Suaeda maritima* (var. macrocarpa [54]) [54]
<22> *Chenopodium album* [58]
<23> *Amaranthus hybridus* [58]
<24> *Brassica oleracea* (italica [58]) [58]
<25> *Raphanus sativus* [58]
<26> *Cucumis sativus* [58]
<27> *Capsicum frutescens* [58]
<28> *Rhodobacter capsulatus* (E1F1 [80]) [80]
<29> *Lemna minor* [58]
<30> *Chlorella fusca* [59]
<31> *Skeletonema costatum* [61, 89]
<32> *Skeletonema tropicum* [61]
<33> *Thalassiosira antarctica* [61]
<34> *Phaeocystis antarctica* [61]
<35> *Dunaliella tertiolecta* [61]
<36> *Amaranthus tricolor* [62]
<37> *Sinapis alba* [63]
<38> *Nicotiana plumbaginifolia* [73]
<39> *Bradyrhizobium sp.* (strain 750 and IM-43B [75]) [75]
<40> *Gossypium hirsutum* [77]
<41> *Arabidopsis thaliana* [79, 88]

3 Reaction and Specificity

Catalyzed reaction

nitrite + NAD^+ + H_2O = nitrate + NADH + H^+ (<2>, random bi bi mechanism [3]; <3>, hexa uni ping pong mechanism [4]; <4>, two-site ping-pong mechanism [12]; <41>, catalytic rate is determined by a combination of rates with no overall rate limiting individual process [79])

Reaction type

oxidation
redox reaction
reduction

Natural substrates and products

S nitrate + NADH <3> (<3>, the enzyme catalyzes the regulated and rate-limiting step in the utilization of inorganic nitrogen by higher plants [84]) (Reversibility: ? <3> [84]) [84]

P nitrite + NAD^+ + H_2O <3> [84]

S Additional information <6, 7, 9, 12, 14, 37, 39> (<9>, induction by nitrate and cytokinin [15]; <12>, the enzyme could have a role in iron assimilation [21]; <14>, key enzyme of nitrogen metabolism in higher plants [24]; <7>, induced by nitrate [25]; <37>, effect of nitrate, ammonium, light and a plastidic factor on the appearance of multiple forms of nitrate reductase [63]; <39>, constitutive nitrate reductase [75]; <6>, in extracts of algae grown under either constant dimlight or light-dark cycle, the activity of nitrate reductase exhibits a daily rhythm, peaking at midday phase as does photosynthesis [87]) [15, 21, 24, 25, 63, 75, 87]

P ?

Substrates and products

S 2,6-dichlorophenolindophenol + NADH <1, 3, 11, 28> (Reversibility: ? <1, 3, 11, 28> [17, 37, 72, 80]) [17, 37, 72, 80]

P ? + NAD^+

S $FADH_2$ + nitrate <1, 7, 10, 20> (Reversibility: ? <1, 7, 10, 20> [50, 57, 72, 76]) [50, 57, 72, 76]

P FAD + nitrite

S $FMNH_2$ + nitrate <1, 3, 8, 18> (Reversibility: ? <1, 8, 18> [5, 14, 51, 64, 72, 86]) [5, 14, 51, 64, 72, 86]

P FMN + nitrite

S NADH + BrO_3^- <3> (Reversibility: ? <3> [37]) [37]

P NAD^+ + ?

S NADH + ClO_3^- <3> (Reversibility: ? <3> [37]) [37]

P NAD^+ + ?

S NADH + IO_3^- <3> (Reversibility: ? <3> [37]) [37]

P NAD^+ + ?

S benzoquinone + NADH <4> (Reversibility: ? <4> [8]) [8]

P benzoquinol + NAD^+

S bromophenol blue + nitrate <17> (Reversibility: ? <17> [42]) [42]

P ? + nitrite

S ferricyanide + NADH <1, 2, 3, 4, 11> (Reversibility: ? <1, 2, 3, 4, 11> [8, 17, 37, 41, 44, 65, 72, 82, 84, 86]) [8, 17, 37, 41, 44, 65, 72, 82, 84, 86]

P ferrocyanide + NAD^+

S ferricytochrome c + NADH <1, 2, 3, 7, 8, 11, 18, 28> (Reversibility: ? <1, 2, 3, 7, 8, 11, 18, 28> [17, 37, 41, 44, 46, 50, 65, 72, 80, 86]) [4, 17, 37, 41, 44, 46, 50, 65, 72, 80, 86]

P ferrocytochrome c + NAD^+

S flavin hydroquinone + nitrate <3> (Reversibility: ? <3> [37]) [37]

P ? + nitrite

S menadione + NADH <4> (Reversibility: ? <4> [8]) [8]

P reduced menadione + NAD^+

S methylene blue + NADH <4> (Reversibility: ? <4> [8]) [8]
P reduced methylene blue + NAD^+
S nitrate + NADH <1-41> (Reversibility: ? <1-41> [1-89]) [1-89]
P nitrite + NAD^+ + H_2O
S nitrate + NADPH <20> (<20>, much more active with NADH [78]; <1, 4, 18>, no activity [5, 8, 86]) (Reversibility: ? <20> [78]) [78]
P nitrite + $NADP^+$ + H_2O
S reduced bromophenol blue + nitrate <28> (Reversibility: ? <28> [80]) [80]
P bromophenol blue + nitrite
S reduced flavin + nitrate <2> (Reversibility: ? <2> [44]) [44]
P ? + nitrite
S reduced methyl viologen + nitrate <1, 2, 3, 5, 8, 10, 16, 18> (Reversibility: ? <1, 2, 3, 5, 8, 10, 16, 18> [5, 9, 26, 31, 37, 44, 56, 57, 65, 72, 80, 86]) [5, 9, 26, 31, 37, 44, 56, 57, 65, 72, 80, 86]
P methyl viologen + nitrite
S Additional information <2, 3, 4> (<4>, no activity with coenzyme Q_{10} and plastoquinone [8]; <2>, partial activities reside on functionally independent domains [44]; <3>, NADH-diaphorase activity [51,64]; <3>, the isolated flavin-containing domain is capable of reducing cytochrome b_5 directly [83]) [8, 44, 51, 64, 83]
P ?

Inhibitors

2-(4'-maleimidylanilino)naphthalene-6-sulfonic acid <3, 11> [82]
ADP <1, 3> (<1>, competitive with ADP [72]; <3>, inhibition of the recombinant FAD domain [83]) [72, 83]
AMP <3> (<3>, inhibition of the recombinant FAD domain [83]) [83]
Ca^{2+} <12> (<12>, 5 mM, 90% inhibition of the low activity form, no activity of high activity form, inhibition is prevented by low concentrations of thiol compounds [81]) [81]
Cl^- <3> (<3>, alters the observed Mo(V) lineshape, mixed-type inhibitor, decreases both NADH:nitrate reductase and reduced methyl viologen:nitrate reductase activities [26]) [26]
Co^{2+} <1> (<1>, 1 mM, strong inhibition [72]) [72]
Cu^{2+} <1, 12> (<12>, potent inhibitor, inhibition can be abolished by prior chelation of the metal by EDTA [18]; <1>, 1 mM, strong inhibition [72]) [18, 72]
Fe^{2+} <12> (<12>, potent inhibitor, inhibition can be abolished by prior chelation of the metal by EDTA [18]) [18]
Fe^{3+} <12> (<12>, less potent inhibitor, effect cannot reversed by EDTA [18]) [18]
Mg^{2+} <12> (<12>, 5 mM, 70% inhibition of the low activity form, no inhibition of high activity form, inhibition is prevented by low concentrations of thiol compounds [81]; <31>, no inhibition [89]) [81]
Mn^{2+} <1, 12> (<1>, 1 mM, 27% inhibition [72]; <12>, less potent inhibitor, effect cannot be reversed by EDTA [18]) [18, 72]

MoO_4^{2-} <1> (<1>, 1 mM, strong inhibition [72]) [72]
NAD^+ <2, 3, 4, 5, 7> (<2>, product inhibition [3]; <4, 7>, competitive towards NADH [12]; <5>, inactivation in presence of thiol compounds is enhanced by cyanide ions and can be reversed by ferricyanide [30]; <3>, inhibition of the recombinant FAD domain [83]) [3, 4, 12, 30, 82, 83]
NADH <3, 5> (<3>, inactivation in a biphasic reaction, immune to inactivation during turnover with nitrate [10]; <5>, conversion of the enzyme to a reduced inactive form, by preincubation with NADH, in absence of nitrate, occurs in presence of either dithiothreitol and/or FAD but not with cysteine. Pretreatment with NADH alone does not inactivate, a nucleophilic agent, i.e. cyanide or superoxide is necessary to inhibit electron transfer by the enzyme to nitrate [33]; <5>, 0.01 mM NADH, in absence of nitrate, 50% loss of activity after 30 min, 0.05 mM nitrate prevents inactivation, 0.001 mM cyanide enhances degree of inactivation. Rapid reactivation after treatment with 0.3 mM ferricyanide or exposure to light, 230 mWatt per cm^2, plus 0.02 mM flavin adenine dinucleotide [48]; <3>, nitrate protects [49]; <3>, *p*-hydroxymercuribenzoate causes the appearance of an FAD-requirement for inactivation by NADH of $FMNH_2$-nitrate reductase [64]) [10, 33, 48, 49, 54]
NADPH <3> (<3>, nitrate protects [49]) [49]
NEM <3, 11, 18> (<18>, inhibition of full and NADH-utilizing partial activities [86]) [82, 86]
NMN <3> (<3>, inhibition of the recombinant FAD domain [83]) [83]
NO_2^- <1> (<1>, 10 mM, 65% inhibition [72]) [72]
NaCl <21> (<21>, uncompetitive for nitrate and NADH [54]) [54]
O-methoxylamine <3> [29]
PCMB <3, 7, 11> (<3>, inactivation is concentration independent [82]) [50, 82]
Pb^{2+} <12> (<12>, less potent inhibitor, effect cannot reversed by EDTA [18]) [18]
VO_3^- <1> (<1>, 1 mM, strong inhibition [72]) [72]
Zn^{2+} <1, 12> (<12>, potent inhibitor, inhibition can be abolished by prior chelation of the metal by EDTA [18]; <1>, 1 mM, strong inhibition [72]) [18, 72]
adenine <3> (<3>, inhibition of the recombinant FAD domain [83]) [83]
adenosine <3> (<3>, inhibition of the recombinant FAD domain [83]) [83]
adenosine 5'-diphosphoribose <2, 3> (<2>, dead-end inhibition [3]; <3>, inhibition of the recombinant FAD domain [83]) [3, 83]
aminooxyacetate <3> [29]
azide <4, 16, 20, 18> (<4, 16, 20>, sodium azide [8, 31, 76]; <4>, 1 mM sodium azide, 90% inhibition of nitrate reduction [8]; <18>, inhibition of full and nitrate-reducing partial activities [86]) [8, 17, 31, 76, 86]
bromophenol blue <17> (<17>, noncompetitive versus NADH [42]) [42]
carbamoyl phosphate <1> (<1>, competitive [72]) [72]
cyanate <11> [17]
cyanide <3, 4, 7, 20, 18> (<3>, mechanism of reactivation of cyanide-inactivated nitrate reductase by flavins in light [20]; <7>, inactivation by simultaneous presence of NADH and low concentrations of cyanide, reactivation by

incubation with ferricyanide or by a short exposure to light in the presence of FAD [32]; <3>, reactivation by incubation with oxidant systems after inactivation by treatment with NADH and cyanide [47]; <20>, sodium azide [76]; <18>, inhibition of full and nitrate-reducing partial activities [86]) [12, 20, 32, 47, 76, 86]
dicoumarol <4> (<4>, competitive towards NADH [8]) [8]
dithiothreitol <3> (<3>, rate of inactivation is increased by NAD^+, but not by $NADP^+$ [49]) [49]
ferricyanide <3> (<3>, inhibition of the recombinant FAD domain [83]) [83]
ferrocytochrome c <3> (<3>, inactivation in a biphasic reaction, immune to inactivation during turnover with nitrate [10]) [10]
hydroxylamine <3, 11, 15> (<3>, interacts with reduced cytochrome b_{557} during catalysis of the enzyme [29]; <15>, NO_3^-, cyanate, carbamoyl phosphate and azide protect from inactivation. Photoreactivation in presence of flavins, early inhibition appears to be competitive versus NO_3^- [27]) [17, 27, 29]
menadione <4> (<4>, inhibits nitrate reductase [8]) [8]
methyl methanethiosulfonate <3, 11> (<3>, inactivation is concentration independent [82]) [82]
nicotinamide <3> (<3>, inhibition of the recombinant FAD domain [83]) [83]
nitrite <2, 4, 7, 11> (<2>, product inhibition [3]; <4,7>, competitive towards nitrate [12]) [3, 12, 17]
p-hydroxymercuribenzoate <4, 3, 16, 17, 18> (<17>, inactivation of NADH:-nitrate reductase activity, no loss of bromphenol blue: nitrate reductase activity [42]; <3>, dissapearance of NADH-diaphorase activity [64]; <18>, inhibition of full and NADH-utilizing partial activities [86]) [12, 31, 42, 64, 86]
phosphate <1> (<1>, 25 mM, increases activity with a nitrate concentration of 2 mM, decreases activity with a nitrate concentration of 0.1 mM [40]) [40]
potassium ferricyanide <5> (<5>, preincubation with potassium ferricyanide inactivates nitrate reductase [22]) [22]
pyridoxal 5'-phosphate <1, 11> (<1>, uncompetitive with nitrate [72]) [17, 72]
thiocyanate <2, 11> (<2>, dead-end inhibition [3]) [3, 17]
Additional information <7, 8, 12, 20> (<20>, purification of a NADH-nitrate reductase inhibitor from young leaves of Glycine max, that causes a reversible inhibition [11]; <8>, roots of seedlings from Oryza sativa contain a substance which inhibits the activity of nitrate reductase when NADH or $FMNH_2$ is used as electron donor [14]; <12>, inhibition by monospecific anti-nitrate reductase rabbit serum [21]; <8>, inhibitor from primary and regenerated roots of nitrate-grown seedlings, main site of action is NADH:cytochrome c reductase component of the nitrate reductase, NADH protects [46]; <7> maize root inactivating enzyme inactivates NADH:cytochrome c reductase at a greater rate than $FADH_2$:nitrate reductase [50]; <19>, maize root inactivating enzyme inactivates pea leaf nitrate reductase [50]; <8>, nitrate reductase inhibitor from root extract of rice seedlings. Inactivation proceeds in two steps: the inhibitor first binds with nitrate reductase to cause a reduction in

both NADH:nitrate reductase and reduced benzyl viologen:nitrate reductase activity. In the second phase, there is a complete inactivation of NADH:nitrate reductase after about 20 min. Reduced benzyl viologen:nitrate reductase activity is not affected by the second phase of inactivation [56]) [11, 14, 21, 46, 50, 56]

Cofactors/prosthetic groups

FAD <1, 2, 3, 12> (<1, 2>, contains FAD [6,7]; <2>, 2-3 mol of FAD per mol of enzyme [7]; <2>, contains a minimum of 2 molecules of FAD [13]; <2, 12>, contains 1 FAD per subunit [28, 65, 71]; <3>, molybdenum, heme and FAD components are localized in distinct domains which are covalently linked by exposed hinge regions [38]; <3>, required for NADH-diaphorase activity [64]; <1>, required for maximal activity, K_m: 8 nM [72]) [6, 7, 13, 28, 38, 64, 65, 71, 72]
NADH <1-41> [1-89]
heme <2, 3, 12> (<2>, prosthetic group [7]; <2>, 2-3 mol of heme per mol of enzyme [7]; <2>, contains a minimum of 2 molecules of heme [13]; <2, 12>, contains 1 heme per subunit [28, 65, 71]; <3>, molybdenum, heme and FAD components are localized in distinct domains which are covalently linked by exposed hinge regions [38]; <3>, enzyme contains a functional heme [70]) [7, 13, 28, 38, 65, 70, 71]
Additional information <11> (<11>, the enzyme is associated with a cytochrome of the b type [17]) [17]

Activating compounds

2-mercaptoethylamine <16> (<16>, 5 mM, 5.6fold stimulation [31]) [31]
EDTA <4> (<4>, reduction of nitrate and menadione requires EDTA [8]) [8]
GSH <16> (<16>, 5 mM, 3.8fold stimulation [31]) [31]
HPO_4^{2-} <1> (<1>, activates [72]) [72]
L-Cys <16> (<16>, 5 mM, 5.6fold stimulation [31]) [31]
L-cysteic acid <16> (<16>, 5 mM, 2fold stimulation [31]) [31]
L-cysteine methyl ester <16> (<16>, 5 mM, 5.6fold stimulation [31]) [31]
calmodulin <36> (<36>, activates [62]) [62]
phosphate <1, 2> (<2>, stimulates [3,41]; <1>, 25 mM, increases activity with a nitrate concentration of 2 mM, decreases activity with a nitrate concentration of 0.1 mM [40]) [3, 40, 41]

Metals, ions

Ca^{2+} <36> (<36>, activates [62]) [62]
Mo^{6+} <2> (<2>, contains one Mo^{6+} per subunit [28]) [28]
molybdenum <1, 2, 3, 12> (<1>, γ subunit which carries molybdenum-containing component of 1000 Da [6]; <2>, enzyme contains molybdenum [7]; <2>, contains a minimum of 2 molecules of molybdenum [13]; <3>, molybdenum, heme and FAD components are localized in distinct domains which are covalently linked by exposed hinge regions. The molybdenum domain appears to be important in the maintenance of subunit interactions in the enzyme complex [38]; <3>, oxidation-reduction midpoint potentials of the molybdenum center: -8 mV for the Mo(VI)/Mo(V) couple and -42 mV for

the Mo(V)/Mo(IV) couple [39]; <2, 12>, contains 1 molybdenum per subunit [65, 71]) [6, 7, 13, 38, 39, 65, 71]
sodium azide <4> (<4>, 1 mM, 50% stimulation of menadione reduction [8]) [8]
Additional information <2> (<2>, increased ionic strength stimulates NADH:nitrate reductase activity [41]) [41]

Specific activity (U/mg)

0.89 <7> [12]
1.747 <10> [57]
1.9 <4> [12]
2 <4> [12]
2.5 <3> [4]
3 <31> [89]
3.37 <6> [87]
4 <16> [31]
6.48 <1> [68]
8 <1> [5]
9.51 <20> (<20>, c2NR [78]) [78]
12.5 <18> [86]
21.8 <5> [34]
23.1 <5> [9]
23.5 <20> [85]
24.8 <3> [66]
32 <41> [88]
37 <1> [72]
51.5 <3> [70]
60-70 <7> [67]
82 <38> [73]
92.68 <2> [7]
95 <12> [71]
100 <12> [21]
103 <3> [69]
Additional information <3, 7, 20, 28> (<3, 7, 20>, in vivo nitrate reductase assay by vacuum-infiltration procedure in leaf sections [53]) [51, 53, 80, 83]

K_m-Value (mM)

0.0026 <10> (FAD) [57]
0.0029 <20> (NADH) [78]
0.003 <7> (NADH) [12]
0.0036 <31> (NADH) [89]
0.0038 <1> (NADH) [72]
0.004 <3, 41> (NADH, <3>, recombinant flavin domain of nitrate reductase C240A mutant [82]; <41>, wild-type recombinant enzyme [88]) [82, 88]
0.006 <3> (NADH, <3>, ionic strength: 200 mM [37]) [37]
0.006 <3> (ferricytochrome c, <3>, ionic strength: 50 mM [37]) [37]
0.007 <3, 4> (NADH, <3>, ionic strength: 50 mM [37]) [8, 37]
0.008 <16> (NADH) [31]

0.0081 <20> (NADH) [76]
0.01 <10> (NADH) [57]
0.011 <3> (NADH, <3>, recombinant flavin domain of nitrate reductase C240G mutant [82]) [82]
0.012 <3> (NADH, <3>, wild-type recombinant flavin domain of nitrate reductase [82]) [82]
0.013 <3> (nitrate, <3>, at ionic strength of 50 mM [26,37]) [26, 37]
0.014 <3> (NADH, <3>, recombinant flavin domain of nitrate reductase C240S mutant [82]) [82]
0.015 <3> (NADH, <3>, recombinant flavin domain of nitrate reductase C17S mutant [82]) [82]
0.015 <3> (NADH, <3>, recombinant flavin domain of nitrate reductase C54S mutant [82]) [82]
0.015 <41> (nitrate, <41>, wild-type recombinant enzyme [88]) [88]
0.016 <3> (NADH, <3>, recombinant flavin domain of nitrate reductase C62S mutant [82]) [82]
0.017 <3> (NADH, <3>, recombinant FAD domain [83]) [83]
0.018 <3> (nitrate, <3>, at ionic strength of 200 mM [26,37]) [26, 37]
0.021 <3> (ferricyanide, <3>, ionic strength: 50 mM [37]) [37]
0.023 <18> (NADH) [86]
0.024 <3> (NADH, <3>, recombinant flavin domain of nitrate reductase C240S mutant [82]) [82]
0.026 <20> (NADPH) [78]
0.027 <3> (ferricyanide, <3>, wild-type recombinant flavin domain of nitrate reductase [82]) [82]
0.029 <3> (ferricytochrome c, <3>, ionic strength: 200 mM [37]) [37]
0.033 <5> (NADH) [9]
0.034 <3> (ferricyanide, <3>, recombinant FAD domain [83]) [83]
0.035 <3> (ferricyanide, <3>, ionic strength: 200 mM [37]) [37]
0.036 <3> (ferricyanide, <3>, recombinant flavin domain of nitrate reductase C54S mutant [82]) [82]
0.04 <16> (nitrate) [31]
0.045 <3> (ferricyanide, <3>, recombinant flavin domain of nitrate reductase C17S mutant [82]) [82]
0.05 <12> (nitrate) [21]
0.052 <28> (NADH, <28>, reaction with nitrate [80]) [80]
0.053 <3> (2,6-dichlorophenolindophenol, <3>, ionic strength: 50 mM [37]) [37]
0.056 <3> (2,6-dichlorophenolindophenol, <3>, ionic strength: 200 mM [37]) [37]
0.056 <3> (flavin hydroquinone, <3>, ionic strength: 50 mM or 200 mM [37]) [37]
0.059 <3> (NADH, <3>, recombinant flavin domain of nitrate reductase C62S mutant [82]) [82]
0.06 <17> (bromophenol blue, <17>, at 10 mM nitrate [42]) [42]
0.064 <18> (nitrate) [86]

0.067 <28> (NADH, <28>, reaction with 2,6-dichlorophenol indophenol [80]) [80]
0.07 <7> (nitrate) [12]
0.09 <1> (nitrate, <1>, in presence of 25 mM phosphate [40]) [40]
0.093 <3> (nitrate, <3>, in 5 mM phosphate buffer [26]) [26]
0.096 <28> (nitrate, <28>, reaction with NADH [80]) [80]
0.11 <20> (nitrate) [76]
0.13 <11> (2,6-dichlorophenolindophenol) [17]
0.13 <28> (nitrate, <28>, reaction with methyl viologen [80]) [80]
0.14 <1> (nitrate, <1>, without phosphate [40]) [40]
0.15 <3> (nitrate) [10]
0.16 <20> (nitrate) [85]
0.19 <20> (nitrate, <20>, reaction with NADH [78]) [78]
0.221 <3> (ClO_3^-, <3>, ionic strength: 50 mM [37]) [37]
0.23 <10> (nitrate) [57]
0.24 <1> (nitrate) [5]
0.27 <1> (nitrate) [72]
0.275 <4> (menadione) [8]
0.295 <31> (nitrate) [89]
0.34 <5> (nitrate) [9]
0.5 <17> (nitrate, <17>, at 0.2 mM bromophenol blue [42]) [42]
0.54 <28> (reduced bromophenol blue) [80]
0.739 <3> (BrO_3^-, <3>, ionic strength: 50 mM [37]) [37]
0.87 <28> (reduced methyl viologen) [80]
10.75 <3> (IO_3^-, <3>, ionic strength: 50 mM [37]) [37]
13 <28> (nitrate, <28>, reaction with bromophenol blue [80]) [80]
80 <28> (ferricytochrome c) [80]
100 <28> (2,6-dichlorophenol indophenol) [80]
Additional information <2, 3> (<3>, K_m-values of the the C-terminal 268 residues corresponding to the flavin-containing domain, amplified and expressed in E. coli [83]) [41, 83]

K_i-Value (mM)

0.013 <4> (dicoumarol, <4>, nitrate reductase assay [8]) [8]
0.018 <1> (carbamoyl phosphate) [72]
0.12 <15> (hydroxylamine) [27]
0.176 <1> (pyridoxal 5'-phosphate) [72]
0.2 <3> (ADP, <3>, inhibition of the recombinant FAD domain [83]) [83]
0.2 <3> (ADPribose, <3>, inhibition of the recombinant FAD domain [83]) [83]
0.3 <17> (Bromophenol Blue) [42]
0.5 <3> (NAD^+, <3>, recombinant flavin domain of nitrate reductase C240G mutant [82]) [82]
0.6 <3> (AMP, <3>, inhibition of the recombinant FAD domain [83]) [83]
0.8 <3> (NAD^+, <3>, recombinant flavin domain of nitrate reductase C240A mutant [82]) [82]

1.2 <3> (NAD^+, <3>, recombinant flavin domain of nitrate reductase C240S mutant [82]) [82]
1.36 <1> (ADP) [72]
1.5 <3> (NAD^+, <3>, recombinant flavin domain of nitrate reductase C54S mutant [82]) [82]
1.9 <3> (NAD^+, <3>, recombinant flavin domain of nitrate reductase wild-type and C17S mutant [82]; <3>, inhibition of the recombinant FAD domain [83]) [82, 83]
2 <3> (NAD^+, <3>, recombinant flavin domain of nitrate reductase C62S mutant [82]) [82]
4.8 <3> (adenosine, <3>, inhibition of the recombinant FAD domain [83]) [83]
15 <3> (Cl^-, <3>, binding to the native enzyme [26]) [26]
23 <3> (ferricyanide, <3>, inhibition of the recombinant FAD domain [83]) [83]
26 <3> (adenine, <3>, inhibition of the recombinant FAD domain [83]) [83]
62 <3> (NMN, <3>, inhibition of the recombinant FAD domain [83]) [83]
122 <3> (nicotinamide, <3>, inhibition of the recombinant FAD domain [83]) [83]
176 <3> (Cl^-, <3>, binding to the enzyme-NO_3^- complex [26]) [26]
Additional information <4, 7> [12]

pH-Optimum

6.3 <12> [21]
6.5 <3, 20> (<3>, NADH:ferricyanide reductase activity, NADH:dichlorophenolindophenol reductase activity and reduced methyl viologen:nitrate reductase activity [37]) [37, 76, 78, 83]
7 <3, 4, 18, 20, 41> (<4>, reaction with NADH and menadione [8]; <3>, NADH:nitrate reductase activity [37]; <20>, in vivo nitrate reduction [53]; <18>, $FMNH_2$-nitrate reductase activity [86]; <41>, wild-type recombinant enzyme [88]) [8, 37, 53, 86, 88]
7.3 <2, 3> (<2,3>, NADH:cytochrome c reductase activity [37,41]) [37, 41]
7.3-7.5 <20> [85]
7.4 <31> [89]
7.5 <1, 3, 4, 7, 16, 18> (<3>, flavin hydroquinone:nitrate reductase activity [37]; <3, 7>, in vivo nitrate reduction [53]; <18>, reduced methyl viologen-nitrate reductase activity [86]) [5, 12, 31, 37, 53, 72, 86]
7.9 <2> (<2>, NADH:nitrate reductase activity [41]) [41]
8.3 <18> [86]
8.5 <18, 28> (<28>, NADH-diaphorase activity [80]; <18>, NADH-ferricyanide reductase activity and NADH-cytochrome c reductase activity [86]) [80, 86]
9 <28> (<28>, NADH-nitrate reductase activity [80]) [80]
9.5 <28> (<28>, methyl viologen-nitrate reductase activity [80]) [80]

pH-Range

6-8.5 <20> (<20>, pH 6.0: about 35% of maximal activity, pH 8.5: about 45% of maximal activity [85]) [85]

Temperature optimum (°C)

14 <31> [89]
15 <31> (<31>, enzyme from cells grown at 5°C, 15°C or at 25°C [61]) [61]
30 <5, 7, 12> [22, 61]

Temperature range (°C)

-10-40 <5> (<5>, -10°C: 8% of maximal activity, 0°C: 23% of maximal activity, 40°C: 19% of maximal activity [22]) [22]

4 Enzyme Structure

Molecular weight

91000 <39> (<39>, 4 active fractions with MW of 91000 Da, 362000 Da, 180000 Da and 720000 Da are detected, gel filtration [75]) [75]
144000 <28> (<28>, non-denaturing pore gradient PAGE [80]) [80]
147000 <28> (<28>, gel filtration [80]) [80]
180000 <39> (<39>, 4 active fractions with MW of 91000 Da, 362000 Da, 180000 Da and 720000 Da are detected, gel filtration [75]) [75]
202000 <5> (<5>, sucrose density gradient centrifugation, gel filtration [34]) [34]
205000 <1> (<1>, non-denaturing PAGE [68]) [68]
220000 <18> (<18>, gel filtration [86]) [86]
221000 <1> (<1>, sucrose density gradient centrifugation, gel filtration [5]) [5]
225000 <1> (<1>, disc gel electrophoresis [72]) [72]
230000 <12> (<12>, disc gel electrophoresis [71]) [71]
270000 <3> (<3>, gel filtration, sucrose density gradient centrifugation [69]) [69]
280000 <2> (<2>, equilibrium sedimentation [7]) [7]
330000 <20> (<20>, gel filtration [76]) [76]
356000 <2> (<2>, sucrose density gradient centrifugation, gel filtration [13]) [13]
360000 <2> (<2>, equilibrium sedimentation [2]) [2]
362000 <39> (<39>, 4 active fractions with MW of 91000 Da, 362000 Da, 180000 Da and 720000 Da are detected, gel filtration [75]) [75]
375000 <2> [65]
440000 <6> (<6>, gel filtration [87]) [87]
720000 <39> (<39>, 4 active fractions with MW of 91000 Da, 362000 Da, 180000 Da and 720000 Da are detected, gel filtration [75]) [75]
Additional information <3> (<3>, the C-terminal 268 residues corresponding to the flavin-containing domain, amplified and expressed in E. coli show a MW of 30000 Da by SDS-PAGE [83]) [83]

Subunits

? <1, 2, 20, 31> (<1>, $\alpha_2\beta_2\gamma_2$molybdenum-containing-component, 2 * 40000 Da FAD-containing α-subunit + 2 * 20000 Da cytochrome b_{557} β subunit + 2 * γ subunit which carries molybdenum-containing component of 1000 Da [6];

<2>, x * 90000, equilibrium sedimentation of enzyme dissociated in 6 M guanidine hydrochloride [7]; <2>, x * 98000, SDS-PAGE [7]; <20>, x * 100000, SDS-PAGE [85]; <31>, x * 110000, SDS-PAGE [89]; <2>, native enzyme is composed of at least 3 subunits [13]) [6, 7, 13, 85, 89]
dimer <1, 3, 12, 18, 28> (<28>, 1 * 46000 + 1 * 95000, SDS-PAGE [80]; <1,18>, 2 * 100000, SDS-PAGE [5, 86]; <1>, 2 * 103000, SDS-PAGE [68]; <1>, 2 * 110000, SDS-PAGE [1]; <3>, 2 * 110000-120000, SDS-PAGE [69]; <12>, 2 * 115000, SDS-PAGE [71]) [1, 5, 68, 69, 71, 80, 86]
tetramer <2, 6> (<2>, 4 * 90000, dihedral symmetry, SDS-PAGE [2]; <2>, 4 * 100000, at low protein concentration, the tetramer dissociates to a fully active dimer, each subunit in the tetramer or dimer can function independently, radiation inactivation analysis [28]; <6>, 4 * 110000, SDS-PAGE [87]) [2, 28, 87]
Additional information <2, 28, 38> (<2>, FAD/NADH-binding domains exposed on the surface of the molecule, a protease-sensitive hinge region which connects the nitrate-reducing and NADH dehydrogenase moieties, the quarternary structure maintains via association sites on the heme/molybdenum domain [65]; <38>, two bands of 107000 Da and 99500 Da are detected by SDS-PAGE [73]; <28>, diaphorase activity is located in the small subunit [80]) [65, 73, 80]

Posttranslational modification

Additional information <2> (<2>, contains little or no carbohydrate [28]; <2>, contains no or little lipid [28]) [28]

5 Isolation/Preparation/Mutation/Application

Source/tissue

callus <7> [60]
cell suspension culture <14> (<14>, suspension culture of the XD cell line. As tobacco XD cell culture progresses though the exponential phase of growth into the stationary phase, due to depletion of nitrate, the overt nitrate reductase activity decreases and the latent form becomes predominant. 9fold activation of the enzyme activity is observed in extracts from late-exponential phase cultures [35]) [35]
cotyledon <4, 12, 25, 26, 37, 40> [12, 18, 21, 58, 63, 71, 77]
embryo <9> [15]
leaf <1, 3, 5, 7, 8, 12, 13, 19, 20, 22, 23, 24, 27, 29, 36, 38> (<5>, from 8-day-old seedlings [9]; <3>, enzyme from leaf and root are different proteins on the basis of their antigenic behaviour [19]; <7>, primary leaf, two enzyme form: NRI and NRII [45]; <1>, 6-day-old primary leaves [52]; <7>, primary leaves [67]; <20>, young first leaves [76]) [1, 4, 5, 9, 11, 14, 19, 20, 25, 26, 30, 32, 33, 38, 40, 45, 47, 48, 50, 52, 53, 55, 58, 62, 67, 69, 70, 72, 73, 76, 78, 81]
root <3, 8, 20> (<3>, enzyme from leaf and root are different proteins on the basis of their antigenic behaviour [19]; <8>, primary and regenerated roots of nitrate-grown seedlings [46]; <20>, root nodules [85]) [19, 37, 46, 85]

root nodule <20> [85]
scutellum <7> [43, 50, 60]
seedling <1, 5, 7, 8> [34, 36, 46, 50, 55, 60, 67]
shoot <1, 5, 7, 21> (<5>, tip [22]; <7>, the activity in roots reaches up to 80% of the shoot activity at 4 d of age [36]; <1>, 4 day old [68]) [6, 22, 34, 36, 54, 68]

Localization

membrane <39> (<39>, constitutive nitrate reductase exists mostly in membranes [75]) [75]

Purification

<1> [5, 68, 72]
<2> [3, 7, 13]
<3> (immunoaffinity chromatography using a monoclonal antibody [66]; removal of nicked subunits by affinity chromatography [70]; the C-terminal 268 residues corresponding to the flavin-containing domain, amplified and expressed in Escherichia coli [83]; mutant enzymes K741R, K741H, K714A, K741E, K741M, K741Q, K741P [84]) [19, 51, 66, 70, 83, 84]
<4> [8, 12]
<5> [9, 34]
<6> [87]
<7> (flavin domain [74]) [12, 32, 67, 69, 74]
<10> (partial [16]) [16, 57]
<12> (NADH:Fe(III)-citrate reductase activity copurifies and has identical electrophoretic mobility [21]) [21]
<16> [31]
<18> [86]
<20> (partial [76]; wild-type and nr1-mutant [78]) [76, 78, 85]
<28> [80]
<31> [89]
<38> (by affinity chromatography using 5'-AMP Sepharose and monoclonal antibodies [73]) [73]
<41> (recombinant enzyme expressed in Pichia pastoris [79,88]) [79, 88]

Cloning

<3> (the C-terminal 268 residues corresponding to the flavin-containing domain, amplified and expressed in Escherichia coli [83]) [83]
<7> (high-level expression of the catalytically active flavin domain in Escherichia coli [74]) [25, 74]
<13> [23]
<14> (two homologous genes coding for nitrate reductase in tobacco [24]) [24]
<41> (expression in Pichia pastoris [88]) [88]

Engineering

C17S <3> (<3>, visible and CD spectra are very similar to that of the wild-type domain, thermal stability is slightly decreased compared to the wild-type domain [82]) [82]

C191A <41> (<41>, enzyme is still produced, but it is inactive [88]) [88]
C191S <41> (<41>, enzyme is still produced, but it is inactive [88]) [88]
C240A <3> (<3>, decrease of diaphorase activity [82]) [82]
C240G <3> (<3>, decrease of diaphorase activity [82]) [82]
C240S <3> (<3>, decrease of diaphorase activity, thermal stability is slightly decreased compared to the wild-type domain, the oxidation reduction midpoint potential for the FAD/$FADH_2$ couple is -219 mV compared to -268 mV in the wild-type domain [82]) [82]
C54S <3> (<3>, visible and CD spectra are very similar to that of the wild-type domain, thermal stability is slightly decreased compared to the wilde-type domain,the oxidation reduction midpoint potential for the FAD/$FADH_2$ couple is -197 mV compared to -268 mV in the wild-type domain [82]) [82]
C62S <3> (<3>, visible and CD spectra are very similar to that of the wild-type domain, thermal stability is decreased compared to the wild-type domain, the oxidation reduction midpoint potential for the FAD/$FADH_2$ couple is -226 mV compared to -268 mV in the wild-type domain [82]) [82]
K714A <3> (<3>, functional flavoprotein which retains FAD as the sole prosthetic group and exhibits spectroscopic properties comparable to that of the wild-type enzyme. Altered NADH:ferricyanide reductase activity with with mutations affecting both turnover number and the K_m-value for NADH. At pH 7.0 the turnover number decreases in the order wild-type, K741R, K741A, K741H, K741E, K741M, K741Q. K_m-value for NADH increases in the same order [84]) [84]
K741E <3> (<3>, functional flavoprotein which retains FAD as the sole prosthetic group and exhibits spectroscopic properties comparable to that of the wild-type enzyme. Altered NADH:ferricyanide reductase activity with with mutations affecting both turnover number and the K_m-value for NADH. At pH 7.0 the turnover number decreases in the order wild-type, K741R, K741A, K741H, K741E, K741M, K741Q. K_m-value for NADH increases in the same order [84]) [84]
K741H <3> (<3>, functional flavoprotein which retains FAD as the sole prosthetic group and exhibits spectroscopic properties comparable to that of the wild-type enzyme. Altered NADH:ferricyanide reductase activity with with mutations affecting both turnover number and the K_m-value for NADH. At pH 7.0 the turnover number decreases in the order wild-type, K741R, K741A, K741H, K741E, K741M, K741Q. K_m-value for NADH increases in the same order [84]) [84]
K741M <3> (<3>, functional flavoprotein which retains FAD as the sole prosthetic group and exhibits spectroscopic properties comparable to that of the wild-type enzyme. Altered NADH:ferricyanide reductase activity with with mutations affecting both turnover number and the K_m-value for NADH. At pH 7.0 the turnover number decreases in the order wild-type, K741R, K741A, K741H, K741E, K741M, K741Q. K_m-value for NADH increases in the same order [84]) [84]
K741P <3> (<3>, functional flavoprotein which retains FAD as the sole prosthetic group and exhibits spectroscopic properties comparable to that of the wild-type enzyme. Altered NADH:ferricyanide reductase activity with with

mutations affecting both turnover number and the K_m-value for NADH. At pH 7.0 the turnover number decreases in the order wild-type, K741R, K741A, K741H, K741E, K741M, K741Q. K_m-value for NADH increases in the same order [84]) [84]
K741Q <3> (<3>, functional flavoprotein which retains FAD as the sole prosthetic group and exhibits spectroscopic properties comparable to that of the wild-type enzyme. Altered NADH:ferricyanide reductase activity with mutations affecting both turnover number and the K_m-value for NADH. At pH 7.0 the turnover number decreases in the order wild-type, K741R, K741A, K741H, K741E, K741M, K741Q. K_m-value for NADH increases in the same order [84]) [84]
K741R <3> (<3>, functional flavoprotein which retains FAD as the sole prosthetic group and exhibits spectroscopic properties comparable to that of the wild-type enzyme. Altered NADH:ferricyanide reductase activity with with mutations affecting both turnover number and the K_m-value for NADH. At pH 7.0 the turnover number decreases in the order wild-type, K741R, K741A, K741H, K741E, K741M, K741Q. K_m-value for NADH increases in the same order [84]) [84]
S266A <41> (<41>, 48% of the activity of wild-type enzyme [88]) [88]

6 Stability

pH-Stability

6.5-8.5 <3> (<3>, 1 h, stable [51]) [51]
7.5 <1, 5> (<5>, highest stability, unstable below [9]; <1>, 0°C, half-life: 42 h [72]) [9, 72]

Temperature stability

0 <1> (<1>, pH 7.5, half-life: 42 h [72]) [72]
10 <5> (<5>, half-life: 94 min [9]) [9]
25 <3, 5> (<5>, half-life is 28 min, half-life in presence of GAD: 70 min [9]; <3>, 20 min, in absence of FAD, stable [51]) [9, 51]
30 <5> (<5>, 30-50% $(NH_4)_2SO_4$, 150 mM potassium phosphate buffer, pH 7.5, 0.01 mM FAD, 1 mM dithiothreitol, half-life: 36.8 min [34]) [34]
30 <5> (<5>, transition temperature [9]) [9]
40 <5, 17> (<5>, $t_{1/2}$: 1 min [9]; <17>, 40 min, 50% loss of nitrate reductase activity, slight increase of bromphenol blue activity [42]) [9, 42]

General stability information

<1>, leupeptin inhibits degradation of NADH-nitrate reductase by thiol-dependent acid endoproteinase in primary leaf extract [52]
<1>, presence of casein in phosphate buffer improves stability at 0°C but not at 30°C [55]
<1>, stabilized at 0°C and at 30°C by buffer containing 0.25 M Tris-HCl, pH 8.5, 3 mM DTT, 0.005 mM FAD, 0.001 mM sodium molybdate and 1 mM EDTA [55]

<2>, incubation of the native enzyme with either trypsin, Staphylococcus aureus V8 protease, or a natural inactivator protease from corn results in loss of NADH:nitrate reductase and NADH:cytochrome c reductase activity, but no loss of methyl viologen:nitrate reductase activity [65]
<3>, 0.03% SDS produces almost complete inactivation of NADH-diaphorase and NADH-nitrate reductase, while FNH_2-nitrate reductase retains 60% of the original activity, FAD has no protecting effect [51]
<3>, 1 M guanidine hydrochloride, in presence of FAD, 80% inactivation of FNH_2-nitrate reductase activity and NADH-nitrate reductase activity, NADH-diaphorase activity is unaffected [51]
<3>, 4 M urea, in presence of FAD, inactivation of $FMNH_2$-nitrate reductase and NADH-nitrate reductase activity, only a slight effect on the NADH-diaphorase activity [51]
<5>, FAD stabilizes at 25°C, at 10°C FAD has no effect on stability [9]
<5>, the decay rate of nitrate reductase activity in crude extracts is due to spontaneous dissociation of the enzyme and to the effects of two decay factors, one present in the 0-30% and the other in the 50-70% saturated $(NH_4)_2SO_4$ fraction of the crude extract [34]
<7>, enzyme in the crude extract is stable for several days at 0°C and for several months at -80°C [67]
<7>, stabilized in vitro by addition of chymostatin to extraction buffer [36]
<7>, the enzyme is very susceptible to inactivation by maize root proteinase and trypsin [43]
<30>, FAD protects from heat inactivation at 45°C [59]
<40>, dilution of a crude extract leads to increasing lability, much more stable in presence of both NADH and nitrate [77]
<40>, imposition of water stress or artificial extension of the dark period leads to significant reduction of nitrate reductase activity, but does not change in vitro nitrate reductase stability [77]
<2, 7>, inactivation by corn root proteinase, comparison of hydrolysis products [45]
<5, 7>, preincubation with NADH stabilizes activity at 0°C and at 25°C [22]

Storage stability

<2>, -20°C, pH 6.9, phosphate buffer, 40% v/v glycerol, stable for more than 2 months [13]
<3>, -20°C, 50 mM MOPS, containing 0.01 mM FAD, 0.1 mM EDTA, pH 7.0, 50% glycerol, stable for several months [26]

References

[1] Min Kuo, T.; Somers, D.A.; Kleinhofs, A.; Warner, R.L.: NADH-nitrate reductase in barley leaves. Identification and amino acid composition of subunit protein. Biochim. Biophys. Acta, **708**, 75-81 (1982)

[2] Howard, W.D.; Solomonson, L.P.: Quaternary structure of assimilatory NADH:nitrate reductase from Chlorella. J. Biol. Chem., **257**, 10243-10250 (1982)
[3] Howard, W.D.; Solomonson, L.P.: Kinetic mechanism of assimilatory NADH:nitrate reductase from Chlorella. J. Biol. Chem., **256**, 12725-12730 (1981)
[4] De La Rosa, F.F.; Palacian, E.: Spinach nitrate reductase. Kinetic studies of NADH-diaphorase. Plant Sci. Lett., **21**, 1-8 (1981)
[5] Kuo, T.; Kleinhofs, A.; Warner, R.L.: Purification and partial characterization of nitrate reductase from barley leaves. Plant Sci. Lett., **17**, 371-381 (1980)
[6] Wray, J.L.; Small, I.S.; Brown, J.: A model for the subunit composition of higher-plant NADH-nitrate reductase. Biochem. Soc. Trans., **7**, 739-741 (1979)
[7] Giri, L.; Ramadoss, C.S.: Physical studies on assimilatory nitrate reductase from Chlorella vulgaris. J. Biol. Chem., **254**, 11703-11712 (1979)
[8] Smarrelli, J.; Campbell, W.H.: NADH dehydrogenase activity of higher plant nitrate reductase (NADH). Plant Sci. Lett., **16**, 139-147 (1979)
[9] Sherrard, J.H.; Dalling, M.J.: In vitro stability of nitrate reductase from wheat leaves. Plant Physiol., **63**, 346-353 (1979)
[10] Maldonado, J.M.; Notton, B.A.; Hewitt, E.J.: Inhibitory effects of ferrocytochrome c on NADH-nitrate reductase activity of spinach (Spinacia oleracea L.). Plant Sci. Lett., **13**, 143-150 (1978)
[11] Jolly, S.O.; Tolbert, N.E.: NADH-nitrate reductase inhibitor from soybean leaves. Plant Physiol., **62**, 197-203 (1978)
[12] Campbell, W.H.; Smarrelli, J.: Purification and kinetics of higher plant NADH:nitrate reductase. Plant Physiol., **61**, 611-616 (1978)
[13] Solomonson, L.P.; Lorimer, G.H.; Hall, R.L.; Borchers, R.; Bailey, J.L.: Reduced nicotinamide adenine dinucleotide-nitrate reductase of Chlorella vulgaris. Purification, prosthetic groups, and molecular properties. J. Biol. Chem., **250**, 4120-4127 (1975)
[14] Kadam, S.S.; Gandhi, A.P.; Sawhney, S.K.; Naik, M.S.: Inhibitor of nitrate reductase in the roots of rice seedlings and its effect on the enzyme activity in the presence of NADH. Biochim. Biophys. Acta, **350**, 162-170 (1974)
[15] Kende, H.; Shen, T.C.: Nitrate reductase in Agrostemma githago. Comparison of the inductive effects of nitrate and cytokinin. Biochim. Biophys. Acta, **286**, 118-125 (1972)
[16] Vennesland, B.; Jetschmann, C.: The nitrate reductase of Chlorella pyrenoidosa. Biochim. Biophys. Acta, **229**, 554-564 (1971)
[17] Solomonson, L.P.; Vennesland, B.: Properties of a nitrate reductase of Chlorella. Biochim. Biophys. Acta, **267**, 544-557 (1972)
[18] Smarrelli, J.; Campbell, W.H.: Heavy metal inactivation and chelator stimulation of higher plant nitrate reductase. Biochim. Biophys. Acta, **742**, 435-445 (1983)
[19] Ferrario, S.; Hirel, B.; Gadal, P.: Immuno chemical characterization of nitrate reductase from spinach leaves and roots. Biochem. Biophys. Res. Commun., **113**, 733-737 (1983)

[20] Jawali, N.; Sane, P.V.: Mechanism of reactivation of cyanide-inactivated nitrate reductase by flavins in light. FEBS Lett., **158**, 213-216 (1983)

[21] Redinbaugh, M.G.; Campbell, W.H.: Reduction of ferric citrate catalyzed by NADH:nitrate reductase. Biochem. Biophys. Res. Commun., **114**, 1182-1188 (1983)

[22] Datta, N.; Rao, L.V.M.; Guha-Mukherjee, S.; Sopory, S. K.: Activation and stabilization of nitrate reductase by NADH in wheat and maize. Phytochemistry, **22**, 821-824 (1983)

[23] Daniel-Vedele, F.; Dorbe, M.F.; Caboche, M.; Rouze, P.: Cloning and analysis of the tomato nitrate reductase-encoding gene: protein domain structure and amino acid homologies in higher plants. Gene, **85**, 371-380 (1989)

[24] Vaucheret, H.; Vincentz, M.; Kronenberger, J.; Caboche, M.; Rouze, P.: Molecular cloning and characterisation of the two homologous genes coding for nitrate reductase in tobacco. Mol. Gen. Genet., **216**, 10-15 (1989)

[25] Gowri, G.; Campbell, W.H.: cDNA clones for corn leaf NADH:nitrate reductase and chloroplast NAD(P)$^+$:glyceraldehyde-3-phosphate dehydrogenase. Plant Physiol., **90**, 792-798 (1989)

[26] Barber, M.J.; Notton, B.A.; Kay, C.J.; Solomonson, L. P.: Chloride inhibition of spinach nitrate reductase. Plant Physiol., **90**, 70-94 (1989)

[27] Balanidin, T.; Fernandez, V.M.; Aparicio, P.J.: Characterization of the reversible inactivation of Ankistrodesmus braunii nitrate reductase by hydroxylamine. Plant Physiol., **82**, 65-70 (1986)

[28] Solomonson, L.P.; McCreery, M.J.: Radiation inactivation of assimilatory NADH:nitrate reductase from Chlorella. Catalytic and physical sizes of functional units. J. Biol. Chem., **261**, 806-810 (1986)

[29] Jawali, N.; Sane, P.V.: Inhibition of the nitrate reductase complex from spinach by oxylamines. Phytochemistry, **23**, 225-228 (1984)

[30] Aryan, A.P.; Wallace, W.; Nicholas, D.J.D.: Effects of NADH and thiol compounds on wheat leaf nitrate reductase. Phytochemistry, **23**, 719-721 (1984)

[31] Smarrelli, J.; Campbell, W.H.: Activation of Thalassiosira pseudonana NADH: nitrate reductase. Phytochemistry, **19**, 1601-1605 (1980)

[32] Echevarria, C.; Maurino, S.G.; Maldonado, J.M.: Reversible inactivation of maize leaf nitrate reductase. Phytochemistry, **23**, 2155-2158 (1984)

[33] Aryan, A.P.; Wallace, W.: Reversible inactivation of wheat leaf nitrate reductase by NADH, involving superoxide ions generated by the oxidation of thiols and FAD. Biochim. Biophys. Acta, **827**, 215-220 (1985)

[34] Jones, P.W.; Mhuimhneachain, M.N.: The activity and stability of wheat nitrate reductase in vitro. Phytochemistry, **24**, 385-392 (1985)

[35] Trinity, P.M.; Filner, P.: Activation and inactivation of NADH nitrate reductase in tobacco XD cells. Phytochemistry, **30**, 69-71 (1991)

[36] Long, D.M.; Oaks, A.: Stabilization of nitrate reductase in maize roots by chymostatin. Plant Physiol., **93**, 846-850 (1990)

[37] Barber, M.J.; Notton, B.A.: Spinach nitrate reductase. Plant Physiol., **93**, 537-540 (1990)

[38] Kubo, Y.; Ogura, N.; Nakagawa, H.: Limited proteolysis of the nitrate reductase from spinach leaves. J. Biol. Chem., **263**, 19684-19689 (1988)

[39] Barber, M.J.; Notton, B.A.; Solomonson, L.P.: Oxidation-reduction midpoint potentials of the molybdenum center in spinach NADH:nitrate reductase. FEBS Lett., **213**, 372-374 (1987)
[40] Oji, Y.; Ryoma, Y.; Wakiuchi, N.; Okamoto, S.: Effect of inorganic orthophosphate on in vitro activity of NADH-nitrate reductase isolated from 2-row barley leaves. Plant Physiol., **83**, 472-474 (1987)
[41] Kay, C.J.; Barber, M.J.: Assimilatory nitrate reductase from Chlorella. Effect of ionic strength and pH on catalytic activity. J. Biol. Chem., **261**, 14125-14129 (1986)
[42] Campbell, W.H.: Properties of bromophenol blue as an electron donor for higher plant NADH:nitrate reductase. Plant Physiol., **82**, 729-732 (1986)
[43] Batt, R.G.; Wallace, W.: A comparison of the effect of trypsin and a maize root proteinase on nitrate reductase and other enzymes from maize. Biochim. Biophys. Acta, **744**, 205-211 (1983)
[44] Solomonson, L.P.; Mc Creery, M.J.; Kay, C.J.; Barber, M.J.: Radiation inactivation analysis of assimilatory NADH:nitrate reductase. Apparent functional sizes of partial activities associated with intact and proteolytically modified enzyme. J. Biol. Chem., **262**, 8934-8939 (1987)
[45] Poulle, M.; Oaks, A.; Bzonek, P.; Goodfellow, V.J.; Solomonson, L.P.: Characterization of nitrate reductases from corn leaves (Zea mays cv W64AxW182E) and Chlorella vulgaris. Plant Physiol., **85**, 375-378 (1987)
[46] Leong, C.C.; Shen, T.C.: Nitrate reductase inhibitor of rice plants. Biochim. Biophys. Acta, **612**, 245-252 (1980)
[47] Maldonado, J.M.; Notton, B.A.; Hewitt, E.J.: Nitrate reductase inhibitor of rice plants. Planta, **156**, 289-294 (1982)
[48] Aryan, A.P.; Batt, R.G.; Wallace, W.: The reactivation of nitrate reductase from spinach (Spinacia oleracea L.) inactivated by NADH and cyanide: effects of peroxidase and associated systems. Plant Physiol., **71**, 582-587 (1983)
[49] Palacian, E.; De La Rosa, F.; Castillo, F.; Gomez-Moreno, C.: Nitrate reductase from Spinacia oleracea. Reversible inactivation by NAD(P)H and by thiols. Arch. Biochem. Biophys., **161**, 441-447 (1974)
[50] Wallace, W.: Effects of a nitrate reductase inactivating enzyme and NAD(P)H on the nitrate reductase from higher plants and Neurospora. Biochim. Biophys. Acta, **377**, 239-250 (1975)
[51] De La Rosa, F.F.; Castillo, F.; Palacian, E.: Effects of denaturing agents on spinach nitrate reductase. Phytochemistry, **16**, 875-879 (1977)
[52] Wray, J.L.; Kirk, D.W.: Inhibition of NADH-nitrate reductase degradation in barley leaf extracts by leupeptin. Plant Sci. Lett., **23**, 207-213 (1981)
[53] Maurino, S.G.; Echevarria, C.; Mejias, J.A.; Vargas, M.A.; Maldonado, J.M.: Properties of the in vivo nitrate reductase assay in maize, soybean, and spinach leaves. J. Plant Physiol., **124**, 123-130 (1986)
[54] Billard, J.P.; Boucaud, J.: Effect of sodium chloride on the nitrate reductase of Suaeda maritima var. macrocarpa. Phytochemistry, **21**, 1225-1228 (1982)
[55] Kuo, T.M.; Warner, R.L.; Kleinhofs, A.: In vitro stability of nitrate reductase from barley leaves. Phytochemistry, **21**, 531-533 (1982)

[56] Leong, C.C.; Shen, T.C.: Action kinetics of the inhibition of nitrate reductase of rice plants. Biochim. Biophys. Acta, **703**, 129-133 (1982)
[57] Schloemer, R.H.; Garrett, R.H.: Partial purification of the NADH-nitrate reductase complex from Chlorella pyrenoidosa. Plant Physiol., **51**, 591-593 (1973)
[58] Hageman, R.H.; Hucklesby, D.P.: Nitrate reductase from higher plants. Methods Enzymol., **23A**, 491-503 (1971)
[59] Zumft, W.G.; Aparicio, P.J.; Paneque, A.; Losada, M.: Structural and functional role of FAD in the NADH-nitrate reducing system from Chlorella. FEBS Lett., **9**, 157-160 (1970)
[60] Sorger, G.; Gooden, D.O.; Earle, E.D.; McKinnon, J.: NADH nitrate reductase and NAD(P)H nitrate reductase in genetic variants and regenerating callus of maize. Plant Physiol., **82**, 473-478 (1986)
[61] Gao, Y.; Smith, G.J.; Alberte, R.S.: Temperature dependence of nitrate reductase activity in marine phytoplankton: Biochemical analysis and ecological implications. J. Phycol., **36**, 304-313 (2000)
[62] Sane, P.V.; Kumar, N.; Baijal, M.; Singh, K.K.; Kochhar, V.K.: Activation of nitrate reductase by calcium and calmodulin. Phytochemistry, **26**, 1289-1291 (1987)
[63] Schuster, C.; Schmidt, S.; Mohr, H.: Effect of nitrate, ammonium, light and a plastidic factor on the appearance of multiple forms of nitrate reductase in mustard (Sinapis alba L.) cotyledons. Planta, **177**, 74-83 (1989)
[64] Castillo, F.; De La Rosa, F.F.; Calero, F.; Palacian, E.: Nitrate reductase from Spinacia oleracea. FAD and the inactivation by NAD(P)H. Biochem. Biophys. Res. Commun., **69**, 277-284 (1976)
[65] Solomonson, L.P.; Barber, M.J.; Robbins, A.P.; Oaks, A.: Functional domains of assimilatory NADH:nitrate reductase from Chlorella. J. Biol. Chem., **261**, 11290-11294 (1986)
[66] Fido, R.J.: Purification of nitrate reductase from spinach (Spinacea oleracea L.) by immunoaffinity chromatography using a monoclonal antibody. Plant Sci., **50**, 111-115 (1987)
[67] Nakagawa, H.; Poulle, M.; Oaks, A.: Characterization of nitrate reductase from corn leaves (Zea mays cv W64A * W182E). Plant Physiol., **75**, 285-289 (1984)
[68] Campbell, J.M.; Wray, J.L.: Purification of barley nitrate reductase and demonstration of nicked subunits. Phytochemistry, **22**, 2375-2382 (1983)
[69] Nakagawa, H.; Yonemura, Y.; Yamamoto, H.; Sato, T.; Ogura, N.; Sato, R.: Spinach nitrate reductase. Purification, molecular weight, and subunit composition. Plant Physiol., **77**, 124-128 (1985)
[70] Fido, R.J.; Notton, B.A.: Spinach nitrate reductase: further purification and removal of "nicked" sub-units by affinity chromatography. Plant Sci. Lett., **37**, 87-91 (1984)
[71] Redinbaugh, M.G.; Campbell, W.H.: Quaternary structure and composition of squash NADH:nitrate reductase. J. Biol. Chem., **260**, 3380-3385 (1985)
[72] Oji, Y.; Mamano, T.; Ryoma, Y.; Miki, Y.; Okamoto, S.: Purification and characterization of NADH-nitrate reductase from leaves of 2-row barley, and its

activity as affected by some metabolites. J. Plant Physiol., **119**, 247-256 (1985)
[73] Moureaux, T.; Leydecker, M.T.; Meyer, C.: Purification of nitrate reductase from Nicotiana plumbaginifolia by affinity chromatography using 5AMP-sepharose and monoclonal antibodies. Eur. J. Biochem., **179**, 617-620 (1989)
[74] Hyde, G.E.; Campbell, W.H.: High-level expression in Escherichia coli of the catalytically active flavin domain of corn leaf NADH:nitrate reductase and its comparison to human NADH:cytochrome B_5 reductase. Biochem. Biophys. Res. Commun., **168**, 1285-1291 (1990)
[75] Chamber-Perez, M.A.; Camacho, M.; de C, A.B.R.; Mercedes-Lucas, M.; Fernandez-Pascual, M.; Manclus, J.J.; Rosario de Felipe, M.: Nitrate reductase isozymes in Bradyrhizobium sp. (Lupinus) bacteroids: localisation, biochemical and kinetic characteristics. J. Plant Physiol., **159**, 525-533 (2002)
[76] Jolly, S.O.; Campbell, W.; Tolbert, N.E.: NADPH- and NADH-nitrate reductases from soybean leaves. Arch. Biochem. Biophys., **174**, 431-439 (1976)
[77] Tischler, C.R.; Purvis, A.C.; Jordan, W.R.: Factors involved in in vitro stabilization of nitrate reductase from cotton (Gossypium hirsutum L.) cotyledons. Plant Physiol., **61**, 714-717 (1978)
[78] Streit, L.; Nelson, R.S.; Harper, J.E.: Nitrate reductases from wild-type and nr1-mutant soybean (Glycine max L. merr.) leaves. Plant Physiol., **78**, 80-84 (1985)
[79] Skipper, L.; Campbell, W.H.; Mertens, J.A.; Lowe, D.J.: Pre-steady-state kinetic analysis of recombinant Arabidopsis NADH:nitrate reductase. Rate-limiting processes in catalysis. J. Biol. Chem., **276**, 26995-27002 (2001)
[80] Blasco, R.; Castillo, F.; Martinez-Luque, M.: The assimilatory nitrate reductase from the phototrophic bacterium, Rhodobacter capsulatus E1F1, is a flavoprotein. FEBS Lett., **414**, 45-49 (1997)
[81] Lim, B.S.; Holm, R.H.: Bis(Dithiolene)molybdenum analogues relevant to the DMSO reductase enzyme family: Synthesis, structures, and oxygen atom transfer reactions and kinetics. J. Am. Chem. Soc., **123**, 1920-1930 (2001)
[82] Trimboli, A.J.; Quinn, G.B.; Smith, E.T.; Barber, M.J.: Thiol modification and site directed mutagenesis of the flavin domain of spinach NADH:nitrate reductase. Arch. Biochem. Biophys., **331**, 117-126 (1996)
[83] Quinn, G.B.; Trimboli, A.J.; Prosser, I.M.; Barber, M.J.: Spectroscopic and kinetic properties of a recombinant form of the flavin domain of spinach NADH:nitrate reductase. Arch. Biochem. Biophys., **327**, 151-160 (1996)
[84] Barber, M.J.; Desai, S.K.; Marohnic, C.C.: Assimilatory nitrate reductase: lysine 741 participates in pyridine nucleotide binding via charge complementarity. Arch. Biochem. Biophys., **394**, 99-110 (2001)
[85] Kanayama, Y.; Kimura, K.; Nakamura, Y.; Ike, T.: Purification and characterization of nitrate reductase from nodule cytosol of soybean plants. Physiol. Plant., **105**, 396-401 (1999)
[86] Nakamura, Y.; Ikawa, T.: Purification and properties of NADH: nitrate reductase from the red alga Porphyra yezoensis. Plant Cell Physiol., **34**, 1239-1249 (1993)

[87] Lopes, P.F.; Oliveira, M.d.C.; Colepicolo, P.: Characterization and daily variation of nitrate reductase in Gracilaria tenuistipitata (Rhodophyta). Biochem. Biophys. Res. Commun., **295**, 50-54 (2002)
[88] Su, W.; Mertens, J.A.; Kanamaru, K.; Campbell, W.H.; Crawford, N.M.: Analysis of wild-type and mutant plant nitrate reductase expressed in the methylotrophic yeast Pichia pastoris. Plant Physiol., **115**, 1135-1143 (1997)
[89] Gao, Y.; Smith, G.J.; Albert, R.S.: Nitrate reductase from the marine diatom Skeletonema costatum. Biochemical and immunological characterization. Plant Physiol., **103**, 1437-1445 (1993)

Nitrate reductase [NAD(P)H] 1.7.1.2

1 Nomenclature

EC number
1.7.1.2

Systematic name
nitrite:NAD(P)$^+$ oxidoreductase

Recommended name
Nitrate reductase [NAD(P)H]

Synonyms
assimilatory NAD(P)H-nitrate reductase
assimilatory nitrate reductase
EC 1.6.6.2 (formerly)
NAD(P)$^+$:nitrate oxidoreductase
NAD(P)H bispecific nitrate reductase
NAD(P)H-nitrate reductase
nitrate reductase (NAD(P)H)
nitrate reductase (reduced nicotinamide adenine dinucleotide (phosphate))
nitrate reductase NAD(P)H
reductase, nitrate (reduced nicotinamide adenine dinucleotide (phosphate))

CAS registry number
9029-27-0

2 Source Organism

<1> *Candida nitratophila* [1, 3]
<2> *Hansenula anomala* (ascomycetous yeast, enzyme exists in 2 interconvertible forms, active and inactive, depending on the state of the molybdenum center [5]) [4-6]
<3> *Sphagnum magellanicum* [2]
<4> *Sphagnum palustre* [2]
<5> *Sphagnum pappilosum* [2]
<6> *Sphagnum squarrosum* [2]
<7> *Sphagnum subsecundum* [2]
<8> *Sphagnum cuspidatum* [2]
<9> *Sphagnum fallox* [2]
<10> *Sphagnum fimbriatum* [2]
<11> *Sphagnum molle* [2]

<12> *Sphagnum rubellum* [2]
<13> *Sphagnum subnitens* [2]
<14> *Chlorella variegata* [7]
<15> *Chlamydomonas reinhardtii* (mutant 104 [8]; mutant 305 [8]; complementation from mutants [8]) [8, 9]
<16> *Glycine max* (contains EC 1.6.6.1 and EC 1.6.6.2 [10]; NADH:nitrate oxidoreductase is more active in young cotyledons, as cotyledons age NAD(P)H:nitrate oxidoreductase becomes more active [11]) [10, 11]
<17> *Rhodotorula glutinis* [12, 20]
<18> *Cyanidium caldarium* (alga) [13]
<19> *Hordeum vulgare* (barley, mutant nar la [14]) [14]
<20> *Monoraphidium braunii* [15]
<21> *Ankistrodesmus braunii* [16-20]

3 Reaction and Specificity

Catalyzed reaction
nitrite + NAD(P)$^+$ + H_2O = nitrate + NAD(P)H + H^+

Reaction type
redox reaction

Natural substrates and products
S NAD(P)H + nitrate <2> (plays a role in first step of nitrate assimilation) [5]

Substrates and products
S NAD(P)H + nitrate <1-21> (other electron donors: combination of dithionite with flavin nucleotides or viologens, <17> [12]; $FMNH_2$, <3-13> [2]; reduced methyl viologen, <6, 17> [2, 12]; also reduction of ClO_3^-, $K_3Fe(CN)_6$, cytochrome c and 2,6-dichlorophenolindophenol with NADPH or NADH, <17> [12]) [1-20]
P NAD(P)$^+$ + nitrite + H_2O <1-21> [1-20]

Inhibitors
NAD(P)H <1, 2, 14, 15, 19> (preincubation with NAD(P)H and cyanide: inactivation, <1> [1]; preincubation with NAD(P)H alone: activation, <1> [1]; hydroxyl radical is involved in in vitro irreversible inactivation, <2> [5]; effect of temperature on inactivation, <2> [5]; FAD enhances inactivation, <2> [5]; in vitro an active diaphorase moiety is required for inactivation by reduced pyridine nucleotides, in vivo the absence of nitrate rather than the presence of ammonium is the triggering event for inactivation, <15> [9]) [1, 5, 7, 9, 14]
azide <17> [12]
cyanate <17> [12]
cyanide <1, 14, 17> (preincubation with NAD(P)H and cyanide: inactivation, <1> [1]) [1, 7, 12]
dithionite <19> [14]

nitrite <17> [12]
p-chloromercuribenzoate <1, 14, 17> (inhibition of diaphorase activity, <14> [7]; NADH or NADPH protect, <14> [7]) [1, 7, 12]
p-hydroxymercuribenzoate <21> [19]

Cofactors/prosthetic groups
FAD <1, 2, 17, 21> (prosthetic group, <21> [17]; essential for NAD(P)H-dependent activity, <21> [17]; maximum activity requires addition of flavin nucleotides, FAD more effective than FMN, <1> [1]; complete reduction of cytochrome by NADPH takes place only when added FAD is present in the enzyme preparation before addition of NADPH, <17> [12]; structure of FAD binding site, <21> [16]; enzyme contains 4 molecules of FAD, <21> [18]; activation, <2> [4]) [1, 4, 12, 16-18]
FMN <1, 2> (maximum activity requires addition of flavin nucleotides, FAD is more effective than FMN, <1> [1]; activation, <2> [4]) [1, 4]
NADH <1, 2, 14, 17, 21> (utilizes both NADH and NADPH, more active with NADH, <1,2> [1, 3, 4]; at high nitrate concentration: more active with NADPH, <17> [12]; at low nitrate concentration: more active with NADH, <17> [12]; NADPH to NADH activity ratio is 1.8, <21> [19]; preincubation with NAD(P)H and CN^- causes inactivation, preincubation with NAD(P)H activates, <1> [1]) [1, 3, 4, 7, 12, 19]
NADPH <1, 2, 14, 17, 21> (utilizes both NADH and NADPH, more active with NADH, <1,2> [1, 3, 4]; at high nitrate concentration: more active with NADPH, <17> [12]; at low nitrate concentration: more active with NADH, <17> [12]; NADPH to NADH activity ratio is 1.8, <21> [19]; preincubation with NAD(P)H and CN^- causes inactivation, preincubation with NAD(P)H alone activates, <1> [1]) [1, 3, 4, 7, 12, 19]
cytochrome b_{557} <1, 17, 21> (contains cytochrome b_{557}, <1, 17> [1, 3, 12]; 1 molecule per subunit, <1> [3]; 4 mol per mol of enzyme <21> [20]; 1 mol per mol of enzyme <17> [20]) [1, 3, 12, 20]
heme <21> (enzyme contains 4 heme groups, <21> [18]; minimum of 4 molecules of heme per molecule of native enzyme, <21> [19]) [18, 19]

Metals, ions
molybdenum <1, 2, 15, 17, 21> (molybdoprotein, <1, 2, 15, 17> [3, 4, 8, 12]; contains one molecule of molybdenum per subunit, <1> [3]; enzyme contains 2 atoms of molybdenum, <21> [18]) [3, 4, 8, 12, 18]

Turnover number (min^{-1})
28180 <21> (NADPH) [19]
Additional information <1, 16> (above 6000, <1> [1]) [1, 10]

Specific activity (U/mg)
0.718 <14> (NADH-nitrate reductase) [7]
0.766 <14> (NADPH-nitrate reductase) [7]
61.25 <21> [19]
72-80 <21> [19]
148.2 <17> [12]
217.4 <1> [1]

K_m-Value (mM)

0.000004 <21> (FAD, as a protector of NADH-cytochrome c reductase activity) [17]
0.0115 <14> (NADH) [7]
0.0145 <14> (NADPH) [7]
0.017 <1> (NADH) [1]
0.02 <17> (NADPH) [12]
0.03 <1> (NADPH) [1]
0.045 <2, 17> (NADPH, <2> [4]; nitrate + NADH, <17> [12]) [4, 12]
0.1 <2> (NADH) [4]
0.11 <1> (nitrate, + NADPH) [1]
0.12 <1> (nitrate, + NADH) [1]
0.125 <17> (nitrate, + NADPH) [12]
0.13 <14> (nitrate, + NADH) [7]
0.14 <14> (nitrate, + NADPH) [7]
0.16 <17> (NADH) [12]
Additional information <15, 17, 19, 21> [8, 14, 20]

pH-Optimum

7 <1, 15> [3, 8]
7.4-7.7 <14> [7]
7.5 <17, 21> [12, 20]
7.8-8.2 <18> [13]

pH-Range

7-9 <18> (at pH 7 and 9: about 60% of activity maximum) [13]
7.1-8.2 <14> (at pH 7.1 and 8.2: about 50% of activity maximum) [7]

4 Enzyme Structure

Molecular weight

230000 <17> (sucrose density gradient centrifugation, gel filtration) [12]
365000 <1> (sedimentation equilibrium centrifugation) [1]
460000-467400 <21> (gel filtration, density gradient centrifugation, disc gel electrophoresis, pore gradient electrophoresis, sucrose-density gradient centrifugation) [18, 19]

Subunits

dimer <17> (2 * 118000, SDS-PAGE) [12]
octamer <21> (8 * 58750-59000, SDS-PAGE) [18, 19]
tetramer <1> (4 * 95000, SDS-PAGE) [1]

5 Isolation/Preparation/Mutation/Application

Source/tissue

cotyledon <16> [11]

Localization

chloroplast <20> (specifically located in the pyrenoid region of chloroplast) [15]
soluble <15, 21> [8, 17]
Additional information <2> (not membrane-bound) [4]

Purification

<1> [1, 3]
<2> [4]
<3> [2]
<4> [2]
<5> [2]
<6> [2]
<7> [2]
<8> [2]
<9> [2]
<10> [2]
<11> [2]
<12> [2]
<13> [2]
<14> [7]
<15> [8]
<16> [10]
<17> [12]
<19> (partial) [14]
<21> [18, 19]

6 Stability

Temperature stability

30 <2> (inactivation) [4]
60 <18> (3 min, complete loss of NADPH-nitrate reductase activity) [13]
Additional information <18> (heat activation of benzyl viologen:nitrate reductase activity) [13]

General stability information

<1>, labile to dilution with protein-free buffer, 2.5 mg bovine serum albumin, stabilizes [1]

Storage stability

<1>, 4°C, 50% loss of activity after 7 days [1]

References

[1] Hipkin, C.R.; Ali, A.H.; Cannons, A.: Structure and properties of assimilatory nitrate reductase from the yeast Candida nitratophila. J. Gen. Microbiol., **132**, 1997-2003 (1986)

[2] Deising, H.: Nitrate reductase from Sphagnum species: isolation, in vitro assays and partial purification. Z. Naturforsch. C, **42**, 653-656 (1987)

[3] Hipkin, C.R.; Kau, D.A.; Cannons, A.C.; Jones, D.H.; Gallon, J.R.; Kay, C.J.; Barber, M.J.; Solomonson, L.P.: Physical and chemical properties of nitrate reductase from Candida nitratophila. Biochem. Soc. Trans., **17**, 928-929 (1989)

[4] Minagawa, N.; Yoshimoto, A.: Assimilatory nitrate reductase of Hansenula anomala: its electron donors and cellular distribution. Agric. Biol. Chem., **47**, 125-127 (1983)

[5] Minagawa, N.; Yoshimoto, A.: The in vivo inactivation and reactivation of assimilatory nitrate reductase in Hansenula anomala. Agric. Biol. Chem., **49**, 2217-2219 (1985)

[6] Minagawa, N.; Yoshimoto, A.: The in vitro irreversible inactivation by NAD(P)H of assimilatory nitrate reductase from Hansenula anomala. Agric. Biol. Chem., **49**, 2213-2215 (1985)

[7] Hipkin, C.R.; Al-Bassam, B.A.; Syrett, P.J.: Pyridine nucleotide specificity and other properties of purified nitrate reductase from Chlorella variegata. Planta, **144**, 137-141 (1979)

[8] Fernandez, E.; Cardenas, J.: In vitro complementation of assimilatory NAD(P)H-nitrate reductase from mutants of Chlamydomonas reinhardii. Biochim. Biophys. Acta, **657**, 1-12 (1981)

[9] Cordoba, F.; Cardenas, J.; Fenandez, E.: Role of the diaphorase moiety on the reversible inactivation of the Chlamydomonas reinhardii nitrate reductase complex. Biochim. Biophys. Acta, **827**, 8-13 (1985)

[10] Streit, L.; Nelson, R.S.; Harper, J.E.: Nitrate reductases from wild-type and nr1-mutant soybean (Glycine max L. merr.) leaves. Plant Physiol., **78**, 80-84 (1985)

[11] Orihuel-Iranzo, B.; Campbell, W.H.: Development of NAD(P)H: and NADH:nitrate reductase activities in soybean cotyledons. Plant Physiol., **65**, 595-599 (1980)

[12] Guerrero, M.G.; Gutierrez, M.: Purification and properties of the NAD(P)H:nitrate reductase of the yeast Rhodotorula glutinis. Biochim. Biophys. Acta, **482**, 272-285 (1977)

[13] Rigano, C.: Studies on nitrate reductase from Cyanidium caldarium. Arch. Mikrobiol., **76**, 265-276 (1971)

[14] Harker, A.R.; Narayanan, K.R.; Warner, R.L.; Kleinhofs, A.: NAD(P)H bispecific nitrate reductase in barley leaves: partial purification and characterization. Phytochemistry, **25**, 1275-1279 (1986)

[15] Lopez-Ruiz, A.; Roldan, J.M.; Verbelen, J.P.; Diez, J.: Nitrate reductase of green algae is located in the pyrenoid. Plant Physiol., **78**, 614-618 (1985)

[16] Marquez, A.J.; De La Rosa, M.A, Vega, J.M.: Studies by affinity chromatography on the NAD(P)H and FAD sites of nitrate reductase from Ankistrodesmus braunii. J. Chromatogr., **235**, 435-443 (1982)
[17] De La Rosa, M.A.; Marquez, A.J.; Vega, J.M.: Dissociation of FAD from the NAD(P)H-nitrate reductase complex from Ankistrodesmus braunii and role of flavin in catalysis. Z. Naturforsch. C, **37**, 24-30 (1982)
[18] De La Rosa, M.A.; Vega, J.M.; Zumft, W.G.: Composition and structure of assimilatory nitrate reductase from Ankistrodesmus braunii. J. Biol. Chem., **256**, 5814-5819 (1981)
[19] De La Rosa, M.A.; Diez, J.; Vega, J.M.; Losada, M.: Purification and properties of assimilatory nitrate reductase [NAD(P)H] from Ankistrodesmus braunii. Eur. J. Biochem., **106**, 249-256 (1980)
[20] Guerrero, M.G.; Vega, J.M.; Losada, M.: The assimilatory nitrate-reducing system and its regulation. Annu. Rev. Plant Physiol., **32**, 169-204 (1981)

Nitrate reductase (NADPH) 1.7.1.3

1 Nomenclature

EC number
1.7.1.3

Systematic name
nitrite:$NADP^+$ oxidoreductase

Recommended name
nitrate reductase (NADPH)

Synonyms
EC 1.6.6.3 (formerly)
NADPH-dependent nitrate reductase
NADPH-nitrate reductase
NADPH2:nitrate oxidoreductase
NADPH:nitrate reductase
assimilatory NADPH-nitrate reductase
assimilatory nitrate reductase
assimilatory reduced nicotinamide adenine dinucleotide phosphate-nitrate reductase
nitrate reductase (NADPH)
nitrate reductase (reduced nicotinamide adenine dinucleotide phosphate)
triphosphopyridine nucleotide-nitrate reductase

CAS registry number
9029-28-1

2 Source Organism

<1> *Aspergillus nidulans* (biA-1 [1]; bi-1 a, translocation-free biotin auxotroph [2]; IFO No. 6083 [5]; biA1, biotin-requiring [6]) [1, 2, 5-7, 13]
<2> *Neurospora crassa* (STA4 [4]; wild type 146 [8]; 5297a [9]; nit-1 mutant [11]; wild-type STA4, nit-1 and nit-3 mutants [14]; wild-type STA4, FGSC 358 and nit-1 mutant, FGSC 262 [15]) [4, 8-12, 14, 15]
<3> *Funaria hygrometrica* (Hedw. [3]) [3]
<4> *Hordeum vulgare* (L. cv. Steptoe nar1a and nar1a:nar7w genotypes [11]) [11]

3 Reaction and Specificity

Catalyzed reaction

nitrite + $NADP^+$ + H_2O = nitrate + NADPH + H^+ (An iron-sulfur molybdenum flavoprotein; <1> random order rapid-equilibrium mechanism, two functional parts: 1. heat labile, FAD and haemoprotein containing, binds NADPH and transfers electrons from NADPH via FAD and perhaps cytochrome b to acceptors such as cytochrome c, 2. heat stable, molybdenum containing, accepts electrons from reduced viologen dyes and transfers them to nitrate [2]; <2> multicenter redox enzyme. Ser920, Arg921 and Arg932 are suggested to be the key enzymes to investigate for a role in determining pyridine nucleotide specificity. Arg932 may be playing a role in binding the adenine ring of NADPH [10]; <2> sulfhydryl groups may participate in the binding of the protein subunits [15])

Reaction type

oxidation
redox reaction
reduction

Natural substrates and products

S nitrate + NADPH <1-4> (<2> first step in the reduction of nitrate to ammonia, biosynthesis of amino acids and other nitrogen-containing cell constituents [4]; <2,4> key enzyme in the assimilation path of nitrate to ammonium. Crude extracts possess endogenous NADPH regenerating systems capable of providing reducing equivalents for effective nitrate reduction in vitro [11]; <2> first step in nitrate assimilation [14]) (Reversibility: ? <1-4> [1-15]) [1-15]

P nitrite + $NADP^+$ + H_2O <1-4> [1-15]

Substrates and products

S chlorate + NADPH <1> (<1> lower affinity than to nitrate, appears to be toxic or its product [2]) (Reversibility: ? <1> [2]) [2]

P chlorite + $NADP^+$ <1> [2]

S nitrate + NADPH <1-4> (<1> some mutants use hypoxanthine as nitrogen source, 4.5-S cytochrome-c reductase activity [1]; <1> two associated activities: cytochrome c reductase and reduced viologen dye:nitrate reductase [2]; <3> associated cytochrome c reductase activity [3]; <1,2> 4 activities: NADPH-nitrate reductase, FADH-nitrate reductase, reduced methyl viologen-nitrate reductase and NADPH-cytochrome c reductase [4,5,12,14]; <1> catalyzes the NADPH-linked reduction of ferricyanide and 2,6-dichlorophenolindophenol, chlorate- and bromate-dependent NADPH oxidation, and FMNH-linked nitrate reduction [5]; <2> 20-fold higher activity with NADPH than with NADH [9]; <2> associated FAD-nitrate reductase and methylviologen-nitrate activity [15]) (Reversibility: ? <1-4> [1-6, 10-15]; ir <2> [9]) [1-6, 9-15]

P nitrite + $NADP^+$ + H_2O <1-4> [1-6, 9-15]

Inhibitors

8-hydroxyquinoline <1, 2> (<1> pH 7.5, 20% inhibition at 0.1 mM [2]; <2> 94% inhibition at 10 mM [4]) [2, 4, 9]
Cu^{2+} <2> (<2> inhibitory [9]) [9]
EDTA <2> [4]
$NADP^+$ <1> (<1> inhibits NADPH oxidation, non-competitive with respect to nitrate [2]) [2]
NADPH <2, 4> (<2,4> elevated concentrations [11]) [11]
azide <1, 2> (<1> pH 7.5, 90% inhibition at 0.01 mM [2]; <2> 96% inhibition at 5 mM [4]) [2, 4, 5, 9]
cyanide <1, 2> (<1> pH 7.5, 100% inhibition at 0.01 mM [2]; <1> 99% inhibition at 10 mM [5]; <2> 1-10 mM inhibits enzyme by 70-90% [8]) [2, 4, 5, 8, 9, 13]
iodoacetamide <1> (<1> pH 7.5, 40% inhibition at 1 mM [2]) [2]
nitrite <1> (<1> competitive with respect to nitrate and non-competitive to NADPH [2]) [2]
o-phenanthroline <2> [9]
p-chloromercuribenzoate <1, 2> (<1> reversion of the inhibition by addition of reduced glutathione [5]; <2> reversed by cysteine or glutathione [9]) [5, 9, 13]
p-hyroxymercuribenzoate <1, 2> (<1> pH 7.5, 85% inhibition at 0.001 mM [2]; <2> reversed by sulfhydryl reagents [4]; <2> at 0.1 mM complete inhibition [15]) [2-4, 15]
phenanthroline hydrate <1> (<1> pH 7.5, 1:10, 20% inhibition at 0.1 mM [2]) [2]
phenylglyoxal <2> (<2> in 0.1 M phosphate, pH 7.3, 4 mM, inactivation after 15 min to 40% and to 20% after 60 min [4]) [4]
potassium chlorate <1> (<1> pH 7.5, slight, 20% inhibition at 5 mM [2]) [2]
potassium ethyl xanthate <2> [9]
sodium nitrite <1> (<1> pH 7.5, slight, 50% inhibition at 5 mM [2]) [2]
thiourea <2> [9]
Additional information <1> (<1> cysteine or dithiothreitol relieve or prevent inhibition. Not inhibited by urea, glutamic acid, aspartic acid and ammonia at 10 mM [2]; <2> protection of inactivation by FAD and restorage by dithiothreitol [4]) [2, 4]

Cofactors/prosthetic groups

FAD <1, 2, 4> (<1,2> stimulated by flavin [2,9]; <2> 1 mol per subunit is assumed [4]; <1> molybdoflavoprotein [5,7]; <1> contains 5.91-7.78 nmol FAD per mg of protein, activation of NADPH-nitrate reductase activity [5]; <1> indispensible role in nitrate reductase complex [7]; <2> flavoprotein, prosthetic group [9]; <1> 1 mol per mol of enzyme [13]) [2, 4, 5, 7, 9-13, 15]
FMN <2> (<2> to a lesser extent than FAD [9]) [9]
cytochrome b_{557} <1, 2> (<2> heme-iron, 2 mol per mol enzyme [4]; <1> involved in intracellular electron transport from NADPH to nitrate [5]) [4, 5, 14, 15]

heme <1> (<2> 2 mol heme iron per mol enzyme protein [14]) [2, 10, 13, 14]
Additional information <1, 2> (<1> NADH: no effect, presence of protoheme IX [5]; <2> cytochrome a or b involved [9]) [5, 9]

Activating compounds

FAD <1> (<1> activates 10-fold by addition of 0.004 mM, stabilizes against heat inactivation [5]) [5]
o-phenanthroline <2> (<2> 40% increase of activity at 2.5 mM [4]) [4]
Additional information <2> (<2> enzyme can be reactivated by molybdenum and dithioerythritol [14]) [14]

Metals, ions

NH_4^+ <3> (<3> represses the intracellular NO_3^- assimilation, which is required for the induction and maintenace of NADPH nitrate reductase, enzyme is regulated by NH_4^+ at the transcriptional level [3]) [3]
molybdenum <1, 2, 4> (<1> molybdenum containing cofactor [1]; <2> 1 mol per mol protein [4]; <1> molybdoflavoprotein [5,7]; <1> part of cofactor [6]; <2> reactivation after cyanide inhibition [8]; <2> 1 mol per mol enzyme protein, activation of enzyme after urea treatment, little stimulation [14]) [1, 2, 4-8, 10-12, 14]
phosphate <1> (stimulates, required for maximal activity [2]) [2]

Turnover number (min^{-1})

1080 <2> (NADPH, <2> R920D/R932S double mutant [10]) [10]
10800 <2> (NADPH, <2> S920D mutant [10]) [10]
16500 <2> (NADH, <2> R932S mutant [10]) [10]
17400 <2> (NADH, <2> wild type [10]) [10]
27000 <2> (NADPH, <2> R932S mutant [10]) [10]
35400 <2> (NADH, <2> R920D/R932S double mutant [10]) [10]
84000 <2> (NADPH, <2> wild type [10]) [10]
102000 <2> (NADH, <2> S920D mutant [10]) [10]

Specific activity (U/mg)

0.004-0.074 <1> (<1> various temperature-sensitive mutants [1]) [1]
0.063 <2> (<2> nit-1 mutant [12]) [12]
0.074 <1> (<1> wild type [1]) [1]
1.24 <2> (<2> wild-type [12]) [12]
2.5 <1> [2]
4-16.1 <1> (<1> at different stages of purification [2]) [2]
8.55 <2> [15]
15.5 <1> [5]
16 <1> [13]
Additional information <2> (<2> one unit is defined as that amount of enzyme which results in the formation of 0.000001 mM of nitrite [9]) [4, 9]

K_m-Value (mM)

0.009 <1> (NADPH, <1> + nitrate, pH 6.5 and + cytochrome c, pH 7.5 [2]) [2]
0.01 <1> (NADPH, <1> + nitrate, pH 7.5 [2]) [2]

0.012-0.015 <3> (nitrate) [3]
0.013 <2> (NADPH, <2> wild type [10]) [10]
0.015 <1> (NADPH, <1> + cytochrome c, pH 8.5 [2]) [2]
0.022-0.05 <2> (NADPH) [12]
0.024 <1> (cytochrome c, <1> + NADPH, pH 8.5 [2]) [2]
0.029 <1> (cytochrome c, <1> + NADPH, pH 7.5 [2]) [2]
0.045 <1> (NADPH, <1> + nitrate [5]) [5]
0.06 <1> (nitrate, <1> + NADPH, pH 7.5 [2]) [2]
0.08 <1> (nitrate, <1> + NADPH, pH 6.5 [2]) [2]
0.2 <1> (nitrate, <1> + NADPH [5]) [5]
0.25 <2> (NADPH, <2> R932S mutant [10]) [10]
0.25 <2> (NADPH, <2> S920D mutant [10]) [10]
0.29 <2> (nitrate) [12]
0.4 <2> (NADH, <2> S920D mutant [10]) [10]
2.3 <2> (NADH, <2> S920D/R932S double mutant [10]) [10]
2.5 <1> ($FADH_2$, <1> + nitrate [5]) [5]
2.6 <2> (NADH, <2> R932S mutant [10]) [10]
3 <1> ($FMNH_2$, <1> + nitrate [5]) [5]
3 <2> (NADH, <2> wild type [10]) [10]
4.2 <2> (NADPH, <2> S920D/R932S double mutant [10]) [10]

K_i-Value (mM)

0.04-0.048 <1> ($NADP^+$, <1> NADPH as substrate in saturating and non-saturating concentrations, competitive [2]) [2]
0.065 <1> ($NADP^+$, <1> nitrate as substrate in non-saturating concentrations, non-competitive [2]) [2]
0.15 <1> (nitrite, <1> NADPH as substrate in non-saturation concentration, non-competitive [2]) [2]
0.18 <1> (nitrite, <1> nitrate as substrate in saturating and non-saturating concentrations, competitive [2]) [2]

pH-Optimum

6 <2> (<2> wild type after his-tag is removed [10]) [10]
7 <2> (<2> his-tagged wild type [10]) [9, 10]
7-8 <1> [5]
7.2 <2> (<2> assay at [11]) [11]
7.3 <2> (<2> assay at [12]) [12]
7.4 <1> (<1> assay at [5]) [5]
7.5 <1, 3> (<1,3> assay at [2,3,10]) [2, 3, 10]
7.5-8 <1> [2]
7.7 <4> (<4> assay at [11]) [11]
Additional information <2> (<2> optima of associated activities [15]) [15]

pH-Range

6-9 <2> [9]

Temperature optimum (°C)

23-28 <2> (<2> assay at [9]) [9]
25 <1, 2> (<1> assay at [2,10,12,13]) [2, 10, 12, 13]

28 <2, 4> (<2,4> assay at [11]) [11]
30 <1, 3> (<1,3> assay at [3,5]) [3, 5]

4 Enzyme Structure

Molecular weight

180000 <1> (<1> gel filtration, sucrose density gradient [5]) [5, 6]
180000-200000 <1> (<1> sucrose density gradient, gel-filtration [2]) [2]
197000 <1> (<1> gel filtration [13]) [13]
204000 <2> (<2> nit-3 enzyme, gel filtration [14]) [14]
230000 <2> (<2> sucrose density gradient, gel filtration [4]) [4]
235000 <2> (<2> sucrose density gradient centrifugation, gel filtration [15]) [15]
272000 <2> (<2> wild-type, gel filtration [14]) [14]
290000 <2> [12]
450000 <3> (<3> native gel electrophoresis [3]) [3]

Subunits

dimer <2> (<2> 1 * 130000 + 1 * 115000, SDS-PAGE, homodimer of 2 * 150000 suggested [4]; <1> 2 * 91000, homodimer, SDS-PAGE, smaller bands are caused by proteolytic cleavage [6]; <2> homodimer, 2 * 145000 [12]; <2> 2 * 97000: nit-3 enzyme, 2 * 132000, wild-type [14]) [4, 6, 12, 14]
multimer <1> [1, 7]
tetramer <1, 3> (<3> 4 * 116000 [3]; <1> 2 * 59000 + 2 * 38000, SDS-PAGE, after heat-treatment all polypeptide chains are 59 kDa [5]) [3, 5]
Additional information <2> (<2> nit-1 mutant enzyme is the apoprotein of nitrate reductase [12,14]; <2> aggregate of two different polypeptide chains: one responsible for transport of electrons fom NADPH to FAD or cytochrome c, nit-1 enzyme. The other transfers electrons from FAD via molybdenum to nitrate, nit-3 enzyme [14]; <2> at least 2 subunits suggested [15]) [12, 15]

5 Isolation/Preparation/Mutation/Application

Source/tissue

chloronema <3> [3]
mycelium <1, 2> [4, 6, 9, 11, 13, 14]

Purification

<1> (ion-exchange, ammonium sulfate precipitation, gel filtration [2]; ammonium sulfate precipitation, ion-exchange, gel filtration, hydroxylapatite column. FAD and EDTA essential in all buffers during purification [5]; quick purification by immunoprecipitation with monospecific anti-nitrate reductase serum, stabilization during purification and protection of proteolytic cleavage by addition of phenylmethylsulphonyl fluoride [6]; salt fractionation, gel filtration, ion-exchange [13]) [2, 5, 6, 13]

<2> (streptomycin sulfate and ammonium sulfate precipitation, ion-exchange, gel filtration, isoelectric focusing, FAD-affinity [4]; ammonium sulfate precipitation [8]; ammonium sulfate precipitation [9]; metal-chelate affinity chromatography for His-tagged proteins, ammonium sulfate fractionation, affinity chromatography [10]; gel filtration [11]; protamine sulfate and ammonium sulfate precipitation, hemoglobin-Sepharose column, Bio-Gel A, FAD-Sepharose affinity column. proteolysis during extensive purification [12]; partially purification of wild-type and nit-3 mutant by streptomycin sulfate precipitation and ammonium sulfate fractionation, of nit-1 mutant by protamine sulfate fractionation, ammonium sulfate precipitation and gel filtration [14]; partially purified [15]) [4, 8, 9, 12, 14, 15]

Cloning

<2> (wild-type and mutants are expressed in Escherichia coli JM109(DE3)-pLysS [10]) [10]

Engineering

R921S <2> (<2> little impact on NADPH and NADH activity, no importance for pyridine nucleotide specificity [10]) [10]

R921T <2> (<2> little impact on NADPH and NADH activity, no importance for pyridine nucleotide specificity [10]) [10]

R932Q <2> (<2> 1/4 wild type NADPH activity is retained, twice as much NADH activity is present as compared to wild type [10]) [10]

R932S <2> (<2> 1/10 wild type NADPH activity is retained, 2/3 of wild type NADH activity [10]) [10]

S920D <2> (<2> important for the enzyme's interaction with the pyridine nucleotide substrates. Mutant retains ~2% of the NADPH activity of the wild type while it has an increased NADH activity, ~15% higher. It is concluded that Ser920 is a ligand involved in binding the 2' phosphate of NADPH in the wild type enzyme [10]) [10]

S920D/R932S <2> (<2> greatest decrease in NADPH activity of all created mutants, shows that Arg932 is a residue interacting with the pyridine nucleotide coenzyme electron donors and that Ser920 and Arg932 have effects on substrate binding and catalytic activity. Both residues may be ligands to the 2'-phosphate of NADPH in the wild type cyt b reductase fragment of nitrate reductase [10]) [10]

Additional information <1, 2> (<1> studies of temperature-sensitive mutations, niaD gene: mutation leads to loss of a 4.5-S cytochrome-c reductase activity, which is a subunit of nitrate reductase. It is suggested that neither the product of the cnxE nor the cnyF genes form part of the nitrate reductase molecule, but some catalytic role in cofactor formation, niaD and cnxH seem to be structural genes [1]; <1> different structural gene (niaD) and cofactor gene (cnx) mutants are analyzed concerning their flavin and molybdenum content [7]; <2> several mutations of recombinant cyt b reductase fragment of nitrate reductase in the region Ser920, Arg921 and Arg932 are created. Conversion from NADPH-specific to virtually NADH-specific cyt b reductase fragment of nitrate reductase [10]; <2> nit-1 mutant, suggested to produce the complete apoenzyme [12]; <2> nit-1 mutant: lacks all activities except

FAD-dependent NADPH:cytochrome c reductase activity, nit-2 mutant: reduced FAD:- and reduced methyl viologen:nitrate reductase activities but lacks the other two activities [14]; <2> nit-3 mutant (FGSC 262): reduced FAD-nitrate reductase and reduced methylviologen-nitrate reductase activities [15]) [1, 7, 10, 12, 14, 15]

6 Stability

Temperature stability

0 <1> (<1> half-life: 36 h [13]) [13]
10 <2, 4> (<2,4> quite stable for at least 24 h [11]) [11]
25 <1> (<1> wild-type half-life: 120 min [1]; <1> half-life: 12 h [13]) [1, 13]
30 <1> (<1> wild-type half-life: 66 min [1]) [1]
35 <1> (<1> wild-type half-life: 18.8 min, various temperature sensitive mutant strains: 3.8-18.7 min [1]) [1]
37 <1> (<1> wild-type half-life: 14.5 min [1]; <1> absence of FAD, rapid loss of activity, decreases to 50% in less than 20 min [5]; <1> half-life: 20 min [13]) [1, 5, 13]
40 <2> (<2> fraction V, pH 7.0, 50% loss of activity within 5 min [9]) [9]
40 <2> (<2> less than 50% loss of activity in 60 min for wild type and S920D mutant [10]) [10]
47 <1> (<1> half-life: 30 sec [13]) [13]
49 <2> (<2> 50% loss of activity after 1.5 min [12]) [12]
50 <2> (<2> fraction V, pH 7.0, loses all of its activity in 5 min [9]) [9]
60 <2> (<2> labile, loss of acitvity after 2 min [15]) [15]

Organic solvent stability

ethanol <2> (<2> mild denaturation [4]) [4]

General stability information

<2>, glutathione stabilizes the enzyme at 1 mM, cysteine as well but not as effective [8]
<3>, loss of activity with ammonium chloride or ammonium nitrate at 10 mM to growing cells, faster decay in washed and resuspended cells [3]

Storage stability

<1>, -80°C, 0.02 mM FAD, several months with only slight loss of the activity [5]
<1>, 4°C, half life of 3-5 days [2]
<2>, -10°C, 0.1 M sodium phosphate buffer, pH 7.3, 0.17 M NaCl, 1 mM dithiothreitol, 5 m M EDTA, 0.5 mM PMSF, 1% ethanol, 30% glycerol [12]
<2>, -10°C, 50 mM sodium phosphate, pH 6.9, 30% glycerol, 0.5 mM EDTA, 0.5 mM dithiothreitol, 0.5 mM phenylmethylsulfonylfluoride, 0.1 mM FAD, stable for at least 6 months [4]
<2>, -15°C, fraction II, most stable, optimal pH: 7.0, 10-20% loss of activity after a month [9]
<2>, -15°C, fraction IV, one week, loses about half of its activity [9]

<2>, -15°C, fraction V, quite unstable, overnight, loses at least half of its activity [9]
<2>, -20°C, stable for 6 months or longer, 5% loss of activty [15]
<2>, -80°C, 25 mM Mops, 0.1 mM EDTA, pH 7.2 [10]
<2>, -80°C, extract, 6 months, without loss of activity [11]
<2>, 4°C or -15°C, fraction III, overnight, 50% loss of activity [9]
<2>, 5°C, urea-treated, 0.1 M potassium phosphate buffer, pH 7.3, 5 mM EDTA, 2 mM dithioerythritol, 0.5 mg/ml or more bovine serum albumin, stable for 18 h without loss of activity [14]
<3>, half life of about 2 h with ammonium sulfate or ammonium chloride as sole N-source [3]

References

[1] MacDonald, D.W.; Cove, D.J.: Studies on temperature-sensitive mutants affecting the assimilatory nitrate reductase of Aspergillus nidulans. Eur. J. Biochem., **47**, 107-110 (1974)
[2] McDonald, D.W.; Coddington, A.: Properties of the assimilatory nitrate reductase from Aspergillus nidulans. Eur. J. Biochem., **46**, 169-178 (1974)
[3] Padidam , M.; Venkateswarlu, K.; Johri, M.M.: Ammonium represses NADPH-nitrate reductase in the moss Funaria hygrometrica. Plant Sci., **75**, 185-194 (1991)
[4] Pan, S.S.; Nason, A.: Purification and characterization of homogeneous assimilatory reduced nicotinamide adenine dinucleotide phosphate-nitrate reductase from Neurospora crassa. Biochim. Biophys. Acta, **523**, 297-313 (1978)
[5] Minagawa, N.; Yoshimoto, A.: Purification and characterization of the assimilatory NADPH-nitrate reductase of Aspergillus nidulans. J. Biochem., **91**, 761-774 (1982)
[6] Cooley, R.N.; Tomsett, A.B.: Determination of the subunit size of NADPH-nitrate reductase from Aspergillus nidulans. Biochim. Biophys. Acta, **831**, 89-93 (1985)
[7] Downey, R.J.: The role of molybdenum in formation of the NADPH-nitrate reductase by Aspergillus nidulans. Biochem. Biophys. Res. Commun., **50**, 920-925 (1973)
[8] Nicholas, D.J.D.; Nason, A.: Molybdenum and nitrate reductase. II. Molybdenum as a constituent of nitrate reductase. J. Biol. Chem., **207**, 353-360 (1954)
[9] Nason, A.; Evans, H.J.: Triphosphopyridine nucleotide-nitrate reductase in Neurospora. J. Biol. Chem., **202**, 655-673 (1953)
[10] Shiraishi, N.; Croy, C.; Kaur, J.; Campbell, W.H.: Engineering of pyridine nucleotide specificity of nitrate reductase: mutagenesis of recombinant cytochrome b reductase fragment of Neurospora crassa NADPH:nitrate reductase. Arch. Biochem. Biophys., **358**, 104-115 (1998)

[11] Savidov, N.A.; Alikulov, Z.A.; Lips, S.H.: Identification of an endogenous NADPH-regenerating system coupled to nitrate reduction in vitro in plant and fungal crude extracts. Plant Sci., **133**, 33-45 (1998)
[12] Horner, R.D.: Purification and comparison of nit-1 and wild-type NADPH:-nitrate reductases of Neurospora crassa. Biochim. Biophys. Acta, **744**, 7-15 (1983)
[13] Downey, R.J.: Characterization of the reduced nicotinamide adenine dinucleotide phosphate-nitrate reductase of Aspergillus nidulans. J. Bacteriol., **105**, 759-768 (1971)
[14] Tachiki, T.; Nason, A.: Preparation and proterties of apoenzyme of nitrate reductases from wild-type and nit-3 mutant of Neurospora crassa. Biochim. Biophys. Acta, **744**, 16-22 (1983)
[15] Antoine, A.D.: Purification and properties of the nitrate reductase isolated from Neurospora crassa mutant nit-3. Kinetics, molecular weight determination, and cytochrome involvement. Biochemistry, **13**, 2289-2294 (1974)

Nitrite reductase [NAD(P)H] 1.7.1.4

1 Nomenclature

EC number
1.7.1.4

Systematic name
ammonium-hydroxide:$NAD(P)^+$ oxidoreductase

Recommended name
nitrite reductase [NAD(P)H]

Synonyms
NAD(P)H:nitrite oxidoreductase
NADH-nitrite oxidoreductase
NADPH-nitrite reductase
assimilitory nitrite reductase
nitrite reductase (reduced nicotinamide adenine dinucleotide (phosphate))
nitrite reductase [$NAD(P)H_2$]

CAS registry number
9029-29-2

2 Source Organism

<1> *Escherichia coli* (strain OR75Ch15 [1]; K12 [2, 4-6]; strain Bn [7]) [1-7, 19]
<2> *Neurospora crassa* [8-13]
<3> *Azotobacter chroococcum* [14]
<4> *Derxia gummosa* [15]
<5> *Azotobacter agile* [16]
<6> *Torulopsis nitratophila* [17]
<7> *Staphylococcus carnosus* [18]
<8> *Candida utilis* [20, 21]
<9> *Bacillus subtilis* [22]

3 Reaction and Specificity

Catalyzed reaction
ammonium hydroxide + 3 $NAD(P)^+$ + H_2O = nitrite + 3 NAD(P)H + 3 H^+

Reaction type

oxidation
redox reaction
reduction

Natural substrates and products

S NAD(P)H + nitrite + H^+ <1, 2, 5, 9> (<5> nitrite assimilation [16]; <1> enzyme is thought to function as a detoxifying mechanism for the nitrite generated by nitrate respiration [19]; <9> functions as both an assimilatory and a dissimilatory enzyme, essential for nitrite respiration during anaerobic growth [22]) (Reversibility: ? <1, 2, 5, 9> [1, 8, 16, 19, 22]) [1, 8, 12, 16, 19, 22]

P $NAD(P)^+$ + NH_3 + H_2O <1, 2, 5, 9> [1, 8, 12, 16, 19, 22]

Substrates and products

S $Fe(CN)_6^{3-}$ + NADH <1> (<1> one-electron reduction [5]) (Reversibility: ? <1> [5]) [5]

P $Fe(CN)_6^{4-}$ + NAD^+ <1> [5]

S NAD(P)H + NH_2OH <1, 2, 5> (<2> NADPH is preferred [9]) (Reversibility: ? <1, 2, 5> [1, 9]) [1, 9, 12, 16]

P $NAD(P)^+$ + NH_3 + H_2O <1, 2, 5> [1, 9, 12, 16]

S NAD(P)H + nitrite + H^+ <1-9> (<1> specific for NADH [7]; <2> NADPH is preferred [9]; <2> NADH is preferred [13]; <3> NADH is preferred, $S_2O_4^{2-}$ can act as electron donor, methyl viologen as electron carrier [14]; <4> NADH is preferred, reduced benzyl viologen or methyl viologen can act as artificial electron donors [15]; <6> specific for NADPH, can utilize reduced benzyl or methyl viologen as electron donor [17]; <8> enzyme can utilize NAD(P)H and methyl viologen as electron donors, NADPH is more effective than NADH [20]; <9> much lower activity in cell extracts with NADPH [22]) (Reversibility: ? <1-9> [1, 8, 14, 15, 16, 17, 18, 20, 22]) [1, 7, 8, 9, 12, 13, 14, 15, 16, 17, 18, 20, 22]

P $NAD(P)^+$ + NH_3 + H_2O <1-9> [1, 7, 8, 9, 12, 13, 14, 15, 16, 17, 18, 20, 22]

S oxidized 2,6-dichlorophenol-indophenol + NADH <1> (<1> one-electron reduction [5]) (Reversibility: ? <1> [5]) [5]

P reduced 2,6-dichlorophenol-indophenol + NAD^+ <1> [5]

S oxidized horse heart cytochrome c + NADH <1> (<1> one-electron reduction [5]) (Reversibility: ? <1> [5]) [5]

P reduced horse heart cytochrome c + NAD^+ <1> [5]

Inhibitors

1,10-phenanthroline <1, 2> [1, 12, 13]

1-cyclohexyl-3-(2-morpholinoethyl)-carbodiimide metho-*p*-toluenesulfonate <4> (<4> preincubation with NO_2^- completely protects [15]) [15]

2,2'-bipyridine <1, 2, 4> (<4> 1 mM, 40% inhibition, no inhibition if enzyme is preincubated with nitrite or NH_2OH prior to adding the inhibitor [15]) [1, 13, 15]

2-heptyl-4-hydroxyquinoline N-oxide <2> [13]

8-hydroxyquinoline <2, 4> (<4> 1 mM, 96% inhibition [15]) [9, 13, 15]

CN^- <1, 2, 3, 4, 5, 7> (<1> 1 mM, complete inhibition of NH_2OH reduction, 12% inhibition of cytochrome c reduction, $K_3Fe(CN)_6$ reduction is not inhibited [5]; <1> 0.02 mM, 88% inhibition [7]; <3> 0.01 mM, 40% inhibition, 0.1 mM, 93% inhibition [14]; <4> 1 mM, 45% inhibition, no inhibition if enzyme is preincubated with nitrite or NH_2OH prior to adding the inhibitor [15]; <7> 1 mM, complete inhibition of activity in cytosolic extracts [18]) [1, 5, 7, 9, 12-16, 18]
CO <2, 3> (<2> almost complete inhibition [9]; <3> competitive vs. nitrite [14]) [9, 14]
EDTA <2> [13]
KCNS <4> (<4> 0.1 mM, 90% inhibition of NH_2OH reductase activity, nitrite reductase activity is not effected [15]) [15]
N-ethylmaleimide <4, 8> (<4> 0.1 mM, 52% inhibition [15]; <8> 1 mM, 64% inhibition after 5 min, reduced glutathione restores activity to 94% [20]) [15, 20]
NAD(P)H <2, 3> (<2> in vitro inactivation after preincubation for 5-10 min in the presence of FAD [11]; <3> inactivation after preincubation, presence of nitrite protects from inactivation [14]) [11, 12, 14]
NAD^+ <1> (<1> hydroxylamine reduction, substrate inhibition at low concentration, mixed inhibition vs. NADH [4]; <1> substrate inhibition at high concentration, required for full activity at low concentrations, maximum at 1 mM [6]) [4, 6]
$NADP^+$ <1> (<1> 0.06 mM, 13% inhibition, 0.27 mM, 43% inhibition [7]) [7]
NH_2OH <4> (<4> competitive vs. NO_2^- [15]) [15]
Na_2S <2> [12]
NaN_3 <1, 3> [1, 14]
SO_3^{2-} <1, 2> (<1> 1 mM, 22% inhibition [1]) [1, 9]
antimycin A <4> (<4> 0.2 mM, 45% inhibition [15]) [15]
arsenite <1, 2> (<1> 10 mM, 24% inhibition [1]; <1> 0.5 mM, 62% inhibition [7]) [1, 7, 9]
atabrine <2> [13]
carbamoyl phosphate <3> [14]
cyanate <3> [14]
dinitrophenol <4> [15]
diquinolyl <2> [13]
hydrazine sulfate <2> (<2> complete inhibition [13]) [13]
hydroxylamine <2, 4> (<4> competitive vs. nitrite [15]) [12, 15]
iodine <4> [15]
iodoacetate <8> (<8> 64% inhibition [20]) [20]
iodosobenzoate <2> [13]
naphthoquinone <2> [13]
nitrite <4, 7> (<4> competitive vs. hydroxylamine [15]; <7> substrate inhibition with increasing concentrations, complete inhibition with 50 mM nitrite [18]) [15, 18]
p-chloromercuribenzoate <1, 2, 4, 5, 8> (<1> 0.02 mM, complete inactivation, partially prevented by 10 mM cysteine [1]; <1> 0.02 mM, complete inhibition of cytochrome c, $K_3Fe(CN)_6$ and NH_2OH reduction [5]; <1>

0.001 mM, 83% inhibition [7]; <2> completely reversed by glutathione [13]; <4> 0.1 mM, 60% inhibition [15]; <5> 0.1 mM, complete inhibition of nitrite reductase activity, approx. 50% reversal by glutathione [16]; <8> 1 mM, 79% inhibition after 5 min, dithiothreitol completely reverses activity, reduced glutathione, L-cysteine and 2-mercaptoethanol reverse activity to 86%, 64% and 82% respectively [20]) [1, 5, 7, 9, 13, 15, 16, 20]
p-hydroxymercuribenzoate <2, 3> (<2> very potent inhibitor, inhibition can be reversed by either cysteine or glutathione [12]; <3> 0.01 mM, 42% inhibition, 0.1 mM, complete inhibition [14]) [12, 14]
Rose Bengal <4> (<4> preincubation with NO_2^- completely protects [15]) [15]
rotenone <4> (<4> 0.1 mM, 56% inhibition [15]) [15]
salicylic acid <2> [13]
sodium diethyldithiocarbamate <2> [13]
sodium disulfite <2> [12]
thiourea <2> [13]

Cofactors/prosthetic groups

FAD <1, 2, 3, 5, 8> (<1> enzyme contains one non-covalently bound FAD molecule [5]; <2> exogenously added FAD is required for activity in vitro [8, 10]; <2> increases activity [13]; <3> stimulation [14]; <5> addition required for maximal activity [16]; <8> addition increases activity [20]) [5, 8, 10, 13, 14, 16, 20]
FMN <8> (<8> 0.01 mM, approx. 2fold increase in activity [20]) [20]
NAD^+ <1> (<1> substrate inhibition at low concentration [4]; <1> substrate inhibition at high concentration, at low concentration required for full activity, maximum at 1 mM [6]) [4, 6]
NADH <1, 2, 3, 4, 7, 8> (<1> enzyme is specific for NADPH, NADH reaction rate approx. 1% of NADPH rate [7]) [1, 2, 3-6, 9, 12-15, 18, 20]
NADPH <1, 2, 3, 4, 7, 8> (<1> enzyme is specific for NADPH, NADH reaction rate approx. 1% of NADPH rate [7]) [7-10, 12, 14, 15, 18, 20]
flavin <1, 2> (<1> 0.4 mol of flavin per mol of enzyme [1]; <1> enzyme contains one non-covalently bound FAD molecule per subunit but no covalently bound flavin [5]; <2> little or no flavin associated with purified protein [8]) [1, 5, 8]
siroheme <1, 2> [5, 8-10]
Additional information <1> (<1> heme is not detectable [1]; <1> FMN is not detectable [5]; <3> FMN has no effect [14]) [1, 5, 14]

Activating compounds

FMN <2, 3> (<2> addition increases activity [13]; <3> no effect [14]) [13, 14]
NAD^+ <1> (<1> activation of hydroxylamine reduction at low concentrations [4]; <1> enzyme from crude extracts and purified enzyme requires NAD^+ for full activity, maximal activation at 1 mM [6]) [4, 6]
pyridoxine <2> (<2> stimulates activity in crude extracts from pyridoxine-deficient felts, no activation of purified enzyme [13]) [13]

Metals, ions

Cu^{2+} <2> (<2> depressed activity in Cu-deficient felt extracts, stimulates activity in extracts from Cu-deficient felts, depresses activity in extracts from control felts [13]) [13]
Fe^{3+} <2> (<2> stimulates activity in Fe-deficient felt extracts [13]) [13]
Mn^{2+} <5> (<5> stimulates hydroxylamine reductase activity [16]) [16, 17]
iron <1, 2> (<1> enzyme may contain a [2Fe-2S]-iron sulfur cluster [3]; <1> enzyme contains 5 Fe atoms per subunit [5]; <2> 9-10 mol Fe per mol of enzyme, heme iron associated with siroheme moieties, nonheme iron involved in iron-sulfur centers [8]) [3, 5, 8]
magnesium <2> (<2> depressed activity in Mg-deficient felt extracts [13]) [13]
zinc <2> (<2> depressed activity in Zn-deficient felt extracts [13]) [13]
Additional information <1, 5> (<5> enzyme contains an essential metal component [16]; <1> no molybdenum detected [5]) [5, 16]

Turnover number (min^{-1})

7800 <2> (NO_2^-) [9]

Specific activity (U/mg)

0.00221 <5> (<5> reduction of NH_2OH [16]) [16]
0.00426 <5> (<5> reduction of NO_2^- [16]) [16]
0.0523 <2> [13]
0.065 <7> (<7> activity in cells grown anaerobically in the presence of 20 mM NO_2^- [18]) [18]
0.079 <7> (<7> activity in cells grown anaerobically in the presence of 20 mM NO_3- [18]) [18]
0.18 <1> [7]
0.35 <7> (<7> activity in cytosolic extracts [18]) [18]
0.422 <4> (<4> reduction of NO_2^- [15]) [15]
0.504 <7> (<7> activity in cytosolic extracts from cells grown anaerobically with 1 mM nitrite [18]) [18]
0.52 <2> (<2> NADH + NO_2^- [12]) [12]
0.73 <2> (<2> NADPH + NO_2^- [12]) [12]
0.868 <4> (<4> reduction of hydroxylamine [15]) [15]
14.6 <9> (<9> activity in extracts from aerobic grown cells [22]) [22]
16.7 <1> [1]
26.9 <2> [9]
33.4 <9> (<9> activity in extracts from anaerobic grown cells [22]) [22]
62.4 <1> [5]
1800 <8> (<8> activity in crude extracts of cells grown in the presence of 50 mM urea [20]) [20]
5900 <8> (<8> activity in crude extracts of cells grown in the presence of 50 mM glutamine [20]) [20]
17300 <8> (<8> activity in crude extracts of cells grown in the presence of 50 mM ammonium sulfate [20]) [20]

24000 <8> (<8> activity in crude extracts of cells grown in the presence of 50 mM asparagine [20]) [20]
60000 <8> (<8> activity in crude extracts of cells grown in the presence of 50 mM potassium nitrate [20]) [20]

K_m-Value (mM)

0.00002 <2> (FAD) [8]
0.000045 <6> (FAD) [17]
0.0001 <2> (FAD) [12]
0.0048 <4> (NO_2^-) [15]
0.005 <1> (NO_2^-) [6]
0.0055 <3> (NO_2^-) [14]
0.0063 <4> (NADH, <4> + NO_2^- [15]) [15]
0.0075 <2> (NO_2^-) [8]
0.01 <2> (NO_2^-, <2> + NADPH or NADH [12]) [12]
0.015 <3> (NADH) [14]
0.015 <2> (NADPH) [8]
0.016 <1> (NADH, <1> in presence of 1 mM NAD^+ and 2 mM NO_2^- [6]) [6]
0.019 <6> (nitrite) [17]
0.05 <2> (FAD) [13]
0.05 <2> (FMN) [13]
0.1 <2> (NO_2^-) [13]
0.15 <4> (NADH, <4> + hydroxylamine [15]) [15]
0.4 <1> (NO_2^-) [7]
0.8 <2> (NO_2^-, <2> + dithionite [12]) [12]
2.5 <2> (hydroxylamine) [12]
5.3 <1, 4> (hydroxylamine) [6, 15]

K_i-Value (mM)

0.000032 <3> (CO) [14,17]
0.002 <2> (*p*-hydroxymercuribenzoate) [12]
0.005 <2> (KCN) [12]
0.52 <4> (NO_2^-) [15]
2 <2> (1,10-phenanthroline) [12]
2 <2> (sodium disulfite) [12]
5 <2> (sodium sulfide) [12]
6 <2> (hydroxylamine) [12]
24 <4> (NH_2OH) [15]

pH-Optimum

7.1 <5> [16]
7.5 <2> [12]
7.5-8.5 <1> [7]
7.6 <2> [13]
8.5 <4> [15]

pH-Range

4-11 <2> (<2> pH 4.0: 40% activity, pH 11.0: 70% activity [13]) [13]
6-8 <5> (<5> pH 6: about 35% activity [16]) [16]
6.5-9.5 <1> (<1> pH 6.5: about 50% activity, pH 9.5: about 15% activity [7]) [7]

4 Enzyme Structure

Molecular weight

67000 <3> (<3> sucrose density gradient centrifugation [14]) [14]
190000 <1> (<1> glycerol density gradient centrifugation [1]) [1]
289400 <2> (<2> deduced from amino acid composition [8]) [8]
290000 <2> (<2> gel filtration, sucrose density gradient centrifugation [12]) [12]

Subunits

dimer <1, 2> (<1> 2 * 88000, SDS-PAGE [1]; <2> 2 * 140000, SDS-PAGE [8]) [1, 8]

5 Isolation/Preparation/Mutation/Application

Source/tissue

mycelium <2> [9]

Localization

cell associated <1> [1, 5, 7]
cytosol <7, 9> (<9> no activity detected in membrane fraction [22]) [18, 22]

Purification

<1> (ammonium sulfate, DEAE-cellulose, DEAE-Sephadex [1]) [1, 5, 7]
<2> [8-13]
<8> [20]

Cloning

<1> (expression in Escherichia coli [2]; expression in Escherichia coli as lacZ transcriptional fusion [19]) [2, 19]

Application

analysis <8> (<8> microplate high-throughput nitrite reductase assay [21]) [21]

6 Stability

Temperature stability

40 <2> (<2> 80% activity is lost after 4 min [12]) [12]
50 <5> (<5> 5 min, complete loss of activity [16]) [16]

General stability information

<1>, dialysis for 15 h against 1 mM cyanide, 1 mM 8-hydroxyquinoline or 1 mM EDTA followed by dialysis against 0.05 M phosphate buffer, no decrease in activity [7]
<1>, nitrite, hydroxylamine and FAD stabilize [5]
<2>, dialysis against water or a variety of buffers deactivates [13]
<5>, stable to dialysis for 6 h against 0.1 M K_2HPO_4 and 1 mM glutathione [16]

Storage stability

<1>, 4°C, 1 mM NO_2^- or 50 mM hydroxylamine, 24 h, 10% loss of activity [5]
<2>, -17°C, overnight, 50% loss of activity [13]
<5>, -15°C, 0.1 M phosphate buffer, pH 7.5, 1 mM glutathione, partially purified enzyme, stable [16]

References

[1] Coleman, K.J.; Cornish-Bowden, A.; Cole, J.A.: Purification and properties of nitrite reductase from Escherichia coli K12. Biochem. J., **175**, 483-493 (1978)
[2] MacDonald, H.; Cole, J.: Molecular cloning and functional analysis of the cysG and nirB genes of Escherichia coli K12, two closely-linked genes required for NADH-dependent nitrite reductase activity. Mol. Gen. Genet., **200**, 328-334 (1985)
[3] Cammack, R.; Jackson, R.H.; Cornish-Bowden, A.; Cole, J.A.: Electron-spin-resonance studies of the NADH-dependent nitrite reductase from Escherichia coli K12. Biochem. J., **207**, 333-339 (1982)
[4] Jackson, R.H.; Cole, J.A.; Cornish-Bowden, A.: The steady-state kinetics of the NADH-dependent nitrite reductase from Escherichia coli K12. Biochem. J., **199**, 171-178 (1981)
[5] Jackson, R.H.; Cornish-Bowden, A.; Cole, J.A.: Prosthetic groups of the NADH-dependent nitrite reductase from Escherichia coli K12. Biochem. J., **193**, 861-867 (1981)
[6] Coleman, K.J.; Cornish-Bowden, A.; Cole, J.A.: Activation of nitrite reductase from Escherichia coli K12 by oxidized nicotinamide-adenine dinucleotide. Biochem. J., **175**, 495-499 (1978)
[7] Lazzarini, R.A.; Atkinson, D.E.: A triphosphopyridine nucleotide-specific nitrite reductase from Escherichia coli. J. Biol. Chem., **236**, 3330-3335 (1961)
[8] Prodouz, K.N.; Garrett, R.H.: Neurospora crassa NAD(P)H-nitrite reductase. Studies on its composition and structure. J. Biol. Chem., **256**, 9711-9717 (1981)
[9] Greenbaum, P.; Prodouz, K.N.; Garrett, R.H.: Preparation and some properties of homogeneous Neurospora crassa assimilatory NADPH-nitrite reductase. Biochim. Biophys. Acta, **526**, 52-64 (1978)

[10] Vega, J.M.; Garrett, R.H.: Siroheme: a prosthetic group of the Neurospora crassa assimilatory nitrite reductase. J. Biol. Chem., **250**, 7980-7989 (1975)

[11] Vega, J.M.; Greenbaum, P.; Garrett, R.H.: Studies on the in vitro inactivation of the Neurospora crassa assimilatory nitrite reductase in the presence of reduced pyridine nucleotides plus flavin. Biochim. Biophys. Acta, **377**, 251-257 (1975)

[12] Lafferty, M.A.; Garrett, R.H.: Purification and properties of the Neurospora crassa assimilatory nitrite reductase. J. Biol. Chem., **249**, 7555-7567 (1974)

[13] Nicholas, D.J.D.; Medina, A.; Jones, O.T.G.: A nitrite reductase from Neurospora crassa. Biochim. Biophys. Acta, **37**, 468-476 (1960)

[14] Vega, J.M.; Guerrero, M.G.; Leadbetter, E.; Losada, M.: Reduced nicotinamide-adenine dinucleotide-nitrite reductase from Azotobacter chroococcum. Biochem. J., **133**, 701-708 (1973)

[15] Wang, R.; Nicholas, D.J.D.: Some properties of nitrite and hydroxylamine reductases from Derxia gummosa. Phytochemistry, **25**, 2463-2469 (1986)

[16] Spencer, D.; Takahashi, H.; Nason, A.: Relationship of nitrite and hydroxylamine reductases to nitrate assimilation and nitrogen fixation in Azotobacter agile. J. Bacteriol., **73**, 553-562 (1957)

[17] Hatefi, Y.; Stiggall, D.L.: Metal-containing flavoprotein dehydrogenases. The Enzymes, 3rd Ed. (Boyer, P.D., ed.), **13**, 175-297 (1975)

[18] Neubauer, H.; Goetz, F.: Physiology and interaction of nitrate and nitrite reduction in Staphylococcus carnosus. J. Bacteriol., **178**, 2005-2009 (1996)

[19] Tyson, K.; Busby, S.; Cole, J.: Catabolite regulation of two Escherichia coli operons encoding nitrite reductases: role of the Cra protein. Arch. Microbiol., **168**, 240-244 (1997)

[20] Sengupta, S.; Shaila, M.S.; Rao, G.R.: In vitro and in vivo regulation of assimilatory nitrite reductase from Candida utilis. Arch. Microbiol., **168**, 215-224 (1997)

[21] McNally, N.; Liu, X.Y.; Choudary, P.V.: A high-throughput assay format for determination of nitrate reductase and nitrite reductase enzyme activities. Appl. Biochem. Biotechnol., **62**, 29-36 (1997)

[22] Nakano, M.M.; Hoffmann, T.; Zhu, Y.; Jahn, D.: Nitrogen and oxygen regulation of Bacillus subtilis nasDEF encoding NADH-dependent nitrite reductase by TnrA and ResDE. J. Bacteriol., **180**, 5344-5350 (1998)

Hyponitrite reductase 1.7.1.5

1 Nomenclature

EC number
1.7.1.5

Systematic name
hydroxylamine:NAD^+ oxidoreductase

Recommended name
hyponitrite reductase

Synonyms
EC 1.6.6.4 (formerly)
EC 1.6.6.6 (formerly)
$NADH_2$:hyponitrite oxidoreductase

CAS registry number
9029-30-5

2 Source Organism

<1> *Neurospora crassa* (wild type 5279a, macroconidial [1]) [1]

3 Reaction and Specificity

Catalyzed reaction
2 hydroxylamine + 2 NAD^+ = hyponitrous acid + 2 NADH + 2 H^+ (a metalloprotein)

Reaction type
oxidation
redox reaction
reduction

Substrates and products
S hyponitrite + NAD^+ <1> (Reversibility: ? <1> [1]) [1]
P hydroxylamine + NADH <1> [1]

Inhibitors
2,2'-diquinolyl <1> [1]
2,6-dinitrophenol <1> [1]
hydrazine sulfate <1> [1]

p-chloromercuribenzoate <1> [1]
potassium cyanide <1> [1]

Cofactors/prosthetic groups
NAD^+ <1> [1]
flavin <1> (<1> flavoprotein [1]) [1]

Metals, ions
Additional information <1> (<1> metalloprotein [1]) [1]

References

[1] Medina, A.; Nicholas, D.J.D.: Hyponitrite reductase in Neurospora. Nature, **179**, 533 - 534 (1957)

Azobenzene reductase 1.7.1.6

1 Nomenclature

EC number
1.7.1.6

Systematic name
N,N-dimethyl-1,4-phenylendiamine, aniline:$NADP^+$ oxidoreductase

Recommended name
azobenzene reductase

Synonyms
EC 1.6.6.7 (formerly)
N,N-dimethyl-4-phenylazoaniline azoreductase
NAD(P)H:1-(4'-sulfophenylazo)-2-naphthol oxidoreductase
$NADPH_2$-dependent azoreductase
NC-reductase
orange I azoreductase <2> [5]
orange II azoreductase
azo dye reductase
azo reductase
azo-dye reductase
azoreductase
dibromopropylaminophenylazobenzoic azoreductase
dimethylaminobenzene reductase
methyl red azoreductase
new Coccine (NC)-reductase
nicotinamide adenine dinucleotide (phosphate) azoreductase
p-aminoazobenzene reductase
p-dimethylaminoazobenzene azoreductase
reductase, azobenzene

CAS registry number
9029-31-6

2 Source Organism

<1> *Rattus norvegicus* [2, 3, 7-10, 12-15, 17, 18]
<2> *Pseudomonas sp.* (strain K24 [5] strain KF46, Orange II azoreductase induced by both Orange II and carboxy-Orange II [4]) [4, 5]
<3> *Bacillus cereus* (T-105 [1]) [1]

<4> *Mus musculus* [11, 18]
<5> *Eubacterium hadrum* [6]
<6> *Eubacterium sp.* [6]
<7> *Clostridium clostridiiforme* [6]
<8> *Butyrivibrio sp.* [6]
<9> *Bacteroides sp.* [6]
<10> *Clostridium paraputrificum* [6]
<11> *Clostridium nexile* [6]
<12> *Clostridium sp.* [6]
<13> *Clostridium perfringens* [6]
<14> *Cavia porcellus* [13, 18]
<15> *Gallus gallus* [13, 18]
<16> *Canis familiaris* [13, 18]
<17> *Microtus pennsylvanicus* (meadow vole [13]) [13]
<18> *Escherichia coli* (azoreductases I and II [16]) [16]
<19> *pigeon* [18]
<20> *hamster* [18]
<21> *Oryctolagus cuniculus* (rabbit [18]) [18]
<22> *Felis catus* [18]
<23> *Capra hircus* [18]
<24> *Ovis aries* [18]
<25> *Macaca mulatta* (rhesus monkey [18]) [18]
<26> *Caulobacter subvibroides* (C7-D, induced by Acid Orange 7 [19]) [19]
<27> *Bacillus sp.* (sp. OY1-2 isolated from soil near a waste-water plant [20]) [20]

3 Reaction and Specificity

Catalyzed reaction

N,N-dimethyl-1,4-phenylenediamine + aniline + $NADP^-$ = 4-(dimethylamino)azobenzene + NADPH + H^+

Reaction type

oxidation
redox reaction
reduction

Natural substrates and products

S NADPH + 4-(dimethylamino)azobenzene <2> (<2> key enzyme of azo dye degradation, selective agent during experimental evolution in continuous cultures [5]) (Reversibility: ? <2> [5]) [5]
P $NADP^+$ + N,N-1,4-phenylenediamine + aniline <2> [5]

Substrates and products

S 1,2-dimethyl-4(*p*-carboxyphenylazo)-5-hydroxybenzene + NADPH <1, 14-17> (Reversibility: ? <1, 14-17> [13]) [13]

P 1,2-dimethyl-4-amino-5-hydroxybenzene + *p*-carboxyphenylamine + $NADP^+$ <1, 14-17> [13]

S 1-(2'-methyl-4'-sulfophenylazo)2-naphthol + NAD(P)H <2> (Reversibility: ? <2> [4]) [4]

P 4-amino-2-methyl-1-benzenesulfonic acid + NAD(P)$^+$ + 1-amino-2-hydroxynaphthalene <2> [4]

S 1-(3'-carboxyphenylazo)-2-naphthol + NAD(P)H <2> (Reversibility: ? <2> [4]) [4]

P 3-amino-1-benzoic acid + NAD(P)$^+$ + 1-amino-2-hydroxynaphthalene <2> [4]

S 1-(3'-phosphophenylazo)-2-naphthol + NAD(P)H <2> (Reversibility: ? <2> [4]) [4]

P 3-amino-1-benzenephosphate + NAD(P)$^+$ + 1-amino-2-hydroxynaphthalene <2> [4]

S 1-(3'-sulfophenylazo)-2-naphthol + NAD(P)H <2> (Reversibility: ? <2> [4]) [4]

P 3-amino-1-benzenesulfonic acid + NAD(P)$^+$ + 1-amino-2-hydroxynaphthalene <2> [4]

S 1-(4'-carboxyphenylazo)-2-hydroxy-6-sulfonaphthalene + NAD(P)H <2> (Reversibility: ? <2> [4]) [4]

P 4-carboxyphenylamine + NAD(P)$^+$ + 1-amino-2-hydroxy-6-sulfonaphthalene <2> [4]

S 1-(4'-carboxyphenylazo)-2-hydroxy-6-sulfonaphthalene + NAD(P)H <2> (Reversibility: ? <2> [4]) [4]

P 4-carboxyphenylamine + NAD(P)$^+$ + 1-amino-2-hydroxy-6-sulfonaphthalene <2> [4]

S 1-(4'-carboxyphenylazo)-2-naphthol + NAD(P)H <2> (Reversibility: ? <2> [4]) [4]

P 4-amino-1-benzoic acid + NAD(P)$^+$ + 1-amino-2-hydroxynaphthalene <2> [4]

S 1-(4'-nitrophenylazo)-2-hydroxy-6-sulfonaphthalene + NAD(P)H <2> (Reversibility: ? <2> [4]) [4]

P 4-nitrophenylamine + NAD(P)$^+$ + 1-amino-2-hydroxy-6-sulfonaphthalene <2> [4]

S 1-(4'-phosphophenylazo)-2-naphthol + NAD(P)H <2> (Reversibility: ? <2> [4]) [4]

P 4-amino-1-benzenephosphate + NAD(P)$^+$ + 1-amino-2-hydroxynaphthalene <2> [4]

S 1-(4'-sulfonaminophenylazo-)2-naphthol + NAD(P)H <2> (Reversibility: ? <2> [4]) [4]

P 4-amino-1-benzenesulfonamine + NAD(P)$^+$ + 1-amino-2-hydroxynaphthalene <2> [4]

S 1-(4'-sulfophenylazo)-2-hydroxy-6-carboxynaphthalene + NAD(P)H <2> (Reversibility: ? <2> [4]) [4]

P 4-sulfophenylamine + NAD(P)$^+$ + 1-amino-2-hydroxy-6-carboxynaphthalene <2> [4]

S 1-(4'-sulfophenylazo)-2-hydroxy-6-sulfonaphthalene + NAD(P)H <2> (Reversibility: ? <2> [4]) [4]

P 4-sulfophenylamine + $NAD(P)^+$ + 1-amino-2-hydroxy-6-sulfonaphthalene <2> [4]

S 1-(4-sulfo-1-naphthylazo)-2-naphthol-3,6-disulfonic acid + NADPH <1, 4, 18> (<1> trivial name amaranth, activity in liver microsomes [8]; <1> activity in liver microsomes, reduces also the dyes Red 2G, Red 10B, Carmoisine and Ponceau 4R [9]) (Reversibility: ? <1, 4, 18> [8, 9, 11, 16]) [8, 9, 11, 16]

P α-naphthylamine-4-sulfonic acid + β-naphthol-3,6-disulfonic acid + $NADP^+$ <1, 4, 18> [8, 9, 11, 16]

S 1-phenylazo-2-hydroxy-6-carboxynaphthalene + NAD(P)H <2> (Reversibility: ? <2> [4]) [4]

P phenylamine + $NAD(P)^+$ + 1-amino-2-hydroxy-6-carboxynaphthalene <2> [4]

S 1-phenylazo-2-hydroxy-6-sulfonaphthalene + NAD(P)H <2> (Reversibility: ? <2> [4]) [4]

P phenylamine + $NAD(P)^+$ + 1-amino-2-hydroxy-6-sulfonaphthalene <2> [4]

S 2'-carboxy-4-N,N-dimethylazobenzene + NAD(P)H <1> (<1> trivial name methyl red [2]) (Reversibility: ? <1> [2]) [2]

P 2-carboxy-1-phenylamine + N,N-dimethyl-1,4-phenylenediamine + $NADP^+$ <1> [2]

S 2,2'-(1,8-dihydroxy-3,6-disulfonaphthalene-2,7-bis-azo)-bis(benzenearsonic acid) + NAD(P)H <1> (<1> reduced to an anion free radical by an outer membrane azoreductase [7]) (Reversibility: ? <1> [7]) [7]

P ?

S 3-methyl-4-(2-chlorophenylhydrazone)-isoxazol-5-one + NAD(P)H <1> (<1> fungicide, trivial name drazoxolon, 90-80% activity with NADH [14]) (Reversibility: ? <1> [14]) [14]

P *o*-chloroaniline + ? <1> [14]

S 4-[(2-hydroxy-1-naphthalenyl)azo]benzene sulfonic acid + NAD(P)H <2, 18> (<2> trivial name Orange II [4]; <18> NADH is more effective [16]) (Reversibility: ? <2, 18> [4, 16]) [4, 16]

P 4-amino-1-benzenesulfonic acid + $NAD(P)^+$ + 1-amino-2-hydroxynaphthalene <2, 18> [4, 16]

S 4-[(4-hydroxy-1-naphthalenyl)azo]benzene sulfonic acid + NAD(P)H <2> (<2> trivial name Orange I [5]) (Reversibility: ? <2> [5]) [5]

P 4-amino-benzenesulfonic acid + 1-amino-4-hydroxynaphthalene + $NADP^+$ <2> [5]

S NADPH + 1,4-benzoquinone <1> (Reversibility: ? <1> [3]) [3]

P $NADP^+$ + 1,4-benzoquinol <1> [3]

S NADPH + 1,4-naphthoquinone <1> (Reversibility: ? <1> [3]) [3]

P $NADP^+$ + 1,4-naphthoquinol <1> [3]

S NADPH + 2,6-dichlorophenolindophenol <1> (Reversibility: ? <1> [3]) [3]

P $NADP^+$ + reduced 2,6-dichlorophenolindophenol <1> [3]

S NADPH + 4-(dimethylamino)azobenzene <1-17> (<1> specific for reduction of methyl red [2]; <1> electron transport protein [3]; <2> Orange I azoreductase: absolute requirement for a hydroxyl group in 4'-position of the naphthol ring of the substrate molecule [5]; <2> Orange II azoreductase: requirement for substrates with a 2-naphthol moiety [4,5]) (Reversibility: ? <1-17> [1-15]) [1-15]
P $NADP^+$ + N,N-1,4-phenylenediamine + aniline <1-17> [1-15]
S NADPH + 9,10-phenanthrenequinone <1> (Reversibility: ? <1> [3]) [3]
P $NADP^+$ + 9,10-phenanthrenequinol <1> [3]
S NADPH + cytochrome c <1> (<1> in the presence of menadione as electron carrier [3]) (Reversibility: ? <1> [3]) [3]
P $NADP^+$ + reduced cytochrome c <1> [3]
S NADPH + potassium ferricyanide <1> (Reversibility: ? <1> [3]) [3]
P $NADP^+$ + potassium ferrocyanide <1> [3]
S Ponceaux SX + NAD(P)H <18> (Reversibility: ? <18> [16]) [16]
P ?
S Rocceline + NADPH <27> (Reversibility: ? <27> [20]) [20]
P ?
S Solar Orange + NADPH <27> (Reversibility: ? <27> [20]) [20]
P ?
S Sumifix Black B + NADPH <27> (Reversibility: ? <27> [20]) [20]
P ?
S acid Orange 7 + NAD(P)H <26> (<26> reduces also Acid Orange 6, 8 and 12, Acid Red 88 and 151, NADH is more effective [19]) (Reversibility: ? <26> [19]) [19]
P ?
S direct blue 15 + NAD(P)H <5-10, 12, 13> (Reversibility: ? <5-10, 12, 13> [6]) [6]
P 3,3'-dimethoxybenzidine + 1-hydroxy-2,8-diamino-3,6-naphthalenedisulfonic acid + $NAD(P)^+$ <5-10, 12, 13> [6]
S new coccine + NADPH <3> (Reversibility: ? <3> [1]) [1]
P ?
S tartrazine + NAD(P)H <18> (Reversibility: ? <18> [16]) [16]
P ?

Inhibitors

4-dimethylaminoazobenzene <1> [15]
Ag^+ <3> (<3> 1 mM, 92% inhibition [1]) [1]
CO <1> (<1> noncompetitive vs. amaranth [8]) [8]
Cu^{2+} <3> (<3> 1 mM, 88% inhibition [1]) [1]
Fe^{2+} <3> (<3> 1 mM, 90% inhibition [1]) [1]
Fe^{3+} <3> (<3> 1 mM, 85% inhibition [1]) [1]
Hg^{2+} <3> (<3> 1 mM, 75% inhibition [1]) [1]
Orange I <2> (<2> substrate inhibition [5]) [5]
Orange II <2> (<2> substrate inhibition at high concentration [4]) [4]
SDS <3> (<3> 1 mM, 50% inhibition [1]) [1]

dicoumarol <1> (<1> 0.0001 mM, 50% inhibition of methyl red reduction, 0.0004 mM, 90% inhibition, competitive vs. NADPH noncompetitve vs. methyl red [3]) [3, 14]
fatty acids <1> (<1> oxidized unsaturated [15]) [15]
Additional information <3> (<3> not inhibited by N-ethylmaleimide, *p*-chloromercuribenzoate and monoiodoacetic acid [1]) [1]

Cofactors/prosthetic groups

FAD <1, 3, 18> (<1, 3> increases activity [1, 8, 9, 12, 13, 14]; <1> one mol of enzyme contains 2 mol of FAD [2]; <1> prosthetic group [15]) [1, 2, 8, 9, 12, 13-15, 16]
NADH <1, 2, 18, 26> [2, 4, 5, 14, 16, 19]
NADPH <1-3, 18, 26> (<3> specific for [1]) [1, 2, 4, 5, 8, 14, 15, 16, 19]
flavin <1, 3> (<3> prosthetic group seems to be flavin [1]; <1> typical flavoprotein absorption spectrum [2]) [1, 2]
riboflavin <1, 18> (<3> no activation [1]; <1> activation [8, 9]) [8, 9, 16]
Additional information <1, 3-13> (<3> no activity with NADH [1]; <5-13> effect of different cofactors on azoreductase activity of intestinal anaerobic bacteria [6]; <1> apparent non-identity of cytochrome c reductase and flavin-dependent azoreductase activity [10]; <4> P-450 type cytochromes are responsible for azoreductase activity [11]) [1, 6, 10, 11]

Activating compounds

FMN <1, 3, 18> (<1,3> increases activity [1, 8, 9]; <1> enzyme contains no FMN [2]) [1, 2, 8, 9, 16]
furazolidone <1> [13]
methylviologen <1> (<1> activation [8]) [8]
nitrofuarantoin <1> [13]
nitrofurazone <1> [13]

Metals, ions

Mg^{2+} <3> (<3> slight activation [1]) [1]
dicoumarol <1> [3, 14]

Specific activity (U/mg)

0.00191 <1> (<1> activity in liver microsomes [10]) [10]
0.0028 <1> (<1> continuous assay for hepatic microsomal azo reductase [12]) [12]
0.0035 <1> (<1> activity in caecal suspension with amaranth [17]) [17]
0.0035 <1> (<1> activity in liver microsomes with amaranth [17]) [17]
0.004 <1> (<1> activity in liver microsomes with Fast Red E [17]) [17]
0.004 <1> (<1> activity in liver microsomes with carmoisine [17]) [17]
0.008 <1> (<1> activity in liver microsomes with Ponceau 4R [17]) [17]
0.011 <1> (<1> activity in liver microsomes after flavin treatment [10]) [10]
0.045 <16> (<16> activity in liver supernatant fraction [13]) [13]
0.05 <18> (<18> azoreductase II, with tartrazine [16]) [16]
0.053 <15> (<15> activity in liver supernatant fraction [13]) [13]
0.054 <1> (<1> activity in caecal suspension with carmoisine [17]) [17]
0.06 <17> (<17> activity in liver supernatant fraction [13]) [13]

0.065 <18> (<18> azoreductase II, with orange II [16]) [16]
0.067 <1> (<1> activity in liver supernatant fraction [13]) [13]
0.07 <18> (<18> azoreductase II, with amaranth [16]) [16]
0.076 <1> (<1> activity in caecal suspension with Fast Red E [17]) [17]
0.087 <14> (<14> activity in liver supernatant fraction [13]) [13]
0.112 <18> (<18> azoreductase I, with orange II [16]) [16]
0.129 <1> (<1> activity in caecal suspension with Ponceau 4R [17]) [17]
0.19 <18> (<18> azoreductase I, with amaranth [16]) [16]
0.212 <18> (<18> azoreductase I, with tartrazine [16]) [16]
0.315 <18> (<18> azoreductase II, with ponceau SX [16]) [16]
0.32 <18> (<18> azoreductase I, with ponceau SX [16]) [16]
0.566 <3> [1]
2.87 <2> [5]
7.6 <27> (<27> recombinant enzyme at 50°C with roccelin [20]) [20]
11.7 <27> (<27> native enzyme at 70°C with roccelin [20]) [20]
17.83 <1> [2]
57 <1> [3]
2868 <2> (<2> orange I azoreductase, strain KF24 [4]) [5]
17830 <2> (<2> orange II azoreductase, strain KF46 [4]) [4]
69570 <26> [19]

K_m-Value (mM)

0.0009 <2> (1-phenylazo-2-hydroxy-6-sulfonaphthalene) [4]
0.001 <2> (1-(4'-sulfophenylazo)-2-naphthol) [4]
0.001 <26> (Acid Orange 7, <26> after induction with Acid Orange 7 [19]) [19]
0.0011 <2> (NADPH) [5]
0.0013 <2> (1-(2'-methyl-4-sulfophenylazo)-2-naphthol) [4]
0.0014 <2> (1-phenylazo-2-hydroxy-6-carboxynaphthalene) [4]
0.0015 <2> (1-(3'-carboxyphenylazo)-2-naphthol) [4]
0.0018 <2> (1-(3'-sulfophenylazo)-2-naphthol) [4]
0.0018 <2> (1-(4'-nitrophenylazo)-2-hydroxy-6-sulfonaphthalene) [4]
0.0024 <2> (1-(4'-carboxyphenylazo)-4-naphthol) [5]
0.0026 <2> (1-(4'-sulfophenylazo)-4-naphthol) [5]
0.003 <26> (Acid Red 88, <26> after induction with Acid Orange 7 [19]) [19]
0.004 <26> (Acid Red 151, <26> after induction with Acid Orange 7 [19]) [19]
0.004 <18> (ponceau SX, <18> azoreductase I [16]) [16]
0.005 <2> (NADPH) [4]
0.005 <18> (ponceau SX, <18> azoreductase II [16]) [16]
0.0052 <2> (1-(4'-phosphophenylazo)-2-naphthol) [4]
0.00645 <1> (CO) [8]
0.008 <2> (1-(4'-carboxyphenylazo)-2-naphthol) [4]
0.009 <26> (Acid Red 88) [19]
0.0105 <2> (1-(3'-phosphophenylazo)-2-naphthol) [4]
0.013 <2> (1-(4'-carboxyphenylazo)-2-hydroxy-6-carboxynaphthalene) [4]
0.0132 <2> (1-(4'-sulfophenylazo)-2-hydroxy-6-carboxynaphthalene) [4]

0.0137 <2> (1-(2'-sulfophenylazo)-2-naphthol) [4]
0.0148 <2> (1-(4'-sulfoaminophenylazo)-2-naphthol) [4]
0.017 <2> (1-(4'-sulfophenylazo)-2-hydroxy-6-sulfonaphthalene) [4]
0.0198 <2> (1-(4'-carboxyphenylazo)-2-hydroxy-6-sulfonaphthalene) [4]
0.034 <1> (amaranth) [9]
0.04 <2> (1-(4'-sulfophenyl-N-methylhydrazo)-4-naphthol) [5]
0.054 <2> (NADH) [5]
0.093 <26> (Acid Orange 8, <26> after induction with Acid Orange 7 [19]) [19]
0.12 <1> (1,2-dimethyl-4-(*p*-carboxyphenylazo)-5-hydroxy-benzene) [13]
0.18 <2> (NADH) [4]
0.204 <2> (1-(4'-sulfophenylazo)-4-anthranol) [5]
0.25 <1> (NADPH) [2]
0.32 <1> (methyl red) [2]
0.4 <1> (NADH) [2]
0.45 <26> (Acid Orange 8) [19]

K_i-Value (mM)

3e-005 <1> (dicoumarol, <1> vs. NADPH [3]) [3]
0.00011 <1> (dicoumarol, <1> vs. methyl red [3]) [3]
0.006 <1> (CO) [8]

pH-Optimum

5-8 <2> (<2> Orange II + NADH [4]) [4]
5.5-6 <1> [2]
6-7.5 <26> [19]
6.2 <2> (<2> Orange I [5]) [5]
6.5 <2> (<2> carboxy-Orange II + NADH [4]) [4]
6.8 <2> (<2> carboxy-Orange I [5]) [5]
7 <3, 18> (<18> azoreductase I and II [16]) [1, 16]

pH-Range

4-7 <1> (<1> pH 4.0: about 25% activity, pH 7.0: about 20% activity [2]) [2]
4-10 <3> (<3> pH 4.0: about 10% activity, pH 10.0: about 20% activity [1]) [1]

Temperature optimum (°C)

40 <3> [1]
41 <2> [5]
45 <2, 18> (<18> azoreductases I and II, with ponceau SX [16]) [4, 16]
50 <27> (<27> recombinant enzyme [20]) [20]
70 <27> (<27> native enzyme [20]) [20]

Temperature range (°C)

5-70 <3> (<3> at 5°C and 70°C: about 55% activity [1]) [1]
20-50 <2> (<2> 20°C: about 25% activity with carboxy-Orange II, about 30% activity with Orange II, 50°C: about 95% activity with carboxy-Orange II, about 80% activity with Orange II [4]) [4]

4 Enzyme Structure

Molecular weight

12000 <18> (<18> azoreductase II, gel filtration [16]) [16]
21500 <2> (<2> strain KF 24, gel filtration, Orange I azoreductase [5]) [5]
32000 <2> (<2> strain KF 46, gel filtration, Orange II azoreductase [4]) [4]
52000 <1> (<1> gel filtration [2]) [2]
53000 <18> (<18> azoreductase I, gel filtration [16]) [16]

Subunits

? <26, 27> (<26> x * 30000, SDS-PAGE [19]; <27> x * 20000, recombinant and native enzyme, SDS-PAGE [20]) [19, 20]
dimer <1, 18> (<1> 2 * 30000, SDS-PAGE [2]; <18> 2 * 28000, azoreductase I, SDS-PAGE [16]) [2, 16]
monomer <2> (<2> 1 * 20200, strain K24, SDS-PAGE, Orange I azoreductase [5]; <2> 1 * 30000, strain KF46, SDS-PAGE, Orange II azoreductase [4]) [4, 5]

5 Isolation/Preparation/Mutation/Application

Source/tissue

cecum <1, 4, 14-16, 19-26> (microbiol azoreductase activity [18]) [18, 19]
intestine <1, 4, 14-16, 19-26> (microbiol azoreductase activity in large intestine [18]) [18, 19]
liver <1, 4, 14-17> [2, 3, 7-9, 11, 13-15]

Localization

cytosol <1> [2]
endoplasmic reticulum <1, 4> (<1,4> i.e. microsomes [8,11]) [8, 11]
extracellular <5-13> [6]
intracellular <3> [1]
mitochondrion <1> (<1> activity higher in crude outer mitochondrial membrane fraction than in mitoplast and intact mitochondria [7]) [7]
soluble <1> [7]

Purification

<1> (partially [14]) [2, 14]
<2> [4]
<2> [5]
<3> [1]
<18> [16]
<26> [19]
<27> (recombinant enzyme [20]) [20]

Cloning

<27> (expression in Escherichia coli [20]) [20]

6 Stability

pH-Stability

6-8 <3> (<3> 5°C, 7 days, stable [1]) [1]

Temperature stability

35 <2> (<2> stable below [4]) [4]
60 <3> (<3> 10 min, stable up to [1]) [1]
80 <3> (<3> 10 min, 45% loss of activity [1]) [1]

Oxidation stability

<5-13>, inactivation by oxygen [6]

Storage stability

<2>, -20°C, 50% glycerol, several months [4]

References

[1] Matsudomi, N.; Kobayashi, K.; Akuta, S.: Purification and some properties of new coccine (NC)-reductase from Bacillus cereus T-105 strain. Agric. Biol. Chem., **41**, 2323-2329 (1977)
[2] Huang, M.T.; Miwa, G.T.; Lu, A.Y.H.: Rat liver cytosolic azoreductase. Purification and characterization. J. Biol. Chem., **254**, 3930-3934 (1979)
[3] Huang, M.T.; Miwa, G.T.; Cronheim, N.; Lu, A.Y.H.: Rat liver cytosolic azoreductase. Electron transport properties and the mechanism of dicumarol inhibition of the purified enzyme. J. Biol. Chem., **254**, 11223-11227 (1979)
[4] Zimmermann, T.; Kulla, H.G.; Leisinger, T.: Properties of purified Orange II azoreductase, the enzyme initiating azo dye degradation by Pseudomonas KF46. Eur. J. Biochem., **129**, 197-203 (1982)
[5] Zimmermann, T.; Gasser, F.; Kulla, H.G.; Leisinger, T.: Comparison of two bacterial azoreductases acquired during adaptation to growth on azo dyes. Arch. Microbiol., **138**, 37-43 (1984)
[6] Rafii, F.; Franklin, W.; Cerniglia, C.E.: Azoreductase activity of anaerobic bacteria isolated from human intestinal microflora. Appl. Environ. Microbiol., **56**, 2146-2151 (1990)
[7] Moreno, S.N.J.; Mason, R.P.; Docampo, R.: Ca^{2+} and Mg^{2+}-enhanced reduction of arsenazo III to its anion free radical metabolite and generation of superoxide anion by an outer mitochondrial membrane azoreductase. J. Biol. Chem., **259**, 14609-14616 (1984)
[8] Fujita, S.; Peisach, J.: The stimulation of microsomal azoreduction by flavins. Biochim. Biophys. Acta, **719**, 178-189 (1982)
[9] Mallett, A.K.; King, L.J.; Walker, R.: A continuous spectrophotometric determination of hepatic microsomal azo reductase activity and its dependence on cytochrome P-450. Biochem. J., **201**, 589-595 (1982)
[10] Mallett, A.K.; Walker, R.; King, L.J.: The apparent non-identity of cytochrome c reductase and flavin-dependent azoreductase activities. Biochem. Soc. Trans., **6**, 1302-1305 (1978)

[11] Fujita, S.; Peisach, J.: Liver microsomal cytochromes P-450 and azoreductase activity. J. Biol. Chem., **253**, 4512-4513 (1978)
[12] Mallett, A.K.; King, L.J.; Walker, R.: A continuous assay for hepatic microsomal azo reductase. Biochem. Soc. Trans., **5**, 1522-1524 (1977)
[13] Smith, E.J.; Van Loon, E.J.: 1,2-Dimethyl-4-(*p*-carboxyphenylazo)-5-hydroxybenzene. A convenient substrate for the measurement of azo reductase. Anal. Biochem., **31**, 315-320 (1969)
[14] Daniel, J.W.: A soluble aerobic reduced nicotinamideadenine dinucleotide (phosphate) azoreductase. Biochem. J., **11**, 19P-20P (1969)
[15] Ketterer, B.; Ross-Mansell, P.; Davidson, H.: The effect of 4-dimethylaminoazobenzene and corn oil on azo-dye reductase in the rat liver. Biochem. J., **107**, 15P-16P (1968)
[16] Ghosh, D.K.; Ghosh, S.; Sadhukhan, P.; Mandal, A.; Chaudhuri, J.: Purification of two azoreductases from Escherichia coli K12. Indian J. Exp. Biol., **31**, 951-954 (1993)
[17] Singh, S.; Das, M.; Khanna, S.K.: Comparative azo reductase activity of red azo dyes through caecal and hepatic microsomal fraction in rats. Indian J. Exp. Biol., **35**, 1016-1018 (1997)
[18] Singh, S.; Das, M.; Khanna, S.K.: Azo reductase activity of microbial population from gastrointestinal tract segments of various animals species. Indian J. Exp. Biol., **36**, 99-103 (1998)
[19] Mazumder, R.; Logan, J.R.; Mikell, A.T., Jr.; Hooper, S.W.: Characteristics and purification of an oxygen insensitive azoreductase from Caulobacter subvibrioides strain C7-D. J. Ind. Microbiol. Biotechnol., **23**, 476-483 (1999)
[20] Suzuki, Y.; Yoda, T.; Ruhul, A.; Sugiura, W.: Molecular cloning and characterization of the gene coding for azoreductase from Bacillus sp. OY1-2 isolated from soil. J. Biol. Chem., **276**, 9059-9065 (2001)

GMP Reductase 1.7.1.7

1 Nomenclature

EC number
1.7.1.7

Systematic name
inosine-5'-phosphate:$NADP^+$ oxidoreductase (aminating)

Recommended name
GMP reductase

Synonyms
EC 1.6.6.8 (formerly)
guanosine 5'-monophosphate oxidoreductase
NADPH:GMP oxidoreductase (deaminating)
guanosine 5'-monophosphate reductase
guanosine monophosphate reductase
guanylate reductase
reductase, guanylate

CAS registry number
9029-32-7

2 Source Organism

<-3> no activity in *Methanocaldococucus* (other name Methanococcus jannaschii [16]) [16]
<-2> no activity in *Mycoplasma genitalium* [16]
<-1> no activity in *Haemophilus influenza* [16]
<1> *Mycoplasma mycoides* [8]
<2> *Homo sapiens* [9, 12, 16]
<3> *Aerobacter aerogenes* [13, 14]
<4> *Escherichia coli* [1, 4, 14]
<5> *Bos taurus* [2]
<6> *Bacillus subtilis* (mutants containing mutations in the salvage pathway of purine synthesis [3]) [3]
<7> *Leishmania donovani* [5]
<8> *Salmonella typhimurium* (enzyme synthesis is induced by 6-diazo-5-oxo-L-norleucine and glutamine starvation [7]) [6, 7, 14]
<9> *Artemia salina* [10, 11]

<10> *Rattus norvegicus* (mRNA increases 30fold in brown adipose tissue within 6 h of cold exposure [15]) [15]
<11> *Homo sapiens* [16]

3 Reaction and Specificity

Catalyzed reaction

inosine 5'-phosphate + NH_3 + $NADP^+$ = guanosine 5'-phosphate + NADPH + H^+ (<2> mechanism: ordered sequential with GMP binding first [9]; <2> first order reaction [12])

Reaction type

oxidation
redox reaction
reduction
reductive deamination

Natural substrates and products

S NADPH + guanosine 5'-phosphate <6, 8, 10> (<6> involved in salvage pathway of purine synthesis [3]; <8> enzyme is involved in interconversion of purine ribonucleotides [7]; <10> enzyme may play a role in brown fat response [15]) (Reversibility: ir <8> [7]; ? <6, 10> [3, 15]) [3, 7, 15]
P $NADP^+$ + inosine 5'-phosphate + NH_3 <6, 8, 10> [3, 7, 15]

Substrates and products

S NADPH + guanosine 5'-phosphate <1-11> (<9> the rate of the reverse reaction is 6% of the forward reaction [11]; <2> arabinosylGMP, 2'-dGMP and 8-azaGMP are reductively deaminated to their corresponding IMP analog at rates 1-2% the rate with GMP [9]; <2> less than 10% of the NADPH rate: thionicotinamide-NADPH, deamino-NADPH, 3-acetylpyrimidine-NADPH [9]; <3, 4, 8> specific for GMP [14]) (Reversibility: ir <2, 3, 4, 8> [7, 12, 14]; r <9> [11]; ? <1, 5, 6, 7, 10, 11> [2, 3, 6, 8, 15, 16]) [1-16]
P $NADP^+$ + inosine 5'-phosphate + NH_3 <1-11> [1-16]

Inhibitors

2'-dXMP <2> [9]
5,5'-dithiobis(2-nitrobenzoate) <2> (<2> 0.66 mM, 33% inhibition after 30 min [12]) [12]
6-chloro-9-β-D-ribofuranosylpurine 5'-phosphate <3> (<3> 0.01 mM, 80% inhibition after 40 min, 1 mM GMP protect [13]) [13]
6-chloropurine ribonucleotide <2> [9]
6-thio-GMP <2, 3> (<3> 0.009 mM, 81% inhibition after 30 min [13]) [9, 13]
6-thio-IMP <2, 3> (<3> 0.32 mM, 75% inhibition after 15 min [13]) [9, 13]
6-thio-XMP <2> [9]
8-aza-7-deaza-XMP <2> [9]
8-aza-XMP <2> [9]
ADP <3> (<3> 2 mM, 85% inhibition [13]) [13]

[5] Spector, T.; Jones, T.E.: Guanosine 5-monophosphate reductase from Leishmania donovani. A possible chemotherapeutic target. Biochem. Pharmacol., **31**, 3891-3897 (1982)
[6] Neuhard, J.; Nygaard, P.: Escherichia coli and Salmonella typhimurium cellular and molecular biology. American Society for Microbiology Washington (Neidhardt, F.C., Ingraham, J.L., Low, K.B., Magasanik, B., Schaechter, M., Umbarger, E., eds.), **1**, 445-473 (1987)
[7] Garber, B.B.; Jochimsen, B.U.; Gots, J.S.: Glutamine and related analogs regulate guanosine monophosphate reductase in Salmonella typhimurium. J. Bacteriol., **143**, 105-111 (1980)
[8] Mitchell, A.; Sin, I.L.; Finch, L.R.: Enzymes of purine metabolism in Mycoplasma mycoides subsp. mycoides. J. Bacteriol., **134**, 706-712 (1978)
[9] Spector, T.; Jones, T.E.; Miller, R.L.: Reaction mechanism and specificity of human GMP reductase. Substrates, inhibitors, activators, and inactivators. J. Biol. Chem., **254**, 2308-2315 (1979)
[10] Renart, M.F.; Renart, J.; Sillero, M.A.G.; Sillero, A.: Guanosine monophosphate reductase from Artemia salina: Inhibition by xanthosine monophosphate and activation by diguanosine tetraphosphate. Biochemistry, **15**, 4962-4966 (1976)
[11] Renart, M.F.; Sillero, A.: GMP reductase in Artemia salina. Biochim. Biophys. Acta, **341**, 178-186 (1974)
[12] Mackenzie, J.J.; Sorensen, L.B.: Guanosine 5-phosphate reductase of human erythrocytes. Biochim. Biophys. Acta, **327**, 282-294 (1973)
[13] Brox, L.W.; Hampton, A.: Inactivation of guanosine 5-phosphate reductase by 6-chloro-, 6-mercapto-, and 2-amino-6-mercapto-9-β-D-ribofuranosylpurine 5-phosphates. Biochemistry, **7**, 398-405 (1968)
[14] Mager, J.; Magasanik, B.: Guanosine 5'-phosphate reductase and its role in the interconversion of purine nucleotides. J. Biol. Chem., **235**, 1474-1478 (1960)
[15] Salvatore, D.; Bartha, T.; Larsen, P.R.: The guanosine monophosphate reductase gene is conserved in rats and its expression increases rapidly in brown adipose tissue during cold exposure. J. Biol. Chem., **273**, 31092-31096 (1998)
[16] Deng, Y.; Wang, Z.; Ying, K.; Gu, S.; Ji, C.; Huang, Y.; Gu, X.; Wang, Y.; Xu, Y.; Li, Y.; Xie, Y.; Mao, Y.: NADPH-dependent GMP reductase isoenzyme of human (GMPR2). Expression, purification, and kinetic properties. Int. J. Biochem. Cell Biol., **34**, 1035-1050 (2002)

1.7.1.8

1 Nomenclature

EC number
1.7.1.8 (deleted 2002)

Recommended name
Recommended Name never specified

Nitroquinoline-N-oxide reductase 1.7.1.9

1 Nomenclature

EC number
1.7.1.9

Systematic name
4-(hydroxyamino)quinoline N-oxide:$NADP^+$ oxidoreductase

Recommended name
nitroquinoline-N-oxide reductase

Synonyms
4-nitroquinoline 1-oxide reductase
4NQO reductase
NAD(P)H2:4-nitroquinoline-N-oxide oxidoreductase

CAS registry number
37256-35-2

2 Source Organism

<1> *Canis familiaris* [1]
<2> *Rattus norvegicus* [2, 3, 5]
<3> *Cavia porcellus* [3]
<4> *hamster* [3]
<5> *Oryctolagus cuniculus* [3]
<6> *Felis catus* [3]
<7> *Mus musculus* [4, 5]

3 Reaction and Specificity

Catalyzed reaction
4-(hydroxyamino)quinoline N-oxide + 2 $NADP^+$ + H_2O = 4-nitroquinoline N-oxide + 2 NAD(P)H + 2 H^+ (<1> 4-(hydroxyamino)quinoline N-oxide may be the active intermediate essential in the carcinogenic process of 4-nitroquinoline N-oxide [1])

Reaction type
oxidation
redox reaction
reduction

Substrates and products

S 4-nitroquinoline N-oxide + 2 NAD(P)H + 2 H^+ <1-7> (<7> strong preference for NADH [4, 5]; <2> NADH is preferred, 15% activity with NADPH [5]) (Reversibility: ? <1-7> [1-5]) [1-5]

P 4-(hydroxyamino)quinoline N-oxide + 2 $NAD(P)^+$ + H_2O <1-7> [1-5]

Inhibitors

Additional information <2, 7> (<7> not inhibited by dicoumarol [4, 5]; <2> not inhibited by 0.015 mM dicoumarol [5]) [4, 5]

Specific activity (U/mg)

0.032 <2> (<2> activity in cheek mucosa [2]) [2]
0.07 <5> (<5> activity in corpus ventriculi and pyloric antrum [3]) [3]
0.084 <2> (<2> activity in tongue mucosa [2]) [2]
0.1 <2> (<2> activity in duodenum [3]) [3]
0.12 <6> (<6> activity in esophagus [3]) [3]
0.15 <2> (<2> activity in ileum and rectum [3]) [3]
0.18 <4> (<4> activity in esophagus and caecum, cofactor NADH [3]) [3]
0.25 <2> (<2> activity in forestomach and colon [3]) [3]
0.3 <1> (<1> activity in cytosol from mucous membrane of esophagus [1]) [1]
0.4 <2> (<2> activity in glandular stomach and caecum [3]) [3]
0.4 <3> (<3> activity in pyloric antrum [3]) [3]
0.5 <2> (<2> activity in esophagus [3]) [3]

K_m-Value (mM)

0.015 <7> (4-nitroquinoline N-oxide) [4]

4 Enzyme Structure

Molecular weight

200000 <7> [4]

5 Isolation/Preparation/Mutation/Application

Localization

cytosol <1, 2> (<1> most of the activity [1]) [1, 5]

Purification

<7> (partially [5]) [5]

References

[1] Suzuki, K.: Distribution of 4-nitroquinoline 1-oxide reductase in the mucosa of canine digestive tract. Gann, **69**, 229-235 (1978)

[2] Booth, D.R.: A relationship found between intra-oral sites of 4NQO reductase activity and chemical carcinogenesis. Cell Tissue Kinet., **23**, 331-340 (1990)
[3] Suzuki, K.: The distribution of 4-nitroquinoline 1-oxide reductase in the mucosa of animal digestive tracts. Nihon Univ. J. Med., **20**, 345-357 (1978)
[4] Stanley, J.S.; York, J.L.; Benson, A.M.: Nitroreductases and glutathione transferases that act on 4-nitroquinoline 1-oxide and their differential induction by butylated hydroxyanisole in mice. Cancer Res., **52**, 58-63 (1992)
[5] Benson, A.M.: Conversion of 4-nitroquinoline 1-oxide (4NQO) to 4-hydroxyaminoquinoline 1-oxide by a dicumarol-resistant hepatic 4NQO nitroreductase in rats and mice. Biochem. Pharmacol., **46**, 1217-1221 (1993)

Hydroxylamine reductase (NADH) 1.7.1.10

1 Nomenclature

EC number
1.7.1.10

Systematic name
ammonium:NAD^+ oxidoreductase

Recommended name
hydroxylamine reductase (NADH)

Synonyms
EC 1.6.6.11 (formerly)
N-hydroxy amine reductase
NADH-hydroxylamine reductase
ammonium dehydrogenase
reductase, hydroxylamine

CAS registry number
9032-06-8

2 Source Organism

<-3> no activity in *fish* [4]
<-2> no activity in *amphibian* [4]
<-1> no activity in *reptile* [4]
<1> *Derxia gummosa* (nitrite and hydroxylamine reductase are associated within the same enzyme [1]) [1]
<2> *Sus scrofa* (microsomal multicomponent enzyme system consisting of NADH, cytochrome b_5, cytochrome b_5 reductase and a third unidentified protein, catalyzes reduction of hydroxylamine and a number of its mono- and disubstituted derivatives [2, 3]) [2-4, 10]
<3> *Rattus norvegicus* (primary hepatocytes [11]) [4-6, 11]
<4> *Mus musculus* [4]
<5> *Ovis aries* [4]
<6> *Oryctolagus cuniculus* [4]
<7> *Neurospora crassa* (3 hydroxylamine reductases: one is identical with sulfite reductase and is absolutely specific for NADPH, accepts hydroxylamine and sulfate as substrates at a common binding site, a second hydroxylamine reductase uses either NADH or NADPH and is stimulated by

FAD, a third hydroxylamine reductase is also able to accept NADH and NADPH [9]) [8, 9]
<8> *Glycine max* (soybean [7]) [7]
<9> *opossum* (very low activity [4]) [4]
<10> *Cavia porcellus* [4]
<11> *hamster* [4]
<12> *Canis familiaris* (2 enzyme forms are present, one may be identical to the multienzyme complex from pig [13]) [4, 13]
<13> *Bos taurus* [4]
<14> *pigeon* [4]
<15> *Gallus gallus* [4]
<16> *Homo sapiens* (HepG2 cells [11]; 2 enzyme forms are present, one may be identical to the multienzyme complex from pig [13]) [10, 11, 13]
<17> *Phormidium uncinatum* (cyanobacterium [12]) [12]

3 Reaction and Specificity

Catalyzed reaction

NH_3 + NAD^+ + H_2O = hydroxylamine + NADH + H^+ (<3-6, 9-15> also acts on hydroxamates [4])

Reaction type

oxidation
redox reaction
reduction

Natural substrates and products

S N-hydroxy-arylamines + NADH <3, 16> (<3, 16> enzyme may be involved in detoxification pathway of carcinogenic aromatic and heterocyclic amines [11]) (Reversibility: ? <3, 16> [11]) [11]
P N-arylamines + NAD^+ + H_2O <3, 16> [11]
S hydroxylamine + NADH <7> (<7> involved in pathway of nitrate reduction [9]) (Reversibility: ? <7> [9]) [9]
P NH_3 + NAD^+ + H_2O <7> [9]

Substrates and products

S N,N-dimethylhydroxylamine + NADH <2> (Reversibility: ? <2> [3]) [3]
P dimethylamine + NAD^+ + H_2O <2> [3]
S N-hydroxy-1-naphthylamine + NADH <3, 16> (<3> activity in liver microsomes [11]) (Reversibility: ? <3, 16> [11]) [11, 11]
P 1-naphthylamine + NAD^+ + H_2O <3, 16> [11]
S N-hydroxy-2-amino-1-methyl-6-phenylimidazo[4,5-b]pyridine + NADH <3, 16> (<3> activity in liver microsomes, primary rat hepatocytes and human HepG2 cells [11]) (Reversibility: ? <3, 16> [11]) [11, 11]
P 2-amino-1-methyl-6-phenylimidazo[4,5-b]pyridine + NAD^+ + H_2O <3, 16> [11]

S N-hydroxy-2-amino-3,8-dimethylimidazo[4,5-f]quinoxaline + NADH <3, 16> (<3> activity in liver microsomes [11]) (Reversibility: ? <3, 16> [11]) [11, 11]

P 2-amino-3-methylimidazo[4,5-f]quinoxaline + NAD^+ + H_2O <3, 16> [11]

S N-hydroxy-2-amino-3,8-dimethylimidazo[4,5-f]quinoxaline + NADH <3, 16> (<3> activity in liver microsomes [11]) (Reversibility: ? <3, 16> [11]) [11, 11]

P 2-amino-3-methylimidazo[4,5-f]quinoxaline + NAD^+ + H_2O <3, 16> [11]

S N-hydroxy-2-amino-α-carboline + NADH <3, 16> (<3> activity in liver microsomes [11]) (Reversibility: ? <3, 16> [11]) [11, 11]

P 2-amino-α-carboline + NAD^+ + H_2O <3, 16> [11]

S N-hydroxy-2-aminofluorene + NADH <3, 16> (<3> activity in liver microsomes [11]) (Reversibility: ? <3, 16> [11]) [11, 11]

P 2-aminofluorene + NAD^+ + H_2O <3, 16> [11]

S N-hydroxy-2-naphthylamine + NADH <3, 16> (<3> activity in liver microsomes [11]) (Reversibility: ? <3, 16> [11]) [11, 11]

P 2-naphthylamine + NAD^+ + H_2O <3, 16> [11]

S N-hydroxy-4,4'-methylenebis(2-chloroaniline) + NADH <3, 16> (<3> activity in liver microsomes [11]) (Reversibility: ? <3, 16> [11]) [11, 11]

P 4,4'-methylenebis(2-chloroaniline) + NAD^+ <3, 16> [11]

S N-hydroxy-4-aminoazobenzene + NADH <3, 16> (<3> activity in liver microsomes [11]) (Reversibility: ? <3, 16> [11]) [11, 11]

P 4-aminoazobenzene + NAD^+ + H_2O <3, 16> [11]

S N-hydroxy-4-aminobiphenyl + NADH <3, 16> (<3> activity in liver microsomes [11]) (Reversibility: ? <3, 16> [11]) [11, 11]

P 4-aminobiphenyl + NAD^+ + H_2O <3, 16> [11]

S N-hydroxy-aniline + NADH <3, 16> (<3> activity in liver microsomes [11]) (Reversibility: ? <3, 16> [11]) [11, 11]

P aniline + NAD^+ + H_2O <3, 16> [11]

S N-hydroxyamphetamine + NADH <2> (Reversibility: ? <2> [3]) [3]

P amphetamine + NAD^+ + H_2O <2> [3]

S N-hydroxydebrisoquine + NADH <2, 16> (<2> reconstituted enzyme complex from liver microsomes consisting of NADH, cytochrome b_5, cytochrome b_5 reductase, a P450 isoenzyme and phosphatidylcholin, 40% activity with NADPH [10]) (Reversibility: ? <2, 16> [10]) [10, 10]

P debrisquine + NAD^+ + H_2O <2, 16> [10]

S N-hydroxypiperidine + NADH <2> (Reversibility: ? <2> [3]) [3]

P piperidine + NAD^+ + H_2O <2> [3]

S N-methyl-N-benzyl-hydroxylamine + NADH <2> (Reversibility: ? <2> [3]) [3]

P N-methylbenzylamine + NAD^+ + H_2O <2> [3]

S N-methyl-N-n-octyl-hydroxylamine + NADH <2> (Reversibility: ? <2> [3]) [3]

P N-methyl-n-octylamine + NAD^+ + H_2O <2> [3]

S N-methylhydroxylamine + NADH <2, 3> (Reversibility: ? <2, 3> [3, 4]) [3, 4]

P methylamine + NAD^+ + H_2O <2, 3> [3, 4]

S NO_2^- + NADH <1, 7> (<1> enzyme has both nitrite reductase and hydroxylamine reductase activity, main function may be nitrite reduction [1]) (Reversibility: ? <1, 7> [1, 8]) [1, 8]

P NH_3 + H_2O + NAD^+ <1, 7> [1, 8]

S anthranillic hydroxamate + NADH <3-6, 9-15> (<3-6, 9-15> activity in liver mitochondria [4]) (Reversibility: ? <3-6, 9-15> [4]) [4]

P anthranilamide + NAD^+ <3-6, 9-15> [4]

S benzamidoxime + NADH <2, 16> (<2> reconstituted enzyme complex from liver microsomes consisting of NADH, cytochrome b_5, cytochrome b_5 reductase, a P450 isoenzyme and phosphatidylcholin, 40% activity with NADPH [10]) (Reversibility: ? <2, 16> [10]) [10, 10]

P benzamidine + NAD^+ <2> [10]

S dinitrobenzene + NADH <8> (Reversibility: ? <8> [7]) [7]

P ?

S guanoxabenz + NADH <2, 16> (<2> reconstituted enzyme complex from liver microsomes consisting of NADH, cytochrome b_5, cytochrome b_5 reductase, a P450 isoenzyme and phosphatidylcholin, 40% activity with NADPH [10]) (Reversibility: ? <2, 16> [10]) [10, 10]

P ?

S hydroxylamine + NADH <1-16> (<3> enzyme also reduces anthranilic hydroxamic acid and nicotinic hydroxamic acid, affinity is about twenty times lower than for hydroxylamine [5]; <3> possible identity of hydroxylamine reductase and the enzyme which reduces anthranilic hydroxamic acid to the amide [6]; <7> no activity with O-methylhydroxylamine, hydrazine, nitrophenylhydroxylamine and salicylhydroxamic acid [8]; <7> no other substances beside nitrite and dinitrobenzene stimulate oxidation of NADPH [8]; <3> no reduction of N,N-diethyl or O-methyl compounds [4]; <3> very low activity with phenylhydroxylamine [4]; <1> 100% activity with NADH, 46% with NADPH, 117% with $FADH_2$, 105% with $FMNH_2$, 220% with methylviologen, 210% with benzyl viologen, 57% with Na_2S2O4 [1]; <2> 30-40% activity with NADPH [2]; <3, 8> NADH is 1.5fold more effective than NADPH [5, 7]; <7> NADPH is 2fold more effective than NADH [9]) (Reversibility: ir <7> [8]; ? <1-6, 9-16> [1-7, 9]) [1-9]

P NH_3 + NAD^+ + H_2O <1-15> [1-9]

S phenylhydroxylamine + NADH <2, 3> (Reversibility: ? <2, 3> [3, 4]) [3, 4]

P phenylamine + NAD^+ + H_2O <3> [3, 4]

S sulphamethoxazole hydroxylamine + NADH <12, 16> (<12, 16> activity in liver microsomes, NADH is preferred [13]) (Reversibility: ? <12, 16> [13]) [13]

P sulfamethoxazole + NADH <12, 16> (<12, 6> antimicrobial for the treatment of gastrointestinal, urinary and respiratory pathogens [13]) [13]

Inhibitors

1-cyclohexyl-3-(2-morpholinoethyl)carbodiimide metho-*p*-toluenesulfonate <1> (<1> 1.5 mM, 68% inhibition, hydroxylamine protects [1]) [1]
2,2'-dipyridyl <1, 7, 8> (<8> 1 mM, 84% inhibition [7]) [1, 7, 8]
8-hydroxyquinoline <1, 7, 8> (<1> 0.1 mM, 50% inhibition of hydroxylamine reductase activity [1]; <8> 1 mM, complete inhibition [7]) [1, 7, 8]
$CaCl_2$ <3> (<3> reactivation by washing [4]) [4]
Cu^{2+} <7> [8]
$HgCl_2$ <3> (<3> 1 mM, complete inhibition [5]) [5, 6]
KCN <3, 7, 8> (<7,8> 0.1 mM, complete inhibition [7,8]) [5, 7, 8]
KCl <3> (<3> reactivation by washing [4]) [4]
N-ethylmaleimide <1, 3> (<3> not inhibited [6]; <1> 0.1 mM, 58% inhibition of hydroxylamine reductase activity [1]; <3> 1 mM, 50% inhibition [5]) [1, 5]
N-methylhydroxylamine <2, 3, 12, 16> (<2> 0.5 mM, complete inhibition [10]) [4, 10, 13]
NADH <1> (<1> inhibits in absence of hydroxylamine [1]) [1]
NaCl <3> (<3> reactivation by washing [4]) [4]
NaN_3 <7, 8> (<8> 1 mM, complete inhibition [7]) [7, 8]
antimycin 1 <1> (<1> 0.1 mM, 26% inhibition of hydroxylamine reductase activity [1]) [1]
bromelain <3> [4]
butanol <3> [4]
deoxycholic acid <3> [4]
dinitrophenol <1> (<1> 0.1 mM, 80% inhibition of hydroxylamine reductase activity [1]) [1]
duponol C <3> [4]
ethanol <3> [4]
hydrazine <8> (<8> 1 mM, complete inhibition [7]) [7]
iodine <1> (<1> 0.5 mM, 80% inhibition, preincubation with hydroxylamine protects [1]) [1]
murine phospholipase <3> [4]
nitrite <1> (<1> nitrite and hydroxylamine reductase activity are associated with the same enzyme, nitrite and hydroxylamine are competitive for the same binding site [1]) [1]
o-phenanthroline <7, 8> (<8> 1 mM, complete inhibition [7]) [7, 8]
p-chloromercuribenzoate <1, 3, 8> (<1> 0.1 mM, 67% inhibition of hydroxylamine reductase activity [1]; <3> 1 mM, complete inhibition [5]; <3> 0.05 mM, complete inhibition, subsequent addition of 3 mM cysteine restores the activity [6]; <8> 0.1 mM, complete inhibition [7]) [1, 5, 6, 7]
p-hydroxymercuribenzoate <3, 12, 16> [5, 13]
papain <3> [4]
phosphate <3> [4-6]
potassium cyanide <1> [1]
potassium ethyl xanthate <7, 8> [7, 8]
potassium thiocyanate <1> (<1> 0.1 mM, 90% inhibition of hydroxylamine reductase activity, nitrite reductase activity is not inhibited [1]) [1]
propanol <3> [4]

Rose Bengal <1> (<1> 0.5 mM, 94% inhibition [1]) [1]
rotenone <1> (<1> 0.1 mM, 52% inhibition of hydroxylamine reductase activity [1]) [1]
salicylaldoxime <7, 8> [7, 8]
sodium arsenate <3> [4, 6]
sodium diethyl dithiocarbamate <7, 8> (<8> 1 mM, complete inhibition [7]) [7, 8]
subtilopeptidase A <3> [4]
sulfite <7> (<7> uncompetitive vs. hydroxylamine [9]) [8, 9]
trypsin <3> [4]
uranyl acetate <3> (<3> ATP, ADP, sodium diphosphate restores activity [4]) [4]
Additional information <2, 12, 16> (<2> not inhibited by 0.1 mM KCN, NaN_3, EDTA, CO and O_2 [3]; <12, 16> not inhibited by azide, CO, dextromethorphan or erythromycin [13]; <12> not inhibited by a nitrogen atmosphere [13]) [3, 13]

Cofactors/prosthetic groups
FAD <7> (<7> stimulates [8]) [8]
NADH <1-17> [1-13]
NADPH <3, 7, 8> (<3, 8> about 65% of the activity with NADH [5, 7]) [5, 7, 8]
Additional information <8> (<8> purified enzyme: absolute requirement for an unidentified, dissociable, heat-stable, organic factor obtained from soybean leaf extract, no flavin requirement [7]) [7]

Metals, ions
Mn^{2+} <8> (<8> 0.1 mM, approx. 6fold stimulation [7]) [7]

Specific activity (U/mg)
0.422 <1> (<1> reduction of nitrite [1]) [1]
0.68 <1> (<1> reduction of hydroxylamine [1]) [1]
Additional information <7> (<7> 360.0 units/mg, 1 unit is defined as that amount of enzyme which causes a rate of change in extinction of 0.001 per minute [8]) [8]
Additional information <8> (<8> 794.1 units/mg, 1 unit is defined as that amount of enzyme which causes a rate of change in extinction of 0.001 per minute [7]) [7]

K_m-Value (mM)
0.0048 <1> (NO_2^-) [1]
0.07 <7> (NADH) [8]
0.1 <7> (NADPH) [8]
0.15 <1> (NADH) [1]
0.5 <3> (hydroxylamine) [6]
3.8 <7> (hydroxylamine) [8]
5.3 <1> (hydroxylamine) [1]

K_i-Value (mM)

0.52 <1> (NO_2^-) [1]
0.7 <7> (sulfite) [9]

pH-Optimum

6.3 <2> [3]
6.6-7.1 <8> [7]

pH-Range

5.4-8.4 <3> (<3> pH 5.4: considerable activity in Tris-maleate buffer, pH 8.4: about 50% activity [5]) [5]
5.5-7.2 <3> (<3> pH 5.5: about 70% of activity maximum, pH 7.2: 50% activity [5]) [5]

Temperature optimum (°C)

37 <2> (<2> assay at [3]) [3]

4 Enzyme Structure

Subunits

Additional information <2> (<2> 3 protein fractions are required to reconstitute NADH-hydroxylamine reductase activity: detergent-extracted cytochrome b_5 and its flavoprotein and a third microsomal protein [2]) [2]

5 Isolation/Preparation/Mutation/Application

Source/tissue

kidney <2> [3]
leaf <8> [7]
liver <2-6, 9, 12, 16> [2-6, 10, 11, 13]

Localization

membrane <3> (<3> firmly attached to mitochondrial membrane [4,6]) [4, 6]
microsome <2, 3, 12, 16> [2, 3, 10, 11, 13]
mitochondrion <2-6, 9> (<3> firmly attached to membrane [4,6]) [4-6]
soluble <8> [7]

Purification

<1> (nitrite and hydroxylamine reductase are associated with the same enzyme [1]) [1]
<7> (partial [8]) [8]
<8> (enzyme not seperated from pyridine nucleotide-nitrite reductase [7]) [7]
<17> [12]

6 Stability

pH-Stability

5.4 <3> (<3> complete loss of activity after brief exposure [6]) [6]
7.5-9.5 <7> (<7> highest stability [8]) [8]
8.4 <3> (<3> no loss of activity after brief exposure [6]) [6]

Temperature stability

27 <3> (<3> stable for more than 1 h [5]) [5]
38 <3> (<3> 1 h, inactivation [5]) [5]
40 <3> (<3> membrane bound enzyme, 10 min, pH 7.4, dithiothreitol, no loss of activity [6]) [6]
50 <3, 7> (<3> membrane bound enzyme, 10 min, pH 7.4, dithiothreitol, 50% loss of activity [6]; <7> 5 min, stable [8]) [6, 8]
50 <8> (<8> complete loss of activity after 5 min [7]) [7]
60 <3> (<3> membrane bound enzyme, pH 7.4, dithiothreitol, complete loss of activity [6]) [6]
Additional information <3> (<3> NADH prevents heat damage [6]) [6]

Oxidation stability

<2>, very unstable in the absence of sulfhydryl protecting agents, stable in the presence of 0.2 mM dithiothreitol [2]

General stability information

<3>, NADH prevents heat damage [6]
<3>, dithiothreitol stabilizes during storage [6]
<3>, freezing partially inactivates membrane bound enzyme [6]
<3>, solubilization with deoxycholate, n-butanol or acetone destroys membrane bound enzyme [6]
<7>, dialysis against 3 mM sodium diphosphate, 1 mM cysteine solution, pH 8, 2 h, less than 50% loss of activity, against 10 mM phosphate, pH 8, 1 h, complete loss of activity [8]
<7>, little or no activity in phosphate buffer unless 1 mM EDTA is included in the reaction mixture [8]

Storage stability

<7>, -15°C, slow loss of activity over several weeks [8]
<7>, 4°C, overnight [8]
<8>, -20°C, 1 month, 60-90% loss of activity [7]
<8>, 4°C, 1 week, no loss in activity [7]

References

[1] Wang, R.; Nicholas, D.J.D.: Some properties of nitrite and hydroxylamine reductases from Derxia gummosa. Phytochemistry, **25**, 2463-2469 (1986)
[2] Kadlubar, F.F.; Ziegler, D.M.: Properties of a NADH-dependent N-hydroxy amine reductase isolated from pig liver microsomes. Arch. Biochem. Biophys., **162**, 83-92 (1974)

[3] Kadlubar, F.F.; Mckee, E.M.; Ziegler, D.M.: Reduced pyridine nucleotide-dependent N-hydroxy amine oxidase and reductase activities of hepatic microsomes. Arch. Biochem. Biophys., **156**, 46-57 (1973)
[4] Bernheim, M.L.C.: The reduction of hydroxylamine and some aryl hydroxamates by liver mitochondria from mammals and birds. Enzymologia, **43**, 167-176 (1972)
[5] Bernheim, M.L.C.; Hochstein, P.: Reduction of hydroxylamine by rat liver mitochondria. Arch. Biochem. Biophys., **124**, 436-442 (1968)
[6] Bernheim, M.L.C.: The hydroxylamine reductase of mitochondria. Arch. Biochem. Biophys., **134**, 408-413 (1969)
[7] Roussos, G.G.; Nason, A.: Pyridine nucleotide-nitrite and -hydroxylamine enzymes from soybean leaves. J. Biol. Chem., **235**, 2997-3007 (1960)
[8] Zucker, M.; Nason, A.: A pyridine nucleotide-hydroxylamine reductase from Neurospora. J. Biol. Chem., **213**, 463-478 (1955)
[9] Siegel, L.M.; Leinweber, F.J.; Monty, K.J.: Characterization of the sulfite and hydroxylamine reductases of Neurospora crassa. J. Biol. Chem., **240**, 2705-2711 (1965)
[10] Clement, B.; Lomb, R.; Woller, W.: Isolation and characterization of the protein components of the liver microsomal O_2-insensitive NADH-benzamidoxime reductase. J. Biol. Chem., **272**, 19615-19620 (1997)
[11] King, R.S.; Teitel, C.H.; Shaddock, J.G.; Casciano, D.A.; Kadlubar, F.F.: Detoxification of carcinogenic aromatic and heterocyclic amines by enzymatic reduction of the N-hydroxy derivative. Cancer Lett., **143**, 167-171 (1999)
[12] Bagchi, S.N.; Shakila, T.M.: Characterization of oxime metabolizing enzymes in a cyanobacterium Phormidium uncinatum, utilizing ethyl acetohydroxamate as sole nitrogen source. Indian J. Exp. Biol., **37**, 990-994 (1999)
[13] Trepanier, L.A.; Miller, J.L.: NADH-dependent reduction of sulphamethoxazole hydroxylamine in dog and human liver microsomes. Xenobiotica, **30**, 1111-1121 (2000)

4-(Dimethylamino)phenylazoxybenzene reductase 1.7.1.11

1 Nomenclature

EC number
1.7.1.11

Systematic name
4-(dimethylamino)phenylazobenzene:NADP$^+$ oxidoreductase

Recommended name
4-(dimethylamino)phenylazoxybenzene reductase

Synonyms
EC 1.6.6.12 (formerly)
N,N'-dimethyl-*p*-aminoazobenzene oxide reductase
NADPH-dependent DMAB N-oxide reductase
dimethylaminoazobenzene N-oxide reductase

CAS registry number
103843-39-6

2 Source Organism

<1> *Rattus norvegicus* [1]

3 Reaction and Specificity

Catalyzed reaction
4-(dimethylamino)phenylazobenzene + NADP$^+$ = 4-(dimethylamino)phenylazoxybenzene + NADPH + H$^+$

Reaction type
oxidation
redox reaction
reduction

Substrates and products
S NADPH + 4-(dimethylamino)phenylazoxybenzene <1> (<1> highly specific for NADPH and 4-(dimethylamino)phenylazoxybenzene [1]) (Reversibility: ? <1> [1]) [1]
P NADP$^+$ + 4-(dimethylamino)phenylazobenzene <1> [1]

Cofactors/prosthetic groups
NADPH <1> (<1> specific for [1]) [1]

Turnover number (min^{-1})
150 <1> (NADPH) [1]

Specific activity (U/mg)
0.27 <1> [1]

K_m-Value (mM)
0.021 <1> (NADPH) [1]
0.7 <1> (4-(dimethylamino)phenylazoxybenzene) [1]

pH-Optimum
6.5 <1> (<1> enzyme assay [1]) [1]

Temperature optimum (°C)
37 <1> (<1> enzyme assay [1]) [1]

4 Enzyme Structure

Molecular weight
370000 <1> (<1> gel filtration [1]) [1]

Subunits
tetramer <1> (<1> 4 * 92000, SDS-PAGE [1]) [1]

5 Isolation/Preparation/Mutation/Application

Source/tissue
heart <1> [1]
kidney <1> [1]
liver <1> [1]
lung <1> [1]
spleen <1> [1]

Localization
cytosol <1> [1]

Purification
<1> [1]

6 Stability

Storage stability
<1>, -20 C, several weeks [1]

References

[1] Lasmet Johnson, P.R.; Ziegler, D.M.: Properties of an N,N-dimethyl-*p*-aminoazobenzene oxide reductase purified from rat liver cytosol. J. Biochem. Toxicol., **1**, 15-27 (1986)

N-Hydroxy-2-acetamidofluorene reductase 1.7.1.12

1 Nomenclature

EC number
1.7.1.12

Systematic name
2-acetamidofluorene:NAD(P)$^+$ oxidoreductase

Recommended name
N-hydroxy-2-acetamidofluorene reductase

Synonyms
EC 1.6.6.13 (formerly)
N-hydroxy-2-acetylaminofluorene reductase
NAD(P)H_2:N-hydroxy-2-acetamidofluorene N-oxidoreductase

CAS registry number
99890-08-1

2 Source Organism

<1> *Oryctolagus cuniculus* [1]

3 Reaction and Specificity

Catalyzed reaction
2-acetamidofluorene + NAD(P)$^+$ + H_2O = N-hydroxy-2-acetamidofluorene + NAD(P)H + H^+

Reaction type
oxidation
redox reaction
reduction

Substrates and products
S N-hydroxy-2-acetamidofluorene + NAD(P)H <1> (<1> other electron donors: cysteine, glutathione, dithiothreitol, 2-mercaptoethanol [1]) (Reversibility: ? <1> [1]) [1]
P 2-acetamidofluorene + NAD(P)$^+$ + H_2O <1> [1]

S N-hydroxy-4-acetylaminobiphenyl + NAD(P)H <1> (<1> much slower reaction compared to N-hydroxy-2-acetamidofluorene [1]) (Reversibility: ? <1> [1]) [1]

P 4-acetylaminobiphenyl + $NAD(P)^+$ + H_2O <1> [1]

Inhibitors
$CuSO_4$ <1> (<1> 1 mM, 52% inhibition [1]) [1]
N-ethylmaleimide <1> (<1> 1 mM, 86% inhibition [1]) [1]
disulfiram <1> (<1> 0.1 mM, 56% inhibition [1]) [1]
p-chloromercuribenzoate <1> (<1> 0.2 mM, 81% inhibition [1]) [1]

Cofactors/prosthetic groups
NADH <1> [1]
NADPH <1> [1]

Activating compounds
2-mercaptoethanol <1> (<1> enhances activity, can replace NADH [1]) [1]
cysteine <1> (<1> enhances activity, can replace NADH [1]) [1]
dithiothreitol <1> (<1> enhances activity, can replace NADH [1]) [1]
glutathione <1> (<1> enhances activity, can replace NADH [1]) [1]

Specific activity (U/mg)
910 <1> [1]

4 Enzyme Structure

Molecular weight
34000 <1> (<1> gel filtration) [1]

Subunits
monomer <1> (<1> 1 * 34000, SDS-PAGE) [1]

5 Isolation/Preparation/Mutation/Application

Source/tissue
liver <1> [1]

Localization
cytosol <1> [1]
microsome <1> (<1> low amounts [1]) [1]

Purification
<1> [1]

References

[1] Kitamura, S.; Tatsumi, K.: Purification of N-hydroxy-2-acetylaminofluorene reductase from rabbit liver cytosol. Biochem. Biophys. Res. Commun., **133**, 67-74 (1985)

Nitrite reductase (NO-forming) 1.7.2.1

1 Nomenclature

EC number
1.7.2.1

Systematic name
nitric-oxide:ferricytochrome-c oxidoreductase

Recommended name
nitrite reductase (NO-forming)

Synonyms
cytochrome cd1
cytochrome oxidase
reductase, nitrite (cytochrome)

CAS registry number
37256-41-0

2 Source Organism

<1> *Rhodopseudomonas sphaeroides* (f. sp. denitrificans [1]) [1, 10]
<2> *Alcaligenes sp.* (NCIB 11015 [2, 3]) [2, 3]
<3> *Micrococcus denitrificans* [4]
<4> *Pseudomonas denitrificans* [5]
<5> *Alcaligenes xylosoxidans* [6, 7, 8, 9]
<6> *Haloarcula marismortui* [7]
<7> *Haloferax denitrificans* [7]
<8> *Alcaligenes cycloclastes* [7]
<9> *Fusarium oxysporum* [7]

3 Reaction and Specificity

Catalyzed reaction
nitric oxide + H_2O + ferricytochrome c = nitrite + ferrocytochrome c + 2 H^+ (<4> a copper protein, cytochrome c-552 or cytochrome c-553 from Pseudomonas denitrificans acts as acceptor [5]; <5> mechanism [9])

Reaction type
oxidation
redox reaction
reduction

Natural substrates and products

S NO_2^- + ferrocytochrome c <3> (<3> role in respiration [4]) (Reversibility: ? <3> [4]) [4]

P NO + H_2O + ferricytochrome c <3> [4]

S NO_2^- + reduced azurin <5> (<5> putative physiological electron donor [8]) (Reversibility: ? <5> [8]) [8]

P NO + oxidized azurin <5> [8]

Substrates and products

S N,N-dimethyl-*p*-phenylenediamine + oxidized benzyl viologen <1> (Reversibility: ? <1> [1]) [1]

P ?

S NH_2OH + $NaNO_2$ <4> (Reversibility: ? <4> [5]) [5]

P N_2O + H_2O <4> [5]

S NO_2^- + $Na_2S_2O_4$ <6> (<6> pysiological electron donor is unknown, no activity with methyl viologen, phenazine methosulfate or N,N,N',N',-tetramethyl-*p*-phenylenediamine [7]) (Reversibility: ? <6> [7]) [7]

P NO + NaS_2O_3 <6> [7]

S NO_2^- + reduced azurin <5> (<5> putative physiological electron donor [8]) (Reversibility: ? <5> [8]) [8]

P NO + oxidized azurin <5> [8]

S O_2^- + H^+ <1> (<1> purified enzyme shows superoxide dismutase activity, approx. one-third that of pure superoxide dismutase [1]) (Reversibility: ? <1> [1]) [1]

P H_2O_2 + O_2 <1> [1]

S ferrocytochrome c + O_2 <3, 4> (<4> in the presence of ascorbate, N,N,N',N-tetramethyl-*p*-phenylenediamine and cytochrome c-553 [5]) (Reversibility: ? <3, 4> [4, 5]) [4, 5]

P ferricytochrome c + H_2O <3, 4> [4, 5]

S nitrite + ferrocytochrome c <1-9> (<1, 3> artificial electron donor: reduced benzyl viologen [1, 4]; <3> artificial electron donors: reduced methyl viologen, phenazine methosulfate and to a lesser extend hydroquinone, highly purified enzyme has cytochrome c oxidase activity [4]; <4> artificial electron donors: thionine, brilliant cresyl blue, methylene blue, 2,6-dichlorophenolindophenol or Pseudomonas stutzeri cytochrome c-552 and 558 [5]) (Reversibility: ? <1-9> [1-10]) [1-10]

P nitric oxide + H_2O + ferricytochrome c <1-9> [1-10]

Inhibitors

EDTA <1, 6> (<1> 1 mM, complete inhibition after preincubation for 10 min [1]; <6> 1 mM, 70% inhibition after 1 h [7]) [1, 7]

KCN <1, 3, 6> (<3> 1 mM, 72% inhibition [4]; <6> 1 mM, complete inhibition [7]) [1, 4, 7]

NH_2OH <3> (<3> 1 mM, 50% inhibition [4]) [4]

NaN_3 <3> (<3> 1 mM, 60% inhibition [4]) [4]

diethyldithiocarbamate <6> (<6> 1 mM, complete inhibition [7]) [7]

urea <1> (<1> 4 M, 60% inhibition [1]) [1]

Additional information <3> (<3> relatively ineffective at 1 mM: CO, bathocuproine, diethyl dithiocarbamate, *o*-phenanthroline and α,α'-dipyridyl [4]) [4]

Cofactors/prosthetic groups

cytochrome <3, 4> (<3> contains cytochrome, probably of the c and a2 types [4]; <4> cytochrome c-552 or cytochrome c-553 from Pseudomonas denitrificans acts as acceptor [5]) [4, 5]

Activating compounds

KCl <6> (<6> 2 M, enzyme is activated by high salt concentrations [7]) [7]
LiCl <6> (<6> 2 M, enzyme is activated by high salt concentrations [7]) [7]
NH_4Cl <6> (<6> 2 M, enzyme is activated by high salt concentrations [7]) [7]
NaCl <6> (<6> 2 M, enzyme is activated by high salt concentrations [7]) [7]
$NaNO_3$ <6> (<6> 2 M, enzyme is activated by high salt concentrations [7]) [7]

Metals, ions

copper <1-9> (<2-4> copper protein [2, 4, 5]; <2> contains 2 type 1 copper atoms per molecule but no other types of copper [3]; <3> 5.5 ng/mg protein [4]; <1> one subunit contains a type 1 copper center, the second subunit a type 2 copper center, activity of copper depleted enzyme can be restored [1]; <5> blue copper enzyme that shows little absorbance in the 460 nm range, purified recombinant nonactivated enzyme contains 1.97 mol copper/mol enzyme, the $CuSO_4$ activated enzyme contains 5.97 mol copper/mol enzyme i.e. 6 copper atoms per trimer, $CuSO_4$ activation restores type 2 copper centers which are the sites of catalysis [6]; <6> blue copper enzyme, 1 type 1 and 1 type 2 Cu atoms per subunit [7]; <7, 8> green copper enzyme [7]; <5, 9> blue copper enzyme [7]; <5> the type 1 Cu center functions as electron acceptor from the physiological electron donor azurin I, the type 2 Cu center is involved in binding and reduction of the substrate [8, 9]) [1-10]

Turnover number (min^{-1})

89 <4> (O_2) [5]
155 <4> (NH_2OH) [5]
581 <4> (NO_2^-) [5]

Specific activity (U/mg)

1.27 <3> [4]
10.8 <5> (<5> recombinant enzyme, without activation, enzyme has only type 1 copper centers [6]) [6]
21.1 <2> [3]
22.7 <2> (<2> enzyme is activated 2.5-4.5fold by freezing at -20°C for 6 h and subsequent thawing [2]) [2]
23.4 <2> [3]
40 <1> (<1> if reduced nitrite is measured [1]) [1]
45.2 <5> (<5> native recombinant enzyme, azurin [8]) [8]
80-90 <1> (<1> if oxidized benzyl viologen is measured [1]) [1]

107 <5> (<5> native recombinant enzyme, dithionite [8]) [8]
117 <2> (<2> after storage at -20°C for 25 h and thawing in a water bath at 20°C [3]) [3]
167.7 <5> (recombinant enzyme, after activation with $CuSO_4$ [6]) [6]
168 <5> (<5> native recombinant enzyme, methyl viologen [8]) [8]
960 <6> (<6> in the presence of 2 M NaCl [7]) [7]

K_m-Value (mM)
0.027 <3> (O_2, <3> cytochrome c oxidase activity [4]) [4]
0.035 <5> (NO_2^-) [6]
0.0357 <1> (nitrite) [1]
0.046 <3> (NO_2^-, <3> reduced cytochrome c as electron donor [4]) [4]
0.046 <3> (cytochrome c) [4]

pH-Optimum
6.5-7 <3> (<3> cytochrome c oxidase activity [4]) [4]
6.7 <3> [4]
6.8-7.4 <4> [5]

Temperature optimum (°C)
30 <3> (<3> assay at [4]) [4]

4 Enzyme Structure

Molecular weight
69000 <8> (<8> gel filtration [7]) [7]
70000 <5> (<5> gel filtration [7]) [7]
79000 <2> [2]
80000 <1> (<1> gel filtration, gradient PAGE [1]) [1]
83300 <9> (<9> gel filtration [7]) [7]
103000 <5> (<5> sedimentation equilibrium [7]) [7]
105000 <8> (<8> sedimentation equlibrium [7]) [7]
127000 <7> (<7> gel filtration [7]) [7]
130000 <3> (<3> gel filtration [4]) [4]
167000 <6> (<6> gel filtration [7]) [7]

Subunits
dimer <1, 2, 7> (<1> 2 * 39000-42000, enzyme composed of 2 nonidentical subunits: one contains Cu^{2+} type 1, the second subunit contains Cu^{2+} type 2, SDS-PAGE [1]; <2> 2 * 37000, SDS-PAGE [3]; <7> 2 * 64000, SDS-PAGE [7]; <9> 2 * 41800, SDS-PAGE [7]) [1, 3, 7]
trimer <5, 6, 8> (<5> recombinant enzyme [6]; <6> 3 * 46000, SDS-PAGE [7]; <5> 3 * 36500, calculated from sequence, crystal structure [7]; <8> 3 * 37000, calculated from sequence, crystal structure [7]) [6, 7]

Posttranslational modification
no glycoprotein <2> [3]

5 Isolation/Preparation/Mutation/Application

Localization

soluble <4> [5]

Purification

<1> (f. sp. denitrificans [1]) [1]
<2> (CM-cellulose, Sephadex g-150, hydroxyapatite [3]) [2, 3]
<3> [4]
<5> (recombinant enzyme [6]) [6]
<5> (recombinant wild-type and D92E and D92N mutant enzyme [8]) [8]
<5> (recombinant wild-type and H139A mutant enzyme [9]) [9]
<6> (butyl-toyopearl, Sepharose CL-6B, Sephacryl S-300, octyl-Sepharose, hydroxyapatite [7]) [7]

Cloning

<1> (expression of wild-type, D129A, D129N, H287A, I289A and I289V mutant enzymes in Escherichia coli [10]) [10]
<5> (expression in Escherichia coli [6]; D92E and D92N mutant enzyme [8]; H139A mutant enzyme [9]) [6, 8, 9]

Engineering

D129A <1> (<1> low activity [10]) [10]
D129N <1> (<1> low activity [10]) [10]
D92E <5> (<5> mutation in type 2 Cu center, very low activity with artificial electron donors methyl viologen and sodium dithionite, 20-30% of wild-type activity with physiological electron donor azurin I [8]) [8]
D92N <5> (<5> mutation in type 2 Cu center, very low activity with artificial electron donors methyl viologen and sodium dithionite, 60-70% of wild-type activity with physiological electron donor azurin I [8]) [8]
H139A <5> (<5> mutation in type 1 Cu center, very low activity with the artificial electron donor methyl viologen, no activity with the physiological electron donor azurin I [9]) [9]
H287A <1> (<1> very low activity [10]) [10]
I289A <1> (<1> activity comparable to wild-type [10]) [10]
I289V <1> (<1> activity comparable to wild-type [10]) [10]
M182T <1> (<1> activity comparable to wild-type [10]) [10]

6 Stability

Temperature stability

70 <1> (<1> activity is not affected after treatment for 10 min, first step in purification [1]) [1]
80 <5> (<5> native and purified enzyme retain 50% activity after 20 min exposure [6]) [6]

Storage stability

<2>, 0°C, 3 weeks, no loss of activity [3]
<3>, -15°C, 4 weeks [4]
<4>, -20°C, more than 1 year, no loss of activity [5]

References

[1] Michalski, W.P.; Nicholas, D.J.D.: Molecular characterization of a copper-containing nitrite reductase from Rhodopseudomonas sphaereoides forma sp. denitrificans. Biochim. Biophys. Acta, **828**, 130-137 (1985)

[2] Masuko, M.; Iwasaki, H.; Sakurai, T.; Suzuki, S.; Nakahara, A.: Effects of freezing on purified nitrite reductase from a denitrifier, Alcaligenes sp. NCIB 11015. J. Biochem., **98**, 1285-1291 (1985)

[3] Masuko, M.; Iwasaki, H.; Sakurai, T.; Suzuki, S.; Nakahara, A.: Characterization of nitrite reductase from a denitrifier, Alcaligenes sp. NCIB 11015. A novel copper protein. J. Biochem., **96**, 447-454 (1984)

[4] Lam, Y.; Nicholas, D.J.D.: A nitrite reductase with cytochrome oxidase activity from Micrococcus denitrificans. Biochim. Biophys. Acta, **180**, 459-472 (1969)

[5] Miyata, M.; Mori, T.: Studies on denitrification. X. The denitrifying enzyme as a nitrite reductase and the electron donating system for denitrification. J. Biochem., **66**, 463-471 (1969)

[6] Prudencio, M.; Eady, R.R.; Sawers, G.: The blue copper-containing nitrite reductase from Alcaligenes xylosoxidans: cloning of the nirA gene and characterization of the recombinant enzyme. J. Bacteriol., **181**, 2323-2329 (1999)

[7] Ichiki, H.; Tanaka, Y.; Mochizuki, K.; Yoshimatsu, K.; Sakurai, T.; Fujiwara, T.: Purification, characterization, and genetic analysis of Cu-containing dissimilatory nitrite reductase from a denitrifying halophilic archaeon, Haloarcula marismortui. J. Bacteriol., **183**, 4149-4156 (2001)

[8] Prudencio, M.; Eady, R.R.; Sawers, G.: Catalytic and spectroscopic analysis of blue copper-containing nitrite reductase mutants altered in the environment of the type 2 copper centre: implications for substrate interaction. Biochem. J., **353**, 259-266 (2001)

[9] Prudencio, M.; Sawers, G.; Fairhurst, S.A.; Yousafzai, F.K.; Eady, R.R.: Alcaligenes xylosoxidans dissimilatory nitrite reductase: alanine substitution of the surface-exposed histidine 139 ligand of the type 1 copper center prevents electron transfer to the catalytic center. Biochemistry, **41**, 3430-3438 (2002)

[10] Zhao, Y.; Lukoyanov, D.A.; Toropov, Y.V.; Wu, K.; Shapleigh, J.P.; Scholes, C.P.: Catalytic function and local proton structure at the type 2 copper of nitrite reductase: the correlation of enzymatic pH dependence, conserved residues, and proton hyperfine structure. Biochemistry, **41**, 7464-7474 (2002)

Nitrite reductase (cytochrome; ammonia-forming) 1.7.2.2

1 Nomenclature

EC number
1.7.2.2

Systematic name
ammonia:ferricytochrome-c oxidoreductase

Recommended name
nitrite reductase (cytochrome; ammonia-forming)

Synonyms
NiR cytochrome c_{552} <6> [6]
ammonia-forming cytochrome c nitrite reductase <5> [5]
cytochrome c NiR <6> [6]
cytochrome c nitrite reductase
cytochrome c_{552} <1> [1]
hexaheme c-type cytochrome <2> [2]
multiheme nitrite reductase

CAS registry number
37256-41-0

2 Source Organism

<1> *Escherichia coli K12* (ST249 [1]; JCB7123, JCB7120, JCB7121 [8]) [1, 2, 8]
<2> *Vibrio fischeri* (formerly Achromobacter fischeri [2]) [2]
<3> *Wolinella succinogenes* [2-5]
<4> *Desulfovibrio desulfuricans* (ATCC 27774 [7]) [2, 7]
<5> *Sulfurospirillum deleyianum* [5]
<6> *Desulfovibrio vulgaris Hildenborough* [6]

3 Reaction and Specificity

Catalyzed reaction
ammonia + 2 H_2O + 6 ferricytochrome c = nitrite + 6 ferrocytochrome c + 7 H^+ (Found as a multiheme cytochrome in many bacteria. The enzyme from E. coli contains five hemes c and requires Ca^{2+}. It also reduces nitric oxide and hydroxylamine to ammonia, and sulfite to sulfide)

Reaction type

oxidation
redox reaction
reduction

Substrates and products

S hydroxylamine + benzyl viologen reduced <1> (Reversibility: ? <1> [1]) [1]
P ? + benzyl viologen oxidized <1> [1]
S nitrite + $FADH_2$ <1> (Reversibility: ? <1> [1]) [1]
P NH_4^+ + FAD <1> [1]
S nitrite + $FMNH_2$ <1> (Reversibility: ? <1> [1]) [1]
P NH_4^+ + FMN <1> [1]
S nitrite + benzyl viologen reduced + H^+ <1> (Reversibility: ? <1> [1]) [1]
P NH_4^+ + benzyl viologen oxidized + H_2O <1> [1]
S sulfite + ? <6> (Reversibility: ? <6> [6]) [6]
P sulfide + ? <6> [6]
S Additional information <1> (<1> not accepted electron donors: NADH, NADPH, ubiquinol-1, menadiol, formate, D-lactate, sn-glycerol-3-phosphate [1]; <1> no nitrate or sulfite reductase activity [1]) [1]
P ?

Inhibitors

CO <4> [7]
$CuCl_2$ <1> (<1> 0.01 mM, 70% inhibition [1]; 0.1 mM, 69% inhibition [1]) [1]
$CuSO_4$ <1> (<1> 0.01 mM, 68% inhibition [1]) [1]
KCN <1> (<1> 0.1 mM, 84% inhibition [1]; 1 mM, 100% [1]) [1]
KNO_3 <1> (<1> 1 mM, 6% inhibition [1]) [1]
NaN_3 <1> (<1> 1 mM, 16% inhibition [1]) [1]
azide <3> [3]
sulfate <3> [3]

Cofactors/prosthetic groups

heme c <1-5> (<1,3> pentaheme c-type cytochrome [3,8]; <2,4> hexaheme protein [1,2]; <5> tetraheme cytochrome [5]) [1, 2, 3, 5, 8]

Activating compounds

NrfE <1> (<1> gene product necessary for the assembly of the holoenzyme [8]) [8]
NrfF <1> (<1> gene product necessary for the assembly of the holoenzyme [8]) [8]
NrfG <1> (<1> gene product necessary for the assembly of the holoenzyme [8]) [8]

Metals, ions

Ca^{2+} <3> (<3> essential constituent discovered in the X-ray structure [3]) [3]
Additional information <1> (<1> no non-heme iron [1]) [1]

Specific activity (U/mg)

36 <1> (<1> $FADH_2$ [1]) [1]
50 <1> (<1> $FMNH_2$ [1]) [1]
345 <4> (<4> nitrite [2]) [2]
682 <2> (<2> nitrite [2]) [2]
685 <6> (<6> NH_4^+ [6]) [6]
708 <3> (<3> nitrite [2]) [2]
810 <3> (<3> nitrite, oligomeric form [5]) [5]
850 <3> (<3> nitrite, monomeric form [5]) [5]
970 <5> (<5> nitrite, oligomeric form [5]) [5]
1050 <5> (<5> nitrite, monomeric form [5]) [5]
2690 <1> (<1> hydroxylamine [1]) [1]
2930 <1> (<1> benzyl viologen [1]) [1]
2930 <1> (<1> nitrite [1]) [1]

K_m-Value (mM)

0.11 <1> (nitrite) [1]
18 <1> (hydroxylamine) [1]

pH-Optimum

7.5 <1> [1]

Temperature range (°C)

10-60 <1> (<1> activity increases up to 60°C [1]) [1]

4 Enzyme Structure

Molecular weight

18000 <6> (<6> SDS-PAGE [6]) [6]
50590 <1> (<1> apoprotein, calculation from DNA sequence [8]) [8]
53590 <1> (<1> MALDI-TOF-MS [8]) [8]
55000 <3> (<3> SDS-PAGE [5]) [5]
55000 <5> (<5> SDS-PAGE [5]) [5]
56000 <6> (<6> SDS-PAGE [6]) [6]
56300 <1> [2]
57000 <2> (<2> SDS-PAGE [2]) [2]
58000 <2> (<2> native enzyme, gel filtration [2]) [2]
63000 <3> [2]
66000 <4> (<4> SDS-PAGE [7]) [2, 7]
69000 <1> (<1> SDS-PAGE [1]) [1]
72000 <1> (<1> native enzyme, gel filtration [1]) [1]
245000 <3, 5> (<3,5> oligomeric form, gel filtration [5]) [5]
760000 <6> (<6> complex form, gel filtration [6]) [6]

Subunits

dimer <3> (<3> crystallization data [3]) [3]
monomer <1> (<1> 1 * 69000, SDS-PAGE [1]) [1]
monomer <2> (<2> 1 * 57000, SDS-PAGE [2]) [2]

monomer <3> (<3> 1 * 55000, SDS-PAGE [5]) [5]
monomer <5> (<5> 1 * 55000, SDS-PAGE [5]) [5]
pentamer <3, 5> (<3, 5> 4 * 55000 + 1 * 22000, gel filtration [5]) [5]
polymer <6> (<6> x * 56000 + x * 18000, gel filtration [6]) [6]

5 Isolation/Preparation/Mutation/Application

Localization
membrane <4> [7]
periplasm <1> [8]

Purification
<1> [1, 8]
<2> [2]
<6> [6]

Crystallization
<3> (<3> Triton X-114 significant improvement in crystal size and quality [4]) [3, 4]

Engineering
K100H <1> (<1> mutagenesis of the hem-binding motif CWSCK results in almost complete loss of formate-dependent nitrite reduction [8]) [8]
K100I <1> (<1> mutagenesis of the hem-binding motif CWSCK results in almost complete loss of formate-dependent nitrite reduction [8]) [8]
K100L <1> (<1> mutagenesis of the hem-binding motif CWSCK results in almost complete loss of formate-dependent nitrite reduction [8]) [8]

References

[1] Kajie, S.; Anraku, Y.: Purification of a hexaheme cytochrome c_{552} from Escherichia coli K 12 and its properties as a nitrite reductase. Eur. J. Biochem., **154**, 457-463 (1986)
[2] Liu, M.C.; Bakel, B.W.; Liu, M.Y.; Dao, T.N.: Purification of Vibrio fischeri nitrite reductase and its characterization as a hexaheme c-type cytochrome. Arch. Biochem. Biophys., **262**, 259-265 (1988)
[3] Einsle, O.; Stach, P.; Messerschmidt, A.; Simon, J.; Kroger, A.; Huber, R.; Kroneck, P.M.H.: Cytochrome c nitrite reductase from Wolinella succinogenes structure at 1.6.ANG. resolution, inhibitor binding, and heme-packing motifs. J. Biol. Chem., **275**, 39608-39616 (2000)
[4] Einsle, O.; Stach, P.; Messerschmidt, A.; Klimmek, O.; Simon, J.; Kroger, A.; Kroneck, P.M.: Crystallization and preliminary X-ray analysis of the membrane-bound cytochrome c nitrite reductase complex (NrfHA) from Wolinella succinogenes. Acta Crystallogr. Sect. D, **58**, 341-342 (2002)
[5] Schumacher, W.; Hole, U.; Kroneck, P.M.: Ammonia-forming cytochrome c nitrite reductase from Sulfurospirillum deleyianum is a tetraheme protein:

new aspects of the molecular composition and spectroscopic properties. Biochem. Biophys. Res. Commun., **205**, 911-916 (1994)
[6] Pereira, I.A.C.; LeGall, J.; Xavier, A.V.; Teixeira, M.: Characterization of a heme c nitrite reductase from a non-ammonifying microorganism, Desulfovibrio vulgaris Hildenborough. Biochim. Biophys. Acta, **1481**, 119-130 (2000)
[7] Costa, C.; Moura, J.J.; Moura, I.; Wang, Y.; Huynh, B.H.: Redox properties of cytochrome c nitrite reductase from Desulfovibrio desulfuricans ATCC 27774. J. Biol. Chem., **271**, 23191-23196 (1996)
[8] Eaves, D.J.; Grove, J.; Staudenmann, W.; James, P.; Poole, R.K.; White, S.A.; Griffiths, I.; Cole, J.A.: Involvement of products of the nrfEFG genes in the covalent attachment of heme c to a novel cysteine-lysine motif in the cytochrome c552 nitrite reductase from Escherichia coli. Mol. Microbiol., **28**, 205-216 (1998)

Trimethylamine-N-oxide reductase (cytochrome c) 1.7.2.3

1 Nomenclature

EC number
1.7.2.3

Systematic name
trimethylamine:cytochrome c oxidoreductase

Recommended name
trimethylamine-N-oxide reductase (cytochrome c)

Synonyms
EC 1.6.6.9 (formerly)
TMAO reductase
TOR
TorA
TorZ
reductase, trimethylamine N-oxide
trimethylamine oxide reductase

CAS registry number
37256-34-1

2 Source Organism

<1> *Escherichia coli* [1, 5, 6, 7, 10, 11]
<2> *Erythrobacter sp.* (strain Och 114 [2]) [2]
<3> *Roseobacter denitrificans* [3]
<4> *Shewanella massilia* (strain MR1 [4]) [4]
<5> *Rhodobacter capsulatus* (wild-type and mutants [9]) [8, 9]

3 Reaction and Specificity

Catalyzed reaction
trimethylamine + 2 (ferricytochrome c)-subunit + H_2O = trimethylamine N-oxide + 2 (ferrocytochrome c)-subunit + 2 H^+

Reaction type
oxidation
redox reaction
reduction

Natural substrates and products

S trimethylamine N-oxide + (ferrocytochrome c)-subunit + H^+ <1, 4, 5> (<1> reaction mechanism [5]; <5> trimethylamine N-oxide reductase and dimethyl sulfoxide reductase are identical enzymes [8]; <1> reduction of trimethylamine N-oxide is catalyzed by at least 2 enzymes: trimethylamine N-oxide reductase and dimethyl sulfoxide reductase [11]) (Reversibility: ? <1, 4, 5> [4, 5, 8, 11]) [4, 5, 8, 11]

P trimethylamine + (ferricytochrome c)-subunit + H_2O

Substrates and products

S 4-methylmorpholine-N-oxide + 2 (ferrocytochrome c)-subunit + H^+ <1> (Reversibility: ? <1> [7]) [7]

P 4-methylmorpholine + (ferrocytochrome c)-subunit + H^+

S biotin sulfoxide + (ferrocytochrome c)-subunit + H^+ <1> (Reversibility: ? <1> [7]) [7]

P ? + (ferrocytochrome c)-subunit + H^+

S bromate + (ferrocytochrome c)-subunit + H^+ <3> (Reversibility: ? <3> [3]) [3]

P ? + (ferrocytochrome c)-subunit + H^+

S dimethylsulfoxide + (ferrocytochrome c)-subunit + H^+ <1, 5> (Reversibility: ? <1, 5> [7, 8, 10]) [7, 8, 10]

P dimethylsulfide + (ferrocytochrome c)-subunit + H^+

S hydroxylamine + (ferrocytochrome c)-subunit + H^+ <3> (Reversibility: ? <3> [3]) [3]

P ? + (ferrocytochrome c)-subunit + H^+

S nicotinic acid N-oxide + (ferrocytochrome c)-subunit + H^+ <3> (Reversibility: ? <3> [3]) [3]

P ? + (ferrocytochrome c)-subunit + H^+

S picoline N-oxide + (ferrocytochrome c)-subunit + H^+ <3> (Reversibility: ? <3> [3]) [3]

P picoline + (ferrocytochrome c)-subunit + H^+

S tetramethylene sulfoxide + (ferrocytochrome c)-subunit + H^+ <1> (Reversibility: ? <1> [7]) [7]

P ? + (ferrocytochrome c)-subunit + H^+

S trimethylamine N-oxide + (ferrocytochrome c)-subunit + H^+ <2, 3, 4, 5> (<2> cytochrome c, cytochrome b-c1 and cytochrome c_{551} can act as electron donors [2];<3> cytochrome c_2, cytochrome c_{552} and methylviologen can serve as electron donors [3]; <5> enzyme forms a functional association with cytochrome c-556 [9]) (Reversibility: ? <2, 3, 4, 5> [2-10]) [2-10]

P trimethylamine + (ferricytochrome c)-subunit + H_2O <2> [2]

Inhibitors

DTT <1> (<1> inhibits cytochrome c formation [1]) [1]

Cofactors/prosthetic groups

cytochrome c <1, 5> (<1> pentahaem c-type cytochrome [1]; <1> c-type cytochrome Tor C is required for electron transfer to terminal enzyme Tor A [5,6]; <5> cytochrome c_{556} [9]) [1, 5, 6, 7, 9]
heme c <3> [3]

Metals, ions

molybdenum <1, 3, 4, 5> [3, 4, 8, 10]

Specific activity (U/mg)

6.8 <1> [10]
11 <2> (<2> without trimethylamine N-oxide, cell free extract, aerobic conditions [2]) [2]
30 <5> [8]
49.5 <3> [3]
70 <2> (<2> 30 mM trimethylamine N-oxide, cell free extract, aerobic conditions [2]) [2]
320 <2> (<2> 30 mM trimethylamine N-oxide, cell free extract, anaerobic conditions [2]) [2]
Additional information <1> (<1> specific activity in extracts after growth on various oxide compounds [11]) [11]

K_m-Value (mM)

0.006 <3> (trimethylamine N-oxide) [3]
0.36 <1> (biotin sulfoxide) [7]
1.4 <1> (4-methylmorpholine N-oxide) [7]
1.44 <1> (trimethylamine N-oxide) [7]
3.37 <1> (tetramethylene sulfoxide) [7]

Temperature optimum (°C)

30 <1> (<1> assay at [10]) [10]

4 Enzyme Structure

Molecular weight

85030 <5> (<5> calculated from amino acid sequence [8]) [8]
85030 <5> (<5> electrospray mass spectroscopy [8]) [8]
172000 <3> (<3> gel filtration [3]) [3]

Subunits

dimer <3> (<3> 2 * 90000, SDS-PAGE [3]) [3]

5 Isolation/Preparation/Mutation/Application

Localization

periplasm <1, 2, 5> [2, 6, 7, 9]

Purification

<1> [5, 6, 7, 10]
<3> [3]
<4> [4]
<5> [8, 9]

Crystallization

<4> (hanging drop vapor diffusion method [4]) [4]
<5> (hanging drop vapor diffusion method [8]) [8]

Cloning

<1> (fusion protein with β-galactosidase [11]) [1, 6, 7, 10, 11]
<4> [4]
<5> [8]

6 Stability

General stability information

<3>, activity is easily lost on dialysis [3]

Storage stability

<3>, -80°C stable for several months [3]

References

[1] Ansaldi, M.; Bordi, C.; Lepelletier, M.; Mejean, V.: TorC apocytochrome negatively autoregulates the trimethylamine N-oxide (TMAO) reductase operon in Escherichia coli. Mol. Microbiol., **33**, 284-295 (1999)

[2] Arata, H.; Serikawa, Y.; Takamiya, K.: Trimethylamine N-oxide respiration by aerobic photosynthetic bacterium, Erythrobacter sp. OCh 114. J. Biochem., **103**, 1011-1015 (1988)

[3] Arata, H.; Shimizu, M.; Takamiya, K.: Purification and properties of trimethylamine N-oxide reductase from aerobic photosynthetic bacterium Roseobacter denitrificans. J. Biochem., **112**, 470-475 (1992)

[4] Czjzek, M.; Dos Santos, J.P.; Pommier, J.; Giordano, G.; Mejean, V.; Haser, R.: Crystal structure of oxidized trimethylamine N-oxide reductase from Shewanella massilia at 2.5 A resolution. J. Mol. Biol., **284**, 435-447 (1998)

[5] Gon, S.; Giudici-Orticoni, M.T.; Mejean, V.; Iobbi-Nivol, C.: Electron transfer and binding of the c-type cytochrome TorC to the trimethylamine N-oxide reductase in Escherichia coli. J. Biol. Chem., **276**, 11545-11551 (2001)

[6] Gon, S.; Jourlin-Castelli, C.; Theraulaz, L.; Mejean, V.: An unsuspected autoregulatory pathway involving apocytochrome TorC and sensor TorS in Escherichia coli. Proc. Natl. Acad. Sci. USA, **98**, 11615-11620 (2001)

[7] Gon, S.; Patte, J.C.; Mejean, V.; Iobbi-Nivol, C.: The torYZ (yecK bisZ) operon encodes a third respiratory trimethylamine N-oxide reductase in Escherichia coli. J. Bacteriol., **182**, 5779-5786 (2000)

[8] Knäblein, J.; Dobbek, H.; Ehlert, S.; Schneider, F.: Isolation, cloning, sequence analysis and X-ray structure of dimethyl sulfoxide/trimethylamine N-oxide reductase from Rhodobacter capsulatus. Biol. Chem., **378**, 293-302 (1997)
[9] McEwan, A.G.; Richardson, D.J.; Hudig, H.; Ferguson, S.J.; Jackson, J.B.: Identification of cytochromes involved in electron transport to trimethylamine N-oxide/dimethyl sulfoxide reductase in Rhodobacter capsulatus. Biochim. Biophys. Acta, **973**, 308-314 (1989)
[10] Mejean, V.; Iobbi-Nivol, C.; Lepelletier, M.; Giordano, G.; Chippaux, M.; Pascal, M.C.: TMAO anaerobic respiration in Escherichia coli: involvement of the tor operon. Mol. Microbiol., **11**, 1169-1179 (1994)
[11] Yamamoto, I.; Hinakura, M.; Seki, S.; Seki, Y.; Kondo, H.: Anaerobic induction of trimethylamine N-oxide reductase and cytochromes by dimethyl sulfoxide in Escherichia coli. Curr. Microbiol., **20**, 245-249 (1990)

Nitroethane oxidase 1.7.3.1

1 Nomenclature

EC number
1.7.3.1

Systematic name
nitroethane:oxygen oxidoreductase

Recommended name
nitroethane oxidase

Synonyms
oxidase, nitroethane

CAS registry number
9029-36-1

2 Source Organism

<1> *Neurospora crassa* [1]

3 Reaction and Specificity

Catalyzed reaction
nitroethane + H_2O + O_2 = acetaldehyde + nitrite + H_2O_2

Reaction type
oxidation
redox reaction
reduction

Natural substrates and products
S nitroethane + H_2O + O_2 <1> (Reversibility: ? <1> [1]) [1]
P acetaldehyde + nitrite + H_2O_2

Substrates and products
S 1-nitropropane + H_2O + O_2 <1> (Reversibility: ? <1> [1]) [1]
P propionaldehyde + nitrite + H_2O_2
S 2-nitropropane + H_2O + O_2 <1> (Reversibility: ? <1> [1]) [1]
P acetone + nitrite + H_2O_2
S 3-nitropropanoic acid + H_2O + O_2 <1> (Reversibility: ? <1> [1]) [1]
P 2-oxopropanoic acid + nitrite + H_2O_2

S nitroethane + H_2O + O_2 <1> (Reversibility: ? <1> [1]) [1]
P acetaldehyde + nitrite + H_2O_2
S Additional information <1> (<1> slight activity, less than 5% of the activity towards nitroethane with: nitromethane, nitroacetic acid, 1-chloro-1-nitropropane, methylene blue also can function as hydrogen acceptor with low reaction rate [1]) [1]
P ?

Inhibitors
$CuSO_4$ <1> [1]
KNO_3 <1> [1]
NH_2OH <1> (<1> slight [1]) [1]
NaCN <1> (<1> slight [1]) [1]
NaCl <1> (<1> slight [1]) [1]
Tris <1> [1]
boric acid <1> [1]
maleic acid <1> [1]

Cofactors/prosthetic groups
Additional information <1> (<1> probably not a hemoprotein [1]) [1]

pH-Optimum
7 <1> [1]

pH-Range
6.2-7.6 <1> (<1> half-maximal activity at pH 6.2 and 7.6 [1]) [1]

5 Isolation/Preparation/Mutation/Application

Source/tissue
mycelium <1> [1]

Purification
<1> (partial [1]) [1]

6 Stability

pH-Stability
5.5 <1> (<1> 1 h, 0°C, complete loss of activity [1]) [1]
11 <1> (<1> 1 h, 0°C, full activity recovered after readjusting the solution to pH 7.0 [1]) [1]

Temperature stability
50 <1> (<1> 10 s, 70% loss of activity [1]) [1]
100 <1> (<1> 2 min, complete loss of activity [1]) [1]

General stability information
<1>, dialysis: stable to [1]

Storage stability

<1>, -20°C, weeks, slow decrease of activity [1]

<1>, in refrigerator for several days [1]

References

[1] Little, H.N.: Oxidation of nitroethane by extracts from Neurospora. J. Biol. Chem., **193**, 347-358 (1951)

Acetylindoxyl oxidase 1.7.3.2

1 Nomenclature

EC number
1.7.3.2

Systematic name
N-acetylindoxyl:oxygen oxidoreductase

Recommended name
acetylindoxyl oxidase

Synonyms
oxidase, acetylindoxyl

CAS registry number
9029-37-2

2 Source Organism

<1> *Zea mays* [1]

3 Reaction and Specificity

Catalyzed reaction
N-acetylindoxyl + O_2 = N-acetylisatin + (?)

Reaction type
oxidation
redox reaction
reduction

Natural substrates and products
S N-acetylindoxyl + O_2 <1> (Reversibility: ? <1> [1]) [1]
P N-acetylisatin + ?

Substrates and products
S N-acetylindoxyl + O_2 <1> (Reversibility: ? <1> [1]) [1]
P N-acetylisatin + ?

Inhibitors
1-naphthol <1> [1]
2-naphthol <1> [1]
8-hydroxyquinoline <1> [1]

caffeic acid <1> [1]
catechol <1> [1]
cyanide <1> [1]
cysteine <1> [1]
diethyl dithiocarbamate <1> [1]
dihydroxyphenylalanine <1> [1]
dioxindole <1> (<1> slight [1]) [1]
glutathione <1> [1]
indole <1> (<1> slight [1]) [1]
indole-3-aldehyde <1> (<1> slight [1]) [1]
m-chlorophenol <1> [1]
m-cresol <1> [1]
o-aminophenol <1> [1]
potassium ethyl xanthate <1> [1]
potassium indoxyl sulfate <1> (<1> slight [1]) [1]
thiourea <1> [1]

pH-Optimum

9 <1> [1]

pH-Range

7.5-9 <1> (<1> increase of activity from pH 7.0 to 9.0 [1]) [1]

5 Isolation/Preparation/Mutation/Application

Source/tissue

shoot <1> [1]

6 Stability

Temperature stability

100 <1> (<1> 15 min, crude preparation, stable [1]) [1]

References

[1] Beevers, H.; French, R.C.: Oxidation of N-acetylindoxyl by an enzyme from plants. Arch. Biochem. Biophys., **50**, 427-439 (1954)

Urate oxidase 1.7.3.3

1 Nomenclature

EC number
1.7.3.3

Systematic name
urate:oxygen oxidoreductase

Recommended name
urate oxidase

Synonyms
N-35
nodule specific uricase
Nodulin 35
Nodulin 35 homolog
non-symbiotic uricase
urate oxidase
oxidase, urate
uric acid oxidase
uricase
uricase II <7> [6, 60]

CAS registry number
9002-12-4

2 Source Organism

<1> *Homo sapiens* [1]
<2> *Oryctolagus cuniculus* [2, 38]
<3> *Candida utilis* [3, 10, 13, 14, 40, 47, 49]
<4> *Rattus norvegicus* [4, 23, 25, 42, 57]
<5> *Sus scrofa* [5, 14, 24, 27-30, 33, 57]
<6> *Bos taurus* [31-37]
<7> *Phaseolus vulgaris* [6, 60, 61]
<8> *Drosophila melanogaster* [7]
<9> *Streptomyces aureofaciens* [8]
<10> *Glycine max* (inoculated with Bradyrhizobium japonicum [9]; thioredoxin fusion protein [55,58]) [9, 12, 51, 55, 58]
<11> *Bacillus fastidiosus* [11, 20, 21, 22, 47, 56]
<12> *Vigna unguiculata* [16]

<13> *Streptomyces cyanogenus* [15]
<14> *Enterobacter cloacae* [17]
<15> *Aspergillus flavus* [18, 47, 62]
<16> *Neurospora crassa* [19]
<17> *Candida tropicalis* [26]
<18> *Phaseolus coccineus* [48]
<19> *Tupaia glis* [38]
<20> *Monkey* [38]
<21> *Zea mays* (maize) [39, 41]
<22> *Ricinus communis* [41]
<23> *Arachis hypogaea* [41]
<24> *Yucca brevifolia* [41]
<25> *Helianthus sp.* [41]
<26> *Carthamus tinctorius* [41]
<27> *Pinus monophylla* [41]
<28> *Acanthamoeba sp.* [43]
<29> *Arthrobacter pascens* [44, 45]
<30> *Dictyostelium discoideum* [46]
<31> *yeast-like fungi* (endosymbiont of Nilaparvata lugens [50]) [50]
<32> *Bacillus sp.* (TB-90 [52]) [52]
<33> *Candida sp.* [53]
<34> *Puccinia recondita* [54]
<35> *Cicer arietinum* [59]
<36> *Triticum aestivum* [59]
<37> *Vicia fabae major* [59, 61]
<38> *Uromyces fabae* [61]
<39> *Uromyces phaseoli* [61]

3 Reaction and Specificity

Catalyzed reaction
urate + O_2 = 5-hydroxyisourate + H_2O_2

Reaction type
oxidation
redox reaction
reduction

Natural substrates and products
S urate + O_2 <1-39> (Reversibility: ? <1-39> [1-61]) [1-61]
P 5-hydroxyisourate + H_2O_2 + CO_2

Substrates and products
S urate + O_2 <1-39> (Reversibility: ? <1-39> [1-61]) [1-61]
P 5-hydroxyisourate + H_2O_2 + CO_2

S Additional information <4, 5> (<4, 5> uricase can function as a voltage-sensitive channel that is highly selective to urate, relative to K^+ and Cl^- [57]) [57]
P ?

Inhibitors

1,3,7-trimethylxanthine <11> (<11> i.e. caffeine, slight [21]) [21]
2,2'-dipyridyl <12, 29> (<29> weak [45]) [16, 45]
2,9-dimethyl-1,10-phenanthroline <29> (<29> neo-cuproin [45]) [45]
2-hydroxypurine <13> [15]
3,7-dimethylxanthine <11> (<11> i.e. theobromine, slight [21]) [21]
3-methyluric acid <11> [21]
5,5'-dithiobis(2-nitrobenzoate) <3> [13]
5-azaorotate <11, 16, 28> [11, 19, 21, 43]
7-methyluric acid <11> [21]
8-azaxanthine <11> [21]
9-methyluric acid <10, 11> (<10> competitive versus urate [51]) [21, 51]
Ag^+ <14> [17]
Ba^{2+} <33> [53]
CN^- <3, 4, 11-16, 28, 29> [11, 15-19, 21, 40, 42-44]
Cd^{2+} <11> [21]
Co^{2+} <11, 12, 14> [16, 17, 21]
Cr^{3+}<11> [21]
Cu^{2+} <3, 4, 6, 11, 12, 14, 29, 33> [11, 13, 16, 17, 21, 32, 40, 42, 44, 45, 53]
EDTA <12, 16> [16, 19]
Fe^{2+} <11> [21]
Fe^{3+} <33> [53]
Hg^{2+} <3, 33> (<11> 1 mM: not [21]) [13, 40, 49, 53]
Mg^{2+} <11, 12> (<12> slight [16]; <11> 1 mM: not [21]) [16, 21]
Mn^{2+} <11, 12> [16, 21]
N-ethylmaleimide <6, 14> [17, 36]
NH_2OH <11> [11]
NH_4^+ <10, 12> (<10> weak [12]) [12, 16]
$Na_2S_2O_4$ <11> [11]
NaN_3 <11> [11]
Ni^{2+} <11> [21]
Pb^{2+} <11> [21]
Zn^{2+} <11> [21]
adenine <10, 12> (<10> weak [12]; <22> not [41]) [12, 16]
allantoic acid <12> (<12> weak [16]) [16]
allantoin <10, 12> (<10> weak [12]) [12, 16]
amelide <6> [32]
arginine <6> (<6> weak [36]) [36]
aspartic acid <3> (<3> slight [10]) [10]
biguanidine salts <6> (<6> inactivation is pH-dependent: slightly inhibitory below pH 10, rapid inactivation at high pH [36]) [36]
cyanurate <6> [32]

dicyandiamide <6> (<6> inactivation is pH-dependent: small below pH 10, rapid increase at high pH [36]) [36]
diethyldithiocarbamic acid <11> [11]
glutamine <10, 12> (<10> weak [12]) [12, 16]
glycerol <4> [42]
glyoxylic acid <12> (<12> weak [16]) [16]
guanidinium salts <6> (<6> inactivation is pH-dependent: slightly inhibitory below pH 10, rapid inactivation at high pH [36]) [36]
guanine <12> [16]
hydroxylamine <15> [18]
hydroxypurines <28> [43]
hypoxanthine <11, 12, 22> (<12, 22> weak [16, 41]) [16, 21, 41]
inosine 5'-monophosphate <12> [16]
iodoacetate <13> (<13> low effect [15]) [15]
neocuproin <35-37> [59]
o-iodosobenzoate <3> [13]
o-phenanthroline <12, 16, 29> (<3> not [40]; <29> weak [45]) [16, 19, 45]
oxonate <7, 34-39> (<7,38,39> 0.025 mM 90% inhibition [61]) [54, 59, 61]
oxopurines <28> [43]
p-chloromercuribenzoate <3, 13, 14> (<13> slight effect [15]) [13, 15, 17, 40]
periodate <15> [18]
phosphate <6> (<6> no inactivation by phosphate, in presence of borate or dithiothreitol [34]) [34]
pyrazinoate <4, 5> (<4,5> channel activity completely disappears [57]) [57]
salicylhydroxamic acid <12> [16]
sucrose <4> [42]
thiourea <16> (<16> slight [19]) [19]
trichloropurine <28, 30> [43, 46]
urate <13, 22, 28> (<22> substrate inhibition: above 0.120 mM [41]; <28> above 0.125 mM [43]; <13> not [15]) [15, 41, 43]
urea <3, 12> (<12> weak [16]) [16, 40]
xanthine <7, 10-13, 22, 34-39> (<10> competitive versus urate, noncompetitive versus O_2 [51]; <7,38,39> 0.05 mM 60-75% inhibition [61]) [15, 16, 21, 41, 51, 54, 59, 61]

Cofactors/prosthetic groups

Additional information <10, 11> (<10, 11> no cofactors [21, 51]) [21, 51]

Metals, ions

copper <3, 5, 11, 13, 29> (<3> neither copper nor iron detected [13]; <13> enzyme contains 0.15 mol of copper [15]; <11> copper : uricase ratio is 1 : 7 [21]; <5> copper : uricase ratio is 1 : 1 [30]; <3> negligible amount of copper [40]; <29> contains 0.2 mol copper per mol enzyme protein [44]) [13, 15, 21, 30, 40, 44]
iron <3, 11-13, 29> (<3> neither copper nor iron detected [13]; <13> enzyme contains less than 0.05 mol of iron [15]; <11> iron : uricase ratio is 1 : 50 [21]; <3> contains nearly one mol per mol of enzyme [40]; <29> contains

0.1 mol iron per mol enzyme [44]; <12, 29> Fe^{3+}: stimulation [16, 4]; <29> slight stimulation [45]) [15-16, 21, 40, 44, 45]

Turnover number (min^{-1})

1878 <10> (uric acid) [55]

Specific activity (U/mg)

1.87 <7> [61]
2.04 <38> [61]
2.55 <4> [25]
3.11 <39> [61]
3.6 <10> [51]
3.74 <7> (<7> cotyledons, 4 d old [60]) [60]
5.29 <32> [52]
5.6 <12> [16]
6.6 <14> [17]
6.85 <10> [12]
8.4 <34> [54]
9.35 <4> (<4> male rats [42]) [42]
10.48 <3> [49]
12.2 <33> [53]
12.3 <13> [15]
13.3 <7> (<7> hypocotyls, 4 d old [60]) [60]
15.3 <4> (<4> female [42]) [42]
15.4 <7> (<7> roots, 4 d old [60]) [60]
1790 <7> (<7> nodules, 21 d old [60]) [60]

K_m-Value (mM)

0.00588 <3> (uric acid) [49]
0.0098 <10> (uric acid) [51]
0.01 <3, 10> (uric acid) [12, 13]
0.014 <32> (uric acid, <32> enzyme without histidine tag [52]) [52]
0.015 <32, 7> (uric acid, <32> enzyme with histidine tag [52]) [52, 61]
0.018 <12> (uric acid) [16]
0.021 <37> (uric acid) [61]
0.029 <12> (O_2) [16]
0.031 <10> (O_2) [12]
0.032 <39> (uric acid) [61]
0.034 <38> (uric acid) [61]
0.035 <34> (uric acid) [54]
0.042 <11, 16> (uric acid, <11> poly(N-acryloylmorpholine)-OSu-uricase [56]) [19, 56]
0.05 <11> (uric acid, <11> native uricase, monomethoxypoly(ethylene glycol) N-leucine-OSu-uricase and branched monomethoxypoly(ethylene glycol) N-leucine-OSu-uricase [56]) [56]
0.061 <10> (O_2) [51]
0.099 <13> (uric acid) [15]
1.5 <14> (uric acid) [17]

pH-Optimum

8-8.5 <13> (<13> borate and phosphate buffer [15]) [15]
8.5 <3, 30, 33> [3, 46, 49, 53]
8.5-9.3 <4> [42]
8.6 <5> (<5> free enzyme [27]) [27]
8.9 <22> [41]
9 <34> [54]
9-10 <12> [16]
9.2 <29> [45]
9.4 <28> [43]
9.5 <5, 10, 11, 13, 14> (<5> immobilized enzyme [27]; <13> glycine buffer [15]) [12, 15, 17, 21, 27]

pH-Range

7-11 <14> (<14> pH 7: about 50% of activity maximum, pH 11: about 40% of activity maximum [17]) [17]
7.4-9.6 <5> (<5> 50% of activity maximum at pH 7.4 and 9.6, free enzyme [27]) [27]
8-9 <39> [61]
8-10.2 <10> (<10> pH 8.0: about 35% of activity maximum, pH 10.2: about 55% of activity maximum [12]) [12]
8-11 <5> (<5> 50% of activity maximum at pH 8 and 11, immobilized enzyme [27]) [27]
8.5-9.1 <38> [61]
8.5-9.5 <33> [53]
8.5-10.5 <11> (<11> 50% of activity maximum at pH 8.5 and 10.5 [21]) [21]
8.6-9.5 <37> [61]
9-9.7 <7> [61]

Temperature optimum (°C)

30 <3, 33> [3, 53]
30-35 <11> [21]
35 <3, 13> (<3> 5 min incubation test [49]) [15, 49]
40 <14> [17]

Temperature range (°C)

20-50 <3, 14> (<3> 50°C: about 60% of activity maximum [49]; <3> 20°C: about 70% of activity maximum [49]; <14> 20°C: about 60% of activity maximum [17]; <14> 50°C: about 50% of activity maximum [17]) [17, 49]

4 Enzyme Structure

Molecular weight

32000-35000 <10> (<10> SDS-PAGE, gel filtration [12]) [12]
32880 <31> (<31> deduced from nucleotide sequence of cDNA [50]) [50]
33270 <4> (<4> determination of nucleotide sequence of cDNA and calculation of corresponding amino acid sequence [4]) [4]

34000 <7> (<7> SDS-PAGE [6]) [6]
35050 <10> (<10> deduced from nucleotide sequence of cDNA [51]) [51]
35060 <10> (<10> MALDI-TOF mass spectrometry [51]) [51]
50000 <12> (<12> gel filtration [16]) [16]
68000 <10> (<10> gel filtration, [12]) [12]
70000-76000 <33> (<33> gel filtration [53]) [53]
100000 <6, 12> (<6> gel filtration [35]; <12> gel filtration [16]) [16, 35]
102000 <3> (<3> polyacrylamide disc electrophoresis [3]) [3]
105000 <14> (<14> gel filtration [17]) [17]
109000 <13> (<13> gel filtration [15]) [15]
114000-128000 <35> (<35> PAGE, gel filtration [59]) [59]
115000-123000 <16> (<16> PAGE, gel filtration [19]) [19]
120000 <3> (<3> gel filtration, equilibrium sedimentation [40,13]) [13, 40]
120000-122000 <37> (<37> PAGE, gel filtration [59]) [59, 61]
120000-140000 <36> (<36> PAGE, gel filtration [59]) [59]
124000 <38, 39> (<38,39> PAGE [61]) [61]
125000 <34> (<34> PAGE [54]) [54]
125000 <5> (<5> short-column meniscus depletion sedimentation equilibrium [30]) [30]
128000 <7, 15> (<7> PAGE [61]) [61, 62]
136300-141600 <10> (<10> ultracentrifugation [55]) [55]
145000-150000 <11> (<11> gel filtration [21]) [21]
230000 <4> (<4> gel filtration [25]) [25]

Subunits

? <4, 7> (<7> x * 34000, SDS-PAGE [6]; <4> x * 33000, SDS-PAGE [23]) [6, 23]
dimer <10, 12> (<10> at pH 8.8 the enzyme associates to form a dimer of about 68000 [12]; <12> at pH 8.8 the enzyme associates to form a dimer of about 10000 [16]) [12, 16]
heterotetramer <11, 34> (<34> $\alpha_2\beta_2$, 2 * 35000 + 2 * 37000, SDS-PAGE [54]; <11> $\alpha_2\beta_2$, 2 * 36000 + 2 * 39000, SDS-PAGE [21]) [21, 54]
hexamer <4> (<4> α_6, 6 * 37000, SDS-PAGE [25]) [25]
monomer <10, 12, 33> (<10> native enzyme at pH 7.5 is a monomer with MW 32000-35000 [12]; <12> native enzyme at pH 7.5 is a monomer with MW 50000 [16]; <33> 1 * 70000-76000, SDS-PAGE [53]) [12, 16, 53]
tetramer <3, 5, 10, 15, 16, 35-37> (<3> α_4, 4 * 30000, SDS-PAGE [13]; <16> α_4, 4 * 33000, SDS-PAGE [19]; <5> α_4, 4 * 32000, SDS-PAGE [30]; <10> α_4, 4 * 35052, calculated from deduced amino acid sequence [55]; <35-37> α_4, 4 * 32000-34000, SDS-PAGE [59]) [13, 19, 30, 55, 59, 62]
trimer <13> (<13> α3, 3 * 32000, SDS-PAGE [15]) [15]

5 Isolation/Preparation/Mutation/Application

Source/tissue

cotyledon <7, 23, 25, 26> [41, 60]
endosperm <22> [41]

hypocotyl <7> [60]
kidney <4, 6> [31, 32, 34, 35, 37, 57]
larva <8> [7]
leaf <7, 18, 35-37> [48, 59, 61]
liver <2, 4, 5, 19, 20> (<19,20> no activity in liver of New World monkeys, low activity in liver of Old World monkeys [38]) [2, 4, 14, 23, 27-30, 38, 42, 57]
malpighian tubule <8> [7]
megagametophyte <27> [41]
perisperm <24> [41]
proximal tubule <4> [57]
root <7> [60]
root nodule <10> [51, 58]
root tip <21> [39]
scutellum <21> [41]
seed <7, 10> [6, 12]
seedling <21-27> (<21-27> fat-degrading tissues [41]) [41]
spore <11> [11]
uredospore <38, 39> [61]

Localization

cell associated <3, 11, 13, 14, 16> [3, 11, 15, 17, 19, 21]
cell membrane <4> [57]
extracellular <11> (<11> soluble [22]) [22]
glyoxysome <21-26> [41]
intracellular <11> [21]
microbody <17> [26]
mitochondrion <6> [35]
peroxisome <4, 10, 18, 21, 28, 30> (<10> limited to large peroxisomes in uninfected cells of root nodules of Glycine max inoculated with Bradyrhizobium japonicum [9]) [9, 25, 39, 43, 46, 48]
soluble <7> [6]

Purification

<3> [3, 10, 13, 49]
<4> [23, 42]
<6> [35]
<10> [12, 51, 55]
<11> [21]
<12> [16]
<13> [15]
<14> [17]
<16> [19]
<29> [45]
<31> [50]
<32> (enzyme synthesized with C-terminal 6-histidine tag [52]) [52]
<33> [53]
<34> [54]

<3, 6> [10, 31, 35]
<35-37> [59]
<7, 38, 39> [61]

Crystallization

<15> (<15> crystallization of large proteins in the presence of polyethylene glycol [62]) [62]

Cloning

<1> (expression in NM 538 [1]) [1]
<2> [2]
<4> [4]
<5> (expression in pPUO1 [5]) [5]
<7> (expression in Escherichia coli MC1061 [6]) [6]
<8> (expression in Escherichia coli RR1 [7]) [7]
<10> (expression in Escherichia coli [51,55,58]) [51, 55, 58]
<31> (expression in Escherichia coli DH5α [50]) [50]
<32> (overexpression in Escherichia coli JM109 [52]) [52]

Engineering

Additional information <10, 11> (<10> thioredoxin urate oxidase fusion protein [55]; <11> uricase covalently linked to monomethoxypoly(ethylene glycol) N-leucine-OSu, branched monomethoxypoly(ethylene glycol) N-leucine-OSu or poly(N-acryloylmorpholine)-OSu last longer in blood [56]) [55, 56]

Application

medicine <32, 33> (<32,33> determining the urate concentration in blood and urine [52,53]) [52, 53]

6 Stability

pH-Stability

6 <14> (<14> 45°C, 30 min, 40% loss of activity [17]) [17]
6-11 <13> (<13> 35°C, 1 h, stable [15]) [15]
6.5-10.5 <5> (<5> 22°C, 30 min, immobilized enzyme, stable [27]) [27]
7-11 <3> (<3> 10 min, stable [49]) [49]
8-9 <14> (<14> 45°C, 30 min, stable [17]) [17]
8-11 <3> (<3> 4°C, 60 days, 35% loss of activity [3]) [3]
10 <14> (<14> 45°C, 30 min, 30% loss of activity [17]) [17]
12 <5> (<5> 22°C, 30 min, 55% loss of activity [27]) [27]

Temperature stability

18-20 <12> (<12> 2 days, 70% loss of activity [16]) [16]
40 <3, 6, 14> (<14> 30 min, stable below [17]; <6> below pH 9.5, 2 h, very slow loss of activity in absence or presence of Cu^{2+} [37]; <3> 10 min, stable below [49]) [17, 37, 49]
45 <33> (<33> after 5 min 50% of initial activity [53]) [53]

50 <11, 13> (<13> pH 7.8, 10 min, stable below [15]; <11> 10 min, stable [21]) [15, 21]
52 <11> (<11> uricase covalently linked to monomethoxypoly(ethylene glycol) N-leucine-OSu-uricase and branched monomethoxypoly(ethylene glycol) N-leucine-OSu-uricase 50% loss of activity [56]) [56]
60 <11, 13, 14> (<13> pH 7.8, 10 min, 50% loss of activity [15]; <14> 30 min, about 40% loss of activity [17]; <11> 10 min, 55% loss of activity [21]) [15, 17, 21]
60-61 <11> (<11> native uricase and poly(N-acryloylmorpholine)-OSu-uricase 50% loss of activity [56]) [56]
Additional information <6> (<6> thermal inactivation rises steeply as $CuSO_4$ concentration rises from 0.025 to 0.175 mM and as the pH of the medium exceedes 9.5 [37]) [37]

Oxidation stability

<22>, sensitive to O_2 [41]

General stability information

<3>, dithiothreitol prevents polymerization and stabilizes throughout purification [13]
<3>, little loss of activity by freeze-drying [49]
<3>, lower stability in solutions of phosphate buffer than in borate buffer [49]
<4>, repeated freezing and thawing has no effect [42]
<4>, urate oxidase from female rat livers is more stable than enzyme from male rat livers [42]
<4>, urea, 4 M, several h without loss of activity [42]
<5>, stability of immobilized enzyme depends on the time of stirring during immobilization and on the quantity of enzyme used [24]
<5>, unusually resistant to SDS [30]
<5>, unusually resistant to guanidinium chloride [30]
<6>, dithiothreitol effect of treatment with dithiothreitol on extraction and purification [35]
<6>, dithiothreitol stabilization [33]
<11>, conjugated uricases are more stable to trypsin digestion [56]
<12>, frozen enzyme retains complete activity [16]
<29>, Cu^{2+} inactivates at low temperature, uric acid prevents inactivation [44]
<29>, Fe^{3+} partially protects against inactivation at low pH or at low ionic strength, stimulates reactivation [44]
<29>, ammonium sulfate protects from inactivation at low pH [44]
<3, 11, 15>, proteolytic digestion by endopeptidases cause rapid loss of activity, exopeptidases have slight effect [47]
<5, 6>, EDTA stabilizes [33]
<5, 6>, borate stabilizes [33]

Storage stability

<4>, 3°C, 0.15% sodium carbonate, several weeks, stable [42]
<10>, 4°C, several weeks [51]

<11>, 0-4°C, crystals in $(NH_4)_2SO_4$ solution, 3 months, stable [21]
<12>, 4°C, 7 days, 40% loss of activity [16]
<13>, 4°C, 20 mM phosphate buffer, pH 7.8, for at least 1 month [15]
<29>, -15°C, 100 mM borate buffer, pH 8.0, containing 100 mM or more ammonium sulfate [45]
<31>, 4°C, 100 mM Tris-HCl buffer, pH 9.5, 90% of activity after 5 days [50]

References

[1] Yeldandi, A.V.; Wang, X.; Alvares, K.; Kumar, S.; Sambasiva Rao, M.; Reddy, J.K.: Human urate oxidase gene: cloning and partial sequence analysis reveal a stop codon within the fifth exon. Biochem. Biophys. Res. Commun., **171**, 641-646 (1990)
[2] Motojima, K.; Goto, S.: Cloning of rabbit uricase cDNA reveals a conserved carboxy-terminal tripeptide in three species. Biochim. Biophys. Acta, **1008**, 116-118 (1989)
[3] Adamek, V.; Kralova, B.; Suchova, M.; Valentova, O.; Demnerova, K.: Purification of microbial uricase. J. Chromatogr., **497**, 268-275 (1989)
[4] Motojima, K.; Kanaya, S.; Goto, S.: Cloning and sequence analysis of cDNA for rat liver uricase. J. Biol. Chem., **263**, 16677-16681 (1988)
[5] Lee, C.C.; Wu, X.; Gibbs, R.A.; Cook, R.G.; Muzny, D.M.; Caskey, C.T.: Generation of cDNA probes directed by amino acid sequence: cloning of urate oxidase. Science, **239**, 1288-1291 (1988)
[6] Sanchez, F.; Campos, F.; Padilla, J.; Bonneville, J.M.; Enriquez, C.; Caput, D.: Purification, cDNA cloning, and developmental expression of the nodule-specific uricase from Phaseolus vulgaris L.. Plant Physiol., **84**, 1143-1147 (1987)
[7] Kral, L.G.; Johnson, D.H.; Burnett, J.B.; Friedman, T.B.: Cloning a cDNA for Drosophila melanogaster urate oxidase. Gene, **45**, 131-137 (1986)
[8] Demnerova, K.; Kralova, B.; Lehejckova, R.; Adamek, V.: Production of uricase by different strains of Streptomyces. Biotechnol. Lett., **8**, 577-578 (1986)
[9] Van den Bosch, K.A.; Newcomb, E.H.: Immunogold localization of nodule-specific uricase in developing soybean root nodules. Planta, **167**, 425-436 (1986)
[10] Kralova, B.; Lehejckova, R.; Demnerova, K.; Dobransky, T.: Extraction of uricase from Candida utilis. Biotechnol. Lett., **8**, 99-102 (1986)
[11] Salas, J.A.; Johnstone, K.; Ellar, D.J.: Role of uricase in the triggering of germination of Bacillus fastidiosus spores. Biochem. J., **229**, 241-249 (1985)
[12] Lucas, K.; Boland, M.J.; Schubert, K.R.: Uricase from soybean root nodules: purification, properties, and comparison with the enzyme from cowpea. Arch. Biochem. Biophys., **226**, 190-197 (1983)
[13] Nishimura, H.; Yoshida, K.; Yokota, Y.; Matsushima, A.; Inada, Y.: Physicochemical properties and states of sulfhydryl groups of uricase from Candida utilis. J. Biochem., **91**, 41-48 (1982)

[14] Chen, R.H.L.; Abuchowski, A.; Van Es, T.; Palczuk, N.C.; Davis, F.F.: Properties of two urate oxidases modified by the covalent attachment of poly(-ethylene glycol). Biochim. Biophys. Acta, **660**, 293-298 (1981)
[15] Ohe, T.; Watanabe, Y.: Purification and properties of urate oxidase from Streptomyces cyanogenus. J. Biochem., **89**, 1769-1776 (1981)
[16] Rainbird, R.M.; Atkins, C.A.: Purification and some properties of urate oxidase from nitrogen-fixing nodules of cowpea. Biochim. Biophys. Acta, **659**, 132-140 (1981)
[17] Machida, Y.; Nakanishi, T.: Purification and properties of uricase from Enterobacter cloacae. Agric. Biol. Chem., **44**, 2811-2815 (1980)
[18] Conley, T.G.; Priest, D.G.: Non-classical inhibition of uricase by cyanide. Biochem. J., **187**, 733-738 (1980)
[19] Wang, L.W.C.; Marzluf, G.A.: Purification and characterization of uricase, a nitrogen-regulated enzyme, from Neurospora crassa. Arch. Biochem. Biophys., **201**, 185-193 (1980)
[20] Bongaerts, G.P.A.; Vogels, G.D.: Mechanism of uricase action. Biochim. Biophys. Acta, **567**, 295-308 (1979)
[21] Bongaerts, G.P.A.; Uitzetter, J.; Brouns, R.; Vogels, G.D.: Uricase of Bacillus fastidiosus. Properties and regulation of synthesis. Biochim. Biophys. Acta, **527**, 348-358 (1978)
[22] Mahler, J.L.: A new bacterial uricase for uric acid determination. Anal. Biochem., **38**, 65-84 (1970)
[23] Watanabe, T.; Suga, T.: A simple purification method for rat liver urate oxidase. Anal. Biochem., **89**, 343-347 (1978)
[24] Johnson, D.B.; Coughlan, M.P.: Studies on the stability of immobilized xanthine oxidase and urate oxidase. Biotechnol. Bioeng., **20**, 1085-1095 (1978)
[25] Antonenkov, V.D.; Pachenko, L.F.: Organization of urate oxidase in peroxisomal nucleoids. FEBS Lett., **88**, 151-154 (1978)
[26] Tanaka, A.; Yamamura, M.; Kawamoto, S.; Fukui, S.: Production of uricase by Candida tropicalis using n-alkane as a substrate. Appl. Environ. Microbiol., **34**, 342-346 (1977)
[27] Hinberg, I.; O'Driscoll, K.F.: Preparation and kinetic properties of gel entrapped urate oxidase. Biotechnol. Bioeng., **17**, 1435-1441 (1975)
[28] Sedor, F.A.; Sander, E.G.: Inhibition of uricase by substituted pyrimidines. Biochem. Biophys. Res. Commun., **75**, 406-413 (1977)
[29] Johnson, D.B.; Coughlan, M.P.: Characterization of immobilized urate oxidase. Biochem. Soc. Trans., **2**, 1362-1363 (1974)
[30] Pitts, O.M.; Priest, D.G.; Fish, W.W.: Uricase, subunit composition and resistance to denaturants. Biochemistry, **13**, 888-892 (1974)
[31] Truscoe, R.: Effect of detergents on extraction and activity of ox-kidney urate oxidase. Enzymologia, **33**, 19-32 (1967)
[32] Truscoe, R.: Effects of some nitrogenous bases on the activity of ox-kidney urate oxidase. Enzymologia, **34**, 337-343 (1968)
[33] Truscoe, R.: Effect of borate on urate oxidase activity. Enzymologia, **34**, 325-336 (1968)

[34] Truscoe, R.: Effect of phosphate on the activity of ox-kidney urate oxidase. Enzymologia, **35**, 19-30 (1968)
[35] James, K.A.C.; Tate, W.P.; Truscoe, R.: Effects of treatment with dithiothreitol on the extraction, activity and purification of ox-kidney urate oxidase. Enzymologia, **37**, 131-152 (1969)
[36] Bentley, K.W.; Truscoe, R.: Effect of guanidine derivates on urate oxidase activity. Enzymologia, **37**, 285-313 (1969)
[37] Turner, J.C.; Truscoe, R.: Effect of copper on urate oxidase. Enzymologia, **43**, 57-70 (1972)
[38] Christen, P.; Peacock, W.C.; Christen, A.E.; Wacker, W.E.C.: Urate oxidase in primate phylogenesis. Eur. J. Biochem., **12**, 3-5 (1970)
[39] Parish, R.W.: Urate oxidase in peroxisomes from maize root tips. Planta, **104**, 247-251 (1972)
[40] Itaya, K.; Fukumoto, J.; Yamamoto, T.: Studies on urate oxidase of Candida utilis. Part II. Some physical and chemical properties of the purified enzyme. Agric. Biol. Chem., **35**, 813-821 (1971)
[41] Theimer, R.R.; Beevers, H.: Uricase and allantoinase in glyoxysomes. Plant Physiol., **47**, 246-251 (1971)
[42] Townsend, D.; Lata, G.F.: Purification of urate oxidase; a sex dependent enzyme from rat liver. Arch. Biochem. Biophys., **135**, 166-172 (1969)
[43] Muller, M.; Moller, K.M.: Urate oxidase and its association with peroxysomes in Acanthamoeba sp.. Eur. J. Biochem., **9**, 424-430 (1969)
[44] Nose, K.; Arima, K.: Studies on bacterial urate:oxygen oxidoreductase II. Observations concerning the properties and components of the active site. Biochim. Biophys. Acta, **151**, 63-69 (1968)
[45] Arima, K.; Nose, K.: Studies on bacterial urate:oxygen oxidoreductase I. Purification and properties of the enzyme. Biochim. Biophys. Acta, **151**, 54-62 (1968)
[46] Parish, R.W.: Mitochondria and peroxisomes from the cellular slime mould Dictyostelium discoideum. Isolation techniques and urate oxidase association with peroxisomes. Eur. J. Biochem., **58**, 523-531 (1975)
[47] Fitzpatrick, D.A.; McGeeney, K.F.: Resistance of urate oxidase to proteolytic digestion. Biochem. Soc. Trans., **3**, 1253-1255 (1975)
[48] Theimer, R.R.; Heidinger, P.: Control of particulate urate oxidase activity in bean roots by external nitrogen supply. Z. Pflanzenphysiol., **73**, 360-370 (1974)
[49] Itaya, K.; Yamamoto, T.; Fukumoto, J.: Studies on yeast uricase. Part I. Purification and some enzymatic properties of yeast uricase. Agric. Biol. Chem., **31**, 1256-1264 (1967)
[50] Hongoh, Y.; Sasaki, T.; Ishikawa, H.: Cloning, sequence analysis and expression in Escherichia coli of the gene encoding a uricase from the yeast-like symbiont of the brown planthopper, Nilaparvata lugens. Insect Biochem. Mol. Biol., **30**, 173-182 (2000)
[51] Kahn, K.; Tipton, P.A.: Kinetic mechanism and cofactor content of soybean root nodule urate oxidase. Biochemistry, **36**, 4731-4738 (1997)

[52] Nishiya, Y.; Hibi, T.; Oda, J.: A purification method of the diagnostic enzyme Bacillus uricase using magnetic beads and non-specific protease. Protein Expr. Purif., **25**, 426. (2002)
[53] Jianguo, L.; Gaoxiang, L.; Hong, L.; Xiukai, Z.: Purification and properties of uricase from Candida sp. and its application in uric acid analysis in serum. Appl. Biochem. Biotechnol., **47**, 57-63 (1994)
[54] Aguilar, M.; Montalbini, P.; Pineda, M.: Urate oxidase from the rust Puccinia recondita is a heterotetramer with two different-sized monomers. Curr. Microbiol., **44**, 257-261 (2002)
[55] Kahn, K.; Tipton, P.A.: Spectroscopic characterization of intermediates in the urate oxidase reaction. Biochemistry, **37**, 11651-11659 (1998)
[56] Schiavon, O.; Caliceti, P.; Ferruti, P.; Veronese, F.M.: Therapeutic proteins: a comparison of chemical and biological properties of uricase conjugated to linear or branched poly(ethylene glycol) and poly(N-acryloylmorpholine). Farmaco, **55**, 264-269 (2000)
[57] Leal-Pinto, E.; London, R.D.; Knorr, B.A.; Abramson, R.G.: Reconstitution of hepatic uricase in planar lipid bilayer reveals a functional organic anion channel. J. Membr. Biol., **146**, 123-132 (1995)
[58] Kahn, K.; Serfozo, P.; Tipton, P.A.: Identification of the true product of the urate oxidase reaction. J. Am. Chem. Soc., **119**, 5435-5442 (1997)
[59] Montalbini, P.; Redondo, J.; Caballero, J.L.; Cardenas, J.; Pineda, M.: Uricase from leaves. Its purification and characterization from three different higher plants. Planta, **202**, 277-283 (1997)
[60] Capote-Mainez, N.; Sanchez, F.: Characterization of the common bean uricase II and its expression in organs other than nodules. Plant Physiol., **115**, 1307-1317 (1997)
[61] Montalbini, P.; Aguilar, M.; Pineda, M.: Isolation and characterization of uricase from bean leaves and its comparison with uredospore enzymes. Plant Sci., **147**, 139-147 (1999)
[62] Vivares, D.; Bonnete, F.: X-ray scattering studies of Aspergillus flavus urate oxidase: towards a better understanding of PEG effects on the crystallization of large proteins. Acta Crystallogr. Sect. D, **58**, 472-479 (2002)

Hydroxylamine oxidase 1.7.3.4

1 Nomenclature

EC number
1.7.3.4

Systematic name
hydroxylamine:oxygen oxidoreductase

Recommended name
hydroxylamine oxidase

Synonyms
HAO
hydroxylamine oxidoreductase
oxidase, hydroxylamine HAO

CAS registry number
9075-43-8

2 Source Organism

<1> *Nitrosomonas europaea* (gene cloned and expressed in Pseudomonas putida [2]; strain ENI-11 [32]) [1-25, 27, 29, 32, 34]
<2> *Pseudomonas sp.* (strain PB16 [26]) [26]
<3> *Paracoccus denitrificans* (GB17 [28]) [28]
<4> *Brocadia anammoxidans* [30]
<5> *Thiosphaera pantotropha* [31]
<6> *Methylococcus capsulatus* [33]

3 Reaction and Specificity

Catalyzed reaction
hydroxylamine + O_2 = nitrite + H_2O

Reaction type
dehydrogenation
oxidation
redox reaction
reduction

Natural substrates and products

S hydroxylamine + O_2 <1> (<1> energy production [1]) (Reversibility: ? <1> [1]) [1]

P nitrite + H_2O <1> [1]

Substrates and products

S hydrazine + phenazine methosulfate <1, 4> (<4> phenazine sulfate and methylthiazolyltetrazolium bromide can serve as electron acceptors [30]) (Reversibility: ? <1, 4> [12, 30]) [12, 30]

P N_2 + reduced phenazine methosulfate <1, 4> [12, 30]

S hydroxylamine + O_2 <1, 2, 3, 4, 5, 6> (<1> in presence of diethyldithiocarbamate, hydroxylamine is oxidized to nitrite, for each mol of nitrite produced approximately 1 mol of diethyldithiocarbamate is oxidized to bis(diethyldithiocarbamoyl)disulfide [12]; <1> phenazine methosulfate, dichlorophenolindophenol, mammalian cytochrome c and 3-(4,5-dimethylthiazolyl-2)-2,5-diphenyltetrazolium chloride can serve as electron acceptors [12]; <1> Nitrosomonas cytochrome c_{554} can act as electron carrier [20]; <1> mammalian cytochrome c can serve as electron acceptor [27]; <4> phenazine sulfate, ferricyanide and methylthiazolyltetrazolium bromide can serve as electron acceptors [30]; <5> enzyme can use periplasmatic redox proteins, cytochrome c_{551} and pseudoazurin as electron acceptors [31]; <6> cytochrome c_{555} is able to accept electrons, but phenazine methosulfate is required for maximum enzyme activity [33]) (Reversibility: ? <1-6> [1, 4, 12, 20, 26, 27, 28, 30, 31, 33]) [1, 4, 12, 20, 26, 27, 28, 30, 31, 33]

P nitrite + H_2O <1, 3, 4> [12, 20, 28, 30]

S pyrogallol + O_2 <1> (Reversibility: ? <1> [11]) [11]

P ?

Inhibitors

2',2'-dipyridyl <1> (<1> 50% inhibition at 0.1 mM [16]) [16]

8-hydroxyquinoline <1> (<1> 50% inhibition at 0.2 mM [16]) [16]

CO <1> (<1> competition of CO and O_2 for P460 [19]) [5, 19]

$Ce(NH_4)_2(NO_3)_6$ <1> (<1> 0.1 mM, stimulates the rate of oxidation of NH_2OH 2.6-fold and inhibits production of HNO_2 by 40% [9]) [9]

$Co(NO_3)_2$ <1> (<1> 0.1 mM, 70% inhibition of the rate of nitrite synthesis [9]) [9]

EDTA <3> (<3> almost complete inhibition at 1 mM [28]) [28]

H_2O_2 <1, 4> (<1> inhibits one step of oxidation of NH_2OH to HNO [9]; <1> irreversible, inactivation most rapid at pH 9-10, overview: substrates which protect against inactivation [14]; <1> only affects P_{460}, c-hemes remain unaffected [23]; <4> binds to cytochrome P_{468} [30]) [9, 14, 23, 30]

KCN <1> (<1> 20% inhibition at 0.5 mM [10]; <1> 50% inhibition at 1 mM [12]; <1> 50% inhibition at 0.07 mM [16]) [10, 16]

Mn^{2+} <1> (<1> complete inhibition of the oxidation of HNO to NO at 0.001 mM, inhibition of HNO_2 synthesis is the same at pH 6, 7, 8 and 9 [9]) [9]

NaCl <3> (<3> 50% inhibition at 25 mM with horse heart cytochrome c as electron acceptor [28]) [28]

NaS <1> (<1> 50% inhibition at 0.3 mM [16]) [16]
azide <1> (<1> 50% inhibition at 3.2 mM [16]) [16]
cyanide <3, 4> (<3> 1 mM [28]; <4> inhibits oxidation of hydroxylamine [30]) [28, 30]
diethyldithiocarbamate <1> (<1> inhibition of nitrate production at 0.1 mM [12]) [12]
hydrazine <1, 4> (<1> 50% inhibition at 0.1 mM [12]; <4> inhibits oxidation of hydroxylamine, 50% inhibition at 0.035 mM, competes with hydroxylamine [30]) [12, 30]
hydroxyethyl-hydrazine <1> [27]
methyl-hydrazine <1> [27]
o-phenanthroline <1> (<1> 50% inhibition at 0.12 mM [16]) [16]
phenazine methosulfate <1> (<1> inhibitory above 0.1 mM [1]) [1]
phenyl hydrazine <1> (<1> N2H3 protects [27]) [27]
sodium dodecylsulfate <4> (<4> 50% inhibition at 3% [30]) [30]

Cofactors/prosthetic groups

heme <1, 4> (<1> contains at least one b- and one c-type cytochrome [15]; <1> heme-like chromophore P_{460} is part of substrate-binding site [3]; <1> resolution of the hemes by redox potentiometry and electron spin resonance spectroscopy [4]; <1> multi-heme enzyme containing at least 5 thermodynamically distinct c-type hemes and the heme-like moiety P460 [4]; <1> contains hemes c_{553}, c_{559} and P_{460} in the ratio 5:2:1, P-460 is the site of electron entry [5]; <1> contains c hemes and the CO-binding heme P_{460} in a 7:1 ratio [6]; <1> resolution of multiple heme centers by electron paramagnetic resonance spectroscopy [7]; <1> enzyme posseses heme c molecules in different states [10]; <1> each α,β subunit of the $(\alpha,\beta)_3$ subunit structure contains 7-8 c-type hemes and one prosthetic group P-460, P-460 is a MW 17000 protein fragment [13]; <1> enzyme consists of one α-β-3 subunit and contains seven-eight c-type hemes and P_{460} [17]; <1> contains cytochrome P_{460} [18]; <1> covalently bound c-heme and P_{460} [22]; <1> 7 c-type hemes and 1 P_{460} [23, 27]; <1> 8 hemes per subunit: a triheme cluster, four hemes that are electronically coupled in two distinguishable pairs and one of these pairs is at the active site, and one heme that is not part of a cluster [24]; <4> 26 hemes per trimer [30]) [3-7, 8, 10, 11, 13, 15, 17, 18, 21, 22, 23, 24, 30]

Activating compounds

8-hydroxyquinoline <6> (<6> 1.5fold stimulation at 0.1 mM [33]) [33]
cyanide <6> (<6> 2fold stimulation at 1 mM [33]) [33]
diethyldithiocarbamate <1> (<1> stimulation of nitrite production at 0.1 mM [12]) [12]

Metals, ions

$Ce(NH_4)_2(NO_3)_6$ <1> (<1> 0.1 mM, stimulates the rate of oxidation of NH_2OH 2.6-fold and inhibits production of HNO_2 by 40% [9]) [9]
Fe <1, 3, 6> (<1> enzyme contains 20 mol of iron but no other metals [11]; <3> non-heme, non iron-sulfur [28]; <6> one iron atom per subunit [33]) [11, 28, 33]

Mn^{2+} <1> (<1> 0.001 mM, stimulation of NH_2OH utilization, stimulation at pH 6 and pH 9 is approximately 40% greater than at pH 7 and pH 8 [9]) [9]
NaCl <3> (<3> activation at 5 to 50 mM with cytochrome c_{550} as electron acceptor [28]) [28]
Additional information <6> (<6> one copper atom per subunit [33]) [33]

Turnover number (min^{-1})
32 <3> (nitrite) [28]
9500 <1> (nitrite) [11]
23000 <1> (hydroxylamine, <1> oxidized [11]) [11]
Additional information <1> (<1> overview: effect of pH on rate of reduction of heme c_{553} of the enzyme by NH_2OH, rate constant for reduction of hemes of the enzyme by dithionite at 2°C and 19°C and by NH_2OH or NH_2NH_2 at 2°C [5]) [5]

Specific activity (U/mg)
0.47 <2> [26]
21 <4> [30]
21 <4> [30]
28 <1> [11]
Additional information <1, 4, 6> (<4> with various electron acceptors and hydroxylamine or hydrazine as substrate [30]) [1, 30, 33]

K_m-Value (mM)
0.01 <5> (cytochrome c_{551}) [31]
0.017 <1> (cytochrome c, <1> from tuna [10]) [10]
0.025 <1> (cytochrome c, <1> from Rhodospirillum rubrum, Saccharomyces oviformis and cow [10]) [10]
0.033 <5> (pseudoazurin) [31]
0.036 <1> (cytochrome c, <1> from horse [10]) [10]
0.04 <1> (cytochrome c, <1> from Chlorobium limicola forma sp. thiosulfatophilum [10]) [10]
0.2 <1> (cytochrome c, <1> from Thiobacillus novellus [10]) [10]
Additional information <4> (<4> K_m-values at different temperatures [30]) [30]

pH-Optimum
9 <2> [26]
9.5 <1> [10]
10.5 <1> (<1> rate of reduction of hemes c_{553} of the enzyme increases with pH [5]) [5]
Additional information <3> (<3> pH optimum for enzyme activity with various electron acceptors and buffers [28]) [28]

Temperature optimum (°C)
25 <1> (<1> assay at [11]) [11]
65 <4> (<4> with hydroxylamine as substrate [30]) [30]

Temperature range (°C)

Additional information <1> (<1> changes in enzyme conformation imposed by changes in solvent, temperature or pressure affect the rates of intramolecular electron transfer from the substrate site to c hemes [3]) [3]

4 Enzyme Structure

Molecular weight

20000 <5> (<5> SDS-PAGE, gel filtration [31]) [31]
38900 <6> (<6> gel filtration [33]) [33]
132000 <2> (<2> gel filtration [26]) [26]
140000 <1> (<1> SDS-PAGE [29]) [29]
175000-180000 <1> (<1> SDS-PAGE, gel filtration [10]) [10]
180000-200000 <1> (<1> SDS-PAGE [8]) [8]
183000 <4> (<4> nondenaturating PAGE [30]) [30]
189000 <1> (<1> calculated from amino acid sequence [11]) [11]
200000 <1> (<1> sedimentation velocity experiments [15]) [15]

Subunits

dimer <2, 6> (<2> 2 * 68000, SDS-PAGE [26]; <6> 2 * 16800, SDS-PAGE [33]) [26, 33]
hexamer <1> (<1> 3 * 63000 + 3 * 11000, $(\alpha,\beta)_3$ subunit structure, SDS-PAGE, 200000 MW enzyme forms a 63000 MW monomer after heme removal, hydroxylamine oxidoreductase probably consists of 3 molecules of monoheme c-type cytochrome with a MW of 11000 and 3 tightly complexed molecules of a catalytically active MW 63000 protein containing 6 c-type hemes and one P_{460} heme, SDS-PAGE [8]) [8]
monomer <5> (<5> 1 * 20000, SDS-PAGE [31]) [31]
oligomer <1> (<1> α_2 or α_3 oligomeric structure with a subunit of 63 kDa, SDS-PAGE [22]; <1> α_2 oligomer [29]) [22, 29]
trimer <4> (<4> 3 * 58000, SDS-PAGE after removal of hemes [30]) [30]

5 Isolation/Preparation/Mutation/Application

Localization

periplasm <5> [31]
soluble <1, 2> [11, 26]

Crystallization

<1> [21]

Cloning

<1> (cloned and expressed in Pseudomonas putida [2]) [2]

Engineering

Additional information <1> (<1> mutants with only one or two of the three copies of the gene for hydroxylamine oxidoreductase have reduced enzyme activity [32]) [32]

6 Stability

pH-Stability

10.5 <1> (<1> enzyme is permanently inactivated above pH 10.5 [5]) [5]

Temperature stability

Additional information <1> (<1> changes in enzyme conformation imposed by changes in solvent, temperature or pressure affect the rates of intramolecular electron transfer from the substrate site to c hemes [3]) [3]

Organic solvent stability

Additional information <1> (<1> changes in enzyme conformation imposed by changes in solvent, temperature or pressure affect the rates of intramolecular electron transfer from the substrate site to c hemes [3]) [3]

General stability information

<1>, relative stable for 72 hours during ammonia starvation [29]

References

[1] Rees, M.K.: Studies of the hydroxylamine metabolism of Nitrosomonas europaea. I. Purification of hydroxylamine oxidase. Biochemistry, **7**, 353-366 (1968)

[2] Tatsuaki, T.; Yusuke, T.; Reiji, T.: Cloning and expression of the hydroxylamine oxidase gene of Nitrosomonas europaea in Pseudomonas putida. Hakko Kogaku Kaishi, **66**, 103-107 (1988)

[3] Balny, C.; Hooper, A.B.: Effect of solvent, pressure and temperature on reaction rates of the multiheme hydroxylamine oxidoreductase. Evidence for conformational change. Eur. J. Biochem., **176**, 273-279 (1988)

[4] Prince, R.C.; Hooper, A.B.: Resolution of the hemes of hydroxylamine oxidoreductase by redox potentiometry and electron spin resonance spectroscopy. Biochemistry, **26**, 970-974 (1987)

[5] Hooper, A.B.; Tran, M.; Balny, C.: Kinetics of reduction by substrate or dithionite and heme-heme electron transfer in the multiheme hydroxylamine oxidoreductase. Eur. J. Biochem., **141**, 565-571 (1984)

[6] Hooper, A.B.; Debey, P.; Andersson, K.K.; Balny, C.: Heme P460 of hydroxylamine oxidoreductase of Nitrosomonas. Reaction with CO and H_2O_2. Eur. J. Biochem., **134**, 83-87 (1983)

[7] Lipscomb, J.D.; Hooper, A.B.: Resolution of multiple heme centers of hydroxylamine oxidoreductase from Nitrosomonas. 1. Electron paramagnetic resonance spectroscopy. Biochemistry, **21**, 3965-3972 (1982)

[8] Terry, K.R.; Hooper, A.B.: Hydroxylamine oxidoreductase: a 20-heme, 200 000 molecular weight cytochrome c with unusual denaturation properties which forms a 63 000 molecular weight monomer after heme removal. Biochemistry, **20**, 7026-7032 (1981)

[9] Hooper, A.B.; Terry, K.R.: Hydroxylamine oxidoreductase of Nitrosomonas. Production of nitric oxide from hydroxylamine. Biochim. Biophys. Acta, **571**, 12-20 (1979)

[10] Yamanaka, T.; Shinra, M.; Takahashi, K.; Shibasaka, M.: Highly purified hydroxylamine oxidoreductase derived from Nitrosomonas europaea. Some physicochemical and enzymatic properties. J. Biochem., **86**, 1101-1108 (1979)

[11] Hooper, A.B.; Maxwell, P.C.; Terry, K.R.: Hydroxylamine oxidoreductase from Nitrosomonas: absorption spectra and content of heme and metal. Biochemistry, **17**, 2984-2989 (1978)

[12] Hooper, A.B.; Terry, K.R.; Maxwell, P.C.: Hydroxylamine oxidoreductase of Nitrosomonas. Oxidation of diethyldithiocarbamate concomitant with stimulation of nitrite synthesis. Biochim. Biophys. Acta, **462**, 141-152 (1977)

[13] Andersson, K.K.; Kent, T.A.; Lipscomb, J.D.; Hooper, A.B.; Munck, E.: Mossbauer, EPR, and optical studies of the P-460 center of hydroxylamine oxidoreductase from Nitrosomonas. A ferrous heme with an unusually large quadrupole splitting. J. Biol. Chem., **259**, 6833-6840 (1984)

[14] Hooper, A.B.; Terry, K.R.: Hydroxylamine oxidoreductase from Nitrosomonas: inactivation by hydrogen peroxide. Biochemistry, **16**, 455-459 (1977)

[15] Rees, M.K.: Studies on the hydroxylamine metabolism of Nitrosomonas europaea. II. Molecular properties of the electron-transport particle, hydroxylamine oxidase. Biochemistry, **7**, 366-372 (1968)

[16] Ritchie, G.A.F.; Nicholas, D.J.D.: The partial characterization of purified nitrite reductase and hydroxylamine oxidase from Nitrosomonas europaea. Biochem. J., **138**, 471-480 (1974)

[17] Nasri, H.; Fischer, J.; Weiss, R.: Synthesis and characterization of five-coordinate high-spin iron(II) porphyrin complexes with unusually large quadrupole splittings. Models for the P_{460} center of hydroxylamine oxidoreductase from Nitrosomonas. J. Am. Chem. Soc., **109**, 2549-2550 (1987)

[18] Numata, M.; Saito, T.; Yamazaki, Y.; Yamanaka, T.: Cytochrome P_{460} of Nitrosomonas europaea: further purification and further characterization. J. Biochem., **108**, 1016-1021 (1990)

[19] Hooper, A.B.; Balny, C.: Reaction of oxygen with hydroxylamine oxidoreductase of Nitrosomonas. FEBS Lett., **144**, 299-303 (1982)

[20] Yamanaka, T.; Sakano, Y.: Oxidation of hydroxylamine to nitrite catalyzed by hydroxylamine oxidoreductase purified from Nitrosomonas europaea. Curr. Microbiol., **4**, 239-244 (1980)

[21] Arciero, D.M.; Golombek, A.; Hendrich, M.P.; Hooper, A.B.: Correlation of optical and EPR signals with the P_{460} heme of hydroxylamine oxidoreductase from Nitrosomonas europaea. Biochemistry, **37**, 523-529 (1998)

[22] Arciero, D.M.; Hooper, A.B.: Hydroxylamine oxidoreductase from Nitrosomonas europaea is a multimer of an octa-heme subunit. J. Biol. Chem., **268**, 14645-14654 (1993)

[23] Collins, M.J.; Arciero, D.M.; Hooper, A.B.: Optical spectropotentiometric resolution of the hemes of hydroxylamine oxidoreductase. Heme quantitation and pH dependence of Em. J. Biol. Chem., **268**, 14655-14662 (1993)
[24] Hendrich, M.P.; Petasis, D.; Arciero, D.M.; Hooper, A.B.: Correlations of structure and electronic properties from EPR spectroscopy of hydroxylamine oxidoreductase. J. Am. Chem. Soc., **123**, 2997-3005 (2001)
[25] Hommes, N.G.; Sayavedra-Soto, L.A.; Arp, D.J.: Mutagenesis of hydroxylamine oxidoreductase in Nitrosomonas europaea by transformation and recombination. J. Bacteriol., **178**, 3710-3714. (1996)
[26] Jetten, M.S.M.; De Bruijn, P.; Kuenen, J.G.: Hydroxylamine metabolism in Pseudomonas PB16: involvement of a novel hydroxylamine oxidoreductase. Antonie Leeuwenhoek, **71**, 69-74 (1997)
[27] Logan, M.S.P.; Hooper, A.B.: Suicide inactivation of hydroxylamine oxidoreductase of nitrosomonas europaea by organohydrazines. Biochemistry, **34**, 9257-9264 (1995)
[28] Moir, J.W.B.; Wehrfritz, J.M.; Spiro, S.; Richardson, D.J.: The biochemical characterization of a novel non-heme-iron hydroxylamine oxidase from Paracoccus denitrificans GB17. Biochem. J., **319**, 823-827 (1996)
[29] Nejidat, A.; Shmuely, H.; Abeliovich, A.: Effect of ammonia starvation on hydroxylamine oxidoreductase activity of Nitrosomonas europaea. J. Biochem., **121**, 957-960. (1997)
[30] Schalk, J.; de Vries, S.; Kuenen, J.G.; Jetten, M.S.: Involvement of a novel hydroxylamine oxidoreductase in anaerobic ammonium oxidation. Biochemistry, **39**, 5405-5412. (2000)
[31] Wehrfritz, J.M.; Reilly, A.; Spiro, S.; Richardson, D.J.: Purification of hydroxylamine oxidase from Thiosphaera pantotropha. Identification of electron acceptors that couple heterotrophic nitrification to aerobic denitrification. FEBS Lett., **335**, 246-250 (1993)
[32] Yamagata, A.; Hirota, R.; Kato, J.; Kuroda, A.; Ikeda, T.; Takiguchi, N.; Ohtake, H.: Mutational analysis of the multicopy hao gene coding for hydroxylamine oxidoreductase in Nitrosomonas sp. strain ENI¯11. Biosci. Biotechnol. Biochem., **64**, 1754-1757. (2000)
[33] Zahn, J.A.; Duncan, C.; DiSpirito, A.A.: Oxidation of hydroxylamine by cytochrome P-460 of the obligate methylotroph Methylococcus capsulatus Bath. J. Bacteriol., **176**, 5879-5887 (1994)
[34] DiSpirito, A.; Taaffe, L.R.; Lipscomb, J.D.; Hooper, A.B.: A 'blue' copper oxidase from Nitrosomonas europaea. Biochim. Biophys. Acta, **827**, 320-326 (1985)

3-Aci-nitropropanoate oxidase 1.7.3.5

1 Nomenclature

EC number
1.7.3.5

Systematic name
3-aci-nitropropanoate:oxygen oxidoreductase

Recommended name
3-aci-nitropropanoate oxidase

Synonyms
propionate-3-nitronate oxidase

CAS registry number
111940-52-4

2 Source Organism

<1> *Penicillium atrovenetum* [1]

3 Reaction and Specificity

Catalyzed reaction
3-aci-nitropropanoate + O_2 = 3-oxopropanoate + nitrite + H_2O_2 (<1> primary products of the enzymatic reaction probably is the nitropropanoate free radical and superoxide [1])

Reaction type
oxidation
redox reaction
reduction

Substrates and products
S 3-aci-nitropropanoate + O_2 <1> (Reversibility: ? <1> [1]) [1]
P 3-oxopropanoate + nitrite + H_2O_2
S butyrate-4-nitronate + O_2 <1> (Reversibility: ? <1> [1]) [1]
P succinate + nitrite + H_2O_2
S Additional information <1> (<1> neither cytochrome c nor $K_3Fe(CN)_6$ serve as oxidants, other substrates tested are oxidized less than 2% the rate of 3-nitropropanoate [1]) [1]
P ?

Cofactors/prosthetic groups

FMN <1> (<1> flavoprotein [1]) [1]

Activating compounds

propane-2-nitronate <1> (<1> increases the enzyme activity over 40fold when added to a reaction mixture that contains propionate-3-nitronate, the stimulation can be partially inhibited by superoxide dismutase [1]) [1]

Turnover number (min^{-1})

21000 <1> (3-nitropropanoate) [1]

Specific activity (U/mg)

160 <1> (<1> butyrate-4-nitronate as substrate [1]) [1]
274 <1> (<1> propionate-3-nitronate as substrate [1]) [1]

pH-Optimum

8.1 <1> (<1> assay at [1]) [1]

Temperature optimum (°C)

25 <1> (<1> assay at [1]) [1]

4 Enzyme Structure

Molecular weight

73000 <1> (<1> gel filtration [1]) [1]

Subunits

dimer <1> (<1> 2 * 38000, SDS-PAGE in presence of mercaptoethanol [1]) [1]

5 Isolation/Preparation/Mutation/Application

Source/tissue

mycelium <1> [1]

Localization

soluble <1> [1]

Purification

<1> (homogeneity [1]) [1]

References

[1] Porter, D.J.T.; Bright, H.J.: Propionate-3-nitronate oxidase from Penicillium atrovenetum is a flavoprotein which initiates the autoxidation of its substrate by O_2. J. Biol. Chem., **262**, 14428-14434 (1987)

Ferredoxin-nitrite reductase 1.7.7.1

1 Nomenclature

EC number
1.7.7.1

Systematic name
ammonia:ferredoxin oxidoreductase

Recommended name
ferredoxin-nitrite reductase

Synonyms
reductase, ferredoxin-nitrite

CAS registry number
37256-44-3

2 Source Organism

<1> *Phormidium laminosum* (cyanobacterium [1]) [1, 2]
<2> *Oryza sativa* (rice [3]) [3, 12]
<3> *Spinacia oleracea* (spinach [4, 35, 37]) [4, 6, 10, 11, 13, 14, 17, 19, 20, 22, 24-26, 28, 31, 35, 37, 38, 39, 40]
<4> *Chlamydomonas reinhardtii* [5, 8, 36, 43]
<5> *Porphyra yezoensis* [7, 11]
<6> *Anacystis nidulans* [9, 10, 11]
<7> *Cucurbita pepo* [10, 11, 18]
<8> *Chlorella fusca* [10, 11, 21, 27]
<9> *Dunaliella tertiolecta* [10]
<10> *Anabaena cylindrica* [10]
<11> *Phaseolus angularis* (bean [15]; red bean [17]) [15, 17, 23]
<12> *Spirulina platensis* (cyanobacterium [16, 17]) [16, 17]
<13> *Phaseolus vulgaris* (french bean [17]) [17]
<14> *Allium tuberosum* (leek [17]) [17]
<15> *Symphytum officinale* [17]
<16> *Zea mays* [17, 34]
<17> *Hordeum vulgare* (barley [17, 30]) [17, 30]
<18> *Spirodela polythiza* [29]
<19> *Candida utilis* [44]
<20> *Helianthus annus* (sunflower [32]) [32]
<21> *Anabaena sp.* (cyanobacterium [33]) [33]

<22> *Lycopersicon esculentum* (tomato, 2 isoforms NiR1 and NiR2 [41]) [41]
<23> *Lotus japonicus* [42]
<24> *Spinacia oleracea* [35]

3 Reaction and Specificity

Catalyzed reaction

ammonia + 2 H_2O + 6 oxidized ferredoxin = nitrite + 6 reduced ferredoxin + 7 H^+ (<3,5-10> mechanism [10,11]; <3> sequential reaction scheme [4])

Reaction type

oxidation
redox reaction
reduction

Natural substrates and products

S nitrite + reduced ferredoxin <1, 3-10> (<3,8> second step of nitrate assimilation [4,21]; <3,6-10> second enzyme of photosynthetic nitrate-reducing system [10]) (Reversibility: ? <1, 3-10> [2, 4, 5, 10, 11, 21]) [2, 4, 5, 10, 11, 21]
P ammonia + oxidized ferredoxin

Substrates and products

S nitrite + reduced ferredoxin <1-17, 19> (<1, 3, 5, 7, 8> other electron donors: flavodoxin [2, 11]; <1, 3-5, 7, 8, 19> other electron donor: methyl viologen [2, 4, 5, 8, 11, 21, 27, 28, 44]; <1,4> other electron donor: benzyl viologen [2, 8]; <1> not: flavin [2]; <1> NADH [2]; <1, 19> NADPH [2, 44]; <3, 5> methyl viologen can serve as a less effective electron donor [7, 24]; <-1, 3, 5, 7, 8> enzyme can also catalyze the reduction of hydroxylamine to ammonia, at lower rate [11, 18, 27]; <8> NADH or NADPH can not serve as electron donors [21]; <4> enzyme forms a covalent adduct with ferredoxin [36]; <3> electrostatic interactions between ferredoxin and nitrite reductase play an important role in the reaction mechanism [40]) (Reversibility: ? <1-17, 19> [1-28, 31, 36, 40, 44]) [1-28, 31, 36, 40, 44]
P ammonia + oxidized ferredoxin

Inhibitors

$(NH_4)_2SO_4$ <3> (<3> slightly inhibitory [25]) [25]
8-hydroxyquinoline <8> (<8> 7% inhibition at 5 mM [27]) [27]
CO <3-5, 7, 8> (<3,5,7,8> enzyme from higher plants and algae [11]; <3> forms a complex with the reduced enzyme [26]) [8, 11, 26, 27]
EDTA <8> [27]
N-acetylsuccinimide <24> (<24> specifically inhibits the ferredoxin binding ability of the enzyme by modifying lysine residues [35]; <3> loss of enzymatic activity when reduced ferredoxin serves as electron donor, but very little effect with methyl viologen as electron donor, ferredoxin protects the enzyme [37]) [35, 37]

N-bromosuccinimide <3> (<3> incubation for 8 hours with 8-fold excess of NBS leads to 80% inhibition of the catalytic activity without effect on substrate binding or other enzyme activities, complex formation with ferredoxin protects the enzyme against inhibition [38,39]; <3> modifies tryptophane and cysteine residues, incubation for 8 hours with 8-fold excess of NBS leads to 50% inhibition of the catalytic activity, incubation for 16 hours with 8-fold excess of NBS leads to 80% inhibition of the catalytic activity [39]) [38, 39]
N-ethylmaleimide <1> (<1> 47% inhibition at 1 mM [2]) [2]
NaN_3 <1> (<1> slight [2]) [2]
SO_3^{2-} <1> (<1> 24% inhibition at 10 mM [2]) [2]
bathophenanthroline <1> (<1> slight [2]) [2]
cyanide <1, 3, 5, 7, 8> (<1> 92% inhibition at 0.1 mM [2]; <3> almost complete inhibition from 0.02 mM to 0.2 mM [25]; <3> competitive inhibition with respect to nitrite [26]; <8> 90% inhibition at 1 mM [27]; <3> 1 mM [28]) [2, 11, 21, 25-28]
hydroxylamine <1> (<1> 34% inhibition at 10 mM [2]) [2]
mersalyl <7> (<7> reversible by glutathione [18]) [11, 18]
o-phenanthroline <8> (<8> 15% inhibition at 5 mM [27]) [27]
p-chloromercuribenzoate <3, 7, 8> (<3> not [28]; <3> nitrite partially protects the inhibition [26]; <8> complete inhibition at 0.1 mM [27]) [18, 21, 26, 27]
p-hydroxymercuribenzoate <1, 3, 5, 7, 8> (<3, 5, 7, 8> enzyme from higher plants and eukaryotic algae inhibited, cyanobacterial enzyme not [11]) [2, 11]
phenyl mercury acetate <7> [18]
phenylglyoxal <3, 24> (<24> inhibits the ability of the enzyme to form a high-affinity complex with ferredoxin by modifying arginine residues [35]; <3> loss of enzymatic activity when reduced ferredoxin serves as electron donor, but very little effect with methyl viologen as electron donor, ferredoxin protects the enzyme [37]) [35, 37]
pyridoxal-5'-phosphate <24> (<24> 75% inhibition of the enzyme with ferredoxin as electron donor after exposure to $NaBH_4$, but no inhibition of the enzyme with methyl viologen as electron donor [35]) [35]
Additional information <4> (<4> anti-glutamate synthase antibodies recognize nitrite reductase, but inhibit only ferredoxin-linked activity and not methyl viologen-linked activity [5]) [5]

Cofactors/prosthetic groups

siroheme <1, 3, 4> (<1, 3> prosthetic group [2, 10, 11, 13, 25, 26]; <4> 0.65 mol of siroheme per mol of enzyme [8]; <3> 1 mol siroheme per mol of enzyme [10,11,26]; <3> 0.93 mol siroheme per mol of enzyme [13]; <3> 2 mol of siroheme per mol of enzyme [25]) [2, 8, 10, 11, 13, 24, 25, 26, 31, 38, 40]

Metals, ions

iron <3-5, 7, 8, 9, 21> (<4> 0.65 molecules of siroheme per enzyme molecule [8]; <3> enzyme contains 1 tetranuclear center (4Fe-4S) and 1 siroheme, minimum of 5 Fe atoms per active enzyme molecule [10,11]; <3> 4.8 Fe atoms per molecule of MW 63000, 0.93 mol siroheme per mol of enzyme

[13]; <7> contains 2 mol of Fe per mol [18]; <3> 7 mol of iron per 86000 MW protein [24];<3> contains about 2 mol of siroheme per mol of enzyme, 3 mol of iron per mol of enzyme (MW 61000), one in siroheme, one (Fe_2-S_2) [26]; <8> 2 atoms of iron probably bound in the chromophoric group [27]) [8, 10, 11, 13, 18, 24-27, 33, 40]

Turnover number (min^{-1})

600 <5> (nitrite) [11]
3240 <8> (nitrite) [11]
5340 <7> (nitrite) [11]
6600 <3> (nitrite) [11]
Additional information <3> (<3> turnover number decreases about 20% after treatment with N-bromosuccinimide [39]) [39]

Specific activity (U/mg)

3.25 <3> [28]
16.06 <19> [44]
51.7 <8> [27]
53.4 <11> [23]
85 <7> [18]
89 <4> [8]
90 <3> (<3> methyl viologen as electron donor [37]) [37]
100 <3> [31]
107.8 <3> [26]
108 <3> [10]
120 <3> (<3> ferredoxin as electron donor [37]) [37]
135 <3> (<3> methyl viologen as electron donor [39]) [39]
140 <3> [20]
181 <2> [3]
188 <3> (<3> ferredoxin as electron donor [39]) [39]
194 <12> [16]
196.8 <3> [19]
207 <3> [13]
301.4 <3> [25]
625 <1> [2]
Additional information <4, 11, 18, 20, 21, 23> (<20> activities under different growth conditions [32]; <21> activities of various mutants [33]; <16> activities in seedlings grown in dark and light [34]; <23> activity in leaves and roots during a day/night cycle [42]; <4> activity during a light/dark cycle [43]) [15, 29, 32, 33, 34, 42, 43]

K_m-Value (mM)

0.006 <3> (ferredoxin) [4]
0.01 <3> (nitrite) [4]
0.02 <12> (ferredoxin) [16]
0.022 <1> (ferredoxin) [2]
0.024 <4> (ferredoxin) [8]
0.025 <5> (ferredoxin) [7]

0.025 <3> (ferredoxin, <3> ferredoxin from Spinacia oleracea [6]) [6]
0.027 <3> (ferredoxin) [24]
0.04 <1> (nitrite) [2]
0.063 <2> (methyl viologen) [3]
0.069 <19> (NADPH) [44]
0.08 <19> (methyl viologen) [44]
0.1 <3> (methyl viologen) [37]
0.2 <12> (nitrite) [16]
0.215 <1> (methyl viologen) [2]
0.28 <3> (nitrite) [6]
0.28 <5> (nitrite) [7]
0.3 <3> (nitrite) [31]
0.36 <2> (nitrite) [3]
0.4 <12> (methyl viologen) [16]
0.91 <4> (methyl viologen) [8]
Additional information <3, 5, 7, 8> [11, 13, 14, 28, 39]

pH-Optimum
7-8 <3, 7-10> [10]
7.1-7.8 <3> [28]
7.3 <1> (<1> phosphate buffer [2]) [2]
7.5 <3, 5, 7, 8, 19> (<19> phosphate buffer [44]) [11, 44]
7.6 <1, 2> (<1> Tris-HCl buffer [2]) [2, 3]
7.7 <3, 5> [6, 7, 37]
7.8 <12> [16]
8 <19> (<19> Tris buffer [44]) [44]

pH-Range
6.8-8.3 <3> (<3> pH 6.8: about 75% of activity maximum, pH 8.3: about 70% of activity maximum [28]) [28]
7.8-8.3 <19> (<19> appreciable amounts of activity at pH 7.8 and pH 8.3 in Tris buffer [44]) [44]

Temperature optimum (°C)
30 <19> [44]
50 <1> (<1> increase of activity up to 50°C [2]) [2]

4 Enzyme Structure

Molecular weight
52000 <12> (<12> gel filtration [16]) [16]
54000 <1> (<1> gel filtration [2]) [2]
59000 <3, 5> (<3,5> SDS-PAGE [6,7]) [6, 7]
59650 <21> (<21> mass spectroscopy [33]) [33]
60000 <2, 3> (<2> SDS-PAGE [3]; <3> gel filtration [26]) [3, 26]
60000-63000 <3> (<3> gel filtration, sedimentation equilibrium [11]) [11, 13, 20, 26]
61000 <2, 3> (<2> gel filtration [3]; <3> SDS-PAGE [31]) [3, 12, 31]

62000 <3, 22> (<3> SDS-PAGE [37]; <22> isoform NiR2, SDS-PAGE [41]) [37, 41]
62690 <3> (<3> amino acid composition [26]) [26]
63000 <3, 4, 5, 7, 8, 22, 23> (<3> gel filtration, SDS-PAGE, sedimentation equilibrium [13]; <8> gel filtration, SDS-PAGE [27]; <22> isoform NiR1, SDS-PAGE [41]; <23> SDS-PAGE [42]; <4> SDS-PAGE [43]) [11, 13, 21, 27, 41, 42, 43]
64830 <23> (<23> calculated from amino acid composition [42]) [42]
68000 <11> (<11> gel filtration, SDS-PAGE [15]) [15]
69000 <1> (<1> native PAGE [1]) [1]
86000 <3, 4> (<3> gel filtration [24]; <4> gel filtration [8]) [8, 24]
100000 <11> (<11> gel filtration [23]) [23]
126000 <19> (<19> gel filtration, native enzyme [44]) [44]
Additional information <3> (<3> the native enyzme with a MW of 85000 can be split into a modified form of a MW of 61000 that retains activity with the non-physiological electron-donor, methyl viologen but loses most of the ferredoxin-linked activity and an 24000 MW coupling protein fragment in which the ferredoxin-binding domain is located [22, 25]) [22, 25]

Subunits

dimer <4, 11, 19> (<4> 1 * 63000 + 1 * 25000, SDS-PAGE [8]; <11> 1 * 64000 + 1 * 35000, SDS-PAGE [23]; <19> 1 * 58000 + 1 * 66000, SDS-PAGE [44]) [8, 23, 44]
monomer <1-3, 8, 11> (<1> 1 * 54000 SDS-PAGE [2]; <2> 1 * 60000 SDS-PAGE [3]; <11> 1 * 68000 SDS-PAGE [15]; <3> SDS-PAGE [13]; <8> 1 * 63000 SDS-PAGE + mercaptoethanol [27]) [1, 2, 3, 11, 13, 15, 27]
Additional information <3> (<3> spinach: the native enyzme with a molecular weight of 85000 can be split into a modified form with a MW of 61000 that retains activity with the non-physiological electron-donor methyl viologen, but loses most of the ferredoxin-linked activity and an 24000 MW coupling protein fragment in which the ferredoxin-binding domain is located [22, 25]) [22, 25]

Posttranslational modification

phosphoprotein <19> (<19> 66 kDa subunit is phosphorylated [44]) [44]

5 Isolation/Preparation/Mutation/Application

Source/tissue

cotyledon <20, 22> (<20> cotyledon of 7 day old seedling [32]; <22> NiR 1 occurs in etiolated cotyledons, NiR2 is predominant in seedlings grown in light [41]) [32, 41]
leaf <2, 3, 16, 24, 23> (<16> 7-days old seedlings [34]) [3, 10, 13, 19, 20, 28, 34, 35, 37, 38, 40, 42]
root <11, 17, 23> (<17> root of 8 day old seedling [30]) [15, 30, 42]
shoot <11> (<11> green [15]; <11> etiolated [15, 23]) [15, 23]
thallus <5> [7]

Localization

chloroplast <1, 3, 6-10, 16> [2, 10, 19, 28, 34, 39, 40]
particle-bound <6> (<6> chlorophyll-containing subcellular particle [11]) [11]
soluble <6> (<6> exception: Anacystis [11]) [11]

Purification

<1> [2]
<2> [3]
<3> (homogeneity [10,20,24,26,31]; 95% purity [40]) [10, 13, 17, 19, 20, 22, 24, 25, 26, 28, 31, 37, 39, 40]
<4> [8]
<5> [7]
<7> [18]
<8> [21, 27]
<11> [15, 17, 23]
<12> [16, 17]
<13> [17]
<14> [17]
<15> [17]
<16> [17]
<17> [17]
<19> (homogeneity [44]) [44]
<21> (homogeneity [33]) [33]

Cloning

<21> [33]
<2, 3> (fusion with β-galactosidase [31]) [6, 12, 31]

Engineering

C514S <3> (<3> almost complete loss of activity [31]) [31]
C518S <3> (<3> almost complete loss of activity [31]) [31]
G513A <3> (<3> increased K_m for nitrite compared to the wild type enzyme [31]) [31]
G513E <3> (<3> loss of most of the activity [31]) [31]
G513V <3> (<3> loss of most of the activity [31]) [31]
G519A <3> (<3> 68% activity of wild type enzyme [31]) [31]
G519E <3> (<3> 8% activity of wild type enzyme [31]) [31]
G519V <3> (<3> 9% activity of wild type enzyme [31]) [31]
P515A <3> (<3> marginal loss of activity [31]) [31]
P515S <3> (<3> marginal loss of activity [31]) [31]
P515T <3> (<3> marginal loss of activity [31]) [31]
Additional information <21> (<21> various mutants that differ in enzymatic activity [33]) [33]

6 Stability

pH-Stability

7 <8> (<8> enzyme in crude extract is rapidly inactivated below pH 7 [27]) [27]

Temperature stability

23 <3> (<3> 1 h, pH 8.0, 75 mM Tris-HCl, stable [14]) [14]
40 <1, 3> (<3> stable [10]; <1> thermostable up to 40°C [2]) [2, 10]
45 <8> (<8> unstable after 5 min [21]) [21]
60 <1, 3> (<1,3> 10 min, more than 95% loss of activity [2,10]) [2, 10]

General stability information

<3>, repetitive freezing and thawing or prolonged dialysis produces a loss of enzyme activity [10]
<8>, unstable in crude extracts and in solutions below pH 7 [27]

Storage stability

<1>, -20°C, several months [2]
<1>, 4°C, 1 week [2]
<3>, -20°C, 50% glycerol, for at least 6 months [10]
<3>, 4°C, 0.1 M potassium phosphate buffer, pH 7.7, 1 week [10]
<8>, -24°C, 30% loss of activity within 4 months [27]
<8>, 4°C, 30% loss of activity within 3 weeks [27]
<11>, -20°C, 10% glycerol, 10 mM 2-mercaptoethanol, stable for several weeks [23]
<12>, 4°C, 200 mM Tris-HCl buffer, pH 7.5, 200 mM NaCl, 10% glycerol, for at least 1 month [16]

References

[1] Arizmendi, J.M.; Serra, J.L.: Composition of the ferredoxin-nitrite reductase from the cyanobacterium Phormidium laminosum. Biochem. Soc. Trans., **18**, 637-638 (1990)

[2] Arizmendi, J.M.; Serra, J.L.: Purification and some properties of the nitrite reductase from the cyanobacterium Phormidium laminosum. Biochim. Biophys. Acta, **1040**, 237-244 (1990)

[3] Ida, S.; Iwagami, K.; Minobe, S.: Purification and characterization of molecular and immunological properties of rice ferredoxin-nitrite reductase. Agric. Biol. Chem., **53**, 2777-2784 (1989)

[4] Mikami, B.; Ida, S.: Spinach ferredoxin-nitrite reductase: characterization of catalytic activity and interaction of the enzyme with substrates. J. Biochem., **105**, 47-50 (1989)

[5] Romero, L.C.; Gotor, C.; Marquez, A.J.; Forde, B.G.; Vega, J.M.: Antigenic similarities between ferredoxin-dependent nitrite reductase and glutamate synthase from Chlamydomonas reinhardtii. Biochim. Biophys. Acta, **957**, 152-157 (1988)

[6] Back, E.; Burkhart, W.; Moyer, M.; Privalle, L.; Rothstein, S.: Isolation of cDNA clones coding for spinach nitrite reductase: complete sequence and nitrate induction. Mol. Gen. Genet., **212**, 20-26 (1988)
[7] Ide, T.; Tamura, G.: Isolation and partial characterization of homogeneous nitrite reductase from a red alga, Porphyra yezoensis. Agric. Biol. Chem., **51**, 3391-3393 (1987)
[8] Romero, L.C.; Galvan, F.; Vega, J.M.: Purification and properties of the siroheme-containing ferredoxin-nitrite reductase from Chlamydomonas reinhardtii. Biochim. Biophys. Acta, **914**, 55-63 (1987)
[9] Herrero, A.; Guerrero, M.G.: Regulation of nitrite reductase in the cyanobacterium Anacystis nidulans. J. Gen. Microbiol., **132**, 2463-2468 (1986)
[10] Vega, J.M.; Cardenas, J.; Losada, M.: Ferredoxin-nitrite reductase. Methods Enzymol., **69**, 255-270 (1980)
[11] Guerrero, M.G.; Vega, J.M.; Losada, M.: The assimilatory nitrate-reducing system and its regulation. Annu. Rev. Plant Physiol., **32**, 169-204 (1981)
[12] Matsui, J.; Takeba, G.; Ida, S.: Molecular cloning and partial amino acid sequence of rice ferredoxin-nitrite reductase. Agric. Biol. Chem., **54**, 3069-3071 (1990)
[13] Ida, S.; Mikami, B.: Spinach ferredoxin-nitrite reductase: a purification procedure and characterization of chemical properties. Biochim. Biophys. Acta, **871**, 167-176 (1986)
[14] Stein Privalle, L.; Privalle, C.T.; Leonardy, N.J.; Kamin, H.: Interactions between spinach ferredoxin-nitrite reductase and its substrates. Evidence for the specificity of ferredoxin. J. Biol. Chem., **260**, 14344-14350 (1985)
[15] Ishiyama, Y.; Shinoda, I.; Fukushima, K.; Tamura, G.: Some properties of ferredoxin-nitrite reductase from green shoots of bean and an immunological comparison with nitrite reductase from roots and etiolated shoots. Plant Sci., **39**, 89-95 (1985)
[16] Yabuki, Y.; Mori, E.; Tamura, G.: Nitrite reductase in the cyanobacterium Spirulina plantensis. Agric. Biol. Chem., **49**, 3061-3062 (1985)
[17] Ishiyama, Y.; Shinoda, I.; Fukushima, K.; Tamura, G.: Immunological comparison of nitrite reductase from different plant sources. Agric. Biol. Chem., **49**, 2225-2226 (1985)
[18] Hucklesby, D.P.; James, D.M.; Banwell, M.J.; Hewitt, E.J.: Properties of nitrite reductase from Cucubita pepo. Phytochemistry, **15**, 599-603 (1976)
[19] Hirasawa, M.; Tamura, G.: Purification of ferredoxin-dependent nitrite reductase from spinach leaves. Agric. Biol. Chem., **43**, 659-661 (1979)
[20] Ida, S.: Purification to homogeneity of spinach nitrite reductase by ferredoxin-sepharose affinity chromatography. J. Biochem., **82**, 915-918 (1977)
[21] Zumft, W.G.; Paneque, A.; Aparicio, P.J.; Losada, M.: Mechanism of nitrate reduction in Chlorella. Biochem. Biophys. Res. Commun., **36**, 980-986 (1969)
[22] Hirasawa, M.; Knaff, D.B.: Interaction of ferredoxin-linked nitrite reductase with ferredoxin. Biochim. Biophys. Acta, **830**, 173-180 (1985)
[23] Ishiyama, Y.; Tamura, G.: Isolation and partial characterization of homogenous nitrite reductase from etiolated bean shoots (Phaseolus angularis W.F. Wight). Plant Sci. Lett., **37**, 251-256 (1985)

[24] Hirasawa-Soga, M.; Tamura, G.: Some properties of ferredoxin-nitrite reductase from Spinacia oleracea. Agric. Biol. Chem., **45**, 1615-1620 (1981)
[25] Hirasawa-Soga, M.; Horie, S.; Tamura, G.: Further characterization of ferredoxin nitrite reductase and the relationship between the enzyme and methyl viologen-dependent nitrite reductase. Agric. Biol. Chem., **46**, 1319-1328 (1982)
[26] Vega, J.M.; Kamin, H.: Spinach nitrite reductase. Purification and properties of a siroheme-containing iron-sulfur enzyme. J. Biol. Chem., **252**, 896-909 (1977)
[27] Zumft, W.G.: Ferredoxin:nitrite oxidoreductase from Chlorella. Purification and properties. Biochim. Biophys. Acta, **276**, 363-375 (1972)
[28] Ramirez, J.M.; Del Campo, F.F.; Paneque, A.; Losada, M.: Ferredoxin-nitrite reductase from spinach. Biochim. Biophys. Acta, **118**, 58-71 (1966)
[29] Appenroth, K.J.; Oelmueller, R.: Regulation of transcript level and nitrite reductase activity by phytochrome and nitrate in turions of Spirodela polyrhiza. Physiol. Plant., **93**, 272-278 (1995)
[30] Aslam, M.; Travis, R.L.; Rains, D.W.; Huffaker, R.C.: Differential effect of ammonium on the induction of nitrate and nitrite reductase activities in roots of barley (Hordeum vulgare) seedlings. Physiol. Plant., **101**, 612-619 (1997)
[31] Bellissimo, D.B.; Privalle, L.S.: Expression of spinach nitrite reductase in Escherichia coli: site-directed mutagenesis of predicted active site amino acids. Arch. Biochem. Biophys., **323**, 155-163 (1995)
[32] Cabello, P.; De La Haba, P.; Gonzalez-Fontes, A.; Maldonado, J.M.: Induction of nitrate reductase, nitrite reductase, and glutamine synthetase isoforms in sunflower cotyledons as affected by nitrate, light, and plastid integrity. Protoplasma, **201**, 1-7 (1998)
[33] Curdt, I.; Singh, B.B.; Jakoby, M.; Hachtel, W.; Bohme, H.: Identification of amino acid residues of nitrite reductase from Anabaena sp. PCC 7120 involved in ferredoxin binding. Biochim. Biophys. Acta, **1543**, 60-68 (2000)
[34] Datta, R.; Sharma, R.: Temporal and spatial regulation of nitrate reductase and nitrite reductase in greening maize leaves. Plant Sci., **144**, 77-83 (1999)
[35] Dose, M.M.; Hirasawa, M.; Kleis-Sanfrancisco, S.; Lew, E.L.; Knaff, D.B.: The ferredoxin-binding site of ferredoxin:nitrite oxidoreductase. Differential chemical modification of the free enzyme and its complex with ferredoxin. Plant Physiol., **114**, 1047-1053 (1997)
[36] Garcia-Sanchez, M.I.; Gotor, C.; Jacquot, E.P.; Stein, M.; Suzuki, A.; Vega, J.M.: Critical residues of Chlamydomonas reinhardtii ferredoxin for interaction with nitrite reductase and glutamate synthase revealed by site-directed mutagenesis. Eur. J. Biochem., **250**, 364-368 (1997)
[37] Hirasawa, M.; de Best, J.H.; Knaff, D.B.: The effect of lysine- and arginine-modifying reagents on spinach ferredoxin:nitrite oxidoreductase. Biochim. Biophys. Acta, **1140**, 304-312 (1993)
[38] Hirasawa, M.; Dose, M.M.; Kleis-SanFrancisco, S.; Hurley, J.K.; Tollin, G.; Knaff, D.B.: A conserved tryptophan at the ferredoxin-binding site of ferredoxin:nitrite oxidoreductase. Arch. Biochem. Biophys., **354**, 95-101 (1998)

[39] Hirasawa, M.; Proske, P.A.; Knaff, D.B.: The role of tryptophan in the reaction catalyzed by spinach ferredoxin-dependent nitrite reductase. Biochim. Biophys. Acta, **1187**, 80-88 (1994)
[40] Hirasawa, M.; Tollin, G.; Salamon, Z.; Knaff, D.B.: Transient kinetic and oxidation-reduction studies of spinach ferredoxin:nitrite oxidoreductase. Biochim. Biophys. Acta, **1185**, 336-345 (1994)
[41] Migge, A.; Carrayol, E.; Hirel, B.; Lohmann, M.; Meya, G.; Becker, T.W.: Two nitrite reductase isoforms are present in tomato cotyledons and are regulated differently by UV-A or UV-B light and during plant development. Planta, **207**, 229-234 (1998)
[42] Orea, A.; Pajuelo, P.; Pajuelo, E.; Marquez, A.J.; Romero, J.M.: Characterisation and expression studies of a root cDNA encoding for ferredoxin-nitrite reductase from Lotus japonicus. Physiol. Plant., **113**, 193-202. (2001)
[43] Pajuelo, E.; Pajuelo, P.; Clemente, M.T.; Marquez, A.J.: Regulation of the expression of ferredoxin-nitrite reductase in synchronous cultures of Chlamydomonas reinhardtii. Biochim. Biophys. Acta, **1249**, 72-78 (1995)
[44] Sengupta, S.; Shaila, M.S.; Rao, G.R.: Purification and characterization of assimilatory nitrite reductase from Candida utilis. Biochem. J., **317**, 147-155 (1996)

Ferredoxin-nitrate reductase 1.7.7.2

1 Nomenclature

EC number
1.7.7.2

Systematic name
nitrite:ferredoxin oxidoreductase

Recommended name
ferredoxin-nitrate reductase

Synonyms
EC 1.6.6.2 (formerly)
assimilatory ferredoxin-nitrate reductase
assimilatory nitrate reductase
nitrate (ferredoxin) reductase
reductase, nitrate (ferredoxin)

CAS registry number
60382-69-6

2 Source Organism

<1> *Plectonema boryanum* [1, 2]
<2> *Phormidium unicatum* (strain CU 1462/7 [3]) [3]
<3> *Nostoc muscorum* (strain 7119 [4]) [4, 7, 10]
<4> *Anacystis nidulans* (strain L 1402-1 [5, 6]; unicellular non-nitrogen fixer [6]) [5-7]
<5> *Anabaena sp.* (strain 7119 [6, 14]; previously classified as Nostoc muscorum 7119, filamentous nitrogen fixer [6]; strains 7120 and M131 [14]) [6, 14]
<6> *Nostoc sp.* (strain PCC 6719, filamentous nitrogen fixer [6]; strains 29105 and 29150 [14]) [6, 14]
<7> *Anabaena cylindrica* [7]
<8> *Azotobacter chroococcum* [7]
<9> *Acinetobacter calcoaceticus* [7]
<10> *Ectothiorhodospira shaposhnikovii* (strain N1, phototrophic, purple sulfur bacterium [8]) [8]
<11> *Clostridium perfringens* [9]
<12> *Rhodopseudomonas capsulata* [7]
<13> *Haloferax mediterranei* (extreme halophilic archaeon [11]) [11]

<14> *Synechococcus sp.* (strain PCC 7942 [12]) [12]
<15> *Anabaena variabilis* (strain ATCC 29413 [13,14]) [13, 14]
<16> *cyanobacteria* (heterocystous nitrogen-fixing cyanobacterial strains [14]) [14]
<17> *Synechocystis sp.* (strain 6803, unicellular cyanobacterium [14]) [14]
<18> *Calothrix sp.* (strains 7101 and 7601 [14]) [14]
<19> *Fischerella muscicola* (strain 1829 [14]) [14]

3 Reaction and Specificity

Catalyzed reaction

nitrite + H_2O + 2 oxidized ferredoxin = nitrate + 2 reduced ferredoxin

Reaction type

oxidation
redox reaction
reduction

Natural substrates and products

S nitrate + reduced ferredoxin <1, 3-7, 11, 13, 14, 16> (<1, 3, 4, 7, 13, 14> assimilatory nitrate reduction [2, 5, 7, 11, 12]; <1> two-electron nitrate reduction [2]; <3> nitrate assimilation, ferredoxin-dependent photosynthetic reduction of nitrate [4]; <4-6, 16> first step in assimilatory reduction of nitrate [6, 14]; <11> nitrate fermentation, ferredoxin is utilized as direct electron donor in nitrate reduction [9]; <3, 4, 7, 14> reduced ferredoxin as physiological electron donor [7, 12]) (Reversibility: ? <1, 3-7, 11, 13, 14, 16> [2, 4-7, 9, 11, 12, 14]) [2, 4-7, 9, 11, 12, 14]
P nitrite + oxidized ferredoxin <3-7, 14, 16> [4, 6, 7, 12, 14]
S nitrate + reduced flavodoxin <4> (<4> flavodoxin can physiologically substitute for ferredoxin under conditions of iron starvation [7]) [7]
P nitrite + oxidized flavodoxin

Substrates and products

S nitrate + reduced ferredoxin <1-19> (<2, 3> coupled ferredoxin-NADPH reductase activity with $NADP^+$-reductase system [3, 10]; <3, 4, 7> enzyme can not accept electrons directly from NAD(P)H, it is dependent on reduced ferredoxin as physiological electron donor [7]; <13> 100 nM dithionite reduced ferredoxin with 94% of the activity with reduced methyl viologen [11]; <11> ferredoxin as electron donor at neutral pH and pH 8, but not at pH 9 [9]; <14> ratio of methyl viologen- to ferredoxin-dependent nitrate reduction is about 27 [12]; <16> depends on reduced ferredoxin for its activity, multimodal kinetic model: a single catalytic site functions in multiple kinetic modes [14]) (Reversibility: ? <1-19> [1-14]) [1-14]
P nitrite + H_2O + oxidized ferredoxin <1-19> [1-14]
S nitrate + reduced methyl viologen <1, 4-6, 10, 11, 13-19> (<1, 11> enzyme also has methyl viologen-linked activity [1, 2, 9]; <4-6, 10, 11, 14,

16> enzyme uses also reduced methyl viologen as reductant [5-10, 12, 14]; <13> reduced methyl viologen is the best electron donor in vitro [11]; <11> nitrite formation 10times higher at pH 9 than at pH 8 [9]) (Reversibility: ? <1, 4-6, 10, 11, 13-19> [11, 2, 5-14]) [1, 2, 5-14]

P nitrite + oxidized methyl viologen <1, 4-6, 10, 11, 13-19> [1, 2, 5-14]

S Additional information <10, 11, 13> (<10> not as electron donors: NADH, NADPH, $FADH_2$, $FMNH_2$, reduced animal cytochrome c, sodium dithionite, methylene blue, phenazine methosulfate [8]; <13> not as electron donors: NADH, NADPH, dithionite, 8% of maximum activity with 1mM $FADH_2$ and less than 7% of maximum activity with 1 mM reduced 2,6-dichlorophenolindophenol or phenazine methosulfate [11]; <11> not as electron donors: NADH, NADPH, $FADH_2$, $FMNH_2$, rubredoxin [9]) [8, 9, 11]

P ?

Inhibitors

EDTA <1, 13> (<1> photoactivated EDTA, irreversible inactivation [1]; <13> partial inhibition, 43% loss of activity [11]) [1, 11]

FAD <1> (<1> photoactivated FAD, irreversible inactivation [1]) [1]

ammonia <3> (<3> after initial slight stimulation rapid and complete inhibition, promotes conversion of ferredoxin-nitrate reductase into its reduced inactive form, metabolic interconversion [4]) [4]

azide <1, 11, 13> (<1> competitive inhibitor [1]; <13> potent inhibitor, 79% loss of activity [11]; <11> 74% inhibition [9]) [1, 9, 11]

cyanide <1, 4, 8, 9, 11, 13> (<1> competitive inhibitor [1]; <4> powerful inhibitor [7]; <13> almost complete inhibition [11]; <11> 10 mM, 87% inhibition [9]) [1, 7, 9, 11]

dithionite <1> (<1> inactivation can be suppressed by cyanide and azide and specifically prevented by superoxide dismutase [1]; <1> converts native enzyme in modified form without ferredoxin-linked activity but retained methyl viologen-linked activity [2]) [1, 2]

light <4> (<4> light- and oxygen-dependent inactivation of enzyme activity in absence but not in presence of any added nitrogen source, inactivation may result from an oxidative modification of enzyme, assay in cell suspension [5]) [5]

oxygen <4> (<4> light- and oxygen-dependent inactivation of enzyme activity in absence but not in presence of any added nitrogen source, inactivation may result from an oxidative modification of enzyme, assay in cell suspension [5]) [5]

p-hydroxymercuribenzoate <4, 8, 9> (<4> powerful inhibitor [7]) [7]

Additional information <4-6, 11, 13> (<4> enzyme synthesis is repressed in presence of ammonia as a result of its metabolism, nitrate is not required as an obligate inducer [6]; <5,6> nitrate plays an active role in enzyme synthesis, ammonia acts as antagonist [6]; <13> not inhibited by potassium chlorate, dithiothreitol, *p*-hydroxymercuribenzoate, sulfite [11]; <11> not inhibited by 0.1 mM *p*-chloromercuribenzoate, enzyme formation is induced by nitrate and inhibited by 0.5 mM tungstate, but recovered by 0.1 mM molybdate [9]) [6, 9, 11]

Cofactors/prosthetic groups

heme c <12> (<12> enzyme contains 1.5 heme molecules of the c-type [7]) [7]
iron-sulfur centre <1, 13> (<1> contains 4 iron and 4 acid-labile sulfur atoms per enzyme molecule, presence of iron-sulfur cluster [2]; <13> enzyme contains Fe-S clusters [11]) [2, 11]
molybdenum cofactor <16> [14]
Additional information <4, 13, 14> (<4> no presence of flavin, heme or cytochrome [7]; <13> no FAD or cytochromes are observed [11]; <14> narB gene with domains that might be involved in binding of the molybdenum cofactor and of an iron-sulfur cluster, but no flavin nucleotide- or heme-binding domains [12]) [7, 11, 12]

Activating compounds

cyanate <8, 9> (<8, 9> stimulates [7]) [7]
Additional information <5, 6, 11, 15> (<5, 6> nitrate plays an active role in enzyme synthesis, ammonia acts as antagonist [6]; <11, 15> enzyme synthesis is induced by nitrate [9, 13]) [6, 9, 13]

Metals, ions

Ca^{2+} <11> (<11> stimulates activity about 30% [9]) [9]
Fe^{2+} <11> (<11> stimulates activity about 50% [9]) [9]
Mg^{2+} <11> (<11> stimulates activity about 30% [9]) [9]
Mn^{2+} <11> (<11> stimulates activity about 50%, stimulates reaction rate and level of nitrite formation [9]) [9]
NaCl <13> (<13> activity depends on salt concentration [11]) [11]
iron <1, 11> (<1, 11> molybdenum iron-sulfur protein [2, 9]; <1> contains 4 iron and 4 acid-labile sulfur atoms per enzyme molecule [2]) [2, 9]
iron-sulfur centre <1, 11, 13> (<1> contains 4 iron and 4 acid-labile sulfur atoms per enzyme molecule, presence of iron-sulfur cluster [2]; <11> molybdo-iron-sulfur protein [9]; <13> enzyme contains Fe-S clusters [11]) [2, 9, 11]
molybdenum <1, 4, 8, 9, 11, 12> (<1> Mo center [1]; <1> enzyme contains 0.95 atom of molybdenum per enzyme molecule [2]; <4, 8, 9, 12> molybdoprotein [7]; <1, 11> molybdenum iron-sulfur protein [2, 9]; <4> essential role in catalytic activity [7]; <12> enzyme contains one molybdenum atom [7]) [1, 2, 7, 9]

Turnover number (min^{-1})

65640 <4> (nitrate, <4> dithionite-reduced methyl viologen as electron donor [7]) [7]

Specific activity (U/mg)

0.2 <4> (<4> in situ assay, cells grown on media containing nitrate [5, 6]) [5, 6]
0.55 <13> [11]
52 <11> [9]
305 <1> (<1> ferredoxin-linked activity [2]) [2]
1020 <1> (<1> methyl viologen-linked activity [2]) [2]
Additional information <2, 4, 10, 12> [3, 7, 8]

K_m-Value (mM)

0.013 <4> (reduced ferredoxin) [7]
0.038 <1> (reduced ferredoxin) [2]
0.066 <13> (reduced methyl viologen) [11]
0.1 <11> (nitrate) [9]
0.69 <4> (nitrate) [7]
0.7-2.1 <14> (nitrate) [12]
0.95 <13> (nitrate) [11]
1.5 <14> (nitrate, <14> reduced methyl viologen as electron donor, recombinant enzyme expressed in Escherichia coli [12]) [12]
2.5 <1> (reduced methyl viologen) [2]
Additional information <5, 6, 15, 17-19> [14]

pH-Optimum

8.5 <1> (<1> anaerobic assay, methyl viologen- and ferredoxin-linked activity [2]) [2]
9 <11, 13> (<11> reduced methyl viologen as electron donor [9]) [9, 11]
9.5-10.5 <14> (<14> reduced methyl viologen as electron donor [12]) [12]
10.2 <1> (<1> aerobic assay, methyl viologen-linked activity [2]) [2]
10.5 <4> (<4> dithionite-reduced methyl viologen as electron donor [7]) [7]

pH-Range

7.3-9.2 <1> (<1> at pH 7.3 and 9.2: 50% of activity maximum, aerobic assay, methyl viologen-linked and ferredoxin-linked activity [2]) [2]
8.3-9.8 <11> (<11> at pH 8.3 and 9.8: about 50% of activity maximum [9]) [9]

Temperature optimum (°C)

30 <1, 3> (<1,3> assay at [2,10]) [2, 10]
37 <11, 14> (<14> reduced methyl viologen as electron donor [12]; <11> assay at [9]) [9, 12]
60 <13> (<13> for NaCl concentrations of 1.3 and 0.9 M [11]) [11]
80 <13> (<13> for NaCl concentrations of 3.1 and 2.2 M [11]) [11]
Additional information <3, 13> (<3> assay at room temperature [4]; <13> strong dependence of activity on the temperature and NaCl concentrations [11]) [4, 11]

4 Enzyme Structure

Molecular weight

75000 <4> [7]
80000 <11> (<11> gel filtration [9]) [9]
85000 <1> (<1> sedimentation equilibrium method [2]) [2]
90000 <11> (<11> ultracentrifugation [9]) [9]
100000 <8, 9> (<8,9> about [7]) [7]
132000 <13> (<13> gel filtration [11]) [11]
185000 <12> [7]

Subunits

? <14> (<14> x * 76000, enzyme expressed in Escherichia coli, SDS-PAGE [12]) [12]
dimer <12, 13> (<12> 2 * 85000 [7]; <13> 1 * 105000 + 1 * 50000, SDS-PAGE [11]) [7, 11]
monomer <1, 4, 11, 16> (<4> 1 * 75000 [7]; <11> 1 * 90000, SDS-PAGE [9]; <1> 1 * 83000, SDS-PAGE [2]) [2, 7, 9, 14]

5 Isolation/Preparation/Mutation/Application

Localization

cytoplasm <11> (<11> enzyme functions in the soluble state in the cytoplasm [9]) [9]
membrane <4> (<4> tightly bound to chlorophyll-containing membrane fractions [7]) [7]
soluble <14> (<14> recombinant enzyme activity is found exclusively in the soluble fraction [12]) [12]
Additional information <3> (<3> activity remains associated with photosynthetic subcellular particles [10]) [10]

Purification

<1> (19000fold purification [2]) [2]
<4> [7]
<11> [9]
<12> [7]
<13> (177fold purification [11]) [11]
<15> (185fold partial purification [14]) [14]
<16> (purified from a few cyanobacterial strains [14]) [14]
<17> (partial purification [14]) [14]

Cloning

<5> (strain 7120: gene encoding nitrate reductase is cloned and its complete amino acid sequence deduced [14]) [14]
<14> (narB structural gene for nitrate reductase is cloned and expressed in Escherichia coli as a 76 kDa polypeptide with 729 amino acids [12]) [12]

6 Stability

Temperature stability

60 <10> (<10> 5 min, crude extract, stable [8]) [8]

Oxidation stability

<1>, reversible inactivation of ferredoxin-linked activity by exposure of dithionite solution of the enzyme to air, but no effect on methyl viologen-linked activity [1]

General stability information

<1>, low ionic strength, 70 h, 0.002 mg protein/ml, 10 mM phosphate buffer or 50 mM Tris-HCl buffer, pH 7.5, unstable [2]
<4>, in vivo stability [5]
<13>, NaCl stabilizes, after 43 h, with 2.7 M, 2.1 M or 0.9 M NaCl: 95% loss of activity, after 90 h, with 4.3 M NaCl: completely stable [11]

Storage stability

<1>, -20°C, in 50 mM Tris-HCl buffer, pH 7.5, 0.5 M NaCl, 1 mM EDTA, 1 month, no loss of activity [2]
<13>, 4°C, 50 mM phosphate buffer, pH 7, 20% glycerol, 2 weeks, no loss of activity [11]

References

[1] Mikami, B.; Ida, S.: Reversible inactivation of ferredoxin-nitrate reductase from the cyanobacterium Plectonema boryanum. The role of superoxide anion and cyanide. Plant Cell Physiol., **27**, 1013-1021 (1986)
[2] Mikami, B.; Ida, S.: Purification and properties of ferredoxin-nitrate reductase from the cyanobacterium Plectonema boryanum. Biochim. Biophys. Acta, **791**, 294-304 (1984)
[3] Nath Bagachi, S.; Singh Chauhan, V.; Palod, A.: Heterotrophy and nitrate metabolism in a cyanobacterium Phormidium uncinatum. Curr. Microbiol., **21**, 53-57 (1990)
[4] Ortega, T.; Rivas, J.; Cardenas, J.; Losada, M.: Metabolic interconversion of ferredoxin-nitrate reductase and NADP reductase of Nostoc muscorum. Biochem. Biophys. Res. Commun., **78**, 185-193 (1977)
[5] Herrero, A.; Flores, E.; Guerrero, M.G.: Regulation of the nitrate reductase level in anacystis nidulans: Activity decay under nitrogen stress. Arch. Biochem. Biophys., **234**, 454-459 (1984)
[6] Herrero, A.; Flores, E.; Guerrero, M.G.: Regulation of nitrate reductase levels in the cyanobacteria Anacystis nidulans, Anabaena sp. strain 7119, and Nostoc sp. strain 6719. J. Bacteriol., **145**, 175-180 (1981)
[7] Guerrero, M.G.; Vega, J.M.; Losada, M.: The assimilatory nitrate-reducing system and its regulation. Annu. Rev. Plant Physiol., **32**, 169-204 (1981)
[8] Malofeeva, I.V.; Kondratieva, E.N.; Rubin, A.B.: Ferredoxin-linked nitrate reductase from the phototrophic bacterium Ectothiorhodospira shaposhnikovii. FEBS Lett., **53**, 188-189 (1975)
[9] Seki-Chiba, S.; Ishimoto, M.: Studies on nitrate reductase of Clostridium perfringens. Purification, some properties, and effect of tungstate on its formation. J. Biochem., **82**, 1663-1671 (1977)
[10] Ortega, T.; Castillo, F.; Cardenas, J.: Photolysis of water coupled to nitrate reduction by Nostoc muscorum subcellular particles. Biochem. Biophys. Res. Commun., **71**, 885-891 (1976)

[11] Martinez-Espinosa, R.M.; Marhuenda-Egea, F.C.; Bonete, M.J.: Assimilatory nitrate reductase from the haloarchaeon Haloferax mediterranei: Purification and characterization. FEMS Microbiol. Lett., **204**, 381-385 (2001)
[12] Rubio, L.M.; Herrero, A.; Flores, E.: A cyanobacterial narB gene encodes a ferredoxin-dependent nitrate reductase. Plant Mol. Biol., **30**, 845-850 (1996)
[13] Sanz, A.P.; Moreno-Vivian, C.; Maldonado, J.M.; Gonzales-Fontes, A.: Effect of a constant supply of different nitrogen sources on protein and carbohydrate content and enzyme activities of Anabaena variabilis cells. Physiol. Plant., **95**, 39-44 (1995)
[14] Nissen, P.; Martin-Nieto, J.: Multimodal kinetics: Cyanobacterial nitrate reductase and other enzyme, transport and binding systems. Physiol. Plant., **104**, 503-511 (1998)

Hydroxylamine reductase 1.7.99.1

1 Nomenclature

EC number
1.7.99.1

Systematic name
ammonia:(acceptor) oxidoreductase

Recommended name
hydroxylamine reductase

Synonyms
hybrid cluster protein <2> (<2> i.e. HCP, formerly termed prismane protein [6]) [6]
reductase, hydroxylamine (acceptor)

CAS registry number
37256-42-1

2 Source Organism

<1> *Aspergillus niger* (iron-chlorin protein also catalyzes methyl viologen-linked reduction of hydroxylamine and that of nitrite at a slower rate [3]) [3]
<2> *Escherichia coli* [6, 10]
<3> *Spinacia oleracea* (hydroxylamine reductase 1, hydroxylamine reductase 2 [5]) [5]
<4> *Pseudomonas aeruginosa* [1]
<5> *Pennisetum purpureum* (low activity [2]) [2]
<6> *Brachiaria mutica* (low activity [2]) [2]
<7> *Chloris gayana* (low activity [2]) [2]
<8> *Panicum maximum* (low activity [2]) [2]
<9> *Cucurbita pepo* [4, 5, 7, 9]
<10> *Triticum aestivum* (wheat, nitrite reductase and hydroxylamine reductase are distinct enzymes [8]) [8]

3 Reaction and Specificity

Catalyzed reaction
ammonia + H_2O + acceptor = hydroxylamine + reduced acceptor

Reaction type

oxidation
redox reaction
reduction

Substrates and products

S hydroxylamine + reduced acceptor <2, 3, 4, 9, 10> (<4> with pyocyanine or reduced methylene blue as electron donor [1]; <9> reduced benzyl viologen as electron donor [9]; <3,9> hydroxylamine reductase 1 reacts with reduced benzyl viologen, ferredoxin and $FMNH_2$ [5,7]; <3,9> $FADH_2$, NADPH [5]; <3> hydroxylamine reductase 2 reacts with reduced benzyl viologen [5]; <9> hydroxylamine reductase 2 reacts with benzyl viologen and dithionite [7]; <10> methyl viologen is more effective than benzyl viologen, flavin nucleotides are also utilized but to a lesser extent [8]) (Reversibility: ? <2, 3, 4, 9, 10> [1, 3, 5-9]) [1, 3, 5, 6, 7, 8, 9]

P ammonia + oxidized acceptor <2> [6]

Inhibitors

CO <9> (<9> with reduced benzyl viologen as electron donor, reductase 1: irreversible inhibition by light, reductase 2 : reversible inhibition by light [4]) [4]
Cu^{2+} <4> (<4> inhibition of dialyzed enzyme at 0.0005 mM [1]) [1]
Zn^{2+} <4> (<4> inhibition of dialyzed enzyme at 0.0005 mM [1]) [1]
atabrine <9> (<9> hydroxylamine reductase 1 [4]) [4]
bathophenanthroline <9> (<9> 50% inhibition of hydroxylamine reductase 2 at 1 mM [4]) [4]
cyanide <9> (<9> 100% inhibition at 5 mM [9]) [9]
guanidinium chloride <9> (<9> 25% inhibition at 2 mM, 30% inhibition at 3 mM [7]) [7]
hydrazine <4, 9> (<9> hydroxylamine reductase 2, slight [4]; <4> competitive inhibition [1]; <9> hydrazine sulfate inhibits 10-15% at 4 mM [9]) [1, 4, 9]
isonicotinic acid <9> (<9> hydroxylamine reductase 2, slight [4]) [4]
nitrite <9> [9]
oxygen <2> (<2> reversible inhibition [6]) [6]
p-choloromercuribenzoate <9> (<9> 47% inhibition at 0.5 mM [9]) [9]
phenylmercuric acetate <9> (<9> 73% inhibition at 0.5 mM [9]) [9]
pyruvate <9> (<9> 20% inhibition at 3.5 mM [9]) [9]
semicarbazide <9> (<9> hydroxylamine reductase 2, slight [4]) [4]

Cofactors/prosthetic groups

FAD <4> (<4> flavoprotein, contains 0.9 nmol FAD per mg protein [1]) [1]

Activating compounds

2,2'-bipyridyl <9> (<9> 10% stimulation at 1 mM [9]) [9]
8-hydroxyquinoline <9> (<9> 5% stimulation at 1 mM [9]) [9]
CN^- <2> (<2> stimulates at concentrations above 5 mM [6]) [6]

Metals, ions

Co^{2+} <4> (<4> can substitute for Mn^{2+}, only 10% as effective [1]) [1]
Mn^{2+} <4> (<4> required for maximal activity [1]) [1]
iron <1, 9> (<1> iron-chlorin protein also catalyzes methyl viologen-linked reduction of hydroxylamine and nitrite at a slower rate [3]; <9> iron containing proteins: nitrite reductase-hydroxylamine reductase 1 and hydroxylamine reductase 2 [4]) [3, 4]

Specific activity (U/mg)

2.87 <4> [1]
92 <2> (<2> at pH 7.5 [6]) [6]
458 <2> (<2> at pH 9 [6]) [6]
Additional information <1, 5, 6, 7, 8, 9> (<9> specific activities after different purification methods [7]) [2, 3, 7]

K_m-Value (mM)

0.0000004 <4> (hydroxylamine) [1]
0.00001 <4> ($FADH_2$) [1]
0.000012 <4> (reduced pyocyanine) [1]
0.000044 <4> (reduced methylene blue) [1]
0.0075 <9> (benzyl viologen) [9]
0.05-0.9 <9> (hydroxylamine) [9]
2.5 <2> (hydroxylamine, <2> at pH 9 [6]) [6]
38.9 <2> (hydroxylamine, <2> at pH 7.5 [6]) [6]

pH-Optimum

7.8 <4, 10> [1, 8]
9 <2> [6]
Additional information <9> (<9> no clear or consistent pH optima observed with hydroxylamine reductase 1 and hydroxylamine reductase 2 [7]) [7]

Temperature optimum (°C)

37 <4> [1]

4 Enzyme Structure

Molecular weight

32000 <3> (<3> hydroxylamine reductase 1, gel filtration [5]) [5]
35000 <9> (<9> hydroxylamine reductase 2, gel filtration [5,7]) [5, 7]
60000 <9> (<9> hydroxylamine reductase 1, gel filtration [5,7]) [5, 7]
65000 <3> (<3> hydroxylamine reductase 2, gel filtration [5]) [5]
280000-320000 <1> (<1> gel filtration and sedimentation [3]) [3]

5 Isolation/Preparation/Mutation/Application

Source/tissue

leaf <3, 5-10> [2, 5, 7, 8]

Localization
chloroplast <10> [8]

Purification
<1> [3]
<2> [6]
<4> [1]
<9> [5, 7]

Crystallization
<1> (gradual concentration for 7 weeks [3]) [3]

6 Stability

Temperature stability
45 <10> (<10> 20 min, 25-38% loss of activity [8]) [8]
55 <9> (<9> 10 min, 25-40% loss of activity [7]) [7]
60 <9> (<9> 10 min, 50-85% loss of activity [7]) [7]

Storage stability
<9>, -18°C, DEAE-cellulose fraction, often stable for several weeks, but unpredictable loss of activity [7]
<10>, 0°C, 65% activity with methyl viologen after 24 hours, 40% activity with benzyl viologen after 24 hours [8]

References

[1] Walker, G.C.; Nicholas, D.J.D.: Hydroxylamine reductase from Pseudomonas aeruginosa. Biochim. Biophys. Acta, **49**, 361-368 (1961)
[2] Venkataramana, S.; Das, V.S.R.: Distribution of nitrogen assimilating enzymes in relation to photosynthesis in certain C4 grasses. Z. Pflanzenphysiol., **105**, 289-296 (1982)
[3] Horie, S.; Watanabe, T.; Nakamura, S.: Isolation, properties, and crystallization of an iron-chlorin protein from Aspergillus niger. J. Biochem., **80**, 579-593 (1976)
[4] Hucklesby, D.P.; Hewitt, E.J.; James, D.M.: Possible active sites in nitrite reductase and hydroxylamine reductases from vegetable marrow (Curcubita pepo L.). Biochem. J., **117**, 30P (1970)
[5] Hewitt, E.J.; Hucklesby, D.P.: Molecular exclusion chromatography of nitrite and hydroxylamine reductases from plants with reference to electron donor systems. Biochem. Biophys. Res. Commun., **25**, 689-693 (1966)
[6] Wolfe, M.T.; Heo, J.; Garavelli, J.S.; Ludden, P.W.: Hydroxylamine reductase activity of the hybrid cluster protein from Escherichia coli. J. Bacteriol., **184**, 5898-5902 (2002)
[7] Hucklesby, D.P.; Hewitt, E.J.: Nitrite and hydroxylamine reduction in higher plants. Fractionation, electron donor and substrate specificity of leaf en-

zymes, principally from vegetable marrow (Cucurbita pepo L.). Biochem. J., **119**, 615-627 (1970)
[8] Sawhney, S.K.; Nicholas, D.J.D.: Nitrite, hydroxylamine and sulphite reductases in wheat leaves. Phytochemistry, **14**, 1499-1503 (1975)
[9] Cresswell, C.F.; Hageman, R.H.; Hewitt, E.J.; Hucklesby, D.P.: The reduction of nitrate, nitrite and hydroxylamine to ammonia by enzymes from Curcubita pepo L. in the presence of reduced benzyl viologen as electron donor. Biochem. J., **94**, 40-53 (1965)
[10] Wolfe, M.T.; Heo, J.; Garavelli, J.S.; Ludden, P.W.: Hydroxylamine reductase activity of the hybrid cluster protein from Escherichia coli. J. Bacteriol., **184**, 5898-5902 (2002)

Nitric-oxide reductase 1.7.99.2

1 Nomenclature

EC number
1.7.99.2 (deleted)

Recommended name
nitric-oxide reductase

Nitrite reductase 1.7.99.3

1 Nomenclature

EC number
1.7.99.3 (deleted, now included in EC 1.7.2.1)

Recommended name
nitrite reductase

Nitrate reductase 1.7.99.4

1 Nomenclature

EC number
1.7.99.4

Systematic name
nitrite:(acceptor) oxidoreductase

Recommended name
nitrate reductase

Synonyms
cytochrome B-NR
NAP <18, 20, 29> (<18, 20, 29> periplasmic respiratory nitrate reductase [35, 37, 40]) [35, 37, 40]
NAR <20> (<20> membrane-bound respiratory nitrate reductase with 3 subunits [37]) [37]
NaR
nitrate reductase (acceptor)
nitrite oxidoreductase <16> [34]
quinol-nitrate oxidoreductase NarGHI <20> [48]
reductase, nitrate (acceptor)
respiratory nitrate reductase
Additional information <5, 15> (<5, 15> the enzymes EC 1.7.99.4 and EC 1.9.6.1 are probably identical, in vivo cytochrome serves as electron donor in the electron transport chain to nitrate [32, 33]) [32, 33]

CAS registry number
37256-45-4

2 Source Organism

<1> *Aerobacter aerogenes* (strain S 45 [1]) [1]
<2> *Micrococcus halodenitrificans* (strain ATCC 13511, moderate halophile, nitrate reductase A [2]) [2]
<3> *Halobacterium sp.* (extreme halophile [3]) [3]
<4> *Klebsiella aerogenes* (strain S 45 [4,28]; identical with Aerobacter aerogenes [4]; 2 forms: nitrate reductase I: ab, nitrate reductase II: abc2 and (abc2)4 [4, 6]) [4, 6, 28]
<5> *Escherichia coli* (strain A1004a, 5-aminolaevulinic acid auxotroph [5]; strain K12 [7, 9, 10, 13, 21, 24]; strain EMG-2 [8]; strain A1002 [9]; strain

RK7 [10]; strain EMG 29 [12]; strain X5119 [13]; several K12 strains, expression of the periplasmic NapABC, identical with nitrate reductase P, and two membrane-bound Nar enzymes, NarGHI, identical with nitrate reductase A, and NarZYV [42]; coexistence of 2 enzyme forms with different subunit compositions [7]; 2 forms of enzyme: monomeric and probably tetrameric form [9]; expression of NAR and NAP under anaerobic growth conditions [37]; NAR [41]) [5, 7-10, 12-14, 21, 24, 26, 31, 32, 37, 41, 42, 45]

<6> *Aeropyrum pernix* (NAR [37]) [37]

<7> *Pseudomonas denitrificans* (strain ATCC 13867 [15]) [15]

<8> *Paracoccus denitrificans* (strain 4, nitrate reductase A [16]; formerly Micrococcus denitrificans [17]; strain NCIB 8944 [19,20]; NAR is predominantly expressed under anaerobic and NAP under aerobic growth conditions [37,41]; strain DSM 65 [41]; NarGHI [42]) [16, 17, 19, 20, 37, 38, 41, 42, 45]

<9> *Rhodobacter sphaeroides* (f.sp. denitrificans, strain IL106 [18,41]; NAP [41]) [18, 37, 41]

<10> *Geobacillus stearothermophilus* (NCA strain 2184, identical with ATCC 12016 [22]) [22]

<11> *Pseudomonas stutzeri* (strain ATCC 14405 [23,46]; formerly Pseudomonas perfectomarina [23]) [23, 46]

<12> *Bacillus licheniformis* (strain S244 [27,28]) [27, 28]

<13> *Thiosphaera pantotropha* (strain LMD82.5, expression of two catalytically distinct forms: a membrane-bound form under anaerobic growth conditions and a periplasmic form mainly under aerobic growth conditions [29]; strain M-6 [44]) [29, 44]

<14> *Rhodobacter capsulatus* (strain KS36 and strains SL113I/II and R507, mutant strains of KS36 [30]) [30]

<15> *Achromobacter fischeri* [32, 33]

<16> *Nitrobacter hamburgensis* [34]

<17> *Enterobacter cloacae* (strain IFO 12935 [25]) [25]

<18> *Desulfovibrio desulfuricans* (strain ATCC 27774 [35]; NAP [37]) [35, 37]

<19> *Haloarcula marismortui* (strain ATCC 43049, halophilic archaeon [36]) [36]

<20> *Paracoccus pantotrophus* (3 nitrate reductases: assimilatory NAS, respiratory NAR and NAP [37]; NAR is predominantly expressed under anaerobic and NAP under aerobic growth conditions [37,41]; NarGHI is expressed during anaerobic growth and NapAB is predominantly expressed under aerobic conditions [48]) [37, 41, 42, 48]

<21> *Pyrobaculum aerophilum* (hyperthermophilic archaeon [38]) [38]

<22> *Pseudomonas aeruginosa* [37]

<23> *Haloferax volcanii* (halophilic archaeon [36,38]) [36, 38]

<24> *Haloferax denitrificans* (halophilic archaeon [36,38]) [36, 38]

<25> *Pseudomonas fluorescens* (strain YT101 [11,39]; only membrane-bound nitrate reductase A, no periplasmic enzyme [39]) [11, 39]

<26> *Geobacter metallireducens* (strain GS15, identical with ATCC 55774 [43]) [43]

<27> *Thermus thermophilus* (NAR [37]; strain HB8, ATCC 27643 [47]) [37, 47]
<28> *bacteria* (distribution of respiratory NAR, NAP and assimilatory NAS in proteobacteria, firmicutes and cyanobacteria [37]) [37]
<29> *Azospirillum brasilense* (strain Sp245, NAP is expressed under oxic and anoxic conditions [40]) [40]
<30> *Ralstonia eutropha* (strain DSM 428, NAR is expressed under anaerobic growth conditions [41]) [41]

3 Reaction and Specificity

Catalyzed reaction

nitrite + acceptor = nitrate + reduced acceptor (<5> mechanism of respiratory-driven proton translocation [8]; <8> mechanism [19]; <5> with reduced viologen dyes: compulsory-order mechanism, Theorell-Chance mechanism, with quinols as electron donor: two-site ping-pong mechanism, enzyme-substitution mechanism [21]; <8,20> electron transfer mechanism of NAR and NAP [37]; <20> electron transfer mechanism of NarGHI [48])

Reaction type

oxidation
redox reaction
reduction

Natural substrates and products

S nitrate + reduced acceptor <1, 2, 4, 5, 7-13, 15, 17-20, 25, 26, 27, 29> (<2, 19> bacterial denitrification [2, 36]; <1, 4, 5, 7, 8, 10-13, 17, 18, 27> nitrate respiration [1, 4-9, 12, 13, 15-17, 20, 22, 23, 25, 27-29, 35, 46, 47]; <5, 8, 26> nitrate dissimilation [43, 45]; <8> respiratory enzyme catalyzes the first step of bacterial denitrification [19]; <13> first step of an aerobic denitrification process in presence of oxygen [29]; <9> first reduction step during denitrification process [41]; <5> catalyzes terminal reaction of a proton-translocating respiratory chain with a two electron transfer from nitrate to nitrite [24]; <5> terminal enzyme in a nitrate-respiring system in vivo, reduced cytochrome b_1 may be the physiological electron donor [26]; <4> physiological function: transfer of electrons from cytochrome $b_5$59 to nitrate [4]; <5> physiological rate of nitrate reduction by the respiratory chain is saturated by 0.04 mM nitrate or less, lower rate compared to the rate obtained with reduced benzyl viologen, enzyme may act as a transmembrane hydrogen carrier [8]; <5> terminal oxidoreduction enzyme of a proton-translocating respiratory chain, two-electron transfer probably from a b-type cytochrome to nitrate [9]; <5> functions as a terminal enzyme of the respiratory chain, when organism is grown anaerobically in presence of nitrate as electron acceptor [12, 13, 21]; <5> transfer of reducing equivalents from reduced ubiquinol via cytochrome b_{556} and nitrate reductase to nitrate [12]; <9> cytochrome c_{552} may be an electron donor for enzyme [18]; <8> γ-subunit catalyzes electron transfer to the α- and β-subunits from physiologically ubiquinol which acts as branch point in the respiratory chain, electron flow from ubiquinol via cytochrome b on the periplasmic side of membrane to enzyme active site at the cytoplasmic side [20]; <5, 15> the enzymes EC 1.7.99.4 and EC 1.9.6.1 are probably identical, in vivo cytochrome serves as electron donor in the electron transport chain to nitrate [32, 33]; <5> in vitro artificial electron donors, e.g. reduced viologen indicators, can replace cytochrome c, reduction of cytochrome b_{556} by ubiquinol which releases two protons, electrons are passed to nitrate reductase and used to reduce NO_3^- to NO_2^- [31]; <5> metabolic role is nitrate respiration, electron transport chain in vivo: from cytochrome b_1 via Mo to NO_3^- [32]; <15> electron transport chain in vivo: Fe^{3+} via specific cytochrome-nitrate reductase to NO_3^- [33]; <20> NAR and NAP are linked to the respiratory electron transport systems, each ultimately taking electrons from the quinol pool, NAP is coupled to quinol oxidation via a membrane anchored tetraheme cytochrome [37]; <29> linked to the respiratory chain at the level of the quinol pool in the cytoplasmic membrane [40]; <5> periplasmic NapABC enzyme is resonsible for nitrate dissimilation and can function in anaerobic respiration but does not constitute a site for generating proton motive force, it may function when low nitrate concentrations limit the bioenergetic efficiency of nitrate respiration via NarGHI enzyme, membrane-bound NarGHI enzyme employs a redox loop to couple quinol oxidation with proton translocation, thereby generating proton motive force for anaerobic respiration [42]; <25> presence of a functional structural gene encoding the respiratory nitrate reductase confers higher rhizosphere competence of maize [11]; <25> genetic and/or functional relationship between the dissimilatory reduction of nitrate and that of nitrite [39]; <27> NarC, a 27 kDa membrane cytochrome c, plays an essential role in the synthesis of active enzyme and for the attachment of enzyme to the membrane, role of NarC as a component of the electron transport chain toward the NR [47]; <20> NarGHI directs electrons from quinol oxidation at the membrane anchor, NarI, to the site of nitrate reduction in the membrane extrinsic Fe-S cluster and Mo-bis-MGD containing dimer, NarGH, it conserves the free energy of the $QH2/NO_3^-$ couple as a transmembrane electrochemical proton gradient in the first step of denitrification pathway, NapAB receives its electrons from the quinol pool via the membrane-anchored protein NapC and dissipates the free energy of the $QH2/NO_3^-$ couple [48]) (Reversibility: ? <1, 2, 4, 5, 7-13, 15, 17-20, 25, 26, 27, 29> [1, 2, 4-9, 11-13, 15-29, 31-33, 35-37, 39-43, 45-48]) [1, 2, 4-9, 11-13, 15-29, 31-33, 35-37, 39-43, 45-48]

P nitrite + acceptor <5> [24]

S nitrite + ferricyanide <16> (<16> nitrification [34]) (Reversibility: ? <16> [34]) [34]

P nitrate + reduced ferricyanide <16> [34]

S Additional information <5> (<5> $FMNH_2$ is a non-physiological reductant [8]) [8]

P ?

Substrates and products

S bromate + reduced acceptor <5, 8> (<5> membrane-bound NAR is able to reduce chlorate and bromate, but not periplasmic NAP [41]) (Reversibility: ? <5, 8> [16, 21]) [16, 21, 41]

P bromite + acceptor

S chlorate + reduced acceptor <2, 4, 5, 7, 8, 13, 19-21, 26> (<7> alternative electron acceptor to nitrate [15]; <8> substrate for purified enzyme, but no substrate in intact cells [20]; <13> membrane-bound enzyme form uses chlorate as substrate, but not periplasmic form [29]; <21> slightly higher rate than with nitrate, but weaker enzyme affinity for chlorate than for nitrate [38]; <5> membrane-bound NAR is able to reduce chlorate and bromate, but not periplasmic NAP [41]; <20> alternative substrate for enzyme [48]) (Reversibility: ? <2, 4, 5, 7, 8, 13, 19-21, 26> [2, 4, 15, 16, 20, 21, 29, 36, 38, 41, 43, 48]) [2, 4, 15, 16, 20, 21, 29, 36, 38, 41, 43, 48]

P chlorite + acceptor <8, 21> [38]

S nitrate + reduced acceptor <1-30> (<1, 2, 4, 5, 7-9, 12, 13, 15, 16, 21, 25, 26, 29> reduced benzyl viologen as electron donor [1, 2, 4-9, 15, 16, 18, 21, 27, 29, 32-34, 38-43, 45]; <2-5, 7, 8,10, 13-17, 19, 20, 26, 27, 29> reduced methyl viologen as electron donor [2-4, 10, 14-16, 19, 20-22, 24-26, 29, 30, 32-34, 36, 40-43, 45, 47, 48]; <5> reduced benzyl viologen with 109% and $FADH_2$ with 8% of the rate obtained with reduced methyl viologen [24]; <2, 5, 8, 11> reduced viologen indicators as electron donors [2, 7, 16, 20, 21, 32, 46]; <5, 8> enzyme prefers viologens, e.g. methyl viologen and benzyl viologen, and phenazines, e.g. safranin T and phenosafranin, as electron donors [45]; <2, 5, 8> $FMNH_2$ as electron donor [2, 5, 8, 9, 16]; <5, 8, 26> $FADH_2$ as electron donor [16, 32, 43]; <13> NADH as electron donor, only membrane-bound enzyme form [29]; <26> NADH as electron donor [43]; <11> reduced cytochrome b as electron donor [46]; <8> cytochrome-containing enzyme uses duroquinol as electron donor, but not cytochrome-free enzyme, reduction rate is 50 times slower compared with reduced methyl viologen as electron donor, ascorbate-phenazine methosulfate as electron donor for cytochrome-containing enzyme [20]; <5> cytochrome-containing holoenzyme: quinols as electron donors, e.g. ubiquinol 1 or duroquinol, but not cytochrome-free $\alpha\beta'$ derivative, obtained by trypsin-cleavage of one subunit, that is fully active with viologen dyes as reductant [21]; <5> electron donors: NADPH, reduced cytochrome b_1, phenosafranine, reduced methylene blue [32]; <5> ferrocytochrome:nitrate reductase [45]; <4> electron donor and acceptor have different binding sites [6]; <5> holoenzyme has two independent active sites for quinol oxidation and nitrate reduction, also reduced viologens and ubiquinol interact with the enzyme at distinct sites [21]; <5> enzyme reacts with $FMNH_2$ at the inner aspect of the cytoplasmic membrane and can catalyze a vectorial reduction of nitrate on the outer aspect with reducing equivalents from the inner aspect [8]; <20> NarGHI: site of nitrate reduction is found in 127 kDa NarG, complex pattern of reversible nitrate concentration-dependent modulation of enzyme activity, catalysis under substrate-limiting conditions occurs via two pathways with distinct kinetic properties reversibly linked by a redox event [48]) (Reversibility: r <16> [34]; ir <8> [19]; ? <1-7, 9-15, 17-30> [1-10, 12-18, 20-33, 35-48]) [1-48]

P nitrite + acceptor <1-30> [1-48]

S nitrite + ferricyanide <16> (<16> nitrite oxidation occurs only with ferricyanide as electron acceptor, but not with oxygen [34]) (Reversibility: ? <16> [34]) [34]

P nitrate + ferrocyanide <16> [34]

S selenate + reduced benzyl viologen <5, 8, 9, 20, 30> (<5, 8, 9, 20, 30> both NAR and NAP can reduce selenate [41]; <5, 8, 9, 20, 30> assay in cell extract [41]; <9> assay with purified enzyme, catalytic efficiency is 300times lower than for nitrate [41]) (Reversibility: ? <5, 8, 9, 20, 30> [41]) [41]

P ?

S tellurite + reduced benzyl viologen <5, 8, 9, 20, 30> (<5, 8, 9, 20, 30> both NAR and NAP can reduce tellurite [41]; <5, 8, 9, 20, 30> assay in cell extract [41]; <9> assay with purified enzyme, catalytic efficiency is 200times lower than for nitrate [41]) (Reversibility: ? <5, 8, 9, 20, 30> [41]) [41]

P ?

S Additional information <2, 4, 5, 7, 8, 15, 18, 19, 25, 26> (<2, 4, 8> not as electron donors: NADH, NADPH [2, 4, 16]; <7> after solubilization and purification enzyme is no longer active with the natural electron donors NADH and formate [15]; <4> not as electron donor: $FADH_2$, $FMNH_2$, menadione [4]; <5> not as electron donors: 2,6-dichlorophenolindophenol and horse heart cytochrome c [24]; <19> not as electron donors: NADH, duroquinol, no reaction with ferredoxin [36]; <5, 15> the enzymes EC 1.7.99.4 and EC 1.9.6.1 are probably identical, in vivo cytochrome serves as electron donor in the electron transport chain to nitrate [32,33]; <5> in vitro artificial electron donors, e.g. reduced viologen indicators, can replace cytochrome c [31]; <18> structure is folded in 4 domains with an α/β-type topology, all are involved in cofactor binding [35]; <25> no enzyme activity in cell extracts when methyl viologen is used as electron donor instead of benzyl viologen [39]; <26> enzyme complex with both nitrate reductase and nitrite reductase activity, but no nitrite reductase activity in its native state, i.e. nondenaturing PAGE, enzyme uses nitrite as substrate when it is denatured and reduced with 2-mercaptoethanol [43]) [2, 4, 9, 15, 16, 24, 31-33, 35, 36, 39, 43]

P ?

Inhibitors

2-n-heptyl-4-hydroxyquinoline N-oxide <5, 8> (<8> only effective as inhibitor with the three-subunit enzyme and duroquinol as substrate [20]; <5> reversible inhibition of the quinol-dependent activity, but not of viologen dye-dependent activity, competitive to quinols [21]) [20, 21, 32]

IO_3^- <7> (<7> 1 mM, 40% inhibition [15]) [15]

$KClO_4$ <7> (<7> 1 mM, 33% inhibition [15]) [15]

NaCl <3, 19> (<3> 0.17 M NaCl + 0.5 mM $MgCl_2$ activates, 4.27 M NaCl + 0.5 mM $MgCl_2$ inhibits [3]; <19> 2 M NaCl activates, but concentrations higher than 2 M inhibit catalytic activity [36]) [3, 36]
ammonia <26> (<26> inhibits at high concentrations, e.g. 3.8 mM [43]) [43]
ammonium sulfate <5> (<5> ammonium sulfate and other ammonium salts inhibit [10]) [10]
azide <2, 4, 5, 8, 12, 13, 15, 20, 21, 26> (<2, 8, 21> competitive inhibition [2, 16, 20, 38]; <2> higher affinity for azide than for nitrate or chlorate [2]; <4, 5> strong inhibitor, competitive to nitrate [6, 8]; <4> uncompetitive to reduced benzyl viologen, completely reversible inhibition [6]; <5> azide-sensitive nitrate-reducing site of enzyme is located on the outer aspect of the cytoplasmic membrane [8]; <8> affinity for azide is one thousand times greater than for nitrate or chlorate [16]; <12> competitive to nitrate, non-competitive to benzyl viologen [27]; <13> membrane-bound enzyme form is highly sensitive, but not periplasmic form [29]; <15> NaN_3 at 30 mM completely inactivates [33]; <5, 8> NarGHI is inhibited by azide [42]; <26> NaN_3: at 1.16 mM: 35% inhibition, but stimulates at high concentrations, e.g. 3.8 mM [43]; <20> inhibitor of NarGH [48]) [2, 6, 8, 10, 16, 20, 27, 29, 32, 33, 38, 42, 43, 48]
bathophenanthroline <4> (<4> uncompetitive inhibition [6]) [6]
bathophenanthroline-sulfonate <4> (<4> uncompetitive inhibition [6]) [6]
bromate <7> (<7> competitive inhibitor, 1 mM, 50% inhibition [15]) [15]
chlorate <4, 7> (<4,7> competitive inhibitor [4, 15]; <7> 1 mM, 50% inhibition [15]) [4, 15]
cyanide <2, 4, 5, 7-9, 21, 26> (<4> mixed non-competitive to nitrate, only partly reversible inhibition, modification of a persulfide group in the vicinity of the Mo atom [6]; <7> 1 mM, 81% inhibition [15]; <8> 1 mM, 90-95% inhibition [16]; <9> 1 mM, about 80% inhibition [18]; <5> reduced enzyme is more sensitive at pH 7.2 than at pH 10.5, at pH 10.5 reversible enzyme-cyanide complexes are formed with reduced enzyme, nitrate protects reduced enzyme for inactivation, oxidized enzyme is less sensitive, kinetics of inactivation [24]; <21> non-competitive inhibition, reduces V_{max} of enzyme by 3fold [38]; <26> NaCN: complete inhibition [43]) [2, 6, 15, 16, 18, 24, 32, 38, 43]
dicoumarol <5> (<5> inhibition can be partially reversed by menadione [32]) [32]
diethyldicarbonate <5> (<5> irreversible inhibition of the quinol-dependent activity, but not of viologen dye-dependent activity [21]) [21]
dithiol <7> (<7> 0.1 mM, 95% inhibition [15]) [15]
dithiothreitol <5> (<5> inactivates at 1 mM [10]) [10]
ferricyanide <5> (<5> at pH 10.5 the as prepared enzyme is inactivated [24]) [24]
hydrogensulfite <15> (<15> 0.1 mM, complete inhibition [33]) [33]
mepacrine <7> (<7> 1 mM, 40% inhibition [15]) [15]
o-phenanthroline <7> (<7> 3 mM, 25% inhibition [15]) [15]

oxygen <5, 13> (<13> inhibition of activity in intact cells in presence of oxygen, membrane-bound form is inhibited by 90% under aerobic conditions [29]) [29, 32]
p-chloromercuribenzoate <4, 7, 15> (<4> above 0.1 mM non-competitive inhibition [4]; <7> 2 mM, 60% inhibition [15]; <15> 0.3 mM, completely inhibits, can be reversed by cysteine or glutathione [33]) [4, 15, 33]
sulfide <4, 5, 15> (<4> more strong inhibition of dithionite-reduced enzyme than of oxidized enzyme, nearly complete restoration of activity by 5 mM dithiothreitol, formation of iron-sulfide complexes [6]; <5> inactivates in absence of cyanide [24]; <15> Na_2S at 0.5 mM inactivates completely [33]) [6, 24, 33]
thiocyanate <4, 7> (<4> non-competitive to nitrate, completely reversible inhibition [6]; <7> 1.5 mM, 75% inhibition [15]) [6, 15]
Additional information <5, 8, 20, 26, 29> (<8> not inhibited by *o*-phenanthroline and EDTA [16]; <29> NAP activity in intact cells is insensitive towards ammonium, no inactivation by oxygen [40]; <5> napF operon-encoded enzyme activity and NapABC is not sensitive to azide [42]; <20> NapA and NapABC is not sensitive to azide [42]; <26> not inhibited by tungstate [43]) [16, 40-43]

Cofactors/prosthetic groups

bactopterin <11> (<11> common cofactor of eubacterial molybdoenzymes [23]) [23]
bis-molybdopterin guanine dinucleotide <18, 20> (<18> one bis-MGD cofactor in a single polypeptide chain of 723 amino acids, extends across the interior of the molecule interacting with residues from all 4 domains, catalytic molybdenum site is coordinated to two MGD cofactors, Cys140 and a water/hydroxo ligand [35]; <20> NAR and NAP bind the bis-MGD cofactor at the active site, NAP: di-oxo Mo(VI) state [37]; <18> NAP binds bis-MGD cofactor, oxo-transfer mechanism of this cofactor, mono-oxo Mo(VI) state [37]; <20> NarGHI and NapAB: Mo-bis-MGD containing, Mo-bis-MGD center is found in 127 kDa NarG [48]) [35, 37, 48]
cytochrome <15> (<15> presence of a bound cytochrome [33]) [33]
cytochrome b <5, 8, 21> (<8> enzyme contains a b-type cytochrome that is reduced by duroquinol and oxidized by nitrate, lack of the γ-subunit is correlated with the lack of cytochrome b, γ-subunit is a b-type cytochrome [20]; <5> in some preparations a γ-subunit is found that is a b-type cytochrome [13]; <5> cytochrome b containing holoenzyme, 0.8-1.7 mol per mol of enzyme [21]; <21> 0.49 mol cytochrome b per mol enzyme complex as cofactor, presence of a chlorate-oxidizable cytochrome b [38]; <5,8> NarGHI enzyme contains diheme cytochrome b-556 [42]) [13, 20, 21, 32, 38, 42]
cytochrome c <9, 16, 20, 26> (<16> c-type cytochrome is present [34]; <9> NapB, β-subunit: diheme cytochrome c [41]; <20> NapABC enzyme contains diheme cytochrome c-552, NapB, and a tetraheme cytochrome c, NapC [42]; <26> enzyme contains a low-midpoint potential, multiheme cytochrome c as the 62 kDa-subunit [43]) [34, 41-43]

heme <20> (<20> NarGHI: 25 kDa NarI is a diheme containing membrane anchor [48]) [48]
heme b <5, 8> (<8> presence of two b-type hemes in enzyme composed of three subunits, α, β and γ, absence of the γ-subunit is correlated with the absence of heme b [19]; <5> an additional 20 kDa subunit is present in heme-containing enzyme [9]) [9, 19]
heme c <20, 26> (<26> enzyme contains heme c in 62 kDa subunit [43]; <20> NapAB with 2 c-type hemes [48]) [43, 48]
iron-sulfur centre <2, 4, 5, 8, 10-12, 16, 18-21> (<2> enzyme contains 2 non-heme Fe atoms and 4 labile sulfide groups in acid medium per enzyme molecule [2]; <4, 11, 16> molybdo-iron-sulfur protein [4, 6, 34, 46]; <5> enzyme contains iron-sulfur centres [21, 31, 35]; <4> nitrate reductase I, MW 260000, abc2: 8 iron-sulfur groups per molecule of enzyme, which participate in electron transfer [6]; <8> one labile sulfide per iron atom [16]; <10> 6 atoms of non-heme iron and 12 mol of labile sulfide in 1 mol of purified enzyme [22]; <5> 16 g atoms of iron and 14 g atoms of acid-labile sulfide per mol of enzyme [24]; <12> 6.9 atoms of non-heme iron and 6.7 atoms of acid-labile sulfide per molecule of enzyme, iron-sulfur groups may participate in enzyme activity [27]; <18> one 4Fe-4S cluster in a single polypeptide chain of 723 amino acids, located near the periphery of the molecule [35]; <19> enzyme contains 4Fe-4S-type clusters as prosthetic cofactor, 4.49 mol Fe and 4.38 mol acid-labile sulfur per mol of the 63 kDa subunit [36]; <5,21> enzyme contains one 3Fe-4S cluster and three 4Fe-4S cluster [38, 41]; <5, 8> NarGHI enzyme contains five iron-sulfur clusters [42]; <20> NapABC enzyme contains one iron-sulfur cluster [42]; <11> 12.5-12.8 non-heme iron atoms and 8.7-12.1 acid-labile sulfur atoms per 172 kDa enzyme [46]; <20> NarGH and NapAB contain Fe-S cluster, 61 kDa NarH contains one 3Fe-4S and three 4Fe-4S clusters [48]) [2, 6, 16, 21, 22, 24, 27, 31, 34-36, 38, 41, 42, 46, 48]
molybdenum cofactor <11, 12, 19> (<12> 0.93 atoms of molybdenum per enzyme molecule, molybdenum seems to be a part of a low-molecular weight peptide, the Mo-cofactor, to which it may be bound by interaction with thiol groups [27]; <19> Mo-molybdopterin complex with Mo(V) as prosthetic cofactor [36]; <11> enzyme contains 1 mol of molybdenum cofactor per mol of enzyme, composed of 1 atom Mo and 1 mol MGD [46]) [27, 36, 46]
molybdopterin <5> (<5> NarG, α-subunit binds a molybdopterin cofactor [41]) [41]
molybdopterin guanine dinucleotide <5, 8, 11, 13, 14, 20, 27, 29> (<29> NapA binds molybdopterin guanine dinucleotide derivative of the molybdenum cofactor [40]; <5,8> NarGHI enzyme contains Mo-molybdopterin guanine dinucleotide [42]; <20> NapABC enzyme contains Mo-molybdopterin guanine dinucleotide [42]; <14> enzyme contains MGD cofactor: covalent attachment of GMP to the terminal phosphate group of molybdopterin via a pyrophosphate link [30]; <13> enzyme contains a GMP conjugate of the molybdopterin cofactor: molybdopterin guanine dinucleotide, molybdenum coordination environment in both Mo(VI) and Mo(IV) oxidation states [44]; <11> molybdopterin guanine dinucleotide is the organic moiety of the molybdenum cofactor, one mol MGD per mol of enzyme [46]; <27> enzyme

contains molybdopterin guanine dinucleotide cofactor [47]) [30, 40, 42, 44, 46, 47]
Additional information <1, 2, 4, 5, 8-10, 15, 19, 29> (<1, 4, 9, 10, 19> enzyme contains no cytochrome [1, 4, 18, 22, 36]; <5> enzyme is free of spectroscopically detectable cytochrome [12]; <5> no b-type cytochrome [7]; <2,5> non-heme enzyme [2, 9]; <2, 8> enzyme contains no FAD and FMN [2, 16]; <5, 15, 19> enzyme contains no flavin [26, 33, 36]; <8> water-soluble α-β-enzyme contains no b-type heme centre consistent with the lack of the γ-subunit [19]; <10> absence of cytochrome b_1 can be attributed to the isolation procedure, it is present up to gel filtration in Sephacryl 200 during purification [22]; <29> napB and napC genes with two and four c-type cytochrome consensus heme-binding motifs [40]) [1, 2, 4, 7, 9, 12, 16, 18, 19, 22, 26, 33, 36, 40]

Activating compounds

deoxycholate <4> (<4> activates [4]) [4]
ferricyanide <5> (<5> enzyme as prepared is slowly activated at pH 7.2 over 250% of the original activity, dependent on ferricyanide concentration, little activation at pH 9.5 [24]) [24]
menadione <5> (<5> stimulates [32]) [32]
p-chloromercuribenzoate <4> (<4> enhances enzyme activity at 0.1 mM and lower concentrations [4]) [4]
sodium azide <26> (<26> at 1.16 mM: 35% inhibition, but stimulates at high concentrations, e.g. 3.8 mM [43]) [43]
Additional information <5> (<5> NarGHI synthesis is maximally induced during anaerobic growth in the presence of nitrate [42]) [42]

Metals, ions

NaCl <3, 19> (<3> 0.17 M NaCl + 0.5 mM $MgCl_2$ activates, 4.27 M NaCl + 0.5 mM $MgCl_2$ inhibits [3]; <19> 2 M NaCl activates, but concentrations higher than 2 M inhibit catalytic activity [36]) [3, 36]
copper <16> [34]
iron <2, 4, 5, 8, 10-12, 15, 16, 19-21, 26> (<12> 6.9 atoms of non-heme iron per molecule of enzyme [27]; <5, 15, 16, 26> contains iron [9, 33, 34, 43]; <5> contains non-heme iron [10, 13]; <2> enzyme contains 2 Fe atoms per enzyme molecule, non-heme iron protein [2]; <4> molybdo-iron-sulfur protein [4,6]; <8> iron content about 8 mol per mol of enzyme, non-heme iron protein, iron plays an essential role in enzyme catalysis [16]; <10> 6 atoms of non-heme iron and 12 mol of labile sulfide in 1 mol of purified enzyme [22]; <5> 16 g atoms of iron per mol of enzyme [24]; <5, 20> enzyme contains iron-sulfur centres [21, 31, 48]; <4> nitrate reductase I and II: 8 iron-sulfur groups, nitrate reductase I: additional 4 tightly bound non-heme iron atoms per enzyme molecule, nitrate reductase II: no tightly bound iron [6]; <19> 4.49 mol Fe and 4.38 mol acid-labile sulfur per mol of the 63 kDa subunit [36]; <5> 40 atoms of bound Fe per mol [26]; <21> 15.4 mol iron per mol enzyme complex as cofactor [38]; <11> contains 12.5-12.8 non-heme iron atoms per 172 kDa enzyme [46]) [2, 4, 6, 9, 10, 13, 16, 22, 24, 26, 27, 31, 33, 34, 36, 38, 43, 46, 48]

iron-sulfur centre <2, 4, 5, 8, 10-12, 16, 18-21> (<2> enzyme contains 2 non-heme Fe atoms and 4 labile sulfide groups in acid medium per enzyme molecule [2]; <4,11,16> molybdo-iron-sulfur protein [4,6,34,46]; <5> enzyme contains iron-sulfur centres [21,31,35]; <4> nitrate reductase I, MW 260000, abc2: 8 iron-sulfur groups per molecule of enzyme, which participate in electron transfer [6]; <8> one labile sulfide per iron atom [16]; <10> 6 atoms of non-heme iron and 12 mol of labile sulfide in 1 mol of purified enzyme [22]; <5> 16 g atoms of iron and 14 g atoms of acid-labile sulfide per mol of enzyme [24]; <12> 6.9 atoms of non-heme iron and 6.7 atoms of acid-labile sulfide per molecule of enzyme, iron-sulfur groups may participate in enzyme activity [27]; <18> one 4Fe-4S cluster in a single polypeptide chain of 723 amino acids, located near the periphery of the molecule [35]; <19> enzyme contains 4Fe-4S-type clusters as prosthetic cofactor, 4.49 mol Fe and 4.38 mol acid-labile sulfur per mol of the 63 kDa subunit [36]; <5,21> enzyme contains one 3Fe-4S cluster and three 4Fe-4S cluster [38,41]; <5,8> NarGHI enzyme contains five iron-sulfur clusters [42]; <20> NapABC enzyme contains one iron-sulfur cluster [42]; <11> 12.5-12.8 non-heme iron atoms and 8.7-12.1 acid-labile sulfur atoms per 172 kDa enzyme [46]; <20> NarGH and NapAB contain Fe-S cluster, 61 kDa NarH contains one 3Fe-4S and three 4Fe-4S clusters [48]) [2, 6, 16, 21, 22, 24, 27, 31, 34-36, 38, 41, 42, 46, 48]
molybdenum <2, 4, 5, 8, 9, 11-13, 16, 18, 19-21, 29> (<2,5> enzyme contains 1 Mo atom per enzyme molecule [2,26]; <5,16> contains molybdenum [9,21,31,32,34]; <5,11,29> molybdenoenzyme [23,13,40]; <4> molybdenum containing iron-sulfur protein [4,6]; <8> traces of molybdenum [16]; <8> molybdenum center [17]; <4> nitrate reductase I, MW 260000, abc2: 0.24 atoms Mo per enzyme molecule, nitrate reductase II, MW 180000, 0.22 atoms Mo per enzyme molecule, Mo participates in electron transfer [6]; <5> 4 molecules of molybdenum per enzyme molecule, 1 Mo is associated with each pair of subunits, large plus small, Mo interacts directly with the substrate [10]; <9> 1 mol molybdenum per mol of enzyme [18]; <5> 0.8 g atom of molybdenum per mol of enzyme [24]; <12> 0.93 atoms of molybdenum per enzyme molecule, molybdenum seems to be a part of a low-molecular weight peptide to which it may be bound by interaction with thiol groups [27]; <18> molybdenum-containing enzyme, catalytic molybdenum site is coordinated to two MGD cofactors, Cys140 and a water/hydroxo ligand [35]; <19> Mo(V) in the Mo-molybdopterin complex, 0.33 mol Mo per mol of the 63 kDa subunit [36]; <21> 0.8 mol Mo per mol enzyme complex as cofactor [38]; <9> NapA, α-subunit is a molybdenum-containing catalytic subunit [41]; <13> molybdenum coordination environment in both Mo(VI) and Mo(IV) oxidation states, conserved Cys-181 of periplasmic enzyme is proposed to be part of Mo-binding site [44]; <11> contains 0.49-0.85 molybdenum atoms per 172 kDa enzyme [46]; <20> NarGH and NapAB: Mo-bis-MGD containing dimers, NarGHI: Mo ion cycles between its 6+ and 4+ oxidation states, mechanism [48]) [2, 4, 6, 9, 10, 13, 16-18, 21, 23, 24, 26, 27, 31, 32, 34-36, 38, 40, 41, 44, 46, 48]
tungsten <21> (<21> 0.06 mol tungsten per mol of enzyme complex [38]) [38]

Additional information <2, 4, 9, 17, 19, 26> (<2> NaCl activates not [2]; <4> no effect on enzyme activity by salts [4]; <9> no iron or cytochrome is found [18]; <17> Fe^{2+} and molybdate in growth medium promotes enzyme production [25]; <19> no tungsten is found [36]; <26> no molybdenum, vanadium, tungsten is detected [43]) [2, 4, 18, 25, 36, 43]

Turnover number (min^{-1})

4260 <19> (nitrate, <19> in absence of NaCl [36]) [36]
4620 <5> (bromate, <5> reduced methyl viologen as electron donor [21]) [21]
6000 <20> (nitrate) [48]
8700 <19> (nitrate, <19> in presence of 2 M NaCl [36]) [36]
19800 <5> (nitrate, <5> reduced benzyl viologen as electron donor [21]) [21]
27300 <5> (nitrate, <5> reduced methyl viologen as electron donor [21]) [21]
42240 <5> (chlorate, <5> reduced methyl viologen as electron donor [21]) [21]
69720 <21> (nitrate, <21> at 75°C [38]) [38]
80880 <21> (chlorate, <21> at 75°C [38]) [38]
Additional information <5> [26]

Specific activity (U/mg)

0.02-0.04 <5> (<5> rate of physiological reduction, lower than that obtained with reduced benzyl viologen [8]) [8]
0.076 <3> [3]
0.1 <16> (<16> nitrite oxidase activity [34]) [34]
0.23 <14> [30]
0.27 <9> (<9> selenate as substrate [41]) [41]
0.342 <29> (<29> methyl viologen as electron donor, aerobic growth conditions, assay in intact cells [40]) [40]
0.39 <26> (<26> reduced methyl viologen [43]) [43]
0.523 <29> (<29> benzyl viologen as electron donor, aerobic growth conditions, assay in intact cells [40]) [40]
0.7-1 <27> [47]
0.85 <7> [15]
1.147 <16> (<16> nitrate reductase activity [34]) [34]
1.407 <29> (<29> benzyl viologen as electron donor, anaerobic growth conditions, assay in intact cells [40]) [40]
1.484 <15> [33]
2.018 <29> (<29> methyl viologen as electron donor, anaerobic growth conditions, assay in intact cells [40]) [40]
2.157 <29> (<29> methyl viologen as electron donor, microaerobic growth conditions, assay in intact cells [40]) [40]
2.264 <29> (<29> benzyl viologen as electron donor, microaerobic growth conditions, assay in intact cells [40]) [40]
2.63 <17> (<17> enzyme in cell-free extract [25]) [25]
8 <20> (<20> NarGH, 30°C, pH 6 [48]) [48]
14.04 <9> [18]
19.3 <12> [27]

21 <1> [1]
24.5 <4> (<4> nitrate reductase I [4]) [4]
39 <9> (<9> nitrate as substrate [41]) [41]
39.1 <4> (<4> nitrate reductase II [4]) [4]
45-80 <5> [21]
50 <8> [20]
62.53 <10> [22]
65.67 <8> [16]
76.19 <5> [10]
80 <5> [24]
84.6 <19> [36]
160 <5> (<5> at 30°C, both enzyme forms, reduced benzyl viologen as reductant [7]) [7]
166.7 <2> (<2> at 37°C [2]) [2]
229 <5> (<5> at 30°C [9]) [9]
326 <21> (<21> nitrate as substrate, at 75°C [38]) [38]
378 <21> (<21> chlorate as substrate, at 75°C [38]) [38]
Additional information <2, 5, 8, 9, 11> [2, 5, 9, 18, 19, 23, 42]

K_m-Value (mM)

0.0002 <9> (reduced benzyl viologen) [18]
0.0019 <5> (nitrate, <5> duroquinol or ubiquinol 1 as electron donor [21]) [21]
0.013 <8> (nitrate, <8> with duroquinol or reduced methyl viologen as electron donor [20]) [20]
0.032 <26> (nitrate) [43]
0.057 <12> (reduced benzyl viologen) [27]
0.058 <21> (nitrate) [38]
0.07 <2> (reduced benzyl viologen, <2> nitrate as substrate [2]) [2]
0.079 <19> (nitrate, <19> in presence of 2 M NaCl [36]) [36]
0.1 <4> (nitrate) [4]
0.11 <12> (nitrate) [27]
0.12 <9> (nitrate) [41]
0.12 <2> (reduced benzyl viologen, <2> chlorate as substrate [2]) [2]
0.14 <21> (chlorate) [38]
0.182 <5> (reduced methyl viologen) [21]
0.2 <5> (nitrate) [8]
0.27 <9> (selenate) [41]
0.283 <8> (nitrate, <8> with reduced methyl viologen as electron donor [20]) [20]
0.33 <5> (nitrate, <5> reduced benzyl viologen as electron donor [21]) [21]
0.42 <5> (nitrate, <5> reduced methyl viologen as electron donor [21]) [21]
0.47 <8> (chlorate, <8> with duroquinol as electron donor [20]) [20]
0.47 <5> (nitrate) [24]
0.51 <5> (nitrate) [32]
0.6 <9> (nitrate) [18]
0.6 <9> (tellurite) [41]

0.67 <7> (nitrate) [15]
0.8 <4> (nitrate, <4> oxidized enzyme [6]) [6]
0.8 <2> (reduced methyl viologen, <2> nitrate as substrate [2]) [2]
0.9 <16> (nitrate, <16> nitrate reduction with reduced methyl or benzyl viologen [34]) [34]
1.3 <2> (nitrate) [2]
1.5 <2> (reduced methyl viologen, <2> chlorate as substrate [2]) [2]
1.6 <4> (reduced benzyl viologen, <4> nitrate as substrate [4]) [4]
3.6 <16> (nitrate) [34]
3.6 <16> (nitrite, <16> nitrite oxidation with ferricyanide [34]) [34]
5 <2> (chlorate) [2]
5 <4> (nitrate, <4> reduced enzyme [6]) [6]
Additional information <5, 8, 15, 19, 20> (<5> apparent K_m for reduction of nitrate by the respiratory chain is 10fold less than that when the reductant is reduced benzyl viologen [8]; <5> much higher K_m for bromate than for nitrate or chlorate as substrate [21]; <19> K_m for nitrate decreases continously with an increase of NaCl [36]) [8, 16, 21, 33, 36, 48]

K_i-Value (mM)

0.00013 <12> (azide, <12> nitrate as substrate [27]) [27]
0.0002 <5> (azide, <5> when assayed spectrophotometrically [8]) [8]
0.0003 <4> (azide, <4> nitrate as substrate [6]) [6]
0.0011 <2> (azide, <2> chlorate as substrate [2]) [2]
0.002 <2> (azide, <2> nitrite as substrate [2]) [2]
0.002-0.005 <5> (azide, <5> when assayed manometrically [8]) [8]
0.01 <4> (azide, <4> reduced benzyl viologen as substrate [6]) [6]
0.0173 <12> (azide, <12> reduced benzyl viologen as substrate [27]) [27]
0.031 <21> (azide) [38]
0.044 <5> (cyanide, <5> at pH 10.5, reduced enzyme [24]) [24]
0.16 <4> (sulfide, <4> nitrate as substrate, reduced enzyme [6]) [6]
0.28 <4> (sulfide, <4> nitrate as substrate, oxidized enzyme [6]) [6]
0.3 <4> (cyanide, <4> nitrate as substrate [6]) [6]
0.5 <4> (bathophenanthroline, <4> nitrate as substrate [6]) [6]
0.55 <8> (azide) [20]
0.8 <4> (bathophenanthroline, <4> reduced benzyl viologen as substrate [6]) [6]
1.3 <4> (bathophenanthroline-sulfonate, <4> nitrate as substrate [6]) [6]
1.3 <4> (chlorate) [4]
1.4 <4> (thiocyanate, <4> nitrate as substrate [6]) [6]
1.7 <7> (chlorate) [15]
2.4 <4> (bathophenanthroline-sulfonate, <4> reduced benzyl viologen as substrate [6]) [6]

pH-Optimum

5.6-5.8 <8> (<8> nitrate as substrate [16]) [16]
6 <20> (<20> maximal activity of NarGH [48]) [48]
6.2-6.4 <8> (<8> chlorate as substrate [16]) [16]
6.3 <2> (<2> nitrate as substrate [2]) [2]

6.4 <2> (<2> chlorate as substrate [2]) [2]
6.5 <21> [38]
6.8 <4, 5> (<4> broad pH-optimum from pH 5.8 to pH 7.1 with maximum at pH 6.8 [4]; <5> assay at [21]) [4, 21]
7 <8, 16, 26> (<8> assay at [16]; <16> nitrate reduction with reduced acceptor [34]) [16, 34, 43]
7-8.5 <15> (<15> same rate between [33]) [33]
7.1 <5, 17> (<17> nitrate as substrate [25]) [25, 26]
8 <7, 16> (<16> nitrite oxidation with ferricyanide [34]) [15, 34]

pH-Range
5-11 <20> (<20> no systematic variations of electroactive film coverage across the pH range [48]) [48]
5.5-10 <15> (<15> sharp drop of activity below pH 5.5 and above pH 10 [33]) [33]

Temperature optimum (°C)
23 <5> (<5> assay at [10]) [10]
25 <5> (<5> assay at [21]) [21]
30 <5, 8, 9, 16, 17, 29> (<5,8,9,16,17,29> assay at [5,9,18,20,24,25,34,40,41]) [5, 9, 18, 20, 24, 25, 34, 40, 41]
37 <2, 8> (<2,8> assay at [2,16]) [2, 16]
50 <26> [43]
56 <3> (<3> presence of 0.5 mM $MgCl_2$ [3]) [3]
73 <3> (<3> presence of 2 M KCl [3]) [3]
75 <17> (<17> optimum of 5 min reaction [25]) [25]
80 <27> (<27> assay at [47]) [47]
85 <3> (<3> presence of 4.27 M NaCl [3]) [3]
95 <21> (<21> highest activity at or above [38]) [38]
Additional information <3> (<3> temperature optimum is a function of both: the concentration and the specific cation present, increasing NaCl and KCl concentrations result in an increase in the maximal activity at higher temperatures [3]) [3]

Temperature range (°C)
Additional information <3> [3]

4 Enzyme Structure

Molecular weight
100000 <15> [33]
106000 <5> [26]
160000 <8> (<8> gel filtration, ultracentifugation [16]) [16]
165000 <2> (<2> mean value, real value between 155000 and 175000, PAGE [2]) [2]
172000 <11> [46]
176000 <12> (<12> analytical ultracentrifugation [27]) [27]

180000 <4, 9> (<4> nitrate reductase II, monomeric form, gel filtration [4]; <9> gel filtration of urea-treated enzyme [18]) [4, 18]
196000 <12> (<12> always present in a monomeric form, PAGE [27]) [27]
200000 <1, 5, 8> (<1> 8.5 S monomeric state, gel filtration [1]; <8> water-soluble α-β-enzyme, gel filtration [19]; <5> in presence of deoxycholate, dissociated form, gel filtration [24]) [1, 19, 24]
208000 <19> (<19> gel filtration [36]) [36]
210000 <10> (<10> PAGE [22]) [22]
214000 <21> (<21> three-subunit complex, calculated as the sum of the MWs of the subunits [38]) [38]
220000 <5> (<5> monomer form of the enzyme, analytical ultracentrifugation in presence of 0.2% deoxycholate [9]) [9]
235000 <5> (<5> minimum value, calculated as the sum of the MWs of the three subunits [21]) [21]
260000 <4> (<4> nitrate reductase I, abc2, monomeric form, gel filtration [4]) [4]
290000 <5> (<5> one of the two enzyme forms with different subunit compositions [7]; <5> gel filtration [7,9]) [7, 9]
320000 <5> (<5> in absence of deoxycholate, associated form, gel filtration [24]) [24]
360000 <16> (<16> nitrite oxidoreductase enzyme complex, gel filtration [34]) [34]
400000 <1> (<1> 13.9 S dimeric state, gel filtration [1]) [1]
620000 <5> (<5> one of the two enzyme forms with different subunit compositions [7]; <5> gel filtration [7,9]) [7, 9]
720000 <5> (<5> gel filtration [10]) [10]
773000 <5> (<5> analytical ultracentrifugation [10]) [10]
880000 <5> (<5> associated, probably tetrameric, form of enzyme [9]) [9]
1000000 <5> [14, 32]
1060000 <4> (<4> nitrate reductase I, $(abc2)_4$, tetrameric form, gel filtration [4]) [4]

Subunits

? <5> (<5> x * 150000 α + x * 67000 β_1 + x * 65000 β_2, molar ratio α:β_1 + β_2 is 1:1, SDS-PAGE [7,9]; <5> an additional 20 kDa subunit is present in heme-containing enzyme [9]; <5> x * 150000 α + x * 59000 β [12]; <5> x * 150000 α + x * 60000 β + x * 19000 γ, α-subunit with catalytic function containing molybdenum cofactor and non-heme iron, β-subunit is possibly involved in membrane binding, γ-subunit is found in some preparations and is a b-type cytochrome [13]; <5> x * 155000 α + x * 60000 β + x * 20000 γ, γ-subunit is a b-type cytochrome, SDS-PAGE [21]; <5> x * 150000 + x * 58000 + x * 42000, SDS-PAGE [24]; <5> x * 150000 α + x * 60000 β + x * 20000 γ, that is the apoprotein of cytochrome c_{556}, α-subunit is involved in catalysis [31]) [7, 9, 12, 13, 21, 24, 31]
dimer <4, 8-10, 12, 20, 24> (<12> α,β, 1 * 150000 + 1 * 57000, SDS-PAGE [27,28]; <4> 1 * 117000 + 1 * 57000, nitrate reductase II, SDS-PAGE [4,6]; <9> α,β, 1 * 120000 + 1 * 60000, α subunit: catalytic subunit, β subunit: a

membrane attachment protein, SDS-PAGE [18]; <8> α,β, 1 * 127000 + 1 * 61000, purified water-soluble enzyme, membrane-bound enzyme contains an additional γ-subunit, SDS-PAGE [19]; <10> α,β, 1 * 150000 + 1 * 44000, SDS-PAGE [22]; <24> 1 * 116000 + 1 * 60000 [36]; <9> α,β, 1 * 91000 + 1 * 17000, α: NapA, molybdenum-containing catalytic subunit, β: NapB, diheme cytochrome c, SDS-PAGE [41]; <20> 1 * 120000 NarG + 1 * 60000 NarH, soluble NarGH, SDS-PAGE [48]) [4, 6, 18, 19, 22, 27, 28, 36, 41, 48]
hexadecamer <4> (<4> 4 * 117000 + 4 * 57000 + 8 * 52000, nitrate reductase II, SDS-PAGE [4]) [4]
homotetramer <19> (<19> 4 * 63000, SDS-PAGE [36]) [36]
octamer <5> (<5> 4 * 142000 + 4 * 58000, subunits are probably associated in form of a double tetrahedron, SDS-PAGE [10]) [10]
pentamer <16> (<16> α_2,β_2,γ, 2 * 116000 + 2 * 65000 + 1 * 32000, nitrite reductase enzyme complex, SDS-PAGE [34]) [34]
tetramer <4, 26> (<4> 1 * 117000 + 1 * 57000 + 2 * 52000, nitrate reductase I with subunit structure abc2, 52 kDa subunit may have structural function, SDS-PAGE [4,6]; <26> 1 * 62000 + 1 * 52000 + 1 * 36000 + 1 * 16000, enzyme complex with nitrate and nitrite reductase activity, 62kDa-subunit is a multiheme cytochrome c and separation of this subunit from the 3 non-heme polypeptides causes a loss of nitrate/nitrite reductase activity, SDS-PAGE [43]) [4, 6, 43]
trimer <5, 8, 20, 21, 23> (<8> α,β,γ, 1 * 127000 + 1 * 61000 + 1 * 21000, SDS-PAGE [20]; <21> 1 * 130000 + 1 * 52000 + 1 * 32000, SDS-PAGE [38]; <23> 1 * 100000 + 1 * 60000 + 1 * 31000 [36,38]; <5> α,β,γ, 1 * 112000-140000 + 1 * 52000-64000 + 1 * 19000-25000, α: NarG, catalytic subunit with a molybdopterin cofactor, β: NarH, with one 3Fe-4S and three 4Fe-4S centers, γ: NarI, membrane diheme b quinol-oxidizing subunit [41]; <20> 1 * 127000 NarG + 1 * 61000 NarH + 1 * 25000 NarI, membrane-anchored NarGHI [48]) [20, 36, 38, 41, 48]
Additional information <20> (<20> NAR is a three-subunit complex, NAP is a two-subunit complex [37]) [37]

Posttranslational modification

no glycoprotein <12> [27]
proteolytic modification <29> (<29> NapB is characterized by a N-terminal signal peptide, cleavage site is predicted to be between two alanine residues at positions 25 and 26 [40]) [40]
Additional information <5> (<5> no proteolytic processing at the N-terminal end except for removal of N-formylmethionine [13]) [13]

5 Isolation/Preparation/Mutation/Application

Localization

cytoplasmic membrane <5, 7-9, 11, 12, 19-21, 27> (<5> β-subunit is located at the internal surface of the cytoplasmic membrane, α-subunit is similarly located at the cytoplasmic face of the membrane [12]; <5,9> bound to cytoplasmic membrane [9,10,18]; <12> located solely on the cytoplasmic surface of the membrane [27]; <7> in cell-membrane fraction [15]; <8> α- and β-subunit of purified water-soluble enzyme are exposed on the cytoplasmic surface of the cytoplasmic membrane, they are peripheral polypeptides that are attached to the membrane [19]; <12> the 2 subunits are localized on the cytoplasmic side of the membrane [28]; <19> membrane-extrinsic protein that combines with the surface of the cytoplasmic membrane by hydrophobic interactions [36]; <20> NAR: three-subunit complex anchored to the cytoplasmic face of the membrane with its active site located in the cytoplasmic compartment [37]; <27> NarC, a cytochrome c, is required for the attachment of the α subunit to the cytoplasmic membrane [47]) [9, 10, 12, 15, 18, 19, 23, 27, 28, 36-38, 47]

integral to plasma membrane <4, 5, 8> (<5> incorporation into the cytoplasmic membrane, enzyme site that oxidizes $FMNH_2$ is located on the cytoplasmic aspect of cytoplasmic membrane [5]; <5> nitrate reductase spans the cytoplasmic membrane and is vectorial, reducing nitrate on the outer aspect of the membrane with 2 H^+ and 2 electrons that have crossed from the inner aspect of the membrane [8]; <5,8> α- and β-subunits are exposed on the cytoplasmic surface of the cytoplasmic membrane, they are peripheral polypeptides that are attached to the membrane presumably via the γ-subunit, which is embedded to the membrane [19,31]; <5> transmembranous location within the cytoplasmic membrane, γ-subunit is located at the periplasmic face and α-, β-subunit at the cytoplasmic face [21]; <4> transmembrane protein of cytoplasmic membrane [28]) [5, 8, 19, 21, 28, 31]

membrane <1, 4, 5, 13, 16, 17, 20, 23, 25, 26> (<1, 4, 5, 16, 17, 23, 25> membrane-bound [1,4,13,24,25,34,36,38,39,41]; <4> solubilized from membrane [4]; <5> in membrane fraction [7]; <13> a periplasmic enzyme is expressed under both aerobic and anaerobic conditions and a membrane-bound enzyme is synthesized only under anaerobic conditions [29]; <5> two membrane-bound proton-translocating enzymes encoded by the operons narGHJI and narZYWV [42]; <26> solubilized from membrane fractions, membrane-associated activity [43]; <20> NarGHI is a membrane-anchored protein, NarGH, liberated from the membrane, retained its enzyme activity [48]) [1, 4, 7, 13, 24, 25, 29, 34, 36, 38, 39, 41-43, 48]

periplasm <5, 8, 9, 13, 14, 18, 20, 29> (<13> a periplasmic enzyme is expressed under both aerobic and anaerobic conditions and a membrane-bound enzyme is synthesized only under anaerobic conditions [29]; <18,20> NAP: two-subunit complex is located in the periplasmic compartment [37]; <5> NapABC enzyme is periplasmic [42]; <20> NapAB [48]) [29, 30, 35, 37, 40-42, 44, 48]

soluble <15, 24, 27> (<24> it may exist an additional enzyme that associates the soluble enzyme with the cytoplasmic membrane [38]; <27> narC:kat mutant with soluble NarG [47]) [33, 38, 47]

Purification

<1> (100fold purification [1]) [1]

<2> (40fold purification [2]) [2]

<3> (partial purification [3]) [3]
<4> (2 forms: nitrate reductase I: 36.6fold purification, nitrate reductase II: 58.4fold purification [4]; 2 forms: nitrate reductase I, nitrate reductase II [6]) [4, 6, 28]
<5> (2 enzyme forms with different subunit compositions [7]; 134fold purification [9]; 112fold purification [10]; strain X5119 [13]; 140fold purification [24]) [7, 9, 10, 12, 13, 21, 24, 26]
<7> (56fold purification [15]) [15]
<8> (50fold purification [16]; partial [19,20]; purified water-soluble enzyme from cytoplasmic membrane contains α- and β-subunit which are released from membrane, but not the unreleased γ-subunit [19]; γ-subunit of enzyme is lost during gel filtration purification, resulting in a cytochrome-free enzyme [20]) [16, 19, 20, 45]
<9> (120fold purification [18]; purification of His-tagged NAP: NapAB-6His [41]) [18, 41]
<10> (53fold purification, $\alpha\beta$ enzyme, cytochrome b_1 is present up to gel filtration in Sephacryl 200 during purification [22]) [22]
<12> (85fold purification [27]) [27, 28]
<13> [44]
<15> [33]
<16> (21fold purification [34]) [34]
<19> (34.3fold purification [36]) [36]
<20> (purification of soluble NarGH [48]) [48]
<21> (137fold purification [38]) [38]
<25> (purification of the catalytic α-subunit of nitrate reductase A [39]) [39]
<26> (partial purification [43]) [43]

Crystallization
<9> [41]
<18> [37]

Cloning
<5> (3 sets of nitrate reductase genes: narGHJI operon encoding NarGHI enzyme, narZYWV operon encoding NarZYV enzyme and napFDAGHBC encoding NapABC enzyme and associated electron transfer components [42]) [42]
<5> (cloning of the narG gene, encoding the large α-subunit gene of enzyme, nucleotide sequence of part of nar DNA and sequence of N-terminal 147 amino acids of the α-subunit [13]) [13]
<27> (NarC, a 27 kDa membrane cytochrome c, is encoded as the first gene of the narCGHJIK1K2 operon for nitrate respiration and plays an essential role in the synthesis of active enzyme and for the attachment of enzyme to the membrane, sequence of narC, NarG is unable to bind to the cytoplasmic membrane in absence of NarC [47]) [47]
<29> (napABC genes, encoding periplasmic nitrate reductase, are isolated and sequenced, NapB has 160 amino acides, NapC with 206 amino acids and a hydrophobic membrane-spanning domain near its N-terminus [40]) [40]

<5, 22> (organisation of the gene clusters of the membrane-bound nitrate reductase, NAR [37]) [37]
<5, 8, 9> (organisation of the gene clusters of the periplasmic nitrate reductase, NAP [37]) [37]

Engineering

Additional information <9, 14, 25, 27, 29> (<29> strain FAJ0164: kanamycin-resistant napA insertion mutant expresses assimilatory nitrate reductase, but no NAP and therefore no methyl viologen- or benzyl viologen-dependent nitrate reductase activity [40]; <9> MS523 mutant strain without synthesis of NAP [41]; <25> strain LP59JG: isogenic mutant deficient in the ability to synthesize the respiratory nitrate reductase, obtained by allelic exchange of a gentamicin resistance gene in the narG gene encoding the catalytic subunit of enzyme [11,39]; <14> strains SL113I and SL113II, mobA mutant strains of KS36, with complete loss of nitrate reductase activity [30]; <27> narC::kat mutants without NR activity expresses constitutively NarG and NarJ, but NarG appears in the soluble fraction instead of associated with the membranes [47]) [11, 30, 39-41, 47]

Application

nutrition <5, 8> (<5,8> supervision of drinking water, enzyme is used for a method for determination of nitrate concentration [45]) [45]

6 Stability

pH-Stability

6.5-10 <17> (<17> stable [25]) [25]

Temperature stability

6 <5> (<5> immobilized enzyme more stable at 6°C than at 23°C [24]) [14]
20 <4> (<4> 24 h, complete loss of activity [4]) [4]
23 <5> (<5> immobilized enzyme more stable at 6°C than at 23°C [14]) [14]
37-45 <21> (<21> 10% of optimal activity [38]) [38]
50 <5, 15, 17> (<5> 1 h, enzyme concentration: 0.1 mg per ml, no inhibition [10]; <15> 10 min, stable [33]; <17> 20 h, stable up to [25]) [10, 25, 33]
60 <5, 15, 17> (<5> after 10 min and 5 h same activity observed, remarkably stable [24]; <15> 10 min, 10-15% loss of activity [33]; <17> 20 h, 15% loss of activity [25]) [24, 25, 33]
70 <15> (<15> 5 min, complete and irreversible loss of activity [33]) [33]
80 <3> (<3> 0.17 M NaCl: 90% loss of activity after 1 min, 0.85 M NaCl: 50% loss of activity after 7 min, 4.27-5.31 M NaCl: no loss of activity after 15 min, high salt concentrations protect enzyme against heat inactivation at 80°C due to a tighter, more stable configuration [3]) [3]
100 <21> (<21> half-life of 1.5 h, half-life within cell-membranes is 6 h, lipid environment stabilizes [38]) [38]

General stability information

<3>, NaCl protects against heat inactivation, 4.27-5.31 M NaCl [3]
<4>, nitrate reductase II, which lacks the 52 kDa subunit, is much more labile than nitrate reductase I, deoxycholate stabilizes nitrate reductase II [4]
<5>, freezing rapidly inactivates enzyme at a concentration of 0.1 mg per ml, 0.1 mM dithiothreitol stabilizes [10]
<5>, immobilized enzyme more labile than free enzyme [14]
<5>, slow freezing inactivates, but rapid freezing in liquid N_2 and thawing at room temperature can be repeated 10 times without effect on enzyme activity, when reduced benzyl viologen is used as reductant [9]
<12>, deoxycholate stabilizes [27]
<15>, stable to prolonged dialysis [33]
<17>, sucrose stabilizes [25]
<19>, stable in absence of NaCl [36]

Storage stability

<4>, 20°C, 0.07 M imidazole buffer, pH 8.1, 24 h, nitrate reductase I: 45% loss of activity, probably due to loss of 52 kDa subunit and conversion to nitrate reductase II, nitrate reductase II: 95% loss of activity [4]
<4>, 4°C, 0.07 M imidazole buffer, pH 8.1, 24 h, nitrate reductase I: 10% loss of activity, nitrate reductase II: 35% loss of activity [4]
<5>, 4°C, phosphate buffer, pH 7.2 or Tris-HCl buffer, pH 8.0, enzyme concentration less than 0.5 mg/ml: half live is 5 days under strict anaerobic conditions and 3 days in air, enzyme concentration above 10 mg/ml: under argon, 20 days, no loss of activity [24]
<5>, 4°C, prolonged storage, stable, when assayed with reduced benzyl viologen as reductant [9]
<5>, liquid N_2 temperature, 6 months, stable [24]
<5>, rapidly frozen in liquid N_2, several months, no loss of activity [21]
<8>, 0°C and 4°C, 0.1 M potassium phosphate buffer, pH 6.5, ammonium sulfate 10% of saturation, several weeks [16]
<8>, 4°C, 3-4 weeks, remains active [20]
<12>, 4°C, aerobically or anaerobically storage, absence of deoxycholate: per 24 h, 30% inactivation, presence of 0.2% deoxycholate: per 24 h, 10% inactivation [27]
<15>, frozen, several months, stable [33]
<17>, 5°C, 0.1 M Tris-HCl buffer, pH 8.8, 10% sucrose, 24 h, 10% loss of activity, without sucrose: 75% loss of activity [25]
<21>, room-temperature, many weeks, stable [38]

References

[1] Van 'T Riet, J.; Planta, R.J.: Purification and some properties of the membrane-bound respiratory nitrate reductase of Aerobacter aerogenes. FEBS Lett., 5, 249-252 (1969)

[2] Rosso, J.P.; Forget, P.; Pichinoty, F.: Bacterial nitrate reductases. Solubilization, purification and properties of the enzyme A of Micrococcus halodenitrificans. Biochim. Biophys. Acta, **321**, 443-455 (1973)
[3] Marquez, E.D.; Brodie, A.F.: The effect of cations on the heat stability of a halophilic nitrate reductase. Biochim. Biophys. Acta, **321**, 84-89 (1973)
[4] Van 'T Riet, J.; Planta, R.J.: Purification, structure and properties of the respiratory nitrate reductase of Klebsiella aerogenes. Biochim. Biophys. Acta, **379**, 81-94 (1975)
[5] Kemp, M.B.; Haddock, B.A.; Garland, P.B.: Synthesis and sidedness of membrane-bound respiratory nitrate reductase (EC 1.7.99.4) in Escherichia coli lacking cytochromes. Biochem. J., **148**, 329-333 (1975)
[6] Van 'T Riet, J.; Van Ee, J.H.; Wever, R.; Van Gelder, B.F.; Planta, R.J.: Characterization of the respiratory nitrate reductase of Klebsiella aerogenes as a molybdenum-containing iron-sulfur enzyme. Biochim. Biophys. Acta, **405**, 306-317 (1975)
[7] Clegg, R.A.: The size of nitrate reductase in Escherichia coli. Biochem. Soc. Trans., **3**, 691-694 (1975)
[8] Garland, P.B.; Downie, J.A.; Haddock, B.A.: Proton translocation and the respiratory nitrate reductase of Escherichia coli. Biochem. J., **152**, 547-559 (1975)
[9] Clegg, R.A.: Purification and some properties of nitrate reductase (EC 1.7.99.4) from Escherichia coli K12. Biochem. J., **153**, 533-541 (1976)
[10] MacGregor, C.H.; Schnaitman, C.A.; Normasell, D.E.; Hodgins, M.G.: Purification and properties of nitrate reductase from Escherichia coli K12. J. Biol. Chem., **249**, 5321-5327 (1974)
[11] Ghiglione, J.F.; Gourbierre, F.; Potier, P.; Philippot, L.; Lensi, R.: Role of respiratory nitrate reductase in ability of Pseudomonas fluorescens YT101 to colonize the rhizosphere of maize. Appl. Environ. Microbiol., **66**, 4012-4016 (2000)
[12] Graham, A.; Boxer, D.H.: Arrangement of respiratory nitrate reductase in the cytoplasmic membrane of Escherichia coli. Location of β subunit. FEBS Lett., **113**, 15-20 (1980)
[13] McPherson, M.J.; Baron, A.J.; Pappin, D.J.C.; Wootton, J.C.: Respiratory nitrate reductase of Escherichia coli. Sequence identification of the large subunit gene. FEBS Lett., **177**, 260-264 (1984)
[14] Schiller, J.G.; Liu, C.C.: Immobilization of nitrate reductase within polyacrylamide gels. Biotechnol. Bioeng., **18**, 1643-1645 (1976)
[15] Radcliffe, B.C.; Nicholas, D.J.D.: Some properties of a nitrate reductase from Pdeudomonas denitrificans. Biochim. Biophys. Acta, **205**, 273-287 (1970)
[16] Forget, P.: The bacterial nitrate reductases. Solubilization, purification and properties of the enzyme from Micrococcus denitrificans. Eur. J. Biochem., **18**, 442-450 (1971)
[17] Turner, N.; Ballard, A.L.; Bray, R.C.; Ferguson, S.: Investigation by electron paramagnetic resonance spectroscopy of the molybdenum centre of respiratory nitrate reductase from Paracoccus denitrificans. Biochem. J., **252**, 925-926 (1988)

[18] Byrne, M.D.; Nicholas, D.J.D.: A membrane-bound dissimilatory nitrate reductase from Rhodobacter sphaeroides f.sp. denitrificans. Biochim. Biophys. Acta, **915**, 120-124 (1987)

[19] Ballard, A.L.; Ferguson, S.J.: Respiratory nitrate reductase from Paracoccus denitrificans. Evidence for two b-type haems in the γ subunit and properties of a water-soluble active enzyme containing α and β subunits. Eur. J. Biochem., **174**, 207-212 (1988)

[20] Craske, A.; Ferguson, S.J.: The respiratory nitrate reductase from Paracoccus denitrificans. Molecular characterisation and kinetic properties. Eur. J. Biochem., **158**, 429-436 (1986)

[21] Morpeth, F.F.; Boxer, D.H.: Kinetic analysis of respiratory nitrate reductase from Escherichia coli K12. Biochemistry, **24**, 40-46 (1985)

[22] Chikwem, J.O.; Downey, R.J.: Purification and characterization of the respiratory nitrate reductase of Bacillus stearothermophilus. Anal. Biochem., **126**, 74-80 (1982)

[23] Kruger, B.; Meyer, O.; Nagel, M.; Andreesen, J.R.; Meincke, M.; Bock, E.; Blumle, S.; Zumft, W.G.: Evidence for the presence of bactopterin in the eubacterial molybdoenzymes nicotinic acid dehydrogenase, nitrite oxidoreductase, and respiratory nitrate reductase. FEMS Microbiol. Lett., **48**, 225-227 (1987)

[24] Adams, M.W.W.; Mortenson, L.E.: The effect of cyanide and ferricyanide on the activity of the dissimilatory nitrate reductase of Escherichia coli. J. Biol. Chem., **257**, 1791-1799 (1982)

[25] Aoki, K.; Shinke, R.; Nishira, H.: Isolation and identification of respiratory nitrate reductase-producing bacteria from soil and production of the enzyme. Agric. Biol. Chem., **45**, 817-822 (1981)

[26] Taniguchi, S.; Itagaki, E.: Nitrate reductase of nitrate respiration type from Escherichia coli. I. Solubilization and purification from the particulate system with molecular characterization as a metalloprotein. Biochim. Biophys. Acta, **44**, 263-279 (1960)

[27] Van 'T Riet, J.; Wientjes, F.B.; Van Doorn,J.; Planta, R.J.: Purification and characterization of the respiratory nitrate reductase of Bacillus licheniformis. Biochim. Biophys. Acta, **576**, 347-360 (1979)

[28] Wientjes, F.B.; Kolk, A.H.J.; Nanninga, N.; Van 'T Riet, J.: Respiratory nitrate reductase: Its localization in the cytoplasmic membrane of Klebsiella aerogenes and Bacillus licheniformis. Eur. J. Biochem., **95**, 61-67 (1979)

[29] Bell, L.C.; Richardson, D.J.; Ferguson, S.J.: Periplasmic and membrane-bound respiratory nitrate reductases in Thiosphaera pantotropha. The periplasmic enzyme catalyzes the first step in aerobic denitrification. FEBS Lett., **265**, 85-87 (1990)

[30] Leimkuhler, S.; Klipp, W.: The molybdenum cofactor biosynthesis protein MobA from Rhodobacter capsulatus is required for the activity of molybdenum enzymes containing MGD, but not for xanthine dehydrogenase harboring the MPT cofactor. FEMS Microbiol. Lett., **174**, 239-246 (1999)

[31] Boxer, D.; Malcolm, A.; Graham, A.: Escherichia coli formate to nitrate respiratory pathway: structural analysis. Biochem. Soc. Trans., **10**, 480-481 (1982)

[32] Nason, A.: Nitrate reductases. The Enzymes, 2nd Ed.(Boyer, P.D., Lardy, H., Myrbäck, K., eds.), **7**, 587-607 (1963)
[33] Sadana, J.C.; McElroy, W.D.: Nitrate reductase from Achromobacter fischeri. Purification and properties: Function of flavines and cytochrome. Arch. Biochem. Biophys., **67**, 16-34 (1957)
[34] Sundermeyer-Klinger, H.; Meyer, W.; Warninghoff, B.; Bock, E.: Membrane-bound nitrite oxidoreductase of Nitrobacter: Evidence for a nitrate reductase system. Arch. Microbiol., **140**, 153-158 (1984)
[35] Dias, J.M.; Than, M.E.; Humm, A.; Huber, R.; Bourenkov, G.P.; Bartunik, H.D.; Bursakov, S.; Calvete, J.; Caldeira, J.; Carneiro, C.; Moura, I.; Romao, M.J.: Crystal structure of the first dissimilatory nitrate reductase at 1.9 A solved by MAD methods. Structure Fold Des., **7**, 65-79 (1999)
[36] Yoshimatsu, K.; Sakurai, T.; Fujiwara, T: Purification and characterization of dissimilatory nitrate reductase from a denitrifying halophilic archaeon, Haloarcula marismortui. FEBS Lett., **470**, 216-220 (2000)
[37] Richardson, D.J.; Berks, B.C.; Russell, D.A.; Spiro, S.; Taylor, C.J.: Functional, biochemical and genetic diversity of prokaryotic nitrate reductases. Cell. Mol. Life Sci., **58**, 165-178 (2001)
[38] Afshar, S.; Johnson, E.; DeVries, S.; Schröder, I.: Properties of a thermostable nitrate reductase from the hyperthermophilic archaeon Pyrobaculum aerophilum. J. Bacteriol., **183**, 5491-5495 (2001)
[39] Ghiglione, J.F.; Philippot, L.; Normand, P.; Lensi, R.; Potier, P.: Disruption of narG, the gene encoding the catalytic subunit of respiratory nitrate reductase, also affects nitrite respiration in Pseudomonas flourescens YT101. J. Bacteriol., **181**, 5099-5102 (1999)
[40] Steenhoudt, O.; Keijers, V.; Okon, Y.; Vanderleyden, J.: Identification and characterization of a periplasmic nitrate reductase in Azospirillum brasilense Sp245. Arch. Microbiol., **175**, 344-352 (2001)
[41] Sabaty, M.; Avazeri, C.; Pignol, D.; Vermeglio, A.: Characterization of the reduction of selenate and tellurite by nitrate reductases. Appl. Environ. Microbiol., **67**, 5122-5126 (2001)
[42] Steward, V.; Lu, Y.; Darwin, A.J.: Periplasmic nitrate reductase (NapABC enzyme) supports anaerobic respiration by Escherichia coli K-12. J. Bacteriol., **184**, 1314-1323 (2002)
[43] Murillo, F.M.; Gugliuzza, T.; Senko, J.; Basu, P.; Stolz, J.F.: A heme-c-containing enzyme complex that exhibits nitrate and nitrite reductase activity from the dissimilatory iron-reducing bacterium Geobacter metallireducens. Arch. Microbiol., **172**, 313-320 (1999)
[44] Benett, B.; Charnock, J.M.; Sears, H.J.; Berks, B.C.; Thomson, A.J.; Ferguson, S.J.; Garner, C.D.; Richardsson, D.J.: Structural investigation of the molybdenum site of the periplasmic nitrate reductase from Thiosphaera pantotropha by X-ray absorption spectroscopy. Biochem. J., **317**, 557-563 (1996)
[45] Strehlitz, B.; Grundig, B.; Vorlop, K.D.; Bartholmes, P.; Kotte, H.; Stottmeister, U.: Artificial electron donors for nitrate and nitrite reductases usable as mediators in amperometric biosensors. Fresenius J. Anal. Chem., **349**, 676-678 (1994)

[46] Frunzke, K.; Heiss, B.; Meyer, O.; Zumft, W.G.: Molybdopterin guanine dinucleotide is the organic moiety of the molybdenum cofactor in respiratory nitrate reductase from Pseudomonas stutzeri. FEMS Microbiol. Lett., **113**, 241-246 (1993)

[47] Zafra, O.; Ramirez, S.; Castan, P.; Moreno, R.; Cava, F.; Valles, C.; Caro, E.; Berenguer, J.: A cytochrome c encoded by the nar operon is required for the synthesis of active respiratory nitrate reductase in Thermus thermophilus. FEBS Lett., **523**, 99-102 (2002)

[48] Anderson, L.J.; Richardson, D.J.; Butt, J.N.: Catalytic protein film voltammetry from a respiratory nitrate reductase provides evidence for complex electrochemical modulation of enzyme activity. Biochemistry, **40**, 11294-11307 (2001)

5,10-Methylenetetrahydrofolate reductase ($FADH_2$) 1.7.99.5

1 Nomenclature

EC number
1.7.99.5

Systematic name
5-methyltetrahydrofolate:(acceptor) oxidoreductase

Recommended name
5,10-methylenetetrahydrofolate reductase ($FADH_2$)

Synonyms
5,10-methylenetetrahydrofolate reductase
5,10-methylenetetrahydropteroylglutamate reductase
5-methyltetrahydrofolate:NAD oxidoreductase
5-methyltetrahydrofolate:NAD^+ oxidoreductase
EC 1.1.1.68 (formerly)
EC 1.1.99.15 (formerly)
MR
MTHFR
$N^{5,10}$-methylenetetrahydrofolate reductase
N^5,N^{10}-methylenetetrahydrofolate reductase
methylenetetrahydrofolate (reduced riboflavin adenine dinucleotide) reductase
methylenetetrahydrofolate reductase
methylenetetrahydrofolic acid reductase
reductase, methylenetetrahydrofolate (reduced riboflavin adenine dinucleotide)

CAS registry number
9028-69-7

2 Source Organism

<1> *Sus scrofa* [1, 2, 8]
<2> *Escherichia coli* (strain 113-3 [2,5]) [2, 5, 8, 14-16]
<3> *Clostridium formicoaceticum* (strain ATCC 23439 [3,4]) [3, 4]
<4> *Rattus norvegicus* (adult male Sprague-Dawley rats, fed a Purina lab chow diet [7]) [6, 7]
<5> *Homo sapiens* (healthy men and women aged 19-63 years [10]) [8-13, 15]
<6> *Caenorhabditis elegans* [8]

<7> *Arabidopsis sp.* (ecotype Columbia, expression in Saccharomyces cerevisiae [14]) [14]
<8> *Zea mays* (expression in Saccharomyces cerevisiae [14]) [14]
<9> *Saccharomyces cerevisiae* (strain DAY4, two isoenzymes, encoded by the genes MET12 and MET13 [17]) [17]
<10> *Schizosaccharomyces pombe* [17]

3 Reaction and Specificity

Catalyzed reaction

5-methyltetrahydrofolate + acceptor = 5,10-methylenetetrahydrofolate + reduced acceptor (<1> mechanism, acceptor is free FAD, stereochemistry [1]; <2> ping-pong bi-bi mechanism, in which $NAD(P)^+$ release precedes the binding of methylenetetrahydrofolate [16])

Reaction type

oxidation
redox reaction
reduction

Natural substrates and products

S 5,10-methylenetetrahydrofolate + $FADH_2$ <1-3> (<1> first step of conversion of 5,10-methylenetetrahydrofolate to methionine [2]; <2> biosynthesis of 5-methyltetrahydrofolate, a donor of methyl groups of methionine [5]; <3> enzyme in pathway of synthesis of acetate from CO_2 via formate and a series of reactions involving tetrahydrofolate and a corrinoid, $FADH_2$ and reduced ferredoxin may serve as natural reductants for 5,10-methylenetetrahydrofolate [4]) (Reversibility: ? <1-4> [2, 4, 5, 7]) [2, 4, 5, 7]
P 5-methyltetrahydrofolate + FAD <3> [4]
S 5,10-methylenetetrahydrofolate + NADH <2, 4, 7, 8> (<4> physiological direction: 5-methyltetrahydrofolate formation, inhibition by S-adenosylmethionine functions as a feedback-type metabolic regulation in vivo [6]; <7,8> folate-mediated one-carbon metabolism [14]; <2> physiological direction is the 5-methyltetrahydrofolate formation by transfer of reducing equivalents from NADH to the enzyme-bound FAD and from reduced FAD to methylenetetrahydrofolate [16]) (Reversibility: r <7, 8> [14]; ? <2, 4> [6, 16]) [6, 14, 16]
P 5-methyltetrahydrofolate + NAD^+ <4, 7, 8> [6, 14, 16]
S 5,10-methylenetetrahydrofolate + NADPH <5, 9> (<5> physiological NADPH-CH_2-H4folate oxidoreductase activity [15]; <9> enzyme encoded by MET13 produces 5-methyltetrahydrofolate used for methylation of homocysteine for methionine biosynthesis in vivo, NADPH is the likely natural reductant [17]) (Reversibility: ? <5, 9> [15, 17]) [15, 17]
P 5-methyltetrahydrofolate + $NADP^+$ <5, 9> (<5> 5-methyltetrahydrofolate is the major methyl donor for the conversion of homocysteine to methionine [15]) [15, 17]

S 5,10-methylenetetrahydrofolate + reduced acceptor <5> (<5> regeneration of methyl group of methionine [8]; <5> homocysteine metabolism [9,11]; <5> supply of the folate needed for the metabolism of homocysteine: 5-methyltetrahydrofolate is required for the methylation of homocysteine to methionine [10]; <5> remethylation of homocysteine to methionine [11,13]; <5> physiological direction is the 5-methyltetrahydrofolate formation [11]) (Reversibility: ? <5> [8-11, 13]) [8-11, 13]

P 5-methyltetrahydrofolate + oxidized acceptor <5> (<5> 5-methyltetrahydrofolate is the main form of circulating folate [11]) [8-11, 13]

S 5,10-methylenetetrahydrofolate + reduced ferredoxin <3> (<3> enzyme is part of synthesis pathway of acetate from CO_2 via a unique tetrahydrofolate-corrinoid pathway [3]; <3> enzyme in pathway of synthesis of acetate from CO_2 via formate and a series of reactions involving tetrahydrofolate and a corrinoid, $FADH_2$ and reduced ferredoxin may serve as natural reductants for 5,10-methylenetetrahydrofolate, electron transfer from reduced ferredoxin via iron-sulfur centers via enzyme-bound FAD to 5,10-methylenetetrahydrofolate [4]; <3> physiological important reaction: enzyme catalyzes reduction of 5,10-methylenetetrahydrofolate with reduced ferredoxin [3,4]) (Reversibility: ? <3> [3, 4]) [3, 4]

P 5-methyltetrahydrofolate + oxidized ferredoxin <3> [3, 4]

Substrates and products

S 5,10-methylene-5,6,7,8-tetrahydropteroylpentaglutamate + reduced acceptor <4> (<4> equilibrium lies far in favor of 5-methyl-5,6,7,8-tetrahydropteroylpentaglutamate formation, pentaglutamate form binds to the same enzyme site as monoglutamate form [7]) (Reversibility: ? <4> [7]) [7]

P (+)-5-methyl-5,6,7,8-tetrahydropteroylpentaglutamate + oxidized acceptor <4> (<4> menadione as acceptor for 5,10-methylene-5,6,7,8-tetrahydropteroylpentaglutamate formation [7]) [7]

S 5,10-methylenetetrahydrofolate + reduced acceptor <1-9> (<4,9> equilibrium lies far in favor of 5-methyltetrahydrofolate formation [7,17]; <1-5> forward reaction: reduced acceptor is $FADH_2$ [1-5,7,10]; <1> forward reaction: NADPH as reduced acceptor causes irreversible reduction of the flavin coenzyme FAD [1]; <3> forward reaction: reduced ferredoxin as reduced acceptor [3,4]; <2,5,9> forward reaction: NADPH as reduced acceptor [8,15,17]; <2,4> forward reaction: NADH as reduced acceptor [6-8,15,16]; <2> prefers NADH as reductant [14]; <7,8> forward reaction: strong preference of NADH over NADPH [14]; <3> reverse reaction: FAD, rubredoxin, benzyl viologen and methylene blue as electron acceptor [3,4]; <1-5,7,8> reverse reaction: menadione as electron acceptor [2-4, 6,7,11,14,16]; <2> under anaerobic conditions and in absence of electron acceptors, the equilibrium lies far to the 5-methyltetrahydrofolate formation, inclusion of menadione or oxygen promotes the oxidation of 5-methyltetrahydrofolate [5]; <4> 5,10-methylene-5,6,7,8-tetrahydropteroylmonoglutamate as substrate [7]) (Reversibility: r <1-4, 7-9> [2-6, 14, 16, 17]; ? <1, 2, 4-6> [1, 7-13, 15]) [1-17]

P 5-methyltetrahydrofolate + oxidized acceptor <1-9> (<4> product is (+)-5-methyl-5,6,7,8-tetrahydropteroylmonoglutamate [7]) [1-17]

S Additional information <1-5, 9> (<2,3> no direct activity with pyridine nucleotides [3-5]; <3> not as electron donor: reduced rubredoxin [3]; <4> reverse reaction: not $NADP^+$ as acceptor [7]; <1> enzyme is allosterically regulated by S-adenosylmethionine, tryptic proteolysis cleaves each subunit in an N-terminal 40 kDa and a C-terminal 37 kDa fragment without effect on catalytic activity, S-adenosylmethionine binds to the 37 kDa fragment, implicating this fragment as the regulatory region [8]; <5> N-terminal region is the catalytic region and contains determinants for binding of FAD, NADPH and methylentetrahydrofolate [8]; <2> catalytic mechanism, Asp-120 and Glu-28 at the flavin active site are relevant to catalysis, Asp-120: located near the enzyme-bound FAD, role in catalysis of folate reduction and in stabilization of the folate intermediate 5-iminium cation, Glu-28: located near N^{10} of the folate, general acid catalyst to aid in 5-iminium cation formation [16]; <9> reverse reaction: menadione and 2,6-dichlorophenolindophenol and $NADP^+$ are not effective as electron acceptor [17]) [3-5, 7, 8, 16, 17]

P ?

Inhibitors

(+)-5-methyl-5,6,7,8-tetrahydropteroylpentaglutamate <4> (<4> inhibits at high concentrations [7]) [7]

S-adenosyl-L-methionine <4, 5> (<4> inhibitory effect at a physiological concentration, dependent on incubation temperature, binds to an allosteric regulatory site different from catalytic site, reversed by S-adenosylhomocysteine [6]; <4> in vivo feedback-inhibition [6,7]; <5> inhibits enzyme activity, which is allosterically regulated by the S-adenosyl-L-methionine:adenosylhomocysteine ratio, with adenosylhomocysteine competing with S-adenosylmethionine for binding to enzyme but not acting as inhibitor [8]; <5> strong, reversible allosteric inhibition, prevented by S-adenosylhomocysteine [15]) [6-8, 11, 15]

menadione <9> (<9> inhibits enzyme activity in yeast extracts [17]) [17]

Additional information <2, 7-9> (<2> vitamin B_{12} and methionine can repress enzyme biosynthesis [5]; <2,7,8> not inhibited by S-adenosylmethionine [14]; <9> gene contains a putative S-adenosylmethionione binding sequence, that is inhibitory in eukaryotic MTHFR [17]) [5, 14, 17]

Cofactors/prosthetic groups

FAD <1-5> (<1-4> flavoprotein, oxidized acceptor in reverse reaction [1,3-5,7,8]; <1> enzyme-bound FAD as coenzyme [1]; <3> 1.7 molecules enzyme-bound FAD per enzyme molecule [3,4]; <2> enzyme-bound FAD [5]; <2> contains non-covalently bound FAD as its cofactor [8]; <1> one equivalent of non-covalently bound FAD per subunit [8]; <5> binds FAD [8]; <5> requires FAD as cofactor [10]; <5> enzyme activity depends on the concentration of its cofactor FAD: maximum activity at 0.05 mM FAD [11]; <2,5> flavoprotein with FAD as cofactor [15,16]; <2> FAD is essential for electron

transfer between NADH and methylenetetrahydrofolate [15]) [1, 3-5, 7, 8, 10, 11, 15, 16]
$FADH_2$ <1-5> (<1-4> flavoprotein, reduced acceptor in forward reaction [1,3-5,7,10]) [1, 3-5, 7, 10]
NADH <2, 4, 7, 8> (<2> enzyme-bound FAD and methylenetetrahydrofolate is reduced by NADH [8]; <7,8> strong preference of NADH over NADPH as reductant [14]) [6-8, 14-16]
NADPH <1, 2, 5> (<1> NADPH-dependent reduction of methylenetetrahydrofolate, NADPH irreversibly reduces the enzyme-bound flavin [1]; <2> enzyme-bound FAD and methylenetetrahydrofolate is reduced by NADPH, but much more slowly than by NADH [8]) [1, 8, 15]
flavin <9> (<9> flavin-dependent enzyme [17]) [17]
Additional information <2, 3> (<2,3> no direct activity with pyridine nucleotides: NADH, NADPH [3-5]) [3-5]

Activating compounds

S-adenosyl-L-methionine <7> (<7> stimulates [14]) [14]

Metals, ions

iron <3> (<3> 15.2 molecules iron per enzyme molecule, enzyme contains iron-sulfur clusters [3,4]) [3, 4]
iron-sulfur centre <3> (<3> enzyme contains 15.2 molecules iron and 19.5 molecules acid-labile sulfur as iron-sulfur clusters [3,4]) [3, 4]
zinc <3> (<3> 2.3 molecules zinc per enzyme molecule [3,4]) [3, 4]

Turnover number (min^{-1})

Additional information <2> [16]

Specific activity (U/mg)

9.4e-005 <9> (<9> wild-type, cell extract [17]) [17]
6.92 <7> (<7> recombinant enzyme [14]) [14]
12.4 <5> (<5> recombinant enzyme [15]) [15]
139 <3> [3, 4]
Additional information <2, 4, 5> (<4> specific activity is greater with the monoglutamate substrate than with the pentaglutamate [7]; <5> values of healthy and homocysteinemic humans before and after heat inactivation, specific activity of thermolabile variant of enzyme in homozygotes is 50% of normal enzyme [11]) [5, 7, 11, 12]

K_m-Value (mM)

0.0005 <2> (5,10-methylenetetrahydrofolate, <2> wild-type enzyme [16]) [16]
0.003 <4> ((+)-5-methyl-5,6,7,8-tetrahydropteroylpentaglutamate) [7]
0.005 <2> (NADH, <2> D120N mutant enzyme [16]) [16]
0.019 <5> (5-methyltetrahydrofolate) [11]
0.02 <2> (NADH, <2> wild-type enzyme [16]) [16]
0.023 <4> ((+)-5-methyltetrahydrofolate) [7]
0.027 <2> (5,10-methylenetetrahydrofolate, <2> D120N mutant enzyme [16]) [16]

0.06-0.08 <4> (5-methyltetrahydrofolate, <4> fresh extract [6]) [6]
0.085 <2> (5-methyltetrahydrofolate, <2> wild-type enzyme [16]) [16]
0.12 <3> (DL-5-methyltetrahydrofolate) [3, 4]
0.16 <2> (5-methyltetrahydrofolate, <2> D120N mutant enzyme [16]) [16]
11.1 <3> (benzyl viologen) [3, 4]

K_i-Value (mM)
0.04-0.051 <4> ((+)-5-methyl-5,6,7,8-tetrahydropteroylpentaglutamate) [7]
0.2 <4> (S-adenosylmethionine) [6]

pH-Optimum
6.3-6.4 <2> (<2> formation of 5-methyltetrahydrofolate [5]) [5]
6.7 <4> (<4> 5-methyltetrahydrofolate and 5-methyl-5,6,7,8-tetrahydropteroylpentaglutamate as substrates [7]) [7]
7.8-8 <3> [4]

pH-Range
6-8 <4> (<4> pH 6: about 70% of activity maximum, pH 8: about 35% of activity maximum with 5-methyl-5,6,7,8-tetrahydropteroylmonoglutamate, about 65% of activity maximum with 5-methyl-5,6,7,8-tetrahydropteroylpentaglutamate [7]) [7]

Temperature optimum (°C)
30 <4> (<4> assay at [6,7]) [6, 7]
37 <1-3, 5, 7-9> (<1-3,5,7-9> assay at [1,4,5,11,14,17]) [1, 4, 5, 11, 14, 17]

4 Enzyme Structure

Molecular weight
124000 <3> (<3> determination at pH 9, α_2,β_2 enzyme form, PAGE [4]) [4]
141000 <7> (<7> gel filtration [14]) [14]
237000 <3> (<3> determination at pH 7.4, α_4,β_4 enzyme form [4]; gel filtration [3,4]) [3, 4]

Subunits
? <8> (<8> x * 66400, calculated from the nucleotide sequence [14]) [14]
dimer <5, 7> (<7> 2 * 66000, SDS-PAGE [14]) [14, 15]
homodimer <1> (<1> 2 * 77000 [8]) [8]
homotetramer <2> (<2> 4 * 33000 [14,15]) [14, 15]
octamer <3> (<3> α_4,β_4, 4 * 26000 + 4 * 35000, SDS-PAGE [3,4]; <3> determined at pH 7.4 [4]) [3, 4]
tetramer <3> (<3> α_2,β_2, 2* 26000 + 2 * 35000, determined at pH 9, SDS-PAGE [4]) [4]

5 Isolation/Preparation/Mutation/Application

Source/tissue

leaf <7, 8> [14]
liver <1, 4> [1, 2, 6-8]
lymphocyte <5> [11]
root <7, 8> [14]
skin fibroblast <5> [12]

Localization

cytosol <7, 8> [14]

Purification

<1> (20fold purification [2]) [1, 2, 8]
<2> (about 100fold purification [5]; purification of wild-type and histidine-tagged D120N and E28Q mutant enzymes [16]) [2, 5, 8, 16]
<3> [3, 4]
<4> (more than 100fold partial purification [7]) [7]
<5> (62fold purification of recombinant enzyme, expressed in Sf9 cells [15]) [15]
<7> (643fold purification of recombinant enzyme, expressed in Saccharomyces cerevisiae [14]) [14]

Crystallization

<2> [16]

Cloning

<2> (expression of the D120N and E28Q mutant plasmids in Escherichia coli strain AB 1909 [16]) [16]
<5> (cDNA sequences specifying methylenetetrahydrofolate reductase [8]) [8]
<5> (expression of recombinant human MTHFR at high levels in Sf9 cells from Spodoptera frugiperda by using a baculovirus expression system [15]) [15]
<7> (two MTHFR genes, two cDNAs encoding functional MTHFRs are cloned, sequenced and expressed in Saccharomyces cerevisiae at high levels [14]) [14]
<8> (two MTHFR genes, a cDNAs encoding functional MTHFR is cloned, sequenced and expressed in Saccharomyces cerevisiae [14]) [14]
<9> (cloning, disruption and expression of the two genes MET12 and MET13 encoding isoenzymes of MTHFR, MET12 is located on chromosome XVI and encodes a 657 amino acids protein, MET13 is located on chromosome VII, encodes a 599 amino acids protein and is responsible for most of the activity in cell, expression in Escherichia coli BL21 [17]) [17]

Engineering

A177V <2> (<2> diminished expression of enzyme in an overexpressing strain, enzyme activity is thermolabile and rapidly lost during purification, it readily loses its flavin cofactor on dilution [8]; <2> enzyme with decreased affinity for its FAD cofactor [15]) [8, 15]

A222V <5> (<5> polymorphism, present in high frequency, homozygous genotype for this polymorphism have reduced enzyme specific activity in fibroblast extracts and increased susceptibility to heat inactivation of enzyme [8]; <5> most frequent genetic cause of mild hyperhomocysteinemia, enzyme with enhanced propensity to dissociate into monomers and to lose its FAD cofactor on dilution, increased thermolability of enzyme activity [15]) [8, 15]
D120N <2> (<2> mutant with 150fold decreased activity in the physiological NADH-CH_2-H4folate oxidoreductase reaction, enzyme is reduced by NADH 30% more rapidly than the wild-type enzyme, it binds methylenetetrahydrofolate in its ring-closed form, but no conversion to the 5-iminium cation, enzyme-bound FAD is more easiliy reduced and more difficult reoxidized than FAD of wild-type enzyme [16]) [16]
E28Q <2> (<2> mutant enzyme is unable to catalyze reduction of methylenetetrahydrofolate and is inactive in the physiological NADH-CH_2-H4folate oxidoreductase reaction, it binds methyltetrahydrofolate, but reduces not the FAD cofactor, 240fold decrease in NADH-menadione oxidoreductase activity [16]) [16]
E429A <5> (<5> A1298C mutation of the MTHRF gene, enzyme with indistinguishable properties from the wild-type [15]) [15]
R157Q <5> (<5> point mutation in gene for methylenetetrahydrofolate reductase with severe deficiency [8]) [8]
T227M <5> (<5> point mutation in gene for methylenetetrahydrofolate reductase with severe deficiency [8]) [8]
Additional information <5, 9> (<5> C677T mutation in the coding region of the gene, which replaces a conserved alanine by valine residue in enzyme with reduced activity, increased risk for cardiovascular disease, homozygosity results in a thermolabile enzyme and homocysteinaemia [9]; <5> C677T mutation of the gene for methylenetetrahydrofolate reductase: thermolabile enzyme with reduced activity, due to the loss of its riboflavin cofactor, homozygous genotype TT is associated with an increase in plasma total homocysteine, 12% of the healthy white population with this polymorphism [10]; <5> homozygous C677T mutation in the MTHFR gene, resulting in a conserved amino acid change from alanine to valine: enzyme with reduced activity and higher thermolability, elevated plasma homocysteine concentrations, but no genetic risk factor for deep-vein thrombosis, irrespective of factor V Leiden genotype [13]; <5> C677T mutation in MTHFR gene is a polymorphism which leads to the substitution of Ala-222 by valine [15]; <5> A1298C mutation of the MTHFR gene, which leads to the substitution of Glu-429 by alanine [15]; <9> mutant strain RRY1 with single disruption of MET13 and strain RRY3 with double disruption of both MET12 and MET13, but not strain RRY2 with single disruption of MET12, result in loss of MTHFR activity and methionine auxothrophy, they can be complemented by overexpression of the human enzyme, but not by overexpression of the Escherichia coli metF gene [17]) [9, 10, 13, 15, 17]

Application

medicine <5> (<5> homocysteinemia: defect in either methylenetetrahydrofolate reductase or methionine synthase [8]; <5> mutation in gene for methylenetetrahydrofolate reductase is a cause of moderate homocysteinaemia, which is a risk factor for arteriosclerosis and thrombosis, patients with MTHFR deficiency have low methionine concentrations in plasma [9]; <5> deficiency and thermolability of enzyme as cause of mild homocysteinemia with premature vascular disease [11]; <5> methylenetetrahydrofolate reductase deficiency causes homocystinuria [12]; <5> methylenetetrahydrofolate reductase deficiency as cause of homocysteinuria [13]; <5> C677T mution of MTHRF gene is the most frequent genetic cause of mild hyperhomocysteinemia, a risk factor for cardiovascular disease [15]) [8, 9, 11, 12]

6 Stability

pH-Stability

7.1-7.6 <3> (<3> at this pH-range enzyme is more stable in Tris-HCl than in triethanolamine-HCl or phosphate buffers [4]) [4]

Temperature stability

40 <5> (<5> 5 min, 17% loss of activity [11]) [11]
46 <5> (<5> 5 min, 53% loss of activity [11]) [11]
49 <5> (<5> 5 min, 70% loss of activity [11]) [11]
Additional information <5> (<5> activity of enzyme from healthy and homocysteinemic humans after heat inactivation, the latter with a more thermolabile enzyme: residual activity in homozygotes is lower than 30% compared with 50% in control subjects [11]; <5> thermolability is enhanced when the FAD cofactor dissociates form enzyme [15]) [11, 15]

Oxidation stability

<3>, oxygen labile, half-life: less than 1 h in aerobic buffer, sodium dithionite prevents inactivation by oxygen [3, 4]

General stability information

<2>, FAD stabilizes [5]
<3>, 0.005 mM FAD stabilizes enzyme during purification procedure [3, 4]
<2, 5>, methyltetrahydrofolate and S-adenosylmethionine protects enzyme from the loss of FAD after dilution [15]

Storage stability

<1>, -15°C, several months, stable [2]
<3>, 10°C, 50 mM Tris-HCl buffer, pH 7.4, 20% glycerol, 2 mM dithionite, in an anaerobic chamber, a few weeks, stable [4]
<3>, frozen, 50 mM Tris-HCl buffer, pH 7.4, 20% glycerol, 2 mM dithionite, longer time, 10-20% loss of activity [4]
<7>, on ice, 3 h, loss of about 50% of activity [14]

References

[1] Vanoni, M.A.; Lee, S.; Floss, H.G.; Matthews, R.G.: Stereochemistry of reduction of methylenetetrahydrofolate to methyltetrahydrofolate catalyzed by pig liver methylenetetrahydrofolate reductase. J. Am. Chem. Soc., **112**, 3987-3992 (1990)

[2] Buchanan, J.M.: Methionine biosynthesis (hog liver). Methods Enzymol., **17B**, 371-378 (1971)

[3] Clark, J.E.; Ljungdahl, L.G.: Purification and properties of 5,10-methylenetetrahydrofolate reductase from Clostridium formicoaceticum. Methods Enzymol., **122**, 392-399 (1986)

[4] Clark, J.E.; Ljungdahl, L.G.: Purification and properties of 5,10-methylenetetrahydrofolate reductase, an iron-sulfur flavoprotein from Clostridium formicoaceticum. J. Biol. Chem., **259**, 10845-10849 (1984)

[5] Katzen, H.M.; Buchanan, J.M.: Enzymatic synthesis of the methyl group of methionine. VIII. Repression – derepression, purification, and properties of 5,10-methylene-tetrahydrofolate reductase from Escherichia coli. J. Biol. Chem., **240**, 825-835 (1965)

[6] Kutzbach, C.; Stokstad, E.L.R.: Feedback inhibition of methylene-tetrahydrofolate reductase in rat liver by S-adenosylmethionine. Biochim. Biophys. Acta, **139**, 217-220 (1967)

[7] Cheng, F.W.; Shane, B.; Stokstad, E.L.R.: Pentaglutamate derivatives of folate as substrates for rat liver tetrahydropteroylglutamate methyltransferase and 5,10-methylenetetrahydrofolate reductase. Can. J. Biochem., **53**, 1020-1027 (1975)

[8] Matthews, R.G.; Sheppard, C.; Goulding, C.: Methylenetetrahydrofolate reductase and methionine synthase: biochemistry and molecular biology. Eur. J. Pediatr., **157**, 54-59 (1998)

[9] Blom, H.J.: Mutated 5,10-methylenetetrahydrofolate reductase and moderate hyperhomocysteinaemia. Eur. J. Pediatr., **157**, 131-134 (1998)

[10] McNulty, H.; McKinley, M.C.; Wilson, B.; McPartlin, J.; Strain, J.J.; Weir, D.G.; Scott, J.M.: Impaired functioning of thermolabile methylenetetrahydrofolate reductase is dependent on riboflavin status: Implications for riboflavin requirements. Am. J. Clin. Nutr., **76**, 436-441 (2002)

[11] Engbersen, A.M.T.; Franken, D.G.; Boers, G.H.J.; Stevens, E.M.B.; Trijbels, F.J.M.; Blom, H.J.: Thermolabile 5,10-methylenetetrahydrofolate reductase as a cause of mild hyperhomocysteinemia. Am. J. Hum. Genet., **56**, 142-150 (1995)

[12] Rosenblatt, D.S.; Lue-Shing, H.; Matiaszuk, N.; Low-Nang, L.; Arzoumanian, A.; Cooper, B.A.: Thermolability of residual methylene-tetrahydrofolate reductase (MR) activity, methionine synthase activity and methyl-cobalamin levels in cultured fibroblasts from patients with MR deficiency. Adv. Exp. Med. Biol., **338**, 703-706 (1993)

[13] Kluijtmans, L.A.J.; den Heijer, M.; Reitsma, P.H.; Heil, S.G.; Blom, H.J.; Rosendaal, F.R.: Thermolabile methylenetetrahydrofolate reductase and factor

V Leiden in the risk of deep-vein thrombosis. Thromb. Haemost., **79**, 254-258 (1998)
[14] Roje, S.; Wang, H.; McNeil, S.D.; Raymond, R.K.; Appling, D.R.; Shachar-Hill, Y.; Bohnert, H.J.; Hanson, A.D.: Isolation, characterization, and functional expression of cDNAs encoding NADH-dependent methylenetetrahydrofolate reductase from higher plants. J. Biol. Chem., **274**, 36089-36096 (1999)
[15] Yamada, K.; Chen, Z.; Rozen, R.; Matthews, R.G.: Effects of common polymorphisms on the properties of recombinant human methylenetetrahydrofolate reductase. Proc. Natl. Acad. Sci. USA, **98**, 14853-14858 (2001)
[16] Trimmer, E.E.; Ballou, D.P.; Ludwig, M.L.; Matthews, R.G.: Folate activation and catalysis in methylenetetrahydrofolate reductase from Escherichia coli: Roles for aspartate 120 and glutamate 28. Biochemistry, **40**, 6216-6226 (2001)
[17] Raymond, R.K.; Kastanos, E.K.; Appling, D.R.: Saccharomyces cerevisiae expresses two genes encoding isoenzymes of methylenetetrahydrofolate reductase. Arch. Biochem. Biophys., **372**, 300-308 (1999)

Nitrous-oxide reductase 1.7.99.6

1 Nomenclature

EC number
1.7.99.6

Systematic name
nitrogen:(acceptor) oxidoreductase (N_2O-forming)

Recommended name
nitrous-oxide reductase

Synonyms
N_2OR
N_2O reductase
nitrous oxide reductase

CAS registry number
55576-44-8

2 Source Organism

<1> *Paracoccus denitrificans* (NCIB 8944 [22]; ATCC 19367 [26,30,31]) [1, 21, 22, 26, 30, 31]
<2> *Paracoccus pantotrophus* (formerly Thiosphaera pantotropha [2,14]; LMD 82.5 [14]) [2, 3, 14]
<3> *Pseudomonas stutzeri* (wild-type, C622D-mutant laccking copper in the CuA-site and N_2OR V a mutant which lacks CuZ [3]) [3, 10, 12, 19]
<4> *Pseudomonas nautica* (strain 617 [4,5]) [4, 5, 17]
<5> *Rhodobacter sphaeroides* (f. sp. denitrificans IL106 [6,7,9]) [6, 7, 9, 23]
<6> *Alcaligenes xylosoxidans* (NCIMB 11015, X-ray scattering data, 20 A resolution [8]) [8]
<7> *Thiobacillus denitrificans* (DSM 807 [11]) [11]
<8> *Wolinella succinogenes* (ATCC 29543 [13]; under anaerobic conditions cytochrome c oxidase activity is observed [18]) [13, 18]
<9> *Flexibacter canadensis* (ATCC 29591 [15]) [15]
<10> *soil organism Is-11* (soil isolate [15]) [15]
<11> *Achromobacter cycloclastes* (ATCC 21921 [16]) [16]
<12> *Rhizobium sp.* (strain 8A55 [20]) [20]
<13> *Pseudomonas perfectomarina* (ATCC 14405 [24,26]) [24, 26, 28]
<14> *Rhodopseudomonas capsulata* (strains N22, Kb1 [25]) [25]
<15> *Rhodopseudomonas spaeroides* (strain 2.4.1 [25]) [25]

<16> *Rhodopseudomonas palustris* (strain PW5 [25]) [25]
<17> *Rhodospirillum rubrum* (stain S1 [25]) [25]
<18> *Pseudomonas denitrificans* (ATCC 13867 [26]) [26]
<19> *Pseudomonas sp.* (strain G59 [27]) [27]
<20> *Pseudomonas fluorescens* [29]

3 Reaction and Specificity

Catalyzed reaction

nitrogen + H_2O + acceptor = nitrous oxide + reduced acceptor

Reaction type

oxidation
redox reaction
reduction

Natural substrates and products

S nitrous oxide + reduced acceptor <1-20> (Reversibility: ? <1-20> [1-31]) [1-31]
P nitrogen + H_2O + acceptor

Substrates and products

S nitrogen + H_2O + acceptor <1> (Reversibility: ? <1> [1]) [1]
P nitrous oxide + reduced acceptor
S nitrous oxide + reduced acceptor <1-20> (<2> methyl viologen as electron donor, activity after incubation with ferricyanide, ascorbate or dithionite [2]; <5> dithionite-reduced benzyl viologen as electron donor [6]; <1, 5, 8, 13, 18> photochemically-reduced benzyl viologen as electron donor [9, 13, 24, 26, 30, 31]; <2,6> dithionite-reduced methyl viologen as electron donor [8, 14]; <1, 11, 13> methyl viologen as electron donor [16, 28, 30]; <1> a mixture of N,N,N',N'-tetramethyl-*p*-phenylenediamine and ascorbate as electron donor [22]; <5> photochemically reduced benzyl viologen and 3,3',5,5'-tetramethylbenzidine as electron donor [23]) (Reversibility: ? <1-20> [1-31]) [1-31]
P nitrogen + H_2O + acceptor

Inhibitors

2-mercaptoethanol <1> (<1> 53% after 30 min at 0.5 mM [21]) [21]
CO <1> [31]
EDTA <1, 5> (<5> complete inhibition at 0.1 mM [9]; <1> 11% at 0.5 mM after 30 min [21]) [9, 21]
EGTA <1> (<1> 36% at 0.5 mM after 30 min [21]) [21]
KCN <1, 2> [14, 31]
N,N,N',N'-tetramethyl-*p*-phenylenediamine <14> (<14> 76% inhibition at 0.1 mM [25]) [25]
Na_2S <9, 10> (<9,10> temporarily inhibition [15]; <9,10> together with sulfide no inhibition [15]) [15]
NaN_3 <1, 2> [14, 31]

Ni^{2+} <5> (<5> 50% inhibition at 0.1 mM [9]) [9]
O_2 <1, 2, 4, 5, 13, 18> (<2> inactivation is faster in the crude extract than solutions of the purified enzyme, acceleration takes place if the enzyme is turning over [2]; <5> turnover dependent inactivation [6]; <4> complete inactivation at 0.25 mg/l, no activity restorage after return to anaerobic conditions [17]; <13> unpurified enzyme is rapidly inactivated at air [24]; <1, 13, 18> inactivation with O_2 is largely reversible by exposure to benzyl viologen cation radical at room temperature for several hours [26]; <1> 50% inactivation after 30 min, 80% regained activity after incubation with reduced benzyl viologen [30]) [2, 6, 17, 24, 26, 30]
$S_2O_4^{2-}$ <11, 13> (<11,13> irreversible formation of inactive blue form of the enzyme [16,28]) [16, 28]
Zn^{2+} <5> (<5> complete inhibition at 0.1 mM [9]) [9]
acetylene <1, 2, 7, 9, 10, 12> (<7> at 10 mM 80% inhibition [11]; <9,10> at 4 kPa [15]; <9,10> together with sulfide no inhibition [15]; <12> at 10% complete inhibition [20]) [11, 14, 15, 20, 30, 31]
methylisonitrile <1> [30]
rotenone <14> (<14> 88% inhibition at 0.025 mM [25]) [25]

Activating compounds

NO_2^- <5> (<5> increases the synthesis of N_2O-reductase [7]) [7]

Metals, ions

Ca <1, 8> (<1> two calcium sites located at the intermonomeric surface [1]; <8> 1.2 atoms per subunit [18]) [1, 18]
Cu <1-8, 13-19> (<1, 2> one dinuclear centre Cu_A and a copper cluster Cu_Z in which four copper ions are litigated by seven histidine imidazoles and a bridging inorganic sufide [1, 2]; <2> Cu_A can exist in two oxidation forms [$Cu^{1.5+}$ – $Cu^{1.5+}$] and [Cu^{1+} – Cu^{1+}] [2]; <2> wild-type 10.5 atoms per protein dimer [3]; <2> wild-type 7.2 atoms per protein dimer [14]; <3> wild-type 9.9 atoms per protein dimer [3]; <4> six copper atoms per monomer arranged in two centers named Cu_A and Cu_Z, 10.7 atoms per dimer [4]; <6> contains 7.1 atoms per dimer [8]; <5> 7.0 atoms per dimer, needs 0.001 mM Cu^{2+} for full activity [9];<3> Cu_A is electron transfer center, Cu_Z is the catalytic center [10]; <11> 4 atoms per subunit [16]; <8> 2.9 atoms per subunit [18]; <1> 4.0 atoms per subunit [21]; <5> 4 atoms per mol [23]; <13> 8 atoms per mol enzyme [24,28]) [1-4, 8-14, 16, 18, 21, 23, 24-28]
Fe <8> (<8> 1 atom per mol [18]) [18]
Ni <5> (<5> 0.76 atoms per mol [23]) [23]
Zn <5> (<5> 2 atoms per mol [23]) [23]

Specific activity (U/mg)

0.05 <7> [11]
1 <2> (<2> ascorbate-reduced phenanzine methosulfate electrode assay [14]) [14]
1.6 <2> (<2> anaerobic form pre-reduced with ascorbate [2]) [2]
2.6 <2> (<2> anaerobic form pre-oxidized with ferricyanide [2]) [2, 14]
2.8 <2> (<2> anaerobic form pre-reduced with dithionite [2]) [2]

3.3 <2> (<2> photoreduced methyl viologen spectrophotometric assay [14]) [14]
3.7 <2> (<2> dithionite-reduced methyl viologen electrode assay [14]) [14]
5.5 <3> (<3> wild-type [3]) [3]
6 <6> [8]
6.9 <2> [3]
8.5 <2> (<2> aerobic form pre-oxidized with ferricyanide [2]) [2]
8.8 <2> (<2> aerobic form pre-reduced with dithionite [2]) [2]
23 <4> (<4> blue-coloured form [4]) [4]
30 <1> [30]
44 <5> (<5> aerobic form [6,9]) [6, 9]
50.4 <5> [23]
55 <4> (<4> purple coloured form [4]) [4]
78 <5> (<5> anaerobic form [6]) [6]
122 <1> (<1> anaerobic form [21]) [21]
160 <8> [18]

K_m-Value (mM)
0.0009 <1> (reduced benzyl viologen) [31]
0.00095 <1> (reduced methyl viologen) [31]
0.004 <8> (benzyl viologen) [18]
0.005 <1> (N_2O) [31]
0.007 <1> (N_2O) [21]
0.0075 <8> (N_2O) [18]

K_i-Value (mM)
0.000025 <2> (KCN) [14]
0.000045 <1> (KCN, <1> non competitive [31]) [31]
0.00035 <1> (NaN_3, <1> non competitive [31]) [31]
0.00064 <2> (NaN_3) [14]
0.0035 <1> (CO, <1> non competitive [31]) [31]
0.028 <1> (acetylene, <1> non competitive [31]) [31]
0.035 <2> (acetylene) [14]

pH-Optimum
8 <1, 2> (<2> above pH 8 with reduced methyl viologen [14]; <1> in presence of N,N,N',N'-tetramethyl-*p*-phenylenediamine and ascorbate [22]) [14, 22, 30]
8.5-9 <5> (<5> with reduced benzyl viologen as electron donor [6]) [6]

pH-Range
6-9.2 <1> (<1> half maximum activity values at pH 6 and pH 9.2 [30]) [30]
9-10 <6> (<6> activity is maximal after incubation at high pH values [8]) [8]

Temperature optimum (°C)
60 <5> [6]

Temperature range (°C)
40-85 <6> (<6> activity is maximal after exposure to high temperatures [8]) [8]

4 Enzyme Structure

Molecular weight

85000 <1> (<1> gel filtration [30]) [30]
87000 <13, 18> (<13,18> gel filtration [26]) [26]
89000 <1> (<1> gel filtration [26]) [26]
95000 <5> (<5> gel filtration [23]) [23]
105000 <5> (<5> gel filtration [5,9]) [5, 9]
115000 <2> (<2> gel filtration [14]) [14]
118000 <13> (<13> gel filtration [24]) [24]
120000 <4, 13, 20> (<4,13> gel filtration [4,28,29]) [4, 28, 29]
130000 <13> (<13> non denaturating PAGE [24]) [24]
134000 <6> (<6> gel filtration [8]) [8]
144000 <1> (<1> gel filtration [21]) [21]
160000 <7> (<7> gel filtration [11]) [11]
162000 <8> (<8> gel filtration [18]) [18]

Subunits

homodimer <1> (<1> 2 * 70000, SDS-PAGE [21]) [21]
homodimer <1, 4> (<1,4> 2 * ?, each monomer is composed of two domains, a C-terminal cupredoxin domain carrying a dinuclear electron entry site CuA, an N-terminal seven-bladed propeller domain with the active center CuZ [1,5]) [1, 5]
homodimer <13> (<13> 2 * 74000, SDS-PAGE [24]) [24]
homodimer <13> (<13> α_2, 2 * 62000, SDS-PAGE [28]) [28]
homodimer <2> (<2> 2 * 66281, ESI mass spectrometry [14]) [14]
homodimer <2> (<2> 2 * 73000, SDS-PAGE [14]) [14]
homodimer <4> (<4> 2 * 65000, SDS-PAGE [4]) [4]
homodimer <4> (<4> 2 * 65373, ESI mass spectrometry [4]) [4]
homodimer <5> (<5> 2 * 66180, MALDI-TOF [6,9]) [6, 9]
homodimer <5> (<5> 2 * 67000, SDS-PAGE [9]) [9]
homodimer <6> (<6> 2 * 66871, ESI mass spectrometry [8]) [8]
homodimer <7> (<7> 2 * 73000, SDS-PAGE [11]) [11]
homodimer <8> (<8> 2 * 88000, SDS-PAGE [18]) [18]
monomer <11> (<11> 1 * 72000, SDS-PAGE [16]) [16]

5 Isolation/Preparation/Mutation/Application

Localization

membrane <7> [11]
periplasm <4, 3, 14> (<4,3> soluble [5,10,12]) [5, 10, 12, 25]

Purification

<1> (purification under anaerobic conditions [21]; partial purification under anaerobic conditions [30]) [21, 30]
<2> (of anaerobically grown cells [14]) [14]

<4> (two forms a purple and a blue form [4]) [4]
<5> (purification under anaerobic conditions [6]; purification aerobically at 4°C [9]) [6, 9]
<6> (purification under anaerobic conditions [8]) [8]
<8> (purification under anaerobic conditions [18]) [18]
<13> (a pink form is isolated when the purification is done aerobically, a purple form when the purification is under anaerobic conditions [24]) [24]

Crystallization
<1> (1.6 A resolution [1]) [1]
<4> (2.4 A resolution, preliminary study [4]; 2.4 A resolution [5]) [4, 5]

Cloning
<3> (expression in Escherichia coli HB101, no activity [19]) [19]

Engineering
C165G <3> (<3> retaines catalytic activity; Cys165 is not available for Cu cooordination [12]) [12]
C622D <3> (<3> no activity, distorted CuA-centre [3]) [3]

Application
degradation <2, 3> (<2,3> plays a critical environmental role in preventing release into the atmosphere of the potent greenhouse gas nitrous oxide [3]) [3]

6 Stability

Temperature stability
25 <13> (<13> 85% inactivation after 20 h [24]) [24]
70 <5> (<5> most of activity lost [6]) [6]

General stability information
<1>, turnover dependent activation, nitrite and fluoride accelerate this process [21]
<2>, an enzyme concentration of at least 0.01 mg per ml is necessary to avoid loss of activity due to dilution [14]
<5>, turnover dependent inactivation, promoted by zinc ions at 0.01 mM [6]

Storage stability
<5>, 4°C, retained most of its activity after aerobic storage for 2 months without any additives [6]
<8>, 4°C, under aerobic conditions, 100 h stable [18]
<8>, 4°C, under aerobic conditions, 5 mM EDTA, 3 h stable [18]

References

[1] Haltia, T.; Brown, K.; Tegoni, M.; Gambillau, C.; Saraste, M.; Mattila, K.; Djinovic-Carugo, K.: The crystal structure of nitrous oxide reductase from Paracoccus denitrificans at 1.6 A resolution. Biochem. J., **369**, 77-88 (2003)

[2] Rasmussen, T.; Berks, B.C.; Butt, J.N.; Thomson, A.J.: Multiple forms of the catalytic centre, Cuz, in the enzyme nitrous oxide reductase from Paracoccus pantotrophus. Biochem. J., **364**, 807-815 (2002)

[3] Rasmussen, T.; Berks, B.C.; Sanders-Loehr, J.; Dooley, D.M.; Zumft, W.G.; Thomson, A.J.: The catalytic center in nitrous oxide reductase, CuZ, is a copper-sulfide cluster. Biochemistry, **39**, 12753-12756 (2000)

[4] Prudencio, M.; Pereira, A.S.; Tavares, P.; Besson, S.; Cabrito, I.; Brown, K.; Samyn, B.; Devreese, B.; Van Beeumen, J.; Rusnak, F.; Fauque, G.; Moura, J.J.G.; Tegoni, M.; Cambillau, C.; Moura, I.: Purification, characterization, and preliminary crystallographic study of copper-containing nitrous oxide reductase from Pseudomonas nautica 617. Biochemistry, **39**, 3899-3907 (2000)

[5] Brown, K.; Tegoni, M.; Prudencio, M.; Pereira, A.S.; Besson, S.; Moura, J.J.; Moura, I.; Cambillau, C.: A novel type of catalytic copper cluster in nitrous oxide reductase. Nat. Struct. Biol., **7**, 191-195 (2000)

[6] Sato, K.; Okubo, A.; Yamazaki, S.: Anaerobic purification and characterization of nitrous oxide reductase from Rhodobacter sphaeroides f. sp. denitrificans IL106. J. Biochem., **125**, 864-868 (1999)

[7] Sabaty, M.; Schwintner, C.; Cahors, S.; Richaud, P.; Vermeglio, A.: Nitrite and nitrous oxide reductase regulation by nitrogen oxides in Rhodobacter sphaeroides f. sp. denitrificans IL106. J. Bacteriol., **181**, 6028-6032 (1999)

[8] Ferretti, S.; Grossmann, J.G.; Hasnain, S.S.; Eady, R.R.; Smith, B.E.: Biochemical characterization and solution structure of nitrous oxide reductase from Alcaligenes xylosoxidans (NCIMB 11015). Eur. J. Biochem., **259**, 651-659 (1999)

[9] Sato, K.; Okubo, A.; Yamazaki, S.: Characterization of a multi-copper enzyme, nitrous oxide reductase, from Rhodobacter sphaeroides f. sp. denitrificans. J. Biochem., **124**, 51-54 (1998)

[10] Farrar, J.A.; Zumft, W.G.; Thomson, A.J.: CuA and CuZ are variants of the electron transfer center in nitrous oxide reductase. Proc. Natl. Acad. Sci. USA, **95**, 9891-9896 (1998)

[11] Hole, U.H.; Vollack, K.U.; Zumft, W.G.; Eisenmann, E.; Siddiqui, R.A.; Friedrich, B.; Kroneck, P.M.H.: Characterization of the membranous denitrification enzymes nitrite reductase (cytochrome cd_1) and copper-containing nitrous oxide reductase from Thiobacillus denitrificans. Arch. Microbiol., **165**, 55-61 (1996)

[12] Dreusch, A.; Riester, J.; Kroneck, P.M.; Zumft, W.G.: Mutation of the conserved Cys165 outside of the CuA domain destabilizes nitrous oxide reductase but maintains its catalytic activity. Evidence for disulfide bridges and a putative protein disulfide isomerase gene. Eur. J. Biochem., **237**, 447-453 (1996)

[13] Zhang, C.s.; Hollocher, T.C.: The reaction of reduced cytochromes c with nitrous oxide reductase of Wolinella succinogenes. Biochim. Biophys. Acta, **1142**, 253-261 (1993)
[14] Berks, B.C.; Baratta, D.; Richardson, D.J.; Ferguson, S.J.: Purification and characterization of a nitrous oxide reductase from Thiosphaera pantotropha. Implications for the mechanism of aerobic nitrous oxide reduction. Eur. J. Biochem., **212**, 467-476 (1993)
[15] Jones, A.M.; Knowles, R.: Denitrification in Flexibacter canadensis. Can. J. Microbiol., **36**, 430-434 (1990)
[16] Hulse, C.L.; Averill, B.A.: Isolation of a high specific activity pink, monomeric nitrous oxide reductase from Achromobacter cycloclastes. Biochem. Biophys. Res. Commun., **166**, 729-735 (1990)
[17] Bonin, P.; Gilewicz, M.; Bertrand, J.C.: Effects of oxygen on each step of denitrification on Pseudomonas nautica. Can. J. Microbiol., **35**, 1061-1064 (1989)
[18] Teraguchi, S.; Hollocher, T.C.: Purification and some characteristics of a cytochrome c-containing nitrous oxide reductase from Wolinella succinogenes. J. Biol. Chem., **264**, 1972-1979 (1989)
[19] Viebrock, A.; Zumft, W.G.: Molecular cloning, heterologous expression, and primary structure of the structural gene for the copper enzyme nitrous oxide reductase from denitrifying Pseudomonas stutzeri. J. Bacteriol., **170**, 4658-4668 (1988)
[20] Coyne, M.S.; Focht, D.D.: Nitrous oxide reduction in nodules: denitrification or N_2 fixation ?. Appl. Environ. Microbiol., **53**, 1168-1170 (1987)
[21] Snyder, S.W.; Hollocher, T.G.: Purification and some characteristics of nitrous oxide reductase from Paracoccus denitrificans. J. Biol. Chem., **262**, 6515-6525 (1987)
[22] Kucera, I.; Boublikova, P.; Dadak, V.: Amperometric assay of activity and pH-optimum of N_2O-reductase of Paracoccus denitrificans. Collect. Czech. Chem. Commun., **49**, 2709-2712 (1984)
[23] Michalski, W.P.; Hein, D.H.; Nicholas, D.J.D.: Purification and characterization of nitrous oxide reductase from Rhodopseudomonas sphaeroides f.sp. denitrificans. Biochim. Biophys. Acta, **872**, 50-60 (1986)
[24] Coyle, C.L.; Zumft, W.G.; Kroneck, P.M.H.; Körner, H.; Jakob, W.: Nitrous oxide reductase from denitrifying Pseudomonas perfectomarina. Purification and properties of a novel multicopper enzyme. Eur. J. Biochem., **153**, 459-467 (1985)
[25] McEwan, A.G.; Greenfield, A.J.; Wetzstein, H.G.; Jackson, J.B.; Ferguson, S.J.: Nitrous oxide reduction by members of the family Rhodospirillaceae and the nitrous oxide reductase of Rhodopseudomonas capsulata. J. Bacteriol., **164**, 823-830 (1985)
[26] Snyder, S.W.; Hollocher, T.C.: Nitrous oxide reductase and the 120,000 MW copper protein of N^2-producing denitrifying bacteria are different entities. Biochem. Biophys. Res. Commun., **119**, 588-592 (1984)
[27] Aida, T.; Hata, S.; Kusunoki, H.: Temporary low oxygen conditions for the formation of nitrate reductase and nitrous oxide reductase by denitrifying Pseudomonas sp. G59. Can. J. Microbiol., **32**, 543-547 (1986)

[28] Zumft, W.G.; Matsubara, T.: A novel kind of multi-copper protein as terminal oxidoreductase of nitrous oxide respiration in Pseudomonas perfectomarinus. FEBS Lett., **148**, 107-112 (1982)
[29] Matsubara, T.; Zumft, W.G.: Identification of a copper protein as part of the nitrous oxide-reducing system in nitrite-respiring (denitrifying) Pseudomonads. Arch. Microbiol., **132**, 322-328 (1982)
[30] Kristjansson, J.K.; Hollocher, T.C.: Partial purification and characterization of nitrous oxide reductase from Paracoccus denitrificans. Curr. Microbiol., **6**, 247-251 (1981)
[31] Kristjansson, J.K.; Hollocher, T.C.: First practical assay for soluble nitrous oxide reductase of denitrifying bacteria and a partial kinetic characterization. J. Biol. Chem., **255**, 704-707 (1980)

Nitric-oxide reductase 1.7.99.7

1 Nomenclature

EC number
1.7.99.7

Systematic name
nitrous-oxide:(acceptor) oxidoreductase (NO-forming)

Recommended name
nitric-oxide reductase

Synonyms
nitrogen oxide reductase

CAS registry number
37256-43-2

2 Source Organism

<1> *Paracoccus denitrificans* (strain 93.11 [3]; strain NCIB 8944 [13,24]; strain Pd 1222 [16]; strain ATCC 35512 formerly Thiosphaera pantotropha [20]; strain ATCC 19367 [28]) [1, 3-5, 7, 9, 10, 13, 16, 20, 22-26, 28-30]
<2> *Synechocystis sp.* (strain PCC6803 [2]) [2, 9]
<3> *Fusarium oxysporum* [6, 10, 11, 14, 15, 17]
<4> *Trichosporon cutaneum* (strain JCM2391, dependent on O_2 supply [8]) [8]
<5> *Pseudomonas stutzeri* (ZoBell strain, ATCC 14405 [21,27]) [9, 10, 21, 27]
<6> *Paracoccus halodenitrificans* [9, 10, 18, 31]
<7> *Achromobacter cycloclastes* (strain ATCC 21921 [28]) [9, 28]
<8> *Ralstonia eutropha* (formerly Alcaligenes eutrophus strain H16 [9,12]; contains two NO reductase enzymes named norB and norZ, that are located the former on the megaplasmid pHG1 and the latter on chromosomal DNA [12,19]) [9, 10, 12, 19]
<9> *fungi* [10]
<10> *Cylindrocarpon tonkinense* (contains two isozymes named cNor1 and cNor2 [17]) [17]
<11> *Alcaligenes faecalis* [10]
<12> *Pseudomonas sp.* [10]
<13> *Bradyrhizobium japonicum* (strain USDA 6 [30]) [10, 30]
<14> *Rhodobacter sphaeroides* [10]
<15> *Pseudomonas aeruginosa* (strain PAO 1 [28]) [10, 28]
<16> *Staphylococcus aureus* [10]

<17> *Geobacillus stearothermophilus* [10]
<18> *Mycobacterium avium* [10]
<19> *Corynebacterium diphtheriae* [10]
<20> *Neisseria meningitidis* [10]
<21> *Neisseria gonorrhoeae* [10]
<22> *Pseudomonas perfectomarina* (strain ATCC 14405 [28]) [28]
<23> *Pseudomonas denitrificans* [32]

3 Reaction and Specificity

Catalyzed reaction

nitrous oxide + acceptor + H_2O = 2 nitric oxide + reduced acceptor

Reaction type

oxidation
redox reaction
reduction

Natural substrates and products

S nitric oxide + reduced acceptor <1-23> (<1> key step in the pathway of denitrification that uses N-oxyanions and N-oxides as respiratory electron acceptors [3]; <1,5,6,11-15> cytochrome c as mediator [10]; <2,8,16-21> quinol-oxidising enzyme [9,10]) (Reversibility: ? <1-23> [1-32]) [1-32]

P nitrous oxide + acceptor + H_2O

Substrates and products

S nitric oxide + reduced acceptor <1-23> (<2> reduced menadione as electron donor [2]; <1> investigation of proton and electron pathways, electron-transfer pathways are similar to those in oxygen-reducing heme-copper oxidases [4]; <1> ascorbate, phenazine methosulfate and horse heart cytochrome c as the electron donor/mediator system [7, 16]; <1, 5-7> as electron donors phenazine methosulfate, N,N,N',N'-tetramethyl-1,4-benzendiamine, N,N,N',N'-tetramethyl-1,4-benzendiamine together with horse cytochrome c [9]; <2, 8, 16-21> quinol-oxidising enzyme [9, 10]; <1, 5, 6, 11-15> cytochrome c as mediator [10, 16]; <3, 9, 10> NAD(P)H and $NAD(P)^+$ as electron donor/acceptor [10, 11, 17]; <8> ascorbate and phenazine methosulfate, NADH, NADH/menadione, NADH/duroquinone, NADH/2,3-dimethoxy-5-methyl-1,4-benzoquinone and N,N,N',N'-tetramethyl-1,4-benzendiamine/ascorbate as electron donors [12]; <1> N,N,N',N'-tetramethyl-1,4-benzendiamine/ascorbate as electron donors [13]; <1> ferrocytochrome c-550 as electron donor [20]; <1> phenazine methosulfate and dithiothreitol as electron donors [22]; <1> ascorbate/phenazine methosulfate, N,N,N',N'-tetramethyl-1,4-benzendiamine/ascorbate, 2,3,5,6-tetramethyl-1,4-benzendiamine/ascorbate and NADH as electron donors [23]; <1> isoascorbate/2,3,5,6-tetramethyl-1,4-benzendiamine as electron donors and horse heart cytochrome c as mediator [25]; <1> NADH, succinate or isoascorbate plus 2,3,5,6-tetramethyl-1,4-benzendiamine as electron donors [26]; <5, 6> ascorbate/phenazine methosulfate as electron donors [27, 31]; <1, 7, 15, 22> ascorbate plus 2,3,5,6-tetramethyl-1,4-benzendiamine as electron donors [28]; <1> NADH, succinate or ascorbate plus 2,3,5,6-tetramethyl-1,4-benzendiamine as electron donors in membrane vesicles [29]; <23> lactate, leuco-forms of thionine, brilliant cresyl blue and methylene blue as electron donors [32]) (Reversibility: ? <1-23> [1-32]) [1-32]

P nitrous oxide + acceptor + H_2O

Inhibitors

KCN <1, 5> (<1> 50% inhibition at 0.3 mM [20]) [20, 27]
NH_2OH <23> (<23> at 5 mM 58% inhibition [32]) [32]
NO <1, 5-7> (<1,5-7> substrate inhibition [9, 13, 16]; <1> reversible inhibition [23]) [9, 13, 16, 23]
Triton X-100 <1, 7, 15, 22> (<1, 15, 22> 5% Triton X-100 results in complete inactivation after 4-5 h [28]; <7> 7.5% Triton X-100 results in complete inactivation after 6 h [28]) [28]
antimycin <1> (<1> 100% inhibition at 0.005 mg per mg vesicle protein, no inhibition with isoascorbate plus 2,3,5,6-tetramethyl-1,4-benzendiamine as electron donors [26]) [26]
antimycin A <1> (<1> 100% inhibition at 0.0075 mg per ml, no inhibition with ascorbate plus 2,3,5,6-tetramethyl-1,4-benzendiamine as electron donors [30]) [30]
elaidic acid <5> [21]
linoleic acid <5> [21]
linolenic acid <5> [21]
myxothiazol <1> (<1> 100% inhibition at 200 ng per mg vesicle protein, no inhibition with isoascorbate plus 2,3,5,6-tetramethyl-1,4-benzendiamine as electron donors [26]) [26]
o-phenanthroline <23> (<23> at 1 mM 60% inhibition [32]) [32]
oleic acid <5> [21]
p-chloromercuribenzoate <23> (<23> at 1 mM 85% inhibition [32]) [32]
palmitoleic acid <5> [21]

Cofactors/prosthetic groups

NADH <1, 3, 4, 8-10> (<4> enzyme activity measured in intact cells [8]) [8, 10-12, 15, 17, 23, 26, 29, 31]
NADPH <4, 9, 10> (<4> enzyme activity measured in intact cells [8]) [8, 10, 11, 17]
heme <1-21> [1, 5-17, 20-27]

Activating compounds

cardiolipin <5> [21]
n-octyl-β-D-glucopyranoside <5> (<5> 340% activity [27]) [27]
n-octyl-β-D-thioglucopyranoside <5> (<5> 340% activity [27]) [27]
phosphatidylglycerol <5> [21]
soybean phospholipids <5> (<5> 340% activity [27]) [27]

Metals, ions

Fe <1-21> (<1,5-7> 2 heme b, 1 heme c and 1 non-heme [1,5,9,16,18]; <1> heme c is six-coordinate, low-spin and in the NorC subunit and NorB contains one six-coordinate, low-spin heme b and a five-coordinate, high-spin heme b_3, which, together with a non-heme iron atom, forms the dinuclear NO reduction site [1,5]; <1> the dinuclear NO reduction site Fe-O-Fe can adopt two different conformations named open and closed, closed is the dominant one [1]; <1> heme c is bound via an Cys-Xxx-Xxx-Cys-His heme-binding motif [5]; <3> preparation of heme-reconstituted enzyme with different types of unnatural heme as 2,4-diethylheme, 2,4-deuteroheme and 2,4-diacetylheme [14]; <1> contains 24.9 nmol heme b and 14.2 nmol heme c per mg protein [20]) [1, 5-17, 20-27]

Turnover number (min^{-1})

12650 <3> (NO, <3> native enzyme [14]) [14]
24000 <10> (NO, <10> NADH as electron acceptor, cNor2 [17]) [17]
31500 <3> (NO, <3> NADH as electron acceptor [17]) [17]
32000 <10> (NO, <10> NADH as electron acceptor, cNor1 [17]) [17]
40000 <10> (NO, <10> NADPH as electron acceptor, cNor2 [17]) [17]

Specific activity (U/mg)

0.0021 <22> (<22> with ascorbate and 2,3,5,6-tetramethyl-1,4-benzoquinone as electron donors [28]) [28]
0.0028 <7> (<7> with ascorbate and 2,3,5,6-tetramethyl-1,4-benzoquinone as electron donors [28]) [28]
0.0033 <1> (<1> with ascorbate and 2,3,5,6-tetramethyl-1,4-benzoquinone as electron donors [28]) [28]
0.004 <15> (<15> with ascorbate and 2,3,5,6-tetramethyl-1,4-benzoquinone as electron donors [28]) [28]
0.101 <2> (<2> with reduced menadione as electron donor [2]) [2]
0.16 <8> (<8> with NADH and 2,3-dimethoxy-5-methyl-1,4-benzoquinone as electron donors [12]) [12]
0.22 <8> (<8> with N,N,N',N'-tetramethyl-1,4-benzendiamine and ascorbate as electron donors [12]) [12]
0.58 <1> [26]
0.78 <8> (<8> with NADH and duroquinone as electron donors [12]) [12]
1.53 <8> (<8> with NADH as electron donor [12]) [12]
2.27 <8> (<8> with NADH and menadione as electron donors [12]) [12]
2.79 <8> (<8> with ascorbate and phenazine methosulfate as electron donors [12]) [12]
10 <1> (<1> at pH 6.5 [23]) [23]
11 <1> [25]
11.8 <5> [27]
22 <1> (<1> at pH 5.0 [23]) [23]
22.2 <6> [18]
25.8 <1> (<1> with ascorbate, phenazine methosulfate and cytochrome c as electron donors [16]) [16]

60 <5> (<5> ascorbate and phenazine methosulfate as electron donors [21]) [21]
245 <4> (<4> with NADPH as electron donor [8]) [8]
295 <4> (<4> with NADH as electron donor [8]) [8]

K_m-Value (mM)

0.00025 <1> (NO) [20]
0.002 <5> (phenazine methosulfate) [27]
0.02 <10> (NO, <10> NADH as electron acceptor, cNor1 [17]) [17]
0.06 <3, 5> (NO, <3> NADH as electron acceptor [17]) [17, 27]
0.15 <10> (NO, <10> NADH and NADPH as electron acceptors, cNor2 [17]) [17]
0.22 <3> (NADH) [17]
0.32 <10> (NADH, <10> cNor1 [17]) [17]
0.32 <10> (NADPH, <10> cNor2 [17]) [17]
0.71 <10> (NADH, <10> cNor2 [17]) [17]

K_i-Value (mM)

0.0065 <5> (oleic acid) [21]
0.0098 <5> (palmitoleic acid) [21]
0.0125 <5> (linoleic acid) [21]
0.0135 <1> (NO) [16]
0.014 <5> (linolenic acid) [21]
0.0235 <5> (elaidic acid) [21]
2.6 <5> (KCN) [27]

pH-Optimum

4.8 <5> [21]
5 <1> [23]
5.2 <5> (<5> membrane-bound activity [27]) [27]
5.3 <6> [31]
6 <1> (<1> in 20 mM bis-Tris propane buffer 8-fold more activity than at pH 8.5 [3]) [3]
7-7.4 <23> [32]

pH-Range

4.8-6 <5> (<5> at pH 6.0 half of the activity of pH 4.8 [21]) [21]

4 Enzyme Structure

Molecular weight

43000 <4> (<4> SDS-PAGE [8]) [8]
69500 <1> (<1> DNA sequence [16]) [16]
75000 <8> (<8> SDS-PAGE [12]) [12]
84500 <8> (<8> amino acid composition [12]) [12]

Subunits

dimer <1, 5, 6> (<1> cytochrome bc complex, subunits named NorB and NorC [1,5,16]; <1,5> NorC is a monoheme membrane-anchored c-type chromosome, NorB is a divergent member of heme-copper oxidases [7,10]; <5> 1 * 17000 + 1 * 38000, SDS-PAGE [9, 21]; <1> 1 * 37000 + 1 * 17000, SDS-PAGE [16]; <6> 1 * 20000 + 1 * 42000, SDS-PAGE [18]; <1> 1 * 34000 + 1 * 15000, SDS-PAGE [20]) [1, 5, 7, 9, 10, 16, 18, 20, 21]

monomer <4, 8, 9> (<4, 8> 1 + 43000, SDS-PAGE [8,9]; <8> no NorC gene upstream NorB, contains an extension including two additional transmembrane spans and a (putative) periplasmic domain [10,12,19]) [8-10, 12, 19]

Posttranslational modification

lipoprotein <5> (<5> contains cardiolipin, phosphatidylglycerol and phosphatidylethanolamine [21]) [21]

5 Isolation/Preparation/Mutation/Application

Localization

membrane <1, 2, 6, 13> [2, 4, 16, 18, 25, 30]

membrane <1, 5-7, 15, 22> (<5> binuclear reaction centre of the enzyme is localised inside the lipidic bilayer of the cytoplasmic membrane [9]; <5> integral membrane protein [10]) [1, 9, 10, 26, 28]

mitochondrion <4> [8]

soluble <9> [10]

Purification

<1> (water-soluble NorC domain expressed in Escherichia coli [5]; using detergent lauryl maltoside [16]) [5, 16, 20, 25]

<3> [15]

<4> [8]

<5> [27]

<6> (with n-dodecyl-β-D-maltoside [18]) [18]

<8> [12]

Crystallization

<3> (in the ferric resting and in the ferrous carbonmonoxy states, at 1.0 and 1.05 A resolution, respectively [6]; at cryogenic temperature -173°C wild-type, S286V- and S286T-mutant [11]) [6, 11]

Cloning

<1> (cycA-promoter induced aerobic expression of the protein dimer in Paracoccus denitrificans, expression in Escherichia coli [7]) [7]

<1> (the water-soluble domain of NorC, cytochrome c type, expressed in Escherichia coli [5]) [5]

<3> (expressed in the soluble fraction of Escherichia coli JM109, modified N-terminus, 92% activity of wild-type [15]) [15]

Engineering

A239K <3> (<3> 6% activity of modified soluble wild-type enzyme [15]) [15]
A242K <3> (<3> 42% activity of modified soluble wild-type enzyme [15]) [15]
E202A <1> (<1> 70-85% activity of wild-type [7]) [7]
G240K <3> (<3> 63% activity of modified soluble wild-type enzyme [15]) [15]
M244K <3> (<3> 60% activity of modified soluble wild-type enzyme [15]) [15]
M247K <3> (<3> 130% activity of modified soluble wild-type enzyme [15]) [15]
N241K <3> (<3> 38% activity of modified soluble wild-type enzyme [15]) [15]
N246K <3> (<3> 45% activity of modified soluble wild-type enzyme [15]) [15]
S286T <3> (<3> 7% activity of wild-type [11]) [11]
S286V <3> (<3> 1% activity of wild-type [11]) [11]
T243H <3> (<3> 8% activity of modified soluble wild-type enzyme [15]) [15]
T243K <3> (<3> 12% activity of modified soluble wild-type enzyme [15]) [15]
T243R <3> (<3> 17% activity of modified soluble wild-type enzyme [15]) [15]

Application

environmental protection <1> (<1> removes cytotoxic nitrous oxide [3]; <4> mantains global environmental homeostasis [8]) [3, 8]

6 Stability

pH-Stability

6.5 <5> (<5> inactivation after 1 h at room temperature [21]) [21]

General stability information

<1>, 0.8% octyl glucoside is needed for solubilzation [23]
<1>, active under aerobic conditions [24]
<1>, after one cycle of freezing to -20°C and thawing stable, after more cycles activity is lost [26]
<1>, solubilized in dodecyl maltoside [25]
<5>, a phospholipid is required for high catalytic activity [21]

Storage stability

<1>, -20°C, stable [23]
<1>, -20°C, stable for 2 months [26]
<1>, 4°C, 2 d, 50% activity, the inclusion of 0.01% 2-phenylethanol, 1 mM benzamidine and 100 nM pepstatin stabilizes [23, 25]

<5>, 17°C, 0.1% laurylpropanediol-3-phosphorylcholine ether or 0.1% dodecylmaltoside, 8 days, 100% activity [21]
<5>, 17°C, Triton X-100, 2 days, 100% activity [21]

References

[1] Pinakoulaki, E.; Gemeinhardt, S.; Saraste, M.; Varotsis, C.: Nitric-oxide reductase. Structure and properties of the catalytic site from resonance Raman scattering. J. Biol. Chem., **277**, 23407-23413 (2002)
[2] Busch, A.; Friedrich, B.; Cramm, R.: Characterization of the norB gene, encoding nitric oxide reductase, in the nondenitrifying cyanobacterium Synechocystis sp. strain PCC6803. Appl. Environ. Microbiol., **68**, 668-672 (2002)
[3] Field, S.J.; Prior, L.; Roldan, M.D.; Cheesman, M.R.; Thomson, A.J.; Spiro, S.; Butt, J.N.; Watmough, N.J.; Richardson, D.J.: Spectral properties of bacterial nitric-oxide reductase: Resolution of pH-dependent forms of the active site heme b_3. J. Biol. Chem., **277**, 20146-20150 (2002)
[4] Hendriks, J.H.M.; Jasaitis, A.; Saraste, M.; Verkhovsky, M.I.: Proton and electron pathways in the bacterial nitric oxide reductase. Biochemistry, **41**, 2331-2340 (2002)
[5] Oubrie, A.; Gemeinhardt, S.; Field, S.; Marritt, S.; Thomson, A.J.; Saraste, M.; Richardson, D.J.: Properties of a soluble domain of subunit C of a bacterial nitric oxide reductase. Biochemistry, **41**, 10858-10865 (2002)
[6] Shimizu, H.; Park, S.Y.; Shiro, Y.; Adachi, S.: X-ray structure of nitric oxide reductase (cytochrome P450nor) at atomic resolution. Acta Crystallogr. Sect. D, **58**, 81-89. (2002)
[7] Butland, G.; Spiro, S.; Watmough, N.J.; Richardson, D.J.: Two conserved glutamates in the bacterial nitric oxide reductase are essential for activity but not assembly of the enzyme. J. Bacteriol., **183**, 189-199 (2001)
[8] Zhang, L.; Takaya, N.; Kitazume, T.; Kondo, T.; Shoun, H.: Purification and cDNA cloning of nitric oxide reductase cytochrome P_{450}nor (CYP55A4) from Trichosporon cutaneum. Eur. J. Biochem., **268**, 3198-3204. (2001)
[9] Koutny, M.: From no-confidence to nitric oxide acknowledgment: a story of bacterial nitric-oxide reductase. Folia Microbiol., **45**, 197-203 (2000)
[10] Hendriks, J.; Oubrie, A.; Castresana, J.; Urbani, A.; Gemeinhardt, S.; Saraste, M.: Nitric oxide reductases in bacteria. Biochim. Biophys. Acta, **1459**, 266-273 (2000)
[11] Shimizu, H.; Obayashi, E.; Gomi, Y.; Arakawa, H.; Park, S.Y.; Nakamura, H.; Adachi, S.I.; Shoun, H.; Shiro, Y.: Proton delivery in NO reduction by fungal nitric-oxide reductase. Cryogenic crystallography, spectroscopy, and kinetics of ferric-NO complexes of wild-type and mutant enzymes. J. Biol. Chem., **275**, 4816-4826 (2000)
[12] Cramm, R.; Pohlmann, A.; Friedrich, B.: Purification and characterization of the single-component nitric oxide reductase from Ralstonia eutropha H16. FEBS Lett., **460**, 6-10 (1999)

[13] Koutny, M.; Kucera, I.: Kinetic analysis of substrate inhibition in nitric oxide reductase of Paracoccus denitrificans. Biochem. Biophys. Res. Commun., **262**, 562-564. (1999)
[14] Singh, U.P.; Obayashi, E.; Takahashi, S.; Iizuka, T.; Shoun, H.; Shiro, Y.: The effects of heme modification on reactivity, ligand binding properties and iron-coordination structures of cytochrome P_{450}nor. Biochim. Biophys. Acta, **1384**, 103-111 (1998)
[15] Okamoto, N.; Tsuruta, K.; Imai, Y.; Tomura, D.; Shoun, H.: Fungal P450nor: expression in Escherichia coli and site-directed mutagenesis at the putative distal region. Arch. Biochem. Biophys., **337**, 338-344 (1997)
[16] Girsch, P.; de Vries, S.: Purification and initial kinetic and spectroscopic characterization of NO reductase from Paracoccus denitrificans. Biochim. Biophys. Acta, **1318**, 202-216 (1997)
[17] Toritsuka, N.; Shoun, H.; Singh, U.P.; Park, S.Y.; Iizuka, T.; Shiro, Y.: Functional and structural comparison of nitric oxide reductases from denitrifying fungi Cylindrocarpon tonkinense and Fusarium oxysporum. Biochim. Biophys. Acta, **1338**, 93-99 (1997)
[18] Sakurai, N.; Sakurai, T.: Isolation and characterization of nitric oxide reductase from Paracoccus halodenitrificans. Biochemistry, **36**, 13809-13815 (1997)
[19] Cramm, R.; Siddiqui, R.A.; Friedrich, B.: Two isofunctional nitric oxide reductases in Alcaligenes eutrophus H16. J. Bacteriol., **179**, 6769-6777 (1997)
[20] Fujiwara, T.; Fukumori, Y.: Cytochrome cb-type nitric oxide reductase with cytochrome c oxidase activity from Paracoccus denitrificans ATCC 35512. J. Bacteriol., **178**, 1866-1871 (1996)
[21] Kastrau, D.H.W.; Heiss, B.; Kroneck, P.M.H.; Zumft, W.G.: Nitric oxide reductase from Pseudomonas stutzeri, a novel cytochrome bc complex. Phospholipid requirement, electron paramagnetic resonance and redox properties. Eur. J. Biochem., **222**, 293-303 (1994)
[22] Turk, T.; Hollocher, T.C.: Oxidation of dithiothreitol during turnover of nitric oxide reductase: evidence for generation of nitroxyl with the enzyme from Paracoccus denitrificans. Biochem. Biophys. Res. Commun., **183**, 983-988 (1992)
[23] Dermastia, M.; Turk, T.; Hollocher, T.C.: Nitric oxide reductase. Purification from Paracoccus denitrificans with use of a single column and some characteristics [published erratum appears in J Biol Chem 1991 Aug 25;266(24):16256]. J. Biol. Chem., **266**, 10899-10905 (1991)
[24] Carr, G.J.; Ferguson, S.J.: Nitric oxide formed by nitrite reductase of Paracoccus denitrificans is sufficiently stable to inhibit cytochrome oxidase activity and is reduced by its reductase under aerobic conditions. Biochim. Biophys. Acta, **1017**, 57-62 (1990)
[25] Carr, G.J.; Ferguson, S.J.: The nitric oxide reductase of Paracoccus denitrificans. Biochem. J., **269**, 423-429 (1990)
[26] Carr, G.J.; Page, M.D.; Ferguson, S.J.: The energy-conserving nitric-oxide-reductase system in Paracoccus denitrificans. Distinction from the nitrite reductase that catalyses synthesis of nitric oxide and evidence from trap-

ping experiments for nitric oxide as a free intermediate during denitrification. Eur. J. Biochem., **179**, 683-692 (1989)
[27] Heiss, B.; Frunzke, K.; Zumft, W.G.: Formation of the N-N bond from nitric oxide by a membrane-bound cytochrome bc complex of nitrate-respiring (denitrifying) Pseudomonas stutzeri. J. Bacteriol., **171**, 3288-3297 (1989)
[28] Shapleigh, J.P.; Davies, K.J.P.; Payne, W.J.: Detergent inhibition of nitric-oxide reductase activity. Biochim. Biophys. Acta, **911**, 334-340 (1987)
[29] Carr, G.J.; Ferguson, S.J.: Nitric oxide reductase of Paracoccus denitrificans. Biochem. Soc. Trans., **16**, 187-188 (1988)
[30] Shapleigh, J.P.; Payne, W.J.: Nitric oxide-dependent proton translocation in various denitrifiers. J. Bacteriol., **163**, 837-840 (1985)
[31] Grant, M.A.; Cronin, S.E.; Hochstein, L.I.: Solubilization and resolution of the membrane-bound nitrite reductase from Paracoccus halodenitrificans into nitrite and nitric oxide reductases. Arch. Microbiol., **140**, 183-186 (1984)
[32] Miyata, M.; Matsubara, T.; Mori, T.: Studies on denitrification. XI. Some properties of nitric oxide reductase. J. Biochem., **66**, 759-765 (1969)

Cysteamine dehydrogenase 1.8.1.1

1 Nomenclature

EC number
1.8.1.1 (deleted)

Recommended name
cysteamine dehydrogenase

Sulfite reductase (NADPH) 1.8.1.2

1 Nomenclature

EC number
1.8.1.2

Systematic name
hydrogen-sulfide:NADP$^+$ oxidoreductase

Recommended name
sulfite reductase (NADPH)

Synonyms
H_2S-NADP oxidoreductase
NADPH-dependent sulfite reductase
NADPH-sulfite reductase
SIR-FP
SIR-HP
SIRHP
desulforubidin
reductase, sulfite (reduced nicotinamide adenine dinucleotide phosphate)

CAS registry number
9029-35-0

2 Source Organism

<1> *Escherichia coli* [1-3, 7, 8, 10-13, 15-17, 24-26, 29-31, 33-35, 37]
<2> *Salmonella typhimurium* [1, 8]
<3> *Saccharomyces cerevisiae* (var. ellipsoideus [9]) [4-6, 8, 9, 18, 20-23, 36]
<4> *Neurospora crassa* [8]
<5> *Saccharomyces bayanus* [9]
<6> *Desulfotomaculum nigrificans* [14]
<7> *Salmonella pullorum* [19]
<8> *Propionibacterium shermanii* [27]
<9> *Desulfosarcina variabilis* [28]
<10> *Actinobacillus pleuropneumoniae* [32]

3 Reaction and Specificity

Catalyzed reaction

hydrogen sulfide + 3 $NADP^+$ + 3 H_2O = sulfite + 3 NADPH + 3 H^+ (<1,2> electron flow sequence: NADPH-FAD-FMN [1]; <1> mechanism of electron transfer [12,13]; <3> ping-pong mechanism [6])

Reaction type

oxidation
redox reaction
reduction

Natural substrates and products

S sulfite + NADPH + H^+ <1, 3> (<1> no physiological nitrite reductase [13]) (Reversibility: ? <1, 3> [13, 18]) [13, 18]
P sulfide + $NADP^+$ + H_2O
S Additional information <1, 3> (<1> can generate tyrosil radicals e.g. for transfer of electrons to the iron center of metR2, small subunit of ribonucletide reductase [31]; <3> by production of the population synchronizer H_2S involved in ultradian metabolic oscillation [36]) [31, 36]
P ?

Substrates and products

S 1,1'-dimethyl-4,4'-bipyridinium dichloride + $NADP^+$ + H_2O <1> (<1> i.e. paraquat, a herbicide [33]) (Reversibility: ? <1> [32]) [33]
P NADPH + oxidized 1,1'-dimethyl-4,4'-bipyridinium dichloride
S $NADP^+$ + reduced methyl viologen + H_2O <1> (Reversibility: ? <1> [13]) [13]
P NADPH + oxidized methyl viologen
S hydroxylamine + NADPH <1> (<1> NADPH can be replaced by reduced methyl viologen [13]) (Reversibility: ? <1> [8, 13, 17, 20]) [8, 13, 17, 20]
P ammonia + $NADP^+$ + H_2O <1> [13, 17]
S nitrite + NADPH <1-4> (<1> NADPH can be replaced by reduced methyl viologen, the latter reaction is catalyzed by hemoprotein subunit alone [7]) (Reversibility: ? <1-4> [1, 7, 8, 13, 17, 18, 20]) [1, 7, 8, 13, 17, 18, 20]
P ammonia + $NADP^+$ + H_2O <1> [13, 17]
S sulfite + NADPH <1-3, 5> (<1,3> NADPH can be replaced by reduced methyl viologen [7, 13, 20]; <3> benzyl viologen [20]; <1-3> reduction of sulfite by reduced methyl viologen is catalyzed by hemoprotein subunit alone [8]; <3,5> sole compound reduced [9]) (Reversibility: ? <1-3, 5> [1, 7, 13, 17, 18, 20]) [1, 7-9, 13, 17, 18, 20]
P sulfide + $NADP^+$ + H_2O <1> [13]
S Additional information <1-3> (<1-3> reactions catalyzed by flavoprotein subunit alone: transfer of electrons from NADPH to cytochrome c, ferricyanide, dichlorphenolindophenol, menadione, FMN, FAD, O_2, [1, 8, 13, 17, 18, 20, 22]; <1-3> transfer of hydrogen from NADPH to 3-acetylpyridineadenine dinucleotide phosphate [1,8,13]; <1-3> reduction of sulfite, nitrite, and hydroxylamine by NADPH requires catalytic activities of both subunits,

reduction of sulfite, nitrite, and hydroxylamine by methyl viologen requires hemoprotein subunit [1,7,8]; <1-3> reactions of FMN depleted enzyme [8]; <3> overview on assay methods [18]) [1, 7, 8, 13, 17, 18, 20]

P ?

Inhibitors

2'-AMP <3> (<3> NADPH-sulfite reductase and NADPH-cytochrome c reductase [20]) [20]
8-hydroxyquinoline <3> [22]
Br^- <1> [13]
CN^- <1-5> (<3> severe inhibition of NADPH-sulfite,-nitrite,-hydroxylamine reductase, no inhibition of methyl viologen-sulfite reductase, small extent of inhibition of NADPH dependent reduction of cytochrome c, ferricyanide, quinones etc. [18, 20]; <3> only when added with NADPH prior to substrate, sulfite prevents NADPH-reduced form from irreversible conversion by CN^- [20]) [7-9, 13, 16-18, 20]
CO <1-5> (<1> binding to siroheme [10]) [7-10, 13, 16, 17, 22]
Cl^- <1> [13]
F^- <1> [13]
$NADP^+$ <1, 3> (<1,3> NADPH-dependent activities [13,17,18,20]; <3> competitive to NADPH, noncompetitive to sulfite [6]) [6, 13, 17, 18, 20]
NO_3^- <1> [13]
SCN^- <1, 3> [13, 22]
arsenite <1, 3> [8, 13, 17, 18, 20]
iodonium diphenyl chloride <1> (<1> acts on NADPH- and FAD-binding domain of flavoprotein component [30]) [30]
low ionic strength <3> (<3> NADPH-dependent activities [4,23]) [4, 23]
mepacrine <3> [22]
o-phenanthroline <3> [22]
p-chloromercuribenzoate <3> (<3> NADPH-dependent activities [18,20,22]) [18, 20, 22]
p-mercuriphenylsulfonate <1> (<1> NADPH-dependent reactions [8, 17]) [8, 13, 17]
sulfate <1> [13]
sulfide <1-3> (<1-3> reduction of sulfite, nitrite, hydroxylamine by reduced methyl viologen [8,18]) [8, 18]

Cofactors/prosthetic groups

flavin <1, 2, 3> (<1,2> 4 FAD and 4 FMN per mol of enzyme [1]; <3> 2 FAD and 2 FMN per mol of enzyme [6]; <1> treatment with *p*-chloromercuriphenylsulfonate causes dissociation of FMN but retetion of FAD and heme [13]; <3> treatment with ammonium sulfate causes dissociation of FMN but retetion of FAD and heme [18,23]) [1, 6, 13, 15, 18, 23]
heme <1, 3, 6, 8, 9> (<1,6,8> siroheme [2,14,27]; <1> tetrahydroporphyrin of isobacteriochlorin type with eight carboxylic side chains [15]; <1> hemoprotein subunit contains one siroheme and one Fe_4S_4 center per polypeptide [7]; <9> 2 fully metallated sirohemes per mol of enzyme, S=9/2 EPR iron-sulfur cluster [28]) [2, 3, 6, 7, 14, 27, 28]

Metals, ions

iron-sulfur cluster <1, 9> (<1> 4Fe-4S, 4 mol per tetramer of hemoprotein subunit, bound to hemoprotein subunit [3]; <1> hemoprotein subunit contains one siroheme and one Fe4S4 center per polypeptide, heme is in high spin Fe^{3+} state [7, 16]; <9> S = 9/2 EPR iron-sulfur cluster [28]) [3, 7, 16, 28]

Turnover number (min^{-1})

75 <1> (NADPH, <1> cosubstrate O_2 [13]) [13]
1800-2400 <1> (sulfite, <1> cosubstrate NADPH [13,17]) [13, 17]
2600-3100 <1> (nitrite, <1> cosubstrate NADPH [13,17]) [13, 17]
3900 <1> (sulfite, <1> cosubstrate reduced methyl viologen [13]) [13]
5900-13000 <1> (hydroxylamine, <1> cosubstrate NADPH [13,17]) [13, 17]
9000 <1> (FAD, <1> cosubstrate NADPH [17]) [17]
9500 <1> (NADPH, <1> cosubstrate 3-acetylpyridineadenine dinucleotide phosphate [13]) [13]
10300 <1> (NADPH, <1> cosubstrate FMN [13]) [13]
19100 <1> (NADPH, <1> cosubstrate cytochrome c [13]) [13]
27700 <1> (NADPH, <1> cosubstrate dichlorphenolindophenol [13]) [13]
28100 <1> (NADPH, <1> cosubstrate ferricyanide [13]) [13]
36300 <1> ($NADP^+$) [13]
36300 <1> (reduced methyl viologen) [13]
Additional information <1, 2> (<1,2> values for reactions catalyzed by holoenzyme, hemoprotein subunit alone, and flavoprotein subunit alone with various electron acceptors [1]) [1]

Specific activity (U/mg)

1.85 <3> [20]
2.73-2.87 <1> [16, 17]
52 <5> [9]
81 <2> [1]
Additional information <3> [22]

K_m-Value (mM)

0.0043-0.0074 <1> (sulfite) [8, 13, 17]
0.005 <1> (NADPH) [8]
0.01 <3> (NADPH) [6]
0.012-0.017 <1, 3> (sulfite) [6, 7, 18, 20]
0.018 <1, 3> (NADPH, <3> reduction of cytochrome c [18]) [17, 18]
0.021 <3> (NADPH, <3> reduction of sulfite [18]) [18]
0.06 <3> (FMN) [22]
0.06 <3> (cytochrome c) [18]
0.063 <3> (FMN) [22]
0.07 <1> (paraquat) [33]
0.08 <3> (FAD) [22]
0.08 <1> (NADPH) [33]
0.083 <3> (FAD) [22]
0.6 <3> (sulfite) [9]

0.8-1.5 <1, 3> (nitrite) [7, 13, 18, 20]
2 <5> (sulfite) [9]
4.5-10.5 <1, 3> (hydroxylamine) [7, 13, 18, 20]

K_i-Value (mM)
0.006-0.009 <1> (cyanide) [17]
0.018 <1> (iodonium diphenyl chloride) [30]
0.08-0.12 <1> (arsenide) [17]

pH-Optimum
7-7.5 <3> (<3> NADPH-sulfite reductase [18,20]) [18, 20]
7.2 <1> (<1> electron donor NADPH [17]) [17]
7.5-7.9 <3> (<3> NADPH-cytochrome c reductase [18,20]) [18, 20]
7.7-8.5 <3> (<3> methyl viologen-sulfite reductase [18]) [18]
7.9 <1> (<1> sulfite [13]) [13]
8.6 <1> (<1> nitrite [13]) [13]
9.5 <1> (<1> hydroxylamine [13]) [13]

4 Enzyme Structure

Molecular weight
200000-208000 <9> (<9> gel filtration [28]) [28]
300000 <3> (<3> gel filtration, sedimentation analysis at low ionic strength, dissociated [5]) [5]
350000 <3> (<3> calculation from FAD and FMN content, sedimentation coefficient [18,20]) [18, 20]
488000 <2> (<2> sedimentation under nondenaturing conditions [1]) [1]
604000 <3> (<3> gel filtration [6]) [6]
650000 <3> (<3> gel filtration, sedimentation analysis [5]) [5]
670000 <1> (<1> sedimentation equilibrium centrifugation, sedimentation and diffusion coefficients [16]) [16]

Subunits
dodecamer <1> (<1> α_8,β_4, 8 * 58000-60000 + 4 * 54000-57000, sedimentation under denaturing conditions [1, 8, 11]; <1> SDS-PAGE [11,25]) [1, 8, 11, 25]
hexamer <9> (<9> $\alpha_2,\beta_2,\gamma_2$, 2 * 50000 + 2 *42000 + 2 * 12500, SDS-PAGE [28]) [28]
tetramer <3> (<3> α_2,β_2, 2 * 116000 + 2 * 167000, SDS-PAGE [6]) [6]
Additional information <1> (<1> dissociation of enzyme by 5 m urea into flavoprotein octamer and hemoprotein monomers [8,11]; <1> α chain is composed of 2 distinct domains, one binding FAD and the other binding FMN, the FMN binding domains cooperate for a head-to-head subunit interaction [34]) [8, 11, 34]

5 Isolation/Preparation/Mutation/Application

Localization

soluble <3> [18, 20]

Purification

<1> (preparation of FMN depleted enzyme [12]) [12, 16, 17]
<2> [1]
<3> (var. ellipsoideus (partial) [9]; <3> preparation of mutant form lacking FAD and hence with no NADPH-sulfite reductase activity but with methyl viologen-sulfite reductase activity [21]) [9, 18, 20, 21, 22]
<5> (partial [9]) [9]

Renaturation

<1> (mixing isolated hemoprotein and flavoprotein in appropriate proportions reconstituts activity [11]) [11]

Crystallization

<1> (hemoprotein subunit [3,26]; <1> X-ray structure of hemoprotein subunit [3]; <1> recombinant monomeric fragment of flavoprotein, having 3 domains and binding FAD and NADPH [35]; <1> FNR-like domain of flavoprotein [37]) [3, 26, 34, 35, 37]

Cloning

<1> (FMN-binding domain of flavoprotein component [29]) [29]
<1> (NADPH- and FAD-binding domain of flavoprotein component [30]) [30]
<1> (α and β subunits [24,25]) [24, 25]
<10> (homology to cysI of Escherichia coli, i.e. α subunit [32]) [32]

Engineering

Additional information <3> (<3> 4 mutants blocked in sulfite reduction, 3 containing only FMN, 1 lacking both flavins [21]) [21]

Application

medicine <10> (<10> potential use of vaccine against Actinobacillus pleuropneumoniae infections [32]) [32]

6 Stability

Temperature stability

25 <7> (<7> stable at [19]) [19]
45 <7> (<7> instable at [19]) [19]
50 <3> (<3> inactivation above [23]) [23]

Organic solvent stability

acetone <3> (<3> inactivation [18]) [18]
ethanol <3> (<3> inactivation [18]) [18]

General stability information

<3>, low ionic strength such as 0.01 M phosphate buffer destroys NADPH-dependent activities [4, 5, 18]

Storage stability

<1>, -15°C, 0.05 M potassium phosphate buffer pH 7.7, 0.1 mM EDTA, 2 years stable [16]
<1>, -15°C, 0.05 M potassium phosphate buffer pH 7.7, 10 mM EDTA, at least 6 months [17]
<1>, -15°C, reduced hemoprotein subunit or reduced flavoprotein subunit [11]
<1>, 4°C, 0.05 M potassium phosphate buffer pH 7.7, 0.1 mM EDTA, 1 month stable [16]
<3>, -20°C, 0.3 M phosphate buffer pH 7.3, 1 year stable stable [18, 20]
<3>, 4°C, 0.3 M phosphate buffer pH 7.3, 1 week stable [18, 20]

References

[1] Ostrowski, J.; Barber, M.J.; Rueger, D.C.; Miller, B. E.; Siegel, L.M.; Kredich, N.M.: Characterization of the flavoprotein moieties of NADPH-sulfite reductase from Salmonella typhimurium and Escherichia coli. Physicochemical and catalytic properties, amino acid sequence deduced from DNA sequence of cysJ, and comparison with NADPH-cytochrome P-450 reductase. J. Biol. Chem., **264**, 15796-15808 (1989)

[2] Young, L.J.; Siegel, L.M.: Activated conformers of Escherichia coli sulfite reductase heme protein subunit [published erratum appears in Biochemistry 1988 Dec 27;27(26):9226]. Biochemistry, **27**, 4991-4999 (1988)

[3] McRee, D.E.; Richardson, D.C.; Richardson, J.S.; Siegel, L.M.: The heme and Fe_4S_4 cluster in the crystallographic structure of Escherichia coli sulfite reductase. J. Biol. Chem., **261**, 10277-10281 (1986)

[4] Kobayashi, K.; Yoshimoto, A.: Studies on yeast sulfite reductase. V. Effects of ionic strength on enzyme activities. Biochim. Biophys. Acta, **709**, 38-45 (1982)

[5] Kobayashi, K.; Yoshimoto, A.: Studies on yeast sulfite reductase. VI. Use of the effects of ionic strength as a probe for enzyme structure and mechanism. Biochim. Biophys. Acta, **709**, 46-52 (1982)

[6] Kobayashi, K.; Yoshimoto, A.: Studies on yeast sulfite reductase. IV. Structure and steady-state kinetics. Biochim. Biophys. Acta, **705**, 348-356 (1982)

[7] Siegel, L.M.; Rueger, D.C.; Barber, M.J.; Krueger, R. J.; Orme-Johnson, N.R.; Orme-Johnson, W.H.: Escherichia coli sulfite reductase hemoprotein subunit. Prosthetic groups, catalytic parameters, and ligand complexes. J. Biol. Chem., **257**, 6343-6350 (1982)

[8] Hatefi, Y.; Stiggall, D.L.: Metal-containing flavoprotein dehydrogenases. The Enzymes, 3rd Ed. (Boyer, P.D., ed.), **13**, 175-297 (1976)

[9] Dott, W.; Truper, H.G.: Sulfite formation by wine yeasts. Arch. Microbiol., **108**, 99-104 (1976)

[10] Murphy, M.J.; Siegel, L.M.; Kamin, H.: Reduced nicotinamide adenine dinucleotide phosphate-sulfite reductase of enterobacteria. VI. The reaction of

carbon monoxide with the Escherichia coli holoenzyme, the hemoprotein, and free siroheme. J. Biol. Chem., **249**, 1610-1614 (1974)
[11] Siegel, L.M.; Davis, P-S.: Reduced nicotinamide adenine dinucleotide phosphate-sulfite reductase of enterobacteria. IV. The Escherichia coli hemoflavoprotein: subunit structure and dissociation into hemoprotein and flavoprotein components. J. Biol. Chem., **249**, 1587-1598 (1974)
[12] Faeder, E.J.; Davis, P.S.; Siegel, L.M.: Reduced nicotinamide adenine dinucleotide phosphate-sulfite reductase of enterobacteria. V. Studies with the Escherichia coli hemoflavoprotein depleted of flavin mononucleotide: distinct roles for the flavin adenine dinucleotide and flavin mononucleotide prosthetic groups in catalysis. J. Biol. Chem., **249**, 1599-1609 (1974)
[13] Siegel, L.M.; Davis, P.S.; Kamin, H.: Reduced nicotinamide adenine dinucleotide phosphate-sulfite reductase of enterobacteria. 3. The Escherichia coli hemoflavoprotein: catalytic parameters and the sequence of electron flow. J. Biol. Chem., **249**, 1572-1586 (1974)
[14] Murphy, M.J.; Siegel, L.M.: Siroheme and sirohydrochlorin. The basis for a new type of porphyrin-related prosthetic group common to both assimilatory and dissimilatory sulfite reductases. J. Biol. Chem., **248**, 6911-6919 (1973)
[15] Murphy, M.J.; Siegel, L.M.; Kamin, H.: Reduced nicotinamide adenine dinucleotide phosphate-sulfite reductase of enterobacteria. II. Identification of a new class of heme prosthetic group: an iron-tetrahydroporphyrin (isobacteriochlorin type) with eight carboxylic acid groups. J. Biol. Chem., **248**, 2801-2814 (1973)
[16] Siegel, L.M.; Murphy, M.J.; Kamin, H.: Reduced nicotinamide adenine dinucleotide phosphate-sulfite reductase of enterobacteria. I. The Escherichia coli hemoflavoprotein: molecular parameters and prosthetic groups. J. Biol. Chem., **248**, 251-264 (1973)
[17] Siegel, L.M.; Kamin, H.: TPNH-Sulfite reductase (Escherichia coli). Methods Enzymol., **17B**, 539-545 (1971)
[18] Yoshimoto, A.; Naiki, N.; Sato, R.: Sulfite reductase (baker's yeast). Methods Enzymol., **17B**, 520-528 (1971)
[19] Hoeksema, W.D.; Schoenhard, D.E.: Characterization of a thermolabile sulfite reductase from Salmonella pullorum. J. Bacteriol., **108**, 154-158 (1971)
[20] Yoshimoto, A.; Sato, R.: Studies on yeast sulfite reductase. I. Purification and characterization. Biochim. Biophys. Acta, **153**, 555-575 (1968)
[21] Yoshimoto, A.; Sato, R.: Studies on yeast sulfite reductase. II. Partial purification and properties of genetically incomplete sulfite reductases. Biochim. Biophys. Acta, **153**, 576-588 (1968)
[22] Prabhakararao, K.; Nicholas, D.J.D.: Sulphite reductase from bakers yeast: a haemoflavoprotein. Biochim. Biophys. Acta, **180**, 253-263 (1969)
[23] Yoshimoto, A.; Sato, R.: Studies on yeast sulfite reductase. 3. Further characterization. Biochim. Biophys. Acta, **220**, 190-205 (1970)
[24] Tei, H.; Murata, K.; Kimura, A.: Molecular cloning of the cys genes (cysC, cysD, cysH, cysI, cysJ, and cysG) responsible for cysteine biosynthesis in Escherichia coli K-12. Biotechnol. Appl. Biochem., **12**, 212-216 (1990)

[25] Li, C.; Peck, H.D.; Przybyla, A.E.: Cloning of the 3-phosphoadenylyl sulfate reductase and sulfite reductase genes from Escherichia coli K-12. Gene, **53**, 227-234 (1987)
[26] McRee, D.E.; Richardson, D.C.: Preliminary X-ray diffraction studies on the hemoprotein subunit of Escherichia coli sulfite reductase. J. Mol. Biol., **154**, 179-180 (1982)
[27] Scott, A.I.; Irwin, A.J.; Siegel, L.M.; Schoolery, J. N.: Sirohydrochlorin. Prosthetic group of sulfite and nitrite reductases and its role in the biosynthesis of vitamin B_{12}. J. Am. Chem. Soc., **100**, 7987-7994 (1978)
[28] Arendsen, A.F.; Verhagen, M.F.J.M.; Wolbert, R.B.G.; Pierik, A.J.; Stams, A.J.M.; Jetten, M.S.M.; Hagen, W.R.: The dissimilatory sulfite reductase from Desulfosarcina variabilis is a desulforubidin containing uncoupled metalated sirohemes and S = 9/2 iron-sulfur clusters. Biochemistry, **32**, 10323-10330 (1993)
[29] Coves, J.; Zeghouf, M.; Macherel, D.; Guigliarelli, B.; Asso, M.; Fontecave, M.: Flavin mononucleotide-binding domain of the flavoprotein component of the sulfite reductase from Escherichia coli. Biochemistry, **36**, 5921-5928 (1997)
[30] Coves, J.; Lebrun, C.; Gervasi, G.; Dalbon, P.; Fontecave, M.: Overexpression of the FAD-binding domain of the sulphite reductase flavoprotein component from Escherichia coli and its inhibition by iodonium diphenyl chloride. Biochem. J., **342**, 465-472. (1999)
[31] Coves, J.; Niviere, V.; Eschenbrenner, M.; Fontecave, M.: NADPH-sulfite reductase from Escherichia coli. A flavin reductase participating in the generation of the free radical of ribonucleotide reductase. J. Biol. Chem., **268**, 18604-18609 (1993)
[32] Willson, P.J.; Gerlach, G.F.; Klashinsky, S.; Potter, A.A.: Cloning and characterization of the gene coding for NADPH-sulfite reductase hemoprotein from Actinobacillus pleuropneumoniae and use of the protein product as a vaccine. Can. J. Vet. Res., **65**, 206-212 (2001)
[33] Gaudu, P.; Fontecave, M.: The NADPH:sulfite reductase of Escherichia coli is a paraquat reductase. Eur. J. Biochem., **226**, 459-463 (1994)
[34] Eschenbrenner, M.; Coves, J.; Fontecave, M.: NADPH-sulfite reductase flavoprotein from Escherichia coli: contribution to the flavin content and subunit interaction. FEBS Lett., **374**, 82-84 (1995)
[35] Gruez, A.; Pignol, D.; Zeghouf, M.; Coves, J.; Fontecave, M.; Ferrer, J.L.; Fontecilla-Camps, J.C.: Four crystal structures of the 60 kDa flavoprotein monomer of the sulfite reductase indicate a disordered flavodoxin-like module. J. Mol. Biol., **299**, 199-212 (2000)
[36] Sohn, H.Y.; Kuriyama, H.: Ultradian metabolic oscillation of Saccharomyces cerevisiae during aerobic continuous culture: hydrogen sulphide, a population synchronizer, is produced by sulphite reductase. Yeast, **18**, 125-135 (2001)
[37] Gruez, A.; Zeghouf, M.; Bertrand, J.; Eschenbrenner, M.; Coves, J.; Fontecave, M.; Pignol, D.; Fontecilla-Camps, J.C.: The FNR-like domain of the Escherichia coli sulfite reductase flavoprotein component: crystallization and preliminary X-ray analysis. Acta Crystallogr. Sect. D, **54**, 135-136 (1998)

Hypotaurine dehydrogenase 1.8.1.3

1 Nomenclature

EC number
1.8.1.3

Systematic name
hypotaurine:NAD^+ oxidoreductase

Recommended name
hypotaurine dehydrogenase

CAS registry number
37256-46-5

2 Source Organism

<1> *Rattus norvegicus* [1]

3 Reaction and Specificity

Catalyzed reaction
hypotaurine + H_2O + NAD^+ = taurine + NADH + H^+ (a molybdohemoprotein)

Reaction type
oxidation
redox reaction
reduction

Natural substrates and products
S hypotaurine + H_2O + NAD^+ <1> (<1> metabolism of cysteine [1]) (Reversibility: ? <1> [1]) [1]
P taurine + NADH <1> [1]

Substrates and products
S hypotaurine + H_2O + NAD^+ <1> (Reversibility: ? <1> [1]) [1]
P taurine + NADH <1> [1]

Inhibitors
KCN <1> (<1> 0.1 mM [1]) [1]
ammonium sulfate <1> (<1> 10 mM [1]) [1]
phosphate <1> (<1> 10 mM [1]) [1]

Cofactors/prosthetic groups

NAD$^+$ <1> [1]
NADP$^+$ <1> (<1> slightly effective [1]) [1]

Activating compounds

2-mercaptoethanolamine <1> (<1> 1 mM [1]) [1]
cysteine <1> (<1> 1 mM [1]) [1]

pH-Optimum

7.5-7.8 <1> [1]

5 Isolation/Preparation/Mutation/Application

Source/tissue

liver <1> [1]

Purification

<1> (ammonium sulfate precipitation, dialysis [1]) [1]

6 Stability

Storage stability

<1>, frozen, 0.1 M Tris-HCl, pH 7.4, 36 h 50% loss of activity, loses total activity within 4 days [1]

References

[1] Sumizu, K.: Oxidation of hypotaurine in rat liver. Biochim. Biophys. Acta, **63**, 210 - 212 (1962)

Dihydrolipoamide dehydrogenase 1.8.1.4

1 Nomenclature

EC number
1.8.1.4

Systematic name
dihydrolipoamide:NAD^+ oxidoreductase

Recommended name
dihydrolipoamide dehydrogenase

Synonyms
DHLDH
DLD <14> [12]
DLDH
E3
E3 component of 2-oxoglutarate dehydrogenase complex
E3 component of acetoin cleaving system
E3 component of α keto acid dehydrogenase complexes
E3 component of pyruvate and 2-oxoglutarate dehydrogenases complexes
E3 component of pyruvate complex
E3 lipoamide dehydrogenase
EC 1.6.4.3 (formerly)
glycine cleavage system L protein
glycine oxidation system L-factor
LDP-Glc
LDP-Val
LPD
LPD-GLC
LPD-VAL
NADH:lipoamide oxidoreductase
ORF-E3
dehydrogenase, lipoamide
dehydrolipoate dehydrogenase
diaphorase
dihydrolipoamide dehydrogenase E3 <3, 7, 8, 9, 10, 13, 14, 15, 16, 17, 18, 19, 20, 21, 22> (<3, 7, 8, 9, 10, 13, 14, 15, 16, 17, 18, 19, 20, 21, 22>, common component of the three 2-oxoacid dehydrogenase complexes oxidizing pyruvate, 2-oxoglutarate, and the branched-chain 2-oxo acids [11]) [11]
dihydrolipoic dehydrogenase
dihydrolipoyl dehydrogenase
lipoamide dehydrogenase (NADH)

lipoamide oxidoreductase (NADH)
lipoamide reductase
lipoamide-dehydrogenase-valine <8, 9> (<8>, is the specific E3 subunit for branched-chain keto acid dehydrogenase [7]) [7]
lipoate dehydrogenase
lipoic acid dehydrogenase
lipoyl dehydrogenase

CAS registry number

9001-18-7

2 Source Organism

<1> *Pelobacter carbinolicus* [1]
<2> *Trypanosoma cruzi* [2, 16]
<3> *Geobacillus stearothermophilus* (dihydrolipoamide dehydrogenase component of the pyruvate dehydrogenase multienzyme complex [3]; dihydrolipoamide dehydrogenase E3 [11]) [3, 11]
<4> *Peptostreptococcus glycinophilus* [4]
<5> *Clostridium cylindrosporum* [4]
<6> *Trypanosoma brucei* [5]
<7> *Escherichia coli* (dihydrolipoamide dehydrogenase E3 [11]; diyhdrolipoamide dehydrogenase component of the pyruvate dehydrogenase complex [28, 37]; K-12 [11, 28, 33]; strain B [33]; strain M191-6 [33]; strain Crookes [33]) [6, 11, 27, 28, 29, 32, 33, 34, 37, 39]
<8> *Pseudomonas putida* (strain PpG2 [7]; dihydrolipoamide dehydrogenase E3 [11]; two lipoamide dehydrogenases: LPD-glc and LPD-val [20]) [7, 11, 20]
<9> *Pseudomonas aeruginosa* (dihydrolipoamide dehydrogenase E3 [11]) [7, 11]
<10> *Bacillus subtilis* (dihydrolipoamide dehydrogenase E3 [11]) [8, 11, 23, 33, 39]
<11> *Eubacterium acidaminophilum* [9]
<12> *Corynebacterium glutamicum* [50]
<13> *Azotobacter vinelandii* (dihydrolipoamide dehydrogenase E3 [11,19]) [11, 13, 19, 33]
<14> *Halobacterium halobium* (dihydrolipoamide dehydrogenase E3 [11]) [11, 12, 18, 21]
<15> *Saccharomyces cerevisiae* (dihydrolipoamide dehydrogenase E3 [11]) [11, 15, 24, 33, 36, 39]
<16> *Saccharomyces carlsbergensis* (dihydrolipoamide dehydrogenase E3 [11]) [11, 35]
<17> *Pisum sativum* (dihydrolipoamide dehydrogenase E3 [11]) [11, 34, 51, 52, 47]
<18> *Ascaris suum* (dihydrolipoamide dehydrogenase E3 [11]) [11, 31]
<19> *Rattus norvegicus* (dihydrolipoamide dehydrogenase E3 [11]) [11, 17, 26, 33]

<20> *Sus scrofa* (dihydrolipoamide dehydrogenase E3 [9,11]) [9, 11, 22, 33, 36, 40]
<21> *Bos taurus* [11, 33]
<22> *Homo sapiens* [11, 33, 43, 44]
<23> *yeast* (dihydrolipoamide dehydrogenase E3 [10]) [10]
<24> *Salmonella typhimurium* [25]
<25> *Pseudomonas fluorescens* [13, 33, 39]
<26> *Azotobacter agilis* [33, 39]
<27> *Thermoplasma acidophilum* [14]
<28> *Halobacterium volcanii* [15, 21]
<29> *Natronobacterium pharaonis* [21]
<30> *Natronobacterium gregoryi* [21]
<31> *Natronococcus occultus* [21]
<32> *Malbranchea pulchella* [30]
<33> *Squalus acanthias* [33]
<34> *Spinacia oleracea* [33, 41]
<35> *Brassica oleracea* [33]
<36> *Saccharomyces oviformis* [33]
<37> *Candida crusei* [33, 38]
<38> *Neurospora crassa* [33, 39]
<39> *Pythium ultimum* [33]
<40> *Phytophthora erythroseptica* [33]
<41> *Leuconostoc mesenteroides* [33]
<42> *Mycobacterium tuberculosis* [33, 46]
<43> *Peptococcus glycinophilus* [33]
<44> *Proteus vulgaris* [33]
<45> *Streptococcus faecalis* [33]
<46> *Serratia marcescens* [33, 39]
<47> *Synechocystis sp.* (PCC6803 [48]) [48]
<48> *Glycine max* [42]
<49> *Streptococcus pneumoniae* [45]
<50> *Streptomyces seoulensis* [49]
<51> *Mycobacterium smegmatis* [47]

3 Reaction and Specificity

Catalyzed reaction

dihydrolipoamide + NAD^+ = lipoamide + NADH + H^+ (<14>, involvement of a reversibly reducible disulfide bond in catalytic mechanism [18]; <7, 42, 50, 51>, ping-pong mechanism [28, 46, 47, 49])

Reaction type

oxidation
redox reaction
reduction

Natural substrates and products

S Additional information <3, 7, 8, 9, 10, 12, 13, 14, 15, 16, 17, 18, 19, 20, 21, 22, 27, 49> (<7>, the most important function of dehydrolipoamide dehydrogenase as a component of the pyruvate dehydrogenase and the 2-oxoglutarate dehydrogenase complex is the implication in the oxidative decarboxylation of pyruvate and 2-oxoglutarate [6]; <27>, the enzyme fulfills its function in the pyruvate, 2-oxoglutarate and branched-chain 2-oxoacid dehydrogenase complexes and in the glycine cleavage system [14]; <8>, LPD-Val is specifically required as the lipoamide dehydrogenase of branched-chain keto acid dehydrogenase, LPD-Glc fulfills all other requirements for lipoamide dehydrogenase [20]; <49>, lack of dihydrolipoamide dehydrogenase results in a deficiency in α-galactoside metabolism and galactose transport [45]; <12>, the enzyme is an essential component of the pyruvate dehydrogenase and 2-oxoglutarate dehydrogenase complexes [50]; <3, 7, 8, 9, 10, 13, 14, 15, 16, 17, 18, 19, 20, 21, 22>, the enzyme is a component of the three 2-oxoacid dehydrogenase complexes oxidizing pyruvate, 2-oxoglutarate, and the branched-chain 2-oxo acids [11]; <17>, the physiological substrates are the dihydrolipoyl domain of the E2 component, dihydrolipoyl acyltransferase, of the 2-oxoacid dehydrogenase multienzyme complexs or the dihydrolipoyl H-protein of the mitochobdrial glycine decarboxylase [51]) [6, 11, 14, 20, 45, 50, 51]

P ?

Substrates and products

S 1,4-benzoquinone + NADH <50, 51> (Reversibility: ? <50, 51> [47, 49]) [47, 49]

P 1,4-benzoquinol + NAD^+

S 2,6-dichlorophenolindophenol + NADH <1, 7, 12, 18, 34, 37, 50> (<1>, 12.3% of the activity with lipoamide [1]; <7>, activity with wild-type enzyme and mutant enzymes C44S and C49S [29]) (Reversibility: ? <1, 7, 12, 18, 34, 37, 50> [1, 23, 29, 31, 38, 41, 49, 50]) [1, 23, 29, 31, 38, 41, 49, 50]

P ? + NAD^+

S 2,6-dimethyl-1,4-benzoquinone + NADH <42> (Reversibility: ? <42> [46]) [46]

P 2,6-dimethyl-1,4-benzoquinol + NAD^+

S 2-(*p*-iodophenyl)-3-*p*-nitrophenyl-5-phenyltetrazolium chloride + NADH <12> (Reversibility: ? <12> [50]) [50]

P ? + NAD^+

S 2-hydroxy-1,4-benzoquinone + NADH <50> (Reversibility: ? <50> [49]) [49]

P 2-hydroxy-1,4-benzoquinol + NAD^+

S 2-methyl-1,4-benzoquinone + NADH <50> (Reversibility: ? <50> [49]) [49]

P 2-methyl-1,4-benzoquinol + NAD^+

S 5-hydroxy-1,4-naphthoquinone + NADH <42> (Reversibility: ? <42> [46]) [46]

P 5-hydroxy-1,4-naphthoquinol + NAD^+

S 5-nitroblue tetrazolium chloride + NADH <1> (<1>, 5.1% of the activity with lipoamide [1]) (Reversibility: ? <1> [1]) [1]
P ? + NAD^+
S O_2 + NADH <7, 42> (<7>, activity with wild-type enzyme and mutant enzymes C44S and C49S [29]) (Reversibility: ? <7, 42> [29, 46]) [29, 46]
P ?
S benzyl viologen + NADH <1> (<1>, 2.9% of the activity with lipoamide [1]) (Reversibility: ? <1> [1]) [1]
P ? + NAD^+
S dihydrolipoamide + NAD^+ <1-51> (<27>, alternate oxidation and reduction of an intrachain disulfide bond [14]; <7>, mutant enzymes C44S and C49S show minute activity [29]) (Reversibility: r <1, 18, 34, 37> [1, 31, 38, 41, 44]; ? <2-19, 20, 21, 23-33, 35, 36, 38-51> [2-52]) [1-52]
P lipoamide + NADH
S dihydrolipoamide + $NADP^+$ <5> (Reversibility: r <5> [4]) [4]
P lipoamide + NADH <5> [4]
S ferricyanide + NADH <1, 7, 20, 34, 50> (<1>, 31.1% of the activity with lipoamide [1]; <7>, activity with wild-type enzyme and mutant enzymes C44S and C49S [29]) (Reversibility: ? <1, 7, 20, 34, 50> [1, 22, 29, 41, 49]) [1, 22, 29, 41, 49]
P ferrocyanide + NAD^+
S hexacyanoferrate + NADH <15> (Reversibility: ? <15> [24]) [24]
P ? + NAD^+
S iodonitrotetrazolium + NADH <15> (Reversibility: ? <15> [24]) [24]
P ? + NAD^+
S lipoamide + 3-acetylpyridine adenine dinucleotide <42> (Reversibility: ? <42> [46]) [46]
P dihydrolipoamide + ?
S lipoamide + NADH <14, 48> (Reversibility: ? <14, 48> [12, 42]) [12, 42]
P dihydrolipoamide + NAD^+
S lipoamide + nicotinamide hypoxanthine dinucleotide <42> (Reversibility: ? <42> [46]) [46]
P dihydrolipoamide + ?
S lipoamide + thio-NADH <42> (Reversibility: ? <42> [46]) [46]
P dihydrolipoamide + thio-NAD^+
S lipoic acid + NADH <14, 42> (Reversibility: ? <14, 42> [12, 46]) [12, 46]
P dihydrolipoic acid + NAD^+
S menadione + NADH <1> (<1>, 13.6% of the activity with lipoamide [1]) (Reversibility: ? <1> [1]) [1]
P ? + NAD^+
S methylene blue + NADH <1, 37> (<1>, 9.4% of the activity with lipoamide [1]) (Reversibility: ? <1, 37> [1, 38]) [1, 38]
P ? + NAD^+
S naphthoquinone + NADH <50, 51> (Reversibility: ? <50, 51> [47, 49]) [47, 49]
P 1,4-naphthoquinol + NAD^+

S thio-NAD^+ + NADH <7, 18, 42> (<7>, activity with wild-type enzyme and mutant enzymes C44S and C49S [29]) (Reversibility: ? <7, 18, 42> [29, 31, 46]) [29, 31, 46]
P thio-NADH + NAD^+
S vitamin K_5 + NADH <34> (Reversibility: ? <34> [41]) [41]
P ?
S Additional information <17, 18, 20, 32, 42> (<20,32>, diaphorase activity [30,40,42]; <18>, NADH:NAD^+ transhydrogenase activity [31]; <42>, reductive half-reaction, hydride transfer from NADH to FAD, is rate limiting when a quinone is the oxidant [46]; <17>, EC 1.8.1.4 is the E3-protein component of the mitochondrial 2-oxoacid dehydrogenase multienzyme complexes and the L-protein component of the glycine decarboxylase system [51]) [30, 31, 40, 46, 50, 51]
P ?

Inhibitors

1,3-bis(2-chloroethyl)-1-nitrourea <2> (<2>, after reduction of the oxidized form of enzyme to the two-electron-reduced state [2]) [2]
2-amino-4-hydroxy-6,7-dimethyl-5,6,7,8-tetrahydropteridine hydrochloride <20> (<20>, inhibition of NADH-lipoamide oxidoreductase activity, no effect on diaphorase activity and transhydrogenase activity [40]) [40]
2-amino-4-hydroxy-6,7-dimethyl-7,8-dihydropteridine <20> (<20>, inhibition of NADH-lipoamide oxidoreductase activity, no effect on diaphorase activity and transhydrogenase activity [40]) [40]
2-amino-4-hydroxy-6-methyl-7,8-dihydropteridine <20> (<20>, inhibition of NADH-lipoamide oxidoreductase activity, no effect on diaphorase activity and transhydrogenase activity [40]) [40]
Cd^{2+} <7> (<7>, in presence of NADH, inhibition is reversed by dithiols and less effectively by monothiols [37]) [37]
NAD^+ <50, 51> (<51>, substrate inhibition. The rate of lipoamide reduction is dependent upon the NAD^+/NADH ratio, the reaction is activated at low ratios and inhibited at high ratios [47]) [47, 49]
NADH <7, 15, 18, 37, 51> (<15,18>, competitive with respect to NAD^+ [24,31]; <17>, chloroplastic enzyme is more susceptible to product inhibition than the mitochondrial enzyme [34]; <51>, substrate inhibition. The rate of lipoamide reduction is dependent upon the NAD^+/NADH ratio, the reaction is activated at low ratios and inhibited at high ratios [47]) [24, 28, 31, 38, 39, 47]
PCMB <34> (<34>, 0.1 mM, 50% inhibition [41]) [41]
arsenite <7, 51> (<7>, in presence of NADH, inhibition is reversed by dithiols and less effectively by monothiols [37]; <51>, reversible inactivation of lipoamide-reducing reaction, no decrease in diaphorase activity [47]; <34>, 0.3 mM, 50% inhibition [41]) [37, 41, 47]
folic acid <20> (<20>, inhibition of NADH-lipoamide oxidoreductase activity, no effect on diaphorase activity and transhydrogenase activity [40]) [40]
iodoacetic acid <30> (<30>, in presence of NADH or dihydrolipoamide [21]) [21]

isobiopterin <20> (<20>, inhibition of NADH-lipoamide oxidoreductase activity, no effect on diaphorase activity and transhydrogenase activity [40]) [40]
p-[(bromoacetyl)-amino]phenyl arsenoxide <7> (<7>, irreversible active site directed inactivation [27]) [27]
p-aminophenyldichloroarsine <2, 14, 27> (<27>, inactivated only in presence of NADH and dihydrolipoamide, no significant loss of activity in absence of NADH and dihydrolipoamide [14]; <2>, in presence of NADH [16]) [14, 16, 18]
Additional information <34> (<34>, no inhibition by iodoacetate [41]) [41]

Cofactors/prosthetic groups

FAD <1, 2, 4, 5, 7, 8, 15, 18, 19, 22, 32, 37, 47, 48, 50> (<1,32>, contains one noncovalently bound FAD per subunit [1,30]; <4, 5, 8, 15, 18, 19, 47, 50>, contains 1 FAD per subunit [4,20,24,25,26,31,48,49]; <7>, the fluorescence of FAD in oxidized wild-type enzyme is markedly temperature dependent, while the fluorescence of FAD in mutants C44S and C49S is independent of temperature [29]; <7>, 2 molecules of FAD per molecule of enzyme [37]; <37>, 0.01882 mM of FAD per mg of protein [38]; <48>, covalently bound to the protein [42]; <22>, wild-type enzyme contains 1 FAD per subunit, mutant enzymes K54E and S53K/K54S have about 25% less bound FAD [44]; <22>, wild-type enzyme contains 1 FAD per subunit, mutant enzyme K37E has about 25% less bound FAD [43]) [1, 2, 4, 20, 24, 25, 26, 28, 29, 30, 31, 37, 38, 42, 43, 44, 48, 49]
NAD$^+$ <1-51> (<1,4>, highly specific for [1,4]; <34>, NADH-oxidation with free lipoic acid is strongly dependent on the addition of NAD$^+$, EDTA, Mg^{2+} and cysteine, the reverse reaction with reduced lipoic acid and NAD$^+$ does not show any requirement for cofactors [41]; <51>, the rate of lipoamide reduction is dependent upon the NAD$^+$/NADH ratio, the reaction is activated at low ratios and inhibited at high ratios [47]) [1-52]
NADH <1, 7, 12, 14, 15, 18, 20, 29, 31, 34, 37, 42, 48, 50, 51> (<1,14,29,31>, highly specific for [1,12,21]; <51>, the rate of lipoamide reduction is dependent upon the NAD$^+$/NADH ratio, the reaction is activated at low ratios and inhibited at high ratios [47]) [1, 12, 21, 22, 23, 24, 29, 31, 38, 41, 42, 46, 47, 49, 50]
NADP$^+$ <5> (<1,4,14,17,29,30,31,34>, no activity [1,4,21,34,41]) [4]
NADPH <5> (<1,4>, no activity [1,4]) [4]
flavin <14> [18]

Activating compounds

EDTA <34> (<34>, NADH-oxidation with free lipoic acid is strongly dependent on the addition of NAD$^+$, EDTA, Mg^{2+} and cysteine, the reverse reaction with reduced lipoic acid and NAD$^+$ does not show any requirement for cofactors [41]) [41]
cysteine <34> (<34>, NADH-oxidation with free lipoic acid is strongly dependent on the addition of NAD$^+$, EDTA, Mg^{2+} and cysteine, the reverse reaction with reduced lipoic acid and NAD$^+$ does not show any requirement for cofactors [41]) [41]

Metals, ions

Mg^{2+} <34> (<34>, NADH-oxidation with free lipoic acid is strongly dependent on the addition of NAD^+, EDTA, Mg^{2+} and cysteine, the reverse reaction with reduced lipoic acid and NAD^+ does not show any requirement for cofactors [41]) [41]
NaCl <14> (<14>, completely inactive in absence of NaCl [18]) [18]

Turnover number (min^{-1})

405 <22> (NAD^+, <22>, mutant enzyme S53K/K54S [44]) [44]
510 <22> (dihydrolipoamide, <22>, mutant enzyme S53K/K54S [44]) [44]
864 <22> (lipoamide, <22>, mutant enzyme S53K/K54S [44]) [44]
1188 <22> (NADH, <22>, mutant enzyme S53K/K54S [44]) [44]
1380 <22> (lipoamide, <22>, mutant enzyme K54E [44]) [44]
1572 <22> (NADH, <22>, mutant enzyme E192Q [44]) [44]
1782 <22> (dihydrolipoamide, <22>, mutant enzyme E192Q [44]) [44]
1800 <22> (NAD^+, <22>, mutant enzyme E192Q [44]) [44]
4197 <50> (lipoic acid) [49]
4878 <22> (lipoamide, <22>, mutant enzyme E192Q [44]) [44]
5277 <50> (lipoamide) [49]
6477 <50> (1,4-benzoquinone) [49]
11400 <17> (R,S-lipoamide) [51]
20400 <22> (NADH, <22>, wild-type enzyme [44]) [44]
20700 <17> (H-protein) [51]
34020 <22> (dihydrolipoamide, <22>, wild-type enzyme [44]) [44]
34440 <22> (NAD^+, <22>, wild-type enzyme [44]) [44]
38940 <22> (lipoamide, <22>, wild-type enzyme [44]) [44]
Additional information <7, 17, 22, 42> (<17>, turnover numbers for small polypeptides derived from proteolytic treatment of the H-protein [51]) [28, 43, 46, 51]

Specific activity (U/mg)

0.012 <27> [14]
0.229 <2> [16]
0.311 <50> [49]
2.43 <51> [47]
3.651 <12> [50]
11 <10> [39]
20 <7> [6]
25 <19> [26]
25.7 <1> [1]
34 <46> [39]
35 <14> [18]
35.7 <25> [39]
40 <26> [39]
62 <17> (<17>, enzyme from chloroplast [34]) [34]
68 <8> [7]
75 <47> [48]
77 <20> [40]

83.06 <48> (<48>, lipoamide reduction [42]) [42]
90.2 <18> [31]
125 <10> [23]
133 <19> [17]
176.8 <38> [39]
214 <4> [4]
295 <17> (<17>, mitochondrial enzyme [34]) [34]
483 <5> [4]
843 <15> [24]
Additional information <7, 32, 37> [28, 30, 38]

K_m-Value (mM)
0.0041 <42> (NADH, <42>, reaction with thio-NAD^+ [46]) [46]
0.0073 <42> (NADH) [46]
0.0084 <42> (NADH, <42>, reaction with O_2 [46]) [46]
0.0085 <18> (NADH, <18>, NADH:2,6-dichlorophenol indophenol reductase activity [31]) [31]
0.009 <18> (NADH, <18>, NADH:NAD^+ transhydrogenase activity [31]) [31]
0.01 <51> (1,4-benzoquinone) [47]
0.01 <18> (NADH, <18>, lipoamide dehydrogenase activity [31]) [31]
0.017 <30> (dihydrolipoamide) [21]
0.0215 <34> (NADH) [41]
0.023 <2> (NADH) [2]
0.025 <7> (dihydrolipoamide) [6]
0.027 <17> (H-protein) [51]
0.029 <48> (ferric leghemoglobin) [42]
0.036 <42> (thio-NAD^+, <42>, reaction with NADH [46]) [46]
0.037 <18> (NAD^+) [31]
0.047 <4> (NADH) [4]
0.05 <51> (1,4-naphthoquinone) [47]
0.05 <22> (dihydrolipoamide, <22>, mutant enzyme S53K/K54S [44]) [44]
0.058 <48> (NADH) [42]
0.071 <22> (NADH, <22>, mutant enzyme E192Q [44]) [44]
0.08 <7> (dihydrolipoamide, <7>, complex-bound dihydrolipoamide dehydrogenase [28]) [28]
0.1-0.11 <17> (NAD^+, <17>, chloroplastic enzyme [34]) [34]
0.11 <42> (thio-NADH) [46]
0.12 <22> (dihydrolipoamide, <22>, mutant enzyme E192Q [44]) [44]
0.13 <22> (NADH, <22>, mutant enzyme S53K/K54S [44]) [44]
0.13 <2> (dihydrolipoamide) [16]
0.15 <7, 15> (NAD^+) [6, 24]
0.15 <1> (NADH) [1]
0.156 <42> (NADH, <42>, reaction with 2,6-dimethyl-1,4-benzoquinone [46]) [46]
0.161 <42> (NADH, <42>, reaction with 5-hydroxy-1,4-naphthoquinone [46]) [46]
0.2 <5> (NAD^+) [4]

0.25 <4> (NAD^+) [4]
0.25 <22> (NAD^+, <22>, mutant enzyme K37E [43]) [43]
0.27 <5> (NADH) [4]
0.28 <2> (NAD^+) [16]
0.28 <7> (dihydrolipoamide, <7>, free dihydrolipoamide dehydrogenase [28]) [28]
0.29 <42> (5-hydroxy-1,4-naphthoquinone) [46]
0.31 <22> (NAD^+, <22>, mutant enzyme E192Q [44]) [44]
0.31 <22> (lipoamide, <22>, mutant enzyme E192Q [44]) [44]
0.32 <22> (NAD^+, <22>, native enzyme and mutant enzyme E457Q [43]) [43]
0.37 <50> (lipoamide) [49]
0.38 <22> (NAD^+, <22>, mutant enzyme H452Q [43]) [43]
0.39 <42> (3-acetylpyridine adenine dinucleotide) [46]
0.4 <50> (1,4-benzoquinone) [49]
0.4 <22> (NAD^+, <22>, wild-type enzyme [44]) [44]
0.4 <7> (NAD^+, <7>, complex-bound dihydrolipoamide dehydrogenase [28]) [28]
0.4 <1> (dihydrolipoamide) [1]
0.48 <15> (dihydrolipoamide) [24]
0.5 <5> (NAD^+) [4]
0.5 <17> (R,S-lipoamide) [51]
0.52 <27> (NAD^+) [14]
0.55 <22> (NAD^+, <22>, mutant enzyme S53K/K54S [44]) [44]
0.56 <42> (lipoamide, <42>, reaction with 3-acetylpyridine adenine dinucleotide [46]) [46]
0.62 <1> (NAD^+) [1]
0.62-0.64 <17> (NAD^+, <17>, mitochondrial enzyme [34]) [34]
0.63 <22> (lipoamide, <22>, mutant enzyme S53K/K54S [44]) [44]
0.69 <22> (dihydrolipoamide, <22>, native enzyme [43]) [43]
0.76 <22> (dihydrolipoamide, <22>, mutant enzyme K37E [43]) [43]
0.83 <5> ($NADP^+$) [4]
0.83 <27> (dihydrolipoamide) [14]
0.86 <42> (2,6-dimethyl-1,4-benzoquinone) [46]
0.88 <22> (dihydrolipoamide, <22>, wild-type enzyme [44]) [44]
1.1 <30> (NAD^+) [21]
1.3 <51> (lipoamide) [47]
1.83 <7> (NAD^+, <7>, free dihydrolipoamide dehydrogenase [28]) [28]
2.1 <34> (lipoic acid) [41]
2.15 <50> (lipoic acid) [49]
2.25 <22> (lipoamide, <22>, wild-type enzyme [44]) [44]
2.6 <42> (nicotinamide hypoxanthine dinucleotide) [46]
2.9 <22> (dihydrolipoamide, <22>, mutant enzyme H457Q [43]) [43]
2.9 <51> (lipoate) [47]
3.38 <48> (lipoamide) [42]
3.7 <1> (lipoamide) [1]
4.7 <51> (dihydrolioate) [47]
5 <2> (lipoamide) [2]

5 <22> (lipoamide, <22>, mutant enzyme K54E [44]) [44]
5.8 <42> (lipoamide, <42>, reaction with thio-NADH [46]) [46]
6.2 <14> (NADH, <14>, reaction with lipoic acid [12]) [12]
6.4 <42> (lipoamide, <42>, reaction with nicotinamide hypoxanthine dinucleotide [46]) [46]
16 <42> (lipoamide, <42>, reaction with NADH [46]) [46]
43.6 <22> (dihydrolipoamide, <22>, mutant enzyme H452Q [43]) [43]
84 <14> (NADH, <14>, reaction with lipoamide [12]) [12]
120 <42> (lipoate) [46]
Additional information <17> (<17>, K_m-values for small polypeptides derived from proteolytic treatment of the H-protein [51]) [51]

K_i-Value (mM)

0.009 <7> (NADH, <7>, free dihydrolipoamide dehydrogenase [28]) [28]
0.016 <7> (NADH, <7>, complex-bound dihydrolipoamide dehydrogenase [28]) [28]
0.02 <17> (NADH, <17>, chloroplastic enzyme [34]) [34]
0.031 <15> (NADH) [24]
0.11 <17> (NADH, <17>, mitochondrial enzyme [34]) [34]

pH-Optimum

6.2 <37> (<37>, NADH-lipoamide oxidoreductase activity [38]) [38]
6.4 <18> (<18>, reduction of lipoamide [31]) [31]
7 <4, 5> (<4,5>, reaction with NAD(P)H and lipoamide [4]) [4]
7-7.5 <12> (<12>, 50 mM phosphate buffer [50]) [50]
7.2 <1> (<1>, NADH-dependent reduction of lipoamide [1]) [1]
7.4 <18> (<18>, reduction of 2,6-dichlorophenol indophenol [31]) [31]
7.5 <18> (<18>, reduction of thio-NAD^+ [31]) [31]
8 <7, 37> (<37>, dihydrolipoamide-NAD^+ oxidoreductase activity [38]) [28, 38]
8.2 <32> [30]
8.5 <15> [24]
9 <4, 5> (<4,5>, reaction with $NAD(P)^+$ and dihydrolipoamide [4]) [4]

pH-Range

5-10 <32> (<32>, pH 5.0: about 10% of maximal activity, pH 7.0-7.5: 60-70% of maximal activity, pH 8.0-8.5: 90% of maximal activity, pH 10.0: 10% of maximal activity in carbonate buffer [30]) [30]
5.5-7.5 <37> (<37>, pH 5.5: about 75% of maximal activity, pH 7.5: about 40% of maximal activity, NADH-lipoamide oxidoreductase activity [38]) [38]
6.5-8.5 <7> (<7>, pH 6.5: about 40% of maximal activity, pH 8.5: about 50% of maximal activity [28]) [28]
7.5-8.8 <37> (<37>, pH 7.5: about 50% of maximal activity, pH 8.8: about 60% of maximal activity, dihydrolipoamide-NAD^+ oxidoreductase activity [38]) [38]

Temperature optimum (°C)

50 <12> [50]

4 Enzyme Structure

Molecular weight

88000 <7> (<7>, gel filtration [6]) [6]
98000 <19> (<19>, gel filtration [26]) [26]
100000 <13> (<13>, equilibrium sedimentation [19]) [19]
102000 <32> (<32>, gel filtration [30]) [30]
105000 <4, 5, 18, 47> (<4,5,18,47>, gel filtration [4,31,48]) [4, 31, 48]
110000 <1, 10, 50> (<1>, non-denaturing gradient PAGE [1]; <10,50>, gel filtration [23,49]) [1, 23, 49]
112000 <14, 29> (<14,29>, gel filtration [21]) [21]
115000 <8> (<8>, LPD-Val, gel filtration [20]) [20]
117000 <15> (<15>, gel filtration [24]) [24]
118000 <27> (<27>, gel filtration [14]) [14]
119000 <28, 30> (<28,30>, gel filtration [21]) [21]
120000 <31> (<31>, gel filtration [21]) [21]
125000 <8> (<8>, LPD-Glc, gel filtration [20]) [20]

Subunits

? <3, 7, 13, 14, 16, 17, 18, 19, 20, 21, 22, 24, 51> (<51>, x * 49342, electrospray mass spectrometry [47]; <17>, x * 49757, mitochondrial enzyme, electrospray mass spectrometry [34]; <20>, x * 49690, calculation from nucleotide sequence [11]; <7>, x * 51274, calculation from nucleotide sequence [11]; <17>, x * 52614, enzyme from chloroplast, electrospray mass spectrometry [34]; <17>, x * 52000, enzyme from chloroplast, SDS-PAGE [34]; <3,18>, x * 54000, SDS-PAGE [11]; <21>, x * 55000, SDS-PAGE [11]; <19>, x * 56000, enzyme from liver enzyme [11]; <13, 19, 24>, x * 56000, SDS-PAGE [11, 25, 26]; <17>, x * 56000, mitochondrial enzyme, SDS-PAGE [34]; <19>, x * 58000, enzyme from heart, SDS-PAGE [11]; <14>, x * 58000, SDS-PAGE [11]; <17>, x * 59000, SDS-PAGE [11]; <16>, x * 61000, SDS-PAGE [35]) [11, 25, 26, 34, 35, 47]

dimer <1, 2, 4, 5, 7, 10, 13, 14, 15, 18, 32, 47, 48, 50> (<7>, x * 46000, SDS-PAGE [6]; <48>, x * 50027, MW only of protein, flavin is reduced during analysis, electrospray MS analysis [42]; <32>, 2 * 51000, SDS-PAGE [30]; <5>, 2 * 52000, SDS-PAGE [4]; <4>, 2 * 53000, SDS-PAGE [4]; <1,18,50>, 2 * 54000, SDS-PAGE [1, 31, 49]; <2, 10, 47>, 2 * 55000, SDS-PAGE [2, 23, 48]; <7,13,15>, 2 * 56000, SDS-PAGE [19,24,28]; <14>, 2 * 58000, SDS-PAGE [18]; <14>, 2 * 60000, SDS-PAGE [12]) [1, 2, 4, 6, 12, 18, 19, 23, 24, 28, 30, 31, 42, 48, 49]

Additional information <17, 23, 51> (<23>, dihydrolipoamide dehydrogenase E3 is directly bound to the core protein E2 of the 2-oxoglutarate dehydrogenase complex, wheras it is bound to the pyruvate dehydrogenase complex through a protein X [10]; <51>, the enzyme oligomerizes to a high-molecular weight species, above 300000 Da, under nondenaturing conditions [47]; <17>, EC 1.8.1.4 is the E3-protein component of the mitochondrial 2-oxoacid dehydrogenase multienzyme complexes and the L-protein component of the glycine decarboxylase system [51]) [10, 47, 51]

5 Isolation/Preparation/Mutation/Application

Source/tissue

bloodstream form <2, 6> [5, 16]
brain <19, 20> [11, 33, 40]
epimastigote <2> [2]
flower <35> [33]
heart <20, 21> [11, 22, 33]
kidney <21> [11, 33]
leaf <17, 34> [11, 33, 34, 41, 51, 52]
liver <19, 21, 22, 33> [11, 17, 26, 33]
muscle <18> [11, 31]
mycelium <32> [30]
nodule <48> (<48>, enzyme form FlbR-2 [42]) [42]
root <17> [34]

Localization

chloroplast <17, 34> [34, 41]
cytoplasm <7> [6]
cytosol <19> (<19>, activity is about one-tenth of that in mitochondria [26]) [26]
mitochondrion <17, 18, 19, 48> [17, 26, 31, 34, 42]
periplasmic space <47> [48]
plasma membrane <2, 6> (<6>, cytoplasmic surface of plasma membrane [5]) [5, 16]
Additional information <19> (<19>, synthesized predominantly on free ribosomes and translocated into mitochondria [26]) [26]

Purification

<1> [1]
<2> [2, 16]
<4> [4]
<5> [4]
<7> (from pyruvate dehydrogenase complex [28]; affinity chromatography on propyllipoamide-glass columns [36]) [6, 28]
<8> [7]
<10> (from pyruvate dehydrogenase complex [23]) [8, 23]
<12> [50]
<13> (from pyruvate dehydrogenase complex [19]) [19]
<14> [12, 18]
<15> (affinity chromatography on propyllipoamide-glass columns [36]) [24, 36]
<17> [34]
<18> [31]

<19> (dissociation of the lipoamide dehydrogenase component from the branched-chain α-keto acid dehydrogenase complex during purification [17]) [17]
<20> (affinity chromatography on propyllipoamide-glass columns [36]) [36, 40]
<22> (wild-type and mutant enzymes K37E, H452Q and E457Q [43]; wild-type and mutant enzymes K54E, S53K54-K53S54 and E192Q [44]) [43, 44]
<24> (from pyruvate dehydrogenase complex [25]) [25]
<32> [30]
<34> [41]
<37> [38]
<42> (recombinant enzyme [46]) [46]
<47> [48]
<48> [42]
<50> [49]
<51> [47]

Crystallization
<17> [52]
<20> (crystallized by hanging-drop vapor-diffusion method [9]) [9]
<23> [10]
<25> (best results obtained by vapour diffusion method, three-dimensional structure at 2.8 A resolution [13]) [13]
<37> (crystallization by dialysis [38]) [38]

Cloning
<3> (dihydrolipoamide dehydrogenase component of the pyruvate dehydrogenase multienzyme complex, expression in Escherichia coli [3]) [3]
<12> (expression in Escherichia coli or in Corynebacterium glutamicum, the cloned gene is expressed in Corynebacterium glutamicum cells harbouring the gene on a plasmid shows 12fold higher specific LPD activity when compared to the wild-type strain [50]) [50]
<15> [15]
<22> (wild-type and mutant enzymes K37E, H452Q and E457Q, overexpression in Escherichia coli [43]; wild-type and mutant enzymes K37E, H452Q and E457Q [43]; wild-type and mutant enzymes K54E, S53K54-K53S54 and E192Q, overexpression in Escherichia coli [44]) [43, 44]
<42> (expression in Escherichia coli [46]) [46]
<48> (overexpression in Escherichia coli [42]) [42]
<50> [49]

Engineering
C44S <7> (<7>, 0.003% of the activity of wild-type enzyme with NAD^+ and dihydrolipoamide. Enzyme is capable to catalyze reactions with NADH as electron donor and ferricyanide, thio-NAD^+, 2,6-dichlorophenol indophenol and O_2 as electron acceptor. The fluorescence of FAD in oxidized wild-type enzyme is markedly temperature dependent, while the fluorescence of FAD in mutants C44S and C49S is independent of temperature [29]) [29]

C49S <7> (<7>, 0.012% of the activity of wild-type enzyme with NAD^+ and dihydrolipoamide. Enzyme is capable to catalyze reactions with NADH as electron donor and ferricyanide, thio-NAD^+, 2,6-dichlorophenol indophenol and O_2 as electron acceptor. The fluorescence of FAD in oxidized wild-type enzyme is markedly temperature dependent, while the fluorescence of FAD in mutants C44S and C49S is independent of temperature [29]) [29]
E192Q <22> (<22>, specific activity is markedly decreased, less than 5% of the wild-type activity, K_m-values for lipoamide and dihydrolipoamide are markedly reduced [44]) [44]
E457Q <22> (<22>, molar ratio of FAD to enzyme is 0.9 compared to 1 for the wild-type enzyme, mutation affects the environment surrounding FAD, decrease in efficiency of electron transfer from the reduced flavin to the oxidized substrate [43]) [43]
H452Q <22> (<22>, molar ratio of FAD to enzyme is 0.94 compared to 1 for the wild-type enzyme, no production of NADH when the enzyme is reduced by dihydrolipoamide, transfer of electrons from the substrate dihydrolipoamide to NAD^+ is extremely low [43]) [43]
K37E <22> (<22>, molar ratio of FAD to enzyme is 0.76 compared to 1 for the wild-type enzyme [43]) [43]
K53R <7> (<7>, spectral and redox properties of FAD in the mutant enzyme as well as the interaction of the flavin with bound NAD^+ are profoundly affected by the mutation, K53R does not catalyze either the dihydrolipoamide-NAD^+ or the NADH-lipoamide reactions except at very low concentrations of reducing substrate. The absorbance spectrum in the visible and near-ultraviolet is little changed from that of wild-type enzyme, in contrast to wild-type enzyme the spectrum of K53R is sensitive to pH. Unlike the wild-type enzyme, the binding of β-NAD^+ to K53R alters the spectrum [32]) [32]
K54E <22> (<22>, about 25% less bound FAD compared to wild-type, specific activity is markedly decreased, less than 5% of the wild-type activity, K_m-value for lipoamide is increased by about twofold, [44]) [44]
S53K/K54S <22> (<22>, about 25% less bound FAD compared to wild-type, specific activity is markedly decreased, less than 5% of the wild-type activity, K_m-values for lipoamide and dihydrolipoamide are markedly reduced. The catalytic rate constant, turnover number/K_m, is significantly lower than wild-type [44]) [44]

6 Stability

Temperature stability

65 <32> (<32>, 170 min, 50% loss of activity [30]) [30]
70 <32> (<32>, 50 min, 50% loss of activity [30]) [30]
75 <32> (<32>, 50% loss of activity after 15 min, 50% loss of activity after 20 min in presence of NAD^+, 50% loss of activity after less than 1 min in presence of dihydrolipoamide [30]) [30]
80 <32> (<32>, 6 min, 50% loss of activity [30]) [30]
86 <25> (<25>, melting temperature [13]) [13]

95 <14, 29, 30, 31> (<14,29,30,31>, 15 min, in presence of 4 M NaCl, enzyme in crude extract, stable [21]) [21]
100 <14, 29, 30, 31> (<14,29,30,31>, inactivated by boiling, enzyme in crude extract [21]) [21]
Additional information <14> [12]

General stability information

<10>, stable to trypsin treatment [23]
<12>, no loss of activity when the enzyme is frozen at -20°C, and thawed three times [50]
<15>, trypsin or papain, treatment with 0.019 mg/ml protease [24]

Storage stability

<1>, stable for months if stored in liquid nitrogen [1]
<4>, -20°C, 6 months, about 10% loss of activity [4]
<5>, -20°C, 6 months, about 60% loss of activity [4]
<10>, -20°C, stable for several months [23]
<12>, 4°C or -20°C, stable for at least 1 month [50]
<18>, -20°C, stable for at least 6 months [31]

References

[1] Oppermann, F.B.; Schmidt, B.; Steinbuchel, A.: Purification and characterization of acetoin:2,6-dichlorophenolindophenol oxidoreductase, dihydrolipoamide dehydrogenase, and dihydrolipoamide acetyltransferase of the Pelobacter carbinolicus acetoin dehydrogenase enzyme system. J. Bacteriol., **173**, 757-767 (1991)
[2] Lohrer, H.; Krauth-Siegel, R.L.: Purification and characterization of lipoamide dehydrogenase from Trypanosoma cruzi. Eur. J. Biochem., **194**, 863-869 (1990)
[3] Borges, A.; Hawkins, C.F.; Packman, L.C.; Perham, R.N.: Cloning and sequence analysis of the genes encoding the dihydrolipoamide acetyltransferase and dihydrolipoamide dehydrogenase components of the pyruvate dehydrogenase multienzyme complex of Bacillus stearothermophilus. Eur. J. Biochem., **194**, 95-102 (1990)
[4] Dietrichs, D.; Andreesen, J.R.: Purification and comparative studies of dihydrolipoamide dehydrogenases from the anaerobic, glycine-utilizing bacteria Peptostreptococcus glycinophilus, Clostridium cylindrosporum, and Clostridium sporogenes. J. Bacteriol., **172**, 243-251 (1990)
[5] Jackman, S.A.; Hough, D.W.; Danson, M.J.; Stevenson, K.J.; Opperdoes, F.R.: Subcellular localisation of dihydrolipoamide dehydrogenase and detection of lipoic acid in bloodstream forms of Trypanosoma brucei. Eur. J. Biochem., **193**, 91-95 (1990)
[6] Richarme, G.: Purification of a new dihydrolipoamide dehydrogenase from Escherichia coli. J. Bacteriol., **171**, 6580-6585 (1989)

[7] Sokatch, J.R.: Purification of branched-chain keto acid dehydrogenase and lipoamide dehydrogenase-valine from Pseudomonas. Methods Enzymol., **166**, 342-350 (1988)
[8] Perham, R.N.; Lowe, P.N.: Isolation and properties of the branched-chain 2-keto acid and pyruvate dehydrogenase multienzyme complex from Bacillus subtilis. Methods Enzymol., **166**, 330-342 (1988)
[9] Toyoda, T.; Kobayashi, R.; Sekiguchi, T.; Koike, K.; Koike, M.; Takenaka, A.: Crystallization and preliminary X-ray analysis of pig E3, lipoamide dehydrogenase. Acta Crystallogr. Sect. D, **54**, 982-985 (1998)
[10] Toyoda, T.; Suzuki, K.; Sekiguchi, T.; Reed, L.J.; Takenaka, A.: Crystal structure of eucaryotic E3, lipoamide dehydrogenase from yeast. J. Biochem., **123**, 668-674 (1998)
[11] Carothers, D.J.; Pons, G.; Patel, M.S.: Dihydrolipoamide dehydrogenase: functional similarities and divergent evolution of the pyridine nucleotide-disulfide oxidoreductases. Arch. Biochem. Biophys., **268**, 409-425 (1989)
[12] Sundquist, A.R.; Fahey, R.C.: The novel disulfide reductase bis-γ-glutamylcystine reductase and dihydrolipoamide dehydrogenase from Halobacterium halobium: purification by immobilized-metal-ion affinity chromatography and properties of the enzymes. J. Bacteriol., **170**, 3459-3467 (1988)
[13] Mattevi, A.; Obmolova, G.; Kalk, K.H.; van Berkel, W.J.; Hol, W.G.: Three-dimensional structure of lipoamide dehydrogenase from Pseudomonas fluorescens at 2.8 A resolution. Analysis of redox and thermostability properties. J. Mol. Biol., **230**, 1200-1215 (1993)
[14] Smith, L.D.; Bungard, S.J.; Danson, M.J.; Hough, D.W.: Cloning and characterization of the gene encoding lipoamide dehydrogenase in Saccharomyces cerevisiae. Biochem. Soc. Trans., **15**, 1097 (1987)
[15] Roy, D.J.; Dawes, I.W.: Cloning and characterization of the gene encoding lipoamide dehydrogenase in Saccharomyces cerevisiae. J. Gen. Microbiol., **133**, 925-933 (1987)
[16] Danson, M.J.; Conroy, K.; McQuattie, A.; Stevenson, K.J.: Dihydrolipoamide dehydrogenase from Trypanosoma brucei. Characterization and cellular location. Biochem. J., **243**, 661-665 (1987)
[17] Ono, K.; Hakozaki, M.; Kimura, A.; Kochi, H.: Purification, resolution, and reconstitution of rat liver branched-chain α-keto acid dehydrogenase complex. J. Biochem., **101**, 19-27 (1987)
[18] Danson, M.J.; McQuattie, A.; Stevenson, K.J.: Dihydrolipoamide dehydrogenase from halophilic archaebacteria: purification and properties of the enzyme from Halobacterium halobium. Biochemistry, **25**, 3880-3884 (1986)
[19] Bosma, H.J.; De Kok, A.; Westphal, A.H.; Veeger, C.: The composition of the pyruvate dehydrogenase complex from Azotobacter vinelandii. Does a unifying model exist for the complexes from gram-negative bacteria?. Eur. J. Biochem., **142**, 541-549 (1984)
[20] Delaney, R.; Burns, G.; Sokatch, J.R.: Relationship of lipoamide dehydrogenases from Pseudomonas putida to other FAD-linked dehydrogenases. FEBS Lett., **168**, 265-270 (1984)

[21] Danson, M.J.; Eisenthal, R.; Hall, S.; Kessell, S.R.; Williams, D.L.: Dihydrolipoamide dehydrogenase from halophilic archaebacteria. Biochem. J., **218**, 811-818 (1984)
[22] Tsai, C.S.; Wand, A.J.; Templeton, D.M.; Weiss, P.M.: Multifunctionality of lipoamide dehydrogenase promotion of electron transferase reaction. Arch. Biochem. Biophys., **225**, 554-561 (1983)
[23] Hodgson, J.A.; Lowe, P.N.; Perham, R.N.: Wild-type and mutant forms of the pyruvate dehydrogenase multienzyme complex from Bacillus subtilis. Biochem. J., **211**, 463-472 (1983)
[24] Heinrich, P.; Ronft, H.; Schartau, W.; Kresze, G.B.: Lipoamide dehydrogenase from bakers yeast. Improved purification and some molecular, kinetic, and immunochemical properties. Hoppe-Seyler's Z. Physiol. Chem., **364**, 41-50 (1983)
[25] Seckler, R.; Binder, R.; Bisswanger, H.: Purification and properties of the pyruvate dehydrogenase complex from Salmonella typhimurium and formation of hybrids with the enzyme complex from Escherichia coli. Biochim. Biophys. Acta, **705**, 210-217 (1982)
[26] Matuda, S.; Saheki, T.: Intracellular distribution and biosynthesis of lipoamide dehydrogenase in rat liver. J. Biochem., **91**, 553-561 (1982)
[27] Adamson, S.R.; Stevenson, K.J.: Inhibition of pyruvate dehydrogenase multienzyme complex from Escherichia coli with a bifunctional arsenoxide: selective inactivation of lipoamide dehydrogenase. Biochemistry, **20**, 3420-3424 (1981)
[28] Schmincke-Ott, E.; Bisswanger, H.: Dihydrolipoamide dehydrogenase component of the pyruvate dehydrogenase complex from Escherichia coli K12. Comparative characterization of the free and the complex-bound component. Eur. J. Biochem., **114**, 413-420 (1981)
[29] Hopkins, N.; Williams, C.H., Jr.: Characterization of lipoamide dehydrogenase from Escherichia coli lacking the redox active disulfide: C44S and C49S. Biochemistry, **34**, 11757-11765 (1995)
[30] McKay, D.J.; Stevenson, K.J.: Lipoamide dehydrogenase from Malbranchea pulchella: isolation and characterization. Biochemistry, **18**, 4702-4707 (1979)
[31] Komuniecki, R.; Saz, H.J.: Purification of lipoamide dehydrogenase from Ascaris muscle mitochondria and its relationship to NADH:NAD^+ transhydrogenase activity. Arch. Biochem. Biophys., **196**, 239-247 (1979)
[32] Maeda-Yorita, K.; Russell, G.C.; Guest, J.R.; Massey, V.; Williams, C.H., Jr.: Modulation of the oxidation-reduction potential of the flavin in lipoamide dehydrogenase from Escherichia coli by alteration of a nearby charged residue, K53R. Biochemistry, **33**, 6213-6220 (1994)
[33] Williams, C.H.: Flavin-containing dehydrogenases. The Enzymes, 3rd Ed. (Boyer, P.D., ed.), **13**, 89-173 (1976)
[34] Conner, M.; Krell, T.; Lindsay, J.G.: Identification and purification of a distinct dihydrolipoamide dehydrogenase from pea chloroplasts. Planta, **200**, 195-202 (1996)

[35] Wais, U.; Gillmann, U.; Ulrich, J.: Isolation and characterisation of pyruvate dehydrogenase complex from brewers yeast. Hoppe-Seyler's Z. Physiol. Chem., **354**, 1378-1388 (1973)
[36] Scouten, W.H.; Torok, F.; Gitomer, W.: Purification of lipoamide dehydrogenase by affinity chromatography on propyllipoamide-glass columns. Biochim. Biophys. Acta, **309**, 521-524 (1973)
[37] Reed, L.J.; Willms, C.R.: Purification and resolution of the pyruvate dehydrogenase complex (Escherichia coli). Methods Enzymol., **9**, 247-265 (1966)
[38] Kawahara, Y.; Misaka, E.; Nakanishi, K.: Purification and properties of lipoamide dehydrogenase from yeast, Candida krusei. J. Biochem., **63**, 77-82 (1963)
[39] Scouten, W.H.; McManus, I.R.: Microbial lipoamide dehydrogenase. Purification and some characteristics of the enzyme derived from selected microorganisms. Biochim. Biophys. Acta, **227**, 248-263 (1971)
[40] Millard, S.A.; Kubose, A.; Gal, E.M.: Brain lipoyl dehydrogenase. Purification, properties, and inhibitors. J. Biol. Chem., **244**, 2511-2515 (1969)
[41] Jacobi, G.; -hlers, U.: Isolation und Eigenschaften von Liponsäuredehydrogenase aus Chloroplasten. Z. Pflanzenphysiol., **58**, 193-206 (1968)
[42] Moran, J.F.; Sun, Z.; Sarath, G.; Arredondo-Peter, R.; James, E.K.; Becana, M.; Klucas, R.V.: Molecular cloning, functional characterization, and subcellular localization of soybean nodule dihydrolipoamide reductase. Plant Physiol., **128**, 300-313 (2002)
[43] Liu, T.C.; Korotchkina, L.G.; Hyatt, S.L.; Vettakkorumakankav, N.N.; Patel, M.S.: Spectroscopic studies of the characterization of recombinant human dihydrolipoamide dehydrogenase and its site-directed mutants. J. Biol. Chem., **270**, 15545-15550 (1995)
[44] Liu, T.C.; Soo Hong, Y.; Korotchkina, L.G.; Vettakkorumakankav, N.N.; Patel, M.S.: Site-directed mutagenesis of human dihydrolipoamide dehydrogenase: role of lysine-54 and glutamate-192 in stabilizing the thiolate-FAD intermediate. Protein Expr. Purif., **16**, 27-39 (1999)
[45] Smith, A.W.; Roche, H.; Trombe, M.C.; Briles, D.E.; Hakansson, A.: Characterization of the dihydrolipoamide dehydrogenase from Streptococcus pneumoniae and its role in pneumococcal infection. Mol. Microbiol., **44**, 431-448 (2002)
[46] Argyrou, A.; Blanchard, J.S.: Mycobacterium tuberculosis lipoamide dehydrogenase is encoded by Rv0462 and not by the lpdA or lpdB genes. Biochemistry, **40**, 11353-11363 (2001)
[47] Marcinkeviciene, J.; Blanchard, J.S.: Catalytic properties of lipoamide dehydrogenase from Mycobacterium smegmatis. Arch. Biochem. Biophys., **340**, 168-176 (1997)
[48] Engels, A.; Kahmann, U.; Ruppel, H.G.; Pistorius, E.K.: Isolation, partial characterization and localization of a dihydrolipoamide dehydrogenase from the cyanobacterium Synechocystis PCC 6803. Biochim. Biophys. Acta, **1340**, 33-44 (1997)
[49] Youn, H.; Kwak, J.; Youn, H.D.; Hah, Y.C.; Kang, S.O.: Lipoamide dehydrogenase from Streptomyces seoulensis: biochemical and genetic properties. Biochim. Biophys. Acta, **1388**, 405-418 (1998)

[50] Schwinde, J.W.; Hertz, P.F.; Sahm, H.; Eikmanns, B.J.; Guyonvarch, A.: Lipoamide dehydrogenase from Corynebacterium glutamicum: molecular and physiological analysis of the lpd gene and characterization of the enzyme. Microbiology, **147**, 2223-2231 (2001)
[51] Neuburger, M.; Polidori, A.M.; Pietre, E.; Faure, M.; Jourdain, A.; Bourguignon, J.; Pucci, B.; Douce, R.: Interaction between the lipoamide-containing H-protein and the lipoamide dehydrogenase (L-protein) of the glycine decarboxylase multienzyme system. Eur. J. Biochem., **267**, 2882-2889 (2000)
[52] Faure, Bourguignon, J.; Neuburger, M.; MacHerel, D.; Sieker, L.; Ober, R.; Kahn, R.; Cohen-Addad, C.; Douce, R.: Interaction between the lipoamide-containing H-protein and the lipoamide dehydrogenase (L-protein) of the glycine decarboxylase multienzyme system 2. Crystal structure of H- and L-proteins. Eur. J. Biochem., **267**, 2890-2898 (2000)

2-Oxopropyl-CoM reductase (carboxylating) 1.8.1.5

1 Nomenclature

EC number
1.8.1.5

Systematic name
2-mercaptoethanesulfonate,acetoacetate:$NADP^+$ oxidoreductase (decarboxylating)

Recommended name
2-oxopropyl-CoM reductase (carboxylating)

Synonyms
NADPH:2-(2-ketopropylthio)ethanesulfonate oxidoreductase/carboxylase
NADPH:2-ketopropyl-coenzyme M oxidoreductase/carboxylase

CAS registry number
244301-63-1

2 Source Organism

<1> *Xanthobacter sp.* (Py2 [1,2]) [1-3]

3 Reaction and Specificity

Catalyzed reaction
2-mercaptoethanesulfonate + acetoacetate + $NADP^+$ = 2-(2-oxopropylthio)-ethanesulfonate + CO_2 + NADPH + H^+

Reaction type
oxidation
redox reaction
reduction

Natural substrates and products
S 2-mercaptoethanesulfonate + acetoacetate + $NADP^+$ <1> (<1> involved in epoxide carboxylation pathway in bacteria, reverse reaction more important [1,2]) (Reversibility: r <1-3> [1-3]) [1-3]
P 2-(2-oxopropylthio)ethanesulfonate + CO_2 + NADPH + H^+ <1> [1-3]

Substrates and products

S 2-(2-oxopropylthio)ethanesulfonate + CO_2 + NADPH + H^+ <1> (Reversibility: r <1> [1-3]) [1-3]
P 2-mercaptoethanesulfonate + acetoacetate + $NADP^+$ <1> [1-3]
S 2-(2-oxopropylthio)ethanesulfonate + NADPH + H^+ <1> (Reversibility: ir <1> [2]) [2]
P acetone + 2-mercaptoethanesulfonate + $NADP^+$ <1> [2]
S 2-mercaptoethanesulfonate + acetoacetate + $NADP^+$ <1> (<1> dual function enzyme, reverse reaction of component II of carboxylation pathway [1]) (Reversibility: r <1> [1-3]) [1-3]
P 2-(2-oxopropylthio)ethanesulfonate + CO_2 + NADPH + H^+ <1> [1-3]
S 3-mercaptopropanesulfonate + acetoacetate + $NADP^+$ <1> (<1> very poor cofactor, 1% of activity [2]) (Reversibility: ? <1> [2]) [2]
P 2-(2-oxopropylthio)ethanesulfonate + CO_2 + NADPH + H^+ <1> [2]
S 3-mercaptopropionate + acetoacetate + $NADP^+$ <1> (<1> poor cofactor, 11% of activity [2]) (Reversibility: ? <1> [2]) [2]
P 3-(2-oxopropylthio)propionate + CO_2 + NADPH + H^+ <1> [2]
S CO_2 + acetoacetate <1> (<1> exchange of C-14 without added cofactors [2]) (Reversibility: ? <1> [2]) [2]
P acetoacetate + CO_2 <1> [2]
S acetoacetate + H^+ <1> (<1> very low rate [2]) (Reversibility: ? <1> [2]) [2]
P acetone + CO_2 <1> [2]

Inhibitors

N-ethylmaleimide <1> [2]

Cofactors/prosthetic groups

2-mercaptoethanesulfonic acid <1> (<1> coenzyme M [1-3]) [1-3]
FAD <1> (<1> enzyme bound [1,2]) [1, 2]

Turnover number (min^{-1})

15.8 <1> (acetoacetate) [2]

Specific activity (U/mg)

0.0041 <1> (<1> rate of decarboxylation without added cofactors [2]) [2]
0.274 <1> (<1> NADPH formation [2]) [2]

K_m-Value (mM)

6.3 <1> (acetoacetate) [2]

4 Enzyme Structure

Molecular weight

57000 <1> (<1> molecular weight of subunits [2,3]) [2, 3]

Subunits

dimer <1> (<1> 2 * 57000 [2,3]) [2, 3]

References

[1] Allen, J.R.; Clark, D.D.; Krum, J.G.; Ensign, S.A.: A role for coenzyme M (2-mercaptoethanesulfonic acid) in a bacterial pathway of aliphatic epoxide carboxylation. Proc. Natl. Acad. Sci. USA, **96**, 8432-8437 (1999)
[2] Clark, D.D.; Allen, J.R.; Ensign, S.A.: Characterization of five catalytic activities associated with the NADPH:2-ketopropyl-coenzyme M [2-(2-ketopropylthio)ethanesulfonate] oxidoreductase/carboxylase of the Xanthobacter strain Py2 epoxide carboxylase system. Biochemistry, **39**, 1294-1304 (2000)
[3] Jang, S.B.; Jeong, M.S.; Clark, D.D.; Ensign, S.A.; Peters, J.W.: Crystallization and preliminary X-ray analysis of a NADPH 2-ketopropyl-coenzyme M oxidoreductase/carboxylase. Acta Crystallogr. Sect. D, **57**, 445-447 (2001)

Cystine reductase 1.8.1.6

1 Nomenclature

EC number
1.8.1.6

Systematic name
cysteine:NAD^+ oxidoreductase

Recommended name
cystine reductase

Synonyms
EC 1.6.4.1 (formerly)
L-cysteine:NAD^+ oxidoreductase
NADH-dependent cystine reductase
$NADH_2$:L-cystine oxidoreductase
cystine reductase (NADH)
cystine reductase (NADH2)
reductase, cystine

CAS registry number
9029-18-9

2 Source Organism

<1> *Pisum sativum* [1]
<2> *Saccharomyces cerevisiae* [1, 3, 4]
<3> *Candida albicans* [1]
<4> *Histoplasma capsulatum* [2]

3 Reaction and Specificity

Catalyzed reaction
2 L-cysteine + NAD^+ = L-cystine + NADH + H^+

Reaction type
oxidation
redox reaction
reduction

Natural substrates and products

S NADH + L-cystine <4> (<4>, the enzyme may provide reduced sulfhydryl groups involved in the transition of mycelium to yeast form [2]) (Reversibility: ? <4> [4]) [2]

P ?

Substrates and products

S NADH + L-cystine <1-4> (Reversibility: ? <1-4> [1-4]) [1-4]

P NAD^+ + L-cysteine <1-4> [1-4]

Inhibitors

p-chloromercuriphenylsulfonic acid <4> (<4>, strong [2]) [2]

Cofactors/prosthetic groups

NADH <1-4> [1-4]

K_m-Value (mM)

0.11 <4> (L-cystine) [2]
0.9 <2> (L-cystine) [4]
Additional information <2> [4]

5 Isolation/Preparation/Mutation/Application

Source/tissue

seed <1> [1]
Additional information <4> (<4>, the enzyme is phase specific it is not present in mycelium appears early in the transition of mycelium to yeast [2]) [2]

References

[1] Romano, A.H.; Nickerson, W.J.: Cystine reductase in the dimorphic fungus Histoplasma capsulatum. J. Biol. Chem., **208**, 409-416 (1954)
[2] Maresca, B.; Jacobson, E.; Medoff, G.; Kobayashi, G.: Cystine reductase in the dimorphic fungus Histoplasma capsulatum. J. Bacteriol., **135**, 987-992 (1978)
[3] Thibert, R.J.; Sarwar, M.; Carroll, J.E.: Simultaneous determination of cysteine and cystine using N-bromosuccinimide: application in an enzymic system - cystine reductase. Mikrochim. Acta, **3**, 615-624 (1969)
[4] Carroll, J.E.; Kosicki, G.W.; Thibert, R.J.: α-Substituted cystines as possible substrates for cystine reductase and L-amino acid oxidase. Biochim. Biophys. Acta, **198**, 601-603 (1970)

Glutathione-disulfide reductase 1.8.1.7

1 Nomenclature

EC number
1.8.1.7

Systematic name
glutathione:NADP$^+$ oxidoreductase

Recommended name
glutathione-disulfide reductase

Synonyms
EC 1.6.4.2 (formerly)
GOR1
GOR2
GR <1, 13, 20, 23, 26, 30, 34, 35, 39> [52, 53, 57, 61-64, 67, 68, 70, 72]
GRase
GSH reductase
GSSG reductase
NADPH-GSSG reductase
NADPH-glutathione reductase
NADPH:oxidized-glutathione oxidoreductase
OBP29
glutathione S-reductase
glutathione disulfide reductase <36> [56]
glutathione reductase
glutathione reductase (NADPH)
reductase, glutathione

CAS registry number
9001-48-3

2 Source Organism

<1> *Homo sapiens* [1, 2, 6, 8, 12, 17-24, 57, 69-71, 73]
<2> *Escherichia coli* (review [1]; strain 011 [41]; strain S33 [31]) [1, 2, 4, 9, 31, 32, 41]
<3> *Chromatium vinosum* [2, 45]
<4> *Plasmodium vinckei* [2]
<5> *Saccharomyces cerevisiae* (baker's yeast [2,3,12,40]) [1-3, 12, 17, 40, 58, 60]
<6> *Penicillium chrysogenum* (review [1]) [1, 2]

<7> *Spinacia oleracea* (spinach [2,12,38,39]) [2, 12, 38, 39]
<8> *mackerel* [2]
<9> *Mus musculus* [2, 37]
<10> *gerbil* [2, 34]
<11> *Oryctolagus cuniculus* [2, 28, 29]
<12> *Sus scrofa* [2, 15, 16, 43]
<13> *Bos taurus* (calf [36]) [2, 5, 36, 62, 63, 71]
<14> *Rattus norvegicus* (review [1]) [1, 2, 5, 30, 33, 44]
<15> *Achromobacter starkeyi* [7]
<16> *Medicago sativa* (alfalfa [10]) [10]
<17> *Onobrychis viciifolia* (sainfoin [10]) [10]
<18> *Ovis aries* [11, 17]
<19> *Phycomyces blakesleeanus* [13]
<20> *Euglena gracilis* (Z [14,72]) [14, 72]
<21> *Pisum sativum* (pea [25,26]) [25, 26]
<22> *Spirulina maxima* (cyanobacterium [27]) [27]
<23> *Anabaena sp.* (strain 7119, filamentous cyanobacterium [35]; PCC 7120, cyanobacterium [61,66]) [35, 61, 66]
<24> *Hemicentrotus pulcherrimus* (sea urchin, review [1]) [1, 46]
<25> *Oryza sativa* (rice [47,48]) [47, 48]
<26> *Triticum aestivum* (2 charge- and mass-separable isoforms in chloroplasts and cytosol [68]) [17, 68]
<27> *Litomosoides carnii* [2]
<28> *Rhodospirillum rubrum* [49]
<29> *Thiobacillus thiooxidans* [42]
<30> *Triticum durum* (durum wheat [67]) [67]
<31> *Bufo bufo* (toad [2]) [2]
<32> *Moniezia expansa* (cestode [50]) [50]
<33> *horse* [51]
<34> *Picea rubens* (red spruce [52]; isozymes from the cold-hardened strains GR-1H and GR-2H and noncold-hardened strains GR1NH and GR-2NH [52]) [52]
<35> *Chlamydomonas reinhardtii* (2 isozymes [53]) [53, 59]
<36> *Plasmodium falsiparum* [54, 56]
<37> *Trypanosoma brucei rhodesiense* (parasitic protozoa [55]) [55]
<38> *Trypanosoma brucei gambiense* (parasitic protozoa [55]) [55]
<39> *Hansenula mrakii* (yeast, IFO 0895 [64]) [64]
<40> *Thiobacillus ferrooxidans* (iron-grown, AP19-3 [65]) [65]

3 Reaction and Specificity

Catalyzed reaction

2 glutathione + $NADP^+$ = glutathione disulfide + NADPH + H^+ (a dimeric flavoprotein (FAD); activity is dependent on a redox-active disulfide in each of the active centres; <1> inhibition mechanism [71]; <1> recombinant glutathione-trypanothione reductase-like enzyme, substrate binding, three di-

mensional structure, complex formation [70]; <1> catalytic cycle [69]; <26> sulfhydryl groups and histidine residues are essential for the catalytic properties of the enzyme [68]; <37, 38> analogous to trypanothione reductase [55]; <35> branching mechanism [59]; <1, 2, 5, 6, 14, 24, 36> mechanism [2, 8, 24, 46, 54, 69, 73]; <1, 5, 7> ping-pong mechanism [12]; <1> GSSG-binding site [2]; <1,23> substrate and cofactor binding site, three-dimensional structure [23,66,69]; <1,11> cofactor binding of glutathione reductase with FAD analogues, reconstitution [24,29]; <13> molecular modeling of enzyme inactivation by peroxynitrite, structure analysis [62]; <1,23> member of pyridine-nucleotide disulfide oxidoreductase family of flavoenzymes [66,69,70])

Reaction type
oxidation
redox reaction
reduction

Natural substrates and products

S GSSG + NADPH <1-15, 18, 19, 20, 21, 23, 24, 26, 31, 32> (<1> special role in cells that do no longer produce macromolecules [22]; <15> specific for GSSG [7]; <31> high GSSG level in dormant stage [2]; <2,6,14> production of substrate for: glutathione peroxidase EC 1.11.1.9, glutathione-homocystine oxidoreductase EC 1.8.4.1, glutathione-protein disulfide oxidoreductase EC 1.8.4.2 [1]; <1-13,32> maintenance of glutathione/GSSG ratio is a protective mechanism for intracellular thiols during growth in atmospheric oxygen [2,50]; <1> maintenance of high levels of GSH in cytoplasm [8]; <1> role in cell division cycle and in stress adaption on cellular level [8]; <14> enzyme plays a key role in nutrient-induced increase in the thiol content of pancreatic-islet cells and this increase itself participates in the coupling of metabolic to secretory events [30]; <13, 14, 21, 23> serves in detoxification [5,25,61]; <1> probably major source of $NADP^+$ for the pentose phosphate pathway in the lens [17]; <23> high intracellular concentration of glutathione is important for functionality of thiol groups of cellular proteins [61]) (Reversibility: r <1, 2> [22, 31]; ir <13, 14, 20, 23> [1, 5, 14, 61]; ? <1-15, 18, 19, 21, 24, 26, 31, 32> [1, 2, 7, 8, 13, 17, 25, 30, 50]) [1, 2, 5, 7, 8, 13, 14, 17, 22, 25, 30, 31, 50, 61]

P glutathione + $NADP^+$

S Additional information <1> (<1> glutathione redox cycle [2]) [2]

P ?

Substrates and products

S GSSG + NADPH <1-40> (<33> ferricyanide and 2,6-dichlorophenolindophenol as electron acceptors [51]; <7> reverse reaction only if very high concentration of $NADP^+$ and glutathione present [39]; <2,5,6,14,24> glutathione or other disulfides, e.g.: bis-L-γ-glutamyl-L-cystinyl-bis-β-alanine [1,2]; <5> slight activity with bis-N,N'-(γ-glutamylcystine) [2]; <5> glutathione-S-sulfonate [2]; <29> methylene blue and L-cystine slowly [42]; <3,7> no activity [38, 45]; <1, 5, 14, 22, 32, 33> mixed disulfide between coenzyme A and glutathione [2,27,40,50,51]; <1> slight activity with D,L-lipoate [2]; <15> 5,5'-dithiobis(2-nitrobenzoic acid) [7]; <12> S-sulfoglutathione and some mixed disulfides, with the exception of the mixed disulfide of coenzyme A and GSH, are poor substrates [16]; <20,26,35> highly specific for GSSG [14,59,68,72]; <36> methylene blue as electron acceptor [54]) (Reversibility: r <1-14, 25-27, 31, 33, 36> [1, 2, 8, 18, 19, 22-24, 31, 32, 48, 51, 54, 68, 69]; ir <1, 2, 13, 14, 20, 23, 35> [2, 4, 5, 14, 57, 59, 61, 66, 67, 73]; ? <1, 2, 5-7, 11-29, 32, 34-40> [1-3, 6, 7, 9-13, 15-17, 20, 21, 25-30,33-47,49,50-56,58,60,62-65,70-72]) [1-73]

P glutathione + $NADP^+$ <1-14, 19, 20, 24-27, 31, 33, 36> [1, 2, 5, 13, 18, 19, 22-24, 31, 32, 48, 51, 54, 68-72]

S ajoene + NADPH <1> (<1> i.e. (E,Z)-4,5,9-trithiadodeca-1,6,11-triene-9-oxide, substrate and inhibitor [73]; <1> also formation of single-electron reduced product and concomitantly superoxide anion radicals [73]) (Reversibility: ? <1> [73]) [73]

P 4,5,9-trithiadodeca-1,6,11-triene + $NADP^+$ <1> [73]

S ajoene + NADPH <1> (<1> i.e. (E,Z)-4,5,9-trithiadodeca-1,6,11-triene-9-oxide, substrate and inhibitor [73]; <1> also formation of single-electron reduced product and concomitantly superoxide anion radicals [73]) (Reversibility: ? <1> [73]) [73]

P 4,8,9,13-tetrathiahexadeca-1,6,1,15-tetraene + $NADP^+$ <1> [73]

S diallyl trisulfide + NADPH <1> (Reversibility: ? <1> [73]) [73]

P ?

S trypanothione + NADPH <1> (<1> recombinant glutathione-trypanothione reductase-like enzyme [70]) (Reversibility: <1> [70]) [70]

P reduced trypanothione + $NADP^+$ <1> [70]

S Additional information <1, 3, 5, 7, 9, 10, 12, 14, 23, 28> (<5> no substrates: dithiolethiones and dithiolones [58]; <1> active with diverse FAD analogues, dependent on orientation of the cofactor [24]; <1> glutathione redox cycle [2]; <12> low transhydrogenase activity with oxidized pyridine nucleotide analogs and diaphorase activity with 2,6-dichlorophenolindophenol as acceptor substrates, NADPH and NADH as donors [16]; <9,14,23> branched mechanism [35,37,44]; <28> bisubstrate mechanism [49]; <7> not: cysteamine [38]; <3> oxidized lipoamide and oxidized lipoic acid [45]) [2, 16, 24, 34, 35, 37, 38, 44, 45, 49, 58]

P ?

Inhibitors

1,2-cyclohexanedione <1> [2]
1,3-bis-(2-chloroethyl)-1-nitrosourea <1, 5, 14, 32> (<14> time- and dose-dependent irreversible inhibition [30]) [2, 30, 50]
1-(2-chloroethyl)-3-(2-hydroxyethyl)-1-nitrosourea <1> [2]
1-chloro-2,4-dinitrobenzene <1> (<1> reversible [2]) [2]
1-fluoro-2,4-dinitrobenzene <5> (<5> reversible [2]) [2]
2,3-butanedione <1> [2]
2,4,6-trinitrobenzenesulfonate <13, 32> (<13> reversible [2]) [2, 50]
2-chloroethylisocyanate <5> [2]
5-nitro-2-furoic acid <32> [50]

5-nitroindole <32> [50]
ADP <3> [45]
AMP <3> [45]
ATP <3> [45]
$AgNO_3$ <3, 29> [42, 45]
$BaCl_2$ <3> [45]
Br^- <28> [49]
Ca^{2+} <3, 15> [7, 45]
Cd^{2+} <1, 8, 39> [2, 64]
Cl^- <3, 28> [45, 49]
Co^{2+} <20, 35, 39> (<35,39> slightly [53,64]; <35> complete inhibition at 1 mM [59]) [14, 53, 59, 64]
Cu^{2+} <3, 5, 15, 20, 21, 35, 39> (<21> root enzyme inhibited, chloroplast enzyme only slightly [26]; <1> no inhibition [1]; <35> complete inhibition at 1 mM [59]) [2, 7, 14, 26, 45, 53, 59, 64]
FAD <1> [6]
FMN <1> [6]
$Fe(NO_3)_2$<3> [45]
Fe^{2+} <21> (<21> root enzyme inhibited, chloroplast enzyme only slightly [26]) [26]
Fe^{3+} <20, 39> [14, 64]
GSSG <32> [50]
Hg^{2+} <14, 15, 20, 29, 35> (<35> complete inhibition at 1 mM [59]) [2, 7, 14, 42, 53, 59]
KCl <33> [51]
KI <3, 33> [45, 51]
KNO_3 <33> [51]
Mg^{2+} <3, 15> [7, 45]
Mn^{2+} <3, 15, 20, 35> (<35> slightly [53]; <35> 40% inhibition [59]) [7, 14, 45, 53, 59]
N,N'-bis(trans-4-hydroxychlorohexyl)-N'-nitrosourea <14> [2]
N-alkylmaleimide <5> [2]
N-ethylmaleimide <3, 5, 7, 15, 20, 21, 23, 26, 28, 35> (<26> chloroplast enzyme, 90% inhibtion at 1 mM, partially reversible by GSSG 0.5 mM [68]; <15> slight [7]; <21> root enzyme is slightly more sensitive than chloroplast enzyme [26]) [1, 7, 14, 26, 35, 39, 45, 49, 59, 68]
NADH <1, 9, 19> (<9> at concentration above 0.3 mM [37]) [13, 19, 37]
$NADP^+$ <1, 5, 7, 9, 14, 20, 30> (<20> 1 mM [14]; <1,7,9,30> competitive against NADPH [19,37,39,67]) [1, 6, 14, 19, 21, 37, 39, 44, 67]
NADPH <1, 9, 19, 26, 34> (<1> protection by $NADP^+$ [71]; <1> slow inactivation in vitro due to inter- or intramolecular disulfide formation [69]; <26> chloroplast enzyme, 50% inhibition at 0.1 mM [68]; <34> reversible reductive inactivation with isozyme GR-1H and slightly with GR-2H [52]; <9> at concentration above 0.3 mM [37]; <1> derivatives [21]; <1> promotes formation of aggregates, reversible by thiols, e.g. glutathione or 2-mercaptoethanol [19]; <21> no inhibition [25]) [1, 13, 19, 21, 37, 52, 68, 69, 71]
NO_3^- <28> [49]

Na^+ <32> [50]
$Na_2S_2O_3$ <33> [51]
NaBr <1> (<1> above 0.2 M [6]) [6]
NaCl <33> [51]
Ni^{2+} <20, 35, 39> (<35> 80% inhibition [59]; <39> slightly [64]) [14, 59, 64]
S-(2,4-dinitrophenyl)-glutathione <1> [2]
SO_4^{2-} <28> [49]
Zn^{2+} <1, 7, 14, 20, 23, 26, 35, 39> (<26> chloroplast enzyme, 94% inhibition at 0.5 mM, reversible and competitive to GSSG [68]; <21> root enzyme inhibited, chloroplast enzyme very slightly [26]; <23> inhibition increased by NADPH [35]; <35> 75% inhibition [59]) [2, 14, 35, 39, 53, 59, 64, 68]
ajoene <1> (<1> i.e. (E,Z)-4,5,9-trithiadodeca-1,6,11-triene-9-oxide, natural compound from garlic, Allium sativum, covalent inhibition, but also substrate [73]; <1> time- and temperature-dependent inhibition, mixed disulfide between active site Cys58 and the inhibitor, modified enzyme shows a markedly increased oxidative activity [73]) [73]
apigenin <13> (<13> flavonone, non-competitive with both NADPH and GSSG, influence on glutathione recognition [63]) [63]
arsenite <1, 5, 14, 15, 23, 27, 35> (<23> inhibition increased by NADPH [35]; <35> slightly [53]) [2, 7, 35, 53]
baicalin <13> (<13> slightly, flavone glycoside, non-competitive with both NADPH and GSSG, influence on glutathione recognition [63]) [63]
benzylselenosulfate <14> [2]
carmustine <1> (<1> irreversible inhibition of enzyme in reduced state [69]) [69]
catechin <13> (<13> slightly, catechin, non-competitive with both NADPH and GSSG, influence on glutathione recognition [63]) [63]
chromate <1-14> (<14> probably irreversible [2]) [2]
chrysin <13> (<13> slightly, flavonone, non-competitive with both NADPH and GSSG, influence on glutathione recognition [63]) [63]
citrate <28> [49]
diethyl dicarbonate <26> (<26> chloroplast enzyme, 100% inhibition at 4 mM, 23% inhibition at 1 mM [68]) [68]
dinitrosated isomers of N,N'-bis[N(2-chloroethyl)-N-carbonyl]cysteamine <14> [2]
epicatechin <13> (<13> slightly, catechin, non-competitive with both NADPH and GSSG, influence on glutathione recognition [63]) [63]
epicatechin gallate <13> (<13> slightly, catechin, non-competitive with both NADPH and GSSG, influence on glutathione recognition [63]) [63]
epigallocatechin <13> (<13> slightly, catechin, non-competitive with both NADPH and GSSG, influence on glutathione recognition [63]) [63]
epigallocatechin gallate <13> (<13> catechin, non-competitive with both NADPH and GSSG, influence on glutathione recognition [63]) [63]
fisetin <13> (<13> flavonol, non-competitive with both NADPH and GSSG, influence on glutathione recognition [63]) [63]
gallic acid <13> (<13> slightly [63]) [63]

glutathione <1, 3, 7, 9, 20, 28, 30> (<30> uncompetitive product inhibition [67]; <20> 1 mM [14]; <1,3,7,9> product inhibition [1,37,39,45]; <28> GSSG, 12 mM, 25% inhibition [49]) [1, 14, 19, 37, 39, 45, 49, 67]
haemin <5> (<5> mediates covalent cross-linking and degradation of the enzyme [2]) [2]
iodide <24> [46]
iodine <5> [2]
iodoacetamide <1, 5, 5> [2, 45]
iodoacetate <5, 7, 15> (<5> in presence but not in absence of reduced coenzyme [3]) [3, 7, 39]
kaempferol <13> (<13> slightly, flavonol, non-competitive with both NADPH and GSSG, influence on glutathione recognition [63]) [63]
melarsen oxide <1, 27, 37, 38> (<1> 2-step process [71]; <37,38> i.e. *p*-(4,6-diamino-s-triazin-2-yl)aminophenylarsenoxide [55]; <37,38> potent inhibitor, reversible by excess of thiols, more sensitive with NADPH [55]) [2, 55, 71]
mercuric sulfate <29> [42]
morin <13> (<13> slightly, flavonol, non-competitive with both NADPH and GSSG, influence on glutathione recognition [63]) [63]
myricetin <13> (<13> flavonol, non-competitive with both NADPH and GSSG, influence on glutathione recognition [63]) [63]
nifurtimox <1, 14> [2, 4]
nitrofurantoin <1-14, 32> (<1> non-competitive [6]) [2, 6, 50]
nitrofurazone <32> [50]
nitrogen mustard <1> [2]
p-chloromercuribenzoate <20, 26, 28, 29, 35, 39> (<26> chloroplast enzyme, 96% inhibition at 1 mM [68]) [14, 42, 49, 59, 64, 68]
p-chloromercuriphenyl sulfonate <5> [1]
p-hydroxymercuribenzoate <2, 3, 10, 11, 14, 21> (<21> chloroplast enzyme is slightly more sensitive than root enzyme [26]) [2, 26, 28, 30, 34, 45]
peroxynitrite <13> (<13> inactivation of enzyme by formation of nitrotyrosine near the catalytic center, 2.5fold increased K_m-value and 1.7fold decreased V_{max}, molecular modeling [62]) [62]
phenyl mercuric acetate <5, 35> (<35> total inhibition [53]) [1, 53]
phenylarsonous acid <5> [2]
phenylglyoxal <1, 5, 35> (<35> slightly [53]) [2, 53]
phenylmethylsulfonylfluoride <26> (<26> i.e. PMSF, chloroplast enzyme, slight inhibition [68]) [68]
phosphate buffer <1> [1]
pyridoxal 5'-phosphate <5> (<5> 70% inactivation, due to specific modification of an ε-amino group lysine residue [60]) [60]
quercetin <13> (<13> flavonol, non-competitive with both NADPH and GSSG, influence on glutathione recognition [63]) [63]
riboflavin <1> [6]
rutin <13> (<13> slightly, flavonol glycoside, non-competitive with both NADPH and GSSG, influence on glutathione recognition [63]) [63]

sulfhydryl reagents <5> (<5> in presence but not in absence of reduced coenzyme [3]) [3]
urea <22> (<22> activation: 0.4-0.6 M, inactivation at higher concentration [27]) [27]
Additional information <1, 9, 5> (<1,9,5> regulation by inactivation in vivo, e.g. by disulfide bridging [2]) [2]

Cofactors/prosthetic groups

FAD <1-14, 18, 19, 23-25, 35-38> (<1,11> properties of glutathione reductase reconstituted with FAD analogues [24,29]; <5> 4 thiol groups per FAD [1]; <2> 5 thiol groups per FAD [1]; <1-14,22-25,28,35-38> FAD enzyme [2, 3, 5, 6, 8, 15, 16, 20-24, 27-29, 34-36, 38, 45, 46, 48, 49, 54-57, 59, 60, 66, 69, 73]; <12-14, 18, 19, 25> 2 mol of FAD per mol of enzyme [5, 11, 13, 15, 16, 36, 44, 47, 48]; <1> amino acid sequence of FAD-binding domain [23]; <10, 13> 1 mol of FAD per mol of subunit [34,36]; <1> activity with diverse FAD analogues [24]) [1-3, 5, 6, 8, 11, 13, 15, 16, 20-24, 28, 29, 34-36, 38, 41, 44-48, 54-57, 59, 60, 66, 69, 70, 73]
NADH <1, 3, 7, 12, 14, 15, 20, 21, 23, 24, 28, 29, 33-36> (<23> 48fold less activity than with NADPH [66]; <33> 8.5% of the activity with NADPH [51]; <12> 200times less effective than NADPH [16]; <1> 20times less activity than with NADPH [2]; <12> activity relative to NADPH: 0.5% [15]; <20> 1.3% activity of that with NADPH [14]; <7,14> 10% activity of that with NADPH [38,44]; <1> 2% activity of that with NADPH [6]; <24> less than 1% activity of that with NADPH [46]; <28> 3% activity of that with NADPH [49]; <1,7,23,29,36> low activity with NADH [18,39,42,56,61]; <15> enzyme can utilize both NADH and NADPH, more active with NADPH [7]; <21> 20times higher activity with NADH than with NADPH [26]; <3> specific for NADH [45]; <1> a single enzyme uses both NADPH and NADH as hydrogen donors [21]) [2, 6, 7, 14, 15, 16, 19, 21, 26, 38, 39, 42, 44-46, 49, 51-54, 56, 61, 66]
NADPH <1-31, 33, 35, 36, 39, 40> (<2> diphosphate group is bound at the C-terminal edge of a parallel stranded β-sheet as is common in nucleotide-binding proteins [4]; <23> 48fold more activity than with NADH [66]; <1> 20times more activity than with NADH [2]; <21> 5% of the activity with NADH, 3fold higher affinity of root enzyme compared to chloroplast enzyme [26]; <1, 3, 15, 23, 36> can utilize both NADPH and NADH, more active with NADPH [2, 7, 56, 61, 69, 72]; <1, 12, 20> best electron donor [14, 15, 19, 72]; <12> 200times more activity than with NADH [16]; <1> a single enzyme uses both NADPH and NADH as hydrogen donors [21]; <5, 10, 11, 22, 25, 35> absolute specificity for NADPH [27, 28, 34, 47, 59, 60]; <1> NADPH promoted aggregation of the enzyme [19]) [1-43, 45-49, 51, 54, 56, 58-67, 69, 70, 72, 73]
deamino-NADPH <33> (<33> 69% of the activity with NADPH [51]) [51]
ferricyanide <33> [51]
flavin <1-14, 22, 25, 33> (<1-14,25,33> flavoprotein [2,5,31,47,51]; <22> 1.06 mol of flavin per mol of subunit [27]) [2, 5, 27, 31, 47, 51]

Additional information <14, 24, 33, 35> (<1> as malaria brings riboflavin deficiency, in effection regions a large portion of enzymes miss FAD, but are as well active [69]; <33> total lack of activity with 3-acetylpyridine NADH [51]; <14,24,35> no activity with FMN [44,46,59]) [44, 46, 51, 59, 69]

Activating compounds

FAD analogues <1, 11> (<1,11> properties of glutathione reductase reconstituted with FAD analogues [24,29]) [24, 29, 71]
KCl <1> [19]
NaBr <1> (<1> non-linear [6]) [6]
NaCl <1> (<1> non-linear [6]) [6]
cysteine <1> (<1> redox active disulfide at the catalytic site [69,70]) [69, 70]
sodium acetate <1> (<1> non-linear [6]) [6]
sodium citrate <1> (<1> non-linear [6]) [6]
sodium phosphate <1> (<1> activates at sub-saturating concentrations, linear [6]) [6, 19]

Metals, ions

KBr <33> (<33> below 0.1 M [51]) [51]
KH_2PO_4 <33> (<33> below 0.1 M [51]) [51]
Mg^{2+} <39> (<39> slight activation [64]) [64]
Mn^{2+} <39> (<39> activates [64]) [64]
NH_4^+ <3> (<3> activates [45]) [45]
NH_4Cl <33> (<33> below 0.1 M [51]) [51]
Na^+ <3> (<3> activates [45]) [45]
NaBr <1> (<1> activates [6]) [6]
NaCl <1> (<1> activates [6]) [6]
NaF <33> (<33> below 0.1 M [51]) [51]
SO_4^{2-} <3> (<3> stimulates [45]) [45]
diphosphate <3> (<3> activates [45]) [45]
phosphate <3, 24> (<3> activates [45]; <24> optimum concentration of potassium phosphate: 0.1 M [46]) [45, 46]
sodium acetate <1> (<1> activates [6]) [6]
sodium citrate <1> (<1> activates [6]) [6]
sodium phosphate <1> (<1> activates [6]) [6]

Turnover number (min^{-1})

9.6 <1> (diallyl trisulfide) [73]
20.4 <1> ((E)-ajoene) [73]
88.2 <1> ((E)-ajoene, <1> + GSH [73]) [73]
156 <1> ((Z)-ajoene) [73]
1250 <1> (trypanothione, <1> recombinant glutathione-trypanothione reductase-like enzyme [70]) [70]
2280 <23> (NADH, <23> recombinant enzyme [61]) [61]
2900 <1> (glutathione disulfide, <1> reoxidation [1]) [1]
12600 <1> (GSSG) [2, 19, 70]
19000 <20> (GSSG) [72]
22000 <23> (NADPH, <23> recombinant enzyme [61]) [61]
Additional information <36> [54]

Specific activity (U/mg)

0.246 <20> [72]
0.331 <29> (<29> purified enzyme [42]) [42]
1.18 <15> [7]
1.41 <2> [4]
2.9 <28> (<28> purified enzyme [49]) [49]
5.03 <32> (<32> purified enzyme [50]) [50]
28.4 <3> [45]
34 <1> (<1> purified enzyme, with NADH [21]) [21]
42-60 <35> (<35> purified enzyme [53]) [53]
61.2 <19> [13]
86.9 <39> (<39> purified enzyme [64]) [64]
96 <26> (<26> purified enzyme [68]) [68]
108 <7> [38]
114 <13> (<13> purified enzyme [71]) [71]
120 <21> (<21> purified enzyme [26]) [26]
125 <12> [43]
140 <5> (<5> purified enzyme [40]) [40]
146 <21> [25]
150 <25> [47]
153 <5> [3]
158 <9> [37]
165 <1> [17]
180 <11> [28]
193.3 <18> [11]
204 <13, 14> (<13> purified enzyme [36]) [5, 36]
207 <1, 14> (<1> purified protein, with NADPH [21]) [21, 44]
221 <12> (<12> purified enzyme [15]) [15]
238 <22> [27]
240 <1> (<1> purified enzyme [18]) [2, 18]
246 <7, 23> (<23> recombinant purified enzyme [61]) [39, 61]
248 <33> [51]
249 <23> (<23> purified enzyme [35]) [35]
250 <1> (<1> purified enzyme [19]) [19]
262 <1> (<1> purified enzyme [71]) [71]
269 <14> (<14> purified enzyme [33]) [5, 33]
273 <17> [10]
281 <16> [10]
320 <24> (<24> purified enzyme [46]) [46]
334 <2> (<2> recombinant additional enzyme in Escherichia coli [9]) [9]
350 <40> (<40> purified enzyme [65]) [65]
361 <2> (<2> purified enzyme [31]) [31]
373.3 <35> (<35> purified enzyme [59]) [59]
505 <2> (<2> purified enzyme [41]) [41]
Additional information <1, 2, 9, 13, 14, 21, 30> (<2,13,14> assay methods [4,5]; <1,9,14> activity decreases with age of the organism [2]; <1> relative

activity and reaction velocity with diverse FAD analogues [24]) [2, 4, 5, 22, 24, 25, 30, 67]

K_m-Value (mM)

0.00052 <3> (FAD) [45]
0.00078 <24> (FAD) [46]
0.0015 <13> (NADPH) [62]
0.0023 <21> (NADPH, <21> root enzyme [26]) [26]
0.003 <7, 21> (NADPH, <7,21> chloroplast enzyme [26,39]) [25, 26, 39]
0.0037 <30> (NADPH) [67]
0.0038 <5> (NADPH) [1, 3]
0.005 <24> (NADPH) [46]
0.0059 <9> (NADPH) [37]
0.007 <35> (NADPH, <35> isozyme GR-1 [53]) [53]
0.008 <1, 11, 14, 28> (NADPH) [2, 5, 19, 28, 49]
0.009 <1, 23, 33> (NADPH, <1> pH 6.8 [21]) [21, 35, 51]
0.0091 <30> (GSSG) [67]
0.0095 <1> (NADPH, <1> 0.03 M phosphate buffer [1]) [1]
0.01 <7, 21> (GSSG, <21> root enzyme [26]) [26, 38]
0.0106 <35> (NADPH) [59]
0.013 <1, 25> (NADPH, <1> 0.3 M phosphate buffer [1]) [1, 48]
0.0143 <39> (NADPH) [64]
0.015 <26> (NADPH, <26> chloroplast enzyme [68]) [68]
0.016 <2> (NADPH) [31]
0.017 <21> (NADH) [25]
0.019 <1> (GSSG, <1> 0.03 M phosphate buffer [1]) [1]
0.02 <19> (NADPH) [13]
0.021 <12, 13, 32> (NADPH) [5, 15, 36, 50]
0.0213 <39> (GSSG) [64]
0.025 <2> (NADPH) [9]
0.026 <40> (NADPH) [65]
0.0263 <14> (GSSG) [5]
0.028 <21> (GSSG, <21> chloroplast enzyme [26]) [26]
0.028 <35> (NADPH, <35> isozyme GR-2 [53]) [53]
0.032 <25> (GSSG) [48]
0.04 <13, 26> (GSSG, <26> chloroplast enzyme [68]) [62, 68]
0.05 <35> (GSSG, <35> isozmes GR-1 and GR-2 [53]) [53]
0.054 <20, 35> (GSSG) [59, 72]
0.055 <5> (GSSG) [1, 3]
0.056 <34> (GSSG, <34> isozyme GR-1H [52]) [52]
0.058 <11, 28> (GSSG) [28, 49]
0.059 <33> (GSSG) [51]
0.062 <21> (GSSG) [25]
0.065 <1> (GSSG) [2, 19, 70]
0.066 <2> (GSSG) [31]
0.07 <2> (GSSG) [9]
0.073 <34> (GSSG, <34> isozyme GR-1NH [52]) [52]

0.076 <12, 32> (GSSG) [15, 50]
0.101 <13, 24> (GSSG) [5, 36, 46]
0.107 <9> (GSSG) [37]
0.121 <22> (GSSG) [27]
0.125 <1> (GSSG, <1> 0.3 M phosphate buffer [1]) [1]
0.125 <40> (NADH) [65]
0.13 <19> (GSSG) [13]
0.131 <34> (GSSG, <34> isozyme GR-2NH [52]) [52]
0.139 <34> (GSSG, <34> isozyme GR-2H [52]) [52]
0.16 <23> (GSSG, <23> recombinant enzyme [61]) [61]
0.16 <1> (diallyl trisulfide) [73]
0.21 <7, 23> (GSSG) [35, 39]
0.22 <21> ($NADP^+$, <21> root and chloroplast enzyme [26]) [26]
0.24 <23> (NADPH, <23> recombinant enzyme [61]) [61]
0.25 <1> (NADH, <1> pH 6.05 [21]) [21]
0.25 <5> (glutathione-S-sulfonate) [2]
0.3 <40> (GSSG) [65]
0.35 <32> (NADH) [50]
0.377 <2> (NADH) [31]
0.38 <1> ((Z)-ajoene) [73]
0.5 <1> (trypanothione, <1> recombinant glutathione-trypanothione reductase-like enzyme [70]) [70]
0.56 <33> (ferricyanide) [51]
0.6 <3> (NADH) [45]
0.67 <1> (NADH, <1> pH 6.8 [21]) [21]
0.68 <33> (NADH) [51]
0.8 <1> ((E)-ajoene) [73]
1 <1> (mixed disulfide of coenzyme A and glutathione) [2]
1.12 <23> (NADH, <23> recombinant enzyme [61]) [61]
3.3 <22> (mixed disulfide of CoA and glutathione) [27]
7 <3> (GSSG) [2, 45]
7.7 <21> (glutathione, <21> chloroplast enzyme [26]) [26]
8.2 <21> (glutathione, <21> root enzyme [26]) [26]
11.6 <5> (bis-N,N'-(γ-glutamylcystine)) [2]
Additional information <1, 2, 5, 12, 16, 17, 20, 23, 33> (<16,17> thermal dependency of K_m [10]; <33> different isozymes, temperature-dependent, homo-glutathione [51]; <23> K_m of wild-type and mutants [66]) [6, 10, 12, 14, 16, 31, 51, 58, 66]

K_i-Value (mM)

0.00033 <14> (Hg^{2+}) [2]
0.0005 <13> (2,4,6-trinitrobenzene-sulfonate, <13> below, at 50% inhibition value, leading to oxidase activity [2]) [2]
0.0008 <1> (Zn^{2+}, <1> at 50% inhibition value [2]) [2]
0.001 <1> (Cd^{2+}, <1> at 50% inhibition value [2]) [2]
0.001 <5> (Cu^{2+}, <5> at 50% inhibition value [2]) [2]

0.001 <1> (nitrofurantoin, <1> at 50% inhibition value, enzyme partially purified from platelets [2]; <1> platelets [6]) [2,6]
0.004 <1> (FMN) [6]
0.005 <1> (Zn^{2+}, <1> at 50% inhibition value [2]) [2]
0.0065 <14> (Zn^{2+}, <14> at 50% inhibition value [2]) [2]
0.0072 <1> ($NADP^+$, <1> with NADH at pH 6.8 [21]) [21]
0.0091 <27> (melarsen oxide, <27> at 50% inhibition value [2]) [2]
0.01 <2> (*p*-hydroxymercuribenzoate) [2]
0.018 <1> ($NADP^+$, <1> with NADH at pH 6.05 [21]) [21]
0.02 <14> (nitrofurantoin, <14> at 50% inhibition value [2]) [2]
0.02 <1> (riboflavin) [6]
0.022 <1> (1-chloro-2,4-dinitrobenzene, <1> at 50% inhibition value [2]) [2]
0.022 <1> (NADPH-X, <1> derivative, pH 6.8, with NADPH [21]) [21]
0.025 <1> (nitrofurantoin, <1> at 50% inhibition value, enzyme from erythrocytes [2]) [2]
0.028 <1> ($NADP^+$, <1> 10 mM sodium phosphate [6]) [6]
0.03 <1> (S-(2,4-dinitrophenyl)-glutathione, <1> at 50% inhibition value [2]) [2]
0.03 <14> (nifurtimox, <14> at 50% inhibition value [2]) [2]
0.04 <1> (nifurtimox, <1> at 50% inhibition value [2]) [2,4]
0.05 <1> (FAD) [6]
0.055 <1> ($NADP^+$, <1> with NADPH at pH 6.8 [21]) [21]
0.06 <1> ($NADP^+$, <1> 0.1 M sodium phosphate [6]) [6]
0.07 <1> ($NADP^+$) [19]
0.25 <8> (Cd^{2+}) [2]
0.25 <1> (melarsen oxide, <1> at 50% inhibition value [2]) [2]
0.35 <14> (chromate, <1> at 50% inhibition value [2]) [2]
0.54 <1> (NADH-X, <1> derivative, pH 6.8, with NADH [21]) [21]
2.2-9 <1> (arsenite, <1> at 50% inhibition value [2]) [2]
3 <27> (arsenite, <27> at 50% inhibition value [2]) [2]

pH-Optimum

4.5 <2> (<2> NADH + GSSG [31]) [31]
5 <12, 32> (<12,32> NADH + GSSG [15,16,50]) [15, 16, 50]
5-5.5 <14> (<14> NADH + GSSG [44]) [44]
5.7 <1> (<1> NADH + GSSG [19]) [19]
6 <1> (<1> NADH + GSSG [21]) [21]
6 <19> (<19> NADH [13]) [13]
6.5 <1, 33, 40> (<1> recombinant glutathione-trypanothione reductase-like enzyme [70]; <33> NADH + glutathione [51]; <40> NADPH + GSSG [65]) [51, 65, 70]
6.6-7.6 <5> [3]
6.8 <1, 36> (<1> broad optimum [1]; <1> NADPH + GSSG [19,21]) [19, 21, 56]
6.8-7.6 <1> [6]
6.9 <11> [28]

7 <3, 5, 12-15, 22, 23, 29, 33> (<5, 12, 14, 23, 33> NADPH + GSSG [1, 15, 16, 44, 51, 61, 66]; <3> NADH + GSSG [45]) [1, 5, 7, 15, 16, 27, 36, 42, 44, 45, 51, 66]
7.2-7.3 <24> [46]
7.4 <32> (<32> NADPH + GSSG [50]) [50]
7.5 <2, 19, 39> (<2,19,39> NADPH + GSSG [13,31,64]) [13, 31, 64]
7.5-8 <30> [67]
7.5-8.2 <28> [49]
7.7 <21> (<21> root enzyme [26]) [26]
7.8 <1, 13> (<1,13> assay at [71]) [71]
7.8-8.6 <1, 5, 7> (<1,5,7> broad optimum [12]) [12]
7.9 <25> [48]
8 <5, 23, 26> (<26> NADPH + GSSG [68]; <23> NADH + GSSG [61,66]; <5> glutathione disulfide oxidation [1]) [1, 61, 66, 68]
8.1 <7> [38]
8.2 <21, 35> (<21> chloroplast enzyme [26]) [26, 59]
8.5-9 <7> [39]
9 <23> [35]

pH-Range

5.2-9 <11> (<11> at pH 5.2 and 9.0 about 20% of activity maximum [28]) [28]
5.9-8 <5> [60]
6.1-9.1 <21> (<21> 50% of maximal activity at pH 6.1 and 9.1, root enzyme [26]) [26]
6.3-9.6 <25> (<25> at pH 6.3-9.6 about 50% of activity maximum [48]) [48]
6.5-9 <30> (<30> 50% activity at pH 6.5 and pH 9.0, 10% activity at pH 5.0 and pH10.0 [67]) [67]
6.6-10.8 <1, 5, 7> (<1,5,7> at saturating NADPH levels [12]) [12]
6.8-8.4 <35> [59]
6.9-9.2 <21> (<21> pH 6.9: about 50% of activity maximum, pH 9.2: about 83% of activity maximum [26]) [26]
7-9.5 <7> (<7> pH 7: about 40% of activity maximum, pH 9.5: about 55% of activity maximum [39]) [39]
7-11 <10> (<10> pH 7: about 13% of activity maximum, pH 11: 36% of activity maximum [34]) [34]
7.3-8.8 <7> (<7> pH 7.3: about 65% of activity maximum, pH 8.6: about 90% of activity maximum [38]) [38]
7.5-9.5 <2> [31]

Temperature optimum (°C)

25 <1, 2, 5, 7, 10, 18, 21, 22, 24, 26, 30> (<1, 2, 5, 7, 10, 18, 21, 22, 24, 26, 30> assay at [1, 3, 4, 12, 17, 25, 27, 34, 46, 58, 60, 67]) [1, 3, 4, 12, 17, 25, 27, 34, 46, 58, 60, 67]
27 <1, 13, 37, 38> (<1,13,37,38> assay at [55,71]) [55, 71]
28 <15> [7]
30 <1, 2, 12, 14, 23, 35> (<1,2,12,14,23,35> assay at [6,16,31,33,53,66]) [6, 16, 31, 33, 53, 66]

37 <18, 32> (<18,32> assay at [11,50]) [11, 50]
40 <40> [65]
45 <26> (<26> exponential increase in activity up to 45°C [68]) [68]
49 <35> [59]
55 <39> [64]

Temperature range (°C)
15-40 <15> (<15> 15°C: about 25% of activity maximum, 40°C: about 50% of activity maximum [7]) [7]
31-54 <35> (<35> 49% activity at 31°C and 68% activity at 54°C [59]) [59]

4 Enzyme Structure

Molecular weight
54000-56000 <35> (<35> gel filtration [59]) [59]
56000 <11> (<11> SDS-PAGE, gel filtration, monomer, enzyme can also be active as dimer [28]) [28]
63000 <28> (<28> gel filtration [49]) [49]
79000 <20> (<20> gel filtration [14]) [14]
94000 <2> (<2> gel filtration [9]) [9]
96000 <1> (<1> apoenzyme without FAD, sedimentation equilibrium [20]) [20]
99800 <22> (<22> Spirulina maxima, pore gradient gel electrophoresis, dimer, enzyme exists predominantly as tetrameric species in equilibrium with a minor dimer fraction [27]) [27]
100000 <1, 2, 13, 19, 40> (<40> gel filtration and PAGE [65]; <13> analytical ultracentrifugation [36]; <1> sedimentation equilibrium [20]) [5, 13, 19, 20, 32, 36, 65, 69]
102000 <24> (<24> gel filtration [46]) [46]
103000 <12> (<12> analytical ultracentrifugation [15,16]) [15, 16]
104000 <23, 25> (<25> gel filtration [47]; <23> PAGE at different acrylamide concentrations, gel filtration [35]) [35, 47]
105000 <2, 9> (<2> gel filtration [4,41]; <9> gel filtration [37]) [4, 37, 41]
106000 <25> (<25> calculation from sedimentation and diffusion coefficients [48]) [48]
107000 <33> (<33> native PAGE [51]) [51]
108000 <11> (<11> ultracentrifugation in glycerol density gradient, dimer, enzyme can also be active also as monomer [28]) [28]
109000 <2> [31]
110000 <30, 36> (<30,36> gel filtration [56,67]) [56, 67]
114000 <32> [50]
116000 <18> (<18> gel filtration [11]) [11]
118000 <5> (<5> equilibrium sedimentation [3]) [3]
120000 <10> (<10> gel filtration [34]) [34]
123000 <39> (<39> gel filtration [64]) [64]
125000 <7, 14> (<7, 14> gel filtration [38, 44]) [38, 44]

127000 <35> (<35> gel filtration [53]) [53]
140000 <16> (<16> native PAGE [10]) [10]
145000 <7> (<7> gel filtration [39]) [39]
149000 <17> (<17> native PAGE [10]) [10]
150000 <26> (<26> chloroplast enzyme, gel filtration [68]) [68]
156000 <21> (<21> gel filtration [26]) [26]
177000 <22> (<22> sedimentation equilibrium, tetramer, enzyme exists predominantly as a tetrameric species in equilibrium with a minor dimer fraction [27]) [27]
Additional information <21> [25]

Subunits

dimer <1, 2, 7, 9-14, 18-20, 22, 23-25, 30, 32, 35, 39, 40> (<39> 2 * 56000, SDS-PAGE [64]; <35> 2 * 66000, SDS-PAGE [53]; <32> 2 * 63000, SDS-PAGE [50]; <14,30> 2 * 60000, SDS-PAGE [44,67]; <24,40> 2 * 52000, SDS-PAGE [46,65]; <2,25> 2 * 52000, SDS-PAGE [41,48]; <1> 2 * 52500 [2,23]; <2> 2 * 55000 [31]; <10> 2 * 53400, SDS-PAGE [34]; <23> 2 * 54000, SDS-PAGE, gel filtration in presence of urea [35]; <18> 2 * 64000, SDS-PAGE [11]; <2> 2 * 49000, SDS-PAGE [9]; <19> 2 * 50200, [13]; <20> 2 * 40000, SDS-PAGE [14]; <12> 2 * 59000, SDS-PAGE [15,16]; <1,20> 2 * 50000, SDS-PAGE [20,72]; <22> 2 * 47000, SDS-PAGE, enzyme exists predominantly as a tetrameric species in equilibrium with a minor dimer fraction [27]; <11> 2 * 56000, SDS-PAGE, enzyme can also be active as monomer [28]; <9> 2 * 55000, SDS-PAGE [37]; <7> 2 * 72000, SDS-PAGE [39]) [2, 5, 9, 11, 13-16, 19, 20, 23, 27, 28, 31, 32, 34-37, 39, 41, 44, 46, 48, 50, 53, 64, 65, 67, 72]
monomer <11, 35> (<35> 1 * 54000-56000, SDS-PAGE [59]; <11> 1 * 56000, SDS-PAGE, enzyme can also be active as dimer [28]) [28, 59]
oligomer <16, 17, 21, 26> (<26> 2 * 56000 + associated polypeptide 1 * 32000, SDS-PAGE [68]; <21> x * 60000, SDS-PAGE [25]; <16> x * 57000 + x * 36000 [10]; <17> x * 57000 + x * 37000, SDS-PAGE [10]) [10, 25, 68]
tetramer <21, 22> (<22> 4 * 47000, SDS-PAGE, enzyme exists predominantly as a tetrameric species in equilibrium with a minor dimer fraction [27]) [26, 27]
Additional information <1> (<1> three-dimensional structure [8]; <1> in absence of thiols the enzyme shows tendency to form aggregates [18]; <1> NADH promotes aggregation [19]; <1> in absence of thiols, glutathione reductase shows a tendency to form tetramers and larger aggregates, catalytically active [20]) [8, 18-20]

5 Isolation/Preparation/Mutation/Application

Source/tissue

brain <18> [11]
egg <24> [46]
erythrocyte <1, 11, 12> [2, 8, 15, 16-23, 43, 70, 71]
germ <26> [17]

grain <30> (<30> mature kernel endosperm, 2 isoforms belonging to different developmental stages [67]) [67]
intestine <13> (<13> mucosa [62,63,71]) [62, 63, 71]
kernel <25> [47, 48]
leaf <7, 16, 17, 21, 26> [2, 10, 26, 31, 38, 39, 68]
lens fiber <1> [2, 17]
leukocyte <1> [2]
liver <9-11, 13-14, 33> [2, 5, 28, 29, 33, 34, 36, 37, 44, 51]
muscle <8> [2]
mycelium <19> [13]
needle <34> [52]
pancreatic islet <14> [30]
platelet <1> [2, 6]
retina <18> [17]
root <21> [26]

Localization

chloroplast <7, 21, 26, 35> (<7,21> cytoplasm-like spaces of chloroplasts [1,2,25]; <7,26> stroma [39,68]; <26,35> isoform dependent [53,68]) [1, 2, 25, 26, 39, 53, 68]
cytoplasm <13, 14, 20, 26, 29, 32, 35> (<20> solely in cytoplasm [14]; <26,35> isoform dependent [53,68]) [2, 5, 6, 14, 33, 42, 50, 53, 68, 72]
mitochondrion <14, 32> (<32> small amount [50]; <14> cytoplasm-like spaces of mitochondria [1,2]) [1, 2, 50]

Purification

<1> (recombinant mutant from Escherichia coli [70]; from lens and erythrocytes [17]; amino acid sequence analysis [23]) [6, 17, 18, 21-23, 70, 71]
<2> (large scale [41]; enzyme expressed from plasmid, small and large scale [9]) [4, 9, 31, 32, 41]
<3> [45]
<5> [3, 17, 40]
<7> [38, 39]
<9> [37]
<10> [34]
<11> [28]
<12> [15, 16, 43]
<13> [5, 36, 71]
<14> [5, 33, 44]
<15> [7]
<16> [10]
<17> [10]
<18> [11, 17]
<19> [13]
<20> [72]
<21> [25, 26]
<22> [27]

<23> (via high-level expression of recombinant enzyme in Escherichia coli [61]; wild-type and recombinant from Escherichia coli [66]) [35, 61, 66]
<24> [46]
<25> [47]
<26> (chloroplast enzyme [68]) [17, 68]
<27> (partially [2]) [2]
<28> [49]
<29> [42]
<30> [67]
<32> [50]
<33> [51]
<34> (from temperature hardened and nonhardened strains [52]) [52]
<35> (partially [53]) [53, 59]
<36> [56]
<39> [64]
<40> [65]

Renaturation

<5> (refolding after treatment with 6 M urea, recovering of full activity [3]) [3]

Crystallization

<1> (recombinant glutathione-trypanothione reductase-like enzyme, hanging-drop vapour diffusion method, 0.1 M potasssium phosphate, pH 8.0, crystal structure [16]; substrate-binding and structure analysis [70]; binding of cofactor NADPH, $NADP^+$ acts as a positive effector in the reduction of non-physiological substrates like 2,4,6-trinitrobenzene [4]; molecular asymmetry of crystals from deionized water and from 2 M ammonium sulfate, exact 2fold molecular axis in the dimer, form A-D, three-dimensional structure [2]; structure analysis overview, ligand binding, active center [8,73]; dialysis for 72 h against 1000 volumes of deionized water with 1 change [18]; binding structure of diverse FAD analogues to the apoenzyme [24]) [2, 8, 18, 21, 22, 24, 69, 70, 73]
<2> (trigonal crystals, high internal symmetry [2]; hanging drop method, form-S grown crystals from salt and form-P grown crystals from PEG-8000, X-ray analysis [32]) [2, 32]
<12> (<12> hanging-drop method at 22°C, size of crystals varies with ammonium sulfate concentration [15]) [15]

Cloning

<1> (mutant enzyme glutathione reductase-trypanothine reductase, expression in Escherichia coli [70]) [70]
<2> (Escherichia coli gene [9]) [9]
<23> (expression in Escherichia coli of a His-tagged high-level expression mutant [61]; expression of wild-type and mutant in Escherichia coli, structure modeling [66]) [61, 66]
<26> (2 isoforms: DNA sequence analysis, multiple gene copies [68]) [68]

Engineering

A34E/R37W <1> (<1> glutathione reductase mutant, activity switches to trypanothione reductase, termed GRTR, 700fold more activity with trypanothione than with glutathione [70]) [70]

Additional information <23> (<23> a mutant lacking a loop involved in ligand binding, showing reduced activity, and a mutant with a modified loop, which is more efficient with NADPH and NADH [66]; introduction of silent random mutations in 5'-region to generate a His-tagged high-level expression clone [61]) [61, 66]

Application

medicine <1, 32, 36> (<1> mimicking of favism by inhibiting the enzyme with specific inhibitors to protect against malaria [69]; <32,36> potential target for chemotherapy [50,56]; <1> potential target of antimalarial and cytostatic agents [57,69]) [50, 56, 57, 69]

6 Stability

pH-Stability

3 <5> (<5> stable [3]) [3]
4-10 <39> [64]
5.6 <7> (<7> 30 min, stable [39]) [39]
5.8-8.4 <20> (<20> 67°C: stable, 78°C: unstable [14]) [14]
6-8 <29> (<29> unstable below and above [42]) [42]
7.5 <2> (<2> unstable below [31]) [31]
7.5-9.5 <2> (<2> 4°C, stable [31]) [31]

Temperature stability

20 <28> (<28> 24 h, stable [49]) [49]
30 <21> (<21> reduced form of the enzyme highly unstable [26]) [26]
37 <9> (<9> 8 h, stable [37]) [37]
40 <39> [64]
42.5 <19> (<19> complete loss of activity above 42°C, 10 min in presence of NADPH [13]) [13]
45 <29, 35> (<29> irreversible denaturation above, complete loss of activity after 40 min [42]; <35> stable [59]) [42, 59]
50 <3, 29> (<29> 10 min, complete loss of activity [42]; <3> 5 min, 60% loss of activity [45]) [42, 45]
55 <19, 22, 28, 35, 19> (10 min, 13% loss of activity [13]; <28> 5 min, no loss of activity [49]; <22> 10 min, pH 7.0, stable [27]; <35> complete loss of activity [59]) [13, 27, 49, 59]
60 <7, 28> (<7> 10 min, complete loss of activity [39]; <28> 3 min, 30% loss of activity [49]) [39, 49]
60-65 <21> (<21> oxidized form stable [26]) [26]
62 <28> (<28> unstable above [49]) [49]
65 <3, 19> (<19> 10 min, complete loss of activity [13]; <3> 5 min, complete loss of activity [45]) [13, 45]

67 <20> (<20> pH 5.8-8.4, stable [14]) [14]
70 <2> (<2> highly stable, significant loss of activity only after 10 min [31]) [31]
75 <9, 11> (<11> 1 h, 25% loss of activity [28]; <9> 10 min, stable [37]) [28, 37]
78 <20> (<20> pH 5. 8-8.4, unstable [14]) [14]
80 <9> (<9> $NADP^+$, GSSG, glutathione and dichloroindophenol protect against thermal inactivation at 80°C [37]) [37]
Additional information <21, 30> (<30> resistant to high temperatures and very resistant to low temperatures [67]; <21> root enzyme more stable than chloroplast enzyme, oxidized form more stable than reduced form [26]) [26, 67]

Oxidation stability

<1>, oxygen promotes enzyme aggregation by oxidation of thiols, that prevent oxidation of the enzyme [19]

General stability information

<2>, FAD stabilizes [31]
<2>, dithiothreitol stabilizes [31]
<2>, glycerol stabilizes [31]
<2>, pure enzyme heated in presence of bovine serum albumin and 0.020 mM NADPH: loss of activity beyond 30°C, significant reduction even at 0°C [31]
<7>, glutathione stabilizes chloroplast enzymes [39]
<7>, partial inactivation after incubation with ATP [38]
<9>, FAD, no stabilization against thermal inactivation at 80°C [37]
<9>, GSSG protects against thermal inactivation at 80°C [37]
<9>, $NADP^+$, stabilizes against thermal inactivation at 80°C [37]
<9>, complete inactivation after incubation with NADH, reversible by addition of GSSG [37]
<9>, dichloroindophenol protects against thermal inactivation at 80°C [37]
<9>, glutathione protects against thermal inactivation at 80°C [37]
<13>, NADPH is not protective against induced inactivation by peroxynitrite [62]
<22>, repeated freezing and thawing, no effect [27]
<22>, urea, activation from 0.4-0.6 M, inactivation at higher concentration [27]
<1-14>, concentrated solution, stable for years [2]
<1-14>, crystals and amorphous protein obtained by precipitation with ammonium sulfate are stable [2]
<1-14>, repeated freezing and thawing, inactivation [2]
<11, 19>, urea, quite stable to [13, 28]
<13, 14>, freezing denaturates [5]
<2, 5>, $NADP^+$ stabilizes [31, 60]
<3, 11>, repeated freezing and thawing, stable [28, 45]
<7, 9>, complete inactivation after incubation with NADPH, reversible by addition of GSSG [37, 39]

Storage stability

<1>, 0-4°C, crystals in buffer, several months [18]
<2>, -18°C, in the dark, 10% glycerol, 5 mM $NADP^+$, for at least 1 year [31]
<2>, -20°C, 20 mM potassium phosphate, pH 7.5, 1 mM EDTA, 10% glycerol, months [32]
<3>, -20°C, 2 years [45]
<5>, -20°C, pH 6.8, 0.05 M potassium phosphate buffer, indefinitely stable [3]
<7>, 4°C, several weeks [39]
<9>, -70°C, pH 7.0, 1 mM EDTA, 1 mM dithiothreitol, 0.020 mM FAD, 5 mM $NADP^+$ [37]
<11>, -20°C, with several freeze-thawing cycles, more than 2 years stable [28]
<13>, in refrigerator, several months [36]
<15>, -20°C, 2-3 months [7]
<19>, 4°C, pH 7.5, 50 mM sodium phosphate buffer, 1 mM 2-mercaptoethanol, 1 mM EDTA [13]
<20>, -20°C, several weeks [14]
<22>, -20°C, 1 year [27]
<23>, 25°C, in the dark [35]
<24>, -28°C, 50 mM phosphate buffer, a few months [46]
<28>, 0°C, 6 days [49]
<32>, 4°C, 50 mM potassium phosphate, pH 7.4, 50% loss of activity after 10 days [50]
<1-14>, 4°C, activation after storage over days, months or years [2]
<1-14>, concentrated solution, years [2]

References

[1] Williams, C.H.: Flavin-containing dehydrogenases. The Enzymes, 3rd Ed. (Boyer, P.D., ed.), **13**, 89-173 (1976)
[2] Schirmer, R.H.; Krauth-Siegel, R.L.; Schulz, G.E.: Glutathione reductase. Coenzymes and cofactors. Glutathione, Chem. Biochem. Med. Aspects Pt. A (Dolphin D, Poulson R, Avromonic O, eds.) John Wiley & Sons, New York, **3**, 553-596 (1989)
[3] Colman, R.F.: Glutathione reductase (yeast). Methods Enzymol., **17**, 500-503 (1971)
[4] Williams, C.H.; Arscott, L.D.: Glutathione reductase (Escherichia coli). Methods Enzymol., **17**, 503-509 (1971)
[5] Carlberg, I.; Mannervik, B.: Glutathione reductase. Methods Enzymol., **113**, 484-490 (1985)
[6] Moroff, G.; Kosow, D.P.: Characterization of human platelet glutathione reductase. Biochim. Biophys. Acta, **527**, 327-336 (1978)
[7] Ruiz-Herrera, J.; Amezcua-Ortega, R.; Trujillo, A.: Purification and properties of a disulfide reductase obtained from Achromobacter starkeyi. J. Biol. Chem., **243**, 4083-4088 (1968)

[8] Schulz, G.E.; Schirmer, R.H.; Pai, E.F.: The three-dimensional structure of glutathione reductase and its substrate complexes at 0.3 nm resolution. Flavins and Flavoproteins (Proc. Int. Symp., 6th, Meeting, Yagi, K., Yamano, T., eds.), 557-567 (1978)

[9] Scrutton, N.S.; Berry, A.; Perham, R.N.: Purification and characterization of glutathione reductase encoded by a cloned and over-expressed gene in Escherichia coli. Biochem. J., **245**, 875-880 (1987)

[10] Kidambi, S.P.; Mahan, J.R.; Matches, A.G.: Purification and thermal dependence of glutathione reductase from two forage legume species. Plant Physiol., **92**, 363-367 (1990)

[11] Acan, N.L.; Tezcan, E.F.: Sheep brain glutathione reductase: purification and general properties. FEBS Lett., **250**, 72-74 (1989)

[12] Wong, K.K.; Vanoni, M.A.; Blanchard, J.S.: Glutathione reductase: solvent equilibrium and kinetic isotope effects. Biochemistry, **27**, 7091-7096 (1988)

[13] Montero, S.; De Arriaga, D.; Soler, J.: Purification and characterization of glutathione reductase from mycelium of Phycomyces blakesleeanus. Biochim. Biophys. Acta, **952**, 56-66 (1988)

[14] Shigeoka, S.; Onishi, T.; Nakano, Y.; Kitaoka, S.: Characterization and physiological function of glutathione reductase in Euglena gracilis Z. Biochem. J., **242**, 511-515 (1987)

[15] Boggaram, V.; Brobjer, T.; Larson, K.; Mannervik, B.: Purification of glutathione reductase from porcine erythrocytes by the use of affinity chromatography on 2,5-ADP-Sepharose 4B and crystallization of the enzyme. Anal. Biochem., **98**, 335-340 (1979)

[16] Boggaram, V.; Larson, K.; Mannervik, B.: Characterization of glutathione reductase from porcine erythrocytes. Biochim. Biophys. Acta, **527**, 337-347 (1978)

[17] Harding, J.J.: Affinity chromatography in the purification of glutathione reductase. J. Chromatogr., **77**, 191-199 (1973)

[18] Worthington, D.J.; Rosemeyer, M.A.: Human glutathione reductase: purification of the crystalline enzyme from erythrocytes. Eur. J. Biochem., **48**, 167-177 (1974)

[19] Worthington, D.J.; Rosemeyer, M.A.: Glutathione reductase from human erythrocytes. Catalytic properties and aggregation. Eur. J. Biochem., **67**, 231-238 (1976)

[20] Worthington, D.J.; Rosemeyer, M.A.: Glutathione reductase from human erythrocytes. Molecular weight, subunit composition and aggregation properties. Eur. J. Biochem., **60**, 459-466 (1975)

[21] Nakashima, K.; Miwa, S.; Yamauchi, K.: Human erythrocyte glutathione reductase. I. Purification and properties. Biochim. Biophys. Acta, **445**, 309-323 (1976)

[22] Krohne-Ehrich, G.; Schirmer, R.H.; Untucht-Grau, R.: Glutathione reductase from human erythrocytes. Eur. J. Biochem., **80**, 65-71 (1971)

[23] Untucht-Grau, R.; Schirmer, R.H.; Schirmer, I.; Krauth-Siegel, R.L.: Glutathione reductase from human erythrocytes: amino-acid sequence of the structurally known FAD-binding domain. Eur. J. Biochem., **120**, 407-419 (1981)

[24] Krauth-Siegel, R.L.; Schirmer, R.H.; Ghisla, S.: FAD analogues as prosthetic groups of human glutathione reductase. Properties of the modified enzyme species and comparisons with the active site structure. Eur. J. Biochem., **148**, 335-344 (1985)
[25] Connell, J.P.; Mullet, J.E.: Pea chloroplast glutathione reductase: purification and characterization. Plant Physiol., **82**, 351-356 (1986)
[26] Bielawski, W.; Joy, K.W.: Properties of glutathione reductase from chloroplasts and roots of pea. Phytochemistry, **25**, 2261-2265 (1986)
[27] Rendon, J.L.; Calcagno, M.; Mendoza-Hernandez, G.; Ondarza, R.N.: Purification, properties, and oligomeric structure of glutathione reductase from the cyanobacterium Spirulina maxima. Arch. Biochem. Biophys., **248**, 215-223 (1986)
[28] Zanetti, G.: Rabbit liver glutathione reductase. Purification and properties. Arch. Biochem. Biophys., **198**, 241-246 (1979)
[29] Zanetti, G.; Beretta, C.; Malandra, D.: Properties of rabbit liver glutathione reductase reconstituted with FAD analogs. Arch. Biochem. Biophys., **244**, 831-837 (1986)
[30] Malaisse, W.J.; Dufrane, S.P.; Mathias, P.C.F.; Carpinelli, A.R.; Malaisse-Lagae, F.; Garcia-Morales, P.; Valverde, I.; Sener, A.: The coupling of metabolic to secretory events in pancreatic islets. The possible role of glutathione reductase. Biochim. Biophys. Acta, **844**, 256-264 (1985)
[31] Mata, A.M.; Pinto, M.C.: Purification by affinity chromatography of glutathione reductase (EC 1.6.4.2) from Escherichia coli and characterization of such enzyme. Z. Naturforsch. C, **39**, 908-915 (1984)
[32] Ermler, U.; Schulz, G.E.: The three-dimensional structure of glutathione reductase from Escherichia coli at 3.0 A resolution. Proteins Struct. Funct. Genet., **9**, 174-179 (1991)
[33] Carlberg, I.; Altmejd, B.; Mannervik, B.: Purification and immunological studies of glutathione reductase from rat liver. Evidence for an antigenic determinant at the nucleotide-binding domain of the enzyme. Biochim. Biophys. Acta, **677**, 146-152 (1981)
[34] Le Trang, N.; Bhargava, K.K.; Cerami, A.: Purification of glutathione reductase from gerbil liver in two steps. Anal. Biochem., **133**, 94-99 (1983)
[35] Serrano, A.; Rivas, J.; Losada, M.: Purification and properties of glutathione reductase from the cyanobacterium Anabaena sp. strain 7119. J. Bacteriol., **158**, 317-324 (1984)
[36] Carlberg, I.; Mannervik, B.: Purification and characterization of glutathione reductase from calf liver. An improved procedure for affinity chromatography on 2,5-ADP-Sepharose 4B. Anal. Biochem., **116**, 531-536 (1981)
[37] Lopez-Barea, J.; Lee, C.Y.: Mouse-liver glutathione reductase. Purification, kinetics, and regulation. Eur. J. Biochem., **98**, 487-499 (1979)
[38] Wirth, E.; Latzko, E.: Partial purification and properties of spinach leaf glutathione reductase. Z. Pflanzenphysiol., **89**, 69-75 (1978)
[39] Halliwell, B.; Foyer, C.H.: Properties and physiological function of a glutathione reductase purified from spinach leaves by affinity chromatography. Planta, **139**, 9-17 (1978)

[40] Carlberg, I.; Mannervik, B.: Purification by affinity chromatography of yeast glutathione reductase, the enzyme responsible for the NADPH-dependent reduction of the mixed disulfide of coenzyme A and glutathione. Biochim. Biophys. Acta, **484**, 268-274 (1977)
[41] Pigiet, V.P.; Conley, R.R.: Purification of thioredoxin, thioredoxin reductase, and glutathione reductase by affinity chromatography. J. Biol. Chem., **252**, 6367-6372 (1977)
[42] Tano, T.; Ishii, K.; Sugio, T.; Imai, K.: A NADPH:glutathione oxidoreductase isolated from Thiobacillus thiooxidans. Agric. Biol. Chem., **40**, 1879-1880 (1976)
[43] Mannervik, B.; Jacobsson, K.; Boggaram, V.: Purification of glutathione reductase from erythrocytes by the use of affinity chromatography on 2,5-ADP-Sepharose 4-B. FEBS Lett., **66**, 221-224 (1976)
[44] Carlberg, I.; Mannervik, B.: Purification and characterization of the flavoenzyme glutathione reductase from rat liver. J. Biol. Chem., **250**, 5475-5480 (1975)
[45] Chung, Y.C.; Hurlbert, R.E.: Purification and properties of the glutathione reductase of Chromatium vinosum. J. Bacteriol., **123**, 203-211 (1975)
[46] Ii, I.; Sakai, H.: Glutathione reductase in the sea urchin egg. I. Purification and general properties. Biochim. Biophys. Acta, **350**, 141-150 (1974)
[47] Ida, S.; Morita, Y.: Studies on respiratory enzymes in rice kernel. Part VII. Purification of an acidic flavoprotein, glutathione reductase, from rice embryos. Agric. Biol. Chem., **35**, 1542-1549 (1971)
[48] Ida, S.; Morita, Y.: Studies on respiratory enzymes in rice kernel. Part VIII. Enzymatic properties and physical and chemical characterization of glutathione reductase from rice embryos. Agric. Biol. Chem., **35**, 1550-1557 (1971)
[49] Boll, M.: Glutathione reductase from Rhodospirillum rubrum. Arch. Mikrobiol., **66**, 374-388 (1969)
[50] McCallum, M.J.; Barrett, J.: The purification and properties of glutathione reductase from the cestode Moniezia expansa. Int. J. Biochem. Cell Biol., **27**, 393-401 (1995)
[51] Garcia-Alfonso, C.; Martinez-Galisteo, E.; Llobell, A.; Barcena, J.A.; Lopez-Barea, J.: Horse-liver glutathione reductase: purification and characterization. Int. J. Biochem., **25**, 61-68 (1993)
[52] Hausladen, A.; Alscher, R.G.: Cold-hardiness-specific glutathione reductase isoenzymes in red spruce. Thermal dependence of kinetic parameters and possible regulatory mechanisms. Plant Physiol., **105**, 215-223 (1994)
[53] Serrano, A.; Llobell, A.: Occurrence of two isoforms of glutathione reductase in the green alga Chlamydomonas reinhardtii. Planta, **190**, 199-205 (1993)
[54] Bohme, C.C.; Arscott, L.D.; Becker, K.; Schirmer, R.H.; Williams, C.H., Jr.: Kinetic characterization of glutathione reductase from the malarial parasite Plasmodium falciparum. Comparison with the human enzyme. J. Biol. Chem., **275**, 37317-37323 (2000)

[55] Cunningham, M.L.; Zvelebil, M.J.J.M.; Fairlamb, A.H.: Mechanism of inhibition of trypanothione reductase and glutathione reductase by trivalent organic arsenicals. Eur. J. Biochem., **221**, 285-295 (1994)

[56] Krauth-Siegel, R.L.; Mueller, J.G.; Lottspeich, F.; Schirmer, R.H.: Glutathione reductase and glutamate dehydrogenase of Plasmodium falciparum, the causative agent of tropical malaria. Eur. J. Biochem., **235**, 345-350 (1996)

[57] Nordhoff, A.; Tziatzios, C.; van den Broek, J.A.; Schott, M.K.; Kalbitzer, H.R.; Becker, K.; Schubert, D.; Schirmer, R.H.: Denaturation and reactivation of dimeric human glutathione reductase–an assay for folding inhibitors. Eur. J. Biochem., **245**, 273-282 (1997)

[58] Levron, B.; Burgot, G.; Burgot, J.L.: On the reduction of dithiolethiones and dithiolylium ions by NADPH and glutathione reductase. Arch. Biochem. Biophys., **382**, 189-194 (2000)

[59] Takeda, T.; Ishikawa, T.; Shigeoka, S.; Hirayama, O.; Mitsunaga, T.: Purification and characterization of glutathione reductase from Chlamydomonas reinhardtii. J. Gen. Microbiol., **139**, 2233-2238 (1993)

[60] Pandey, A.; Iyengar, L.; Katiyar, S.S.: Modification of an essential amino group of glutathione reductase from yeast by pyridoxal 5'-phosphate. J. Enzyme Inhib., **12**, 143-154 (1997)

[61] Jiang, F.; Mannervik, B.: Optimized heterologous expression of glutathione reductase from cyanobacterium Anabaena PCC 7120 and characterization of the recombinant protein. Protein Expr. Purif., **15**, 92-98 (1999)

[62] Francescutti, D.; Baldwin, J.; Lee, L.; Mutus, B.: Peroxynitrite modification of glutathione reductase: modeling studies and kinetic evidence suggest the modification of tyrosines at the glutathione disulfide binding site. Protein Eng., **9**, 189-194 (1996)

[63] Iio, M.; Kawaguchi, H.; Sakota, Y.; Otonari, J.; Nitahara, H.: Effects of polyphenols, including flavonoids, on glutathione S-transferase and glutathione reductase. Biosci. Biotechnol. Biochem., **57**, 1678-1680 (1993)

[64] Miki, T.; Tsujimoto, Y.; Miyabe, S.; Sugiyama, K.i.; Izawa, S.; Inoue, Y.; Kimura, A.: Oxidative stress response in yeast: purification and characterization of glutathione reductase from Hansenula mrakii. Biosci. Biotechnol. Biochem., **60**, 1207-1209 (1996)

[65] Sugio, T.; Matsugi, S.; Tanaka, T.; Fujita, M.; Tano, T.: Purification and some properties of glutathione reductase from iron-oxidizing bacterium Thiobacillus ferrooxidans. Biosci. Biotechnol. Biochem., **59**, 1568-1570 (1995)

[66] Danielson, U.H.; Jiang, F.; Hansson, L.O.; Mannervik, B.: Probing the kinetic mechanism and coenzyme specificity of glutathione reductase from the Cyanobacterium Anabaena PCC 7120 by redesign of the pyridine-nucleotide-binding site. Biochemistry, **38**, 9254-9263 (1999)

[67] De Lamotte, F.; Vianey-Liaud, N.; Duviau, M.P.; Kobrehel, K.: Glutathione reductase in wheat grain. 1. Isolation and characterization. J. Agric. Food Chem., **48**, 4978-4983 (2000)

[68] Lascano, H.R.; Casano, L.M.; Melchiorre, M.N.; Trippi, V.S.: Biochemical and molecular characterisation of wheat chloroplastic glutathione reductase. Biol. Plant., **44**, 509-516 (2001)

[69] Krauth-Siegel, R.L.; Jockers-Scheruebl, M.C.; Becker, K.; Schirmer, R.H.: NADPH-dependent disulphide reductases. Biochem. Soc. Trans., **17**, 315-317 (1989)

[70] Stoll, V.S.; Simpson, S.J.; Krauth-Siegel, R.L.; Walsh, C.T.; Pai, E.F.: Glutathione reductase turned into trypanothione reductase: structural analysis of an engineered change in substrate specificity. Biochemistry, **36**, 6437-6447 (1997)

[71] Cunningham, M.L.; Zvelebil, M.J.J.M.; Fairlamb, A.H.: Mechanism of inhibition of trypanothione reductase and glutathione reductase by trivalent organic arsenicals. Eur. J. Biochem., **221**, 285-295 (1994)

[72] Montrichard, F.; Le Guen, F.; Laval-Martin, D.L.; Davioud-Charvet, E.: Evidence for the co-existence of glutathione reductase and trypanothione reductase in the non-trypanosomatid Euglenozoa: Euglena gracilis Z. FEBS Lett., **442**, 29-33 (1999)

[73] Gallwitz, H.; Bonse, S.; Martinez-Cruz, A.; Schlichting, I.; Schumacher, K.; Krauth-Siegel, R.L.: Ajoene is an inhibitor and subversive substrate of human glutathione reductase and Trypanosoma cruzi trypanothione reductase: crystallographic, kinetic, and spectroscopic studies. J. Med. Chem., **42**, 364-372 (1999)

Protein-disulfide reductase 1.8.1.8

1 Nomenclature

EC number
1.8.1.8

Systematic name
protein-dithiol:NAD(P)$^+$ oxidoreductase

Recommended name
protein-disulfide reductase

Synonyms
EC 1.6.4.4 (formerly)
NAD(P)H:protein-disulfide oxidoreductase
disulfide reductase
insulin-glutathione transhydrogenase
protein disulfide reductase
protein-disulfide reductase (NAD(P)H)
reductase, protein disulfide

CAS registry number
9029-19-0

2 Source Organism

<1> *Pisum sativum* [1]
<2> *Avena sativa* [1]
<3> *Vicia faba* [1]
<4> *Triticum aestivum* [1]
<5> *Hordeum vulgare* [1]
<6> *Zea mays* [1]

3 Reaction and Specificity

Catalyzed reaction
protein dithiol + NAD(P)$^+$ = protein disulfide + NAD(P)H + H$^+$

Reaction type
oxidation
redox reaction
reduction

Substrates and products

S NAD(P)H + protein disulfide <1-6> (<1>, reaction rate with NADPH is approximately twice that with NADH [1]) (Reversibility: ? <1-6> [1]) [1]

P $NAD(P)^+$ + protein dithiol <1-6> [1]

Inhibitors

$AgNO_3$ <1> (<1>, 0.009 mM, 66% inhibition [1]) [1]
$CdSO_4$ <1> (<1>, 0.005 mM, 91% inhibition [1]) [1]
$CuSO_4$ <1> (<1>, 0.015 mM, complete inhibition [1]) [1]
$HgCl_2$ <1> (<1>, 0.002 mM, 92% inhibition [1]) [1]
$ZnSO_4$ <1> (<1>, 0.003 mM, 82% inhibition [1]) [1]
arsenite <1> (<1>, 0.15 mM, complete inhibition [1]) [1]
idoacetate <1> (<1>, 0.054 mM, 57% inhibition [1]) [1]
p-chloromercuribenzoate <1> (<1>, 0.054 mM, 88% inhibition [1]) [1]

Cofactors/prosthetic groups

NADH <1> (<1>, reaction rate with NADPH is approximately twice of that with NADH [1]) [1]
NADPH <1> (<1>, reaction rate with NADPH is approximately twice of that with NADH [1]) [1]

Specific activity (U/mg)

Additional information <1> [1]

K_m-Value (mM)

0.002 <1> (NADPH) [1]

pH-Optimum

6.9-7.4 <1> [1]

pH-Range

6.2-7.8 <1> (<1>, pH 6.2: about 50% of activity maximum, pH 7.8: about 20% of activity maximum [1]) [1]

5 Isolation/Preparation/Mutation/Application

Source/tissue

root <1, 5, 6> [1]
seed <1-6> [1]
shoot <1, 3-6> [1]

Purification

<1> [1]

6 Stability

pH-Stability

5.5 <1> (<1>, 5°C or 22°C, irreversible inactivation [1]) [1]

Temperature stability

60 <1> (<1>, 10 min, stable [1]) [1]

70 <1> (<1>, 10 min, complete inactivation [1]) [1]

Storage stability

<1-6>, -15°C, 8 weeks, stable [1]

References

[1] Hatch, M.D.; Turner, J.F.: A protein disulfide reductase from pea seeds. Biochem. J., **76**, 556-562 (1960)

Thioredoxin-disulfide reductase 1.8.1.9

1 Nomenclature

EC number
1.8.1.9

Systematic name
thioredoxin:$NADP^+$ oxidoreductase

Recommended name
thioredoxin-disulfide reductase

Synonyms
EC 1.6.4.5 (formerly)
GSP35
general stress protein 35
NADP-thioredoxin reductase
NADPH-thioredoxin reductase
$NADPH_2$:oxidized thioredoxin oxidoreductase
NTR <5, 6> [9, 12]
TR <1-3, 11, 12> [1, 3, 5, 7, 31, 38, 42, 48, 49]
TRase <7> [13]
TrR <2> [32, 47]
TrxR <2, 3, 11, 12, 16, 17> [39, 44-46, 51]
TrxT <16> [35]
reductase, thioredoxin
thioredoxin reductase (NADPH)
Additional information <1, 16> (<1,16> 2 classes of thioredoxin reductases: high molecular weight or large proteins e.g. in mammalia, Plasmodium falsiparum and Caenorhabditis elegans, and low molecular weight proteins e.g. in bacteria [35,40,49]) [35, 40, 49]

CAS registry number
9074-14-0

2 Source Organism

<1> *Bos taurus* (<1> cytosolic and mitochondrial isoforms [49]) [1, 19, 21, 33, 41, 49]
<2> *Escherichia coli* (enzyme with active site mutation [7]; overproducing strain, trxB gene in a pUC vector [2]) [1, 2, 5, 7, 10, 11, 15, 19, 23, 25, 28, 30, 32, 43, 45, 47-50]

<3> *Homo sapiens* (3 different isoforms TR1, TR2, TR3 [42]; HeLa cells [20,36,51]; cell line JPX9 [42]) [3, 5, 6, 8, 14, 20, 31, 36-38, 41, 42, 44, 51]
<4> *Corynebacterium nephridii* [4]
<5> *Spinacia oleracea* (spinach [9]) [9]
<6> *Daucus carota* [12]
<7> *Rhodobacter sphaeroides* (Y [13]) [13]
<8> *Cavia porcellus* (guinea pig [14]) [14]
<9> *Saccharomyces cerevisiae* (involvement of thioredoxin reductase in the dimethyl sulphoxide reductase system [16]) [16, 26]
<10> *Chromatium vinosum* [17]
<11> *Rattus norvegicus* (3 isoforms TR1, TR2, TR3 [42]) [18, 27, 39, 42]
<12> *Mus musculus* (3 isoforms TR1, TR2, TR3 [42]) [19, 42, 44]
<13> *Triticum aestivum* [22, 34]
<14> *Euglena gracilis* [24]
<15> *Arabidopsis thaliana* (EMBL accession numbers Z23109, Z23108 [29]) [29]
<16> *Plasmodium falciparum* (recombinant [44]) [35, 40, 44]
<17> *Drosophila melanogaster* (3 isoforms TrxR1, TrxR2, TrxR3, encoded by 3 genes [46]; isoform TR1 and TR2 [42]; recombinant [44]) [42, 44, 46]
<18> *Mycobacterium leprae* [50]
<19> *Homo sapiens* [42]
<20> *Rattus norvegicus* (RN5CATOMT) [42]
<21> *Mus musculus* [42]
<22> *Mus musculus* [42]
<23> *Drosophila melanogaster* [42]

3 Reaction and Specificity

Catalyzed reaction

thioredoxin + $NADP^+$ = thioredoxin disulfide + NADPH + H^+ (A flavoprotein (FAD); <3> conformational change upon reduction with NADPH or tris-(2-carboxyethyl)phosphine [51]; <2> 2 conformations possible, termed FR and FO conformation, which differ in their fluorescence spectroscopic behaviour, their accessibility for inhibitors and in the efficiency of electron transfer to FAD, involving position 138 in the wild-type and the mutant C138S [45]; <2,3,16> electron transfer in the enzyme complex of apoenzyme, FAD and thioredoxin with NADPH [32,38,40,43,45]; <3> three-dimensional structure [31]; <3,8> UV-light radical reduction mechanism [14]; <2,3> inhibition mechanism [6,15,38]; <2,16> catalytic mechanism [7,11,30,35,43]; <2,16,18> oxidation-reduction cycle of thioredoxin [2,23,25,30,35,50])

Reaction type

oxidation
redox reaction
reduction

Natural substrates and products

S thioredoxin + NADP$^+$ <2, 3, 8, 11, 12, 16, 17> (<17> defense against oxidative stress [46]; <16> detoxification of hydrogen peroxide, protection of the cell against oxidative damage [35]; <3,8> radical reduction, prevention of cells from UV-generated free radical caused damage on the skin [14]; <3> major anti-oxidant in keratinocytes, melanocytes, melanoma cells [6]; <2> metabolic function of thioredoxin reductase-thioredoxin system: supplies reducing equivalents for a wide variety of acceptors, e.g. : ribonucleotide reductase, nonspecific protein disulfide reductase, methionine sulfoxide reductase, D-proline reductase [11]; <12> reduction of free radicals at the surface of the epidermis, enzyme may play a role in physiology of pancreatic β-cells [19]) (Reversibility: ? <2, 3, 8, 11, 12, 16, 17> [6, 11, 14, 19, 27, 31, 35, 46]) [6, 11, 14, 19, 27, 31, 35, 46]

P thioredoxin disulfide + NADPH

S Additional information <11> (<11> ribonucleotide reductase, thioredoxin and thioredoxin reductase constitute a system necessary for the biosynthesis of deesoxyribonucleotides [27]) [27]

P ?

Substrates and products

S 1,4-naphthoquinone + NADPH <18> (Reversibility: ? <18> [50]) [50]

P ?

S 5,5'-dithiobis(2-nitrobenzoic acid) + NADPH <1-3, 6, 7, 11, 12, 16-18> (<18> native thioredoxin-thioredoxin reductase fusion protein [50]; <1-3, 6, 7, 11, 12, 16-18> i.e. DTNB [3, 5, 8, 10, 12, 13, 15, 18-20, 35, 36, 37, 39-41, 43, 44, 46, 49-51]; <2, 7, 12> requires thioredoxin for reduction of DTNB [5, 13, 15, 19]; <1-3, 11, 16-18> coupled assay [15, 18, 35, 36, 37, 39-41, 43, 44, 46, 49-51]) (Reversibility: ? <1-3, 6, 7, 11, 12, 16-18> [3, 5, 8, 10, 12, 13, 15, 18-20, 35, 36, 37, 39-41, 43, 44, 46, 49-51]) [3, 5, 8, 10, 12, 13, 15, 18-20, 35-37, 39-41, 43, 44, 46, 49-51]

P 2-nitro-5-thiobenzoate + NADP$^+$ <1, 17> [46, 49]

S GSSG + NADPH <3> (<1> no activity [49]; <3> in presence of methylselenol [44]) (Reversibility: r <3> [44]; ? <3> [37]) [37, 44]

P GSH + NADP$^+$

S H_2O_2 + NADPH <3> (Reversibility: ? <3> [51]) [51]

P ?

S L-cysteine + NADPH <11> (Reversibility: ? <11> [18]) [18]

P L-cystine + NADP$^+$

S alloxan + NADPH <1, 2, 11, 12> (Reversibility: ? <1, 2, 11, 12> [18, 19, 21]) [18, 19, 21]

P ?

S dithiothreitol + NADPH <3, 5, 14> (<1> no activity [49]) (Reversibility: ? <3, 5, 14> [8, 9, 24]) [8, 9, 24]

P ?

S insulin + NADPH <13> (Reversibility: ? <13> [34]) [34]

P insulin + NADP$^+$ (<13> insulin with reduced disulfides [34])

S menadione + NADPH <3, 11> (Reversibility: ? <3, 11> [18, 37]) [18, 37]

P ?

S methylseleninate + H_2O_2 <3, 11> (<3> in presence of methylseleninate [44]; <11> only wild-type enzyme [39]; <11> addition of selenocysteine increases the activity by 20% [39]) (Reversibility: ? <3, 11> [39, 44]) [39, 44]

P ?

S methylseleninate + NADPH <3> (<3> also utilizes glutathione instead of NADPH [44]) (Reversibility: ir <3> [44]) [44]

P CH_3SeH + $NADP^+$ <3> [44]

S oxidized lipoamide + NADPH <11> (Reversibility: ? <11> [18]) [18]

P lipoamide disulfide + $NADP^+$

S oxidized lipoate + NADPH <3> (<1> no activity with DL-α-lipoate [49]) (Reversibility: ? <3> [37]) [37]

P ?

S protein disulfide isomerase + NADPH <1> (Reversibility: r <1> [1]) [1]

P protein disulfide isomerase + $NADP^+$ <1> (<1> protein disulfide isomerase with reduced disulfides [1]) [1]

S protein disulfide isomerase like protein 1 + NADPH <2> (<2> protein disulfide isomerase like protein 1 from rat liver containing a thioredoxin domain [48]) (Reversibility: ? <2> [48]) [48]

P protein disulfide isomerase like protein 1 + $NADP^+$ <2> (<2> protein disulfide isomerase like protein 1 with reduced disulfides, coupled assay with insulin [48]) [48]

S protein disulfide isomerase like protein 2 + NADPH <2> (<2> protein disulfide isomerase like protein 2 from rat liver containing a thioredoxin domain [48]) (Reversibility: ? <2> [48]) [48]

P protein disulfide isomerase like protein 2 + $NADP^+$ <2> (<2> protein disulfide isomerase like protein 2 with reduced disulfides, coupled assay with insulin [48]) [48]

S seleninate + NADPH <3> (Reversibility: ? <3> [44]) [44]

P ?

S selenocysteine + NADPH <1> (Reversibility: ? <1> [49]) [49]

P selenocystine + $NADP^+$

S thioredoxin + 3-acetylpyridine adenine dinucleotide <2> (<2> wild-type enzyme, mutant enzyme C135S and thioredoxin in subunit complex C135-C32S with the enzyme [32]) (Reversibility: ? <2> [32]) [32]

P thioredoxin disulfide + reduced 3-acetylpyridine adenine dinucleotide <2> [32]

S thioredoxin + 5,5'-dithiobis(2-nitrobenzoic acid) + NADPH <2, 3, 5, 9, 10, 13, 15> (<2, 3, 5, 9, 10, 13, 15> i.e. DTNB [8, 9, 17, 22, 23, 26, 29, 31, 34]) (Reversibility: ir <2, 10> [17, 23]; ? <3, 5, 9, 13, 15> [8, 9, 22, 26, 29, 31, 34]) [8, 9, 17, 22, 23, 26, 29, 31, 34]

P thioredoxin disulfide + 2(2-nitro-5-mercaptobenzoic acid) <2, 10> [17, 23]

S thioredoxin + $NADP^+$ <1-18> (<18> native thioredoxin-thioredoxin reductase fusion protein [50]; <2> substrate e.g. thioredoxin disulfide from phage T4 [25]; <2> NADH in the standard assay, wild-type and chimeric mutants with and without amino acid exchanges [47]; <4> coupled assay with ribonucleotide reductase or methionine sulfoxide reductase from E. coli, thioredoxin-2 [4]; <5> coupled assay with dithiothreitol [9]; <2> wide variety of electron acceptors [11]; <3> coupled assay with nitroxide reductase [14]; <3> coupled assay with adenosine 3'-phosphate 5'-phosphosulfate reductase [20]; <1-3, 5, 6, 9, 11, 13, 16-18> coupled assay with DTNB [5, 7-10, 12, 20, 22, 26, 34-41, 43, 44, 46, 47, 49-51]; <1-3, 11, 13, 16> coupled assay, measurement of NADPH oxidation in presence of insulin and thioredoxin [31, 34-37, 39-41, 47]) (Reversibility: r <1-3, 7, 12, 16, 17> [1, 7, 13, 15, 25, 38, 44]; ir <1, 2, 10-12> [2, 17-19, 21, 23, 28]; ? <1-6, 9, 13, 14, 16-18> [4-6, 8-12, 14, 16, 22, 24, 26, 27, 30-37, 39-41, 43, 46, 47, 49-51]) [1-41, 43, 44, 46, 47, 49-51]

P thioredoxin disulfide + NADPH <1-3, 7, 10-12, 17> [1, 2, 7, 13, 15, 17-19, 21, 23, 25, 28, 31, 46]

S thioredoxin + insulin disulfide <1-3, 7, 11, 13, 16> (<1> no activity [49]; <2,3,7,11,13,16> in presence of NADPH, coupled assay [8,13,23,31,34-36,39,48]) (Reversibility: ir <2, 11> [18, 23, 28]; r <1-3, 7> [13, 21, 25, 38, 48]; ? <3, 11, 13, 16> [8, 22, 31, 34-37, 39]) [8, 13, 18, 21-23, 25, 28, 31, 34-39, 48, 49]

P thioredoxin disulfide + insulin <2, 3, 11> [18, 21, 23, 28, 38, 48]

S thioredoxin + tert-butyl-hydroperoxide <3> (<3> in presence of methylseleninate, coupled assay [44]) (Reversibility: ? <3> [44]) [44]

P ?

S Additional information <1-3, 7, 11, 13, 16> (<2, 3, 13, 16> redox system for electron transfer in the complex of apoenzyme, FAD and thioredoxin with NADPH or other electron acceptors [30-35]; <2,13> slowly reduces other proteins, e.g. insulin, lipoate and ribonuclease [22, 28]; <2> reduction of thioredoxin by NADPH is virtually complete, equilibrium constant is 48 at pH 7 [11]; <1, 11> specific for NADPH, B side of nicotinamide ring [1, 27]) [1, 11, 13, 22, 27, 28, 30-35]

P ?

Inhibitors

1-chloro-2,4-dinitrobenzene <1, 3> (<1> mitochondrial isoform [49]; <1,3> i.e. DNCB [38,41,49]; <1,3> irreversible, with NADPH, alkylation of the active site cysteine disulfide, strong increase in oxidation activity of the enzyme against NADPH [38,41]) [38, 41, 49]

1-fluoro-2,4-dinitrobenzene <1, 3> (<1,3> irreversible, with NADPH, alkylation of the active site cysteine disulfide, strong increase in oxidation activity of the enzyme against NADPH [41]) [41]

13-cis-retinoic acid <3> [6]

3-bromopropionate <3> [37]

4-vinylpyridine <1, 3> (<1,3> irreversible [41]) [41]

5,5'-dithiobis(2-nitrobenzoate) <3> (<3> above 0.1 mM [20]) [20]

Ca^{2+} <1-3> (<3> non-reversible by EDTA [3]) [3, 5, 8, 49]

$CdCl_2$ <2> [28]

Cu^{2+} <1> (<1> strong inhibition [49]) [49]

Fe^{2+} <1> (<1> strong inhibition [49]) [49]
Mg^{2+} <1> (<1> slightly [49]) [49]
Mn^{2+} <1> (<1> strong inhibition [49]) [49]
N-ethylmaleimide <2> (<2> reaction only with the reduced enzyme [15]) [15]
$NADP^+$ <2> (<2> product inhibition [11]) [11]
Zn^{2+} <1> (<1> strong inhibition [49]) [49]
arsenite <2, 3, 6, 10, 11> [12, 17, 18, 28, 37]
auranofin <3> [44]
iodoacetamide <18> [50]
iodoacetate <1, 3> (<1,3> irreversible [41]) [37, 41]
p-chloromercuribenzoate <9> (<9> with NADPH [26]) [26]
p-mercuribenzoate <2> [28]
phenyl mercuric acetate <2> (<2> stabilizes enzyme in one of two possible conformations [45]) [45]
Additional information <1, 3> (<1,3> no inhibition with dinitrohalobenzene analogues: 1,4-dichlorobenzene, 1-chloro-4-nitrobenzene, 1-chloro-3,4-dinitrobenzene, 1-chloro-2,5-dinitrobenzene [41]) [41]

Cofactors/prosthetic groups

3-acetylpyridine adenine dinucleotide <2> [32]
FAD <2, 3, 5-7, 9-13, 15-18> (<2> wild-type and chimeric mutants partly from Salmonella typhimurium AhpF protein [47]; <2,5,7,9> 2 FAD per dimer [9,11,13,26,30]) [2, 7-9, 11-13, 15, 17, 18, 20, 23, 25, 26-32, 34-47, 50, 51]
NADH <2> (<2> wild-type and mutants [47]) [47]
$NADP^+$ <2, 5, 6> [9, 12, 30]
NADPH <1-18> (<2> wild-type and mutants [47]; <2,11,13,15> specific for [11,27,29,34]; <7> strictly dependent on NADPH, inactive with NADH [13]) [1-50]

Activating compounds

EDTA <1> (<1> 4% activity remaining without [49]) [49]
KCN <2> [28]
cysteine <2, 3, 7, 11, 13, 15-17> (<1> C-terminal redox active selenocysteine motif [49]; <3,16,17> C-terminal redox active disulfide [44,46]; <2> 2 pairs of cysteine residues, C32, C35 and C135, C138 [43]; <16> at the active center, C88 and C93 [40]; <16> 1 pair of redox active cysteines at the N-terminal active center and 1 pair of selenenylsulfide at the C-terminus, crucial for the reduction of thioredoxin [35]; <2> 1 redox-active disulfide per subunit [30]; <2,3,7,11,13,15-17> part of the active site [13,15,18,22,29,36,38,39,44]) [13, 15, 18, 22, 29, 30, 34-36, 38-40, 43, 44, 46]
phosphate <2, 3> (<2> 2fold activation [11]; <3> under growth conditions [51]) [11, 51]

Metals, ions

selenocysteine <1, 3, 11, 12, 19-22> (<3> heparin affinity depends on the selenium content, binding can be induced by reduction of the enzyme with NADPH or tris-(2-carboxyethyl)phosphine [51]; <3> isoform 1 contains 1 selenium per subunit, the mitochondrial isoform 2 contains 0.5 selenium per subunit [51]; <1,11> encoded by TGA stop codon [39,49]; <1,3,11> contains selenocysteine as part of the redox active selenosulfide in the active center [36,37,39,49,51]) [36, 37, 39, 42, 49, 51]

Turnover number (min^{-1})

9.6 <16> (DTNB, <16> mutant enzyme H509A [40]) [40]
14 <16> (DTNB, <16> mutant enzyme H509Q [40]) [40]
14.6 <11> (thioredoxin, <11> SeC498C mutant enzyme, human thioredoxin [39]) [39]
110 <3> (rat thioredoxin, <3> insulin coupled assay [37]) [37]
143 <2> (thioredoxin, <2> mutant C136S [7]) [7]
275 <16> (DTNB, <16> wild-type enzyme [40]) [40]
335 <3> (thioredoxin, <3> insulin coupled assay method [37]) [37]
620 <2> (thioredoxin-CAC, <2> mutant, pH 8.0 [2]) [2]
620 <2> (thioredoxin-R, <2> mutant, pH 8.0 [2]) [2]
720 <17> (methylseleninate, <17> recombinant protein [44]) [44]
790 <2> (thioredoxin, <2> mutant C139S [7]) [7]
840 <12> (methylseleninate) [44]
1320 <2> (thioredoxin, <2> wild-type [7]) [7]
1360 <2> (thioredoxin K36E, <2> mutant, pH 8.0 [2]) [2]
1360 <2> (thioredoxin P34S, <2> mutant, pH 8.0 [2]) [2]
1365 <2> (thioredoxin, <2> wild-type, pH 8.0 [2]) [2]
1380 <3> (methylseleninate) [44]
1770 <3> (DTNB) [37]
1860 <16> (methylseleninate) [44]
1980 <2> (FAD) [30]
2000 <9> (NADPH) [26]
2500 <11> (thioredoxin, <11> wild-type enzyme, human thioredoxin [39]) [39]
3000 <11> (thioredoxin) [18]
3018 <2> (DTNB, <2> chimeric enzyme mutant, partly from Salmonella typhimurium AhpF protein [47]) [47]
4000 <11> (DTNB) [18]
10000 <1> (alloxan) [21]
30000 <1> (protein disulfide-isomerase) [1]
61800 <1> (thioredoxin, <1> human thioredoxin [49]) [49]
72000 <1> (thioredoxin) [49]
78000 <1> (thioredoxin, <1> thioredoxin from E. coli [49]) [49]
96000 <1> (selenocysteine) [49]
Additional information <2> (<2> wild-type and recombinant chimeric mutants in a coupled assay [47]; <2> transhydrogenase activity of mutants and wild-type enzyme with thioredoxin and FAD [32]) [11, 30, 32, 47]

Specific activity (U/mg)

0.2 <16> (<16> purified double mutant C535S/C88A as homodimer, insulin coupled assay [35]) [35]
0.22 <16> (<16> mutant H509Q, coupled assay with DTNB [40]) [40]

0.3 <16> (<16> purified mutant C535S as homodimer, insulin coupled assay [35]) [35]
0.4 <16> (<16> homodimer of mutant C88A, insulin coupled assay [35]; <16> purified double mutant C535S/C88A as heterodimer with the wild-type, insulin coupled assay [35]) [35]
0.595 <11> (<11> purified enzyme [27]) [27]
1.5 <16> (<16> purified mutant C535A as heterodimer with the wild-type, insulin coupled assay [35]) [35]
2.2 <16> (<16> purified mutant C88A as heterodimer with the wild-type, insulin coupled assay [35]) [35]
3.9 <16> (<16> purified wild-type enzyme, insulin coupled assay [35]) [35]
4.3 <16> (<16> mutant H509A, coupled assay with DTNB [40]) [40]
4.3 <16> (<16> wild-type enzyme, coupled assay with DTNB [40]) [40]
4.6 <10> (<10> purified enzyme, coupled assay with DTNB [17]) [17]
13.9 <1> (<1> mitochondrial isoform [49]) [49]
20 <5> [9]
22 <1, 2> (<1,2> purified enzyme, coupled assay with DTNB [49]) [49]
31 <3> (<3> DTNB coupled assay, purified enzyme [37]) [37]
33.8 <3> (<3> purified enzyme [31]) [31]
35 <11> (<11> purified enzyme, coupled assay with DTNB [18]) [18]
42 <3> (<3> purified enzyme, coupled assay with DTNB [20]) [20]
58 <2> (<2> purified enzyme [28]) [28]
67 <2> (<2> purified enzyme [23]) [23]
103 <9> (<9> purified enzyme [26]) [26]
Additional information <1-5, 7, 9, 11, 12, 14, 16> (<2> wild-type and chimeric mutants with and without amino acid exchanges [47]; <11> several mutants with different substrates [39]; <16> quantification by DTNB coupled assay method [35]; <1> enzyme quantification by immunochemical method, investigation of cross-reactivity of specific antibodies with enzymes from other species, calf tissue overview [33]; <2> 3 different assay methods [2]; <4> coupled assay [4]) [2, 4, 8, 9, 13, 16, 19, 24, 25, 27, 28, 33, 35, 39, 47]

K_m-Value (mM)
0.0004 <13> (NADPH, <13> wild-type [34]) [34]
0.0004 <11> (thioredoxin, <11> SeC498C mutant, human thioredoxin [39]) [39]
0.00047 <4> (thioredoxin C-2) [4]
0.00076 <4, 13> (thioredoxin, <13> assay with DTNB instead of NADPH [34]; <4> substrate: thioredoxin from E. coli [4]) [4, 34]
0.0008 <2> (NADPH, <2> at 4°C [11]) [11]
0.00088 <2> (thioredoxin, <2> mutant C136S [7]) [7]
0.001 <17> (NADPH, <17> cytosolic and mitochondrial TrxR1 isoform [46]) [46]
0.0011 <15> (thioredoxin, <15> recombinant enzyme [29]) [29]
0.0012 <2> (NADPH, <2> at 25°C [11]) [11]
0.00125 <2> (thioredoxin, <2> wild-type [7]) [7]
0.0017 <2> (thioredoxin, <2> mutant C139S [7]; <2> at 4°C [11]) [7, 11]

0.002 <2> (thioredoxin, <2> substrate wild-type, thioredoxin P34S, pH 8.0 [2]) [2]
0.0025 <11> (reductase, <11> thioredoxin from rat, bovine substrate [18]) [18]
0.0025 <11> (thioredoxin, <11> substrate: thioredoxin from calf and rat [18]) [18]
0.0028 <1, 2> (thioredoxin, <2> at 25°C [11]) [11, 49]
0.003 <5> (NADPH) [9]
0.003 <7, 13> (thioredoxin) [13, 22]
0.0032 <13> (NADPH) [22]
0.0033 <11> (human thioredoxin, <11> wild-type enzyme [39]) [39]
0.0033 <7> (reductase) [13]
0.0037 <3> (thioredoxin, <3> thioredoxin from rat [37]) [37]
0.0044 <9> (thioredoxin-II) [26]
0.006 <11> (NADPH) [18]
0.006 <1> (thioredoxin, <1> human thioredoxin [49]) [49]
0.0067 <2> (thioredoxin K36R, <2> substrate mutant, pH 8.0 [2]) [2]
0.007 <17> (thioredoxin, <17> cytosolic TrxR1 isoform [46]) [46]
0.0144 <7> (thioredoxin, <7> substrate: thioredoxin from E. coli [13]) [13]
0.018 <3> (methylseleninate) [44]
0.019 <17> (thioredoxin, <17> mitochondrial TrxR1 isoform [46]) [46]
0.034 <3> (thioredoxin, <3> thioredoxin from E. coli [37]) [37]
0.035 <1> (protein disulfide-isomerase) [1]
0.036 <13> (thioredoxin, <13> wild-type, thioredoxin from E. coli [34]) [34]
0.045 <11> (thioredoxin, <11> substrate: thioredoxin from E. coli [18]) [18]
0.05 <16> (DTNB, <16> mutant enzyme H509Q [40]) [40]
0.07 <1, 2> (thioredoxin, <1> E. coli thioredoxin [49]) [21, 49]
0.116 <3> (DTNB) [37]
0.125 <2> (thioredoxin-R, <2> substrate mutant, thioredoxin-CAC, pH 8.0 [2]) [2]
0.147 <16> (DTNB, <16> mutant enzyme H509A [40]) [40]
0.212 <16> (DTNB, <16> wild-type enzyme [40]) [40]
0.33 <1> (alloxan) [21]
0.33 <1> (thioredoxin) [21]
0.38 <17> (DTNB, <17> cytosolic TrxR1 isoform [46]) [46]
0.41 <17> (DTNB, <17> mitochondrial TrxR1 isoform [46]) [46]
0.6 <13> (NADH, <13> wild-type [34]) [34]
0.62 <1> (selenocysteine) [49]
0.66 <11> (DTNB) [18]
2.5 <11> (H_2O_2, <11> wild-type enzyme, K_m-value can be reduced by addition of selenocysteine [39]) [39]
Additional information <2, 3, 15> (<15> additionally to the Arabidopsis thaliana thioredoxin, several other thioredoxins have been tested, K_m-values [29]; <2> K_m-values are pH-dependent [2]) [2, 29, 31, 32]

K_i-Value (mM)

0.015 <2> ($NADP^+$, <2> at 25°C [11]) [11]

pH-Optimum

6.8 <9> (<9> assay at [26]) [26]
7 <1-3, 11, 18> (<18> wild-type fusion protein [50]; <11> wild-type enzyme [39]; <1-3> assay at [21,28,37,38]) [21, 28, 37-39, 50]
7.1 <17> (<17> cytosolic and mitochondrial TrxR1 isoform [46]) [46]
7.4 <16> (<16> assay at [40]) [40]
7.7 <2> (<2> about [11]) [11]
8 <1, 3> (<1,3> assay at [41]) [41]
9 <11, 18> (<18> around, mutant enzyme, broad optimum [50]; <11> SeC498C mutant enzyme, broad optimum [39]) [39, 50]

pH-Range

4.2-10.5 <11> (<11> wild-type and mutant [39]) [39]
5.5-9 <18> (<18> wild-type fusion protein [50]) [50]
6.5-10 <18> (<18> mutant enzyme, 50% activity at pH 6.5 [50]) [50]

Temperature optimum (°C)

20 <1, 3, 16> (<1,3,16> assay at [38,40,41]) [38, 40, 41]
25 <1-3, 9, 13, 17> (<1-3,9,13,17> assay at [8,15,21,22,26,46]) [8, 15, 21, 22, 26, 46]
37 <1> (<1> assay at [49]) [49]
40 <3> (<3> reduction of insulin [31]) [31]
52 <3> (<3> reduction of DTNB [31]) [31]

4 Enzyme Structure

Molecular weight

53200 <17> (<17> cytosolic TrxR1 isoform, deduced from amino acid sequence [46]) [46]
60000 <10> (<10> gel filtration [17]) [17]
62000 <6> (<6> gel filtration [12]) [12]
63700 <17> (<17> mitochondrial TrxR1 isoform, deduced from amino acid sequence [46]) [46]
65000 <13> (<13> gel filtration [22]) [22]
65800 <2> (<2> gel filtration [10]) [10]
66000 <2> (<2> sedimentation equilibrium analysis [11]) [11]
68000 <2, 5, 7> (<2> sedimentation equilibrium, ultracentrifugation [28]; <5> gel filtration [9]; <7> native PAGE [13]) [9, 13, 28]
73000-75000 <2> (<2> amino acid analysis based on 2 mol FAD per molecule of enzyme [11]) [11]
75000 <9> (<9> gel filtration [26]) [26]
100000 <1> (<1> gel filtration [49]) [49]
105000 <3> (<3> gel filtration [20]) [20]
110000 <1, 3> (<1,3> gel filtration [21,37]) [21, 37]
116000 <11> (<11> gel filtration [18]) [18]
120000 <1, 3> [41]

130000-160000 <3> (<3> sucrose density gradient centrifugation and gel filtration [31]) [31]
185000 <14> (<14> gel filtration [24]) [24]
Additional information <14> [24]

Subunits

? <3, 13, 14> (<13> x * 35000, SDS-PAGE [34]; <3> x * 58000, melanoma metastases, SDS-PAGE [8]) [8, 24, 34]
dimer <1-3, 5-7, 9-11, 15, 16, 19-22> (<3> 2 * 54800, HeLa cell enzyme, MALDI [51]; <1> 2 * 56000, SDS-PAGE [49]; <11,21> 2 * 67000, isoform TR1, SDS-PAGE [42]; <3,11,12,19-22> 2 * 55000-57000, isoforms TR1 and TR3, SDS-PAGE [42]; <3> 2 * 57000, SDS-PAGE [36,37]; <3> 2 * 60000, SDS-PAGE [31]; <15> 2 * 35000-35328, SDS-PAGE and amino acid analysis [29]; <2> 2 * 32000, SDS-PAGE [10]; <5> 2 * 33000, SDS-PAGE [9]; <6,7> 2 * 34000, SDS-PAGE [12,13]; <10> 2 * 33500, SDS-PAGE [17]; <3,11> 2 * 58000, SDS-PAGE [18,20]; <9> 2 * 38000, SDS-PAGE [26]; <2> 2 * 35300 [30]) [9, 12, 13, 17, 18, 20, 26, 29-31, 35-37, 42, 49, 51]
Additional information <16> (<16> intersubunit organisation [35]) [35]

Posttranslational modification

Additional information <3> (<3> no glycosyl groups on the protein [37]) [37]

5 Isolation/Preparation/Mutation/Application

Source/tissue

HeLa cell <3> (<3> high and low affinity form with respect to heparin [36, 51]; <3> heparin affinity depends on the selenium content [51]) [20, 36, 51]
adrenal gland <1> (<1> cortex [49]) [49]
callus <6> [12]
keratinocyte <3, 8> [5, 14]
leaf <5, 13> [9, 34]
liver <1, 11, 21> (<21> isoform TR1, normal and tumor [42]) [18, 27, 33, 42]
lung <3> (<3> adenocarcinoma cell line NCI-H441 [37]; <3> adenocarcinoma cells, high and low affinity form with respect to heparin [36]) [36, 37]
melanoma cell <3> (<3> metastatic melanotic [3,5,6,8]; <3> amelanotic [3]) [3, 5, 6, 8]
pancreatic islet <12> [19]
placenta <3> [31, 41, 44]
prostate gland <11> (<11> isoform TR1 [42]) [42]
seed <13> (<13> aleurone layer, scutellum, starchy endosperm [34]) [34]
seedling <13> (<13> root and shoot [34]) [34]
thymus <1> [1, 21, 41]

Localization

chloroplast <5> [9]
cytoplasmic membrane <3> (<3> associated [8]) [8, 14]

cytosol <1, 3, 11, 12, 15, 17, 23> (<3,17> isoform TrxR1 [46,51]; <11,21,23> isoform TR1 [42]) [29, 42, 46, 49, 51]
extracellular <3, 8> [14]
mitochondrion <1, 3, 11, 17, 22, 23> (<3> isoform TrxR2 [51]; <17> isoform TrxR3 and TrxR1 [46]; <23> isoform TR2 [42]; <11,20,22> isoform TR3 [42]) [42, 46, 49, 51]

Purification
<1> (mitochondrial isoform [49]) [21, 33, 41, 49]
<2> (wild-type from plasmid [50]; wild-type and mutant enzymes [43]; wild-type from overexpressing cell line, mutants [2,7]; recombinant enzyme-thioredoxin subunit complex C135-C32 and mutant S135-S32 [32]) [2, 7, 23, 28, 30, 43, 50]
<4> [4]
<5> [9]
<6> (partial [12]) [12]
<7> [13]
<9> [16, 26]
<10> [17]
<13> (recombinant, wild-type and mutant C53S [34]) [34]
<14> [24]
<15> (recombinant from Escherichia coli [29]) [29]
<16> (recombinant His-tagged wild-type and mutant enzymes from Escherichia coli [35]; recombinant wild-type and mutants from Escherichia coli [40]) [35, 40]
<17> (recombinant wild-type cytosolic and mitochondrial isoforms and mutants from Escherichia coli [46]) [46]
<18> (recombinant wild-type fusion protein and truncated mutant [50]) [50]
<11, 20> [18, 27, 42]
<12, 21, 22> [42]
<3, 19> (heparin affinity depends on the selenium content, binding can be induced by reduction of the enzyme with NADPH or tris-(2-carboxyethyl)-phosphine [51]; HeLa cells, partially [20]) [6, 8, 14, 20, 31, 37, 41, 42, 51]

Cloning
<1> (amino acid and DNA sequence analysis [49]) [49]
<2> (subunit complex of enzyme via C135 with thioredoxin C32, mutant complex S135-S32, expression from plasmid [32,50]; trxB gene, cloning of active site mutants, expression from plasmids [2]; overexpression of wild-type and mutant enzymes from plasmids [7,10,47]; trxB gene, DNA sequence analysis [10]) [2, 7, 10, 32, 47, 50]
<11> (overexpression of antisense mutant in Escherichia coli, expression of wild-type in COS-7 cells, amino acid sequence analysis [39]) [39]
<13> (expression of wild-type as His-tagged protein and mutant C53S in Escherichia coli, DNA sequence analysis [34]) [34]
<15> (strong overexpression in Escherichia coli, DNA sequence analysis [29]) [29]

<16> (expression of wild-type and mutants in Escherichia coli [40]; expression of His-tagged or Strep-tagged wild-type and mutants in Escherichia coli, DNA sequence analysis [35]) [35, 40]
<17> (expression of cytosolic and mitochondrial isoform as His-tagged enzymes in Escherichia coli [46]) [46]
<18> (expression of wild-type fusion protein and truncated mutant [50]) [50]
<12, 21, 22> (nucleotide sequence analysis and comparison [42]) [42]

Engineering

C135S <2> (<2> fluorescence spectroscopic investigation of the interaction with the flavin group [43]; <2> exchange of 1 cysteine in the active center disulfide, 1 cysteine at position 138 is remaining [43]) [43]
C135S/C32S <2> (<2> via the active disulfide centers a subunit complex of tightly bound enzyme, C135 and C138, and thioredoxin, C32 and C35, is formed, exchange of one cysteine for one serine in each protein by site-directed mutagenesis, conformation analysis [32]) [32]
C135S/C35S <2> (<2> fluorescence spectroscopic investigation of the interaction with the flavin group [43]) [43]
C136S <2> (<2> site-directed mutagenesis, reduced activity [7]) [7]
C138S <2> (<2> fluorescence spectroscopic investigation of the interaction with the flavin group [43]; <2> exchange of 1 cysteine in the active center disulfide, 1 cysteine at position 135 is remaining, very low activity [43]) [43, 45]
C138S/C35S <2> (<2> fluorescence spectroscopic investigation of the interaction with the flavin group [43]) [43]
C139S <2> (<2> site-directed mutagenesis, reduced activity [7]) [7]
C32S <2> (<2> fluorescence spectroscopic investigation of the interaction with the flavin group [43]; <2> exchange of 1 cysteine in the active center disulfide, 1 cysteine at position 35 is remaining, low activity [43]) [43]
C35S <2> (<2> fluorescence spectroscopic investigation of the interaction with the flavin group [43]; <2> exchange of 1 cysteine in the active center disulfide, 1 cysteine at position 32 is remaining [43]) [43]
C535S <2, 16> (<16> construction of a homodimer and heterodimer, the latter containing 1 mutant and 1 wild-type subunit, activity is reduced by 56 and 92%, respectively [35]; <2> site-directed mutagenesis, changed conformation [32]) [32, 35]
C535S/C88A <16> (<16> double mutant, construction of a homodimer and a heterodimer, the latter containing 1 mutant and 1 wild-type subunit, activity is reduced by 89 and 95%, respectively [35]) [35]
C53S <13> (<13> site-directed mutagenesis, no activity since the redox cycle system is abolished [34]) [34]
C73S <3> (<3> recombinant, His-tagged [44]) [44]
C88A <16> (<16> site-directed mutagenesis, no activity [40]; <16> construction of a homodimer and a heterodimer, the latter containing 1 mutant and 1 wild-type subunit, activity is reduced by 4 and 90% [35]) [35, 40]
C88S <16> (<16> site-directed mutagenesis, no activity [40]) [40]
C93A <16> (<16> site-directed mutagenesis, no activity [40]) [40]

H509A <16> (<16> site-directed mutagenesis, reduced activity [40]) [40]
H509Q <16> (<16> site-directed mutagenesis, reduced activity [40]) [40]
SeC498C <11> (<11> antisense technique, exchange in the catalytic active selenosulfide at the C-terminus, resulting in higher pH-optimum, 100fold lower turnover number, 10fold lower K_m-value, no activity with H_2O_2 [39]) [39]
SeC498G <11> (<11> antisense technique, reduced activity [39]) [39]
SeC498S <11> (<11> antisense technique, reduced activity [39]) [39]
Additional information <2, 11, 17, 18> (<18> thioredoxin is cut off the native fusion protein thioredoxin-thioredoxin reductase resulting in enhanced activity [50]; <2> trxB gene is combined to chimeric enzyme NT-TrR by exchanging the N-terminus of Escherichia coli with the N-terminus of Salmonella typhimurium AhpF gene protein, which encodes a protein with about 35% homology in the N-terminal region, 2 other mutants are constructed in the same way but with double mutation C129S/C132S and C342S/C345S, the first in the Escherichia coli part and the latter in the Salmonella part of the chimeric protein, activity corressponding to organism wild-type giving the N-terminal part, except for C342S/C345S chimeric mutant who has no activity [47]; <17> His-tagged transcript specific mutants, varied N-terminus, 1 null-mutant [46]; <11> truncated enzyme mutant without catalytic active C-terminus [39]; <2> transposon mutagenesis [10]) [10, 39, 46, 47, 50]

Application

medicine <3, 16> (<3> target for immunostimulating effect of 1-chloro-2,4-dinitrobenzene as agent in AIDS treatment [38]; <16> target for a chemotherapeutic intervention of malaria [35]) [35, 38]

6 Stability

Temperature stability

60 <3, 15> (<3> 1 h, no loss of activity [31]; <15> 5 min, active [29]) [29, 31]
65 <2, 3> (<3> above, irreversible loss of activity, no displacement of tightly bound FAD [31]; <2> 7 min, active [28]) [28, 31]
70 <13> (<13> 10 min, complete loss of activity [22]) [22]

Oxidation stability

<3>, the reduced enzyme is very labile to O_2, decrease of selenium content and activity after exposure to air, protection by NADPH, $NADP^+$ and tris-(2-carboxyethyl)phosphine [51]

General stability information

<1>, 4% activity remaining without EDTA [49]
<3>, tightly bound FAD is not replaced during heat treatment or guanidinium hydrochloride [31]

Storage stability

<11>, -20°C, several months [27]

References

[1] Lundström, J.; Holmgren, A.: Protein disulfide-isomerase is a substrate for thioredoxin reductase and has thioredoxin-like activity. J. Biol. Chem., **265**, 9114-9120 (1990)
[2] Gleason, F.K.; Lim, C.J.; Gerami-Nejad, M.; Fuchs, J. A.: Characterization of Escherichia coli thioredoxins with altered active site residues. Biochemistry, **29**, 3701-3709 (1990)
[3] Schallreuter, K.U.; Wood, J.M.: Calcium regulates thioredoxin reductase in human metastatic melanoma. Biochim. Biophys. Acta, **997**, 242-247 (1989)
[4] McFarlan, S.C.; Hogenkamp, H.P.C.; Eccleston, E.D.; Howard, J.B.; Fuchs, J.A.: Purification, characterization and revised amino acid sequence of a second thioredoxin from Corynebacterium nephridii [published erratum appears in Eur J Biochem 1989 Apr 1;180(3):603]. Eur. J. Biochem., **179**, 389-398 (1989)
[5] Schallreuter, K.U.; Pittelkow, M.R.; Wood, J.M.: EF-Hands calcium binding regulates the thioredoxin reductase/thioredoxin electron transfer in human keratinocytes. Biochem. Biophys. Res. Commun., **162**, 1311-1316 (1989)
[6] Schallreuter, K.U.; Wood, J.M.: The stereospecific suicide inhibition of human melanoma thioredoxin reductase by 13-cis-retinoic acid. Biochem. Biophys. Res. Commun., **160**, 573-579 (1989)
[7] Prongay, A.J.; Engelke, D.R.; Williams, C.H.: Characterization of two active site mutations of thioredoxin reductase from Escherichia coli [published erratum appears in J Biol Chem 1989 Jul 15;264(20):12113]. J. Biol. Chem., **264**, 2656-2664 (1989)
[8] Schallreuter, K.U.; Wood, J.M.: The activity and purification of membrane-associated thioredoxin reductase from human metastatic melanotic melanoma. Biochim. Biophys. Acta, **967**, 103-109 (1988)
[9] Florencio, F.J.; Yee, B.C.; Johnson, T.C.; Buchanan, B.: An NADP/thioredoxin system in leaves: purification and characterization of NADP-thioredoxin reductase and thioredoxin h from spinach. Arch. Biochem. Biophys., **266**, 496-507 (1988)
[10] Russel, M.; Model, P.: Direct cloning of the trxB gene that encodes thioredoxin reductase. J. Bacteriol., **163**, 238-242 (1985)
[11] Williams, C.H.: Flavin-containing dehydrogenases. The Enzymes, 3rd Ed. (Boyer, P.D., ed.), **13**, 89-173 (1976)
[12] Johnson, T.C.; Cao, R.Q.; Kung, J.E.; Buchanan, B.B.: Thioredoxin and NADP-thioredoxin reductase from cultured carrot cells. Planta, **171**, 321-331 (1987)
[13] Clement-Metral, J.D.; Höög, J.O.; Holmgren, A.: Characterization of the thioredoxin system in the facultative phototroph Rhodobacter sphaeroides Y [published erratum appears in Eur J Biochem 1987 May 4;164(3):713]. Eur. J. Biochem., **161**, 119-126 (1986)
[14] Schallreuter, K.U.; Wood, J.M.: The role of thioredoxin reductase in the reduction of free radicals at the surface of the epidermis. Biochem. Biophys. Res. Commun., **136**, 630-637 (1986)

[15] O'Donnell, M.E.; Williams, C.H.: Reaction of both active site thiols of reduced thioredoxin reductase with N-ethylmaleimide. Biochemistry, **24**, 7617-7621 (1985)
[16] Gibson, R.M.; Large, P.J.: The involvement of thioredoxin and thioredoxin reductase in the dimethyl sulphoxide reductase system of Saccharomyces cerevisiae. FEMS Microbiol. Lett., **26**, 89-94 (1985)
[17] Johnson, T.C.; Crawford, N.A.; Buchanan, B.B.: Thioredoxin system of the photosynthetic anaerobe Chromatium vinosum. J. Bacteriol., **158**, 1061-1069 (1984)
[18] Luthman, M.; Holmgren, A.: Rat liver thioredoxin and thioredoxin reductase: purification and characterization. Biochemistry, **21**, 6628-6633 (1982)
[19] Grankvist, K.; Holmgren, A.; Luthman, M.; Täljedal, I.B.: Thioredoxin and thioredoxin reductase in pancreatic islets may participate in diabetogenic free-radical production. Biochem. Biophys. Res. Commun., **107**, 1412-1418 (1982)
[20] Lik-Shing Tsang, M.; Weatherbee, J.A.: Thioredoxin, glutaredoxin, and thioredoxin reductase from cultured HeLa cells. Proc. Natl. Acad. Sci. USA, **78**, 7478-7482 (1981)
[21] Holmgren, A.; Lyckeborg, C.: Enzymatic reduction of alloxan by thioredoxin and NADPH-thioredoxin reductase. Proc. Natl. Acad. Sci. USA, **77**, 5149-5152 (1980)
[22] Suske, G.; Wagner, W.; Follmann, H.: NADPH-dependent thioredoxin reductase and a new thioredoxin from wheat. Z. Naturforsch. C, **34**, 214-221 (1979)
[23] Pigiet, V.P.; Conley, R.R.: Purification of thioredoxin, thioredoxin reductase, and glutathione reductase by affinity chromatography. J. Biol. Chem., **252**, 6367-6372 (1977)
[24] Munavalli, S.; Parker, D.V.; Hamilton, F.D.: Identification of NADPH-thioredoxin reductase system in Euglena gracillis. Proc. Natl. Acad. Sci. USA, **72**, 4233-4237 (1975)
[25] Berglund, O.; Holmgren, A.: Thioredoxin reductase-mediated hydrogen transfer from Escherichia coli thioredoxin-$(SH)_2$ to phage T4 thioredoxin-S2. J. Biol. Chem., **250**, 2778-2782 (1975)
[26] Speranza, M.L.; Ronchi, S.; Minchiotti, L.: Purification and characterization of yeast thioredoxin reductase. Biochim. Biophys. Acta, **327**, 274-281 (1973)
[27] Larsson, A.: Thioredoxin reductase from rat liver. Eur. J. Biochem., **35**, 346-349 (1973)
[28] Moore, E.C.; Reichard, P.; Thelander, L.: Enzymatic synthesis of desoxyribonucleotides. V. Purification and properties of thioredoxin reductase from Escherichia coli B. J. Biol. Chem., **239**, 3445-3452 (1964)
[29] Jacquot, J.P.; Rivera-Madrid, R.; Marinho, P.; Kollarova, M.; Le Marechal, P.; Miginiac-Maslow, M.; Meyer, Y.: Arabidopsis thaliana NADPH thioredoxin reductase. cDNA characterization and expression of the recombinant protein in Escherichia coli. J. Mol. Biol., **235**, 1357-1363 (1994)
[30] Lennon, B.W.; Williams, C.H., Jr.: Enzyme-monitored turnover of Escherichia coli thioredoxin reductase: insights for catalysis. Biochemistry, **35**, 4704-4712 (1996)

[31] Oblong, J.E.; Gasdaska, P.Y.; Sherrill, K.; Powis, G.: Purification of human thioredoxin reductase: properties and characterization by absorption and circular dichroism spectroscopy. Biochemistry, **32**, 7271-7277 (1993)
[32] Wang, P.F.; Veine, D.M.; Ahn, S.H.; Williams, C.H., Jr.: A stable mixed disulfide between thioredoxin reductase and its substrate, thioredoxin: preparation and characterization. Biochemistry, **35**, 4812-4819 (1996)
[33] Martinez-Galisteo, E.; Padilla, C.A.; Holmgren, A.; Barcena, J.A.: Characterization of mammalian thioredoxin reductase, thioredoxin and glutaredoxin by immunochemical methods. Comp. Biochem. Physiol. B, **111**, 17-25 (1995)
[34] Serrato, A.J.; Perez-Ruiz, J.M.; Cejudo, F.J.: Cloning of thioredoxin h reductase and characterization of the thioredoxin reductase-thioredoxin h system from wheat. Biochem. J., **367**, 491-497 (2002)
[35] Krnajski, Z.; Gilberger, T.W.; Walter, R.D.; Muller, S.: Intersubunit interactions in Plasmodium falciparum thioredoxin reductase. J. Biol. Chem., **275**, 40874-40878 (2000)
[36] Liu, S.Y.; Stadtman, T.: Heparin-binding properties of selenium-containing thioredoxin reductase from HeLa cells and human lung adenocarcinoma cells. Proc. Natl. Acad. Sci. USA, **94**, 6138-6141 (1997)
[37] Tamura, T.; Stadtman, T.C.: A new selenoprotein from human lung adenocarcinoma cells: purification, properties, and thioredoxin reductase activity. Proc. Natl. Acad. Sci. USA, **93**, 1006-1011 (1996)
[38] Arner, E.S.J.; Bjoernstedt, M.; Holmgren, A.: 1-Chloro-2,4-dinitrobenzene is an irreversible inhibitor of human thioredoxin reductase. Loss of thioredoxin disulfide reductase activity is accompanied by a large increase in NADPH oxidase activity. J. Biol. Chem., **270**, 3479-3482 (1995)
[39] Zhong, L.; Holmgren, A.: Essential role of selenium in the catalytic activities of mammalian thioredoxin reductase revealed by characterization of recombinant enzymes with selenocysteine mutations. J. Biol. Chem., **275**, 18121-18128 (2000)
[40] Gilberger, T.W.; Walter, R.D.; Muller, S.: Identification and characterization of the functional amino acids at the active site of the large thioredoxin reductase from Plasmodium falciparum. J. Biol. Chem., **272**, 29584-29589 (1997)
[41] Nordberg, J.; Zhong, L.; Holmgren, A.; Arner, E.S.J.: Mammalian thioredoxin reductase is irreversibly inhibited by dinitrohalobenzenes by alkylation of both the redox active selenocysteine and its neighboring cysteine residue. J. Biol. Chem., **273**, 10835-10842 (1998)
[42] Sun, Q.A.; Zappacosta, F.; Factor, V.M.; Wirth, P.J.; Hatfield, D.L.; Gladyshev, V.N.: Heterogeneity within animal thioredoxin reductases. Evidence for alternative first exon splicing. J. Biol. Chem., **276**, 3106-3114 (2001)
[43] Veine, D.M.; Mulrooney, S.B.; Wang, P.F.; Williams, C.H., Jr.: Formation and properties of mixed disulfides between thioredoxin reductase from Escherichia coli and thioredoxin: evidence that cysteine-138 functions to initiate dithiol-disulfide interchange and to accept the reducing equivalent from reduced flavin. Protein Sci., 7, 1441-1450 (1998)

[44] Gromer, S.; Gross, J.H.: Methylseleninate is a substrate rather than an inhibitor of mammalian thioredoxin reductase: implications for the antitumor effects of selenium. J. Biol. Chem., **277**, 9701-9706 (2002)
[45] Van den Berg, P.A.W.; Mulrooney, S.B.; Gobets, B.; van Stokkum, I.H.M.; van Hoek, A.; Williams, C.H., Jr.; Visser, A.J.W.G.: Exploring the conformational equilibrium of E. coli thioredoxin reductase: characterization of two catalytically important states by ultrafast flavin fluorescence spectroscopy. Protein Sci., **10**, 2037-2049 (2001)
[46] Missirlis, F.; Ulschmid, J.K.; Hirosawa-Takamori, M.; Gronke, S.; Schafer, U.; Becker, K.; Phillips, J.P.; Jackle, H.: Mitochondrial and cytoplasmic thioredoxin reductase variants encoded by a single Drosophila gene are both essential for viability. J. Biol. Chem., **277**, 11521-11526 (2002)
[47] Reynolds, C.M.; Poole, L.B.: Attachment of the N-terminal domain of Salmonella typhimurium AhpF to Escherichia coli thioredoxin reductase confers AhpC reductase activity but does not affect thioredoxin reductase activity. Biochemistry, **39**, 8859-8869 (2000)
[48] Lundstroem-Ljung, J.; Birnbach, U.; Rupp, K.; Soeling, H.D.; Holmgren, A.: Two resident ER-proteins, CaBP1 and CaBP2, with thioredoxin domains, are substrates for thioredoxin reductase: comparison with protein disulfide isomerase. FEBS Lett., **357**, 305-308 (1995)
[49] Watabe, S.; Makino, Y.; Ogawa, K.; Hiroi, T.; Yamamoto, Y.; Takahashi, S.Y.: Mitochondrial thioredoxin reductase in bovine adrenal cortex its purification, properties, nucleotide/amino acid sequences, and identification of selenocysteine. Eur. J. Biochem., **264**, 74-84 (1999)
[50] Wang, P.F.; Marcinkeviciene, J.; Williams, C.H., Jr.; Blanchard, J.S.: Thioredoxin reductase-thioredoxin fusion enzyme from Mycobacterium leprae: comparison with the separately expressed thioredoxin reductase. Biochemistry, **37**, 16378-16389 (1998)
[51] Gorlatov, S.N.; Stadtman, T.C.: Human selenium-dependent thioredoxin reductase from HeLa cells: properties of forms with differing heparin affinities. Arch. Biochem. Biophys., **369**, 133-142 (1999)

CoA-Glutathione reductase 1.8.1.10

1 Nomenclature

EC number
1.8.1.10

Systematic name
glutathione:$NADP^+$ oxidoreductase (CoA-acylating)

Recommended name
CoA-glutathione reductase

Synonyms
CoA disulfide reductase <4> [7]
CoADR <4> [7]
CoASSG reductase
EC 1.6.4.6 (formerly)
NADPH-dependent coenzyme A-SS-glutathione reductase
$NADPH_2$:CoA-glutathione oxidoreductase
coenzyme A disulfide-glutathione reductase
coenzyme A glutathione disulfide reductase
reductase, coenzyme A-glutathione disulfide
Additional information <4> (<4> subfamily of pyridine nucleotide-disulfide oxidoreductase [7]) [7]

CAS registry number
37256-33-0

2 Source Organism

<1> *Rattus norvegicus* (male Sprague-Dawley rats [3]; male albino, normal and alloxan diabetic rats, no change of enzyme activity and CoASSG content in diabetic rats [5]) [1, 3-5]
<2> *Escherichia coli* [2]
<3> *Saccharomyces cerevisiae* (strain ATCC 1946 [6]) [6]
<4> *Staphylococcus aureus* (strain 8325-4 [7]) [7]

3 Reaction and Specificity

Catalyzed reaction
CoA + glutathione + $NADP^+$ = CoA-glutathione + NADPH + H^+

Reaction type

coenzyme A transfer
oxidation
redox reaction
reduction

Natural substrates and products

S CoA disulfide + NADPH <4> (<4> central role in thiol metabolism [7]) (Reversibility: ? <4> [7]) [7]
P CoA + $NADP^+$ <4> [7]
S CoA-glutathione + NADPH <1, 2> (<2> metabolism of CoA-glutathione disulfide, role may be to replenish the CoA pool from CoAS-SG formed either chemically or enzymatically during cell growth, accumulation of CoAS-SG in vivo in spite of high enzyme levels indicates involvement of a control mechanism [2]; <1> may be important in metabolism of CoASSG [5]) (Reversibility: ? <1, 2> [2, 3, 5]) [2, 3, 5]
P glutathione + CoA + $NADP^+$ <1> [3, 5]

Substrates and products

S CoA disulfide + NADPH <4> (<4> substrate CoASSCoA, enzyme forms during catalysis a stable mixed disulfide intermediate with CoA [7]) (Reversibility: ? <4> [7]) [7]
P CoA + $NADP^+$ <4> [7]
S CoA-glutathione + NADPH <1-3> (<2> substrate coenzyme A-glutathione mixed disulfide, CoAS-SG, is capable of binding one molecule of Fe^{3+} [2]; <1> high substrate specificity [4]) (Reversibility: ir <1-3> [2, 3, 6]; ? <1> [1, 4, 5]) [1-6]
P glutathione + CoA + $NADP^+$ <1-3> [1-6]
S Additional information <1, 3> (<1> no reduction of disulfides such as cystine, cystamine, panthetine, and insulin [4]; <3> no reduction of CoASSCys, GSSCys or CysSSCys [6]) [4, 6]
P ?

Inhibitors

phosphate <3> (<3> 25 mM, partially inhibits [6]) [6]

Cofactors/prosthetic groups

FAD <4> (<4> flavoprotein, gene with two FAD-binding regions [7]) [7]
NADPH <1-4> (<1-4> NADPH-dependent [2-7]; <4> gene with one NADPH-binding region [7]) [1-7]

Turnover number (min^{-1})

66000 <4> (CoA disulfide) [7]

Specific activity (U/mg)

0.0007 <1> (<1> alloxan diabetic rats, liver extract [5]) [5]
0.00085 <1> (<1> normal rats, liver extract [5]) [5]
Additional information <1-4> (<2> specific activity of enzyme in crude extract increases during growth from early log phase into stationary phase and during a shift from aerobic to anaerobic growth [2]) [2, 4, 6, 7]

K_m-Value (mM)

0.0019 <4> (NADPH, <4> recombinant enzyme expressed in E. coli [7]) [7]
0.014 <4> (CoA disulfide, <4> recombinant enzyme expressed in E. coli [7]) [7]
0.2 <3> (CoA-glutathione, <3> at pH 5.5 and a fixed value of 110 nmol NADPH [6]) [6]
0.23 <1> (CoA-glutathione, <1> at pH 5.75 and a fixed value of 110 nmol NADPH [4]) [4]

pH-Optimum

5.5 <3> [6]
5.8 <1> [4]

Temperature optimum (°C)

24 <3> (<3> assay at [6]) [6]
25 <1> (<1> assay at [5]) [5]

4 Enzyme Structure

Molecular weight

42500 <1> (<1> gel filtration, one protein with CoASSG reductase and GSSG reductase activities, EC 1.6.4.2 [4]) [4]
90000 <4> (<4> native molecular weight [7]) [7]
108000 <3> (<3> sucrose gradient ultracentrifugation [6]) [6]

Subunits

dimer <4> (<4> 2 * 49200, calculated from the nucleotide sequence [7]) [7]

5 Isolation/Preparation/Mutation/Application

Source/tissue

adipose tissue <1> [1]
brain <1> [1]
heart <1> [1]
intestine <1> [1]
kidney <1> [1]
liver <1> [1, 3-5]
lung <1> [1]
muscle <1> [1]
spleen <1> [1]
testis <1> [1]

Localization

cytosol <1> [3]

Purification

<1> (not separated from GSSG reductase activity, EC 1.6.4.2 [3,4]; 289fold purification [4]) [3, 4]
<3> (143fold partial purification, separated from GSSG reductase activity, EC 1.6.4.2 [6]) [6]
<4> (purification of recombinant CoADR, expressed in Escherichia coli [7]) [7]

Cloning

<4> (cdr gene encoding coenzyme A glutathione disulfide reductase is cloned, sequenced and overexpressed in Escherichia coli BL21 using the pET expression system, gene encodes a 438-amino acid polypeptide [7]) [7]

Application

medicine <4> (<4> potential point of intervention in the treatment of staphylococcal infections [7]) [7]
pharmacology <4> (<4> anti-staphylococcal agent, possible target for the design of selective inhibitors that would interrupt the thiol metabolism of the human pathogen Staphylococcus aureus [7]) [7]

6 Stability

Storage stability

<1>, 4°C, dialyzed extract, 16 hours, stable [5]

References

[1] Acuna, R.; Vargas, E.; Ondarza, R.N.: Distribution of GSSG- and CoASSG-reductase in rat tissues. Temas Bioquim. Actual. (Pina, E., Pena, A., Chagoya de Sanchez, V., eds.), **93**, 355-359 (1978)
[2] Loewen, P.C.: Identification of a coenzyme A–glutathione disulfide (DSI), a modified coenzyme A disulfide (DSII), and a NADPH-dependent coenzyme A–glutathione disulfide reductase in E. coli. Can. J. Biochem., **55**, 1019-1027 (1977)
[3] Eriksson, S.; Guthenberg, C.; Mannervik, B.: The nature of the enzymatic reduction of the mixed disulfide of coenzyme A and glutathione. FEBS Lett., **39**, 296-300 (1974)
[4] Ondarza, R.N.; Escamilla, E.; Gutierrez, J.; de la Chica, G.: CoAS-Sglutathione and GSSG reductases from rat liver. Two disulfide oxidoreductase activities in one protein entity. Biochim. Biophys. Acta, **341**, 162-171 (1974)
[5] Dyar, R.E.; Wilken, D.R.: Rat liver levels of coenzyme A-glutathione and of enzymes in its metabolism. Arch. Biochem. Biophys., **153**, 619-626 (1972)
[6] Ondarza, R.N.; Abney, R.; Lopez-Colome, A.M.: Characterization of a NADPH-dependent coenzyme A-SS-glutathione reductase from yeast. Biochim. Biophys. Acta, **191**, 239-248 (1969)
[7] delCardayre, S.B.; Davies, J.E.: Staphylococcus aureus coenzyme A disulfide reductase, a new subfamily of pyridine nucleotide-disulfide oxidoreductase. J. Biol. Chem., **273**, 5752-5757 (1998)

Asparagusate reductase 1.8.1.11

1 Nomenclature

EC number
1.8.1.11

Systematic name
3-mercapto-2-mercaptomethylpropanoate:NAD^+ oxidoreductase

Recommended name
asparagusate reductase

Synonyms
EC 1.6.4.7 (formerly)
$NADH_2$:asparagusate oxidoreductase
asparagusate dehydrogenase
asparagusate reductase (NADH)
asparagusate reductase ($NADH_2$)
asparagusic dehydrogenase

CAS registry number
56126-52-4

2 Source Organism

<1> *Asparagus officinalis* (2 asparagusate dehydrogenases with lipoyl dehydrogenase activity [1-4]; etiolated or green, activity levels are higher in etiolated than in green shoots [2] [2]; etiolated [4]) [1-4]

3 Reaction and Specificity

Catalyzed reaction
3-mercapto-2-mercaptomethylpropanoate + NAD^+ = asparagusate + NADH + H^+ (also acts on lipoate; <1> exhibits lipoyl dehydrogenase activity [1-3]; <1> exhibits diaphorase activity [2,3])

Reaction type
oxidation
redox reaction
reduction

Natural substrates and products

S asparagusate + NADH <1> (<1> essential component of pyruvate dehydrogenase complex [1,2]; <1> participates in enzymatic pyruvate and α-ketoglutarate dehydrogenation [4]) (Reversibility: r <1> [1-4]) [1-4]
P dihydroasparagusate + NAD^+ <1> [1-4]

Substrates and products

S $Fe(CN)_6^{3-}$ + NADH <1> (Reversibility: ? <1> [1-3]) [1-3]
P $Fe(CN)_6^{2-}$ + NAD^+ <1> [1-3]
S asparagusic acid + NADH <1> (<1> reverse reaction very slow [2,3]) (Reversibility: r <1> [1-4]) [1-4]
P dihydroasparagusate + NAD^+ <1> [1-4]
S lipoic acid + NADH <1> (<1> reverse reaction very slow [2,3]) (Reversibility: r <1> [1-4]) [1-4]
P dihydrolipoic acid + NAD^+ <1> [1-4]

Inhibitors

FAD <1> (<1> strongly inhibits the two reductase activities [3]) [3]
Hg^{2+} <1> (<1> inhibitory [2,3]) [2, 3]
N-ethylmaleimide <1> [2, 3]
NAD^+ <1> (<1> inactivator during asparagusic acid reduction [4]) [4]
NADH <1> (<1> above 0.1 mM [1,2]; <1> high concentrations, increase of inhibitory effect by addition of arsenite [3]) [1-4]
arsenite <1> (<1> low inhibition [2,3]) [2, 3]
asparagusic acid <1> (<1> inhibits lipoyl dehydrogenase activity [2-4]) [2-4]
cetyltrimethyl ammonium bromide <1> (<1> affects lipoyl dehydrogenase activity [3]) [3]
iodoacetamide <1> (<1> low inhibition [2,3]) [2, 3]
p-chloromercuribenzoic acid <1> [2, 3]
Additional information <1> (<1> inhibitors indicate an involvement of protein disulfide linkage or thiol group in the catalytic site [2,3]) [2, 3]

Cofactors/prosthetic groups

FAD <1> (<1> flavoprotein, approximately 1 mol of FAD per mol of protein [1-3]) [1-3]
NAD^+ <1> (<1> up to 0.2 mM, stimulates the reduction activity if NADH concentration is 0.2 mM [2,3]) [2, 3]
NADH <1> (<1> 0.1 mM optimal concentration for maximal activity [1-3]; <1> no replacement of NADH by NADPH [2,3]) [1-3]

Activating compounds

NAD^+ <1> (<1> reduction of lipoic acid is activated [3,4]) [3, 4]
Tween 80 <1> (<1> drastic activation of lipoyl dehydrogenase activity, no activation of asparagusate dehydrogenase activity, enzyme II [3]) [3]
lecithin <1> (<1> drastic activation of lipoyl dehydrogenase activity, no activation of asparagusate dehydrogenase activity, enzyme II [3]) [3]
sodium dodecylsulfate <1> (<1> treatment of mitochondria with at 0.5% for 10-20 min, maximal solubilization of enzyme [4]) [4]

Specific activity (U/mg)

0.064 <1> (<1> enzyme/fraction I towards asparagusic acid [4]) [4]
0.08323 <1> (<1> enzyme/fraction II towards asparagusic acid [4]) [4]
0.1631 <1> (<1> enzyme/fraction I towards lipoic acid [4]) [4]
0.213 <1> (<1> enzyme/fraction II towards lipoic acid [4]) [4]
0.215 <1> (<1> enzyme/fraction II towards asparagusic acid [1-3]) [1-3]
0.425 <1> (<1> enzyme/fraction I towards asparagusic acid [1-3]) [1-3]
0.5464 <1> (<1> enzyme/fraction II towards lipoic acid [2,3]) [2, 3]
1.082 <1> (<1> enzyme/fraction I towards lipoic acid [2,3]) [2, 3]

K_m-Value (mM)

0.8 <1> ($K_3Fe(CN)_6$, <1> enzyme II [1-3]) [1-3]
0.9 <1> ($K_3Fe(CN)_6$, <1> enzyme I [1-3]) [1-3]
3 <1> (lipoic acid) [1-3]
3-3.3 <1> (lipoate, <1> enzyme I and II [3]) [3]
20 <1> (asparagusic acid) [1-3]
20-21.5 <1> (asparagusate, <1> enzyme I and II [3]) [3]

pH-Optimum

5.25 <1> (<1> for $K_3Fe(CN)_6$ reduction [1-3]) [1-3]
5.9 <1> (<1> asparagusic and lipoic acid reduction [1,3]) [1-3]
5.95 <1> (<1> assay at [4]) [4]

Temperature optimum (°C)

25 <1> (<1> assay at [1-4]) [1-4]
40-45 <1> [2, 3]

4 Enzyme Structure

Molecular weight

110000-111000 <1> (<1> enzyme I and II, sedimentation equilibrium [1,2]; <1> ultracentrifugation [3]) [1-3]
112000 <1> (<1> enzyme I and II, gel filtration [1-3]; <1> calculation of amino acid content [2,3]) [1-3]

5 Isolation/Preparation/Mutation/Application

Source/tissue

shoot <1> [1-4]
shoot apex <1> (<1> around, higher levels than other regions [2]) [2]

Localization

mitochondrion <1> [2-4]
soluble <1> [3]

Purification

<1> (gel filtration, ion-exchange, ultracentrifugation, copurification of type I and II [1,3]; <1> gel filtration, ion-exchange, copurification of enzyme I and II [2]; SDS-extraction, gel filtration, ion-exchange [4]) [1-4]

6 Stability

Temperature stability

50 <1> (<1> 5 min, activity completely retained [1-3]) [1-3]
50 <1> (<1> above, gradual loss of lipoyl dehydrogenase activity, complete loss at 90°C [1-3]) [1-3]
60-70 <1> (<1> 5 min, 1.4-fold activation of asparagusate dehydrogenase activity [1-3]) [1-3]

General stability information

<1>, EDTA ion stabilizes the enzyme [1, 2, 4]
<1>, PO_4^{3-} stabilizes the enzyme [1, 2, 4]

Storage stability

<1>, -80°C, 67 mM sodium phosphate, pH 7.0, no loss of activity for at least 1 month [1, 2, 4]
<1>, 4°C, 67 mM sodium phosphate, pH 7.0, very unstable [1, 2]
<1>, 4°C, 67 mM sodium phosphate, pH 7.0, very unstable, complete loss of activity in two weeks [4]

References

[1] Yanagawa, H.: Asparagusate reductase. Methods Enzymol., **143**, 516-521 (1987)
[2] Yanagawa, H.: Asparagusate dehydrogenase and lipoyl deydrogenase from Asparagus. Methods Enzymol., **62**, 172-181 (1979)
[3] Yanagawa, H.; Egami, F.: Asparagusate dehydrogenases and lipoyl dehydrogenase from Asparagus mitochondria. Physical, chemical, and enzymatic properties. J. Biol. Chem., **251**, 3637-3644 (1976)
[4] Yanagawa, H.; Egami, F.: Asparagusate dehydrogenases and lipoyl dehydrogenase from Asparagus mitochondria. Biochim. Biophys. Acta, **384**, 342-352 (1975)

Trypanothione-disulfide reductase 1.8.1.12

1 Nomenclature

EC number
1.8.1.12

Systematic name
trypanothione:$NADP^+$ oxidoreductase

Recommended name
trypanothione-disulfide reductase

Synonyms
EC 1.6.4.8 (formerly)
N^1,N^8-bis(glutathionyl)spermidine reductase
$NADPH_2$:trypanothione oxidoreductase
TR <1, 2, 5-8> [12-28]
trypanothione disulfide reductase
trypanothione reductase

CAS registry number
102210-35-5

2 Source Organism

<1> *Crithidia fasciculata* (insect-parasitic trypanosomatid [1-4,6,8,10]) [1-4, 6, 8, 10, 17, 19, 23]
<2> *Leishmania donovani* [1, 18, 20, 24, 28]
<3> *Leishmania aethiopica* [1]
<4> *Trypanosoma brucei* [1, 3, 24]
<5> *Trypanosoma cruzi* (Silvio strain [21]) [1, 2, 4-6, 9, 11, 12, 14-16, 18, 21, 24-27, 29, 30]
<6> *Trypanosoma congolense* [1, 2, 4, 7, 12, 13, 23]
<7> *Homo sapiens* [16]
<8> *Euglena gracilis* (Z [22]) [22]
<9> *Leishmania infantum* [24]

3 Reaction and Specificity

Catalyzed reaction

trypanothione + $NADP^+$ = trypanothione disulfide + NADPH + H^+ (trypanothione disulfide is the oxidized form of N^1,N^8-bis(glutathionyl)-spermidine from the insect-parasitic trypanosomatid Crithidia fasciculata. The enzyme from Crithidia fasciculata is a flavoprotein (FAD), whose activity is dependent on a redox-active cystine at the active centre.; <1,5,6> mechanistic scheme for NAD(P)H:disulfide oxidoreductases [4]; <5> mechanistic reaction scheme, structure-function relation [26,27]; <6> hydride ion transfer reaction of the pro-S-hydrogen of NADPH to N^5 of FAD, redox potentials [13]; <5> substrate binding mechanism, complex structure analysis, protease-like catalytic triade and electronic induced fit [14]; <1> neither the α-carboxy nor the α-amino function of the L-γ-glutamyl group is essential for activity, can be replaced by uncharged benzyloxycarbonyl or tert-butyloxycarbonyl groups [23]; <6> substrate binding structure, structural model [23]; <5> molecular modeling of inhibitor binding using X-ray data [24])

Reaction type

oxidation
redox reaction
reduction

Natural substrates and products

S trypanothione disulfide + NADPH <1, 2, 4-6, 9> (<1> protection against oxidative stress [19]; <2> scheme for trypanothione-dependent H_2O_2 metabolism [18]; <2> metabolizes H_2O_2, recombinant and wild-type cells [18]; <6> physiological function is to protect the organism from oxidative stress and to assist in maintaining low intracellular levels of hydrogen peroxide [13]) (Reversibility: ? <1, 2, 4-6, 9> [3, 6, 10, 12, 13, 18, 21, 23, 24]) [3, 6, 10, 12, 13, 18, 19, 23, 24]

P trypanothione + $NADP^+$ <1, 2, 5, 6> [6, 10, 12, 13, 18]

Substrates and products

S 4-aminobutyryl-L-cysteinylglycyl-3-methylaminopropylamide disulfide + NADPH <1> (<1> substrate analogue [23]) (Reversibility: ? <1> [23]) [23]

P ?

S L-β-aspartyl-L-cysteinylglycyl-3-methylaminopropylamide disulfide + NADPH <1> (<1> substrate analogue [23]) (Reversibility: ? <1> [23]) [23]

P ?

S L-cysteinylglycyl-3-methylaminopropylamide disulfide + NADPH <1> (<1> substrate analogue [23]) (Reversibility: ? <1> [23]) [23]

P ?

S L-γ-glutamyl-L-cysteinylglycyl-3-methylaminopropylamide disulfide + NADPH <1> (<1> substrate analogue [23]) (Reversibility: ? <1> [23]) [23]

P ?
S ajoene + NADPH <5> (<5> i.e. (E,Z)-4,5,9-trithiadodeca-1,6,11-triene-9-oxide, substrate and inhibitor [27]) (Reversibility: ? <5> [27]) [27]
P ?
S benzyloxycarbonyl-L-cysteinylglycyl-3-methylaminopropylamide disulfide + NADPH <1> (<1> substrate analogue [23]) (Reversibility: ? <1> [23]) [23]
P ?
S deazatrypanothione + NADPH <6> (<6> substrate analogue [2]) (Reversibility: ? <6> [2]) [2]
P ?
S glutathionylspermidine disulfide + NADPH <1, 2, 5, 8> (Reversibility: r <1, 5> [5, 8, 9]; ? <2, 8> [20, 22, 24]) [5, 8, 9, 20, 22, 24]
P 2-glutathionylspermidine + $NADP^+$ <1, 5, 8> [5, 8, 9, 22]
S tert-butyloxycarbonyl-L-cysteinylglycyl-3-methylaminopropylamide disulfide + NADPH <1> (<1> substrate analogue [23]) (Reversibility: ? <1> [23]) [23]
P ?
S trypanothione disulfide + NADPH <1-9> (<7> recombinant glutathione-trypanothione reductase-like enzyme [16]; <1> highly specific for [10]; <1,2,4> trypanothione is the oxidized form of N^1,N^8-bis-(glutathionyl)-spermidine, the latter form a stable adduct with melarsen oxide [3,20]) (Reversibility: ir <2, 6> [13, 21]; r <1-6> [1-12]; ? <1, 2, 5-9> [14, 15-19, 22-29]) [1-29]
P trypanothione + $NADP^+$ <1-8> [1-29]
S trypanothione disulfide + β-thio-NADPH <6> (<6> wild-type and mutant enzymes [13]) (Reversibility: ? <6> [13]) [13]
P trypanothione + β-thio-$NADP^+$
S trypanothione disulfide + reduced 3-acetylpyridine adenine dinucleotide <6> (<6> wild-type and mutant enzymes [13]) (Reversibility: ? <6> [13]) [13]
P trypanothione + oxidized 3-acetylpyridine adenine dinucleotide
S Additional information <1, 5> (<5> serve as subversive substrates in redox cycling activity: 3,3'-[polyaminobis(carbonylalkyl)]bis(1,4-naphthoquinone) derivatives, inhibitory and substrate activity depending on spacer length [24]; <1> glutaryl-L-cysteinylglycyl-3-methylaminopropylamide disulfide is no substrate [23]; <5> wild-type enzyme shows transhydrogenase activity between NADPH and thio-$NADP^+$ [21]; <5> recombinant mutants C53S, C53A, C58S: no activity with trypanothione disulfide and NADPH, but showing transhydrogenase activity between NADPH and thio-$NADP^+$ [21]; <5> reduces 5,5'-dithiobis(2-nitrobenzamides) [15]; <5> chinifur increases the oxidation of the enzyme 40fold [12]; <1,5> naphthoquinones and nitrofuran derivatives are reduced, inhibition of the enzyme [6]; <5> reduces inhibitory nitrofurans, e.g. chinifur, nifuroxamide, nifuroxime and nifurprazine, and naphthoqinone derivatives [12]; <1> reduces several trypanothione derivatives, e.g. N^1-monoglutathionyl-

spermidine, mixed disulfide of N^1-monoglutathionylspermidine and glutathione [8]) [6, 8, 12, 15, 21, 23, 24]

P ?

Inhibitors

(4'-chloro-2,2':6',2'-terpyridine)platinum(II) amine complex <5> (<5> irreversible, altering specifically the Cys52 residue in the active center, increased oxidase activity [25]) [25]
(4-picoline) (4'-*p*-bromophenyl-2,2':6',2'-terpyridine)platinum(II) complex <5> (<5> irreversible, altering specifically the Cys52 residue in the active center, increased oxidase activity [25]) [25]
(6-benzyloxycarbonylamino-6-[[[methyl-(2-propylamino-ethyl)-carbamoyl]-methyl]-carbamoyl]-1-[[(3-propylamino-propylcarbamoyl)-methyl]-carbamoyl]-hex-3-enyl)-carbamic acid benzyl ester <5> [11]
(6-benzyloxycarbonylamino-6-[[[methyl-(2-propylamino-ethyl)-carbamoyl]-methyl]-carbamoyl]-1-[[(3-propylamino-propylcarbamoyl)-methyl]-carbamoyl]-hexyl)-carbamic acid benzyl ester <5> [11]
1,3-bis(2-chloroethyl)-1-nitrosourea <5> [5]
2,3-bis(3-(2-amidinohydrazono)-butyl)-1,4-naphthoquinone dihydrochloride <5> [5]
2-(5-nitro-2-furanylmethylidene)-N,N'-(1,4-piperazinediylbis(1,3-propanediyl))bishydrazinecarboximidamide tetrahydrobromide <5> [5]
3,3'-[polyaminobis(carbonylalkyl)]bis(1,4-naphthoquinone) derivatives <5> [24]
5,5'-dithiobis(N-[3-(4-methylpiperazin-1-yl)propyl]-2-nitrobenzamide) <5> (<5> recombinant enzyme [15]) [15]
6-chlorotacrine <5> (<5> competitive inhibition [26]) [26]
9-amino-1,2,3,4-tetrahydroacridine <5> (<5> i.e. tacrine, competitive inhibition [26]) [26]
9-aminoacridine derivatives <5> (<5> competitive inhibition [26]) [26]
9-thioacridine derivatives <5> (<5> mixed-type inhibition [26]) [26]
H_2O_2 <2> (<2> wild-type and recombinant enzyme overexpressing cells [18]) [18]
N^1,N^{12}-bis(hihydrocaffeoyl)-spermine <1> (<1> i.e. kukoamine A, mixed inhibitor [19]) [19]
$NADP^+$ <6> [13]
NADPH <1> (<1> NADPH inhibits the enzyme at higher concentration and during longer incubation times, enzyme can be protected by $NADP^+$ [17]) [17]
NaCl <2> (<2> below 5% remaining activity at 1 M [20]) [20]
[6-benzyloxycarbonylamino-1,6-bis-(dimethylcarbamoylmethyl-carbamoyl)-hex-3-enyl]-carbamic acid benzyl ester <5> [11]
[6-benzyloxycarbonylamino-1,6-bis-(dimethylcarbamoylmethyl-carbamoyl)-hexyl]-carbamic acid benzyl ester <5> [11]
ajoene <5> (<5> i.e. (E,Z)-4,5,9-trithiadodeca-1,6,11-triene-9-oxide, natural compound from garlic, Allium sativum, covalent inhibition, but also substrate [27]; <5> time- and temperature-dependent inhibition [27]) [27]

antimony chloride <2> (<2> assay containing 10% vol. methanol [20]) [20]
arsenite <1> (<1> + NADPH, slightly [17]) [17]
bisbenzylisoquinoline alkaloids <5> (<5> e.g. cepharanthine, (-)curine, daphnoline, antioquine, limacine, cycleanine [29]) [29]
chinifur <5, 6> (<5> increases oxidation of the enzyme [12]) [12]
clomipramine <5> (<5> recombinant enzyme [15]) [15, 30]
cymelarsen <1> (<1> + NADPH [17]) [17]
gentian violet <2> (<2> wild-type and recombinant enzyme overexpressing cells [18]) [18]
iodoacetamide <1, 5> (<1,5> inhibition only of the enzyme with reduced cysteine residues in the catalytic center by alkylation [9,10]) [9, 10]
juglone <2, 4, 5> [24]
juglone derivatives <2, 4, 5> [24]
lunarine <5> (<5> specific [14]) [14]
melarsen oxide <1> (<1> inhibition is increased in presence of NADPH, inhibition is abolished by melarsen-trypanothione disulfide adduct [17]) [17]
melarsen-trypanothione disulfide adduct <1, 2, 4> (<1,2,4> i.e. adduct of *p*-(4,6-diamino-S-triazinyl-2-yl)aminophenyl arsenoxide with N^1,N^8-bis-(glutathionyl)spermidine [3,20]; <1> no inhibition [17]) [3, 20]
melarsoprol <1> (<1> + NADPH, slightly [17]) [17]
menadione <5> [12, 24]
menadione derivatives <2, 4, 5, 9> [24]
mepacrine <5> (<5> i.e. quinacrine, competitive inhibition [26]) [26]
nifuraxide <5> (<5> increases oxidation of the enzyme [12]) [12]
nifuroxime <5> (<5> increases oxidation of the enzyme [12]) [12]
nifurprazine <5> (<5> increases oxidation of the enzyme [12]) [12]
nifurtimox <2, 5> (<5> weak inhibitor [4]; <2> wild-type and recombinant enzyme overexpressing cells [18]) [4-6, 12, 18]
nitrofurazone <2> (<2> wild-type and recombinant enzyme overexpressing cells [18]) [18]
oxidized 3-acetylpyridine adenine nucleotide <6> (<6> wild-type and mutant E201D and E201Q [13]) [13]
phenothiazines <5> [30]
phenylarsenoxide <1> (<1> + NADPH [17]) [17]
plumbagin <4, 5> [12, 24]
plumbagin derivatives <2, 4, 5, 9> [24]
potassium antimony tartrate <2> (<2> assay containing 10% vol. methanol [20]) [20]
spermidine derivatives <5> (<5> derivatives of 2-amino-diphenylsulfide + phenothiazine [30]) [30]
spermine derivatives <5> (<5> derivatives of 2-amino-diphenylsulfide + phenothiazine [30]) [30]
thio-$NADP^+$ <6> (<6> wild-type and mutant E201D and E201Q [13]) [13]
triostam <2> (<2> assay containing 10% vol. methanol [20]) [20]
Additional information <5> (<5> antiparasitic drugs designed using three-dimensional structure [11]; <5> increasing ionic strength leads to inhibition, e.g. with NaCl, KCl, $(NH_4)_2SO_4$, HEPES [5]) [5, 11]

Cofactors/prosthetic groups

3-acetylpyridine adenine dinucleotide <6> [13]
FAD <1-9> [1-30]
NADH <5, 6> (<5> low activity with NADH [9]; <6> 7% activity of that with NADPH [2]; <8> not [22]) [2, 9]
$NADP^+$ <1-6> [1, 2]
NADPH <1-9> (<8> strictly dependent on [22]; <5,6> enzyme also utilizes NADH, but less efficiently [2,9]) [1-30]
β-thio-NADPH <5, 6> [13, 21]

Activating compounds

Additional information <1-6> (<2> activity is highly dependent on ionic strength, optimum activity at 40 mM NaCl [20]; <1-6> activity is dependent on a redox-active cysteine at the active centre [1-15,17-28]) [1-15, 17-28]

Turnover number (min^{-1})

11.4-28.2 <5> (thio-NADPH, <5> recombinant mutants C53S, C53A, C58S [21]) [21]
12.6-30.6 <5> (NADPH, <5> recombinant mutants C53S, C53A, C58S [21]) [21]
40.2 <5> ((E,Z)-ajoene) [27]
174 <6> (9,10-phenanthrenequinone) [13]
288 <6> (1,4-benzoquinone) [13]
306 <6> (trypanothione disulfide, <6> mutant E201Q [13]) [13]
318 <6> (trypanothione disulfide, <6> mutant E201D [13]) [13]
379.8 <5> (NADPH, <5> recombinant wild-type [21]) [21]
421.2 <5> (thio-NADPH, <5> recombinant wild-type [21]) [21]
564 <6> (9,10-phenanthrenequinone, <6> mutant E201Q [13]) [13]
816 <6> (1,4-benzoquinone, <6> mutant E201D [13]) [13]
1250 <7> (trypanothione disulfide, <7> recombinant glutathione-trypanothione reductase-like enzyme [16]) [16]
1308 <1> (4-aminobutyryl-L-cysteinylglycyl-3-methylaminopropylamide disulfide) [23]
1470 <6> (1,4-benzoquinone, <6> mutant E201Q [13]) [13]
1740 <6> (9,10-phenanthrenequinone, <6> mutant E201D [13]) [13]
3192 <1> (L-cysteinylglycyl-3-methylaminopropylamide disulfide) [23]
3678 <1> (L-β-aspartyl-L-cysteinylglycyl-3-methylaminopropylamide disulfide) [23]
5000 <6> (deazatrypanothione disulfide) [2]
5124 <5> (glutathionylspermidine disulfide, <5> recombinant wild-type [21]) [21]
6060 <1> (tert-butyloxycarbonyl-L-cysteinylglycyl-3-methylaminopropylamide disulfide) [23]
6486 <5> (trypanothione disulfide, <5> recombinant wild-type [21]) [21]
6540 <1> (benzyloxycarbonyl-L-cysteinylglycyl-3-methylaminopropylamide disulfide) [23]
7320 <6> (trypanothione disulfide) [13]

7500 <5> (5,5'-dithiobis(N-[3-(4-methylpiperazin-1-yl)propyl]-2-nitrobenzamide), <5> recombinant enzyme [15]) [15]
7500 <5> (5,5'-dithiobis(N-[3-(dimethylamino)propyl]-2-nitrobenzamide), <5> recombinant enzyme [15]) [15]
9600 <6> (trypanothione disulfide) [2]
10760 <2> (trypanothione disulfide) [20]
14200 <5> (trypanothione disulfide, <5> dependent on assay conditions [9]) [9]
14400 <5> (trypanothione disulfide, <5> recombinant enzyme [15]) [15]
14500 <8> (glutathionylspermidine disulfide) [22]
17000 <8> (trypanothione disulfide) [22]
31000 <1> (trypanothione disulfide) [10]
49080 <1> (L-γ-glutamyl-L-cysteinylglycyl-3-methylaminopropylamide disulfide) [23]

Specific activity (U/mg)

0.06 <2> [18]
0.067 <8> (<8> crude extract [22]) [22]
0.18 <5> (<5> recombinant mutant C58S: transhydrogenase activity between NADPH and thio-$NADP^+$ [21]) [21]
0.2 <2> (<2> crude extract of wild-type cells [20]) [20]
0.21 <5> (<5> recombinant mutant C53A: transhydrogenase activity between NADPH and thio-$NADP^+$ [21]) [21]
0.22 <5> [18]
0.362 <2> [28]
0.44 <5> (<5> recombinant mutant C53S: transhydrogenase activity between NADPH and thio-$NADP^+$ [21]) [21]
0.55 <2> (<2> crude extract cells overexpressing the recombinant enzyme [18]) [18]
2.1 <2> (<2> crude extract cells overexpressing the recombinant enzyme [20]) [20]
2.15 <2> (<2> recombinant overexpressed in Trypanosoma cruzi cells [18]) [18]
6.49 <5> (<5> recombinant wild-type: transhydrogenase activity between NADPH and thio-$NADP^+$ [21]) [21]
28.9 <6> (<6> purified enzyme [2]) [2]
112.7 <2> (<2> purified enzyme [20]) [20]
137 <5> (<5> recombinant purified enzyme [15]) [15]
143.2 <5> (<5> purified recombinant enzyme [21]) [21]
247 <1> (<1> purified enzyme [17]) [17]
272 <8> (<8> purified enzyme [22]) [22]
284 <1, 5> (<1,5> purified enzyme [9,10]) [9, 10]
Additional information <2> (<2> several replacement clones [28]) [28]

K_m-Value (mM)

0.0012 <6> (NADPH) [13]
0.0015-0.002 <6> (NADPH, <6> mutant E201D and E201Q [13]) [13]
0.0023-0.0066 <6> (β-thio-NADPH, <6> mutant E201D and E201Q [13]) [13]

0.003 <6> (3-acetylpyridine adenine dinucleotide, <6> mutant E201D and E201Q [13]) [13]
0.0039 <6> (3-acetylpyridine adenine dinucleotide) [13]
0.004-0.015 <5> (thio-NADPH, <5> recombinant mutants C53S, C53A, C58S [21]) [21]
0.005-0.007 <1, 5, 6> (NADPH) [2, 5, 9, 10]
0.007-0.014 <5> (NADPH, <5> recombinant mutants C53S, C53A, C58S [21]) [21]
0.009 <2> (NADPH) [20]
0.0113 <6> (β-thio-NADPH) [13]
0.017 <5> (NADPH, <5> recombinant wild-type [21]) [21]
0.018 <5> (trypanothione disulfide) [12, 16]
0.018-0.14 <5> (trypanothione disulfide, <5> dependent on buffer system [5]) [5]
0.022-0.275 <1, 5> (glutathionylspermidine disulfide, <5> dependent on buffer system [5]) [5, 8, 9]
0.024 <5> (trypanothione disulfide, <5> recombinant wild-type [21]) [21]
0.031 <6, 8> (trypanothione disulfide) [2, 22]
0.035 <5> (5,5'-dithiobis(N-[3-(dimethylamino)propyl]-2-nitrobenzamide), <5> recombinant enzyme [15]) [15]
0.036 <2> (trypanothione disulfide) [20]
0.045 <5> (trypanothione disulfide, <5> recombinant enzyme [15]) [2, 9, 15]
0.046 <5> (glutathionylspermidine disulfide, <5> recombinant wild-type [21]) [21]
0.05 <5> (trypanothione disulfide, <5> with cepharanthine [29]) [29]
0.051 <1> (trypanothione disulfide) [2, 8]
0.052 <1> (benzyloxycarbonyl-L-cysteinylglycyl-3-methylaminopropylamide disulfide) [23]
0.053 <1> (trypanothione disulfide) [3, 10]
0.058 <4> (trypanothione disulfide) [3]
0.091 <8> (glutathionylspermidine disulfide) [22]
0.096 <6> (NADH) [2]
0.099 <5> (trypanothione disulfide, <5> with cycleanine [29]) [29]
0.115 <2> (glutathionylspermidine disulfide) [20]
0.12 <6> (deazatrypanothione disulfide) [2]
0.225 <1> (L-γ-glutamyl-L-cysteinylglycyl-3-methylaminopropylamide disulfide) [23]
0.251 <1> (tert-butyloxycarbonyl-L-cysteinylglycyl-3-methylaminopropylamide disulfide) [23]
0.275 <5> (glutathionylspermidine disulfide) [9]
0.3 <5> (5,5'-dithiobis(N-[3-(4-methylpiperazin-1-yl)propyl]-2-nitrobenzamide), <5> recombinant enzyme [15]) [15]
0.448 <1> (4-aminobutyryl-L-cysteinylglycyl-3-methylaminopropylamide disulfide) [23]
0.5 <7> (trypanothione disulfide, <7> recombinant glutathione-trypanothione reductase-like enzyme [16]) [16]

0.693 <1> (L-β-aspartyl-L-cysteinylglycyl-3-methylaminopropylamide disulfide) [23]
0.968 <1> (L-cysteinylglycyl-3-methylaminopropylamide disulfide) [23]
2 <5> ((E,Z)-ajoene, <5> above, with NADPH [27]) [27]
Additional information <1, 5> (<5> K_m dependent on assay conditions [9]; <1> K_m of several trypanothione disulfide derivatives, e.g. N^1-monoglutathionylspermidine, mixed disulfide of N^1-monoglutathionylspermidine and glutathione [8]) [8, 9]

K_i-Value (mM)

0.0017 <6> (3-acetylpyridine adenine nucleotide, <6> mutant E201Q [13]) [13]
0.0018 <1> (N^1,N^{12}-bis(hihydrocaffeoyl)-spermine, <1> i.e. kukoamine A [19]) [19]
0.0033 <6> (3-acetylpyridine adenine nucleotide, <6> mutant E201D [13]) [13]
0.0045 <6> (chinifur) [12]
0.0054 <6> (thio-$NADP^+$, <6> mutant E201D, slopes [13]) [13]
0.007-0.043 <5> (9-aminoacridine derivatives) [26]
0.0075 <1> (N^1,N^8-bis(dihydrocaffeoyl)spermidine) [19]
0.0076 <5> (cepharanthine) [29]
0.0084 <6> (3-acetylpyridine adenine nucleotide) [13]
0.0086 <5> (clomipramine, <5> recombinant enzyme, with 5,5'-dithiobis(N-[3-(dimethylamino)propyl]-2-nitrobenzamide) [15]) [15]
0.009 <2,4> (melarsen-trypanothione disulfide-adduct) [3,20]
0.01 <5> (phenothiazines) [30]
0.011 <6> ($NADP^+$, <6> wild-type and mutant E201D [13]) [13]
0.011 <6> (thio-$NADP^+$, <6> mutant E201Q [13]) [13]
0.011-0.607 <1> (acylspermines and acylspermidines) [19]
0.014 <6> ($NADP^+$, <6> mutant E201Q [13]) [13]
0.016 <1> (melarsen oxide, <1> with melarsen-trypanothione disulfide-adduct [17]) [17]
0.0172 <1> (melarsen oxide, <1> with NADPH [17]) [17]
0.019 <5> (mepacrine) [26]
0.021-0.037 <5> (9-thioacridine derivatives) [26]
0.03 <5> ((6-benzyloxycarbonylamino-6-[[[methyl-(2-propylamino-ethyl)-carbamoyl]-methyl]-carbamoyl]-1-[[(3-propylamino-propylcarbamoyl)-methyl]-carbamoyl]-hex-3-enyl)-carbamic acid benzyl ester) [11]
0.036 <1> (melarsen-trypanothione disulfide-adduct) [3]
0.04 <5> (nifurtimox) [5]
0.048 <5> ((6-benzyloxycarbonylamino-6-[[[methyl-(2-propylamino-ethyl)-carbamoyl]-methyl]-carbamoyl]-1-[[(3-propylamino-propylcarbamoyl)-methyl]-carbamoyl]-hexyl)-carbamic acid benzyl ester) [11]
0.051 <6> (thio-$NADP^+$, <6> mutant E201D, intercepts [13]) [13]
0.074 <5> ([6-benzyloxycarbonylamino-1,6-bis-(dimethylcarbamoylmethyl-carbamoyl)-hex-3-enyl]-carbamic acid benzyl ester) [11]
0.08 <5> (6-chlorotacrine) [26]

0.091 <5> ([6-benzyloxycarbonylamino-1,6-bis-(dimethylcarbamoylmethyl-carbamoyl)-hexyl]-carbamic acid benzyl ester) [11]
0.14 <5> (9-amino-1,2,3,4-tetrahydroacridine) [26]
0.144 <5> (lunarine) [14]
0.15 <5> (nifurtimox) [4]
0.194 <5> (cycleanine) [29]
Additional information <5> (<5> derivatives of 2-amino-diphenylsulfide + phenothiazine [30]) [30]

pH-Optimum

6.5 <7> (<7> recombinant glutathione-trypanothione reductase-like enzyme [16]) [16]
7.25 <5> (<5> assay at [15,30]) [15, 30]
7.25-7.75 <5> [16]
7.4 <2> (<2> additional 2 minor optima at pH 6.6 and pH 8.2 [20]) [20]
7.5 <5> (<5> glutathionylspermidine disulfide + NADPH, additional 2 minor optima at pH 7.0 and pH 8.0 [5]) [5]
7.5-8 <1> (<1> trypanothione disulfide + NADPH [10]) [10]
7.8 <1, 5> (<1,5> assay at [6,17]) [6, 8, 17]

pH-Range

6.5-9 <2, 5> [5, 20]

Temperature optimum (°C)

21 <5> (<5> assay at [30]) [30]
25 <5> (<5> assay at [12,15,21,26]) [12, 15, 21, 26]
27 <1, 5> (<1,5> assay at [6,8,17]) [6, 8, 17]

4 Enzyme Structure

Molecular weight

100000 <5> (<5> gel filtration [9]) [9]
106000 <2> (<2> gel filtration [20]) [20]
110000 <1, 6> (<6> recombinant enzyme, gel filtration [2]; <1> enzyme cross-linked with dimethyl suberimidate, SDS-PAGE [10]) [2, 10]

Subunits

? <1-5, 8> (<8> x * 54000, SDS-PAGE [22]; <1-5> x * 50000-55000, SDS-PAGE, immunostaining [1]) [1, 22]
dimer <1, 2, 5, 6> (<2> 2 * 50900, SDS-PAGE [20]; <1> 1 * 53800, SDS-PAGE [10]; <5> 2 * 50000, SDS-PAGE [2,9]; <1,6> 2 * 54000, SDS-PAGE [2]; <6> 2 * 53443, DNA sequence [7]) [2, 4, 7, 9, 10, 20]

5 Isolation/Preparation/Mutation/Application

Purification

<1> [10, 17, 20]
<2> (recombinant overexpression from plasmid in Leishmania donovani and in Trypanosoma cruzi [18]; recombinant overexpression of the enzyme from plasmid [20]) [18, 20]
<5> (recombinant wild-type and mutants C53S, C53A, C58S from Escherichia coli [21]) [9, 21]
<6> (recombinant from Escherichia coli [2]; wild-type and mutants recombinant from Escherichia coli [13]) [2, 13]
<7> (recombinant mutant from Escherichia coli [16]) [16]
<8> [22]

Crystallization

<5> (hanging drop method in presence of 2 M $(NH_4)_2SO_4$ at 4°C [9]; substrate binding and crystal structure in complex with trypanothione disulfide, molecular modeling with potential inhibitors [14]) [9, 14]
<7> (recombinant glutathione-trypanothione reductase-like enzyme, hanging-drop vapour diffusion method, 0.1 M potasssium phosphate, pH 8.0 [16]; substrate-binding and structure analysis [16]) [16]

Cloning

<2> (replacement clones of tyrA, cloning of replacement vector, several genotypes [28]; single copy gene [1]; cloned into expression vector and expressed in wild-type cells [20]; cloned into expression vector and overexpressed in Leishmania donovani and in Trypanosoma cruzi [18]) [1, 18, 20, 28]
<5> (overexpression of wild-type and mutants C53S, C53A, C58S in Escherichia coli [21]; overexpression in Escherichia coli [15,16]) [15, 16, 21]
<6> (expression of wild-type and mutant enzymes in Escherichia coli [13]; expression in Escherichia coli, sequence analysis [2]; DNA sequence analysis [7]) [2, 7, 13]
<7> (mutant enzyme glutathione reductase-trypanothine reductase, expression in Escherichia coli [16]) [16]

Engineering

A34E/R37W <7> (<7> glutathione reductase mutant, activity switches to trypanothione reductase, termed GRTR, 700fold more activity with trypanothione disulfide than with glutathione [16]) [16]
C53A <5> (<5> site-directed mutagenesis, without catalytic active cysteine residue in the active center, no activity with trypanothione disulfide and NADPH, but showing transhydrogenase activity between NADPH and thio-$NADP^+$ [21]) [21]
C53S <5> (<5> site-directed mutagenesis, without catalytic active cysteine residue in the active center, no activity with trypanothione disulfide and NADPH, but showing transhydrogenase activity between NADPH and thio-$NADP^+$ [21]) [21]

C58S <5> (<5> site-directed mutagenesis, without catalytic active cysteine residue in the active center, no activity with trypanothione disulfide and NADPH, but showing transhydrogenase activity between NADPH and thio-$NADP^+$ [21]) [21]
E201D <6> (<6> by site-directed mutagenesis, 5% reductive activity and 50fold increased oxidative activity both compared to wild-type, enhanced quinone reductase activity [13]) [13]
E201Q <6> (<6> by site-directed mutagenesis, 5% reductive activity and 50fold increased oxidative activity both compared to wild-type, enhanced quinone reductase activity [13]) [13]

Application

medicine <1, 2, 4-6> (<5> target enzyme for chemotherapy against Chagas' disease [29]; <2> target for rational approach to design new anti-leishmanial chemotherapeutic agents [20,24]; <1,5,6> potential target for drug design against trypanosomiasis [4,15,19,24,26] <5> nitrofuran derivatives show antiparasitic effect on the trypanocidal drug design target trypanothione reductase [12]; <1,5> inhibition of the enzyme by naphthoquinone and nitrofuran derivatives abolish the cell penetration by the parasite trypomas stigotes, alternate approach to chemotherapy of trypanosomiasis and leishmaniasis, Chagas' disease [6]; <4> enzyme substrate trypanothione disulfide is the primary target for arsenical drugs against african trypanosomes [3]; <5,6> good target for structure/function studies and trypanicidal inhibitor design [2,14,24]) [2-4, 6, 12, 14, 20, 24, 26, 29]

6 Stability

General stability information

<1>, loss of 39% activity after incubation with NADPH at 25°C [17]

References

[1] Taylor, M.C.; Chapman, C.J.; Kelly, J.M.; Fairlamb, A.H.; Miles, M.A.: Cloning by expression of a cDNA corresponding to Leishmania donovani trypanothione reductase. Biochem. Soc. Trans., **17**, 579-580 (1989)
[2] Sullivan, F.X.; Shames, S.L.; Walsh, C.T.: Expression of Trypanosoma congolense trypanothione reductase in Escherichia coli: overproduction, purification, and characterization. Biochemistry, **28**, 4986-4992 (1989)
[3] Fairlamb, A.H.; Henderson, G.B.; Cerami, A.: Trypanothione is the primary target for arsenical drugs against African trypanosomes. Proc. Natl. Acad. Sci. USA, **86**, 2607-2611 (1989)
[4] Krauth-Siegel, R.L.; Jockers-Scheruebl, M.C.; Becker, K.; Schirmer, R.H.: NADPH-dependent disulphide reductases. Biochem. Soc. Trans., **17**, 315-317 (1989)

[5] Jockers-Scheruebl, M.C.; Schirmer, R.H.; Krauth-Siegel, R.L.: Trypanothione reductase from Trypanosoma cruzi. Catalytic properties of the enzyme and inhibition studies with trypanocidal compounds. Eur. J. Biochem., **180**, 267-272 (1989)
[6] Henderson, G.B.; Ulrich, P.; Fairlamb, A.H.; Rosenberg, I.; Pereira, M.; Sela, M.; Cerami, A.: Subversive substrates for the enzyme trypanothione disulfide reductase: alternative approach to chemotherapy of Chagas disease. Proc. Natl. Acad. Sci. USA, **85**, 5374-5378 (1988)
[7] Shames, S.L.; Kimmel, B.E.; Peoples, O.P.; Agabian, N.; Walsh, C.T.: Trypanothione reductase of Trypanosoma congolense: gene isolation, primary sequence determination, and comparison to glutathione reductase. Biochemistry, **27**, 5014-5019 (1988)
[8] Henderson, G.B.; Fairlamb, A.H.; Ulrich, P.; Cerami, A.: Substrate specificity of the flavoprotein trypanothione disulfide reductase from Crithidia fasciculata. Biochemistry, **26**, 3023-3027 (1987)
[9] Krauth-Siegel, R.L.; Enders, B.; Henderson, G.B.; Fairlamb, A.H.; Schirmer, R.H.: Trypanothione reductase from Trypanosoma cruzi. Purification and characterization of the crystalline enzyme. Eur. J. Biochem., **164**, 123-128 (1987)
[10] Shames, S.L.; Fairlamb, A.H.; Cerami, A.; Walsh, C.T.: Purification and characterization of trypanothione reductase from Crithidia fasciculata, a newly discovered member of the family of disulfide-containing flavoprotein reductases. Biochemistry, **25**, 3519-3526 (1986)
[11] Garrard, E.A.; Borman, E.C.; Cook, B.N.; Pike, E.J.; Alberg, D.G.: Inhibition of trypanothione reductase by substrate analogues. Org. Lett., **2**, 3639-3642 (2000)
[12] Blumenstiel, K.; Schoneck, R.; Yardley, V.; Croft, S.L.; Krauth-Siegel, R.L.: Nitrofuran drugs as common subversive substrates of Trypanosoma cruzi lipoamide dehydrogenase and trypanothione reductase. Biochem. Pharmacol., **58**, 1791-1799 (1999)
[13] Zheng, R.; Cenas, N.; Blanchard, J.S.: Catalytic and potentiometric characterization of E201D and E201Q mutants of Trypanosoma congolense trypanothione reductase. Biochemistry, **34**, 12697-12703 (1995)
[14] Bond, C.S.; Zhang, Y.; Berriman, M.; Cunningham, M.L.; Fairlamb, A.H.; Hunter, W.N.: Crystal structure of Trypanosoma cruzi trypanothione reductase in complex with trypanothione, and the structure-based discovery of new natural product inhibitors. Structure, **7**, 81-89 (1999)
[15] Davioud-Charvet, E.; Becker, K.; Landry, V.; Gromer, S.; Loge, C.; Sergheraert, C.: Synthesis of 5,5'-dithiobis(2-nitrobenzamides) as alternative substrates for trypanothione reductase and thioredoxin reductase: a microtiter colorimetric assay for inhibitor screening. Anal. Biochem., **268**, 1-8 (1999)
[16] Stoll, V.S.; Simpson, S.J.; Krauth-Siegel, R.L.; Walsh, C.T.; Pai, E.F.: Glutathione reductase turned into trypanothione reductase: structural analysis of an engineered change in substrate specificity. Biochemistry, **36**, 6437-6447 (1997)

[17] Cunningham, M.L.; Zvelebil, M.J.J.M.; Fairlamb, A.H.: Mechanism of inhibition of trypanothione reductase and glutathione reductase by trivalent organic arsenicals. Eur. J. Biochem., **221**, 285-295 (1994)
[18] Kelly, J.M.; Taylor, M.C.; Smith, K.; Hunter, K.J.; Fairlamb, A.H.: Phenotype of recombinant Leishmania donovani and Trypanosoma cruzi which overexpress trypanothione reductase. Sensitivity towards agents that are thought to induce oxidative stress. Eur. J. Biochem., **218**, 29-37 (1993)
[19] Ponasik, J.A.; Strickland, C.; Faerman, C.; Savvides, S.; Karplus, P.A.; Ganem, B.: Kukoamine A and other hydrophobic acylpolyamines: potent and selective inhibitors of Crithidia fasciculata trypanothione reductase. Biochem. J., **311**, 371-375 (1995)
[20] Cunningham, M.L.; Fairlamb, A.H.: Trypanothione reductase from Leishmania donovani. Purification, characterization and inhibition by trivalent antimonials. Eur. J. Biochem., **230**, 460-468 (1995)
[21] Borges, A.; Cunningham, M.L.; Tovar, J.; Fairlamb, A.H.: Site-directed mutagenesis of the redox-active cysteines of Trypanosoma cruzi trypanothione reductase. Eur. J. Biochem., **228**, 745-752 (1995)
[22] Montrichard, F.; Le Guen, F.; Laval-Martin, D.L.; Davioud-Charvet, E.: Evidence for the co-existence of glutathione reductase and trypanothione reductase in the non-trypanosomatid Euglenozoa: Euglena gracilis Z. FEBS Lett., **442**, 29-33 (1999)
[23] El-Waer, A.F.; Smith, K.; McKie, J.H.; Benson, T.; Fairlamb, A.H.; Douglas, K.T.: The glutamyl binding site of trypanothione reductase from Crithidia fasciculata: enzyme kinetic properties of γ-glutamyl-modified substrate analogs. Biochim. Biophys. Acta, **1203**, 93-98 (1993)
[24] Salmon-Chemin, L.; Buisine, E.; Yardley, V.; Kohler, S.; Debreu, M.A.; Landry, V.; Sergheraert, C.; Croft, S.L.; Krauth-Siegel, R.L.; Davioud-Charvet, E.: 2- And 3-substituted 1,4-naphthoquinone derivatives as subversive substrates of trypanothione reductase and lipoamide dehydrogenase from Trypanosoma cruzi: Synthesis and correlation between redox cycling activities and in vitro cytotoxicity. J. Med. Chem., **44**, 548-565 (2001)
[25] Bonse, S.; Richards, J.M.; Ross, S.A.; Lowe, G.; Krauth-Siegel, R.L.: (2,2':6',2'-Terpyridine)platinum(II) complexes are irreversible inhibitors of Trypanosoma cruzi trypanothione reductase but not of human glutathione reductase. J. Med. Chem., **43**, 4812-4821 (2000)
[26] Bonse, S.; Santelli-Rouvier, C.; Barbe, J.; Krauth-Siegel, R.L.: Inhibition of Trypanosoma cruzi trypanothione reductase by acridines: kinetic studies and structure-activity relationships. J. Med. Chem., **42**, 5448-5454 (1999)
[27] Gallwitz, H.; Bonse, S.; Martinez-Cruz, A.; Schlichting, I.; Schumacher, K.; Krauth-Siegel, R.L.: Ajoene is an inhibitor and subversive substrate of human glutathione reductase and Trypanosoma cruzi trypanothione reductase: crystallographic, kinetic, and spectroscopic studies. J. Med. Chem., **42**, 364-372 (1999)
[28] Tovar, J.; Wilkinson, S.; Mottram, J.C.; Fairlamb, A.H.: Evidence that trypanothione reductase is an essential enzyme in Leishmania by targeted replacement of the tryA gene locus. Mol. Microbiol., **29**, 653-660 (1998)

[29] Fournet, A.; Inchausti, A.; Yaluff, G.; Royas De Arias, A.; Guinaudeau, H.; Bruneton, J.; Breidenbach, M.A.; Karplus, P.A.; Faerman, C.H.: Trypanocidal bisbenzylisoquinoline alkaloids are inhibitors of trypanothione reductase. J. Enzyme Inhib., **13**, 1-9 (1998)
[30] Bonnet, B.; Soullez, D.; Davioud-Charvet, E.; Landry, V.; Horvath, D.; Sergheraert, C.: New spermine and spermidine derivatives as potent inhibitors of Trypanosoma cruzi trypanothione reductase. Bioorg. Med. Chem., **5**, 1249-1256 (1997)

Bis-γ-glutamylcystine reductase 1.8.1.13

1 Nomenclature

EC number
1.8.1.13

Systematic name
γ-glutamylcysteine:NADP$^+$ oxidoreductase

Recommended name
bis-γ-glutamylcystine reductase

Synonyms
EC 1.6.4.9 (created 1992 as EC 1.6.4.9, transferred 2002 to EC 1.8.1.13)
GCR <1> [1, 2]
NADPH$_2$:bis-γ-glutamylcysteine oxidoreductase

CAS registry number
117056-54-9

2 Source Organism

<1> *Halobacterium halobium* [1, 2]

3 Reaction and Specificity

Catalyzed reaction
2 γ-glutamylcysteine + NADP$^+$ = bis-γ-glutamylcystine + NADPH + H$^+$ (highly specific. Not identical with EC 1.8.1.7 glutathione-disulfide reductase or EC 1.8.1.14 CoA-disulfide reductase)

Reaction type
oxidation
redox reaction
reduction

Natural substrates and products
S bis-γ-glutamylcystine + NADPH <1> (Reversibility: ? <1> [1, 2]) [1, 2]
P γ-glutamylcysteine + NADP$^+$

Substrates and products
S 5,5'-dithiobis(2-nitrobenzoic acid) + NADPH <1> (Reversibility: ? <1> [2]) [2]

P 4-nitrobenzenethiol + $NADP^+$
S bis-γ-glutamylcystine + NADPH <1> (Reversibility: ? <1> [1, 2]) [1, 2]
P γ-glutamylcysteine + $NADP^+$
S Additional information <1> (<1> lipoic acid disulfide and dihydrolipoamide are no substrates [1,2]; <1> no activity detected with GSSG [1,2]) [1, 2]
P ?

Inhibitors
AsO_2^- <1> (<1> 25 mM arsenite, relative activity 46% [2]) [2]
Cu^{2+} <1> (<1> relative activity less than 5% [2]) [2]
Hg^{2+} <1> (<1> relative activity less than 5% [2]) [2]
Zn^{2+} <1> (<1> relative activity less than 44% [2]) [2]

Cofactors/prosthetic groups
FAD <1> (<1> flavoprotein, 1 mol FAD/mol subunit [2]) [1, 2]
NADH <1> (<1> less effective than NADPH [1]) [1]
NADPH <1> [1]

Turnover number (min^{-1})
1700 <1> (FAD) [2]

Specific activity (U/mg)
25 <1> [1]
28 <1> [2]

K_m-Value (mM)
0.29 <1> (NADPH) [2]
0.81 <1> (bis-γ-glutamylcystine) [2]

pH-Optimum
7.5 <1> [2]

pH-Range
6.3-8.2 <1> (<1> about half-maximal activity at pH 6.3 and 8.2 [2]) [2]

4 Enzyme Structure

Molecular weight
122000 <1> (<1> gel filtration [1]) [1, 2]

Subunits
dimer <1> (<1> 2 * 61000, SDS-PAGE [1,2]) [1, 2]

5 Isolation/Preparation/Mutation/Application

Localization
cytoplasm <1> [2]

Purification

<1> [1, 2]

6 Stability

Temperature stability

60 <1> (<1> retains full activity after incubation in high-and intermediate-ionic-strength buffers, inactivated after incubation in low-ionic strength buffers [1]) [1]
80 <1> (<1> inactivation in buffers of each of the 3 ionic strengths, low-ionic, intermediate-ionic and high-ionic [1]) [1]

General stability information

<1>, high-ionic-strength buffer above 2 M stabilizes during purification and storage [1]
<1>, high-ionic-strength buffer above 3.5 M increases thermal stability [1]
<1>, low-ionic-strength buffer below 1.0 M leads to rapid denaturation [2]

Storage stability

<1>, 20°C, stable to prolonged storage at room temperature in high-ionic-strength buffer above 3.5 M [1]

References

[1] Sundquist, A.R.; Fahey, R.C.: The novel disulfide reductase bis-γ-glutamylcystine reductase and dihydrolipoamide dehydrogenase from Halobacterium halobium: Purification by immobilized-metal-ion affinity chromatography and properties of the enzymes. J. Bacteriol., **170**, 3459-3467 (1988)
[2] Sundquist, A.R.; Fahey, R.C.: The function of γ-glutamylcysteine and bis-γ-glutamylcystine reductase in Halobacterium halobium. J. Biol. Chem., **264**, 719-725 (1989)

CoA-Disulfide reductase 1.8.1.14

1 Nomenclature

EC number
1.8.1.14

Systematic name
CoA:NAD^+ oxidoreductase

Recommended name
CoA-disulfide reductase

Synonyms
CoA-disulfide reductase (NADH)
CoA-disulfide reductase ($NADH_2$)
$NADH_2$:CoA-disulfide oxidoreductase
NADH:CoA-disulfide oxidoreductase

CAS registry number
206770-55-0

2 Source Organism

<1> *Bacillus megaterium* (QM) [1]

3 Reaction and Specificity

Catalyzed reaction
2 CoA + NAD^+ = CoA-disulfide + NADH + H^+

Reaction type
oxidation
redox reaction
reduction

Substrates and products
S NADH + CoA-disulfide <1> (Reversibility: ? <1> [1]) [1]
P NAD^+ + CoA <1> [1]

Cofactors/prosthetic groups
NADH <1> [1]

5 Isolation/Preparation/Mutation/Application

Source/tissue

spore <1> (<1>, at highest level in dormant spores [1]) [1]

References

[1] Setlow, B.; Setlow, P.: Levels of acetyl coenzyme A, reduced and oxidized coenzyme A, and coenzyme A in disulfide linkage to protein in dormant germinated spores and growing and sporulating cells of Bacillus megaterium. J. Bacteriol., **132**, 444-452 (1979)

Mycothione reductase 1.8.1.15

1 Nomenclature

EC number
1.8.1.15

Systematic name
mycothiol:NAD(P)$^+$ oxidoreductase

Recommended name
mycothione reductase

Synonyms
mycothiol-disulfide reductase

2 Source Organism

<1> *Mycobacterium tuberculosis* [1-3]

3 Reaction and Specificity

Catalyzed reaction
2 mycothiol + NAD(P)$^+$ = mycothione + NAD(P)H + H$^+$

Reaction type
oxidation
redox reaction
reduction

Natural substrates and products
S mycothione + NADPH <1> (Reversibility: ? <1> [1]) [1]
P mycothiol + NADP$^+$

Substrates and products
S 2,6-dimethylbenzoquinone + NADPH <1> (Reversibility: ? <1> [2]) [2]
P ?
S 2-methyl-1,4-naphthoquinone + α-NADPH <1> (Reversibility: ? <1> [2]) [2]
P ?
S 5,8-dihydroxy-1,4-naphthoquinone + NADPH <1> (Reversibility: ? <1> [2]) [2]
P ?

S 5-hydroxy-1,4-naphthoquinone + NADPH <1> (Reversibility: ? <1> [2]) [2]
P ?
S DTNB + NADPH <1> (Reversibility: ? <1> [2]) [2]
P ?
S des-myo-inositol mycothione + NADH <1> (Reversibility: ? <1> [2]) [2]
P ?
S des-myo-inositol mycothione + reduced β-nicotinamide hypoxanthine dinucleotide <1> (Reversibility: ? <1> [2]) [2]
P ?
S des-myo-inositol mycothione + reduced β-nicotinamide hypoxanthine dinucleotide phosphate <1, 1> (Reversibility: ? <1> [2]) [2]
P ?
S des-myo-inositol mycothione + thio-NADH <1> (Reversibility: ? <1> [2]) [2]
P ?
S des-myo-inositol mycothione + thio-NADPH <1> (Reversibility: ? <1> [2]) [2]
P ?
S desmyoinositol mycothione + NADPH <1> (Reversibility: ? <1> [2]) [2]
P ?
S mycothione + NADPH <1> (Reversibility: ? <1> [1, 2]) [1, 2]
P mycothiol + $NADP^+$

Cofactors/prosthetic groups

FAD <1> (1 mol FAD per monomer [2]) [2]
NADPH <1> [2]

Turnover number (min^{-1})

1800 <1> (NADH) [3]
7740 <1> (NADPH) [3]

Specific activity (U/mg)

3.63 <1> [2]

K_m-Value (mM)

0.001 <1> (thio-NADPH, <1> des-myo-inositol mycothione as second substrate [2]) [2]
0.008 <1> (β-NADPH, <1> des-myo-inositol mycothione as second substrate [2]) [2]
0.023 <1> (reduced β-nicotinamide hypoxanthine dinucleotide phosphate, <1> des-myo-inositol mycothione as second substrate [2]) [2]
0.043 <1> (NADH, <1> des-myo-inositol mycothione as second substrate [2]) [2]
0.055 <1> (α-NADPH, <1> des-myo-inositol mycothione as second substrate [2]) [2]
0.073 <1> (mycothione) [2]
0.09 <1> (thio-NADH, <1> des-myo-inositol mycothione as second substrate [2]) [2]

0.2 <1> (reduced β-nicotinamide hypoxanthine dinucleotide, <1> des-myo-inositol mycothione as second substrate [2]) [2]
0.24 <1> (2-methyl-1,4-naphthoquinone) [2]
0.34 <1> (5,8-dihydroxy-1,4-naphthoquinone) [2]
0.4 <1> (des-myo-inositol mycothione) [1]
0.51 <1> (des-myo-inositol mycothione) [2]
0.54 <1> (5-hydroxy-1,4-naphthoquinone) [2]
3.3 <1> (DTNB) [2]
4 <1> (2,6-dimethylbenzoquinone) [2]

4 Enzyme Structure

Molecular weight
100000 <1> (<1> gel filtration [2]) [2]

Subunits
homodimer <1> (<1> α_2, 2 * 50000, SDS-PAGE [2]) [2]

5 Isolation/Preparation/Mutation/Application

Purification
<1> [2, 3]

Cloning
<1> (expression in Mycobacterium smegmatis pyrF [1]) [1-3]

References

[1] Patel, M.P.; Blanchard, J.S.: Synthesis of des-myo-inositol mycothiol and demonstration of a mycobacterial specific reductase activity. J. Am. Chem. Soc., **120**, 11538-11539 (1998)
[2] Patel, M.P.; Blanchard, J.S.: Expression, purification, and characterization of Mycobacterium tuberculosis mycothione reductase. Biochemistry, **38**, 11827-11833 (1999)
[3] Patel, M.P.; Blanchard, J.S.: Mycobacterium tuberkulosis mycothione reductase: pH dependence of the kinetic parameters and kinetic isotope effects. Biochemistry, **40**, 5119-5126 (2001)

Sulfite dehydrogenase 1.8.2.1

1 Nomenclature

EC number
1.8.2.1

Systematic name
sulfite:ferricytochrome-c oxidoreductase

Recommended name
sulfite dehydrogenase

Synonyms
dehydrogenase, sulfite
sulfite cytochrome c reductase
sulfite oxidase
sulfite-cytochrome c oxidoreductase
sulphite:cytochrome c oxidoreductase

CAS registry number
37256-47-6

2 Source Organism

<1> *Thiobacillus versutus* (partial [3]) [1, 2, 3, 4]
<2> *Thiobacillus novellus* [5, 6, 8, 9, 12, 14]
<3> *Rhodotorula sp.* (enzyme complex that posseses sulfite oxidase activity and thiosulfate-oxidizing activity [7]) [7]
<4> *Thiobacillus denitrificans* [9]
<5> *Thiobacillus thioparus* [10, 16]
<6> *Thiobacillus ferrooxidans* [11]
<7> *Paracoccus pantotrophus* (GB17 [13]) [13]
<8> *Paracoccus denitrificans* [15]
<9> *Starkeya novella* [16]
<10> *Thiobacillus thiooxidans* (JCM7814 [17]) [17]

3 Reaction and Specificity

Catalyzed reaction
sulfite + 2 ferricytochrome c + H_2O = sulfate + 2 ferrocytochrome c (<2>, ping-pong mechanism [14])

Reaction type

oxidation
redox reaction
reduction

Natural substrates and products

S sulfite + ferricytochrome c + H_2O <1, 2, 8> (<1>, the enzyme is involved in oxidation of thiosulfate [1]; <1>, the enzyme is a component of a thiosulfate-oxidizing system [3]; <2>, oxidative phosphorylation is coupled to sulfite oxidation with a low P/O ratio [6]; <8>, enzyme is essential for growth with thiosulfate, essential role in lithotrophic sulfur oxidation [15]) (Reversibility: ? <1, 2, 8> [1, 3, 6, 15]) [1, 3, 6, 15]

P sulfate + ferrocytochrome c + H^+

Substrates and products

S sulfite + ferricyanide + H_2O <2, 3, 5, 6, 10> (<2>, 3-5% of the activity with ferricytochrome [8]; <2>, enzyme is involved in thiosulfate oxidation [9]) (Reversibility: ? <2, 3, 5, 6, 10> [6, 7, 8, 9, 10, 11, 17]) [6, 7, 8, 9, 10, 11, 17]

P sulfate + ferrocyanide + H^+

S sulfite + ferricytochrome c + H_2O <1-10> (<2>, specific for sulfite [6]; <5,6>, horse heart cytochrome or yeast cytochrome [10,11]) (Reversibility: ? <1-10> [1-17]) [1-17]

P sulfate + ferrocytochrome c + H^+

Inhibitors

1,10-phenanthroline <3, 5> (<3>, 5 mM, 30% inhibition [7]; <5>, 5 mM, complete inhibition [10]) [7, 10]

2,2'-bipyridyl <3, 5> (<3>, 5 mM, 50% inhibition [7]; <5>, 5 mM, complete inhibition [10]) [7, 10]

AMP <2, 3, 5> (<3>, 0.01 mM, 50% inhibition [7]; <2>, 0.01 mM, 25% inhibition, activity with ferricyanide [6]) [6, 7, 10]

Cl^- <5> (<5>, non-competitive [10]) [10]

EDTA <2, 3, 5> (<2>, 0.005 mM, 20% inhibition [6]; <3>, 10 mM Na_2EDTA, 30% inhibition [7]; <5>, 4 mM complete inhibition [10]) [6, 7, 10]

$HgCl_2$ <2, 3, 5, 10> (<2>, 0.667 mM, 56% inhibition [6]; <3>, 0.25 mM, complete inhibition [7]; <5>, 10 mM, complete inhibition [10]; <10>, 1 mM, 91% inhibition [17]) [6, 7, 10, 17]

KCN <1> (<1>, 2 mM, complete inhibition after 10 min [1]) [1]

KCl <5> [10]

NEM <2, 3, 5, 10> (<3>, 1 mM, complete inhibition [7]; <2>, 0.667 mM, 30% loss of activity [6]; <5>, 1 mM, 32% inhibition [10]; <10>, 1 mM, 53% inhibition [17]) [6, 7, 10, 17]

$NaAsO_2$ <2> (<2>, 31% inhibition [6]) [6]

NaCl <2, 5> (<2>, competitive with respect to sulfite [6]; <5>, noncompetitive [10]) [6, 10]

NaF <2> (<2>, 0.01 mM, 17.5% inhibition [6]) [6]

Ni^{2+} <7> (<7>, 1 mM, 32% inhibition [13]) [13]

PCMB <5, 10> (<5>, 1 mM, complete inhibition [10]; <10>, 1 mM, 47% loss of activity [17]) [10, 17]
SO_4^{2-} <2, 7> (<2>, product inhibition, mixed-type non-competitive with respect to sulfite [14]) [13, 14]
Triton X-100 <3> (<3>, 1.0%, 15% inhibition [7]) [7]
Zn^{2+} <7> (<7>, 1 mM, 27% inhibition [13]) [13]
arsenite <5, 10> (<5>, 1 mM, 83% inhibition [10]; <10>, 1 mM, complete inhibition [17]) [10, 17]
monoiodoacetate <10> (<10>, 1 mM, 19% inhibition [17]) [17]
p-hydroxymercuribenzoate <1, 2, 3> (<1>, 1 mM, 75% inhibition after 15 min [1]; <2>, 0.667 mM, 86% loss of activity, inhibition is completely reversed by GSH [6]; <3>, 1 mM, complete inhibition [7]) [1, 6, 7]
phosphate <1, 5, 7> (<1>, noncompetitive [1]; <5>, 50 mM, 90% inhibition, uncompetitive with respect to sulfite [10]) [1, 10, 13]
sodium deoxycholate <3> (<3>, 1.0%, 80% inhibition [7]) [7]
Additional information <2> (<2>, $CaCl_2$, $MgSO_4$, $ZnSO_4$, $CuSO_4$, $FeCl_3$, $NiCl_2$ and Na_2MoO_4 inhibits the reaction by 70% at 0.267 mM, this effect is apparently due to the non-enzymatic oxidation of sulfite with air [6]) [6]

Cofactors/prosthetic groups

cytochrome c_{551} <1, 2> (<1>, intimately associated with the enzyme [1]; <2>, MW 23000 Da, necessary for the enzyme as an integral part [12]) [1, 12]
cytochrome c_{550} <2> (<2>, enzyme reacts exclusively with cytochrome c_{550} [14]) [5, 14]
heme <2, 5, 7, 9> (<2>, enzyme contains heme [8]; <7> the 38815 Da subunit consists of a polypeptide and two covalently bound hemes, 3.53 mol of heme per mol of enzyme [13]; <2>, the enzyme contains 0.8 heme groups per holoenzyme [14]; <9>, contains heme [16]) [8, 13, 14, 16]

Activating compounds

Triton X-100 <10> (<10>, increases activity [17]) [17]
phosphate <3> (<3>, maximal activity at final substrate concentrations of 0.02-0.05 M at pH 7.8 [7]) [7]

Metals, ions

Mg^{2+} <7> (<7>, 1 mM, 150% increase in activity [13]) [13]
Mn^{2+} <7> (<7>, 1 mM, 127% increase in activity [13]) [13]
iron <5, 7> (<5>, enzyme contains nonheme iron [10]; <7>, enzyme contains 3.17 mol of iron per mol of enzyme [13]) [10, 13]
molybdenum <2, 7, 9> (<2,9>, enzyme contains molybdenum [8,16]; <7>, the 43897 Da subunit contains the 455 Da molybdenum cofactor, 1.3 mol of molybdenum per mol of enzyme [13]; <2>, contains 1.2 Mo atoms per holoenzyme [14]) [8, 13, 14, 16]

Specific activity (U/mg)

0.049 <7> [13]
0.69 <5> [10]
21 <10> [17]
21.1 <1> [2]

128 <1> [1]
139.5 <2> (<2>, reaction with ferricyanide [6]) [6]
327.9 <2> [14]
Additional information <2> [8]

K_m-Value (mM)
0.0018 <2> (ferricytochrome c) [12]
0.002 <2> (sulfite, <2>, pH 6.5, reaction with cytochrome c [6]) [6]
0.0025 <2> (cytochrome c_{550}) [14]
0.004 <2> (cytochrome c) [14]
0.007 <7> (sulfite) [13]
0.01 <2> (sulfite, <2>, reaction with yeast ferricytochrome c [12]) [12]
0.013 <5> (cytochrome c) [10]
0.014 <1> (sulfite) [1]
0.019 <7> (cytochrome c) [13]
0.02 <2> (ferricyanide, <2>, pH 8.0 [6]) [6]
0.027 <2, 9> (sulfite) [14, 16]
0.04 <2> (sulfite, <2>, pH 8.0, reaction with cytochrome c [6]) [6]
0.06 <3> (sulfite, <3>, reaction with cytochrome c [7]) [7]
0.088 <5> (sulfite) [10, 16]
0.25 <6> (ferricyanide) [11]
0.42 <3> (ferricyanide) [7]
0.54 <6> (sulfite, <6>, reaction with ferricyanide [11]) [11]
0.58 <6> (ferricyanide, <6>, reaction with ferricyanide or cytochrome c [11]) [11]
0.58 <6> (sulfite, <6>, reaction with cytochrome c [11]) [11]
1 <3> (sulfite, <3>, reaction with ferricyanide [7]) [7]
1.95 <10> (sulfite) [17]

K_i-Value (mM)
4.5 <2> (NaCl) [6]
12 <1> (phosphate) [1]
38 <5> (Cl^-) [10]
45 <7> (phosphate) [13]
Additional information <2> [14]

pH-Optimum
6.5 <7> [13]
7.5 <2, 10> [12, 17]
7.8 <3> (<3>, reaction with native cytochrome c [7]) [7]
8 <1, 2, 5, 9> (<2>, potassium phosphate buffer [6]; <5>, with Tricine-HCl or Tricine-NaOH buffer [10]) [1, 6, 10, 16]
8.5 <2> [8]

pH-Range
7.3-9.5 <1> (<1>, 90% of maximal activity at pH 7.3 and 9.5 [1]) [1]

Temperature optimum (°C)
20 <10> [17]

30 <3> [7]
65 <9> [16]

4 Enzyme Structure

Molecular weight

40000 <2> (<2>, gel filtration [8]) [8]
41500 <6> (<6>, nondenaturing PAGE [11]) [11]
46000 <2> (<2>, gel filtration [14]) [14]
54000 <5> (<5>, sucrose density gradient method [10,16]) [10, 16]
190000 <7> (<7>, native gradient PAGE [13]) [13]
400000 <10> (<10>, gel filtration [17]) [17]

Subunits

? <10> (<10>, x * 62000 + x * 70000 + x * 74000, SDS-PAGE [17]) [17]
heterodimer <9> (<9>, 2 * 40600 + 2 * 8800 [16]) [16]
monomer <1, 2> (<2>, 1 * 40000, SDS-PAGE [8]; <1>, 1 * 44000, SDS-PAGE [1]) [1, 8]
tetramer <7> (<7>, 2 * 43897 + 2 * 38815, the 43897 Da subunit contains the 455 Da molybdenum cofactor, the 38815 Da subunit consists of a polypeptide and two covalently bound hemes, electrospray ionization mass spectrometry [13]; <7>, $\alpha_2\beta_2$, 2 * 47000 + 2 * 50000, SDS-PAGE [13]) [13]

5 Isolation/Preparation/Mutation/Application

Localization

membrane <1, 10> (<1>, the enzyme is a component of the membrane-associated thiosulfate-oxidizing complex [4]; <10>, bound to [17]) [4, 17]
periplasm <2, 7, 9> [13, 14, 16]
soluble <5> [16]

Purification

<1> [1, 2]
<2> [6, 8, 9, 12, 14]
<5> [10]
<6> (partial [11]) [11]
<7> [13]
<10> [17]

Cloning

<2> [14]
<8> [15]

6 Stability

pH-Stability

7 <10> (<10>, 20°C, 1 h, stable [17]) [17]
9 <10> (<10>, 20°C, 1 h, about 20% loss of activity [17]) [17]

Temperature stability

3 <6> (<6>, 8 h, stable [11]) [11]
20 <1> (<1>, 1 week, about 35% loss of activity [1]) [1]
30 <10> (<10>, 1 h, stable below [17]) [17]
40 <10> (<10>, 1 h, about 20% loss of activity [17]) [17]
50 <10> (<10>, 1 h, complete loss of activity [17]) [17]
55 <2, 3, 5> (<2>, complete loss of activity after 5 min, 25% loss of activity after 5 min in presence of 0.05 M sodium sulfite [6]; <3>, 5 min, 60% loss of activity [7]; <5>, 9 min, 95% loss of activity [10]) [6, 7, 10]
58 <2> (<2>, rapid inactivation of oxidized enzyme, substrate-reduced enzyme remains active for more than 1 h [8]) [8]
60 <2, 3, 5> (<2>, 1 min, complete loss of activity [6]; <3>, 7 min, 70% loss of activity [7]; <5>, 6 min, 95% loss of activity [10]) [6, 7, 10]
100 <3, 5, 6> (<3>, boiling of the purified enzyme complex destroys activity completely [7]; <5>, 1 min, complete loss of activity [10]; <6>, 15 min, complete loss of activity [11]) [7, 10, 11]

Organic solvent stability

ethanol <2> (<2>, 50%, significant loss of activity [6]) [6]

General stability information

<2>, proteolytic treatment with trypsin at room temperature for up to 5 h at trypsin to enzyme ratios of 1/20 and 1/1 does not affect the activity of the enzyme [8]
<2>, stable to repeated freezing and thawing [6]
<3>, treatment with 1-2 mg trypsin per 1.5 mg of protein or 0.1-0.2 mg of pronase or 1.5 mg of protein at 25°C causes decline of activity [7]
<5, 6>, thawing and refreezing has no effect [10, 11]

Storage stability

<1>, -20°C, no loss of activity after several months [1]
<1>, 4°C, 20% loss of activity after 1 month [1]
<2>, -20°C, retains high activity for more than 6 months [6]
<3>, -20°C, in presence of 20% glycerol stable for several days, without glycerol most of the activity is lost after 48 h [7]
<5>, -20°C, stable for several months [10]
<5>, 4°C, 10-20% loss of activity after 6 h at room temperature, 80% loss of activity after 5 days [10]

References

[1] Lu, W.P.; Kelly, D.P.: Properties and role of sulphite:cytochrome c oxidoreductase purified from Thiobacillus versutus (A2). J. Gen. Microbiol., **130**, 1683-1692 (1984)
[2] Lu, W.P.; Kelly, D.P.: Purification and some properties of two principal enzymes of the thiosulphate-oxidizing multi-enzyme system from Thiobacillus A2. J. Gen. Microbiol., **129**, 3549-3564 (1983)
[3] Lu, W.P.; Kelly, D.P.: Partial purification and resolution of a thiosulfate-oxidizing system from Thiobacillus A2. J. Gen. Microbiol., **129**, 1673-1681 (1983)
[4] Oh, J.K.; Suzuki, I.: Resolution of a membrane-associated thiosulfate-oxidizing complex of Thiobacillus novellus. J. Gen. Microbiol., **99**, 413-423 (1977)
[5] Yamanaka, T.: A comparative study on the redox reactions of cytochrome c with certain enzymes. J. Biochem., **77**, 493-499 (1975)
[6] Charles, A.M.; Suzuki, I.: Purification and properties of sulfite:cytochrome c oxidoreductase from Thiobacillus novellus. Biochim. Biophys. Acta, **128**, 522-534 (1966)
[7] Kurek, E.J.: Properties of an enzymatic complex active in sulfite and thiosulfate oxidation by Rhodotorula sp.. Arch. Microbiol., **143**, 277-282 (1985)
[8] Toghrol, F.; Southerland, W.M.: Purification of Thiobacillus novellus sulfite oxidase. Evidence for the presence of heme and molybdenum. J. Biol. Chem., **258**, 6762-6766 (1983)
[9] Charles, A.M.; Suzuki, I.: Mechanism of thiosulfate oxidation by Thiobacillus novellus. Biochim. Biophys. Acta, **128**, 510-521 (1966)
[10] Lyric, R.M.; Suzuki, I.: Enzymes involved in the metabolism of thiosulfate by Thiobacillus thioparus. I. Survey of enzymes and properties of sulfite: cytochrome c oxidoreductase. Can. J. Biochem., **48**, 334-343 (1970)
[11] Vestal, J..R.; Lundgren, D.G.: The sulfite oxidase of Thiobacillus ferrooxidans (Ferrobacillus ferrooxidans). Can. J. Biochem., **49**, 1125-1130 (1971)
[12] Yamanaka, T.; Yoshioka, T.; Kimura, K.: Purification of sulphite-cytochrome c reductase of Thiobacillus novellus and reconstitution of its sulphite oxidase system with the purified constituents. Plant Cell Physiol., **22**, 613-622 (1981)
[13] Quentmeier, A.; Kraft, R.; Kostka, S.; Klockenkamper, R.; Friedrich, C.G.: Characterization of a new type of sulfite dehydrogenase from Paracoccus pantotrophus GB17. Arch. Microbiol., **173**, 117-125 (2000)
[14] Kappler, U.; Bennett, B.; Rethmeier, J.; Schwarz, G.; Deutzmann, R.; McEwan, A.G.; Dahl, C.: Sulfite:cytochrome c oxidoreductase from Thiobacillus novellus. Purification, characterization, and molecular biology of a heterodimeric member of the sulfite oxidase family. J. Biol. Chem., **275**, 13202-13212 (2000)
[15] Wodara, C.; Bardischewsky, F.; Friedrich, C.G.: Cloning and characterization of sulfite dehydrogenase, two c-type cytochromes, and a flavoprotein

of Paracoccus denitrificans GB17: essential role of sulfite dehydrogenase in lithotrophic sulfur oxidation. J. Bacteriol., **179**, 5014-5023 (1997)
[16] Kappler, U.; Dahl, C.: Enzymology and molecular biology of prokaryotic sulfite oxidation. FEMS Microbiol. Lett., **203**, 1-9 (2001)
[17] Nakamura, K.; Yoshikawa, H.; Okubo, S.; Kurosawa, H.; Amano, Y.: Purification and properties of membrane-bound sulfite dehydrogenase from Thiobacillus thiooxidans JCM7814. Biosci. Biotechnol. Biochem., **59**, 11-15 (1995)

Thiosulfate dehydrogenase 1.8.2.2

1 Nomenclature

EC number
1.8.2.2

Systematic name
thiosulfate:ferricytochrome-c oxidoreductase

Recommended name
thiosulfate dehydrogenase

Synonyms
enzymes, thiosulfate-oxidizing
oxidase, thiosulfate
tetrathionate synthase
thiosulfate oxidase
thiosulfate-acceptor oxidoreductase
thiosulfate-oxidizing enzyme

CAS registry number
9076-88-4

2 Source Organism

<1> *Thiobacillus thioparus* [8, 12]
<2> *Ferrobacillus ferrooxidans* [8, 13]
<3> *Thiobacillus sp.* (X [9]; W5 [16]) [9, 16]
<4> *Thiobacillus tepidarius* [1, 6, 15]
<5> *Rhodotorula sp.* [2]
<6> *Rhodopseudomonas globiformis* [3]
<7> *Pseudomonas aeruginosa* [4]
<8> *Chromatium vinosum* [5]
<9> *heterotrophic bacterium* (two bacteria designated A-50 and C-3, isolated from percolation units containing garden soil and elemental sulfur, the enzyme system in C-3 is constitutive, but in A-50 it is induced by thiosulfate or tetrathionate [7]) [7]
<10> *Pseudomonas sp.* (16B, marine heterotroph [10]) [10]
<11> *Alcaligenes sp.* [11]
<12> *Thiobacillus acidophilus* [14, 15]
<13> *Thiobacillus neapolitanus* [15]
<14> *Acidithiobacillus thiooxidans* (JCM7814 [17]) [17]

3 Reaction and Specificity

Catalyzed reaction

2 thiosulfate + 2 ferricytochrome c = tetrathionate + 2 ferrocytochrome c

Reaction type

oxidation
redox reaction
reduction

Natural substrates and products

S thiosulfate + cytochrome c <1, 2, 4> (Reversibility: ? <1, 2, 4> [1, 8]) [1, 8]

P tetrathionate + reduced cytochrome c <1, 2, 4> [1, 8]

Substrates and products

S thiosulfate + oxidized acceptor <1-12, 14> (<1,4-7,9-11> ferricyanide or native cytochrome c act as electron acceptors [1-4,7,8,10,11]; <8,12> ferricyanide is a good electron acceptor [5,14]; <2,3,5,6,10,14> the activity does not couple with mammalian cytochrome c as electron acceptor [2,3,10,13,16,17]; <1,8> horse cytochrome c is a poor electron acceptor [5,12]; <1> cytochrome c of yeast is a poor electron acceptor [12]; <1,3-11> cytochrome c acts as electron acceptor [1-7,9-12]; <8> cytochrome c of yeast and cytochrome c of tuna act as electron acceptors [5]; <1,3,7-9,12> mammalian cytochrome c acts as electron acceptor [4,5,7-9,14]; <10> only native cytochrome c acts as electron acceptor [10]; <3> cytochrome $c_{553.5}$ acts as electron acceptor [9]; <3> cytochrome c_{557} becames reduced when mixed with thiosulfate, enzyme and a trace of cytochrome $c_{553.5}$, the omission of either the enzyme or cytochrome $c_{553.5}$ prevents this reduction [9]; <1-12,14> ferricyanide acts as electron acceptor [1-5,7-12,13,14,16,17]; <2> the enzyme is active only with ferricyanide, not with native or mammalian cytochrome c [8,13]; <3,11> strictly specific for thiosulfate [11,16]; <8> the enzyme rapidly reduces high-potential non-heme iron protein in the presence of thiosulfate [5]; <9> A-50: once induced to oxidize thiosulfate, also reduces tetrathionate to thiosulfate in the presence of an electron donor such as lactate, the possibility cannot be excluded, that a separate tetrathionate-reducing enzyme is induced [7]; <1> ferricyanide is a much better electron acceptor than cytochrome c [12]; <12> Wurster's blue can also serve as electron acceptor [14]; <12> phenazine methosulfate and phenazine ethosulfate assayed in the presence of dichlorophenolindophenol and Thialhalivibrio versutus cytochrome c_{550} support low, but significant enzyme activities [14]) (Reversibility: ir <2, 3> [9, 13]; r <9> [7]; ? <1, 3, 4, 5-8, 10-12, 14> [1-8, 10-12, 14, 16, 17]) [1-14, 16, 17]

P tetrathionate + reduced acceptor <1-12, 14> [1-14, 16, 17]

S Additional information <1, 3, 8, 11, 12> (<8> no activity with: cytochrome c_{552}, cytochrome c', cytochrome c_{553} [5]; <1,3,11,12> NAD^+, methylene blue, 2,6-dichlorophenol indophenol [11,12,14,16]; <1,12> no activity with: $NADP^+$, O_2 [12,14]; <11> no activity with FAD [11]; <12> no activity with: pyrrolo-quinoline quinone, methyl- or benzyl viologen, methylthiazol tetrazolium bromide [14]; <3> no activity with: benzyl viologen, 2,6-dichloroindophenol, 2,6-dichloroindophenol with phenazine methosulfate, 2,6-dichloroindophenol with phenazine ethosulfate, horse heart cytochrome c with phenazine methosulfate, horse heart cytochrome c with phenazine ethosulfate [16]) [5, 11, 12, 14, 16]

P ?

Inhibitors

2,2'-bipyridyl <1> (<5> not inhibitory at 5 mM [2]; <9> not inhibitory [7]; <1> 50% inhibition at 1 mM, a 60 min incubation period is necessary to potentiate the inhibition [12]) [12]

EDTA <1, 7, 11> (<7> 13% inhibition at 1 mM, 3% inhibition at 0.1 mM [4]; <11> 18% inhibition at 1 mM and 46% inhibition at 20 mM [11]; <1> 44% inhibition at 10 mM, a 15 min incubation period is necessary to potentiate the inhibition [12]; <3,12> not inhibitory [14,16]) [4, 11, 12]

$HgCl_2$ <1, 2, 5, 7, 11> (<5> 100% inhibition at 0.5 mM, 45% inhibition at 0.25 mM and 15% inhibition at 0.1 mM [2]; <7> 37% inhibition at 0.1 mM, 13% inhibition at 0.01 mM [4]; <11> 100% inhibition at 1 mM [11]; <1> 100% inhibition at 0.5 mm [12]; <2> 73.3% inhibition at 1 mM [13]) [2, 4, 11-13]

N-ethylmaleimide <1, 5, 7, 11> (<5> 12% inhibition at 1 mM [2]; <7> 40% inhibition at 1 mM, 15% inhibition at 0.1 mM [4]; <3,9,12> not inhibitory [7,14,16]; <11> 78% inhibition at 1 mM [11]; <1> 100% inhibition at 1 mM, a 15 min incubation period is necessary to potentiate the inhibition [12]) [2, 4, 11, 12]

Na_2SeO_3 <3, 12> (<12> 39% inhibition at 0.5 mM [14]; <3> it inhibits the oxidation of thiosulfate by whole cells [16]) [14, 16]

Na_2SeO_4 <12> (<12> 51% inhibition at 5 mM [14]) [14]

SO_3^{2-} <1-3, 5, 7, 8, 9, 11, 12> (<7> 75% inhibition at 1 mM, 48% inhibition at 0.1 mM, 33% inhibition at 0.01 mM [4]; <8> 50% inhibition at 1.7 mM [5]; <1> strongly inhibits, the inhibition is time-dependent and irreversible, 50% inhibition at 0.005 mM in 5 min [8]; <11> 62% inhibition at 3 mM [11]; <1> 50% inhibition at 0.005 mM , time-dependent inhibition, irreversible [12]; <12> 100% inhibition at 0.5 mM [14]; <3> inhibits at concentrations above 0.05 mM, the sulfite inhibition of thiosulfate oxidation by whole cells is very similar to the inhibition pattern of the enzyme [16]) [2, 4, 5, 7, 8, 11, 12, 14, 16]

SO_4^{2-} <7> (<7> 33% inhibition at 1 mM, 15% inhibition at 0.1 mM, 7% inhibition at 0.01 mM [4]) [4]

azide <2, 9, 12> (<9> strongly inhibits thiosulfate oxidation by A-50 extracts [7]; <3> not inhibitory at 10 mM [9]; <2> 60% inhibition at 10 mM [13]; <12> 63% inhibition at 0.5 mM [14]) [7, 13, 14]

cupferron <9> (<9> strongly inhibits thiosulfate oxidation by A-50 extracts [7]) [7]

cyanide <2, 8, 9> (<8> 50% inhibition at 1.3 mM [5]; <9> strongly inhibits thiosulfate oxidation by A-50 extracts [7]; <2> 28.9% inhibition at 10 mM [13]; <3> not inhibitory [16]) [5, 7, 13]
deoxycholate <5> [2]
diethyldithiocarbamate <9> (<9> slightly inhibits [7]) [7]
dithionate <7> (<7> 18% inhibition at 1 mM, 15% inhibition at 0.1 mM and 0.01 mM [4]) [4]
o-phenantroline <5> (<5> 17% inhibition at 5 mM [2]) [2]
p-chloromercuribenzoate <1, 7, 11, 12> (<7> 47% inhibition at 1 mM, 7% inhibition at 0.1 mM [4]; <9> not inhibitory [7]; <11> 90% inhibition at 1 mM [11]; <1> 72.8% inhibition at 1 mM [12]; <12> 32% inhibition at 0.1 mM [14]) [4, 11, 12, 14]
p-hydroxymercuribenzoate <5> (<5> 50% inhibition at 1 mM, 40% inhibition at 0.5 mM and 10% inhibition at 0.1 mM [2]) [2]
thiocyanate <4> (<4> slightly inhibits [6]) [6]

Cofactors/prosthetic groups

heme c <3, 12, 14> (<12> 5.3 mole per mol of native enzyme, made up of two different c_{553} hemes, present in both subunit types [14,15]; <3> 4 mol per mol of enzyme, present in both subunit types [16]) [14-17]

Metals, ions

iron <1> (<1> enzyme contains 2 mol of non-heme iron [12]) [12]

Turnover number (min^{-1})

138000 <3> (thiosulfate) [16]
510000 <12> (thiosulfate, <12> at pH 7 [14]) [14]
15000000.0 <12> (thiosulfate, <12> at least, at pH 3 [14]) [14]

Specific activity (U/mg)

0.06 <14> [17]
1 <4> (<4> cytoplasm [1]) [1]
2 <4> (<4> membrane [1]) [1]
9.13 <6> (<6> partially purified enzyme [3]) [3]
9.6 <4> (<4> periplasm [1]) [1]
11.38 <10> (<10> purified enzyme [10]) [10]
146.1 <1> [12]
316 <4> (<4> partially purified enzyme [1]) [1]
1149 <3> (<3> purified enzyme [16]) [16]
4925 <12> (<12> purified enzyme [14]) [14]
6010 <11> [11]
Additional information <2, 3, 11> (<11> specific activities in crude cell-free extracts of mercaptosuccinate-grown Alcaligenes under a variety of nutritional conditions [11]) [9, 11, 13]

K_m-Value (mM)

0.004 <4> (thiosulfate, <4> with cytochrome c as electron acceptor [1]) [1]
0.005 <1> (thiosulfate, <1> cytochrome c as electron acceptor [12]) [12]
0.1 <1> (thiosulfate, <1> ferricyanide as electron acceptor [12]) [12]

0.11 <4> (thiosulfate, <4> with ferricyanide as electron acceptor [1]) [1]
0.12 <3> (thiosulfate, <3> at pH 7.5 [16]) [16]
0.13 <8> (high-potential non-heme iron protein) [5]
0.14 <5> (thiosulfate, <5> with native cytochrome c as electron acceptor [2]) [2]
0.16 <5> (thiosulfate, <5> with ferricyanide as electron acceptor [2]) [2]
0.34 <5> (ferricyanide) [2]
0.54 <12> (thiosulfate, <12> at pH 7 [14]) [14]
0.62 <6> (thiosulfate) [3]
0.67 <7> (thiosulfate) [4]
0.81 <14> (thiosulfate) [17]
0.9 <6> (ferricyanide) [3]
0.9 <2> (thiosulfate) [13]
1.1 <7> (ferricyanide) [4]
1.57 <10> (thiosulfate, <10> pH 6.1-6.3, 30°C [10]) [10]
3.3 <8> (ferricyanide) [5]
Additional information <1, 2, 4, 7-10> (<10> temperature dependency of K_m [10]) [4-6, 7, 10, 12, 13]

K_i-Value (mM)

0.03 <3> (Na_2SeO_3, <3> it is possible that selenite blocks access to the thiosulfate dehydrogenase or has an inhibiting effect on the respiratory chain itself [16]) [16]
0.3 <3> (SO_3^{2-}, <3> sulfite resembles the sulfonate part of thiosulfate and might, therefore, block the active site of the enzyme [16]) [16]

pH-Optimum

3 <12> (<12> purified enzyme [14]) [14]
3.5 <14> [17]
4-6 <6> [3]
4.5 <4, 13> [15]
5 <1-3, 11> (<1,11> below, using ferricyanide as electron acceptor [11,12]; <3> the rate of reduction of cytochrome $c_{553.5}$ at pH 5 is about 10% of the rate at pH 7 [9]) [9, 11, 12, 13]
5.2 <9> (<9> thiosulfate + ferricyanide [7]) [7]
5.5 <3> [16]
5.5-6.5 <7> [4]
6.5 <1> (<1> broad, with cytochrome c as electron acceptor [12]) [12]
7 <3> (<3> for the reduction of cytochrome $c_{553.5}$ [9]) [9]
7.2-7.6 <11> (<11> cytochrome c as electron acceptor [11]) [11]
7.8 <5> [2]
Additional information <10> (<10> no definite pH-optimum [10]) [10]

pH-Range

4.5-9 <7> [4]
5-9 <11> [11]
5.5-8.8 <1> (<1> little change in activity between [12]) [12]

Temperature optimum (°C)

30 <5, 10> [2, 10]
30-42 <7> [4]
35 <14> [17]
40 <11> [11]

Temperature range (°C)

22-48 <11> (<11> about 50% of activity maximum at 22°C and 48°C [11]) [11]

4 Enzyme Structure

Molecular weight

27000 <14> (<14> gel filtration [17]) [17]
90000 <11> (<11> gel filtration [11]) [11]
102000 <12> (<12> gel filtration [14]) [14, 15]
115000 <1> (<1> sucrose density gradient method [12]) [12]
120000 <3> (<3> gel filtration [16]) [16]
138000 <4> (<4> gel filtration [1]) [1, 15]
180000 <6> (<6> gel filtration [3]) [3]

Subunits

heterotetramer <3, 12> (<12> 2 * 24000 + 2 * 20000, SDS-PAGE, both subunits contain c_{553}-type heme [14,15]; <3> 2 * 33000 + 2 * 27000, SDS-PAGE [16]) [14-16]
monomer <14> (<14> 1 * 27900, SDS-PAGE [17]) [17]
trimer <4> (<4> 3 * 45000, SDS-PAGE [1,15]) [1, 15]

5 Isolation/Preparation/Mutation/Application

Localization

periplasm <3, 4, 14> (<3,4> at least 73% of the enzyme activity is recovered in the periplasmic fraction [1,16]) [1, 6, 15-17]

Purification

<1> (using sonication, column chromatography on DEAE-cellulose, pH-adjustment to 4.5 and calcium phosphate gel treatment [12]) [12]
<2> (using heat-treatment, 5-17% ethanol precipitation and CM-cellulose treatment [13]) [13]
<3> (using chromatography on Phenyl-Sepharose column and Mono Q column, gel filtration on Superdex 200 and column chromatography on Phenyl Superose [16]) [16]
<4> (partial, using ion exchange chromatography, gel filtration and hydrophobic interaction chromatography [1,15]) [1, 15]

<6> (partial, using ultracentrifugation, ammonium sulfate fractionation, column chromatography on DEAE-cellulose and gel filtration on Sephacryl S-200 [3]) [3]
<7> (partial, using ultracentrifugation, ultrafiltration, column chromatography on DEAE-Sephadex A-25 and gel filtration on Sephadex S-200 [4]) [4]
<8> (partial [5]) [5]
<10> (partial, using ion-exchange chromatography and calcium phosphate gel treatment [10]) [10]
<11> (using $MnCl_2$ precipitation, treatment with calcium phosphate gel, and successive column chromatography on DEAE-Sephadex and Sephadex G-150 [11]) [11]
<12> (using ammonium sulfate precipitation, hydrophobic interaction chromatography, anion-exchange chromatography and gel filtration [14,15]) [14, 15]
<14> (using chromatography on Phenyl-Sepharose column, CM-cellulose column and hydroxylapatite column [17]) [17]

6 Stability

pH-Stability

2 <12> (<12> unstable below [14]) [14]
4 <2> (<2> unstable below [13]) [13]
5-7 <14> [17]

Temperature stability

0 <12> (<12> 20% decrease of activity in 2 h and 60% decrease of activity in 7 h [14]) [14]
4 <1> (<1> 6 h, 10% loss of activity, 50% loss of activity after 5 days [12]) [12]
25 <11> (<11> 30 min, 7% loss of activity [11]) [11]
30 <12> (<12> 50% decrease of activity in 2 h [14]) [14]
45 <3, 11, 14> (<11> 30 min, complete loss of activity [11]; <3> 3 min, pH 7, complete loss of activity [9]; <14> stable up to [17]) [9, 11, 17]
55 <1> (<1> 5 min, stable [12]) [12]
55 <5> (<5> 30% decrease of activity after 5 min [2]) [2]
60 <5> (<5> 50% decrease of activity after 7 min [2]) [2]
60 <1, 3, 9> (<9> 10 min, complete loss of activity, crude extract [7]; <1> 5 min, 50% loss of activity [12]; <3> enzyme activity completely destroyed after 3 min at pH 7 [9]) [7, 9, 12]
100 <1, 2, 5, 8, 10> (<1,8> 1 min, complete loss of activity [5,12]; <5> after 2 min the activity is completely destroyed [2]; <10> after 1 min the activity is completely destroyed [10]; <2> complete loss of activity [13]) [2, 5, 10, 12, 13]

General stability information

<4>, lability of enzyme during purification and storage at -20°C [1]

Storage stability

<1>, -20°C, the ability of the enzyme to couple with cytochrome c decreases during storage [12]
<1>, 4°C, 10% loss of activity after 6 h and 50% loss of activity after 5 days [12]
<1>, frozen, stable, indefinitely as measured by ferricyanide reduction [12]
<2>, -20°C, little loss of activity after 2 months [13]
<2>, diluted enzyme loses 10-15% of the activity each time it is frozen and thawed [13]
<3>, -25°C, 0.02 M phosphate, 40% loss of activity after 2 months [9]
<4>, 20-30% loss of activity after freezing at -20°C and thawing [1]
<5>, -20°C, the activity of cell-free extracts is stable for several days in the presence of 20% v/v glycerol, without glycerol most of the activity is lost after 2 days [2]
<5>, -20°C, the activity of whole cells is stable for 3 months [2]
<5>, 4°C, cell-free extracts lose most of the activity after 2 days even in the presence of glycerol [2]
<5>, the activity in cell-free extracts is lost very rapidly after freezing and thawing [2]
<7>, 4°C, 50% loss of activity after 3 days, 75% loss of activity after 14 days [4]
<9>, -20°C, crude extract, 50% loss of activity after 2 months [7]
<10>, -20°C, 36% loss of activity after 9 days [10]
<10>, 4°C, stable, 2 days [10]
<11>, 4°C, purified enzyme, 10% loss of activity after 24 h, completely inactive after 7 days [11]
<11>, frozen, 0.05 M phosphate buffer, pH 6.2, stable for at least 90 days [11]
<12>, frozen, stable for at least 3 months [14]

References

[1] Lu, W.P.; Kelly, D.P.: Cellular location and partial purification of the 'thiosulphate-oxidizing enzyme' and the 'trithionate hydrolase' from Thiobacillus tepidarius. J. Gen. Microbiol., **134**, 877-885 (1988)
[2] Kurek, E.J.: Properties of an enzymatic complex active in sulfite and thiosulfate oxidation by Rhodotorula sp.. Arch. Microbiol., **143**, 277-282 (1985)
[3] Then, J.; Truper, H.G.: The role of thiosulfate in sulfur metabolism of Rhodopseudomonas globiformis. Arch. Microbiol., **130**, 143-146 (1981)
[4] Schook, L.B.; Berk, R.S.: Partial purification and characterization of thiosulfate oxidase from Pseudomonas aeruginosa. J. Bacteriol., **140**, 306-308 (1979)
[5] Fukumori, Y.; Yamanaka, T.: A high potential nonheme iron protein (HiPIP)-linked, thiosulfate-oxidizing enzyme derived from Chromatium vinosum. Curr. Microbiol., **3**, 117-120 (1979)

[6] Lu, W.P.; Kelly, D.P.: Kinetic and energetic aspects of inorganic sulphur compound oxidation by Thiobacillus tepidarius. J. Gen. Microbiol., **134**, 865-876 (1988)
[7] Trudinger, P.A.: Metabolism of thiosulfate and tetrathionate by heterotrophic bacteria from soil. J. Bacteriol., **93**, 550-559 (1967)
[8] Suzuki, I.: Mechanisms of inorganic oxidation and energy coupling. Annu. Rev. Microbiol., **28**, 85-101 (1974)
[9] Trudinger, P.A.: Thiosulfate oxidation and cytochromes in Thiobacillus X. Biochem. J., **78**, 680-686 (1961)
[10] Tuttle, J.H.; Schwartz, J.H.; Whited, G.M.: Some properties of thiosulfate-oxidizing enzyme from marine heterotroph 16B. Appl. Environ. Microbiol., **46**, 438-445 (1983)
[11] Hall, M.R.; Berk, R.S.: Thiosulfate oxidase from an Alcaligenes grown on mercaptosuccinate. Can. J. Microbiol., **18**, 235-245 (1972)
[12] Lyric, R.M.; Suzuki, I.: Enzymes involved in the metabolism of thiosulfate by Thiobacillus thioparus. 3. Properties of thiosulfate-oxidizing enzyme and proposed pathway of thiosulfate oxidation. Can. J. Biochem., **48**, 355-363 (1970)
[13] Silver, M.; Lundgren, D.G.: The thiosulfate-oxidizing enzyme of Ferrobacillus ferrooxidans (Thiobacillus ferrooxidans). Can. J. Biochem., **46**, 1215-1220 (1968)
[14] Meulenberg, R.; Pronk, J.T.; Hazeu, W.; van Dijken, J.P.; Frank, J.; Bos, P.; Kuenen, J.G.: Purification and partial characterization of thiosulfate dehydrogenase from Thiobacillus acidophilus. J. Gen. Microbiol., **139**, 2033-2039 (1993)
[15] Kelly, D.P.; Wood, A.P.: Enzymes involved in microbiological oxidation of thiosulfate and polythionates. Methods Enzymol., **243**, 501-510 (1994)
[16] Visser, J.M.; De Jong, G.A.H.; Robertson, L.A.; Kuenen, J.G.: Purification and characterization of a periplasmic thiosulfate dehydrogenase from the obligately autotrophic Thiobacillus sp. W5. Arch. Microbiol., **166**, 372-378 (1996)
[17] Nakamura, K.; Nakamura, M.; Yoshikawa, H.; Amano, Y.: Purification and properties of thiosulfate dehydrogenase from Acidithiobacillus thiooxidans JCM7814. Biosci. Biotechnol. Biochem., **65**, 102-108 (2001)

1.8.2.3

1 Nomenclature

EC number

1.8.2.3 (deleted)

Recommended name

Recommended Name never specified

Sulfite oxidase 1.8.3.1

1 Nomenclature

EC number
1.8.3.1

Systematic name
sulfite:oxygen oxidoreductase

Recommended name
sulfite oxidase

Synonyms
oxidase, sulfite
sulphite oxidase cytochrome b_9

CAS registry number
9029-38-3

2 Source Organism

<1> *Vertebrates* (immunochemical comparison [7]) [6, 7]
<2> *Gallus gallus* [1-5, 13, 20-22, 25, 29, 30, 35, 36, 45, 46, 48, 50, 51]
<3> *Rattus norvegicus* [7-10, 12-21, 23, 24, 28, 31-34, 39, 41, 49]
<4> *Mus musculus* [12]
<5> *Merluccius productus* (pacific hake [11]) [11]
<6> *Bos taurus* [20, 26-28]
<7> *Oryctolagus cuniculus* [20]
<8> *Sus scrofa* [20]
<9> *Canis familiaris* [28]
<10> *Malva sylvestris* [37]
<11> *Spinacia oleracea* [38]
<12> *Homo sapiens* (expression in Escherichia coli [47]) [40-42, 44, 47]
<13> *Arabidopsis thaliana* [43]
<14> *Nicotiana tabacum* [43]

3 Reaction and Specificity

Catalyzed reaction
sulfite + O_2 + H_2O = sulfate + H_2O_2 (<2> kinetics and proposed mechanism [30,46]; <2,3> mechanism and kinetics of electron transfer [35,39,48]; <2,6> ping-pong mechanism [25,26]; <2> X-ray absorption spectroscopy of oxida-

tion states [29]; <3> this direct reduction of O_2 is prevented completely in presence of cytochrome c [19]; <2> intramolecular electron transfer and effect of solution viscosity [45])

Reaction type

redox reaction

Natural substrates and products

S sulfite + cytochrome c <1-3, 5, 12> (<3> natural acceptor [24]; <2> catalytic cycle [30]; <1,5> detoxification [6,11]; <12> genetic deficiency results in neurological abnormities [40]) (Reversibility: ? <1-3, 5, 12> [6, 11, 24, 30, 40]) [6, 11, 24, 30, 40]

P sulfate + reduced cytochrome c

Substrates and products

S sulfite + H_2O + A <2-4, 6, 9-11> (<2-4,6> A: electron acceptor, i.e. O_2, cytochrome c, $K_3[Fe(CN)_6]$, 2,6-dichloroindophenol, methylene blue, highly specific for sulfite as electron donor [12,19,24-26]; <3> H_2O_2 acceptor only when respiratory chain is inhibited [19]; <3> artificial A: tetramethylphenylenediamine, 2,6-dichloroindophenol, methylene blue [39]) (Reversibility: ? <2-4, 6, 9-11> [12, 19, 24-26, 28, 37, 38]) [12, 19, 24-26, 28, 37, 38, 39]

P SO_4^{2-} + AH_2 (reduced acceptor, i.e. H_2O_2, ferrocytochrome c, $K_4[Fe(CN)_6]$, reduced 2,6-dichloroindophenol, reduced methylene blue)

Inhibitors

1-ethyl-3-(3-dimethylaminopropyl)carbodiimide hydrochloride <2> (<2> EDC [4]) [4]

2,6-dichloroindophenol <6> (<6> inhibition of O_2 consumption [26]) [26]

Ag^+ <3, 6, 9> [28]

CN^- <2, 3-8> (<5> more than 10 mM [11]; <3,4> profound at low O_2 concentration, not at high O_2 concentration [12]; <3> mechanism of inactivation [33]) [2, 11, 12, 20, 26, 33]

Hg^{2+} <3, 6, 9> [28]

K_2HPO_4 <2> (<2> at 26 mM 50% inhibition if cytochrome c or ferricyanide is electron acceptor [22]) [22]

K_2SO_4 <2> (<2> at 22 mM 50% inhibition if cytochrome c or ferricyanide is electron acceptor [22]) [22]

KCl <2> (<2> at 95 mM 50% inhibition if cytochrome c or ferricyanide is electron acceptor [22]) [22]

KF <2> (<2> at 72 mM 50% inhibition if cytochrome c or ferricyanide is electron acceptor [22]) [22]

KNCS <2> (<2> at 57 mM 50% inhibition if cytochrome c or ferricyanide is electron acceptor [22]) [22]

KNO_3 <2> (<2> at 78 mM 50% inhibition if cytochrome c or ferricyanide is electron acceptor [22]) [22]

N-bromosuccinimide <2, 3> (<2,3> 94% inhibition at 0.1 mM [21]) [21]

N-cyclohexyl-N'-[2-(N-methylmorpholino)-ethyl]carbodiimide *p*-toluene sulfonate <2> (<2> CMC [3]) [3]

N-ethyl-5-phenylisoxazolium-3'-sulfonate <2> (<2> Woodward's reagent K [4]) [4]
NaCl <2, 13> (<2> at 100 mM 50% inhibition if cytochrome c or ferricyanide is electron acceptor [22]; <13> 50% inhibition at 70 mM, in Tris/HCl 20 mM, pH 8.5 [43]) [43, 22]
Tris/HCl <13> (<13> 50% inhibition at 90 mM, pH 8.5 [43]) [43]
arsenate <12> (<12> 100 mM, EPR spectra [42]) [42]
arsenite <2, 3, 6-8> [20, 21, 26]
cytochrome c <6> (<6> inhibition of O_2 consumption [26]) [26]
diethylpyrocarbonate <2> (<2> modifies ten His per enzyme molecule [4]) [4]
ferricyanide <6> (<6> inhibition of O_2 consumption [26]; <2> irreversible inactivation of molybdenum center [21]) [21, 26]
heavy metal ions <3, 5, 6, 9> [11, 28]
mannitol <6> (<6> only with O_2 as electron acceptor [26]) [26]
methylene blue <2, 3> (<2,3> 70% inhibition at 0.4 mM [21]) [21]
p-chloromercuribenzoate <3, 6, 9> [28]
phosphate <12> (<12> 100 mM, EPR spectra [42]) [42]
potassium nitrate <13> (<13> 50% inhibition at 1 mM, in Tris/HCl 20 mM, pH 8.5 [43]) [43]
potassium phosphate <13> (<13> 50% inhibition at 30 mM, pH 8.5 [43]) [43]
sulfate <5> [11]
Additional information <3, 6> (<3,6> not inhibitory: CN^- [17,28]; <3> not inhibitory: NaN_3 at 0.5 mM, NaCN at 0.5 mM [24]) [17, 24, 28]

Cofactors/prosthetic groups
heme <2, 3, 5-8, 12> [2, 11, 20, 25, 27, 40]
Additional information <13> (<13> no heme domain [43]) [43]

Activating compounds
cytochrome c <3, 4> (<3,4> at lower oxygen concentrations [12]) [12]
di-(carboxamidomethyl)molybdopterin <2, 3> [5, 13, 14]
glycine <6, 9> [28]
sodium deoxycholate <3> (<3> at 0.04%, acceptor: cytochrome c [24]) [24]

Metals, ions
Mo <2, 3, 5-8, 12, 13> (<2,3> molybdohemoprotein, structure of molybdopterin [5,13,14]; <13> molybdopterin [43] <2,3> X-ray absorption spectroscopy of oxidation states [29,49]; <12> proposed structure of molybdenum center [41]; <2> modeling of active site, kinetics [50]) [5, 11, 20, 23, 25, 29, 40, 41, 43, 49, 50]

Turnover number (min^{-1})
5100 <2> (cytochrome c) [3]

Specific activity (U/mg)
35.4 <2> [2]
75 <3> [17]
90 <1, 3> [7]

772 <14> [43]
Additional information <1-3, 5-8> (<1,3> comparison of activity in different cells and tissues [6,28]; <3,5,6> activity in different organelles [11,23,28]) [6, 11, 20, 23, 28]

K_m-Value (mM)
0.00092 <2> (cytochrome c) [3]
0.002 <1-3> (cytochrome c) [7, 12, 16, 20, 25]
0.006 <3> (cytochrome c) [17]
0.015 <3, 4> (O_2) [12]
0.034 <13> (sulfite) [43]
0.051 <14> (sulfite) [43]
0.17 <3> (sulfite) [17]
0.39 <4> (sulfite) [12]
0.39 <6> (sulfite) [20]
5.33 <10> (sulfite) [37]
Additional information <2, 3, 6, 9> (<2> kinetic studies [30]; <2> variation of K_M with pH [46]) [28, 30, 46]

pH-Optimum
7.5 <3, 4> [12]
7.6 <11> [38]
8.5 <6, 9> [28]
8.6 <6> (<6> with O_2 as acceptor [20]) [20, 26]
8.7 <5> [11]
9 <1, 3> [7]

pH-Range
7 <5> (<5> below: less than 50% of maximal activity [11]) [11]
7-8.5 <11> [38]

Temperature range (°C)
45-55 <11> [38]

4 Enzyme Structure

Molecular weight
27000 <10> (<10> gel filtration [37]) [37]
40000 <13> (<13> gel filtration, predicted amino acid sequence [43]) [43]
115000-120000 <3, 5, 6> (<5> gel filtration [11]; <3> sedimentation equilibrium centrifugation [16]; <6> sedimentation equilibrium centrifugation, gel filtration [27]) [11, 16, 27]

Subunits
? <2, 14> (<2> ? * 55000, SDS-PAGE, heterogenous behaviour in sedimentation equilibrium experiments [25]; <14> ? * 46000, SDS-PAGE [43]) [25, 43]
dimer <2, 3, 5, 6> (<2> 2 * 50545, amino acid sequence [1]; <5> 2 * 55000-60000, SDS-PAGE [11]; <3> 2 * 60000, SDS-PAGE [16]; <3> 56000 + 46500,

SDS-PAGE [7]; <3> SDS-PAGE [23]; <6> 2 * 55000, SDS-PAGE [27]) [1, 11, 16, 23, 25, 27]
monomer <13> (<13> 1 * 40000, gel filtration, predicted amino acid sequence [43]) [43]
Additional information <3> (<3> after tryptic cleavage: 9500 Da heme-containing fragment, gel filtration with guanidine-HCl, 47400 Da molybdenum-containing fragment, SDS-PAGE [18]; <3> study of domains [16]; <3> study on precursor and processing [8-10]) [8, 18]

5 Isolation/Preparation/Mutation/Application

Source/tissue

brain <3, 6, 9> (low activity [28]) [28]
heart <3, 6, 9> [28]
kidney <3, 6, 9> [28]
leaf <10> [37]
lung <3, 6, 9> (low activity [28]) [28]
spleen <3, 6, 9> (low activity [28]) [28]
testis <3, 6, 9> (low activity [28]) [28]
Additional information <1-3, 5, 7-8> (<1,3> activity in tissues and cells [6,28]; <5> distribution in fish organs [11]) [6, 11, 14, 20, 23, 28]

Localization

chloroplast <11> [38]
cytosol <3> (<3> precursor enzyme [8-10]) [8-10, 17, 34]
microsome <3> (<3> 10% of total activity, inside vesicles [17]) [17]
mitochondrion <2, 3, 5-9> (<3> intermembrane space [8,9,19,24]) [3, 8, 11, 17, 20, 23, 24, 28]
peroxisome <13, 14> [43]
thylakoid <11> [38]
Additional information <3> (<3> subcellular localisation [17,23,24]) [17, 23, 24]

Purification

<2> (<2> 70-77% pure [2]) [2, 3, 25, 36]
<5> (<5> partial [11]) [11]
<6> (<6> 75% pure [26]) [26-28]
<10> [37]
<11> (<11> partial [38]) [38]
<12> (<12> recombinant protein from Escherichia coli [47]) [47]
<13> [43]
<14> [43]
<1-3, 6-9> (<3> isolation of heme-and molybdenum-containing fragments [18]) [7, 17, 18, 20, 23, 28]

Renaturation

<3> [31, 32]
<12> (<12> in vitro insertion of molybdopterin into aposulfite oxidase and conversion of molybdopterin into molybdenum cofactor [44]) [44]

Crystallization

<3> (<3> isolated molybdenum- and heme-fragments [18]) [18]

Cloning

<13> [43]

Engineering

C207S <12> (<12> C207 essential for enzyme activity, probably as ligand of Mo [41]) [41]
C207S <3> (<3> C207 essential for enzyme activity [41]) [41]
C242S <3> (<3> silent mutation [41]) [41]
C260S <3> (<3> silent mutation [41]) [41]
C451S <3> (<3> silent mutation [41]) [41]
R160Q <12> (<12> sulfite-oxidase deficient patient [40]) [40]
Additional information <12> (<12> optimized expression in Escherichia coli, untagged and His-tagged enzyme, expression in presence of tungstate [47]) [47]

Application

analysis <2> (<2> electroimmobilisation into polypyrrole film, use for amperometric detection of sulfite [51]) [51]

6 Stability

pH-Stability

7-9.5 <3, 6, 9> (<3,6,9> quite stable, but unstable below pH 7.0 [28]) [28]

Temperature stability

25 <5> (<5> 10 min stable [11]) [11]
50 <5> (<5> rapid inactivation [11]) [11]
52 <6> (<6> inactivation, protection by sulfate [26]) [26]
54 <3> (<3> reduced form has higher stability than oxidized form [15]) [15]

General stability information

<3, 6, 9>, trypsin inactivates [28]

Storage stability

<2>, -80°C, several weeks [4]

References

[1] Neame, P.J.; Barber, M.J.: Conserved domains in molybdenum hydroxylases. The amino acid sequence of chicken hepatic sulfite oxidase. J. Biol. Chem., **264**, 20894-20901 (1989)

[2] Kipke, C.A.; Enemark, J.H.; Sunde, R.A.: Purification of prosthetically intact sulfite oxidase from chicken liver using a modified procedure. Arch. Biochem. Biophys., **270**, 383-390 (1989)

[3] Ritzman, M.; Bosshard, H.R.: Sulfite oxidase from chicken liver. Further characterization of the role of carboxyl groups in the reaction with cytochrome c. Eur. J. Biochem., **172**, 377-381 (1988)

[4] Ritzmann, M.; Bosshard, H.R.: Sulfite oxidase from chicken liver. The role of imidazole and carboxyl groups for the reaction with cytochrome c. Eur. J. Biochem., **159**, 493-497 (1986)

[5] Kramer, S.P.; Johnson, J.L.; Ribeiro, A.A.; Millington, D.S.; Rajagopalan, K.V.: The structure of the molybdenum cofactor. Characterization of di-(carboxamidomethyl)molybdopterin from sulfite oxidase and xanthine oxidase. J. Biol. Chem., **262**, 16357-16363 (1987)

[6] Beck-Speier, I.; Hinze, H.; Hozer, H.: Effect of sulfite on the energy metabolism of mammalian tissues in correlation to sulfite oxidase activity. Biochim. Biophys. Acta, **841**, 81-89 (1985)

[7] Kuwahara, T.; Yoshimoto, I.; Ito, A.: Comparative immunochemical studies of sulfite oxidases of vertebrate livers. J. Biochem., **92**, 1925-1931 (1982)

[8] Ono, H.; Ito, A.: Evidence for participation of the inner membrane in the import of sulfite oxidase into the intermembrane space of liver mitochondria. Biochem. Biophys. Res. Commun., **107**, 258-264 (1982)

[9] Mihara, K.; Omura, T.; Harano, T.; Brenner, S.; Fleischer, S.; Rajagopalan, K.V.; Blobel, G.: Rat liver L-glutamate dehydrogenase, malate dehydrogenase, D-β-hydroxybutyrate dehydrogenase, and sulfite oxidase are each synthesized as larger precursors by cytoplasmic free polysomes. J. Biol. Chem., **257**, 3355-3358 (1982)

[10] Ono, H.; Ito, A.: Biosynthesis of sulfite oxidase in rat liver - presence of a large precursor form of the enzyme in cytosol. J. Biochem., **91**, 117-123 (1982)

[11] Onoue, Y.: Sulfite oxidase from Merluccius productus. Biochim. Biophys. Acta, **615**, 48-58 (1980)

[12] Shibuya, A.; Horie, S.: Studies on the composition of the mitochondrial sulfite oxidase system. J. Biochem., **87**, 1773-1784 (1980)

[13] Johnson, J.L.; Hainline, B.E.; Rajagopalan, K.V.: Characterization of the molybdenum cofactor of sulfite oxidase, xanthine, oxidase, and nitrate reductase. Identification of a pteridine as a structural component. J. Biol. Chem., **255**, 1783-1786 (1980)

[14] Berg, J.M.; Hodgson, K.O.; Cramer, S.P.; Corbin, J.L.; Elsberry, A.; Pariyadath, N.; Stiefel, E.I.: Structural results relevant to the molybdenum sites in xanthine oxidase and sulfite oxidase. Crystal structures of MoO_2L,

L=$(SCH_2CH_2)_2NCH_2CH_2X$ with X=SCH_3, $N(CH_2)_2$. J. Am. Chem. Soc., **101**, 2774-2776 (1979)
[15] Southerland, W.M.; Rajagopalan, K.V.: Domain interactions in oxidized and reduced forms of rat liver sulfite oxidase. J. Biol. Chem., **253**, 8753-8758 (1978)
[16] Southerland, W.M.; Winge, D.R.; Rajagopalan, K.V.: The domains of rat liver sulfite oxidase. Proteolytic separation and characterization. J. Biol. Chem., **253**, 8747-8752 (1978)
[17] Ito, A.; Kuwahara, T.; Mitsunari, Y.; Omura, T.: Distribution of hepatic sulfite oxidase among subcellular organelles and its intraorganelle localization. J. Biochem., **81**, 1531-1541 (1977)
[18] Johnson, J.L.; Rajagopalan, K.V.: Tryptic cleavage of rat liver sulfite oxidase. Isolation and characterization of molybdenum and heme domains. J. Biol. Chem., **252**, 2017-2025 (1977)
[19] Oshino, N.; Chance, B.: The properties of sulfite oxidation in perfused rat liver; interaction of sulfite oxidase with the mitochondrial respiratory chain. Arch. Biochem. Biophys., **170**, 514-528 (1975)
[20] Bray, R.C.: Molybdenum iron-sulfur flavin hydroxylases and related enzymes. The Enzymes, 3rd Ed. (Boyer, P.D., ed.), **12**, 299-419 (1975)
[21] Kessler, D.L.; Rajagopalan, K.V.: Hepatic sulfite oxidase. Identification of the molybdenum center as the site of irreversible inactivation by ferricyanide. Biochim. Biophys. Acta, **370**, 399-409 (1974)
[22] Kessler, D.L.; Rajagopalan, K.V.: Hepatic sulfite oxidase. Effect of anions on interaction with cytochrome c. Biochim. Biophys. Acta, **370**, 389-398 (1974)
[23] Kessler, D.L.; Johnson, J.L.; Cohen, H.J.; Rajagopalan, K.V.: Visualization of hepatic sulfite oxidase in crude tissue preparations by electron paramagnetic resonance spectroscopy. Biochim. Biophys. Acta, **334**, 86-96 (1974)
[24] Cohen, H.J.: Hepatic sulfite oxidase. Congruency in mitochondria of prosthetic groups and activity. J. Biol. Chem., **247**, 7759-7766 (1972)
[25] Kessler, D.L.; Rajagopalan, K.V.: Purification and properties of sulfite oxidase from chicken liver. Presence of molybdenum in sulfite oxidase from diverse sources. J. Biol. Chem., **247**, 6566-6573 (1972)
[26] Cohen, H.J.; Fridovich, I.: Hepatic sulfite oxidase. Purification and properties. J. Biol. Chem., **246**, 359-366 (1971)
[27] Cohen, H.J.; Fridovich, I.: Hepatic sulfite oxidase. The nature and function of the heme prosthetic groups. J. Biol. Chem., **246**, 367-373 (1971)
[28] MacLeod, R.M.; Farkas, W.; Fridovich, I.; Handler, P.: Purification and properties of hepatic sulfite oxidase. J. Biol. Chem., **236**, 1841-1846 (1961)
[29] George, G.N.; Kipke, C.A.; Prince, R.C.; Sunde, R.A.; Enemark, J.H.; Cramer, S.P.: Structure of the active site of sulfite oxidase. X-ray absorption spectroscopy of the Mo(IV), Mo(V), and Mo(VI) oxidation states. Biochemistry, **28**, 5075-5080 (1989)
[30] Kipke, C.A.; Cusanovich, M.A.; Tolin, G.; Sunde, R.A.; Enemark, J.H.: Chicken liver sulfite oxidase. Kinetics of reduction by laser-photoreduced flavins and intramolecular electron transfer. Biochemistry, **27**, 2918-2926 (1988)

[31] Jones, H.P.; Johnson, J.L.; Rajagopalan, K.V.: In vitro reconstitution of demolybdosulfite oxidase by molybdate. J. Biol. Chem., **252**, 4988-4993 (1977)
[32] Johnson, J.L.; Jones, H.P.; Rajagopalan, K.V.: In vitro reconstitution of demolybdosulfite oxidase by a molybdenum cofactor from rat liver and other sources. J. Biol. Chem., **252**, 4994-5003 (1977)
[33] Coughlan, M.P.; Johnson, J.L.; Rajagopalan, K.V.: Mechanisms of inactivation of molybdoenzymes by cyanide. J. Biol. Chem., **255**, 2694-2699 (1980)
[34] Johnson, J.L.: Rat liver "cytochrome b_9" is sulfite oxidase. J. Biol. Chem., **256**, 9824-9825 (1981)
[35] Sullivan, E.P., Jr.; Hazzard, J.T.; Tollin, G.; Enemark, J.H.: Electron transfer in sulfite oxidase: Effects of pH and anions on transient kinetics. Biochemistry, **32**, 12465-12470 (1993)
[36] Ratnam, K.; Brody, M.S.; Hille, R.: Purification of xanthine dehydrogenase and sulfite oxidase from chicken liver. Prep. Biochem. Biotechnol., **26**, 143-154. (1996)
[37] Ganai, B.A.; Masood, A.; Baig, M.A.: Isolation, purification and partial characterization of sulfite oxidase from Malva sylvestris. Phytochemistry, **45**, 879-880 (1997)
[38] Jolivet, P.; Bergeron, E.; Meunier, J.C.: Evidence for sulfite oxidase activity in spinach leaves. Phytochemistry, **40**, 667-672 (1995)
[39] Coury, L.A., Jr.; Yang, L.; Murray, R.W.: Electrochemical study of the rate of activation of the molybdoheme protein sulfite oxidase by organic electron acceptors. Anal. Chem., **65**, 242-246 (1993)
[40] Garrett, R.M.; Johnson, J.L.; Graf, T.N.; Feigenbaum, A.; Rajagopalan, K.V.: Human sulfite oxidase R160Q: identification of the mutation in a sulfite oxidase-deficient patient and expression characterization of the mutant enzyme. Proc. Natl. Acad. Sci. USA, **95**, 6394-6398 (1998)
[41] Garrett, R.M.; Rajagopalan, K.V.: Site-directed mutagenesis of recombinant sulfite oxidase: identification of cysteine 207 as a ligand of molybdenum. J. Biol. Chem., **271**, 7387-7391. (1996)
[42] George, G.N.; Garrett, R.M.; Graf, T.; Prince, R.C.; Rajagopalan, K.V.: Interaction of arsenate with the molybdenum site of sulfite oxidase. J. Am. Chem. Soc., **120**, 4522-4523 (1998)
[43] Eilers, T.; Schwarz, G.; Brinkmann, H.; Witt, C.; Richter, T.; Nieder, J.; Koch, B.; Hille, R.; Hansch, R.; Mendel, R.R.: Identification and biochemical characterization of Arabidopsis thaliana sulfite oxidase: a new player in plant sulfur metabolism. J. Biol. Chem., **276**, 46989-46994 (2001)
[44] Leimkuhler, S.; Rajagopalan, K.V.: In vitro incorporation of nascent molybdenum cofactor into human sulfite oxidase. J. Biol. Chem., **276**, 1837-1844. (2001)
[45] Feng, C.; Kedia, R.V.; Hazzard, J.T.; Hurley, J.K.; Tollin, G.; Enemark, J.H.: Effect of solution viscosity on intramolecular electron transfer in sulfite oxidase. Biochemistry, **41**, 5816-5821 (2002)
[46] Brody, M.S.; Hille, R.: The kinetic behavior of chicken liver sulfite oxidase. Biochemistry, **38**, 6668-6677. (1999)

[47] Temple, C.A.; Graf, T.N.; Rajagopalan, K.V.: Optimization of expression of human sulfite oxidase and its molybdenum domain. Arch. Biochem. Biophys., **383**, 281-287. (2000)
[48] Pacheco, A.; Hazzard, J.T.; Tollin, G.; Enemark, J.H.: The pH dependence of intramolecular electron transfer rates in sulfite oxidase at high and low anion concentrations. J. Biol. Inorg. Chem., **4**, 390-401 (1999)
[49] Cramer, S.P.; Gray, H.B.; Rajagopalan, K.V.: The molybdenum site of sulfite oxidase. Structural information from X-ray absorption spectroscopy. J. Am. Chem. Soc., **101**, 2772-2774 (1979)
[50] Das, S.K.; Chaudhury, P.K.; Biswas, D.; Sarkar, S.: Modeling for the active site of sulfite oxidase: synthesis, sharacterization, and reactivity of $[MoVIO_2(mnt)_2]^{2-}$ (mnt2- = 1,2-dicyanoethylenedithiolate). J. Am. Chem. Soc., **116**, 9061-9070 (1994)
[51] Adeloju, S.B.; Barisci, J.N.; Wallace, G.G.: Electroimmobilization of sulfite oxidase into a polypyrrole film and its utilization for flow amperometric detection of sulfite. Anal. Chim. Acta, **332**, 145-153 (1996)

Thiol oxidase 1.8.3.2

1 Nomenclature

EC number
1.8.3.2

Systematic name
thiol:oxygen oxidoreductase

Recommended name
thiol oxidase

Synonyms
DTT-oxidase
ERv2p
oxidase, thiol
sulfhydryl oxidase
sulfhydryl oxidase SOx-3
thiooxidase

CAS registry number
9029-39-4

2 Source Organism

<1> *Bos taurus* [1, 3, 4, 6, 7, 8, 10, 12, 15, 18, 20]
<2> *Rattus norvegicus* [2, 7, 9, 11, 13, 18, 24]
<3> *Sus scrofa* [5, 7, 20]
<4> *Homo sapiens* [7, 19, 20]
<5> *Capra hircus* [7]
<6> *Mesocricetus auratus* [14]
<7> *Magnaporthe grisea* [16, 17]
<8> *Polyporus versicolor* [16]
<9> *Gallus gallus* [20, 22, 25, 27]
<10> *Saccharomyces cerevisiae* [21, 23, 28]
<11> *Cavia porcellus* [26]

3 Reaction and Specificity

Catalyzed reaction

4 R'C(R)SH + O_2 = 2 R'C(R)S-S(R)CR' + 2 H_2O (<1>, ter bi substituted mechanism, O_2 binds first [6])

Reaction type

oxidation
redox reaction
reduction

Natural substrates and products

S Additional information <4, 6, 9, 10, 11> (<6>, enzyme appears to protect sperm structure and function against damage by endogeneous sulfhydryls [14]; <4>, the enzyme may provide a crucial switch for the regulation of receptor-Ck-dependent mevalonate pathway [19]; <10>, essential function of the mitochondrial sulfhydryl oxidase Erv1p/ALR in the maturation of cytosolic but not of mitochondrial Fe-S proteins [21]; <9>, possible role for oxidase in protein secretory pathway [22]; <10>, Erv2p functions in the generation of microsomal disulfide bonds acting in parallel with Ero1p, the essential FAD-dependent oxidase of protein disulfide isomerase [23]; <11>, sulfhydryl oxidase Sox-3 can be implicated in the negative cell cycle control [26]; <9>, the enzyme may play an important role in the introduction of disulfide bridges in egg white proteins [27]) [14, 19, 21, 22, 23, 26, 27]
P ?

Substrates and products

S 2-mercaptobenzothiazole + O_2 <7> (Reversibility: ? <7> [16]) [16]
P ? + H_2O
S 2-mercaptoethanol + O_2 <2, 6, 9> (<2>, 3.7% of the activity with dithiothreitol [2,13];) (Reversibility: ? <2, 6, 9> [13, 14, 25]) [2, 13, 14, 25]
P ? + H_2O
S 2-nitro-5-thiobenzoic acid + O_2 <1> (Reversibility: ? <1> [12]) [12]
P ? + H_2O
S 5,5'-dithiobis(2-nitrobenzoic acid) + O_2 <1, 2> (Reversibility: ? <1, 2> [1, 2, 13]) [1, 2, 13]
P ? + H_2O
S 5-amino-2-benzimidazolethiol + O_2 <7> (Reversibility: ? <7> [16]) [16]
P ? + H_2O
S D-Cys + O_2 <1> (Reversibility: ? <1> [4, 6, 10, 12]) [4, 6, 10, 12]
P ? + H_2O
S D-penicillamine + O_2 <2> (<2>, 33% of the activity with dithiothreitol [2]; <2>, 33% of the activity with dithiothreitol [13]) (Reversibility: ? <2> [2, 13]) [2, 13]
P ? + H_2O
S GSH + O_2 + O_2 <1, 2, 3, 4, 6, 9> (<2>, 0.7% of the activity with dithiothreitol [2]; <2>, 0.7% of the activity with dithiothreitol [13]; <6>, 13.5% of the activity with dithiothreitol [14]; <7>, no activity [16]) (Reversibil-

ity: ? <1, 2, 3, 4, 6, 9> [1, 2, 3, 4, 5, 6, 7, 8, 9, 10, 12, 13, 14, 15, 18, 22, 25, 27]) [1, 2, 3, 4, 5, 6, 7, 8, 9, 10, 12, 13, 14, 15, 18, 22, 25, 27]

P GSSG + H_2O <1, 2, 3, 4> (<9>, production of H_2O_2 [27]) [1, 2, 3, 4, 5, 6, 7, 8, 9, 10, 12]

S Gly-Gly-L-Cys + O_2 <1> (Reversibility: ? <1> [1]) [1]

P ? + H_2O

S L-Cys + O_2 <1, 2, 3, 9> (<2>, 17% of the activity with dithiothreitol [2]; <2>, 17% of the activity with dithiothreitol [13]; <6>, 20.3% of the activity with dithiothreitol [14]; <7>, no activity [16]) (Reversibility: ? <1, 2, 3, 9> [2, 4, 5, 6, 10, 12, 13, 14, 18, 25]) [2, 4, 5, 6, 10, 12, 13, 14, 18, 25]

P ? + H_2O

S N-acetyl-EAQCGTS + O_2 <9> (Reversibility: ? <9> [22]) [22]

P ? + H_2O

S N-acetylcysteine + O_2 <1, 2, 3> (<2>, 4.6% of the activity with dithiothreitol [2]; <2>, 4.6% of the activity with dithiothreitol [13]; <6>, 12.6% of the activity with dithiothreitol [14]) (Reversibility: ? <1, 2, 3> [1, 2, 4, 5, 6, 12, 18]) [1, 2, 4, 5, 6, 12, 18]

P ? + H_2O

S cysteamine + O_2 <1, 3> (Reversibility: ? <1, 3> [4, 5, 6, 12]) [4, 5, 6, 12]

P ? + H_2O

S diethyldithiocarbamate + O_2 <7> (Reversibility: ? <7> [16]) [16]

P tetraethylthiuram disulfide + H_2O <7> [16]

S dithioacetate + O_2 <7> (Reversibility: ? <7> [16]) [16]

P ? + H_2O

S dithioerythritol + O_2 <2> (Reversibility: ? <2> [2, 13]) [2, 13]

P ? + H_2O

S dithiooxalate + O_2 <7> (Reversibility: ? <7> [16]) [16]

P ? + H_2O

S dithiothreitol + O_2 <2, 3, 4, 6, 9> (<9>, anaerobically, the ferricenium ion is a facile alternative electron acceptor [25]) (Reversibility: ? <2, 3, 4, 6, 9> [2, 5, 13, 14, 18, 19, 20, 24, 25, 27]) [2, 5, 13, 14, 18, 19, 20, 24, 25, 27]

P ? + H_2O <2> (<2,9>, production of H_2O_2 [24,25]) [24, 25]

S ergothioneine + O_2 <7> (Reversibility: ? <7> [16]) [16]

P ? + H_2O

S *p*-bromophenylthiopseudourea + O_2 <7> (Reversibility: ? <7> [16]) [16]

P ? + H_2O

S reduced aldolase + O_2 <9> (Reversibility: ? <9> [22]) [22]

P aldolase + H_2O

S reduced insulin A chain + O_2 <9> (Reversibility: ? <9> [22]) [22]

P insulin A chain + H_2O

S reduced insulin B chain + O_2 <9> (Reversibility: ? <9> [22]) [22]

P insulin B chain + H_2O

S reduced lysozyme + O_2 <9, 10> (Reversibility: ? <9, 10> [22, 28]) [22, 28]

P lysozyme + H_2O

S reduced ovalbumin + O_2 <9> (Reversibility: ? <9> [22]) [22]

P ovalbumin + H_2O

S reduced pyruvate kinase + O_2 <9> (Reversibility: ? <9> [22]) [22]

P pyruvate kinase + H_2O
S reduced riboflavin-binding protein + O_2 <9> (Reversibility: ? <9> [22]) [22]
P riboflavin-binding protein + H_2O
S reduced ribunuclease + O_2 <9> (Reversibility: ? <9> [22]) [22]
P renatured ribonuclease + H_2O <9> [22]
S reductively denatured ribonuclease A + O_2 <1, 2, 9> (<1>, reductively denatured pancreatic ribonuclease A [1]) (Reversibility: ? <1, 2, 9> [1, 2, 4, 13, 15, 27]) [1, 2, 4, 13, 15, 27]
P renatured ribonuclease + H_2O <9> (<9>, production of H_2O_2 [27]) [27]
S thioacetate + O_2 <7> (Reversibility: ? <7> [16]) [16]
P ? + H_2O
S thioglycolate + O_2 <7> (Reversibility: ? <7> [16]) [16]
P ? + H_2O
S thioglycolate + O_2 <6> (<6>, 11.1% of the activity with dithiothreitol [14]) (Reversibility: ? <6> [14]) [14]
P ? + H_2O
S thiohistidine + O_2 <7> (Reversibility: ? <7> [16]) [16]
P ? + H_2O
S thiophenol + O_2 <7> (Reversibility: r <7> [17]; ? <7> [16]) [16, 17]
P diphenyl disulfide + H_2O <7> [17]
S Additional information <1> (<1>, enzyme does not catalyze thiol-disulfide interchange [12]) [12]
P ?

Inhibitors

2,2'-dipyridyl <7> (<7>, 0.25 mM, 34% inhibition [17]) [17]
2-mercaptoethanol <7> (<7>, 1 mM, 58% inhibition [17]) [17]
8-hydroxyquinoline <7> (<7>, 0.1 mM, 50% inhibition [17]) [17]
Cys <7> [16]
EDTA <1, 2> (<2>, 5 mM, 93% inhibition [9]; <1>, milk enzyme [18]; <2>, inhibition of enzyme from kidney and small intestine, no inhibition of enzyme from skin and seminal vesicles [18]) [9, 15, 18]
GSSG <1, 7> (<1>, substrate inhibition above 0.8 mM [15]) [15, 16]
H_2O_2 <2, 3> (<2>, 0.1 mM, 89% inhibition [9]) [5, 9]
KCN <2> [18]
L-(αS,5S)-α-amino-3-chloro-4,5-dihydro-5-isoxazoleacetic acid <2, 3> [5, 18]
NEM <2, 3> (<2>, 1 mM, 86% inhibition [9]) [5, 9]
azide <6, 7> [14, 16]
bathocuproine disulfonate <2> (<2>, 0.5 mM, 89% inhibition [9]) [9, 18]
cyanide <7> (<7>, 20 mM, 38% inhibition [17]) [16, 17]
diazabicyclooctane <1> [12]
diethyldithiocarbamate <2> [18]
dithiothreitol <2> (<2>, at high concentrations [2]) [2]
guanine <1> [12]
iodoacetamide <2, 9> (<2>, moderate [13]) [2, 13, 25]

iodoacetic acid <2> (<2>, moderate [13]) [2, 13]
o-dianisidine <1> [12]
o-phenylenediamine <1> [12]
ribonuclease <1> (<1>, substrate inhibition above 0.04 mM [4]) [4]

Cofactors/prosthetic groups

FAD <2, 9, 10, 11> (<9>, contains FAD [27]; <2>, enzyme from seminal vesicles [18]; <2,9>, dependent on [22,24]; <10>, enzyme is linked to FAD [23]; <10>, one molecule of FAD per subunit [28]; <10>, firmly attached but nut covalently linked to the protein [28]) [18, 22, 23, 24, 25, 26, 27, 28]

Activating compounds

Triton X-100 <2> (<2>, 0-0.1% concentration, at most 30% activation [13]) [13]

Metals, ions

copper <2, 3> (<2>, native activity is increased 2fold [2]; <3>, contains 1 mol copper per subunit, in a binuclear complex [5]; <2>, involved in catalysis [9]; <2>, enzyme from kindney and small intestine [18]) [2, 5, 9, 18]
iron <1> (<1>, contains enzyme-bound iron [12]; <1>, iron appears to be required for enzymatic activity [15]) [12, 15, 18]

Turnover number (min^{-1})

1033 <9> (dithiothreitol) [25]
1215 <9> (2-mercaptoethanol) [25]
1275 <9> (L-Cys) [25]
1385 <9> (GSH) [25]
Additional information <10> [28]

Specific activity (U/mg)

9.515 <2> [13]
10 <1> [1]
24 <1> [3]
103.8 <1> [15]
Additional information <1, 2, 3, 4, 7, 9> (<1,3,4,9>, a continuous fluorescence assay for sulfhydryl oxidase [20]) [2, 5, 17, 20]

K_m-Value (mM)

0.0174 <1> (reduced ribonuclease, <1>, corresponds to a sulfhydryl concentration of 0.14 mM [4]) [4]
0.09 <1> (GSH) [15]
0.11 <9> (reduced lysozyme, <9>, expressed per thiol basis [22]) [22]
0.115 <9> (reduced ribonuclease, <9>, expressed per thiol basis [22]) [22]
0.15 <9> (dithiothreitol) [25]
0.16 <9> (reduced aldolase, <9>, expressed per thiol basis [22]) [22]
0.215 <9> (reduced insulin A chain, <9>, expressed per thiol basis [22]) [22]
0.23 <9> (reduced riboflavin-binding protein, <9>, expressed per thiol basis [22]) [22]
0.25 <2> (O_2) [18]

0.3 <1> (GSH) [12]
0.3 <9> (reduced insulin B chain, <9>, expressed per thiol basis [22]) [22]
0.33 <9> (reduced ovalbumin, <9>, expressed per thiol basis [22]) [22]
0.34 <1> (GSH) [10]
0.41 <7> (thiophenol) [17]
0.42 <3> (L-Cys) [5]
0.5 <2> (GSH) [18]
0.51 <3> (dithiothreitol) [5]
0.73 <3> (GSH) [5]
0.77 <1> (O_2) [6]
0.8 <1> (L-Cys) [10, 12]
0.805 <1> (L-Cys) [4, 6]
0.966 <1> (L-Cys) [4, 6]
1.13 <1> (N-acetyl-L-Cys) [4, 6]
1.25 <1> (cysteamine) [4, 6]
1.25 <9> (reduced pyruvate kinase, <9>, expressed per thiol basis [22]) [22]
1.33 <1> (D-Cys) [10, 12]
1.6 <3> (cysteamine) [5]
1.72 <9> (N-acetyl-EAQCGTS, <9>, expressed per thiol basis [22]) [22]
2.35 <1> (N-acetyl-L-Cys) [1]
3.16 <1> (GSH) [6]
3.3 <3> (N-acetylcysteine) [5]
3.85 <1> (N-acetyl-L-Cys) [10, 12]
6.31 <1> (Gly-Gly-L-Cys) [1]
8.16 <1> (GSH) [1]
10.9 <9> (L-Cys) [25]
20 <9> (GSH) [22, 25]
30 <1> (cysteamine) [10, 12]
54 <9> (2-mercaptoethanol) [25]
100 <1> (2-nitro-5-thiobenzoic acid) [12]

pH-Optimum

6.5 <7> [17]
6.8-7 <1> [15]
7.4 <2> [18]
7.5 <9> [25]
8-8.2 <2> (<2>, oxidation of 5,5'-dithiobis(2-nitrobenzoic acid) and reactivation of ribonuclease [2]) [2]

pH-Range

5-7.5 <7> (<7>, pH 5.0: about 40% of maximal activity, pH 7.5: about 60% of maximal activity [17]) [17]
5-11 <9> (<9>, pH 5.0: about 40% of maximal activity, pH 11: about 70% of maximal activity [25]) [25]
6-7 <7> (<7>, rate of oxidation is considerable less at pH 7.0 than at pH 6.0 [16]) [16]

Temperature optimum (°C)
35 <1> [15]

4 Enzyme Structure

Molecular weight
47000 <4> (<4>, gel filtration [19]) [19]
66000 <2> (<2>, gel filtration [2,13]) [2, 13]
120000 <3> (<3>, gel filtration [5]) [5]
202000 <9> (<9>, gel filtration [25]) [25]

Subunits
? <1, 2> (<2>, x * 64000, SDS-PAGE [24]; <1>, x * 89000, SDS-PAGE [12,15]; <1>, x * 90000, may exist in an aggregated molecular form, SDS-PAGE [1]) [1, 12, 15, 24]
dimer <3, 9> (<3>, 2 * 70000, also aggregates to larger multimers, SDS-PAGE [5]; <9>, 2 * 93000, SDS-PAGE [25]) [5, 25]
monomer <2> (<2>, 1 * 66000, SDS-PAGE [2,13]) [2, 13]

Posttranslational modification
glycoprotein <1, 2, 9> (<1>, contains 11% carbohydrate [15]; <2>, two potential sites for N-glycosylation. One of them is used and the 64000 Da purified protein is transformed to 61000 Da by the action of endoglycosidase F [24]) [15, 24, 25]

5 Isolation/Preparation/Mutation/Application

Source/tissue
cauda epididymis <6> [14]
egg <9> (<9>, egg white [20,22,25,26]) [20, 22, 25, 26]
endometrial cell line <11> (<11>, endometrial cells. Enzyme level increases during a serum depletion-induced quiescence, decreases when cells enter the G1 phase after serum stimulation, and is restored during the S and G2/M phases [26]) [26]
epididymis <2> (<2>, high activity [24]) [24]
heart <2> (<2>, small amounts [24]) [24]
kidney <2, 3> (<1,3>, cortex [5,8]; <2>, small amounts [24]) [5, 7, 8, 11, 18, 20, 24]
liver <2> (<2>, small amounts [24]) [24]
lung <2> (<2>, small amounts [24]) [24]
mammary gland <2> [7]
milk <1, 2, 3, 4, 5> [1, 3, 4, 6, 7, 8, 10, 12, 15, 18]
mycelium <8> [16]
pancreas <1, 2> [1, 7]
platelet <4> [19]
prostate gland <2> (<2>, small amounts [24]) [24]

seminal vesicle <2> (<2>, fluid, high activity [24]) [18, 20, 24]
skin <2> [2, 13, 18]
small intestine <2> (<2>, epithelium [9]) [9, 18]
spleen <2> (<2>, small amounts [24]) [24]
tear <4> [20]
testis <2> (<2>, small amounts [24]) [24]
uterine endometrium <11> [26]
uterus <11> (<11>, epithelial cells [26]) [26]

Localization

cytosol <2> [18]
membrane <1> [8]
microsome <10> [23]
mitochondrion <10> (<10>, intermembrane space [21]) [21]
plasma membrane <2, 3> (<2>, associated to basal-lateral region of the plasma membrane [9,11,18]) [5, 9, 11, 18]
vesicular fraction <1> [3, 4, 18]

Purification

<1> (transient covalent affinity chromatography with cysteinyl-CPG-glass [3]) [1, 3, 12, 15]
<2> [2, 13, 24]
<3> [5]
<4> [19]
<9> [25]
<10> [28]

Cloning

<2> (expression in Chineses hamster ovary epithelium cells [24]) [24]
<10> (expression in Escherichia coli [23,28]) [23, 28]

Application

nutrition <1> (<1>, flavor modification of ultra-high temperature milk [10]; <1>, elimination of cooked flavour in ultra-high temperature commercially sterile milk, may have other applications for flavour modification [12]) [10, 12]

6 Stability

Temperature stability

60 <2, 7> (<2>, 15 min, 85% loss of activity. 30 min, complete loss of activity [2,13]; <7>, 5 min, stable [17]) [2, 13, 17]
100 <7> (<7>, 5 min, complete inactivation [16]; <7>, destroyed by boiling [17]) [16, 17]
Additional information <2> (<2>, dithiothreitol, 1 mM, increases the stability to heating [2,13]) [2, 13]

General stability information

<1>, crosslinked immobilized enzyme is very stable to urea [12]

<1>, immobilized enzyme, 10% ethanol, 0.01% H_2O_2, or in 0.02% sodium azide, loss of only small amounts of activity [12]

Storage stability

<2>, 4°C, stable for at least 2 weeks [13]

<7>, 4°C, stable for weeks [16]

<9>, -20°C, stable for months [25]

References

[1] Clare, D.A.; Pinnix, I.B.; Lecce, J.G.; Horton, H.R.: Purification and properties of sulfhydryl oxidase from bovine pancreas. Arch. Biochem. Biophys., **265**, 351-361 (1988)

[2] Goldsmith, L.A.: Sulfhydryl oxidase from rat skin. Methods Enzymol., **143**, 510-515 (1987)

[3] Swaisgood, H.E.; Horton, H.R.: Sulfhydryl oxidase from milk. Methods Enzymol., **143**, 504-510 (1987)

[4] Janolino, V.G.; Swaisgood, H.E.: Sulfhydryl oxidase-catalyzed formation of disulfide bonds in reduced ribonuclease. Arch. Biochem. Biophys., **258**, 265-271 (1987)

[5] Lash, L.H.; Jones, D.P.: Purification and properties of the membranal thiol oxidase from porcine kidney. Arch. Biochem. Biophys., **247**, 120-130 (1986)

[6] Sliwkowski, M.X.; Swaisgood, H.E.; Clare, D.A.; Horton, H.R.: Kinetic mechanism and specificity of bovine milk sulphydryl oxidase. Biochem. J., **220**, 51-55 (1984)

[7] Clare, D.A.; Horton, H.R.; Stabel, T.J.; Swaisgood, H. E.; Lecce, J.G.: Tissue distribution of mammalian sulfhydryl oxidase. Arch. Biochem. Biophys., **230**, 138-145 (1984)

[8] Schmelzer, C.H.; Phillips, C.; Swaisgood, H.E.; Horton, H.R.: Immunological similarity of milk sulfhydryl oxidase and kidney glutathione oxidase. Arch. Biochem. Biophys., **228**, 681-685 (1984)

[9] Lash, L.H.; Jones, D.P.: Characterization of the membrane-associated thiol oxidase activity of rat small-intestinal epithelium. Arch. Biochem. Biophys., **225**, 344-352 (1983)

[10] Swaisgood, H.E.; Sliwkowski, M.X.; Skudder, P.J.; Janolino, V.G.: Sulfhydryl oxidase: characterization and application for flavor modification of UHT milk. Util. Enzymes Technol. Aliment., Symp. Int. (Dupuy, P., ed.), Tech. Doc. Lavoisier Paris, 229-235 (1982)

[11] Lash, L.H.; Jones, D.P.: Localization of the membrane-associated thiol oxidase of rat kidney to the basal-lateral plasma membrane. Biochem. J., **203**, 371-376 (1982)

[12] Swaisgood, H.E.: Sulphydryl oxidase: properties and application. Enzyme Microb. Technol., **2**, 265-272 (1986)

[13] Takamori, K.; Thorpe, J.M.; Goldsmith, L.A.: Skin sulfhydryl oxidase. Purification and some properties. Biochim. Biophys. Acta, **615**, 309-323 (1980)
[14] Chang, T.S.K.; Morton, B.: Epididymal sulfhydryl oxidase: a sperm-protective enzyme from the male reproductive tract. Biochem. Biophys. Res. Commun., **66**, 309-315 (1975)
[15] Janolino, V.G.; Swaisgood, H.E.: Isolation and characterization of sulfhydryl oxidase from bovine milk. J. Biol. Chem., **250**, 2532-2538 (1975)
[16] Neufeld, H.A.; Green, L.F.; Latterell, F.M.; Weintraub, R.L.: Thiooxidase, a new sulfhydryl-oxidizing enzyme from Piricularia oryzae and Polyporus versicolor. J. Biol. Chem., **232**, 1093-1099 (1958)
[17] Aurbach, G.D.; Jakoby, W.B.: The multiple functions of thiooxidase. J. Biol. Chem., **237**, 565-568 (1962)
[18] Lash, L.H.; Jones, D.P.; Orrenius, S.: The renal thiol (glutathione) oxidase. Subcellular localization and properties. Biochim. Biophys. Acta, **779**, 191-200 (1984)
[19] Kaul, D.; Dhawan, V.; Kaur, M.: Evidence and nature of a novel thiol-oxidase in human platelets. Mol. Cell. Biochem., **159**, 81-84 (1996)
[20] Raje, S.; Glynn, N.M.; Thorpe, C.: A continuous fluorescence assay for sulfhydryl oxidase. Anal. Biochem., **307**, 266-272 (2002)
[21] Lange, H.; Lisowsky, T.; Gerber, J.; Muhlenhoff, U.; Kispal, G.; Lill, R.: An essential function of the mitochondrial sulfhydryl oxidase Erv1p/ALR in the maturation of cytosolic Fe/S proteins. EMBO Rep., **2**, 715-720 (2001)
[22] Hoober, K.L.; Sheasley, S.L.; Gilbert, H.F.; Thorpe, C.: Sulfhydryl oxidase from egg white. A facile catalyst for disulfide bond formation in proteins and peptides. J. Biol. Chem., **274**, 22147-22150 (1999)
[23] Gerber, J.; Muhlenhoff, U.; Hofhaus, G.; Lill, R.; Lisowsky, T.: Yeast Erv2p is the first microsomal FAD-linked sulfhydryl oxidase of the Erv1p/Alrp protein family. J. Biol. Chem., **276**, 23486-23491 (2001)
[24] Benayoun, B.; Esnard-Feve, A.; Castella, S.; Courty, Y.; Esnard, F.: Rat seminal vesicle FAD-dependent sulfhydryl oxidase. Biochemical characterization and molecular cloning of a member of the new sulfhydryl oxidase/quiescin Q_6 gene family. J. Biol. Chem., **276**, 13830-13837 (2001)
[25] Hoober, K.L.; Joneja, B.; White, H.B.; Thorpe, C.: A sulfhydryl oxidase from chicken egg white. J. Biol. Chem., **271**, 30510-30516 (1996)
[26] Musard, J.F.; Sallot, M.; Dulieu, P.; Fraichard, A.; Ordener, C.; Remy-Martin, J.P.; Jouvenot, M.; Adami, P.: Identification and expression of a new sulfhydryl oxidase SOx-3 during the cell cycle and the estrus cycle in uterine cells. Biochem. Biophys. Res. Commun., **287**, 83-91 (2001)
[27] Hoober, K.L.; Thorpe, C.: Egg white sulfhydryl oxidase: kinetic mechanism of the catalysis of disulfide bond formation. Biochemistry, **38**, 3211-3217 (1999)
[28] Lee, J.E.; Hofhaus, G.; Lisowsky, T.: Erv1p from Saccharomyces cerevisiae is FAD-linked sulfhydryl oxidase. FEBS Lett., **477**, 62-66 (2000)

Glutathione oxidase 1.8.3.3

1 Nomenclature

EC number
1.8.3.3

Systematic name
glutathione:oxygen oxidoreductase

Recommended name
glutathione oxidase

Synonyms
GSH oxidase]
GSH sulfhydryl oxidase
glutathione sulfhydryl oxidase
oxidase, glutathione
sulfhydryl oxidase

CAS registry number
55467-56-6

2 Source Organism

<1> *Rattus norvegicus* [1]
<2> *Penicillium sp.* (overview [2]; K-6-5 [3]) [2-5]
<3> *Fusarium sp.* (overview [2]) [2]
<4> *Aspergillus sp.* (overview [2]) [2]

3 Reaction and Specificity

Catalyzed reaction
2 glutathione + O_2 = glutathione disulfide + H_2O_2 (a FAD flavoprotein, acts, more slowly, also on L-cysteine and several other thiols)

Reaction type
oxidation
redox reaction
reduction

Natural substrates and products

S glutathione + O_2 <2> (<2> measurement of glutathione in vivo via microelectrodes with immobilized enzyme [4]; <2> measurement of H_2O_2 in vivo via microelectrodes with immobilized enzyme [5]) (Reversibility: ? <2> [3-5]) [3-5]
P glutathione disulfide + H_2O_2 <2> [4, 5]

Substrates and products

S 2-mercaptoethanol + O_2 <1, 2> (<2> slowly [3]; <2> 19% of the activity with glutathione [5]) (Reversibility: r <1, 2> [3]; ? <2> [5]) [1, 3, 5]
P mercaptoethanol disulfide
S D-cysteine + O_2 <2> (<2> slowly [3]; <2> low activity [5]) (Reversibility: ? <2> [3, 5]) [3, 5]
P D-cystine
S L-cysteine + O_2 <1, 2> (<2> low activity [5]) (Reversibility: ? <1, 2> [1, 3, 5]) [1, 3, 5]
P L-cystine
S L-cysteine ethyl ester + O_2 <2> (<2> low activity [5]) (Reversibility: ? <2> [5]) [5]
P L-cystine diethyl ester
S L-cysteine methyl ester + O_2 <2> (<2> slowly [3]) (Reversibility: ? <2> [3]) [3]
P L-cystine dimethyl ester
S N-acetyl-L-cysteine + O_2 <2> (<2> slowly [3]; <2> low activity [5]) (Reversibility: ? <2> [3, 5]) [3, 5]
P N,N'-diacetyl-L-cystine
S S-methylglutathione + O_2 <2> (<2> enzyme immobilized on microelectrode sensor, low activity [4]) (Reversibility: ? <2> [4]) [4]
P ?
S cysteamine + O_2 <2> (<2> slowly [3]) (Reversibility: ? <2> [3]) [3]
P ?
S cysteine-glycine + O_2 <2> (<2> enzyme immobilized on microelectrode sensor [4]) (Reversibility: ? <2> [4]) [4]
P ?
S dithiothreitol + O_2 <1, 2> (<2> low activity [3]; <2> 38% of the activity with glutathione [5]) (Reversibility: ? <1, 2> [1, 3, 5]) [1, 3, 5]
P ?
S glutathione + O_2 <1-4> (<2> enzyme immobilized on microsensor [4,5]) (Reversibility: ? <1-4> [1-5]) [1-5]
P glutathione disulfide + H_2O_2 <2> [4, 5]
S reduced RNase A + O_2 <1, 2> (Reversibility: ? <1, 2> [1, 3]) [1, 3]
P ?
S thiophenol + O_2 <2> (<2> slowly [3]) (Reversibility: ? <2> [3]) [3]
P diphenyl sulfide

S Additional information <1, 2> (<2> D-penicilamine and coenzyme A are no substrates [5]; <1> not: D,L-homocysteine [1]; <2> no other amio acids but D- and L-cysteine serve as substrate [3]) [1, 3, 5]

P ?

Inhibitors

$CuSO_4$ <2> (<2> slight inhibition [3]) [3]
KCl <2> (<2> slight inhibition [3]) [3]
KI <2> (<2> slight inhibition [3]) [3]
$MnSO_4$ <2> (<2> slight inhibition [3]) [3]
$NiCl_2$ <2> (<2> slight inhibition [3]) [3]
$ZnSO_4$ <2> (<2> complete inhibition at 1 mM [3]) [3]
thiols <1> (<1> at high concentrations above 20 mM [1]) [1]
Additional information <1, 2> (<1,2> EDTA had no effect on oxidation reaction [1,3]) [1, 3]

Cofactors/prosthetic groups

FAD <1-4> (<1-4> flavoprotein [1-4]; <2> contains 2 mol of FAD per mol of enzyme [3]; <1> 1 mol FAD per mol of enzyme [1]) [1-4]

Activating compounds

EDTA <2> (<2> activates slightly [3]) [3]

Metals, ions

$AlCl_2$ <2> (<2> activates slightly [3]) [3]
$CaCl_2$ <2> (<2> activates slightly [3]) [3]
$FeCl_2$ <2> (<2> activates slightly [3]) [3]

Specific activity (U/mg)

272 <2> [3]

K_m-Value (mM)

0.69 <2> (glutathione) [3]
0.7 <1> (dithiothreitol) [1]
1.7 <2> (glutathione, <2> immobilized enzyme on sensor surface out of cellulose acetate [4]) [4]
3.6 <2> (L-cysteine) [3]
4.4 <1> (glutathione) [1]
6.7 <2> (dithiothreitol) [3]
9.1 <1> (L-cysteine) [1]
32 <1> (2-mercaptoethanol) [1]

pH-Optimum

7 <1> (<1> substrate dithioerythritol [1]) [1]
7-7.8 <2> [3]

pH-Range

3.8-11 <2> (<2> enzyme immobilized on microsensor surface out of cellulose acetate [4]) [4]
5-10 <1, 2> (<1,2> minimal activity below pH 5.0, active at least up to pH 10.0 [1,3]) [1, 3]

6.2-8.8 <2> (<2> sharp decline in activity below pH 6.2 and above pH 8.8 [3]) [3]

Temperature optimum (°C)

25 <1, 2> (<1> assay at [1]; <2> assay with immobilized enzyme on microsensor [5]) [1, 5]
30 <2> (<2> assay at [3]) [3]

4 Enzyme Structure

Molecular weight

66000 <1> (<1> gel filtration [1]) [1]
95000 <2> (<2> gel filtration [3]) [3]

Subunits

dimer <2> (<2> 2 * 47000, SDS-PAGE [3]) [3]
monomer <1> (<1> 1 * 66000, SDS-PAGE [1]) [1]

5 Isolation/Preparation/Mutation/Application

Source/tissue

seminal vesicle <1> [1]

Localization

extracellular <2-4> [2, 3]
intracellular <2> [3]

Purification

<2> [3]

Application

analysis <2> (<2> development of a glutathione micro-enzyme sensor for amperometric detection in vivo, immobilization of enzyme on microelectrode tips [4]; <2> development of amperometric glutathione electrodes with immobilized enzyme for detection of H_2O_2 [5]) [4, 5]

6 Stability

pH-Stability

5.2-8.6 <2> (<2> 45°C, 30 min [3]) [3]

Temperature stability

55 <2> (<2> 30 min, no loss of activity [3]) [3]
60 <1> (<1> 3.5 min, 50% loss of activity [1]) [1]
75 <2> (<2> 30 min, complete loss of activity [3]) [3]

General stability information

<2>, after 36 measurements with enzyme immobilized on microelectrode sensor loss of 30% activity [4]

<2>, after 50 measurements with enzyme immobilized on microelectrode sensor loss of 50% activity [5]

Storage stability

<2>, -20°C, 0.1 M, potassium phosphate buffer, pH 7.4, 6 months [3]

<2>, 4°C, pH 6.5, phosphate buffer, immobilized enzyme on microelectrode sensor, loss of 20% activity, 60 days [5]

<2>, 4°C, phosphate buffered saline, immobilized enzyme on microelectrode sensor, loss of 20% activity, 68 days [4]

References

[1] Ostrowski, M.C.; Kistler, W.S.: Properties of a flavoprotein sulfhydryl oxidase from rat seminal vesicle secretion. Biochemistry, **19**, 2639-2645 (1980)

[2] Kusakabe, H.; Midorikawa, Y.; Kuninaka, A.; Yoshino, H.: Distribution of extracellular oxygen related enzymes in molds. Agric. Biol. Chem., **47**, 1385-1387 (1983)

[3] Kusakaba, H.; Kuninaka, A.; Yoshino, H.: Purification and properties of a new enzyme, glutathione oxidase from Penicillium sp. K-6-5. Agric. Biol. Chem., **46**, 2057-2067 (1982)

[4] Huang, Z.; Villarta-Snow, R.L.; Lubrano, G.J.; Guilbault, G.G.: Glutathione amperometric enzyme microsensor. Anal. Lett., **27**, 263-271 (1994)

[5] Compagnone, D.; Federici, G.; Scarciglia, L.; Palleschi, G.: Amperometric glutathione electrodes. Biosens. Bioelectron., **8**, 257-263 (1993)

Methanethiol oxidase 1.8.3.4

1 Nomenclature

EC number
1.8.3.4

Systematic name
methanethiol:oxygen oxidoreductase

Recommended name
methanethiol oxidase

Synonyms
MM-oxidase
MT-oxidase
methyl mercaptan oxidase
methylmercaptan oxidase
oxidase, methyl mercaptan

CAS registry number
112821-28-0

2 Source Organism

<1> *Hyphomicrobium sp.* (strain EG) [1]
<2> *Thiobacillus thioparus* (strain TK-m [2]; E6 [3]) [2, 3]
<3> *Rhodococcus rhodocrous* [4]

3 Reaction and Specificity

Catalyzed reaction
methanethiol + O_2 + H_2O = formaldehyde + H_2S + H_2O_2

Reaction type
oxidation
redox reaction
reduction

Natural substrates and products
S methanethiol + O_2 + H_2O <2, 3> (<2>, enzyme in oxidation pathway of dimethyl disulfide [3]; <3>, enzyme can be induced by methylmercaptan [4]) (Reversibility: ? <2, 3> [3, 4]) [3, 4]
P formaldehyde + H_2S + H_2O_2

Substrates and products

S ethyl mercaptan + O_2 + H_2O <1, 2> (Reversibility: ? <1, 2> [1, 2]) [1, 2]
P acetaldehyde + H_2S + H_2O_2
S methanethiol + O_2 + H_2O <1-3> (Reversibility: ? <1-3> [1-4]) [1-4]
P formaldehyde + H_2S + H_2O_2 <1-3> [1-4]
S sulfide + O_2 <1> (<2>, no activity [2]) (Reversibility: ? <1> [1]) [1]
P sulfur + H_2O_2 <1> [1]

Inhibitors

$(NH_4)_2SO_4$ <1> [1, 2]
Cu^{2+} <2> [2]
$CuSO_4$ <2> [2]
K_2SO_4 <3> (<3>, inhibition at concentration below 200 mM, original activity is recovered at 200 mM [4]) [4]
NEM <2> [2]
NH_4Cl <1> [1, 2]
NaCl <3> (<3>, inhibition at concentration below 200 mM, original activity is recovered at 200 mM [4]) [4]
cyanide <1, 2> [1, 2]
ethyl mercaptan <1, 2> [1, 2]
methanethiol <1, 2> (<1>, above 0.014 mM [1]) [1-3]
methanol <3> [4]
methylamine <1> [1]
sulfide <1> (<1>, non-competitive inhibition of methyl mercaptan oxidation [1]; <2>, no inhibition [2]) [1]

Specific activity (U/mg)

1.27 <2> [2]
2.13 <1> [1]
Additional information <3> [4]

K_m-Value (mM)

0.005-0.01 <1> (methanethiol) [1]
0.0097 <2> (methanethiol) [3]
0.018 <1> (ethyl mercaptan) [1]
0.0313 <2> (methanethiol) [2]
0.06 <1> (sulfide) [1]

K_i-Value (mM)

0.008 <2> (methanethiol) [3]
0.04 <1> (cyanide) [1]
0.042 <1> (methanethiol) [1]
0.09 <1> (sulfide) [1]
1.4 <1> (methylamine) [1]
4 <1> (NH_4^+) [1]

pH-Optimum

8-9.6 <2> [2]
8.2 <1> [1]

Temperature optimum (°C)

60 <3> [4]

4 Enzyme Structure

Molecular weight

29000 <2> (<2>, gel filtration [2]) [2]
40000-50000 <1> (<1>, gel filtration [1]) [1]

Subunits

? <3> (<3>, x * 64500, SDS-PAGE [4]) [4]
monomer <1, 2> (<1>, 1 * 49000, SDS-PAGE [1]; <2>, 1 * 40000, SDS-PAGE [2]) [1, 2]

5 Isolation/Preparation/Mutation/Application

Purification

<1> [1]
<2> [2]
<3> [4]

6 Stability

Temperature stability

50 <1> (<1>, 10 min, stable [1]) [1]
65 <1> (<1>, 10 min, complete loss of activity [1]) [1]

Storage stability

<1>, -20°C, stable for months [1]

References

[1] Suylen, G.M.H.; Large, P.J.; van Dijken, J.P.; Kuenen, J.G.: Methyl mercaptan oxidase, a key enzyme in the metabolism of methylated sulphur compounds by Hyphomicrobium EG. J. Gen. Microbiol., **133**, 2989-2997 (1987)

[2] Gould, W.D.; Kanagawa, T.: Purification and properties of methyl mercaptan oxidase from Thiobacillus thioparus TK-m. J. Gen. Microbiol., **138**, 217-221 (1992)

[3] Smith, N.A.; Kelly, D.P.: Mechanism of oxidation of dimethyl disulphide by Thiobacillus thioparus. J. Gen. Microbiol., **134**, 3031-3039 (1988)

[4] Kim, S.J.; Shin, H.J.; Kim, Y.C.; Yang, J.W.: Isolation and purification of methyl mercaptan oxidase from Rhodococcus rhodochrous for mercaptan detection. Biotechnol. Bioprocess Eng., **5**, 465-468 (2000)

Prenylcysteine oxidase 1.8.3.5

1 Nomenclature

EC number
1.8.3.5

Systematic name
S-prenyl-L-cysteine:oxygen oxidoreductase

Recommended name
prenylcysteine oxidase

Synonyms
EC 4.4.1.18 (formerly)
PCLY
prenylcysteine lyase

CAS registry number
196717-99-4

2 Source Organism

<1> *Mus musculus* (mouse, wild type and knockout mutants [1]) [1]
<2> *Homo sapiens* [2]

3 Reaction and Specificity

Catalyzed reaction
an S-prenyl-L-cysteine + O_2 + H_2O = a prenal + L-cysteine + H_2O_2

Reaction type
oxidation
redox reaction
reduction

Natural substrates and products

S S-prenyl-L-cysteine + O_2 + H_2O <1, 2> (<1> physiologic role in cleaving prenylcysteines in mammals, cleaves the thioether bond of prenylcysteines to yield free cysteine and the aldehyde of the isoprenoid lipid [1]; <2> kinetic mechanism, enzyme transfers the proS hydride of the farnesylcysteine to FAD [2]) (Reversibility: ? <1, 2> [1, 2]) [1, 2]

P prenal + L-cysteine + H_2O_2

Substrates and products

S S-prenyl-L-cysteine + O_2 + H_2O <1, 2> (Reversibility: ? <1, 2> [1, 2]) [1, 2]
P prenal + L-cysteine + H_2O_2 <1, 2> [1, 2]
S farnesyl-L-cysteine + O_2 + H_2O <1, 2> (Reversibility: ? <1, 2> [1, 2]) [1, 2]
P farnesal + L-cysteine + H_2O_2 <2> [2]
S geranylgeranyl-L-cysteine + O_2 + H_2O <1> (Reversibility: ? <1> [1]) [1]
P geranylgeranal + L-cysteine + H_2O_2

Inhibitors

farnesal <2> (<2> non-competitive versus farnesyl-L-cysteine [2]) [2]
farnesol <2> (<2> non-competitive versus farnesyl-L-cysteine [2]) [2]

Cofactors/prosthetic groups

FAD <2> [2]

Turnover number (min^{-1})

0.008 <2> (farnesyl-L-cysteine) [2]

K_m-Value (mM)

0.003 <2> (farnesyl-L-cysteine) [2]
0.05 <2> (O_2, <2> K_m below [2]) [2]

pH-Optimum

7.4-7.7 <2> (<2> assay at, depending on type of assay [2]) [2]
7.7 <1> (<1> assay at [1]) [1]

Temperature optimum (°C)

25 <2> (<2> assay at [2]) [2]
37 <1> (<1> assay at [1]) [1]

5 Isolation/Preparation/Mutation/Application

Source/tissue

brain <1> [1]
heart <1> [1]
kidney <1> [1]
liver <1> [1]

Cloning

<1> [1]

References

[1] Beigneux, A.; Withycombe, S.K.; Digits, J.A.; Tschantz, W.R.; Weinbaum, C.A.; Griffey, S.M.; Bergo, M.; Casey, P.J.; Young, S.G.: Prenylcysteine lyase deficiency in mice results in the accumulation of farnesylcysteine and ger-

anylgeranylcysteine in brain and liver. J. Biol. Chem., **277**, 38358-38363 (2002)
[2] Digits, J.A.; Pyun, H.J.; Coates, R.M.; Casey, P.J.: Stereospecificity and kinetic mechanism of human prenylcysteine lyase, an unusual thioether oxidase. J. Biol. Chem., **277**, 41086-41093 (2002)

Glutathione-homocystine transhydrogenase 1.8.4.1

1 Nomenclature

EC number
1.8.4.1

Systematic name
glutathione:homocystine oxidoreductase

Recommended name
glutathione-homocystine transhydrogenase

Synonyms
transhydrogenase, glutathione-homocystine

CAS registry number
9029-40-7

2 Source Organism

<1> *Bos taurus* [1]

3 Reaction and Specificity

Catalyzed reaction
2 glutathione + homocystine = glutathione disulfide + 2 homocysteine (the reactions catalyzed by this enzyme and by other in this subclass may be similar to those catalyzed by EC 2.5.1.18)

Reaction type
oxidation
redox reaction
reduction

Natural substrates and products
S GSH + homocystine <1> (Reversibility: r <1> [1]) [1]
P GSSG + homocysteine <1> [1]

Substrates and products
S GSH + homocystine <1> (Reversibility: r <1> [1]) [1]
P GSSG + homocysteine <1> [1]

Temperature optimum (°C)

37 <1> (<1> assay at [1]) [1]

5 Isolation/Preparation/Mutation/Application

Source/tissue

liver <1> [1]

References

[1] Racker, E.: Glutathione-homocystine transhydrogenase. J. Biol. Chem., **217**, 867-874 (1955)

Protein-disulfide reductase (glutathione) 1.8.4.2

1 Nomenclature

EC number
1.8.4.2

Systematic name
glutathione:protein-disulfide oxidoreductase

Recommended name
protein-disulfide reductase (glutathione)

Synonyms
GIT <1-4> [1, 3, 33]
GSH-insulin transhydrogenase
disulphide interchange enzyme <1, 3> [19, 35]
glutathione-insulin transhydrogenase
glutathione-protein disulfide oxidoreductase
insulin reductase
protein disulfide reductase (glutathione)
protein disulfide transhydrogenase
protein-disulfide interchange enzyme
protein-disulfide isomerase/oxidoreductase
reductase, protein disulfide (glutathione)
thiol-disulfide oxidoreductase <5> [48]
thiol-protein disulphide oxidoreductase
thiol:protein-disulfide oxidoreductase
Additional information <1-5> (<5> thiol-disulfide oxidoreductase of the thioredoxin family [48]; <5> DsbA protein family including periplasmic protein disulfide: thiol oxidoreductases with high homology in amino acid sequences [44,46]; <1-4> thiol:protein-disulfide oxidoreductase EC 1.8.4.2 and thiol:protein-disulfide isomerase EC 5.3.4.1 are immunological identical [1,2,18]; <1-4> the two activities, cleavage and formation of protein-disulfide bonds, present alternate activities of the same enzyme [1,2,18,36]; <1,2> glutathione-insulin transhydrogenase EC 1.8.4.2 and protein disulfide-isomerase EC 5.3.4.1 activities are not both catalyzed by a single enzyme species [13,30,41,43]; <4> protein-disulfide isomerase EC 5.3.4.1 and protein-disulfide oxidoreductase EC 1.8.4.2 are identical and form the glutathione-insulin transhydrogenase [3]) [1-3, 13, 18, 30, 36, 41, 43, 44, 46, 48]

CAS registry number
9082-53-5

2 Source Organism

<1> *Bos taurus* (ox [23,41]; steer [31]) [1, 2, 8, 10, 16-20, 22-31, 36, 38-43]
<2> *Rattus norvegicus* (Wistar rat [15]; sand rat, i.e. Psammomys obesus [15]) [1-9, 11-15, 21, 23, 34, 38]
<3> *Mus musculus* [1, 15, 33, 35]
<4> *Homo sapiens* (macrophage migration inhibitory factor is a cytokine and pituitary hormone [54]) [1-3, 32, 37, 45, 54]
<5> *Escherichia coli* (DsbG gene [52]; different forms of protein thiol:disulfide oxidoreductases with either oxidizing, gene DsbA, reducing, gene DsbB, or isomerising, gene DsbC and DsbD, activity exist in Escherichia coli [44]; in periplasm: DsbA and DsbB enzymes [44]; thiol-disulfide oxidoreductase like activity from gene DnaJ [49]) [44, 46, 48, 49, 52]
<6> *Vibrio cholerae* (DsbA homologous [44,46]; tcpG gene [46]) [44, 46]
<7> *Haemophilus influenzae* (DsbA homologous [44,46]; Por gene [46]) [44, 46]
<8> *Shigella flexneri* (DsbA homologous [44]) [44]
<9> *Salmonella typhimurium* (DsbA homologous [44]) [44]
<10> *Erwinia chrysanthemi* (DsbA homologous [44]) [44]
<11> *Erwinia carotovora* (DsbA homologous [44]) [44]
<12> *Cricetulus sp.* (chinese hamster [45]) [45]
<13> *Yersinia pestis* (DsbA homologous [44]) [44]
<14> *Klebsiella sp.* (DsbA homologous [44]) [44]
<15> *Pseudomonas syringae* (DsbA homologous [44]) [44]
<16> *Pseudomonas aeruginosa* (DsbA homologous [44]) [44]
<17> *Legionella pneumophila* (DsbA homologous [44]) [44]
<18> *Enterobacter amnigenes* (DsbA homologous [44]) [44]
<19> *Brassica napus* [47]
<20> *Bacillus brevis* (gene bdb [50]) [50]
<21> *Bacillus subtilis* (genes bdbA, bdbB in BdbAB operon [53]; gene bdbC and bdbD in BdbDC operon [51,53]) [51, 53]
<22> *Salmonella enteritidis* (DsbA homologous [44]) [44]

3 Reaction and Specificity

Catalyzed reaction

2 glutathione + protein-disulfide = glutathione disulfide + protein-dithiol (<5> zinc finger protein with several CXXC motifs [49]; <19> mechanism [47]; <4,5,20> C-X-X-C motif in active site for disulfide bonding, either oxidizing, reducing and isomerising activity [44,46,48-50,54]; <4> insulin inducible [32]; <2> high similarity between enzymes of different species and sources, respectively [7]; <1> random mechanism [42]; <1,2,4> random and allosteric mechanism via multiple distinct reaction sites [2])

Reaction type

oxidation
redox reaction
reduction

Natural substrates and products

- **S** GSH + protein disulfide <1-5> (<5> disulfide bonding step in folding pathway of many periplasmic and outer membrane proteins with structural disulfide bonds [44,46]; <2> initial step in sequential insulin degradation [5,12]; <1-4> physiological function [1]; <1-4> modulation of enzymatic activity from latent to active form and vice versa [1]; <1-4> feedback control via insulin in the liver [1,3]; <1-4> reductive degradation and assembly of proteins [1]; <3> enzyme plays a role in formation of intramonomer bonds common to all immunoglobulin molecules [35]; <3> synthesis of protein disulfide bond [35]; <2> enzyme not directly involved in the subcellular processing of receptor-bound internalized insulin [4]; <3> major contributor to the inactivation of oxytoxin by lactating mammary gland [33]) (Reversibility: r <5> [44, 46]; ? <1-4> [1, 3-5, 12, 21, 32, 33, 35, 45]) [1, 3-5, 12, 21, 32, 33, 35, 44-46]
- **P** GSSG + protein-dithiol
- **S** Additional information <21> (<21> bdbC and bdbD catalyze the formation of disulfide bonds that are essential for the DNA binding and uptake machinery [51,53]) [51, 53]
- **P** ?

Substrates and products

- **S** 2,3-dimercaptopropanol + protein disulfide <1-4> (Reversibility: ? <1-4> [1, 42]) [1, 42]
- **P** 2,3-dimercaptopropanol disulfide + protein-dithiol
- **S** 2-mercaptoethanol + protein disulfide <1, 2> (Reversibility: ? <1, 2> [6, 16, 39, 42]) [6, 16, 39, 42]
- **P** 2-mercaptoethanol disulfide + protein-dithiol
- **S** GSH + insulin <1, 4, 5> (<1-4> proinsulin [1,2]) (Reversibility: ? <1, 4, 5> [39, 49, 54]) [1-42, 49, 54]
- **P** GSSG + reduced insulin chain A and B
- **S** GSH + oxytoxin <1-4> (Reversibility: ? <1-4> [1, 2, 39]) [1, 2, 39]
- **P** ?
- **S** GSH + protein disulfide <1-5> (<5> protein disulfide: enzyme itself [52]; <5> protein disulfide: thioredoxin [48]; <5> very rapid disulfide interchange reaction [44]; <1> protein disulfide: e.g. of alcohol dehydrogenase, hexokinase, fructose-1,6-diphosphatase, malate dehydrogenase, glyceraldehyde phosphate dehydrogenase, glycerol phosphate dehydrogenase [10]; <1> prolactin is a poor substrate [31]) (Reversibility: r <1, 5> [27, 42, 44, 48, 52]; ? <1-4> [1-26, 28-41, 45]) [1-42, 44, 45, 48, 52]
- **P** GSSG + protein-dithiol <1-5> [1-42, 52]
- **S** GSH + vasopressin <1-4> (Reversibility: ? <1-4> [1, 2, 39]) [1, 2, 39]
- **P** ?
- **S** GSH + vasotocin <1> (Reversibility: ? <1> [39]) [39]

P ?

S GSSG + reduced ribonuclease <1, 5> (Reversibility: r <5> [49]; ? <1> [10, 39, 42]) [10, 39, 42, 49]

P GSH + oxidized ribonuclease <5> [49]

S L-cysteine + protein disulfide <1, 2> (Reversibility: ? <1, 2> [6, 16, 19, 39]) [6, 16, 19, 39]

P cystine + protein-dithiol

S α-lipoate + protein disulfide <4> (Reversibility: ? <4> [45]) [45]

P ?

S cysteamine + protein disulfide <1-4> (<4> low activity [45]) (Reversibility: ? <1-4> [1, 16, 45]) [1, 16, 45]

P ?

S dihydrolipoamide + protein disulfide <2> (Reversibility: ? <2> [6]) [6]

P ?

S dihydrolipoate + protein disulfide <1, 2> (Reversibility: ? <1, 2> [6, 42]) [6, 42]

P ?

S dihydrolipoic acid + protein disulfide <1-4> (Reversibility: ? <1-4> [1]) [1]

P ?

S dithiothreitol + protein disulfide <1, 2, 4, 5, 19> (<4> low activity [45]; <19> protein disulfide: insulin or thioredoxin [47]) (Reversibility: r <5, 19> [47, 52]; ? <1, 2, 4> [6, 39, 42, 45]) [6, 39, 42, 45, 47, 52]

P ?

S hydroxyethyl disulfide + protein disulfide <4> (Reversibility: ? <4> [45]) [45]

P ?

S thioglycolic acid + protein disulfide <2> (Reversibility: ? <2> [15]) [15]

P ?

S Additional information <1-5> (<5> redox reaction between different Dsn proteins [52]; <5> DsbB protein re-oxidizes the reduced DsbA protein [44,46]; <1-4> enzyme also catalyzes reactivation and folding of protein containing incorrectly paired disulfide bond, e.g.: scrambled ribonuclease [1,2,39]; <1-4> scrambled lysozyme [1,2]; <1,2,4> scrambled trypsin inhibitor and proinsulin [2]; <1-4> immunoglobulin IgM and IgG [1,2]; <1,2> reduction of ricin and other plant thiols [23]; <1> reduction of choleragen [25]; <1-4> thiol:protein-disulfide oxidoreductase EC 1.8.4.2 and thiol:protein-disulfide isomerase EC 5.3.4.1 are immunological identical [1,2,18]; <1-4> the two activities, cleavage and formation of protein-disulfide bonds, present alternate activities of the same enzyme [1,2,18,36]; <1,2> glutathione-insulin transhydrogenase EC 1.8.4.2 and protein disulfide-isomerase EC 5.3.4.1 activities are not both catalyzed by a single enzyme species [13,30,41,43]) [1, 2, 13, 18, 23, 25, 30, 36, 39, 41, 43, 44, 46, 52]

P ?

Inhibitors

1,3-bis(2-chloroethyl)-1-nitrosurea <4, 12> [45]
Ca^{2+} <1-4> (<1-4> slight [1,39]; <19> no inhibition [47]) [1, 39, 40]
Cu^{2+} <1-4> [1, 31, 39]
EDTA <2, 5> (<2> 50% inhibition at 5 mM [6]; <5> oxidation activity [49]) [6, 49]
GSSG <1> (<1> product inhibition [42]) [42]
Hg^{2+} <1-4> [1, 31, 39, 40]
Mg^{2+} <1> (<1> slight inhibition [39]; <19> no inhibition [47]) [39]
N-acetyl-L-tyrosine ethyl ester <1> [31]
N-ethylmaleimide <1> (<1> treatment with thiol prior to incubation with GSH and substrate: no inhibition, preincubation with GSH and thiol reagent: inhibition [22]) [22]
Ni^{2+} <1-4> (<1-4> slight [1]) [1]
S-sulfonated A-chain or B-chain of insulin <1> (<1> product inhibition [42]) [42]
Se^{2+} <1-4> [1, 40]
Triton X-100 <2> (<2> inhibitory for the solubilized enzyme [2]) [2]
Zn^{2+} <1-4> (<1> slight inhibition [39]) [1, 39]
aprotinin <1-4> [1]
arsenite <4, 12> [45]
bacitracin <1-4> (<2> antibiotic [21]) [1, 21]
cysteine <2> (<2> at high concentration [6]) [6]
deoxycholate <1, 2> (<2> inhibitory for the solubilized enzyme [2]) [2, 39]
glucagon <1> [28]
guanidinium hydrochloride <5> (<5> destroys zinc finger thiols of DnaJ protein [49]) [49]
insulin analogs <1> (<1> competitive inhibition of insulin degradation, overview [28]) [28]
iodoacetamide <1> (<1> treatment with thiol prior to incubation with GSH and substrate: no inhibition, preincubation with GSH and thiol reagent: inhibition [22]) [22]
iodoacetate <1, 2> (<1> treatment with thiol prior to incubation with GSH and substrate: no inhibition, preincubation with GSH and thiol reagent: inhibition [22]) [7, 22]
lysolecithin <1-4> [1, 2]
lysophosphatidylcholine <1, 2> [38]
oxytoxin <1> [28]
phenylarsine oxide <4, 12> (<4> complete inhibition at 1 mM [45]) [45]
phosphatidic acid <1-4> [1, 2, 38]
ribonuclease <1> [31]
scrambled forms of ribonuclease and lysozyme <1> (<1> inhibition of insulin degradation [28]) [28]
selenium oxide <4> [45]
vasopressin <1> [28]
Additional information <1-5> (<5> carboxylation of thiol groups abolish reduction activity [49]; <1> overview product inhibition [42]; <1,2> enzyme

preincubated with toxins, e.g. ricin and Viscum album lectin, enhances their inhibitory effects against protein synthesis [23]; <1-4> overview [1]) [1, 23, 42, 49]

Activating compounds

EDTA <1-4> (<1-4> and other chelating agents activate [1,40]; <3> activates [35]; <1> activation negated by metal ions [40]; <2> 5 mM: inhibition [6]) [1, 35, 40]
Triton X-100 <2> (<2> activating the enzyme membrane-bound in the microsomal fraction [2]) [2]
chelating agents <1-4> (<1-4> e.g. EDTA, EGTA, 8-hydroxyquinoline, 1,10-phenanthroline, activation [1,40]) [1, 40]
cysteine disulfide <4> (<4> cysteine residues in the redox active catalytic center are essential for conformation stability [54]) [54]
deoxycholate <2> (<2> activating the enzyme membrane-bound in the microsomal fraction [2]) [2]
histidine <1> (<1> activates, activation negated by metal ions [40]) [40]
phosphatidylethanolamine <1-4> (<1-4> activates at low concentration [1,3,8]) [1, 3, 8]
phospholipids <1, 2> (<1,2> slight activation by all phospholipids tested except for phosphatidic acid and phosphatidylserine, highest activation by phosphatidylethanolamine [38]) [38]
Additional information <1-4> (<1-4> overview [1]) [1]

Metals, ions

Cs^{+} <2> (<2> 0.3 M: activates [6]) [6]
Mn^{2+} <2> (<2> 1 mM: activates [6]) [6]
Zn^{2+} <5> (<5> essential for activity [49]) [49]
phosphate <1> (<1> enhances activity [36]) [36]

Specific activity (U/mg)

0.00046-0.0006 <2> (<2> insulin-treated and normal rats [4]) [4]
0.00081 <1> [30]
0.0035 <4> (<4> A549 carcinoma cells, substrates GSH, cystamine or cystine [45]) [45]
0.0075 <4> (<4> A549 carcinoma cells, substrate dithiothreitol [45]) [45]
0.01 <1> (<1> purified enzyme [41]) [41]
0.03 <4> (<4> A549 carcinoma cells, substrate hydroxyethyl disulfide [45]) [45]
0.067 <4> (<4> A549 carcinoma cells, substrate lipoate [45]) [45]
4.1 <4> [32]
6 <1> (<1> highest activity of all bovine tissues [19]) [19]
9.81 <2> [15]
83 <2> [9]
458 <2> (<2> purified enzyme [11]) [11]
758 <1> (<1> purified enzyme [8]) [8]
3273 <2> (<2> purified enzyme [7]) [7]
6000 <1> (<1> purified enzyme [16]) [16]

16000 <1> (<1> purified enzyme [20]) [20]
Additional information <1-5, 19> (<5> reductase activity after several treatments [49]; <2> tissue distribution [8]; <2> hepatic enzyme activity, active and latent, during growth phases and during liver regeneration [5]; <1-4> assay methods [1,13,24]; <1> specific activity in bovine tissues [19]; <19> fluorometric assay method [47]; <4> specific activities of different human carcinoma cell lines with different thiol and disulfide substrates [45]) [1, 5, 7, 8, 13, 19, 21, 24, 31, 33, 36, 41, 45, 47, 49]

K_m-Value (mM)

0.003-0.012 <2> (insulin, <2> enzymes from different rat tissues [8]) [8]
0.004 <2> (insulin) [21]
0.004-0.0078 <4> (insulin) [37]
0.01 <1-4> (insulin) [1]
0.014 <2> (insulin) [15]
0.016 <2> (insulin) [5]
0.021 <2> (insulin) [7]
0.029 <2> (insulin) [13]
0.031 <2> (insulin) [11]
0.13 <2> (reduced dithiothreitol) [6]
0.14 <2> (dihydrolipoamide) [6]
0.2 <2> (2-mercaptoethanol) [6]
0.2 <2> (dihydrolipoate) [6]
0.27 <2> (cysteine) [6]
0.31 <2> (GSH) [6]
0.8-1.7 <2> (GSH, <2> enzyme from different rat tissues [8]) [8]
1.28 <2> (GSH) [15]
1.43 <2> (GSH) [11]
Additional information <1-4> (<1> comparison of K_m values in different assay methods [24]; <1> K_m values for insulin at various fixed thiol concentrations and for various thiol compounds at fixed insulin concentrations [42]; <2> K_m in different tissues for GSH and insulin [8]) [1, 6-8, 31, 24, 36, 42]

K_i-Value (mM)

Additional information <1> (<1> product inhibition, overview [42]) [42]

pH-Optimum

7 <2> [11]
7.2 <2> [15]
7.5 <1, 2> (<2> GSH [6]) [6, 16, 39]
7.5-8.5 <1> [31]
7.8 <1> [36]
7.9 <19> (<19> thioredoxin [47]) [47]
8 <3> [35]
9 <2> (<2> DL-dihydrolipoate [6]) [6]

pH-Range

6.2-8.4 <19> [47]
6.2-10.2 <1> (<1> pH 6.2: about 10% of activity maximum, pH 10.2: about 20% of activity maximum [39]) [39]
6.5-8.5 <1> [16]

Temperature optimum (°C)

25 <2> (<2> assay at [11]) [11]
37 <1, 2> (<1,2> assay at [8,9,14,22,24,27]) [8, 9, 14, 22, 24, 27]

4 Enzyme Structure

Molecular weight

20000 <5> (<5> DsbB gene product [44]) [44]
21100 <5> (<5> DsbA gene product [44,46]) [44, 46]
37000 <1> (<1> gel filtration [30]) [30]
50000-55000 <2> (<2> gel filtration [11]) [11]
54000 <1> (<1> amino acid sequence determination [17]) [17]
58000 <1> (<1> gel filtration [22]) [22]
60000 <1-3> (<3> SDS-PAGE [35]; <2> SDS-PAGE, a 120000 MW protein also found [11]; <1> SDS-PAGE, polyacrylamide disc-gel electrophoresis [26]; <1> sedimentation equilibrium centrifugation [29]) [11, 26, 29, 35]
63000 <2> (<2> gel filtration [7]) [7]
92000 <1> (<1> gel filtration [29]) [29]
120000 <1, 2> (<2> SDS-PAGE, MW 60000 determined for the main component [11]; <1> MW 180000 also found in traces, monomer has MW 60000 [27]) [11, 27]
150000-200000 <1> (<1> gel filtration [16]) [16]
Additional information <2> (<2> 3 molecular forms: MW 56000 is the major form, MW 51000 is active and only present in spleen, possibly a proteolytic product of 56000 MW protein, MW 67000 is possibly a precursor of 56000 MW protein [1]) [1]

Subunits

? <4, 5> (<4> ? * 12000, SDS-PAGE [54]; <5> ? * 25700, SDS-PAGE [52]) [52, 54]
dimer <1, 5> (<5> 2 * 41000, SDS-PAGE [49]; <1> 2 * 60000, SDS-PAGE, dimerization after prolonged storage at -20°C, freeze-thawing or heating at 60°C, monomers held together by an intermolecular disulfide bond [27]) [27, 49]
monomer <1, 5> (<5> 1 * 21100 [44]; <1> 1 * 60000, SDS-PAGE [27]; <1> 1 * 62500, SDS-PAGE [29]) [27, 29, 44]
trimer <1> (<1> enzyme contains 3 amino-terminal residues, therefore might be composed of 3 polypeptide chains or subunits [17]) [17]

Posttranslational modification

glycoprotein <1> (<1> bovine pancreas: 1.6% carbohydrate [17]; <1> bovine liver: 12% carbohydrate [29]) [17, 29, 39]
phospholipoprotein <2> (<2> part of complex [9]) [1, 9]

5 Isolation/Preparation/Mutation/Application

Source/tissue

adipose tissue <2> [1, 3, 8, 19]
aorta <1> [19]
brain <2> [1, 3, 8, 19]
diaphragm <2> [1, 8]
erythrocyte <4> [1]
heart <1-4> [1, 2, 8, 19]
insulinoma <4> (<4>, glutathione-insulin transhydrogenase is present in an inactive state as a divalent metal ion complex that can be activated by EDTA and/or GSH [37]) [37, 45]
intestine <2> [1, 8, 19]
kidney <1-4> [1, 3, 8, 19]
lens <2> [1]
leukocyte <4> [1]
liver <1-4> [1, 3-9, 11-14, 16, 18-21, 23, 25-27, 29-31, 38, 41, 43]
lung <2, 4> (<4> A549 carcinoma [45]) [1, 3, 8, 19, 45]
lymph node <1> [19]
lymphoid tissue <3> [35]
macrophage <4> [54]
mammary gland <3, 4> (<4> MCF7 carcinoma [45]) [33, 45]
ovary <12> [45]
pancreas <1-4> (<2,3> pancreatic islets [15,34]) [1, 8, 10, 15, 17, 19, 22, 28, 34, 38, 40, 42]
parotid gland <1> [19]
placenta <2> [1, 5]
prostate gland <4> (<4> DU145 carcinoma [45]) [45]
retina <2> [1]
rumen <1> [19]
skeletal muscle <2> [1, 3, 8, 19]
skin fibroblast <4> [32]
spinal cord <1> [19]
spleen <2> [1, 3, 8, 19]
testis <2> [1, 3, 8, 19]
thymus <2> [1, 8]
Additional information <1-4> (<4> higher activity in mononuclear cells than in polymorphonuclear cells [1]; <1,2> relative activities of enzyme in various tissues [8,19]) [1, 8, 19]

Localization

endoplasmic reticulum <2> [1, 4]
extracellular <20, 21> (<20> cell periphery [50]) [50, 51]
membrane <2, 5> (<5> outer cytoplasmic membrane, protein termini localised in the periplasm [44]; <2> bound to [1]; <2> plasma membrane [12]; <2> weakly bound to membrane phospholipid components [14]) [1, 9, 12-14, 44]
microsome <2> (<2> majority of the enzyme located at [1]; <2> weakly bound on cisternal surface of microsomes via phospholipids [1,14]) [1, 9, 11, 13, 14, 38]
periplasm <5> (<5> DsbG protein [52]; <5> DsbA protein soluble, DsbB protein is membrane intrinsic [44]) [44, 46, 52]
secretory granule <2> (<2> of α and β cells of islets of Langerhans [1]) [1]
zymogen granule <2> (<2> of acinar cells [1]) [1]

Purification

<1> (partially [26,30,41]) [8, 10, 16, 17, 20, 22, 26, 27, 29-31, 38, 39, 41]
<2> [7, 11, 21]
<3> [35]
<4> (recombinant wild-type and mutant enzymes [54]) [54]
<5> (recombinant enzyme [49]; recombinant enzyme from periplasm [52]) [49, 52]
<20> (recombinant enzyme [50]) [50]

Cloning

<4> (expression from plasmid in bacterial expression system, amino acid analysis [54]) [54]
<5> (expression from plasmid in Escherichia coli into the periplasm [52]; expression in Escherichia coli from plasmid [49]) [49, 52]
<20> (overexpression from plasmid in Bacillus brevis, DNA sequence analysis [50]) [50]
<21> (systematic gene function analysis programm, expression of diverse constructs in Escherichia coli and Bacillus subtilis [51]) [51]
<2, 4> (characterization of cDNA for human glutathione-insulin transhydrogenase, amino acid sequence analysis, 94% homology to rat amino acid sequence [3]) [3]

Engineering

C57S <4> (<4> site-directed mutagenesis, reduced activity [54]) [54]
C57S/C60S <4> (<4> site-directed mutagenesis, no activity [54]) [54]
C60S <4> (<4> site-directed mutagenesis, no activity [54]) [54]
Additional information <5> (<5> DsbA and DsbB insertion mutants are sensitive to dithiothreitol and benzylpenicillin or Cd^{2+}, Hg^{2+} and Zn^{2+}, pleitropic phenotype [44]) [44]

Application

medicine <5> (<5> DsbA protein is a target for treatment of pathogenic bacteria [44]) [44]

6 Stability

Temperature stability

37 <2> (<2> phosphate buffer, 5 h without loss of activity [6]) [6]
60 <1> (<1> 30 min, 30% loss of activity [31]; <1> 30 min, 50% loss of activity [39]) [31, 39]
65 <1> (<1> 30 min, 90% loss of activity [39]) [39]
80 <1> (<1> 30 min, about 60% loss of activity [31]) [31]

General stability information

<1>, GSH, 10 mM, enhances heat stability [39]
<1>, dimerization after prolonged storage at -20°C, freeze-thawing or heating at 60°C [27]
<1>, insulin, 0.087 mM, decreases heat stability [39]
<2>, EDTA 5mM, 37°C, phosphate buffer, total loss of activity within 30-60 min, can partially be prevented by NADPH [6]
<2>, metal ions required for activity and maintenance of the proper conformation of the enzyme [6]
<3>, highly susceptible to proteolytic attack [35]
<1, 2>, EDTA stabilizes during purification [11, 39]

Storage stability

<1>, -25°C, pH 7.5, phosphate buffer, months without loss of activity [16]
<3>, 4°C, 90% of the enzyme is converted from its native form with MW 60000 to a fragment of MW 40000-45000 [35]

References

[1] Varandani, P.T.: Glutathione-insulin transhydrogenase (protein-disulfide interchange enzyme). Coenzymes and cofactors, Glutathione, Chem. Biochem. Med. Aspects Pt. A (Dolphin D, Poulson R, Avromonic O, eds.) John Wiley & Sons, New York, **3**, 753-765 (1989)

[2] Varandani, P.T.: Mechanistic and structural aspects of glutathione-insulin transhydogenase (protein-disulfide interchange enzyme). Dev. Biochem., **1**, 29-42 (1978)

[3] Morris, J.I.; Varandani, P.T.: Characterization of a cDNA for human glutathione-insulin transhydrogenase (protein-disulfide isomerase/oxidoreductase). Biochim. Biophys. Acta, **949**, 169-180 (1988)

[4] Chowdhary, B.K.; Smith, G.D.; Mahler, R.; Peters, T.J.: Studies on the subcellular localization and role of glutathione-insulin transhydrogenase in rat liver. Biosci. Rep., **3**, 323-329 (1983)

[5] Hern, E.P.; Varandani, P.T.: Nonlatent and latent hepatic glutathione-insulin transhydrogenase activity during perinatal development and liver regeneration in rats and in rat placenta. Biochem. Biophys. Res. Commun., **116**, 909-915 (1983)

[6] Spolter, P.D.; Vogel, J.M.: Rat-liver thiol: protein-disulphide oxidoreductase. Biochim. Biophys. Acta, **167**, 525-537 (1968)

[7] Varandani, P.T.: Insulin degradation. I. Purification and properties of glutathione-insulin transhydrogenase of rat liver. Biochim. Biophys. Acta, **286**, 126-135 (1972)

[8] Chandler, M.L.; Varandani, P.T.: Insulin degradation. II. The widespread distribution of glutathione-insulin transhydrogenase in the tissues of the rat. Biochim. Biophys. Acta, **286**, 136-145 (1972)

[9] Varandani, P.T.: Insulin degradation. V. Unmasking of glutathione-insulin transhydrogenase in rat liver microsomal membrane. Biochim. Biophys. Acta, **304**, 642-659 (1973)

[10] Chandler, M.L.; Varandani, P.T.: Insulin degradation. 8. Protein thiols as cosubstrates for glutathione-insulin transhydrogenase. Biochim. Biophys. Acta, **320**, 258-266 (1973)

[11] Ansorge, S.; Bohley, P.; Kirschke, H.; Langner, J.; Wiederanders, B.; Hanson, H.: Metabolism of insulin and glucagon. Glutathione-insulin transhydrogenase from microsomes of rat liver. Eur. J. Biochem., **32**, 27-35 (1973)

[12] Varandani, P.T.: Insulin degradation. X. Identification of insulin degrading activity of rat liver plasma membrane as glutathione-insulin transhydrogenase. Biochem. Biophys. Res. Commun., **55**, 689-696 (1973)

[13] Ibbetson, A.L.; Freedman, R.B.: Thiol-protein disulphide oxidoreductases. Assay of microsomal membrane-bound glutathione-insulin transhydrogenase and comparison with protein disulphide-isomerase. Biochem. J., **159**, 377-384 (1976)

[14] Hern, E.P.; Varandani, P.T.: Topology of glutathione-insulin transhydrogenase in rat liver microsomes. Biochim. Biophys. Acta, **732**, 170-178 (1983)

[15] Kohnert, K.D.; Hahn, H.J.; Zuhlke, H.; Schmidt, S.; Fiedler, H.: Breakdown of exogenous insulin by Langerhans islets of the pancreas in vitro. Biochim. Biophys. Acta, **338**, 68-77 (1974)

[16] Schneider, F.; Schauer, R.; Martini, O.; Hahn, J.: Reversibility of glutathione-insulin-transhydrogenation (protein disulfide-reductase reaction). Hoppe-Seyler's Z. Physiol. Chem., **348**, 391-400 (1967)

[17] Varandani, P.T.: Insulin degradation. XII. The amino acid composition, amino-terminal, and carbohydrate content of beef pancreatic glutathione-insulin transhydrogenase. Biochim. Biophys. Acta, **371**, 577-581 (1974)

[18] Bjelland, S.; Wallevik, K.; Kroll, J.; Dixon, J.E.; Morin, J.E.; Freedman, R.B.; Lambert, N.; Varandani, P.T.; Nafz, M.A.: Immunological identity between bovine preparations of thiol:protein-disulphide oxidoreductase, glutathione-insulin transhydrogenase and protein-disulphide isomerase. Biochim. Biophys. Acta, **747**, 197-199 (1983)

[19] Bjelland, S.: Tissue distribution and molecular heterogeneity of bovine thiol:protein-disulphide oxidoreductase (disulphide interchange enzyme). Comp. Biochem. Physiol. B, **87**, 907-914 (1987)

[20] Bjelland, S.; Foltmann, B.; Wallevik, K.: Purification of thiol:protein-disulfide oxidoreductase from bovine liver. Anal. Biochem., **142**, 463-466 (1984)

[21] Roth, R.A.: Bacitracin: an inhibitor of the insulin degrading activity of glutathione-insulin transhydrogenase. Biochem. Biophys. Res. Commun., **98**, 431-438 (1981)
[22] Varandani, P.T.; Plumley, H.: Mechanism of action of glutathione-insulin transhydrogenase. Presence of a functional sulfhydryl group for activity. Biochim. Biophys. Acta, **151**, 273-275 (1968)
[23] Barbieri, L.; Batelli, M.G.; Stirpe, F.: Reduction of ricin and other plant toxins by thiol:protein disulfide oxidoreductases. Arch. Biochem. Biophys., **216**, 380-383 (1982)
[24] Chandler, M.L.; Varandani, P.T.: Insulin degradation. XV. Use of different assay methods for the study of mechanism of action of glutathione-insulin transhydrogenase. Biochim. Biophys. Acta, **397**, 307-317 (1975)
[25] Moss, J.; Stanley, S.J.; Morin, J.E.; Dixon, J.E.: Activation of choleragen by thiol: protein disulfide oxidoreductase. J. Biol. Chem., **255**, 11085-11087 (1980)
[26] Pace, M.; Pietta, P.G.; Fiorino, A.; Pocaterra, E.; Dixon, J.E.: Nonspecific reaction of a thiol: protein disulfide oxidoreductase with the disulfide bonds of insulin. Experientia, **41**, 1332-1335 (1985)
[27] Pace, M.; Dixon, J.E.: The nature of the multiple forms of bovine thiol:protein disulfide oxidoreductase. Int. J. Pept. Protein Res., **14**, 409-413 (1979)
[28] Varandani, P.T.; Nafz, M.A.; Chandler, M.L.: Interaction of insulin analogs, glucagon, growth hormone, vasopressin, oxytocin, and scrambled forms of ribonuclease and lysozyme with glutathione-insulin transhydrogenase (thiol: protein-disulfide oxidoreductase): dependence upon conformation. Biochemistry, **14**, 2115-2120 (1975)
[29] Carmichael, D.F.; Morin, J.E.; Dixon, J.E.: Purification and characterization of a thiol:protein disulfide oxidoreductase from bovine liver. J. Biol. Chem., **252**, 7163-7167 (1977)
[30] Hillson, D.A.; Freedman, R.B.: Bovine liver thiol-protein disulphide oxidoreductases. An alternative method for differential purification and resolution of protein disulphide-isomerase and glutathione-insulin transhydrogenase. Biochem. J., **191**, 389-393 (1980)
[31] Tomizawa, H.H.: Glutathione-insulin transhydrogenase (steer liver). Methods Enzymol., **17B**, 515-519 (1971)
[32] Morin, J.E.; Dixon, J.E.; Chang, P.P.; Moss, J.: Identification of thiol:protein disulfide oxidoreductase activity in cultured human fibroblasts: dependence of enzyme activity on growth conditions. Biochem. Biophys. Res. Commun., **111**, 872-877 (1983)
[33] Ferrier, B.M.; Hendrie, J.M.; Cardy, C.A.: Glutathione-insulin transhydrogenase activity in pregnant mouse mammary gland: hormonal influences studied in tissue culture. Can. J. Biochem., **55**, 340-345 (1977)
[34] Kohnert, K.D.; Jahr, H.; Schmidt, S.; Hahn, H.J.; Zuhlke, H.: Demonstration of insulin degradation by thiol-protein disulfide oxidoreductase (glutathione-insulin transhydrogenase) and proteinases of pancreatic islets. Biochim. Biophys. Acta, **422**, 254-259 (1976)
[35] Roth, R.A.; Koshland, M.E.: Role of disulfide interchange enzyme in immunoglobulin synthesis. Biochemistry, **20**, 6594-6599 (1981)

[36] Lambert, N.; Freedman, R.B.: Kinetics and specificity of homogeneous protein disulphide-isomerase in protein disulphide isomerization and in thiol-protein-disulphide oxidoreduction. Biochem. J., **213**, 235-243 (1983)
[37] Varandani, P.T.: Insulin degradation in insulinoma: evidence for the occurrence of an inactive form of glutathione-insulin transhydrogenase and for the absence of insulin A and B chains degrading protease(s). Biochem. Biophys. Res. Commun., **60**, 1119-1126 (1974)
[38] Varandani, P.T.; Nafz, M.A.: Interaction of glutathione-insulin transhydrogenase (disulfide interchange enzyme) with phospholipids. Biochim. Biophys. Acta, **438**, 358-369 (1976)
[39] Morin, J.E.; Carmichael, C.F.; Dixon, J.E.: Characterization, kinetics and comparative properties of thiol:protein disulfide oxidoreductase. Arch. Biochem. Biophys., **189**, 354-363 (1978)
[40] Varandani, P.T.; Nafz, M.A.: Inhibition of glutathione-insulin transhydrogenase by metal ions and activation by histidine and other chelating agents. Biochim. Biophys. Acta, **832**, 7-13 (1985)
[41] Hawkins, H.C.; Freedman, R.B.: Thiol-protein disulphide oxidoreductases. Differences between protein disulphide-isomerase and glutathione-insulin transhydrogenase activities in ox liver. Biochem. J., **159**, 385-393 (1976)
[42] Chandler, M.L.; Varandani, P.T.: Kinetic analysis of the mechanism of insulin degradation by glutathione-insulin transhydrogenase (thiol: protein-disulfide oxidoreductase). Biochemistry, **14**, 2107-2115 (1975)
[43] Hillson, D.A.; Freedman, R.B.: Resolution of protein disulphide-isomerase and glutathione-insulin transhydrogenase activities by covalent chromatography. Biochem. J., **191**, 373-388 (1980)
[44] Fabianek, R.A.; Hennecke, H.; Thony-Meyer, L.: Periplasmic protein thiol:-disulfide oxidoreductases of Escherichia coli. FEMS Microbiol. Rev., **24**, 303-316 (2000)
[45] Biaglow, J.E.; Donahue, J.; Tuttle, S.; Held, K.; Chrestensen, C.; Mieyal, J.: A method for measuring disulfide reduction by cultured mammalian cells: relative contributions of glutathione-dependent and glutathione-independent mechanisms. Anal. Biochem., **281**, 77-86 (2000)
[46] Loferer, H.; Hennecke, H.: Protein disulfide oxidoreductases in bacteria. Trends Biochem. Sci., **19**, 169-171 (1994)
[47] Heuck, A.P.; Wolosiuk, R.A.: Di-fluoresceinthiocarbamyl-insulin: A fluorescent substrate for the assay of protein disulfide oxidoreductase activity. Anal. Biochem., **248**, 94-101 (1997)
[48] Moessner, E.; Iwai, H.; Glockshuber, R.: Influence of the pKa value of the buried, active-site cysteine on the redox properties of thioredoxin-like oxidoreductases. FEBS Lett., **477**, 21-26 (2000)
[49] Tang, W.; Wang, C.C.: Zinc fingers and thiol-disulfide oxidoreductase activities of chaperone DnaJ. Biochemistry, **40**, 14985-14994 (2001)
[50] Ishihara, T.; Tomita, H.; Hasegawa, Y.; Tsukagoshi, N.; Yamagata, H.; Udaka, S.: Cloning and characterization of the gene for a protein thiol-disulfide oxidoreductase in Bacillus brevis. J. Bacteriol., **177**, 745-749 (1995)
[51] Meima, R.; Eschevins, C.; Fillinger, S.; Bolhuis, A.; Hamoen, L.W.; Dorenbos, R.; Quax, W.J.; Van Dijl, J.M.; Provvedi, R.; Chen, I.; Dubnau, D.; Bron,

S.: The bdbDC operon of Bacillus subtilis encodes thiol-disulfide oxidoreductases required for competence development. J. Biol. Chem., **277**, 6994-7001 (2002)
[52] Van Straaten, M.; Missiakas, D.; Raina, S.; Darby, N.J.: The functional properties of DsbG, a thiol-disulfide oxidoreductase from the periplasm of Escherichia coli. FEBS Lett., **428**, 255-258 (1998)
[53] Dorenbos, R.; Stein, T.; Kabel, J.; Bruand, C.; Bolhuis, A.; Bron, S.; Quax, W.J.; Van Dijl, J.M.: Thiol-disulfide oxidoreductases are essential for the production of the lantibiotic sublancin 168. J. Biol. Chem., **277**, 16682-16688 (2002)
[54] Kleemann, R.; Rorsman, H.; Rosengren, E.; Mischke, R.; Mai, N.T.; Bernhagen, J.: Dissection of the enzymatic and immunologic functions of macrophage migration inhibitory factor. Full immunologic activity of N-terminally truncated mutants. Eur. J. Biochem., **267**, 7183-7192 (2000)

Glutathione-CoA-glutathione transhydrogenase 1.8.4.3

1 Nomenclature

EC number
1.8.4.3

Systematic name
coenzyme A:glutathione-disulfide oxidoreductase

Recommended name
glutathione-CoA-glutathione transhydrogenase

Synonyms
glutathione coenzyme A-glutathione transhydrogenase
glutathione-coenzyme A glutathione disulfide transhydrogenase
glutathione:coenzyme A-glutathione transhydrogenase
transhydrogenase, glutathione-coenzyme A glutathione disulfide

CAS registry number
37256-48-7

2 Source Organism

<1> *Bos taurus* [1]
<2> *Rattus norvegicus* [1, 2]

3 Reaction and Specificity

Catalyzed reaction
CoA + glutathione disulfide = CoA-glutathione + glutathione

Reaction type
oxidation
redox reaction
reduction

Natural substrates and products
S CoA-glutathione + GSH <1, 2> (Reversibility: r <1, 2> [1]) [1]
P CoA + GSSG

Substrates and products
S CoA-glutathione + GSH <1, 2> (Reversibility: r <1, 2> [1, 2]) [1, 2]
P CoA + GSSG

S cysteine-glutathione + GSH <1, 2> (Reversibility: ? <1, 2> [1]) [1]
P cysteine + GSSG
S pantetheine-glutathione + GSH <1> (Reversibility: ? <1> [1]) [1]
P pantetheine + GSSG
S thiolethanolamine-glutathione + GSH <1> (Reversibility: ? <1> [1]) [1]
P thioethanolamine + GSSG
S Additional information <1, 2> (<1,2> catalyzes GSH-GSSG exchange reaction [1]; <1,2> possible functions: 1. salvage of the coenzyme form of pantothenic acid which might be lost to the cell as CoA-glutathione, 2. maintenance of the free CoA-SH at cellular level [1]) [1]
P ?

Specific activity (U/mg)
0.513 <1> [1]
12000 <2> (<2> transhydrogenase of normal rat [2]) [2]
13300 <2> (<2> transhydrogenase of diabetic rat [2]) [2]

K_m-Value (mM)
0.045 <1> (CoA-glutathione) [1]
0.33 <1> (GSH) [1]

pH-Optimum
8.2 <1> [1]

pH-Range
7-9 <1> (<1> pH 7: about 30% of activity maximum, pH 9: about 75% of activity maximum [1]) [1]

Temperature optimum (°C)
25 <1, 2> (<1,2> assay at [1]) [1]

4 Enzyme Structure

Molecular weight
120000 <1> (<1> gel filtration [1]) [1]

5 Isolation/Preparation/Mutation/Application

Source/tissue
brain <2> [1]
heart <2> (<2> low activity [1]) [1]
kidney <1, 2> [1]
liver <2> [1, 2]
lung <2> [1]
muscle <2> [1]
pancreas <2> [1]

Purification

<1> [1]

6 Stability

General stability information

<1>, inactivation during storage is largely reversed by glutathione [1]

Storage stability

<1>, -12°C, 6 weeks, stable [1]
<1>, -12°C, 7 months, 70% loss of activity [1]

References

[1] Chang, S.H.; Wilken, D.R.: Participation of the unsymmetrical disulfide of coenzyme A and glutathione in an enzymatic sulfhydryl-disulfide interchange. J. Biol. Chem., **241**, 4251-4260 (1966)
[2] Dyar, R.E.; Wilken, D.R.: Rat liver levels of coenzyme A-glutathione and of enzymes in its metabolism. Arch. Biochem. Biophys., **153**, 619-626 (1972)

Glutathione-cystine transhydrogenase 1.8.4.4

1 Nomenclature

EC number
1.8.4.4

Systematic name
glutathione:cystine oxidoreductase

Recommended name
glutathione-cystine transhydrogenase

Synonyms
GSH-cystine transhydrogenase
NADPH-dependent GSH-cystine transhydrogenase
transhydrogenase, glutathione-cystine

CAS registry number
37256-49-8

2 Source Organism

<1> *Saccharomyces cerevisiae* [1-3]
<2> *Rattus norvegicus* [4-6]

3 Reaction and Specificity

Catalyzed reaction
2 glutathione + cystine = oxidized glutathione + 2 cysteine

Reaction type
oxidation
redox reaction
reduction

Natural substrates and products
S cystine + GSH <1, 2> (Reversibility: ? <1, 2> [1-6]) [1-6]
P cysteine + GSSG

Substrates and products
S D-cystine + GSH <1> (Reversibility: ? <1> [1]) [1]
P D-cysteine + GSSG
S GSSG + GSH <1> (Reversibility: ? <1> [3]) [3]

P GSSG + GSH
S L-cystine + GSH <1> (Reversibility: ? <1> [1]) [1]
P L-cysteine + GSSG
S L-cystine diamide + GSH <1> (Reversibility: ? <1> [1]) [1]
P ?
S L-cystinyldiglycine + GSH <1> (Reversibility: ? <1> [1]) [1]
P L-Cys-Gly-Gly + GSSG
S β-hydroxyethyl disulfide + GSH <1> (Reversibility: ? <1> [1]) [1]
P ?
S cystine + GSH <1, 2> (Reversibility: ? <1, 2> [1-6]) [1-6]
P cysteine + GSSG
S diacetyl-L-cystine + GSH <1> (Reversibility: ? <1> [1]) [1]
P ?
S dimethyl disulfide + GSH <1> (Reversibility: ? <1> [3]) [3]
P ?
S dithiodiglycolic acid + GSH <1> (Reversibility: ? <1> [3]) [3]
P ?
S homocystine + GSH <1> (Reversibility: ? <1> [1, 3]) [1, 3]
P ?
S lipoic acid oxidized + GSH <1> (Reversibility: ? <1> [3]) [3]
P ?
S Additional information <1> (<1> at low substrate concentration the reaction rate is greater with L-cystine than with any of the other substances [1-3]; <1> inactive with: insulin and other proteins [2]; <1> inactive with albumin and kreatin [3]) [1-3]
P ?

Inhibitors

Cd^{2+} <1> [3]
Co^{2+} <1> [3]
Cu^{2+} <1> [3]
Hg^{2+} <1> [3]
Mn^{2+} <1> [3]
MoO_3 <1> [3]
Na_2HAsO_4 <1> [3]
Zn^{2+} <1> [3]
iodoacetate <1> [3]

Activating compounds

EDTA <1> [3]

Specific activity (U/mg)

61.6 <1> (<1> β-hydroxyethyldisulfide [1,2]) [1, 2]

K_m-Value (mM)

0.23 <2> (cystine) [6]
1.67 <2> (glutathione) [6]

pH-Optimum
8.6 <1> (<1> pH 7.8 chosen for the assay to achieve smaller nonenzymatic rate [1,2]) [1, 2]

pH-Range
6-8.5 <1> (<1> pH 6.0: 10% of activity maximum, pH 8.5: activity maximum [1]) [1]

Temperature optimum (°C)
23 <1> (<1> assay at [1]) [1]

4 Enzyme Structure

Molecular weight
150000 <1> (<1> gel filtration [1]) [1]

5 Isolation/Preparation/Mutation/Application

Source/tissue
intestinal mucosa <2> [5]
kidney <2> [4]
liver <2> [4]
small intestine <2> [4, 6]
Additional information <2> (<2> no activity in brush border [5]) [5]

Localization
cell associated <1> [1, 2]
soluble <2> (<2> supernatant fraction [5,6]) [5, 6]

Purification
<1> [1-3]

6 Stability

Temperature stability
4 <2> (<2> 24 h, complete inactivation [6]) [6]
54 <1> (<1> crude state, quite labile to heat above 54°C [2]) [2]
60 <2> (<2> denaturation, crude state [6]) [6]

General stability information
<1>, EDTA stabilizes [1]
<1>, ammonium sulfate, 1 M, complete stabilization, 6 days at 20°C [1]
<1>, cystine, alone: no stabilization, in addition to glutathione: increases stabilizing effect [1]
<1>, ethylene glycol, 10%, stabilizes [1, 2]
<1>, glutathione stabilizes, stabilizing effect increased by cystine [1, 2]
<1>, glycerol, no stabilization [1]

Storage stability

<1>, -20°C, pH 5.8, 1 mM glutathione, several months [1, 2]
<2>, 4°C, 24 h, complete inactivation, crude state [6]

References

[1] Nagai, S.; Black, S.: A thiol-disulfide transhydrogenase from yeast. J. Biol. Chem., **243**, 1942-1947 (1968)
[2] Nagai, S.: Thiol-disulfide transhydrogenase (yeast). Methods Enzymol., **17B**, 510-515 (1971)
[3] Minoda, Y.; Kurane, R.; Yamada, K.: Thiol-disulfide transhydrogenase from baker's yeast and a new method for the direct assay of an enzyme-catalyzed thiol-disulfide interchange activity. Agric. Biol. Chem., **37**, 2511-2516 (1973)
[4] Wendell, P.L.: Distribution of glutathione reductase and detection of glutathione-cystine transhydrogenase in rat tissues. Biochim. Biophys. Acta, **159**, 179-181 (1968)
[5] States, B.; Segal, S.: Distribution of glutathione-cystine transhydrogenase activity in subcellular fractions of rat intestinal mucosa. Biochem. J., **113**, 443-444 (1969)
[6] States, B.; Segal, S.: Interrelationship of glutathione-cystine transhydrogenase and glutathione reductase in developing rat intestine. Biochem. J., **132**, 623-631 (1973)

Methionine-S-oxide reductase 1.8.4.5

1 Nomenclature

EC number
1.8.4.5

Systematic name
L-methionine:thioredoxin-disulfide S-oxidoreductase

Recommended name
methionine-S-oxide reductase

Synonyms
acetylmethionine sulfoxide reductase
methionine S-oxide reductase
methionine sulfoxide reductase
peptide methionine sulfoxide reductase
reductase, methionine sulfoxide

CAS registry number
70248-65-6

2 Source Organism

<1> *Escherichia coli* [1, 4, 6, 10]
<2> *Saccharomyces cerevisiae* [2, 9]
<3> *yeast* (enzyme system consisting of 3 separable proteins which catalyze the specific reduction of L(-)-methionine sulfoxide to methionine by NADPH [3]; 2 isomers, specific for S-form or R-form of substrate, MsrA resp. MsrB [5]) [3, 5]
<4> *Mus musculus* (2 isomers, specific for S-form or R-form of substrate, MsrA resp. MsrB [5]) [5, 7, 9]
<5> *Xenopus sp.* [8]
<6> *Drosophila melanogaster* [9]
<7> *Saccharomyces cerevisiae* (cosmid 9379) [11]
<8> *Bos taurus* [11]

3 Reaction and Specificity

Catalyzed reaction

L-methionine + thioredoxin disulfide = L-methionine S-oxide + thioredoxin (<1> ping-pong mechanism with sulfenic acid intermediate [10])

Reaction type

oxidation
redox reaction
reduction

Natural substrates and products

S L-methionine S-oxide + reduced thioredoxin <1, 3-6> (<1> methionine sulfoxide can support the growth of a methionine auxotroph [4]; <3,4,6> involved in response to oxidative stress [5,7,9]; <1> required for cadmium resistance, repair of oxidized calmodulin [6]; <5> possibly involved in regulation of protein activities [8]) (Reversibility: ? <1, 3, 4, 6> [4-7, 9]; r <5> [8]) [4-9]
P L-methionine + oxidized thioredoxin

Substrates and products

S L-methionine S-oxide + reduced thioredoxin <1-5> (<4,6> SelR is specific for methionine-R-sulfoxide [9]) (Reversibility: r <1, 2> [1, 2]) [1-9]
P L-methionine + oxidized thioredoxin
S dabsyl methionine-R-sulfoxide + reduced thioredoxin <4, 6> (Reversibility: ? <4, 6> [9]) [9]
P dabsyl methionine + oxidized thioredoxin
S dimethylsulfoxide oxide + reduced thioredoxin <2> (Reversibility: ? <2> [2]) [2]
P ? + oxidized thioredoxin
S Additional information <1, 3> (<1,2> reductase catalyzes reduction of methionine sulfoxide to methionine, no reduction of protein bound sulfoxide [1, 2]; <1> in the reverse reaction, dithiothreitol but not mercaptoethanol or glutathione may replace thioredoxin [1,2]; <3> enzyme system consisting of 3 separable proteins which catalyze the specific reduction of L(-)-methionine sulfoxide to methionine by NADPH [3]; <1> thioredoxin may be replaced by NADPH [4]; <7,8> assay and determination of methionine sulfoxide [11]) [1-4, 11]
P ?

Inhibitors

arsenite <2, 3> [2, 3]
dimethylsulfoxide <2> [2]
formaldehyde <2> [2]
iodoacetamide <2, 3> [2, 3]
methyl ethyl sulfoxide <2> [2]

Metals, ions

Se <4> (<4> recombinant protein from E. coli [9]) [9]
Zn <4, 6> (<4,6> recombinant protein from E. coli [9]) [9]

Turnover number (min^{-1})

0.6 <1> (methionine S-oxide, <1> MsrB isomer [6]) [6]
20 <1> (methionine S-oxide, <1> MsrA isomer [6]) [6]

Specific activity (U/mg)

Additional information <1> [1]

K_m-Value (mM)

0.0003 <1> (methionine sulfoxide) [1]

4 Enzyme Structure

Subunits

? <1> (<1> ? * 21000, SDS-PAGE [1]) [1]

5 Isolation/Preparation/Mutation/Application

Source/tissue

oocyte <5> [8]

Purification

<1> [1]
<7> (<7> recombinant protein from Escherichia coli [11]) [11]
<8> (<8> recombinant protein from Escherichia coli [11]) [11]

Cloning

<1> (<1> gene of MsrB isomer [6]) [6]

Engineering

Additional information <1, 3, 4> (<1> mutants with truncated N- or C-termini [10]; <3> mutants lacking/overexpressing MsrA gene for reduction of S-isomer, sensitivity to oxidative stress depends on intracellular level of MsrA product [5]; <4> knock-out mice without MsrA activity specific for reduction of S-isomer, sensitivity to oxidative stress depends on level of MsrA product [5]; <4> knock-out mice without MsrB activity, sensitivity to oxidative stress depends on level of MsrA product [7]) [5, 7, 10]

6 Stability

Temperature stability

90 <1> (<1> 2 min, 50% loss of activity [1]) [1]

References

[1] Ejiri, S.; Weissbach, H.; Brot, N.: The purification of methionine sulfoxide reductase from Escherichia coli. Anal. Biochem., **102**, 393-398 (1980)
[2] Gibson, R.M.; Large, P.J.: The methionine sulphoxide reductase activity of the yeast dimethyl sulphoxide reductase system. FEMS Microbiol. Lett., **26**, 95-99 (1985)
[3] Black, S.; Harte, E.M.; Hudson, B.; Wartolofsky, L.: A specific enzymatic reduction of L(-)methionine sulfoxide and a related nonspecific reduction of disulfides. J. Biol. Chem., **235**, 2910-2916 (1960)
[4] Ejiri, S.; Weissbach, H.; Brot, N.: Reduction of methionine sulfoxide to methionine by Escherichia coli. J. Bacteriol., **139**, 161-164 (1979)
[5] Stadtman, E.R.; Moskovitz, J.; Berlett, B.S.; Levine, R.L.: Cyclic oxidation and reduction of protein methionine residues is an important antioxidant mechanism. Mol. Cell. Biochem., **234-235**, 3-9 (2002)
[6] Grimaud, R.; Ezraty, B.; Mitchell, J.K.; Lafitte, D.; Briand, C.; Derrick, P.J.; Barras, F.: Repair of oxidized proteins: identification of a new methionine sulfoxide reductase. J. Biol. Chem., **276**, 48915-48920 (2001)
[7] Moskovitz, J.; Bar-Noy, S.; Williams, W.M.; Requena, J.; Berlett, B.S.; Stadtman, E.R.: Methionine sulfoxide reductase (MsrA) is a regulator of antioxidant defense and lifespan in mammals. Proc. Natl. Acad. Sci. USA, **98**, 12920-12925 (2001)
[8] Ciorba, M.A.; Heinemann, S.H.; Weissbach, H.; Brot, N.; Hoshi, T.: Modulation of potassium channel function by methionine oxidation and reduction. Proc. Natl. Acad. Sci. USA, **94**, 9932-9937 (1997)
[9] Kryukov, G.V.; Kumar, R.A.; Koc, A.; Sun, Z.; Gladyshev, V.N.: Selenoprotein R is a zinc-containing stereo-specific methionine sulfoxide reductase. Proc. Natl. Acad. Sci. USA, **99**, 4245-4250 (2002)
[10] Boschi-Muller, S.; Azza, S.; Branlant, G.: E. coli methionine sulfoxide reductase with a truncated N terminus or C terminus, or both, retains the ability to reduce methionine sulfoxide. Protein Sci., **10**, 2272-2279 (2001)
[11] Moskovitz, J.; Berlett, B.S.; Poston, J.M.; Stadtman, E.R.: Methionine sulfoxide reductase in antioxidant defense. Methods Enzymol., **300**, 239-244 (1999)

Protein-methionine-S-oxide reductase 1.8.4.6

1 Nomenclature

EC number
1.8.4.6

Systematic name
protein-L-methionine:thioredoxin-disulfide S-oxidoreductase

Recommended name
protein-methionine-S-oxide reductase

Synonyms
fruit-ripening protein E4
K05_orf151
peptide Met(O) reductase
protein-methionine-S-oxide reductase
met(O)-peptide reductase
methionine sulfoxide peptide reductase
peptide methionine sulfoxide reductase
protein (methionine sulfoxide) reductase
reductase, methionine sulfoxide (protein)

CAS registry number
78206-57-2

2 Source Organism

<1> *Escherichia coli* (the central and the C-terminus domains are separate proteins [24]) [1, 3-6, 8, 10, 11, 15, 16, 24]
<2> *Homo sapiens* (HeLa cells [4]; the central and the C-terminus domains are separate proteins [24]) [1-5, 15, 23, 24]
<3> *Bos taurus* [2, 10, 12, 14, 15, 20, 21]
<4> *Rattus norvegicus* [4, 15, 18]
<5> *Euglena gracilis* [4]
<6> *Tetrahymena pyriformis* [4]
<7> *Spinacia oleracea* (spinach [4]) [4]
<8> *Gossypium hirsutum* (cotton [7]) [7]
<9> *Pisum sativum* (pea [7]) [7]
<10> *Triticum aestivum* (wheat [7]) [7]
<11> *Solanum tuberosum* (potato [7]) [7]
<12> *Streptococcus pneumoniae* [8, 15]

<13> *Neisseria gonorrhoeae* [8, 15]
<14> *Saccharomyces cerevisiae* [9, 15]
<15> *Saccharomyces cerevisiae* [10]
<16> *Arabidopsis sp.* (ecotype Columbia [13]) [13, 15]
<17> *Erwinia chrysanthemi* [15]
<18> *Mycobacterium tuberculosis* [16]
<19> *Mycobacterium bovis* (bacillus Calmette-Guerin [16]) [16]
<20> *Mycobacterium microti* [16]
<21> *Mycoplasma genitalium* [17]
<22> *Mycoplasma pneumoniae* [17]
<23> *Ochrobactrum anthropi* [19]
<24> *Neisseria meningitidis* [22]
<25> *Neisseria gonorrhoeae* (fusion of the central and C terminus reductase domains [24]) [24]
<26> *Neisseria meningiditis* (fusion of the central and C terminus reductase domains [24]) [24]
<27> *Haemophilus influenzae* (fusion of the central and C terminus reductase domains [24]) [24]
<28> *Streptococcus pneumoniae* (fusion of the central and C terminus reductase domains [24]) [24]
<29> *Helicobacter pylori* (fusion of the central and C terminus reductase domains [24]) [24]
<30> *Streptococcus gordonii* (fusion of the central and C terminus reductase domains [24]) [24]
<31> *Treponema pallidum* (has the domain order switched, with the C-terminus domain at the N-terminus [24]) [24]
<32> *Arabidopsis thaliana* (lacks the central domain and has tandem C-terminus domains [24]) [24]

3 Reaction and Specificity

Catalyzed reaction

protein L-methionine + thioredoxin disulfide = protein L-methionine S-oxide + thioredoxin

Reaction type

oxidation
redox reaction
reduction

Natural substrates and products

S protein L-methionine S-oxide + reduced thioredoxin <1-4, 12-14, 16, 17, 21, 23> (<1> biological reducing system consists of NADPH, thioredoxin and thioredoxin reductase, enzymatic reduction can restore biological activity of an inactive oxidized protein or peptide [1,5,6]; <1> e.g. N-formyl-L-methionyl-sulfoxide-L-leucyl-L-phenylalanine [5]; <1> α-1-proteinase inhibitor [6]; <1,12,13> the enzyme is responsible for the production of functional adhesins on the surfaces of the bacterial pathogens [8]; <1> the enzyme is implicated in the defense of organisms against oxidative stress and in the regulation of processes involving peptide methionine oxidation/reduction [11]) (Reversibility: ? <1-4, 12-14, 16, 17, 21, 23> [1, 5, 6, 8, 11, 15, 17, 18, 19, 21, 23]) [1, 5, 6, 8, 11, 15, 17, 18, 19, 21, 23]

P protein L-methionine + oxidized thioredoxin <1-4, 12-14, 16, 17, 21, 23> [1, 5, 6, 8, 11, 15, 17, 18, 19, 21, 23]

S Additional information <1-4, 12-14, 16, 17, 21> (<12> the enzyme is involved in bacterial adherence: decrease of the ability to bind to lung epithelial cells and human vein epithelial cells observed in a mutant [8, 15]; <1> the enzyme is involved in bacterial adherence: an enzyme mutant decreases type 1 fimbriae-mediated mannose-dependent agglutination of erythrocytes [8, 15]; <17> the enzyme is involved in bacterial adherence: an enzyme mutant of the plant pathogen not only is more sensitive to oxidative stress, but has decreased motility and loses its ability to cause systemic invasion of whole plants [15]; <21> a mutant, constructed through homologous recombination, displays markedly reduced adherence to sheep erythrocytes, is incapable of growing in hamsters and exhibits hypersensitivity to hydrogen peroxide when compared to wild-type, the enzyme is important for the maintenance of cytoadherence and virulence potential [17]; <4> the down-regulation of the enzyme can contribute to the accumulation of oxidized protein that is associated with the aging process [18]; <3> the enzyme may be important in increasing the lifespan in other organisms including humans [21]) [8, 15, 17, 18, 21]

P ?

Substrates and products

S DL-methionine-R,S-sulfoxide + reduced acceptor <24> (<24> mixture of 4 diastereomers, the substrate is quantitatively reduced by the fused subdomain, only 50% of the mixture of diasteromers is reduced either by the central subdomain or the C terminus subdomain, only L-methionine-S-sulfoxide and D-methionine-S-sulfoxide are quantitatively reduced by the central subdomain, D-methionine-R-sulfoxide and L-methionine-R-sulfoxide are only reduced by the C terminus subdomain, reduced acceptor is dithiothreitol or thioredoxin [22]) (Reversibility: ? <24> [22]) [22]

P DL-methionine + oxidized acceptor <24> [22]

S N-(9-fluorenylmethoxycarbonyl)-methionine sulfoxide + reduced acceptor <8-11> (<8-11> reduced acceptor is dithiothreitol [7]) (Reversibility: ? <8-11> [7]) [7]

P N-(9-fluorenylmethoxycarbonyl)-methionine + oxidized acceptor <8-11> [7]

S N-acetyl-L-methionine sulfoxide + reduced acceptor <1-3, 12-14, 16, 18, 23> (<1,2> good substrate [1]; <2,3> complete requirement for a dithiol [2]; <3> the reduced acceptor is either dithiothreitol or thioredoxin [12]; <23> crude extracts containing the methionine sulfoxide reductase-glutathione S-transferase fusion protein, as well as the purified enzyme are able to reduce N-acetyl-L methionine sulfoxide to N-acetyl-L-methionine [19]) (Reversibility: ? <1-3, 12-14, 16, 18, 23> [1-3, 8, 9, 12, 13, 15, 16, 19]) [1-3, 8, 9, 12, 13, 15, 16, 19]

P N-acetyl-L-methionine + oxidized acceptor <1-3, 12-14, 16, 18, 23> [1-3, 8, 9, 12, 13, 15, 16, 19]

S N-formyl-L-methionyl-sulfoxide-L-leucyl-L-phenylalanine + reduced acceptor <1, 2> (<1,2> reduced acceptor is dithiothreitol [5]) (Reversibility: ? <1, 2> [1, 5]) [1, 5]

P N-formyl-L-methionyl-L-leucyl-L-phenylalanine + oxidized acceptor <1, 2> (<1,2> the product of the enzymatic reduction shows chemotactic activity [1,5]) [1, 5]

S dabsyl-methionine sulfoxide + reduced acceptor <1, 3, 4, 14, 15> (<1,3,4,15> reduced acceptor is dithiothreitol [10,18]) (Reversibility: ? <1, 3, 4, 14, 15> [9, 10, 18]) [9, 10, 18]

P dabsyl-methionine + oxidized acceptor <1, 3, 4, 14, 15> [9, 10, 18]

S methionine sulfoxide-enkephalin + reduced acceptor <1> (<1> reduced acceptor is dithiothreitol [4]) (Reversibility: ? <1> [4]) [4]

P methionine-enkephalin + oxidized acceptor <1> [4]

S methionine-R-sulfoxide + reduced acceptor <1, 2, 25> (Reversibility: ? <1, 2, 25> [24]) [24]

P methionine + oxidized acceptor <1, 2, 25> [24]

S methionine-S-sulfoxide + reduced acceptor <25> (Reversibility: ? <25> [24]) [24]

P methionine + oxidized acceptor <25> [24]

S oxidized α-1-antitrypsin + reduced acceptor <1, 2, 16> (<1> other name for substrate: α-1 proteinase inhibitor, reduced acceptor is dithiothreitol [6]) (Reversibility: ? <1, 2, 16> [1, 6, 13, 15]) [1, 6, 13, 15]

P α-1-antitrypsin + oxidized acceptor <1, 2, 16> [1, 6, 13, 15]

S protein L-methionine S-oxide + reduced thioredoxin <1-7, 14, 15, 18> (<1,2,14,18> dithiothreitol can replace reduced thioredoxin [1,4,9,16]; <1> dithiothreitol can not be replaced by 2-mercaptoethanol [4]; <1-3> peptide L-methionine S-oxide can replace protein L-methionine, e.g. E. coli ribosomal protein [1,2,4]; <1,3,15> the enzyme has a high specificity toward L-methionine S-sulfoxide [10,15]; <1> E. coli lacking the enzyme are hypersensitive to killing not only by hydrogen peroxide, but also by nitrite and S-nitrosoglutathione [16]) (Reversibility: ? <1-7, 14, 15, 18> [1, 2, 4-6, 9-11, 15, 16]) [1, 2, 4-6, 10, 11, 15, 16]

P protein L-methionine + oxidized thioredoxin <1-7, 14, 15, 18> (<1-3> or peptide L-methionine, or oxidized dithiothreitol [1,2,4-6]) [1, 2, 4-6, 9-11, 15, 16]

S Additional information <23> (<23> the enzyme is also able to reduce free D- and L-methionine sulfoxide, as well as S-(-)-methyl *p*-tolyl sulfoxide, tetramethylene sulfoxide, D-methionine sulfoxide, L-methionine sulfoxide, dimethylsulfoxide, diphenyl sulfoxide and L-ethionine sulfoxide [19]) [19]

P ?

Inhibitors

H_2O_2 <3> (<3> inactivates [12]; <23> no effect [19]) [12]
iodoacetamide <1, 3, 15> [10]
methionine sulfoxide-L12 <1, 2> (<1> 0.1 mM, less than 20% inhibition [4]; <1,2> inhibits the N-acetyl-L-methionine sulfoxide reduction [3]) [3, 4]

Cofactors/prosthetic groups

Additional information <3> (<3> the enzyme activity is independent of cofactors [12]) [12]

Activating compounds

4-chlorophenol <23> (<23> induces the enzyme [19]) [19]
KCl <1, 3, 15> (<1,3,15> increases the specific activity of the enzyme in a linear manner [10]) [10]
Na_2SO_4 <1, 3, 15> (<1,3,15> increases the specific activity of the enzyme in a linear manner [10]) [10]
NaCl <1, 3, 15> (<1,3,15> increases the specific activity of the enzyme in a linear manner [10]; <23> no effect [19]) [10]
NaF <1, 3, 15> (<1,3,15> increases the specific activity of the enzyme in a linear manner [10]) [10]
phenol <23> (<23> induces the enzyme [19]) [19]

Metals, ions

Additional information <3> (<3> the enzyme activity is independent of bound metal [12]) [12]

Turnover number (min^{-1})

12 <24> (DL-methionine-R,S-sulfoxide, <24> C terminus subdomain [22]) [22]
168 <24> (DL-methionine-R,S-sulfoxide, <24> fused subdomain [22]) [22]
222 <24> (DL-methionine-R,S-sulfoxide, <24> central subdomain [22]) [22]

Specific activity (U/mg)

0.000015 <14> (<14> extract of yeast parent strain, substrate: L-methionine sulfoxide [9]) [9]
0.000068 <14> (<14> extract of yeast parent strain, substrate: dabsyl-methionine sulfoxide [9]) [9]
0.000086 <4> (<4> liver, 26 months old rat [18]) [18]
0.000145 <4> (<4> brain, 26 months old rat [18]) [18]
0.000158 <4> (<4> kidney, 26 months old rat [18]) [18]
0.000166 <4> (<4> brain, 9 months old rat [18]) [18]
0.000181 <4> (<4> liver, 9 months old rat [18]) [18]
0.000186 <4> (<4> brain, 24 months old rat [18]) [18]
0.000187 <4> (<4> brain, 18 months old rat [18]) [18]
0.000206 <4> (<4> liver, 24 months old rat [18]) [18]
0.000214 <4> (<4> liver, 18 months old rat [18]) [18]
0.00025 <4> (<4> kidney, 18 months old rat [18]) [18]
0.00027 <4> (<4> kidney, 24 months old rat [18]) [18]
0.000296 <4> (<4> kidney, 9 months old rat [18]) [18]

0.00426 <14> (<14> recombinant yeast methionine sulfoxide reductase, substrate:L-methionine sulfoxide [9]) [9]
0.043 <14> (<14> recombinant yeast methionine sulfoxide reductase, substrate:dabsyl-methionine sulfoxide [9]) [9]
0.075 <25> (<25> with methionine-R-sulfoxide as substrate, full-length PILB protein consisting of three domains, an N-terminal domain, a central domain and a C-terminal domain [24]) [24]
0.083 <25> (<25> with methionine-S-sulfoxide as substrate, full-length PILB protein consisting of three domains, an N-terminal domain, a central domain and a C-terminal domain [24]) [24]
0.085 <1> (<1> with methionine-R-sulfoxide as substrate, C-terminal domain [24]) [24]
0.092 <2> (<2> with methionine-R-sulfoxide as substrate, C-terminal domain [24]) [24]
0.113 <25> (<25> with methionine-R-sulfoxide as substrate, C-terminal domain [24]) [24]
0.15 <25> (<25> with methionine-S-sulfoxide as substrate, central domain [24]) [24]
0.47 <2> [1]
1.88 <1> [1]
Additional information <1, 8-11, 23, 24> (<8> stress treatment: water deficits, loss of enzyme activity is found in response to water-deficit stress [7]; <9> stress treatment: water deficits or high temperature, large loss of enzyme activity is found in response to water-deficit stress or high temperature stress [7]; <10> stress treatment: water-deficits, a large increase of enzyme activity is found in response to water-deficit stress [7]; <11> stress treatment: no irrigation, the enzyme activity does not differ between the unstressed and stressed treatments [7]; <1> results of methionine sulfoxide reductase activity restoration in a mutant of Escherichia coli by methionine sulfoxide reductase from Mycobacterium tuberculosis or Escherichia coli [16]) [7, 16, 19, 22]

K_m-Value (mM)

0.033 <3> (dabsyl-methionine sulfoxide, <15> recombinant enzyme [10]) [10]
0.034 <24> (thioredoxin, <24> C-terminus subdomain [22]) [22]
0.044 <15> (dabsyl-methionine sulfoxide, <15> recombinant enzyme [10]) [10]
0.075 <24> (thioredoxin, <24> central subdomain [22]) [22]
0.12 <1> (dabsyl-methionine sulfoxide, <15> recombinant enzyme [10]) [10]
9 <24> (DL-methionine-R,S-sulfoxide, <24> central subdomain or fused subdomain [22]) [22]
56 <24> (DL-methionine-R,S-sulfoxide, <24> C-terminus subdomain [22]) [22]
Additional information <24> (<24> kinetic parameters for the enzyme activity of wild-type central subdomain, C-terminus subdomain and fused subdomain with a series of substrates [22]) [22]

pH-Optimum

6.9 <1, 2, 25> (<1,2,25> assay at [24]) [24]
7.4 <1-4, 23> (<1-4,23> assay at [1-4,6,12,18,19]) [1-4, 6, 12, 18, 19]
7.5 <1, 3, 8-11, 14, 15> (<1,3,8-11,14,15> assay at [5,7,9,10]) [5, 7, 9, 10]

Temperature optimum (°C)

37 <1-4, 14, 15, 25> (<1-4,14,15,25> assay at [1-6,9,10,12,18,24]) [1-6, 9, 10, 12, 18, 24]

4 Enzyme Structure

Molecular weight

18000 <2> (<2> gel filtration [1]) [1]
18000-20000 <1> (<1> gel filtration [4]) [4]
21000 <18-20> (<18-20> immunoblot, single polypeptide [16]) [16]
21140 <14> (<14> amino acid sequence analysis, calculated value [9]) [9]
22000 <1> (<1> immunoblot, single polypeptide [16]) [16]
23000 <1> [11]
23000-24000 <1> (<1> 1 major band, electrophoresis under nondenaturing conditions [1]) [1]
25000-35000 <1, 2, 4> [15]
28120 <4> (<4> analytical centrifugation [18]) [18]
58000 <13> (<13> predicted molecular mass [8,15]) [8, 15]

Subunits

? <4, 18> (<18> x * 27000, SDS-PAGE, recombinant enzyme [16]; <4> x * 30000, SDS-PAGE, recombinant enzyme [18]) [16, 18]
monomer <4> (<4> 1 * 28115, analytical centrifugation, in solution [18]) [18]

5 Isolation/Preparation/Mutation/Application

Source/tissue

bone marrow <2, 4> [15]
brain <4> [4, 18]
cauline leaf <16> (<16> high levels of mRNA encoding plastid-targeted isoform, the plastid-targeted protein distribution largely but not exactly mirrors that of its mRNA [13]) [13]
cerebellum <2, 4> [15]
heart <2, 4> [15]
kidney <2, 4> (<4> the enzyme mRNA level in fetal kidney is only about 14% of the adult value [15]) [4, 15, 18]
leaf <16> (<16> high levels of mRNA encoding cytosolic isoform, the cytosolic protein distribution is not exactly the same as that of the corresponding mRNA [13]) [13]

lens <2, 3> (<2,3> high activity in outer epithelial layer with decreasing activity in the inner layer [2]) [2]
leukocyte <2> [1]
liver <2, 4> (<4> enzyme mRNA levels are about the same in fetal and adult tissue [15]) [4, 15, 18]
lung <4> [4]
macrophage <4> [15]
neutrophil <2> (<2> prepared from the peripheral blood of normal volunteers [3]) [3, 5]
renal medulla <4> (<4> in kidney [15]) [15]
retinal pigment epithelium <4> [15]
root <16> (<16> high levels of mRNA encoding cytosolic isoform, the cytosolic protein distribution is not exactly the same as that of the corresponding mRNA [13]) [13]
rosette <16> (<16> high levels of mRNA encoding plastid-targeted isoform, the plastid-targeted protein distribution largely but not exactly mirrors that of its mRNA [13]) [13]

Localization

chloroplast <16> (<16> plastid-targeted isoform [13]) [13]
mitochondrion <2> [23]
ribosome <1> [4]
soluble <2, 3, 16> (<16> cytosolic isoform [13]) [2, 13]

Purification

<1> (using ammonium sulfate precipitation, column chromatography on DE-52, ultrogel filtration and ammonium sulfate extraction [1]; partial, using sonication, ammonium sulfate precipitation, column chromatography on DE-52, ultrogel and DEAE-Sephadex [4]) [1, 4]
<2> (partial, using ammonium sulfate fractionation, column chromatography on DEAE-cellulose and ultrogel AcA-44 chromatography [1]) [1]
<4> (recombinant enzyme, using Ni^{2+}-chelation chromatography [18]) [18]

Crystallization

<1> (structure determined at 3 A resolution by multiple wavelength anomalus dispersion method for the selenomethionine-substituted enzyme, and refined to 1.9 A resolution for the native enzyme, the protein is folded into an α/β roll and contains a large proportion of coils, among the three cysteine residues involved in the catalytic mechanism, Cys-51 is positioned at the N terminus of an α helix, in a solvent-exposed area composed of highly conserved amino acids, the two others Cys-198 and Cys-206, are located in the C-terminal coil [11]) [11]
<3> (crystals obtained by the vapor diffusion method, presentation of the structures of two modifications of the enzyme, the first structure is a complex with dithiothreitol, determined at 1.6 A resolution by multiwavelength anomalous dispersion, the second modification includes an additional 10 residues at the C-terminus, determined by molecular replacement to 1.7 A resolution. The two structures identify the catalytic residues, there are three

cysteine residues located in the vicinity of the active site, conformational changes in a glycine-rich C-terminal tail appear to allow all three thiols to come together and to participate in catalysis. The structures support a unique, thiol-disulfide exchange mechanism that relies upon an essential cysteine as a nucleophile and additional conserved residues that interact with the oxygen atom of the sulfoxide moiety. The key players in enzyme methionine sulfoxide catalysis are seen to be Cys-72, Phe-73, and Trp-74 of the Gly-Cys-Phe-Trp-Gly motif, the residues that bind the oxygen atom of the sulfoxide moiety, Tyr-103, Glu-115 and Tyr-155, and possibly the substrate contact residue Asp-150 [14]; the structures of thioredoxin and peptide methionine sulfoxide reductase exhibit resemblance in their $\alpha\beta$ core regions and the active cysteines in the two proteins occupy equivalent positions downstream of a central β-strand and at the N-terminus of an α-helix, identification of a enzyme subfamily that contains an active CxxC motif positioned similarly to the catalytic redox active CxxC motif in thioredoxin reductase [20]) [14, 20]
<25> (a three-wavelength MAD data set collected on a crystal cryoprotected in paratone, the overall fold of the C-terminus subdomain revealed by the 1.85 A crystal structure shows no resemblance to the previously determined central subdomain structures from other organisms, despite the lack of homology, the active sites show approximate mirror symmetry, in each case conserved amino acid motifs mediate the stereo-specific recognition and reduction of the substrate [24]) [24]

Cloning

<1> (<1> transformation of Escherichia coli with plasmid of Mycobacterium tuberculosis [16]) [16]
<3> (expression in Escherichia coli [12,14,15]; overexpression in Drosophila melanogaster, expression predominantly in nervous system markedly extends the lifespan of the fruit fly Drosophila melanogaster [21]) [12, 14, 15, 21]
<4> (cloning, sequencing and overexpression of the recombinant enzyme in Escherichia coli [18]) [18]
<12> (expression in Escherichia coli, msrA is the gene homologous to the enzyme [8,15]) [8, 15]
<13> (expression in Escherichia coli, msrA is the gene homologous to the enzyme [8,15]) [8, 15]
<14> (msrA is the gene homologous to the enzyme, which encodes a protein of 184 amino acids, protein expressed in Escherichia coli, interruption of the yeast gene results in a null mutant, which totally losses its cellular methionine sulfoxide reductase activity [9,15]) [9, 15]
<15> (transformation of the nicked DNA vector harboring the desired mutation into Escherichia coli [10]) [10]
<16> (characterization of two members of a gene family from Arabidopsis that encode, cytosolic and plastid-targeted isoforms of the enzyme, respectively, overexpression of both proteins in Escherichia coli, analysis of the Arabidopsis database indicates that there are probably at least two plastid-tar-

geted-enzyme like genes and three cytosolic enzyme-like genes in the Arabidopsis genome [13,15]) [13, 15]
<18> (production of recombinant enzyme, overexpression of the recombinant enzyme in Escherichia coli [16]) [16]
<22> (overexpression in Escherichia coli, the amino acid sequence of Mycoplasma pneumoniae is 79% identical to the enzyme sequence of Mycoplasma genitalium [17]) [17]
<23> (identification of a gene homologous to the enzyme, the gene is located just downstream of a gluthatione S-transferase gene, overexpression of the hypothetical protein in Escherichia coli as a functional 51 kDa glutathione S-transferase fusion protein [19]) [19]
<24> (truncated PILB variants and their mutants overexpressed in Escherichia coli [22]) [22]

Engineering

C107/218/227S <3> (<3> a low level of residual activity is observed for the mutant with dithiothreitol, the mutation has a much greater effect in the physiologically relevant, thioredoxin-based assay [12]) [12]
C107/218S <3> (<3> a low level of residual activity is observed for the mutant with either dithiothreitol or thioredoxin as reduced acceptors [12]) [12]
C107/227S <3> (<3> the mutation has no significant effect on activity with dithiothreitol as reduced acceptor, but a low level of activity is observed in the physiologically relevant, thioredoxin-based assay [12]) [12]
C107S <3> (<3> the mutation has no significant effect on activity [12]) [12]
C206S <24> (<24> mutation in central subdomain, the mutation leads to complete loss of activity [22]) [22]
C218/227S <3> (<3> a low level of residual activity is observed for the mutant with dithiothreitol as reduced acceptor, the mutation has a much greater effect in the physiologically relevant, thioredoxin-based assay [12]) [12]
C218S <3> (<3> a low level of residual activity is observed for the mutant with either dithiothreitol or thioredoxin as reduced acceptors, the observed lower specific activity of the mutant in the presence of thioredoxin suggests that this residue may facilitate catalysis by relaying the reducing equivalents from thioredoxin through thiol-disulfide exchange [12]) [12]
C227S <3> (<3> the mutation causes a small reduction in activity in the dithiothreitol-based assay but has a much greater effect in the physiologically relevant, thioredoxin-based assay, the observed lower specific activity of the mutant in the presence of thioredoxin suggests that this residue may facilitate catalysis by relaying the reducing equivalents from thioredoxin through thiol-disulfide exchange [12]) [12]
C25S <15> (<15> the mutation leads to complete loss of the ability of the enzyme to catalyze reduction of either L-methionine S-sulfoxide or dabsylmethionine sulfoxide [10]) [10]
C348S <24> (<24> mutation in central subdomain, activity is observed only with dithiothreitol as reduced acceptor, no activity with thioredoxin as reduced acceptor [22]) [22]

C439S <24> (<24> mutation in C-terminus subdomain, activity is observed only with dithiothreitol as reduced acceptor, no activity with thioredoxin as reduced acceptor [22]) [22]
C494S <24> (<24> mutation in C-terminus subdomain, the mutation leads to complete loss of activity [22]) [22]
C72/107/227S <3> (<3> the mutation leads to complete loss of the ability of the enzyme to catalyze reduction of N-acetyl-L-methionine sulfoxide with either dithiothreitol or thioredoxin as reduced acceptors, Cys-72 is essential for activity [12]) [12]
C72/218S <3> (<3> the mutation leads to complete loss of the ability of the enzyme to catalyze reduction of N-acetyl-L-methionine sulfoxide with either dithiothreitol or thioredoxin as reduced acceptors, Cys-72 is essential for activity [12]) [12]
C72S <3> (<3> the mutation leads to complete loss of the ability of the enzyme to catalyze reduction of N-acetyl-L-methionine sulfoxide with either dithiothreitol or thioredoxin as reduced acceptors, Cys-72 is essential for activity [12,15]) [12, 15]
F26A <15> (<15> the mutation leads to complete loss of the ability of the enzyme to catalyze reduction of either L-methionine S-sulfoxide or dabsyl-methionine sulfoxide [10]) [10]
F26H <15> (<15> the mutation leads to complete loss of the ability of the enzyme to catalyze reduction of either L-methionine S-sulfoxide or dabsyl-methionine sulfoxide [10]) [10]
G24A <15> (<15> the mutation results in 63% loss of enzyme activity [10]) [10]
G28A <15> (<15> the mutation results in 83% loss of enzyme activity [10]) [10]
W27A <15> (<15> the mutation leads to complete loss of the ability of the enzyme to catalyze reduction of either L-methionine S-sulfoxide or dabsyl-methionine sulfoxide [10]) [10]

6 Stability

Storage stability

<16>, -80°C, both cytosolic and plastid-targeted isoforms retain more than 80% of activity, at least 6 months [13]

References

[1] Brot, N.; Fliss, H.; Coleman, T.; Weissbach, H.: Enzymatic reduction of methionine sulfoxide residues in proteins and peptides. Methods Enzymol., **107**, 352-360 (1984)
[2] Spector, A.; Scotto, R.; Weissbach, H.; Brot, N.: Lens methionine sulfoxide reductase. Biochem. Biophys. Res. Commun., **108**, 429-436 (1982)

[3] Brot, N.; Werth, J.; Koster, D.; Weissbach, H.: Reduction of N-acetyl methionine sulfoxide: a simple assay for peptide methionine sulfoxide reductase. Anal. Biochem., **122**, 291-294 (1982)
[4] Brot, N.; Weissbach, L.; Werth, J.; Weissbach, H.: Enzymatic reduction of protein-bound methionine sulfoxide. Proc. Natl. Acad. Sci. USA, **78**, 2155-2158 (1981)
[5] Fliss, H.; Vasanthakumar, G.; Schiffmann, E.; Weissbach, H.; Brot, N.: Enzymatic reduction of oxidized chemotactic peptide N-formyl-L-methionyl-sulfoxide-L-leucyl-L-phenylalanine. Biochem. Biophys. Res. Commun., **109**, 194-201 (1982)
[6] Abrams, W.R.; Weinbaum, G.; Weissbach, L.; Weissbach, H.; Brot, N.: Enzymatic reduction of oxidized α-1-proteinase inhibitor restores biological activity. Proc. Natl. Acad. Sci. USA, **78**, 7483-7486 (1981)
[7] Ferguson, D.L.; Burke, J.J.: Methionyl sulfoxide content and protein-methionine-S-oxide reductase activity in response to water deficits or high temperature. Physiol. Plant., **90**, 253-258 (1994)
[8] Wizemann, T.M.; Moskovitz, J.; Pearce, B.J.; Cundell, D.; Arvidson, C.G.; So, M.; Weissbach, H.; Brot, N.; Masure, H.R.: Peptide methionine-sulfoxide reductase contributes to the maintenance of adhesions in three major pathogens. Proc. Natl. Acad. Sci. USA, **93**, 7985-7990 (1996)
[9] Moskovitz, J.; Berlett, B.S.; Poston, J.M.; Stadtman, E.R.: The yeast peptide-methionine sulfoxide reductase functions as an antioxidant in vivo. Proc. Natl. Acad. Sci. USA, **94**, 9585-9589 (1997)
[10] Moskovitz, J.; Poston, J.M.; Berlett, B.S.; Nosworthy, N.J.; Szczepanowski, R.; Stadtman, E.R.: Identification and characterization of a putative active site for peptide methionine sulfoxide reductase (MsrA) and its substrate stereospecificity. J. Biol. Chem., **275**, 14167-14172 (2000)
[11] Tete-Favier, F.; Cobessi, D.; Boschi-Muller, S.; Azza, S.; Branlant, G.; Aubry, A.: Crystal structure of the Escherichia coli peptide methionine sulfoxide reductase at 1.9 ANG resolution. Structure, **8**, 1167-1178 (2000)
[12] Lowther, W.T.; Brot, N.; Weissbach, H.; Honek, J.F.; Matthews, B.W.: Thiol-disulfide exchange is involved in the catalytic mechanism of peptide methionine sulfoxide reductase. Proc. Natl. Acad. Sci. USA, **97**, 6463-6468 (2000)
[13] Sadanandom, A.; Poghosyan, Z.; Fairbairn, D.J.; Murphy, D.J.: Differential regulation of plastidial and cytosolic isoforms of peptide methionine sulfoxide reductase in Arabidopsis. Plant Physiol., **123**, 255-263 (2000)
[14] Lowther, W.T.; Brot, N.; Weissbach, H.; Matthews, B.W.: Structure and mechanism of peptide methionine sulfoxide reductase, an "anti-oxidation" enzyme. Biochemistry, **39**, 13307-13312 (2000)
[15] Brot, N.; Weissbach, H.: Peptide methionine sulfoxide reductase: biochemistry and physiological role. Biopolymers, **55**, 288-296 (2001)
[16] St. John, G.; Brot, N.; Ruan, J.; Erdjument-Bromage, H.; Tempst, P.; Weissbach, H.; Nathan, C.: Peptide methionine sulfoxide reductase from Escherichia coli and Mycobacterium tuberculosis protects bacteria against oxidative damage from reactive nitrogen intermediates. Proc. Natl. Acad. Sci. USA, **98**, 9901-9906 (2001)

[17] Dhandayuthapani, S.; Blaylock, M.W.; Bebear, C.M.; Rasmussen, W.G.; Baseman, J.B.: Peptide methionine sulfoxide reductase (MsrA) is a virulence determinant in Mycoplasma genitalium. J. Bacteriol., **183**, 5645-5650 (2001)
[18] Petropoulos, I.; Mary, J.; Perichon, M.; Friguet, B.: Rat peptide methionine sulphoxide reductase: cloning of the cDNA, and down-regulation of gene expression and enzyme activity during aging. Biochem. J., **355**, 819-825 (2001)
[19] Tamburro, A.; Allocati, N.; Masulli, M.; Rotilio, D.; Di Ilio, C.; Favaloro, B.: Bacterial peptide methionine sulphoxide reductase: co-induction with glutathione S-transferase during chemical stress conditions. Biochem. J., **360**, 675-681 (2001)
[20] Gladyshev, V.N.: Thioredoxin and peptide methionine sulfoxide reductase: convergence of similar structure and function in distinct structural folds. Proteins, **46**, 149-152 (2002)
[21] Ruan, H.; Tang, X.D.; Chen, M.L.; Joiner, M.A.; Sun, G.; Brot, N.; Weissbach, H.; Heinemann, S.H.; Iverson, L.; Wu, C.F.; Hoshi, T.: High-quality life extension by the enzyme peptide methionine sulfoxide reductase. Proc. Natl. Acad. Sci. USA, **99**, 2748-2753 (2002)
[22] Olry, A.; Boschi-Muller, S.; Marraud, M.; Sanglier-Cianferani, S.; Van Dorsselear, A.; Branlant, G.: Characterization of the methionine sulfoxide reductase activities of PILB, a probable virulence factor from Neisseria meningitidis. J. Biol. Chem., **277**, 12016-12022 (2002)
[23] Hansel, A.; Kuschel, L.; Hehl, S.; Lemke, C.; Agricola, H.J.; Hoshi, T.; Heinemann, S.H.: Mitochondrial targeting of the human peptide methionine sulfoxide reductase (MSRA), an enzyme involved in the repair of oxidized proteins. FASEB J., **16**, 911-913 (2002)
[24] Lowther, W.T.; Weissbach, H.; Etienne, F.; Brot, N.; Matthews, B.W.: The mirrored methionine sulfoxide reductase of Neisseria gonorrhoeae pilB. Nature Struct. Biol., **9**, 348-352 (2002)

Enzyme-thiol transhydrogenase (glutathione-disulfide) 1.8.4.7

1 Nomenclature

EC number
1.8.4.7

Systematic name
[xanthine-dehydrogenase]:glutathione-disulfide S-oxidoreductase

Recommended name
enzyme-thiol transhydrogenase (glutathione-disulfide)

Synonyms
[xanthine-dehydrogenase]:oxidized-glutathione S-oxidoreductase
enzyme-thiol transhydrogenase (oxidized glutathione)
glutathione-dependent thiol:disulfide oxidoreductase
thiol:disulphide oxidoreductase

CAS registry number
85030-79-1 (formerly 12345-67-8)

2 Source Organism

<1> *Rattus norvegicus* (newborn and adults [1]; Wistar [2]) [1, 2]

3 Reaction and Specificity

Catalyzed reaction
[xanthine dehydrogenase] + glutathione disulfide = [xanthine oxidase] + 2 glutathione (converts EC 1.1.1.204 xanthine dehydrogenase into EC 1.1.3.22 xanthine oxidase in presence of oxidized glutathione. Also reduces the disulfide bond of ricin. Not inhibited by Cu^{2+} or thiol reagents; <1> thioltransferase, GSH:insulin transhydrogenase and ricin-disulfide-bond reductase activity [1]; <1> thioltransferase activity with cysteine as substrate. Mainly D-form of enzyme exists [2])

Reaction type
oxidation
redox reaction
reduction

Substrates and products

S [xanthine dehydrogenase] + oxidized glutathione <1> (<1> conversion into oxidase by treatment with proteolytic enzymes, storage at -20°C/-25°C or preincubation with either thiol reagents or whole liver homogenate. Can act in the opposite direction in the reduction of the disulfide bond of insulin and ricin [1]; reverse conversion possible by dithioerythritol, not obtained in presence of GSH [2]) (Reversibility: r <1> [1, 2]) [1, 2]

P [xanthine oxidase] + glutathione <1> [1, 2]

S Additional information <1> (<1> other disulfide compounds are either inactive or far less active than oxidized glutathione, reduction of disulfide bonds in ricin, thioltransferase activity, GSH:insulin transhydrogenase activity [1]) [1]

P ?

Inhibitors

Additional information <1> (<1> insensitive against 1 mM $CuSO_4$ and 1 mM N-ethylmaleimide at 37°C for 30 min [1]) [1]

Specific activity (U/mg)

Additional information <1> (<1> 372 U/mg, other definition of units [1]; <1> 3750 U/mg, 1 unit is defined as the amount of the enzyme producing a 1% reduction of the ration NADH formed/uric acid formed, corrected for non-enzymic conversion [2]) [1, 2]

pH-Optimum

7-8.5 <1> [2]
8.1 <1> (<1> assay at [1,2]) [1, 2]

Temperature optimum (°C)

37 <1> (<1> assay at [1,2]) [1, 2]

4 Enzyme Structure

Molecular weight

43000 <1> (<1> gel filtration [1]) [1]

Subunits

monomer <1> (<1> 1 * 40000, SDS-PAGE [1]) [1]

5 Isolation/Preparation/Mutation/Application

Source/tissue

brain <1> [1]
heart <1> [1]
kidney <1> [1]
liver <1> (<1> highest amount of activity [1]) [1, 2]

lung <1> [1]
skeletal muscle <1> [1]
skin <1> [1]
spleen <1> [1]

Localization
cytosol <1> [1]

Purification
<1> (ion-exchange, gel filtration [1]; ammonium sulfate precipitation, gel filtration, ion-exchange [2]) [1, 2]

6 Stability

Storage stability
<1>, -25°C [2]

References

[1] Battelli, M.G.; Lorenzoni, E.: Purification and properties of a new glutathione-dependent thiol:disulphide oxidoreductase from rat liver. Biochem. J., **207**, 133-138 (1982)
[2] Battelli, M.G.: Enzymic conversion of rat liver xanthine oxidase from dehydrogenase (D form) to oxidase (O form). FEBS Lett., **113**, 47-51 (1980)

Phosphoadenylyl-sulfate reductase (thioredoxin) 1.8.4.8

1 Nomenclature

EC number
1.8.4.8

Systematic name
adenosine 3',5'-bisphosphate,sulfite:thioredoxin-disulfide oxidoreductase (3'-phosphoadenosine-5'-phosphosulfate-forming)

Recommended name
phosphoadenylyl-sulfate reductase (thioredoxin)

Synonyms
3'-phosphoadenosine 5'-phosphosulfate reductase
3'-phosphoadenylylsulfate reductase
adenosine 3'-phosphate 5'-phosphosulfate reductase
adenylyl phosphosulfate reductase
PAPS reductase
PAPS reductase, thioredoxin dependent
PAPS reductase, thioredoxin-dependent
PAPS sulfotransferase
PAdoPS reductase
phospho-adenylylsulfate (PAPS) reductase
phospho-adenylylsulfate reductase
reductase, 3'-phosphoadenosine 5'-phosphosulfate
thioredoxin: 3'-phospho-adenylylsulfate reductase
thioredoxin:adenosine 3'-phosphate 5'-phosphosulfate reductase

CAS registry number
9068-63-7

2 Source Organism

<1> *Saccharomyces cerevisiae* [1]
<2> *Escherichia coli* (K12 [6]; recombinant enzyme [4]) [2-4, 6]
<3> *Synechococcus sp.* (PCC7942) [5]

3 Reaction and Specificity

Catalyzed reaction

adenosine 3',5'-bisphosphate + sulfite + thioredoxin disulfide = 3'-phosphoadenylyl sulfate + thioredoxin (<1>, from the kinetic data it is concluded that the enzyme follows an ordered mechanism with thioredoxin as the first substrate followed by 5-phosphoadenosine 3-phosphosulfate as the second [1]; <2>, special ping-pong mechanism with 5-phosphoadenosine 3-phosphosulfate reacting with the reduced enzyme isomer in a Theorell-Chance mechanism [4])

Reaction type

oxidation
reduction

Natural substrates and products

S 5-phosphoadenosine 3-phosphosulfate + reduced thioredoxin <2> (<2>, enzyme is involved in sulfur metabolism [3]; <2>, essential step in the biosynthesis of Cys in E. coli. Blocking of the enzyme inhibits cell growth [2]) [2, 3]
P ?

Substrates and products

S 5-phosphoadenosine 3-phosphosulfate + reduced thioredoxin <1-3> [1-5]
P adenosine-3',5'-bisphosphate + oxidized thioredoxin + sulfite <1, 2> [1, 4]

Inhibitors

N-(iodoacetyl-aminoethyl)-5-N'-naphthylamine-sulfonic acid <2> [4]
N-bromosuccinimide <2> [4]
adenosine 3',5'-bisphosphate <2> (<2>, competitive with respect to 5-phosphoadenosine 3-phosphosulfate) [4]
p-chloromercuribenzoate <2> [4]

Specific activity (U/mg)

44.13 <1> [1]
99 <2> [4]

K_m-Value (mM)

0.0006 <1> (thioredoxin from Saccharomyces cerevisiae) [1]
0.0014 <1> (thioredoxin from E. coli) [1]
0.01 <2> (5-phosphoadenosine 3-phosphosulfate) [4]
0.019 <1> (5-phosphoadenosine 3-phosphosulfate, with thioredoxin from Saccharomyces cerevisiae as cosubstrate, <1>) [1]
0.021 <1> (5-phosphoadenosine 3-phosphosulfate, with thioredoxin from E. coli as cosubstrate, <1>) [1]
0.023 <2> (reduced thioredoxin) [4]

4 Enzyme Structure

Molecular weight

44000 <2> (<2>, reduction of the enzyme by dithiothreitol results in a shift of the apparent MW to 62000 without formation of an enzyme-thioredoxin complex, HPLC gel filtration) [6]

Subunits

? <3> (<3>, x * 22635, calculation from nucleotide sequence [5]) [5]
dimer <2> (<2>, homodimer, each monomer consisting of a six-stranded β-sheet surrounded by α-helices, crystallographic date [2]; <2>, 2 * 27927, calculation from nucleotide sequence [6]) [2, 6]

5 Isolation/Preparation/Mutation/Application

Purification

<1> [1]

Crystallization

<2> (crystal structure at 2.0 A. The enzyme shows striking similarity to the structure of the ATP pyrophosphatase domain of GMP synthetase, indicating that both enzyme families have evolved from a common ancestral nucleotide-binding fold [2]) [2, 3]

Cloning

<2> [6]

Engineering

C239S <2> (mutant enzymes Cys239Ser and Tyr209Phe are inactive) [4]
Q160L <2> (mutant enzyme Gln160Leu is as active as the wild-type enzyme) [4]
Y209F <2> (mutant enzymes Cys239Ser and Tyr209Phe are inactive) [4]

References

[1] Schwenn, J.D.; Krone, F.A.; Husmann, K.: Yeast PAPS reductase: properties and requirements of the purified enzyme. Arch. Microbiol., **150**, 313-319 (1988)

[2] Savage, H.; Montoya, G.; Svensson, C.; Schwenn, J.D.; Sinning, I.: Crystal structure of phosphoadenylyl sulphate (PAPS) reductase: a new family of adenine nucleotide α hydrolases. Structure, **5**, 895-906 (1997)

[3] Montoya, G.; Svensson, C.; Savage, H.; Schwenn, J.D.; Sinning, I.: Crystallization and preliminary X-ray diffraction studies of phospho-adenylylsulfate (PAPS) reductase from E. coli. Acta Crystallogr. Sect. D, **54**, 281-283 (1998)

[4] Berendt, U.; Haverkamp, T.; Prior, A.; Schwenn, J.D.: Reaction mechanism of thioredoxin:3'-phospho-adenylylsulfate reductase investigated by site-directed mutagenesis. Eur. J. Biochem., **233**, 347-356 (1995)
[5] Niehaus, A.; Gisselmann, G.; Schwenn, J.D.: Primary structure of Synechococcus PCC 7942 PAPS reductase gene. Plant Mol. Biol., **20**, 1179-1183 (1992)
[6] Krone, A.; Westphal, G.; Schwenn, J.D.: Characterisation of the gene cysH and of its product phospho-adenylylsulphate reductase from Escherichia coli. Mol. Gen. Genet., **225**, 314-319 (1991)

Adenylyl-sulfate reductase (glutathione) 1.8.4.9

1 Nomenclature

EC number
1.8.4.9

Systematic name
AMP,sulfite:glutathione-disulfide oxidoreductase (adenosine-5'-phosphosulfate-forming)

Recommended name
adenylyl-sulfate reductase (glutathione)

Synonyms
3'-phosphoadenosine-5'-phosphosulfate reductase homolog 19
3'-phosphoadenosine-5'-phosphosulfate reductase homolog 26
3'-phosphoadenosine-5'-phosphosulfate reductase homolog 43
5'-adenylylsulfate reductase
APR
APS reductase
PAPS reductase homolog 19
PAPS reductase homolog 26
PAPS reductase homolog 43
Prh-19
Prh-26
Prh-43
plant-type 5'-adenylylsulfate reductase

CAS registry number
9027-75-2

2 Source Organism

<1> *Arabidopsis thaliana* [1, 2, 4]
<2> *Catharanthus roseus* [3]
<3> *Enteromorpha intestinalis* [4]
<4> *Brassica juncea* (indian mustard [4]) [4]
<5> *Tetraselmis sp.* [4]
<6> *Dunaliella salina* [4]
<7> *Thalassiosira weissflogii* [4]
<8> *Thalassiosira oceanica* [4]
<9> *Isochrysis galbana* [4]

<10> *Emiliania huxleyi* [4]
<11> *Heterocapsa triquetra* [4]
<12> *Lemna minor* [5]

3 Reaction and Specificity

Catalyzed reaction

AMP + sulfite + glutathione disulfide = adenylyl sulfate + 2 glutathione

Reaction type

oxidation
redox reaction
reduction

Substrates and products

S 3'-phosphoadenosine-5'-phosphosulfate + glutathione <1, 3-12> (<1> 50fold lower activity than with 5'-adenylylsulfate [1]; <3-11,12> no activity with 3'-phosphoadenosine-5'-phosphosulfate [4,5]) (Reversibility: ? <1> [1, 2]) [1, 2, 4, 5]
P adenosine 3',5'-diphosphate + sulfite + oxidized glutathione
S 5'-adenylylsulfate + dithioerythritol <2, 12> (Reversibility: ? <2, 12> [3, 5]) [3, 5]
P adenosine 5'-monophosphate + sulfite + oxidized dithioerythritol
S 5'-adenylylsulfate + dithiothreitol <1, 2> (Reversibility: ? <1, 2> [1, 3]) [1, 3]
P adenosine 5'-monophosphate + sulfite + oxidized dithiothreitol
S 5'-adenylylsulfate + glutathione <1-12> (Reversibility: ? <1-12> [1-5]) [1-5]
P adenosine monophosphate + sulfite + oxidized glutathione

Inhibitors

5'-adenylylsulfate <12> (<12> at 0.12 mM [5]) [5]
adenosine-5'-phosphate <1, 12> (<12> competitive inhibition, Ki of 0.34 mM with 10 mM glutathione [5]) [1, 5]
glutathione <12> (<12> at 60 mM [5]) [5]

Activating compounds

$(NH_4)_2SO_4$ <1> (<1> less effective than Na_2SO_4 [1]) [1]
DTT <1> (<1> DTT can replace glutathione as reductant, 5 mM DTT increases activity 40fold [1]) [1]
Na_2SO_4 <1> [1]
sulfate <2> (<2> optimal concentration between 350 and 500 mM [3]) [3]

Turnover number (min^{-1})

126 <2> (5'-adenylylsulfate, <2> with glutathione as reductant [3]) [3]
204 <2> (5'-adenylylsulfate, <2> with DTT as reductant [3]) [3]
3360 <12> (5'-adenylylsulfate) [5]

Specific activity (U/mg)

0.000002 <1> (<1> with 3'-phosphoadenosine 5'-phosphosulfate as substrate and 5 mM DTT, recombinant protein expressed in Escherichia coli [2]) [2]
0.000007 <1> (<1> with 3'-phosphoadenosine 5'-phosphosulfate as substrate and 5 mM DTT, recombinant protein expressed in Escherichia coli [1]) [1]
0.00022 <1> (<1> with 5'-adenylylsulfate as substrate and 5 mM DTT, recombinant protein expressed in Escherichia coli [1]) [1]
0.0005-0.005 <11> (<11> crude cell lysate [4]) [4]
0.00059 <1> (<1> with 5'-adenylylsulfate as substrate and 5 mM DTT, recombinant protein expressed in Escherichia coli [2]) [2]
0.001 <4> (<4> crude cell lysate from young leaves [4]) [4]
0.0013 <1> (<1> crude cell lysate from young leaves [4]) [4]
0.004-0.02 <3> (<3> crude cell lysate [4]) [4]
0.013 <1> (<1> after purification of His-tagged protein with nickel-affinity chromatography [2]) [2]
0.02-0.025 <12> (<12> crude extract, plants treated with 10 mM NH_4Cl [5]) [5]
0.04 <3> (<3> purified recombinant enzyme [4]) [4]
0.05-0.1 <6> (<6> crude cell lysate [4]) [4]
0.08-0.13 <8> (<8> crude cell lysate [4]) [4]
0.1-0.25 <9> (<3> crude cell lysate [4]) [4]
0.1-0.4 <10> (<10> crude cell lysate [4]) [4]
0.1-0.5 <5> (<5> crude cell lysate [4]) [4]
0.2 <7> (<7> crude cell lysate [4]) [4]

K_m-Value (mM)

0.0025 <2> (5'-adenylylsulfate, <2> with glutathione as reductant [3]) [3]
0.0038 <2> (5'-adenylylsulfate, <2> with DTT as reductant [3]) [3]
0.0065 <12> (5'-adenylylsulfate, <12> with 10 mM glutathione as reductant [5]) [5]
0.0065 <3> (5'-adenylylsulfate, <3> with glutathione as reductant [4]) [4]
0.0122 <12> (5'-adenylylsulfate, <12> with 40 mM glutathione as reductant [5]) [5]
0.4 <2> (DTT, <2> with 5'-adenylylsulfate as substrate [3]) [3]
1.39 <3> (glutathione, <3> with 5'-adenylylsulfate as substrate [4]) [4]
3 <2> (glutathione, <2> with 5'-adenylylsulfate as substrate [3]) [3]
16.8 <12> (glutathione, <12> with 0.04 mM adenylylsulfate [5]) [5]

pH-Optimum

8.5 <1> (<1> recombinant protein, expressed in E. coli [1]) [1]
8.8 <2> (<2> recombinant protein, expressed in E. coli [3]) [3]

Temperature optimum (°C)

20-25 <3> (<3> purified, recombinant enzyme, E_a 10.1 kcal per mol between 0°C and 20°C [4]) [4]

4 Enzyme Structure

Molecular weight

43000 <1> (<1> mature protein without chloroplast transit peptide, calculated from cDNA sequence [2]) [2]
43000 <12> (<12> SDS-PAGE [5]) [5]
43640 <12> (<12> calculated from cDNA sequence [5]) [5]
45000 <2> (<2> mature protein without chloroplast transit peptide, calculated from cDNA sequence [3]) [3]
45600 <1> (<1> clone APR2, calculated from cDNA sequence [1]) [1]
45700 <3> (<3> mature protein without chloroplast transit peptide, calculated from cDNA sequence [4]) [4]
47300 <1> (<1> clone APR3, calculated from cDNA sequence [1]) [1]
50450 <1> (<1> clone Prh-43, calculated from cDNA sequence [2]) [2]
50500 <1> (<1> clone Prh-26, calculated from cDNA sequence [2]) [2]
50800 <1> (<1> clone APR1, calculated from cDNA sequence [1]) [1]
51300 <1> (<1> clone Prh-19, calculated from cDNA sequence [2]) [2]
51390 <2> (<2> calculated from cDNA sequence [3]) [3]
91300 <12> (<12> estimation by gel filtration with Superose 12 [5]) [5]

Subunits

dimer <12> (<12> α_2, 2 * 43000, comparison of native mass after gel filtration with denatured after SDS-PAGE [5]) [5]

5 Isolation/Preparation/Mutation/Application

Localization

chloroplast <1, 2, 12> (<1,12> sequence contains chloroplast transit peptide [1,2,5]; <2> precursor imported in pea chloroplast in vitro [3]) [1, 2, 3, 5]

Purification

<1> (purification of recombinant 6 x His-tagged protein with Nickel-affinity chromatography [2]) [2]
<12> (purification to homogeneity [5]) [5]

Cloning

<1> [1, 2]
<2> [3]
<3> [4]
<12> [5]

6 Stability

General stability information

<12>, enzyme unstable during purification and storage, addition of 5 mM 5'-adenosine monophosphate stabilizes activity [5]

References

[1] Setya, A.; Murillo, M.; Leustek, T.: Sulfate reduction in higher plants: Molecular evidence for a novel 5'-adenylylsulfate reductase. Proc. Natl. Acad. Sci. USA, **93**, 13383-13388 (1996)
[2] Gutierrez-Marcos, J.F.; Roberts, M.A.; Campbell, E.I.; Wray, J.L.: Three members of a novel small gene-family from Arabidopsis thaliana able to complement functionally an Escherichia coli mutant defective in PAPS reductase activity encode proteins with a thioredoxin-like domain and "APS reductase" activity. Proc. Natl. Acad. Sci. USA, **93**, 13377-13382 (1996)
[3] Prior, A.; Uhrig, J.F.; Heins, L.; Wiesmann, A.; Lillig, C.H.; Stoltze, C.; Soll, J.; Schwenn, J.D.: Structural and kinetic properties of adenylyl sulfate reductase from Catharanthus roseus cell cultures. Biochim. Biophys. Acta, **1430**, 25-38 (1999)
[4] Gao, Y.; Schofield, O.M.E.; Leustek, T.: Characterization of sulfate assimilation in marine algae focusing on the enzyme 5'-adenylylsulfate reductase. Plant Physiol., **123**, 1087-1096 (2000)
[5] Suter, M.; von Ballmoos, P.; Kopriva, S.; Op den Camp, R.; Schaller, J.; Kuhlemeier, C.; Schurmann, P.; Brunold, C.: Adenosine 5'-phosphosulfate sulfotransferase and adenosine 5'-phosphosulfate reductase are identical enzymes. J. Biol. Chem., **275**, 930-936 (2000)

Adenylyl-sulfate reductase (thioredoxin) 1.8.4.10

1 Nomenclature

EC number
1.8.4.10

Systematic name
AMP,sulfite:thioredoxin-disulfide oxidoreductase (adenosine-5'-phosphosulfate-forming)

Recommended name
adenylyl-sulfate reductase (thioredoxin)

Synonyms
thioredoxin-dependent 5'-adenylylsulfate reductase

CAS registry number
9027-75-2

2 Source Organism

<1> *Rhizobium meliloti* [2]
<2> *Allochromatium vinosum* [4]
<3> *Pseudomonas aeruginosa* [1]
<4> *Mycobacterium tuberculosis* [3]

3 Reaction and Specificity

Catalyzed reaction
AMP + sulfite + thioredoxin disulfide = 5'-adenylyl sulfate + thioredoxin (Uses adenylyl sulfate, not phosphoadenylyl sulfate, distinguishing this enzyme from EC 1.8.4.8, phosphoadenylyl-sulfate reductase (thioredoxin). Uses thioredoxin as electron donor, not glutathione or other donors, distinguishing it from EC 1.8.4.9 [adenylyl-sulfate reductase (glutathione)] and EC 1.8.99.2 (adenylyl-sulfate reductase))

Reaction type
oxidation
redox reaction
reduction

Natural substrates and products

S AMP + sulfite + thioredoxin disulfide <1, 2> (Reversibility: ? <1, 2> [2, 4]) [2, 4]

P 5'-adenylyl sulfate + thioredoxin

Substrates and products

S AMP + sulfite + thioredoxin disulfide <1, 2> (Reversibility: ? <1, 2> [2, 4]) [2, 4]

P 5'-adenylyl sulfate + thioredoxin

K_i-Value (mM)

1 <7> (adenosine-5'-phosphosulfate) [2]

References

[1] Bick, J.A.; Dennis, J.J.; Zylstra, G.J.; Nowack, J.; Leustek, T.: Identification of a new class of 5'-adenylylsulfate (APS) reductases from sulfate-assimilating bacteria. J. Bacteriol., **182**, 135-142 (2000)

[2] Abola, A.P.; Willits, M.G.; Wang, R.C.; Long, S.R.: Reduction of adenosine-5'-phosphosulfate instead of 3'-phosphoadenosine-5'-phosphosulfate in cysteine biosynthesis by Rhizobium meliloti and other members of the family Rhizobiaceae. J. Bacteriol., **181**, 5280-5287 (1999)

[3] Williams, S.J.; Senaratne, R.H.; Mougous, J.D.; Riley, L.W.; Bertozzi, C.R.: 5'-Adenosinephosphosulfate lies at a metabolic branchpoint in mycobacteria. J. Biol. Chem., **277**, 32606-32615 (2002)

[4] Neumann, S.; Wynen, A.; Truper, H.G.; Dahl, C.: Characterization of the cys gene locus from Allochromatium vinosum indicates an unusual sulfate assimilation pathway. Mol. Biol. Rep., **27**, 27-33 (2000)

Glutathione dehydrogenase (ascorbate) 1.8.5.1

1 Nomenclature

EC number
1.8.5.1

Systematic name
glutathione:dehydroascorbate oxidoreductase

Recommended name
glutathione dehydrogenase (ascorbate)

Synonyms
DHA reductase
DHA-R
DasA reductase
GDOR
GSH-DHAR
GSH:DHA-oxidoreductase
dehydroascorbate reductase
dehydroascorbic acid reductase
dehydroascorbic reductase
dehydrogenase, glutathione (ascorbate)
glutathione dehydroascorbate reductase
glutathione-dependent dehydroascorate reductase
glutathione:dehydroascorbic acid oxidoreductase

CAS registry number
9026-38-4

2 Source Organism

<1> *Triticum aestivum* [1, 3, 4, 10]
<2> *Cavia porcellus* [2]
<3> *Euglena gracilis* [5]
<4> *Spinacia oleracea* [6, 8, 9, 12, 18]
<5> *Homo sapiens* [7, 11, 17]
<6> *Solanum tuberosum* [13]
<7> *Rattus norvegicus* [14, 15, 21, 22]
<8> *Pleurotus ostreatus* [16]
<9> *Oryza sativa* [19]
<10> *Sus scrofa* [20]

3 Reaction and Specificity

Catalyzed reaction

2 glutathione + dehydroascorbate = glutathione disulfide + ascorbate (<10>, catalytic mechanism [20])

Reaction type

oxidation
redox reaction
reduction

Natural substrates and products

S glutathione + dehydroascorbate <2, 7, 9> (<2>, vitamin C-conserving mechanism [2]; <7>, regenerates ascorbate after it is oxidized during normal aerobic metabolism [14,15]; <9>, the enzyme is critical for maintenance of an appropriate level of ascorbate in plant cells [19]) [2, 14, 15, 19]
P GSSG + ascorbate <2> [2]

Substrates and products

S GSH + 1,2,3-trioxocyclopentane <1> (Reversibility: ? <1> [4]) [4]
P GSSG + ?
S GSH + dehydroascorbate <1-10> (<1>, L-threo-dehydroascorbate is the best and D-threo-dehydroascorbate is the worst substrate of the four dehydroascorbate stereoisomers [4]; <1,8>, specific for glutathione as hydrogen donor [4,16]; <1>, L-threo-diastereomer is reduced faster than the L-erythro-dehydroascorbate and D-erythro-dehydroascorbate [10]) (Reversibility: ? <1-10> [1-22]) [1-22]
P GSSH + ascorbate <1, 2> [1, 2]
S GSH + isodehydroascorbate <7> (Reversibility: ? <7> [22]) [22]
P GSSG + isoascorbate
S L-Cys + dehydroascorbate <7> (<7>, 8% of the activity with GSH [22]) (Reversibility: ? <7> [22]) [22]
P ? + ascorbate
S L-acetylcysteine + dehydroascorbate <7> (<7>, 4% of the activity with GSH [22]) (Reversibility: ? <7> [22]) [22]
P N,N'-diacetyl-L-cystine + ascorbate
S Additional information <1> (<1>, L-Cys-L-Gly is not active as hydrogen donor [4]; <10>, glutathione-dependent dehydroascorbate reductase activity of thioltransferase (glutaredoxin) [20]) [4, 20]
P ?

Inhibitors

Ca^{2+} <8> (<8>, 5 mM, 34% loss of activity [16]) [16]
Cd^{2+} <8> (<8>, 5 mM $CdCl_2$, complete loss of activity [16]) [16]
Co^{2+} <3> [5]
Cu^{2+} <4, 8> (<8>, 0.1 mM $CuSO_4$, 43% loss of activity [16]; <4>, 0.1 mM $CuSO_4$, 98% inhibition [6]) [6, 16]
Fe^{2+} <8> (<8>, 0.1 mM $FeSO_4$, 34% loss of activity [16]) [16]

Fe^{3+} <3> (<3>, 1 mM, complete inhibition [5]) [5]
$HgCl_2$ <6, 8> (<8>, 5 mM $HgCl_2$, 94% loss of activity [16]) [13, 16]
Mn^{2+} <8> (<8>, 5 mM $MnSO_4$, 82% loss of activity [16]) [16]
NEM <1, 3, 4, 8> (<3>, 0.5 mM, reversed by thiol reagents [5]; <4>, 1 mM, 47% inhibition [6]; <8>, 1 mM, 23% loss of activity [16]) [5, 6, 10, 16]
PCMB <3, 8> (<3>, 0.05 mM, reversed by thiol reagents [5]; <8>, 1 mM, 32% loss of activity [16]) [5, 16]
Zn^{2+} <3, 8> (<3>, 1 mM, complete inhibition [5]; <8>, 5 mM $ZnSO_4$, 50% loss of activity [16]) [5, 16]
dehydroascorbate <4> (<4>, at high concentrations [6]) [6]
imidazole <8> (<8>, 1 mM, 14% loss of activity [16]) [16]
iodoacetamide <10> [20]
iodoacetic acid <1, 4, 8, 9> (<4>, 1 mM, 77% inhibition [6]; <8>, 1 mM, 26% loss of activity [16]; <9>, 0.1 mM, 48% loss of activity. 1.0 mM, 91% loss of activity [19]) [6, 10, 16, 18, 19]
mersalyl <6> [13]
p-chloromercuriphenylsulfonate <6> [13]
p-hydroxymercuribenzoate <4> (<4>, 1 mM, 70% inhibition [6]) [6]
sodium azide <8> (<8>, 1 mM, 32% loss of activity [16]) [16]

Cofactors/prosthetic groups

flavin <8> (<8>, enzyme shows an unusual flavin peak, enzyme might form a flavin adduct [16]) [16]

Activating compounds

2-mercaptoethanol <3, 9> (<3>, 1 mM, stimulates about 2.1fold [5]; <9>, 5 mM, enhances the activity by 13% [19]) [5, 19]
dithiothreitol <3> (<3>, 1 mM, stimulates about 2.7fold [5]) [5]

Turnover number (min^{-1})

316 <5> (dehydroascorbate) [17]
Additional information <4, 10> [18, 20]

Specific activity (U/mg)

0.7645 <7> [21]
1.31 <7> [21]
1.77 <5> [17]
7.68 <3> [5]
9.3 <8> [16]
49.1 <9> [19]
360 <4> (<4>, mitochondrial enzyme [18]) [18]
400 <4> (<4>, plastid enzyme [18]) [18]
508 <1> [10]
Additional information <4, 6> (<4>, spectrophotometric assay [12]) [6, 12, 13]

K_m-Value (mM)

0.06 <4> (dehydroascorbate, <4>, recombinant enzyme form DHAR-1 [18]) [18]
0.07 <4> (dehydroascorbate, <4>, enzyme form DHAR-a [18]) [18]

0.08 <4> (dehydroascorbate, <4>, enzyme form DHAR-b [18]) [18]
0.21 <5> (dehydroascorbate) [17]
0.245 <7> (dehydroascorbate) [22]
0.26 <3> (dehydroascorbate) [5]
0.32 <10> (dehydroascorbate, <10>, glutathione-dependent dehydroascorbate reductase activity of thioltransferase (glutaredoxin) [20]) [20]
0.34 <4> (dehydroascorbate) [6]
0.35 <9> (dehydroascorbate) [19]
0.39 <6> (dehydroascorbate) [13]
0.51 <10> (dehydroascorbate, <10>, glutathione-dependent dehydroascorbate reductase activity of thioltransferase (glutaredoxin), C25S mutant [20]) [20]
0.58 <1> (dehydroascorbate) [10]
0.7 <8> (GSH) [16]
0.84 <9> (GSH) [19]
0.85 <3> (glutathione) [5]
1.1 <4> (GSH, <4>, enzyme form DHAR-a and recombinant enzyme form DHAR-a [18]) [18]
1.3 <5> (dehydroascorbate) [7]
2.5 <4> (GSH, <4>, enzyme form DHAR-b [18]) [18]
2.5 <8> (dehydroascorbate) [16]
2.8 <7> (GSH) [22]
3.5 <5> (GSH) [17]
3.7 <10> (GSH, <10>, glutathione-dependent dehydroascorbate reductase activity of thioltransferase (glutaredoxin) [20]) [20]
3.7 <4> (GSH, <4>, pH 6.8, in presence of 0.5 mM dehydroascorbate [8,12]) [8, 12]
3.8 <5> (GSH) [7]
4.32 <4> (GSH, <4>, pH 6.3, in presence of 0.3 mM dehydroascorbate [8,12]) [8, 12]
4.35 <6> (GSH) [13]
4.43 <4> (GSH) [6]
5.2 <10> (GSH, <10>, glutathione-dependent dehydroascorbate reductase activity of thioltransferase (glutaredoxin), C25S mutant [20]) [20]

pH-Optimum

7 <1, 3> (<1>, pH 7.0 and higher [10]) [5, 10]
7.4 <5> [7]
7.5 <4, 8> [6, 8, 16]
7.5-8 <7> [22]
7.7 <4> (<4>, enzyme form DHAR-b [18]) [18]
7.9 <4> (<4>, enzyme form DHAR-a [18]) [18]
8 <6> (<6>, at 25°C [13]) [13]
8.2 <9> [19]

pH-Range

6.5-8.5 <8> (<8>, pH 6.5: about 25% of maximal activity, pH 8.5: about 60% of maximal activity [16]) [16]

Temperature optimum (°C)

38 <3> [5]
38-45 <8> [16]

4 Enzyme Structure

Molecular weight

23000 <6> (<6>, gel filtration [13]) [13]
25000 <4> (<4>, gel filtration [6,8]) [6, 8]
27000 <9> (<9>, gel filtration [19]) [19]
28000 <3> (<3>, gel filtration [5]) [5]
48700 <7> (<7>, gel filtration [22]) [22]
86000 <8> (<8>, gel filtration [16]) [16]

Subunits

? <1, 7> (<1>, x * 24200, SDS-PAGE [10]; <7>, x * 31000, SDS-PAGE [21,22]) [10, 21, 22]
monomer <4, 5, 6, 8, 9> (<6>, 1 * 23000, SDS-PAGE [13]; <4>, 1 * 25000, enzyme form DHAR-b, SDS-PAGE [18]; <9>, 1 * 26000, SDS-PAGE [19]; <4>, 1 * 26000, enzyme form DHAR-a, SDS-PAGE [18]; <5>, 1 * 32000, SDS-PAGE [17]; <8>, 1 * 86000, SDS-PAGE [16]) [13, 16, 17, 18, 19]

5 Isolation/Preparation/Mutation/Application

Source/tissue

adrenal gland <2, 7> [2, 21]
blood vessel <7> (<7>, basal membrane [14]) [14]
brain <2, 7> (<7>, widespread in the gray matter, not found in the white matter [15]) [2, 14, 15]
central nervous system <7> [15]
cerebellum <7> (<7>, myelin sheet [14]) [14, 15]
colon <7> [21]
corpus striatum <7> (<7>, myelin sheet [14]) [14]
erythrocyte <2, 5> [2, 11, 17]
fibroblast <5> [7]
grain <1> (<1>, germinating [1]) [1, 3]
heart <7> [21]
hippocampus <7> [15]
ileum <7> [21]
jejunum <7> [21]
kidney <7> [21]
leaf <4> [6, 8, 12, 18]

liver <2, 7, 10> [2, 20, 21, 22]
lung <7> [21]
lymphocyte <5> [7]
monocyte <5> [7]
mycelium <8> [16]
myelin sheath <7> (<7>, of striatum and cerebellum [14]) [14]
neuron <7> [14]
neutrophil <5> [7]
olfactory cortex <7> [15]
pancreas <7> [21]
perivascular astrocyte <7> [14]
seed <1> [4, 10]
skeletal muscle <7> [21]
stomach <2> [2]
submandibular gland <7> [21]
testis <7> [21]
thyroid gland <7> [21]
tuber <6> [13]

Localization

basal plasma membrane <7> (<7>, of blood vessels [14]) [14]
chloroplast <4> (<4>, stroma [18]) [9, 18]
cytosol <3, 5, 6, 7, 8> (<7>, of neurons, perinuclear position [14]; <7>, liver enzyme shows exclusively cytosolic location [21]) [5, 7, 13, 14, 15, 16, 21]
nucleus <7> (<7>, of neurons [14]) [14]
plastid <4> [18]
Additional information <4> (<4>, two major dehydroascorbate reductases exist in spinach leaves. One form, DHAR-1, originates in chloroplast, the other, DHAR-b, occurs in a subcellular compartment other than chloroplast [18]) [18]

Purification

<1> [10]
<3> (partial [5]) [5]
<4> [8, 18]
<5> [17]
<6> [13]
<7> [21, 22]
<8> [16]
<9> [19]

Crystallization

<4> (<4>, recombinant enzyme produced in E. coli, crystallized by hanging-drop vapour-diffusion method [9]) [9]

Cloning

<4> (enzyme form DHAR-a, expression in Escherichia coli [18]) [18]

Engineering

C25S <10> (<10>, has equivalent specificity constants for dehydroascorbate and GSH, but may have a different catalytic mechanism [20]) [20]

6 Stability

pH-Stability

6.5 <8> (<8>, highest stability at, unstable under acidic and highly alkaline conditions [16]) [16]

Temperature stability

42 <3> (<3>, pH 7.0-8.0, stable up to [5]) [5]
50 <4, 6, 7, 8> (<4>, 10 min, 35% loss of activity [6]; <6>, 7 min, stable up to [13]; <8>, pH 6.2, 1 h, stable below [16]; <7>, 5 min, 50% loss of activity [22]) [6, 13, 16, 22]
60 <4> (<4>, 10 min, complete inactivation [6]) [6]
65 <8> (<8>, complete loss of activity above [16]) [16]
70 <9> (<9>, 10 min, 12% loss of activity [19]) [19]
75 <7> (<7>, 5 min, 90% loss of activity [22]) [22]
80 <4, 9> (<4>, 10 min, 81% loss of activity of enzyme form DHAR-a, 97% loss of activity of enzyme form DHAR-b [18]; <9>, 10 min, 10% loss of activity [19]) [18, 19]

Storage stability

<4>, -20°C, stable for at least 1 month [8]
<4>, 4°C, stable for at least 1 week [18]
<6>, 4°C, in presence of 2-mercaptoethanol, stable for several days [13]

References

[1] Redman, D.G.: Dehydroascorbic acid reductase activity in germinating wheat grains. Chem. Ind., **10**, 414-415 (1974)
[2] Grimble, R.F.; Hughes, R.E.: The glutathione: dehydroascorbate oxidoreductase activity of guinea-pigs from two different age groups. Life Sci., **7**, 383-386 (1968)
[3] Kahnt, W.D.; Mundy, V.; Grosch, W.: Determination of GSH-DH activity (E.C. 1.8.5.1) presence of the enzyme in different wheat varieties. Z. Lebensm. Unters. Forsch., **158**, 77-82 (1975)
[4] Walther, C.; Grosch, W.: Substrate specificity of the glutathione dehydrogenase (dehydroascorbate reductase) from wheat flour. J. Cereal Sci., **5**, 299-305 (1987)
[5] Shigeoka,, S.; Yasumoto, R.; Onishi, T.; Nakano, Y.; Kitaoka, S.: Properties of monodehydroascorbate reductase and dehydroascorbate reductase and their participation in the regeneration of ascorbate in Euglena gracilis. J. Gen. Microbiol., **133**, 227-232 (1987)

[6] Foyer, C.H.; Halliwell, B.: Purification and properties of dehydroascorbate reductase from spinach leaves. Phytochemistry, **16**, 1347-1350 (1977)
[7] Bigley, R.; Riddle, M.; Layman, D.; Stankova, L.: Human cell dehydroascorbate reductase. Kinetic and functional properties. Biochim. Biophys. Acta, **659**, 15-22 (1981)
[8] Stahl, R.L.; Liebes, L.F.; Silber, R.: Glutathione dehydrogenase (ascorbate). Methods Enzymol., **122**, 10-12 (1986)
[9] Mizohata, E.; Kumei, M.; Matsumura, H.; Shimaoka, T.; Miyake, C.; Inoue, T.; Yokota, A.; Kai, Y.: Crystallization and preliminary X-ray diffraction analysis of glutathione-dependent dehydroascorbate reductase from spinach chloroplasts. Acta Crystallogr. Sect. D, **57**, 1726-1728 (2001)
[10] Boeck, D.; Grosch, W.: Glutathatione-dehydrogenase of wheat flour. Purification and properties. Z. Lebensm. Unters. Forsch., **162**, 243-251 (1976)
[11] Basu, S.; Som, S.; Deb, S.; Mukherjee, D.; Chatterjee, I.B.: Dehydroascorbic acid reduction in human erythrocytes. Biochem. Biophys. Res. Commun., **90**, 1335-1340 (1979)
[12] Stahl, R.L.; Liebes, L.F.; Farber, C.M.; Silber, R.: A spectrophotometric assay for dehydroascorbate reductase. Anal. Biochem., **131**, 341-344 (1983)
[13] Dipierro, S.; Borraccino, G.: Dehydroascorbate reductase from potato tubers. Phytochemistry, **30**, 427-429 (1991)
[14] Fornai, F.; Piaggi, S.; Gesi, M.; Saviozzi, M.; Lenzi, P.; Paparelli, A.; Casini, A.F.: Subcellular localization of a glutathione-dependent dehydroascorbate reductase within specific rat brain regions. Neuroscience, **104**, 15-31 (2001)
[15] Fornai, F.; Saviozzi, M.; Piaggi, S.; Gesi, M.; Corsini, G.U.; Malvaldi, G.; Casini, A.F.: Localization of a glutathione-dependent dehydroascorbate reductase within the central nervous system of the rat. Neuroscience, **94**, 937-948 (1999)
[16] Kim, Y.R.; Kang, S.O.: Purification and characterization of dehydroascorbate reductase from Pleurotus ostreatus. J. Microbiol., **36**, 164-170 (1998)
[17] Xu, D.P.; Washburn, M.P.; Sun, G.P.; Wells, W.W.: Purification and characterization of a glutathione dependent dehydroascorbate reductase from human erythrocytes. Biochem. Biophys. Res. Commun., **221**, 117-121 (1996)
[18] Shimaoka, T.; Yokota, A.; Miyake, C.: Purification and characterization of chloroplast dehydroascorbate reductase from spinach leaves. Plant Cell Physiol., **41**, 1110-1118 (2000)
[19] Kato, Y.; Urano, J.i.; Maki, Y.; Ushimaru, T.: Purification and characterization of dehydroascorbate reductase from rice. Plant Cell Physiol., **38**, 173-178 (1997)
[20] Washburn, M.P.; Wells, W.W.: The catalytic mechanism of the glutathione-dependent dehydroascorbate reductase activity of thioltransferase (glutaredoxin). Biochemistry, **38**, 268-274. (1999)
[21] Paolicchi, A.; Pezzini, A.; Saviozzi, M.; Piaggi, S.; Andreuccetti, M.; Chieli, E.; Malvaldi, G.; Casini, A.F.: Localization of a GSH-dependent dehydroascorbate reductase in rat tissues and subcellular fractions. Arch. Biochem. Biophys., **333**, 489-495 (1996)
[22] Maellaro, E.; Del Bello, B.; Sugherini, L.; Comporti, M.; Casini, A.F.: Purification and characterization of glutathione-dependent dehydroascorbate reductase from rat liver. Methods Enzymol., **279**, 30-35 (1997)

Nitrate-ester reductase 1.8.6.1

1 Nomenclature

EC number
1.8.6.1 (deleted, included in EC 2.5.1.18)

Recommended name
nitrate-ester reductase

Sulfite reductase (ferredoxin) 1.8.7.1

1 Nomenclature

EC number
1.8.7.1

Systematic name
hydrogen-sulfide:ferredoxin oxidoreductase

Recommended name
sulfite reductase (ferredoxin)

Synonyms
desulphoviridin
ferredoxin-sulfite reductase

CAS registry number
37256-50-1

2 Source Organism

<1> *Desulfovibrio gigas* (strain No. 9332 [10]) [10]
<2> *Hordeum vulgare* (var. Sekitori No. 2 [11]) [11]
<3> *Clostridium pasteurianum* (strain W5 [14]) [14]
<4> *Porphyra yezoensis* [1]
<5> *Brassica chinensis* (var. Komatsuna [2]) [2]
<6> *Spirulina platensis* (strain OU-1 [3,6]) [3, 6]
<7> *Spinacia oleracea* (cv. Michinoku No. 2 [4]) [4, 7, 8, 9, 12, 13, 17]
<8> *Pisum sativum* (cv. Fruebusch [5]) [5, 8, 21]
<9> *Allium tuberosum* [15, 16]
<10> *Brassica rapa* (fresh roots [17]; cv. Hakutaka [18]) [17, 18]
<11> *Zea mays* (cv. Golden Cross Bantam T51 [20]) [19, 20]
<12> *Glycine max* [21]

3 Reaction and Specificity

Catalyzed reaction
hydrogen sulfide + 3 oxidized ferredoxin + 3 H_2O = sulfite + 3 reduced ferredoxin (<12> extracted protein probably dual function as a major constituent of the organellar nucleoid and as the redox enzyme ferredoxin-sulfite reductase [21])

Reaction type

oxidation
redox reaction
reduction

Natural substrates and products

S sulfite + reduced ferredoxin <1-11> (<5> assimilatory sulfate reduction [2]) (Reversibility: ? <1-11> [1-5, 9, 10, 11, 12, 14, 16-19]) [1-5, 9, 10, 11, 12, 14, 16-19]
P hydrogen sulfide + oxidized ferredoxin + H_2O

Substrates and products

S S-sulfoglutathione + reduced ferredoxin <6> (Reversibility: ? <6> [6]) [6]
P glutathione persulfide + oxidized ferredoxin
S sulfite + reduced ferredoxin <1-11> (<3,7> reduced methyl viologen as electron donor also [9,13,14]; <3> ferredoxin can be partially replaced by benzyl viologen [14]; <11> three isoforms of ferredoxin: ferredoxin I, II and III, ferredoxin I and II in leaves, ferredoxin III in roots, ferredoxin I and III shows a comparable ability to donate electrons to enzyme [20]) (Reversibility: ? <1-11> [1-5, 9-14, 16-20]) [1-5, 9-14, 16-20]
P hydrogen sulfide + oxidized ferredoxin + H_2O
S Additional information <9> (<9> enzyme activites of shoot and root increase by sulfate limitation in the early stage of growth [15]) [15]
P ?

Inhibitors

azide <6> [6]
cyanide <6, 10> (<10> 1 mM, added to the reduced enzyme [18]) [6, 17, 18]

Cofactors/prosthetic groups

siroheme <1, 4, 5, 6, 7, 9, 10> (<7> one per enzyme subunit, in the high-spin Fe^{3+} state in the reduced enzyme [8]; <7> 1 mol per mol subunit, in the high-spin Fe^{3+} state in the reduced enzyme [9]; <1> in the high-spin Fe^{3+} state [10]) [1, 2, 3, 4, 8, 9, 10, 16, 17, 18]

Metals, ions

4Fe-4S <1, 7> (<7> one per enzyme subunit [8]; <7> 1 mol per mol subunit [9]; <1> two clusters per molecule [10]) [8, 9, 10]

Turnover number (min^{-1})

144 <10> (sulfite, <10> root enzyme [18]) [18]
216 <10> (ferredoxin, <10> leaf enzyme [18]) [18]
360 <10> (ferredoxin, <10> root enzyme [18]) [18]
546 <10> (sulfite, <10> leaf enzyme [18]) [18]

Specific activity (U/mg)

0.0242 <7> [13]
0.0252 <3> [14]
5.09 <7> (<7> sulfite as substrate [12]) [12]
5.3 <7> (<7> with reduced ferredoxin as the electron donor [7]) [7]

8.9 <7> (<7> 69K enzyme [9]) [9]
10.3 <7> (<7> 63K enzyme [9]) [9]
44.6 <7> (<7> sulfide [4]) [4]
45.4 <6> (<6> ferredoxin [6]) [6]
49 <6> [3]
61.1 <5> [2]
100 <4> [1]
103 <10> [17]
156 <10> [18]

K_m-Value (mM)

0.0017 <11> (ferredoxin, <11> wild type ferredoxin [19]) [19]
0.002 <11> (ferredoxin, <11> D66N-D67N ferredoxin [19]) [19]
0.0028 <10> (ferredoxin, <10> root enzyme [18]) [18]
0.0042 <11> (ferredoxin, <11> E93Q ferredoxin [19]) [19]
0.006 <7> (reduced methyl viologen) [7]
0.011 <7> (sulfite, <7> 63K enzyme, reduced methyl viologen as electron donor [9]) [9]
0.012 <7> (sulfite, <7> 69K enzyme, reduced methyl viologen as electron donor [9]) [9]
0.013 <10> (ferredoxin, <10> leaf enzyme [18]) [18]
0.019 <7> (ferredoxin, <7> 69K enzyme [9]) [9]
0.021 <7> (ferredoxin, <7> 63K enzyme [9]) [9]
0.021 <7> (sulfite, <7> 63K enzyme, ferredoxin als electron donor [9]) [9]
0.025 <7> (reduced ferredoxin) [7]
0.025 <7> (sulfite, <7> 69K enzyme, ferredoxin as electron donor [9]) [9]
0.028 <6> (sulfite) [3, 6]
0.076 <10> (ferredoxin) [17]
0.077 <10> (sulfite) [17]
0.2 <10> (sulfite, <10> root enzyme [18]) [18]
0.22 <10> (sulfite, <10> leaf enzyme [18]) [18]
0.6 <6> (S-sulfoglutathione) [6]

pH-Optimum

7 <3> (<3> phosphate buffer [14]) [14]
7-8.8 <6> (<6> Tris-phosphate buffer [6]) [6]
7.7 <10> (<10> potassium phosphate buffer [17]) [17]
7.8 <10> (<10> potassium phosphate buffer [18]) [18]
8-8.5 <2> (<2> Tris-HCl buffer [11]) [11]

Temperature optimum (°C)

22 <7> (<7> enzyme assay, reduced methyl viologen as electron donor [9]) [9]
25 <6> (<6> enzyme assay [3,6]) [3, 6]
30 <10> (<10> enzyme assay [17]) [17]
37 <2, 7> (<2,7> enzyme assay [11,12]) [11, 12]

4 Enzyme Structure

Molecular weight

13600 <7> (<7> [4]) [4]
27000 <7> (<7> gel filtration [4]) [4]
65000 <9> (<9> three isoenzymes, gel filtration [16]) [16]
66000 <10> (<10> in presence of 200 mM NaCl, gel filtration [18]) [18]
68000 <10> (<10> in presence of 200 mM NaCl, gel filtration [17]) [17]
70000 <4, 5> (<4,5> 100 mM potassium sulfate, gel filtration [1,2]) [1, 2]
112000-119000 <7> (<7> 63K enzyme, equilibrium sedimentation, gel filtration [9]) [9]
120000 <6> (<6> 100 mM potassium sulfate, gel filtration [3]; <6> gel filtration [6]) [3, 6]
134000 <7> (<7> 69K enzyme, equilibrium sedimentation [9]) [9]
137000 <7> (<7> high ionic strength, gel filtration, at low ionic strength enzyme co-migrates with ferredoxin [7]) [7]

Subunits

? <8, 12> (<8,12> ? * 68000, probably partially processed to a 63000 Da species, SDS-PAGE [21]) [21]
dimer <6> (<6> 2 * 63000, SDS-PAGE [3]) [3, 6]
dimer <7> (<7> 2 * 63000, 63K form of the enzyme, 50 mM phosphate buffer pH 7.7, 20°C, SDS-PAGE [9]) [9]
dimer <7> (<7> 2 * 69000, 69K form of the enzyme, probably partially processed to a 63000 Da species, 50 mM phosphate buffer pH 7.7, 20°C, SDS-PAGE [9]) [9]
monomer <4, 10> (<4,10> 1 * 65000, SDS-PAGE [1,17]) [1, 17]
monomer <5, 9, 10> (<5,9,10> 1 * 64000, SDS-PAGE [2,16,18]) [2, 16, 18]
tetramer <7> (<7> 4 * 136000, dimer of two dimers, in the presence of 200 mM NaCl, gel filtration, 4 * 71000, SDS-PAGE [4]) [4]

5 Isolation/Preparation/Mutation/Application

Source/tissue

bundle sheath cell <11> (<11> ferredoxin I and ferredoxin II [20]) [20]
cell culture <12> [21]
leaf <5, 7, 9, 10, 11> [2, 4, 7, 12, 13, 16, 18, 19]
mesophyll <11> (<11> ferredoxin I and ferredoxin II [20]) [20]
root <2, 10> [11, 17, 18]
seedling <8, 9, 11> (<8> 5 day old, root, shoot [5]; <9> root, shoot [15]; <11> primary leaves, third leaves, root [20]) [5, 15, 20, 21]
thallus <4> (<4> fresh [1]) [1]

Localization

chloroplast nucleoid <8, 12> [21]
proplastid <8> (<8> in seedling roots [5]) [5]
thylakoid <7> (<7> grana in chloroplasts [12]) [12]

Purification

<4> [1]
<5> [2]
<6> [3]
<7> (partially, crude leaves extract [12]) [4, 7, 9, 12, 13]
<10> [17, 18]
<12> [21]

6 Stability

Temperature stability

50 <7> (<7> no activity loss after heating for 5 h at this temperature, activity loss after incubation at 60°C for 30 min [13]) [13]
60 <10> (<10> 91% remaining activity of leaf enzyme after heating for 5 min, root enzyme lower heat stability [18]) [18]

General stability information

<7>, stable in 0.01 M phosphate buffer, pH 7.7 [13]

Storage stability

<7>, -20°C, stable for 3 months [13]

References

[1] Koguchi, O.; Tamura, G.: Isolation and partial characterization of homogeneous ferredoxin-sulfite reductase from a red alga, Porphyra yezoensis. Agric. Biol. Chem., **53**, 1653-1662 (1989)
[2] Koguchi, O.; Takahashi, H.; Tamura, G.: Purification and partial characterization of ferredoxin-sulfite reductase from rape leaves. Agric. Biol. Chem., **52**, 1867-1868 (1988)
[3] Koguchi, O.; Tamura, G.: Ferredoxin-sulfite reductase from a cyanobacterium, Spirulina platensis. Agric. Biol. Chem., **52**, 373-380 (1988)
[4] Aketagawa, J.; Tamura, G.: Ferredoxin-sulfite reductase from spinach. Agric. Biol. Chem., **44**, 2371-2378 (1980)
[5] Brunold, C.; Suter, M.: Localization of enzymes of assimilatory sulfate reduction in pea roots. Planta, **179**, 228-234 (1989)
[6] Koguchi, O.; Tamura, G.: Reduction of S-sulfoglutathione by cyanobacterial ferredoxin-sulfite reductase. Agric. Biol. Chem., **53**, 783-788 (1989)
[7] Hirasawa, M.; Boyer, J.M.; Gray, K.A.; Davis, D.J.; Knaff, D.B.: The interaction of ferredoxin-linked sulfite reductase with ferredoxin. FEBS Lett., **221**, 343-348 (1987)
[8] Krueger, R.J.; Siegel, L.M.: Evidence for siroheme-Fe4S4 interaction in spinach ferredoxin-sulfite reductase. Biochemistry, **21**, 2905-2909 (1982)
[9] Krueger, R.J.; Siegel, L.M.: Spinach siroheme enzymes: Isolation and characterization of ferredoxin-sulfite reductase and comparison of properties with ferredoxin-nitrite reductase. Biochemistry, **21**, 2892-2904 (1982)

[10] Hall, M.H.; Prince, R.H.; Cammack, R.: EPR spectroscopy of the iron-sulphur cluster and sirohaem in the dissimilatory sulphite reductase (desulphoviridin) from Desulphovibrio gigas. Biochim. Biophys. Acta, **581**, 27-33 (1979)
[11] Tamura, G.; Hosoi, T.: The occurrence of ferredoxin-sulfite reductase in barley roots. Agric. Biol. Chem., **43**, 1601-1602 (1979)
[12] Tamura, G.; Hosoi, T.; Aketagawa, J.: Ferredoxin-dependent sulfite reductase fron spinach leaves. Agric. Biol. Chem., **42**, 2165-2167 (1978)
[13] Asada, K.; Tamura, G.; Bandurski, R.S.: Methyl viologen- and ferredoxin-linked sulfite reductase (spinach). Methods Enzymol., **17B**, 528-539 (1971)
[14] Laishley, L.E.J.; Lin, P.M.; Peck, H.D.: A ferredoxin-linked sulfite reductase from Clostridium pasteurianum. Can. J. Microbiol., **17**, 889-895 (1971)
[15] Takahashi, S.; Yip, W.C.; Matsugami, T.; Tamura, G.: Effect of sulfur and nitrogen nutrition on derepression of ferredoxin-sulfite reductase in leek seedlings. J. Plant Res., **109**, 363-368 (1996)
[16] Takahashi, S.; Yoshida, Y.; Tamura, G.: Purification and characterization of ferredoxin-sulfite reductases from leek (Allium tuberosum) leaves. J. Plant Res., **109**, 45-52 (1996)
[17] Takahashi, S.; Sakata, T.; Tamura, G.: Ferredoxin-linked sulfite reductase from turnip roots. Biosci. Biotechnol. Biochem., **60**, 142-144 (1996)
[18] Takahashi, S.; Yip, W.C.; Tamura, G.: Purification and characterization of ferredoxin-sulfite reductase from turnip (Brassica rapa) leaves and comparison of properties with ferredoxin-sulfite reductase from turnip roots. Biosci. Biotechnol. Biochem., **61**, 1486-1490 (1997)
[19] Akashi, T.; Matsumura, T.; Ideguchi, T.; Iwakiri, K.I.; Kawakatsu, T.; Taniguchi, I.; Hase, T.: Comparison of the electrostatic binding sites on the surface of ferredoxin for two ferredoxin-dependent enzymes, ferredoxin-$NADP^+$ reductase and sulfite reductase. J. Biol. Chem., **274**, 29399-29405 (1999)
[20] Yonekura-Sakakibara, K.; Onda, Y.; Ashikari, T.; Tanaka, Y.; Kusumi, T.; Hase, T.: Analysis of reductant supply systems for ferredoxin-dependent sulfite reductase in photosynthetic and nonphotosynthetic organs of maize. Plant Physiol., **122**, 887-894 (2000)
[21] Chi-Ham, C.L.; Keaton, M.A.; Cannon, G.C.; Heinhorst, S.: The DNA-compacting protein DCP68 from soybean chloroplasts is ferredoxin:sulfite reductase and co-localizes with the organellar nucleoid. Plant Mol. Biol., **49**, 621-631 (2002)

Sulfite reductase 1.8.99.1

1 Nomenclature

EC number
1.8.99.1

Systematic name
hydrogen-sulfide:(acceptor) oxidoreductase

Recommended name
sulfite reductase

Synonyms
ASiR <9, 12, 14> [9, 11]
SIR <18> [14]
aSiR <14> [13]
assimilatory sulfite reductase
assimilatory-type sulfite reductase
reductase, sulfite
sulphite reductase <13, 14> [6, 7]
Additional information <9> (<9> a third type of sulfite reductase of low molecular range in sulfate-reducing bacteria, termed assimilatory because of reduction of sulfite to sulfide with high fidelity [13]) [13]

CAS registry number
37256-51-2

2 Source Organism

<1> *Porphyra yezoensis* (red algae [1]) [1]
<2> *Chlorobium sp.* (strain PM. and T [2]) [2]
<3> *Rhodopseudomonas viridis* [2]
<4> *Rhodomicrobium vannielii* [2]
<5> *Rhodospirillum rubrum* [2]
<6> *Rhodopseudomonas gelatinosa* [2]
<7> *Rhodopseudomonas palustris* [2]
<8> *Chromatium sp.* (strain D [2]) [2]
<9> *Desulfovibrio vulgaris* (Hildenborough [11,12,15]) [3, 4, 11-13, 15]
<10> *Aspergillus nidulans* [1, 5]
<11> *Allium odoratum* [1]
<12> *Spinacia oleracea* [1, 9, 13]
<13> *Saccharomyces cerevisiae* (commercial wine yeast [6]) [1, 6]

<14> *Escherichia coli* (complex of flavoprotein component and haemoprotein component [7,8,13]) [1, 4, 7-9, 13]
<15> *Clostridium pasteurianum* [10]
<16> *Methanosarcina barkeri* [12]
<17> *Desulfuromonas acetoxidans* [12]
<18> *Arabidopsis thaliana* [14]
<19> *Synechococcus sp.* (strain PCC7942) [14]

3 Reaction and Specificity

Catalyzed reaction

hydrogen sulfide + acceptor + 3 H_2O = sulfite + reduced acceptor (an iron-protein. no products: trithionate, thiosulfate. A stochiometry of six molecules of reduced methyl viologen per molecule of sulfide formed is found. No products: trithionate, thiosulfate; <9,12,14,16,17> mechanism [12,13]; <10> inducible by sulfate [5]; <15> cysteine represses the expression in vivo [10]; <9,14> mechanism, structure and kinetic of ligand-binding by siroheme [13,15])

Reaction type

oxidation
redox reaction
reduction

Natural substrates and products

S sulfite + reduced acceptor <1-18> (<18> key enzyme in sulfate-assimilation and integration of sulfur into cysteine [14]; <14> physiological acceptor: e.g. cytochrome P_{450} or sulphite reductase haemoprotein [7]; <1,9-14> intermediate step of assimiliatory reduction of sulfate to the SH level [1,3-5,13]; <2-9> assimilatory sulfate reduction during biosynthesis of cell material [2,11]; <9,16,17> involved in the assimilatory synthesis of sulfur-containing compounds [12]) (Reversibility: ? <1-18> [1-15]) [1-15]
P sulfide + oxidizied acceptor

Substrates and products

S hydroxylamine + reduced methyl viologen <9, 10, 12, 14> (Reversibility: ? <9, 10, 12, 14> [5, 9, 11, 13]) [5, 9, 11, 13]
P ?
S nitrite + reduced methyl viologen <9, 12, 14> (<9,12,14> higher reactivity than with sulfite [13]) (Reversibility: ? <9, 12, 14> [9, 11, 13]) [9, 11, 13]
P NH_4^+ + oxidized methyl viologen <14> [13]
S sulfite + reduced acceptor <1-18> (<15> ferredoxin as electron acceptor [10]; <14> riboflavin, 3-acetylpyridine adenine dinucleotide phosphate, or ferricyanide as electron acceptors [7]; <9> specific [3]; <1-17> reduced methyl viologen as electron donor [1-3,9,10,12]; <9> e.g. methyl viologen, sole artificial electron donor for Desulfovibrio vulgaris enzyme [3];

<1,10> electron donor benzyl viologen [1,5]; <1> electron donors: phenazine methosulfate, janus green, methylene blue, nile blue, and toluidine blue [1]) (Reversibility: r <9> [15]; ? <1-18> [1-14]) [1-15]

P sulfide + oxidized acceptor <1-18> [1-15]

S Additional information <10> (<10> no substrates: neutral red, phenosafranine, janus green, nile blue, methylene blue, toluidine blue, thionine, phenazine methosulfate, FMN, FAD, riboflavin, cytochrome c_3, ferredoxin, NAD(P)H [5]) [5]

P ?

Inhibitors

8-hydroxyquinoline <10> (<10> slight inhibition [5]) [5]
EDTA <1, 10> (<10> slight inhibition [5]) [1, 5]
KCN <1, 10> [1, 5]
N-ethylmaleimide <10> (<10> slight inhibition [5]) [5]
Na_2S <10> (<10> concentration above 1 mM [5]) [5]
Tris-HCl <15> (<15> partially inhibited the enzyme [10]) [10]
arsenite <1, 10> [1, 5]
diethyldithiocarbamate <10> (<10> slight inhibition [5]) [5]
iodonium diphenyl chloride <14> (<14> covalent phenylation of FAD cofactor [7]) [7]
p-chloromercuribenzoate <1> (<10> no inhibition [5]) [1]
Additional information <9, 10, 12, 14> (<9,12,14> inhibition via reduced cofactors [13]; <10> not inhibitory: thiourea, NH_2NH_2, atabrin, $CuSO_4$, NaN_3 [5]) [5, 13]

Cofactors/prosthetic groups

3-acetylpyridine adenine dinucleotide <14> [7]
FAD <14> (<14> electron transfer cascade: NADPH reduces FAD, which reduces FMN, $FMNH_2$ reduces the physiological acceptor, e.g. cytochrome P_{450} or the sulphite reductase haemoprotein, in a catalytic cycle [7]) [7, 13]
FMN <14> (<14> electron transfer cascade: NADPH reduces FAD, which reduces FMN, $FMNH_2$ reduces the physiological acceptor, e.g. cytochrome P_{450} or the sulphite reductase haemoprotein, in a catalytic cycle [7]) [7, 13]
NADPH <14> (<14> electron transfer cascade: NADPH reduces FAD, which reduces FMN, $FMNH_2$ reduces the physiological acceptor, e.g. cytochrome P_{450} or the sulphite reductase haemoprotein, in a catalytic cycle [7]) [7]
ferredoxin <12, 15> [10, 13]
iron tetrahydroporphyrin <9, 12, 14, 16, 17> (<9,12,14,16,17> i.e. siroheme [12,13,15]) [12, 13, 15]
Additional information <1-8, 10> (<1-8,10> not NAD(P)H as electron donor [1,2,5]; <1> not reduced ferredoxin as electron donor [1]) [1, 2, 5]

Activating compounds

Additional information <9, 12, 14> (<9,12,14> activation via reduced cofactors [13]; <9> activation of low-molecular weight assmilatory sulfite reductases may also be related to exchange of a heme axial ligand [13]) [13]

Metals, ions

Cu <1> (<1> contains Cu [1]) [1]
Fe <1, 9, 12, 14, 16-18> (<9> comparison of reduction potentials of siroheme and siroheme-[4Fe-4S] cluster, pH-dependency [15]; <9,12,14> overview about α-β subunit structure, topology and iron bound, a siroheme is coupled covalently to an 4Fe-4S-cluster [13]; <9,16,17> iron tetrahydroporphyrin prosthetic group, termed siroheme, in addition to nonheme iron in 4Fe-4S clusters, 4.7 iron atoms per mole of enzyme and 4.6 mole of acid-labile sulfur atoms per mole of enzyme [12]; <9> 4.9 iron-centers per mole of enzyme and 5.1 sulfide per mole of enzyme [11]; <14> 0.7 4Fe4S cluster per haemoprotein component [8]; <1,9,14,16,17> contains Fe [1,7,12]; <9> 1 4Fe-4S center per mol of enzyme [4]; <18> 4Fe-4S-cluster [14]; <9> [4Fe-4S]-siroheme prosthetic center [15]) [1, 4, 7, 8, 11-15]
Mg^{2+} <13> (<13> essentiell for activity [6]) [6]
Additional information <1> (<1> no Co, Mn, Mo, Zn [1]) [1]

Turnover number (min^{-1})

11.4 <9> (sulfite, <9> recombinant enzyme [11]) [11]
12 <9, 14> (sulfite) [9, 11]
27 <14> (nitrite) [9]
30.6 <12> (sulfite) [9]
78 <9> (sulfite, <9> electrons per heme per sec [13]) [13]
264 <12> (nitrite) [9]
660 <12> (sulfite, <12> electrons per heme per sec [13]) [13]
1200 <9> (nitrite) [11]
1380 <9> (nitrite, <9> recombinant enzyme [11]) [11]
2598 <14> (sulfite, <14> electrons per heme per sec [13]; <14> hemoprotein component [13]) [13]
4860 <12> (hydroxylamine) [9]
5700 <12> (nitrite) [13]
5898 <14> (nitrite, <14> hemoprotein component [13]; <14> electrons per heme per sec [13]) [13]
6000 <14> (hydroxylamine) [9]
7200 <9> (nitrite, <9> electrons per heme per sec [13]) [13]
34800 <12> (hydroxylamine, <12> electrons per heme per sec [13]) [13]
45000 <14> (hydroxylamine, <14> electrons per heme per sec [13]; <14> hemoprotein component [13]) [13]
138000 <9> (hydroxylamine) [11]
144000 <9> (hydroxylamine, <9> recombinant enzyme [11]) [11]
276000 <9> (hydroxylamine, <9> electrons per heme per sec [13]) [13]

Specific activity (U/mg)

0.00074 <3> [2]
0.0026 <2> (<2> strain T [2]) [2]
0.0037 <6, 7> [2]
0.0042 <8> [2]
0.0047 <5> [2]
0.0065 <4> [2]

0.007 <2> (<2> strain P.M. [2]) [2]
0.0125 <10> [5]
0.9 <9> [3]
1.48 <1> [1]
4.98 <10> [5]
6.5 <14> (<14> native holoenzyme [8]) [8]
11.5 <14> (<14> purified flavoprotein component [7]) [7]
17.5 <10> (<10> sulfate-induced wild-type [5]) [5]
Additional information <13> (<13> activity on different assimilable nitrogen sources [6]) [6]

K_m-Value (mM)
0.01 <12, 14> (sulfite) [9, 13]
0.025 <10> (sulfite) [5]
0.05 <9> (sulfite) [11, 13]
0.055 <9> (sulfite, <9> recombinant enzyme [11]) [11]
0.066 <14> (riboflavin, <14> with NADPH in electron transfer cycle [7]) [7]
0.088 <14> (NADPH, <14> with riboflavin in electron transfer cycle [7]) [7]
0.19 <14> (3-acetylpyridine adenine dinucleotide, <14> with NADPH in electron transfer cycle [7]) [7]
0.65 <1> (sulfite) [1]
1.5 <14> (nitrite) [9, 13]
3.5 <12> (nitrite) [9, 13]
4.2 <9> (nitrite, <9> recombinant enzyme [11]) [11]
4.7 <9> (nitrite) [11, 13]
10.5 <14> (hydroxylamine) [9, 13]
14 <9, 12> (hydroxylamine) [9, 11, 13]
18 <9> (hydroxylamine, <9> recombinant enzyme [11]) [11]
20 <10> (hydroxylamine) [5]

K_i-Value (mM)
0.0185 <14> (iodonium diphenyl chloride) [7]

pH-Optimum
7.2-7.8 <10> (<10> Tris-HCl buffer [5]) [5]
7.5 <9, 15> [10, 15]
7.5-8 <1> [1]
8 <10> (<10> above, phosphate buffer [5]) [5]
Additional information <9> (<9> comparison of pH-dependency of reduction potentials of siroheme and siroheme-[4Fe-4S] cluster [15]) [15]

Temperature optimum (°C)
25 <9> (<9> assay at [15]) [15]
30 <13> (<13> assay at [6]) [6]
37 <9, 16, 17> (<9,16,17> assay at [12]) [12]

4 Enzyme Structure

Molecular weight

23500 <9> (<9> gel filtration [15]) [15]
26800 <9> (<9> sedimentation equilibrium [3]) [3]
27200 <9> (<9> amino acid composition [4,12]) [4, 12]
84000-87000 <1> (<1> gel filtration [1]) [1]
780000 <14> (<14> sulfite reductase both flavoprotein and haemoprotein component, gel filtration [8]) [8]
Additional information <9, 12, 14> (<9,12,14> overview [13]) [13]

Subunits

monomer <9> (<9> 1 * 23500, SDS-PAGE [15]; <9> 1* 26800, SDS-PAGE [3]; <9> 1 * 27200, SDS-PAGE [4,12]) [3, 4, 12, 15]
octamer <14> (<14> 8 * 63000, SDS-PAGE [8]) [7, 8]
Additional information <9, 12, 14> (<9,12,14> overview about α-β subunit structure and ligand binding [13]; <14> sulphite reductase contains a β-haemoprotein and an α-flavoprotein component, the latter composed of 8 α-chains including FMN-, NADPH-, and FAD-binding domain [7,8]; <14> the holoenzyme has $(\alpha\text{-}\beta)_8$-structure, spectroscopic and titration experiments [8]) [7, 8, 13]

5 Isolation/Preparation/Mutation/Application

Source/tissue

leaf <12, 18> [1, 14]
mycelium <10> [5]
root <18> [14]
thallus <1> [1]

Localization

soluble <2-9, 14> [2, 8, 12]

Purification

<1> [1]
<9> (cytochromes and a minor flavoprotein present in the sulfite reductase-containing fraction are removed [12]; siroheme component is extracted from the enzyme preparation and further purified [12]) [3, 4, 12]
<10> [5]
<14> (NADPH- and FAD-binding sulfite reductase flavoprotein component, recombinant from Escherichia coli [7]; sulfite reductase haemoprotein component [8]) [7, 8]

Crystallization

<14> (flavoprotein component [7]; overview structure and the complex binding of substrate sulfite or nitrite by siroheme cofactor [13]) [7, 13]

Cloning

<9> (gene expression in Desulfovibrio hosts using broad-range plasmid pDSK519, 50fold increase to native expression, recombinant and native enzymes are indistiguishable [11]) [11]
<14> (CysJ gene, encoding NADPH- and FAD-binding sulfite reductase flavoprotein component, overexpression in Escherichia coli [7]; CysG gene, encoding sulfite reductase haemoprotein component, overexpression in Escherichia coli [8]) [7, 8]
<18> (functional expression of single copy sir gene in transgenic Arabidopsis thaliana plants, identification of promotor, DNA sequence analysis [14]) [14]

Engineering

Additional information <10, 14> (<14> SiR-FP60, N-terminal truncated sulfite reductase flavoprotein component behaves as a monomer and no longer polymerizes to an octamer, but retaines full catalytic activity [8]; <10> naturally occuring mutant strains with blocks in the assimilatory sulfate reduction pathway [5]) [5, 8]

Application

agriculture <13> (<13> possible target to minimize H_2S production of yeast during wine fermentation [6]) [6]
steel industry <14> (<14> enzyme inhibition helps against the corrosion of steel and contamination by excess production of sulfide from sulfate-reducing bacteria [13]) [13]

6 Stability

Temperature stability

0 <13> (<13> half-life 1 h [6]) [6]
25 <13> (<13> half-life 4 h [6]) [6]
50 <10> (<10> 5 min, 50% loss of activity [5]) [5]

General stability information

<9>, 0.5 M pyridine stabilizes the purified siroheme component [12]
<13>, EGTA stabilizes [6]
<13>, glycol stabilizes and reduces the cold lability at 30% concentration [6]

Storage stability

<1>, -13.5°C, 6 months, no loss of activity [1]
<9>, -80°C, 0.05 M Tris-HCl buffer, pH 7.6, several months [4]
<9>, -80°C, 0.05 M Tris-HCl, 6 months [12]
<10>, -20°C, several months [5]
<13>, -20°C, 30% glycerol, loss of 85% activity in 24 h, therefore storage of extracts is impractical [6]

References

[1] Saito, E.; Tamura, G.: Studies on the sulfite reducing system of algae. Part II. Purification and properties of the reduced methyl viologen-linked sulfite reductase from a red algae, Porphyra yezoensis. Agric. Biol. Chem., **35**, 491-500 (1971)
[2] Peck, H.D.; Tedro, S.; Kamen, M.D.: Sulfite reductase activity in extracts of various photosynthetic bacteria. Proc. Natl. Acad. Sci. USA, **71**, 2404-2406 (1974)
[3] Lee, J.P.; LeGall, J.; Peck, H.D.: Isolation of assimilatory- and dissimilatory-type sulfite reductases from Desulfovibrio vulgaris. J. Bacteriol., **115**, 529-542 (1973)
[4] Huynh, B.H.; Kang, L.; DerVatanian, D.V.; Peck, H.D.; LeGall, J.: Characterization of a sulfite reductase from Desulfovibrio vulgaris. Evidence for the presence of a low-spin siroheme and an exchange-coupled siroheme-[4Fe-4S] unit. J. Biol. Chem., **259**, 15373-15376 (1984)
[5] Yoshimoto, A.; Nakamura, T.; Sato, R.: Isolation from Aspergillus nidulans, of a protein catalyzing the reduction of sulfite by reduced viologen dyes. J. Biochem., **62**, 756-766 (1967)
[6] Jiranek, V.; Langridge, P.; Henschke, P.A.: Determination of sulfite reductase activity and its response to assimilable nitrogen status in a commercial Saccharomyces cerevisiae wine yeast. J. Appl. Bacteriol., **81**, 329-336 (1996)
[7] Coves, J.; Lebrun, C.; Gervasi, G.; Dalbon, P.; Fontecave, M.: Overexpression of the FAD-binding domain of the sulphite reductase flavoprotein component from Escherichia coli and its inhibition by iodonium diphenyl chloride. Biochem. J., **342**, 465-472 (1999)
[8] Zeghouf, M.; Fontecave, M.; Coves, J.: A simplified functional version of the Escherichia coli sulfite reductase. J. Biol. Chem., **275**, 37651-37656 (2000)
[9] Wolfe, B.M.; Lui, S.M.; Cowan, J.A.: Desulfoviridin, a multimeric-dissimilatory sulfite reductase from Desulfovibrio vulgaris (Hildenborough). Purification, characterization, kinetics and EPR studies. Eur. J. Biochem., **223**, 79-89 (1994)
[10] Harrison, G.; Curle, C.; Laishley, E.J.: Purification and characterization of an inducible dissimilatory type sulfite reductase from Clostridium pasteurianum. Arch. Microbiol., **138**, 72-78 (1984)
[11] Tan, J.; Soriano, A.; Lui, S.M.; Cowan, J.A.: Functional expression and characterization of the assimilatory-type sulfite reductase from Desulfovibrio vulgaris (Hildenborough). Arch. Biochem. Biophys., **312**, 516-523 (1994)
[12] Moura, I.; Lino, A.R.: Low-spin sulfite reductases. Methods Enzymol., **243**, 296-303 (1994)
[13] Crane, B.R.; Getzoff, E.D.: The relationship between structure and function for the sulfite reductases. Curr. Opin. Struct. Biol., **6**, 744-756 (1996)
[14] Bork, C.; Schwenn, J.D.; Hell, R.: Isolation and characterization of a gene for assimilatory sulfite reductase from Arabidopsis thaliana. Gene, **212**, 147-153 (1998)

[15] Lui, S.M.; Cowan, J.A.: Direct reversible protein electrochemistry at a pyrolytic graphite electrode. Characterization of the redox thermodynamics of the Fe_4S_4-siroheme prosthetic center in the hexameric dissimilatory sulfite reductase and the monomeric assimilatory sulfite reductase from Desulfovibrio vulgaris (Hildenborough). Systematic pH titration experiments and implications for active site chemistry. J. Am. Chem. Soc., **116**, 11538-11549 (1994)

Adenylyl-sulfate reductase 1.8.99.2

1 Nomenclature

EC number
1.8.99.2

Systematic name
AMP, sulfite:(acceptor) oxidoreductase (adenosine-5'-phosphosulfate-forming)

Recommended name
adenylyl-sulfate reductase

Synonyms
AMP,sulfite:flavin oxidoreductase
APS reductase
APS-reductase
ATAPR2 <46> [23]
$AdoPSO_4$ reductase
EAPR <47> [24]
LMAPR <45> [23]
adenosine 5'-phosphosulfate reductase
adenosine phosphosulfate reductase
reductase, adenylylsulfate

CAS registry number
9027-75-2

2 Source Organism

<1> *Escherichia coli* [1, 22]
<2> *Aerobacter aerogenes* (weak activity [1]) [1]
<3> *Desulfovibrio desulfuricans* (G100A, ATCC27774, API, ATCC 13541, Norway [2]; weak activity [2]; strain 8301 (Norway), strain 8307, strain 8380 [17]) [1, 2, 17, 20, 27, 33]
<4> *Clostridium nigrificans* [1]
<5> *Vibrio cholinicus* [1]
<6> *Thiobacillus thioparus* [1, 13, 32]
<7> *Thiobacillus thiooxidans* [1, 2]
<8> *Thiobacillus denitrificans* [1, 4, 27, 32, 34]
<9> *Chromatium sp.* (weak activity [1]) [1]

<10> *Desulfovibrio vulgaris* (strain Miyazaki [14]; strain 8303, strain 8305 [17]; strain Hildenborough [27]; strain Marburg [11]; strain Miyazaki [30]) [1, 2, 5, 7, 11, 14, 15, 17, 18, 27, 30, 33]
<11> *Thiocapsa roseopersicina* [1, 19]
<12> *Catharanthus roseus* [31]
<13> *Desulfovibrio gigas* [2, 3, 5, 10, 17, 33]
<14> *Desulfovibrio salexigens* [2, 17, 33]
<15> *Desulfovibrio multispirans* [2, 33]
<16> *Desulfotomaculum orientis* [2, 17]
<17> *Desulfotomaculum ruminis* [2, 17]
<18> *Desulfotomaculum nigrificans* (strain 8395, strain 8351 [17]) [2, 17]
<19> *Desulfotomaculum propionicus* [2]
<20> *Desulfotomaculum variabilis* [2]
<21> *Chromatium vinosum* (strain D [12]) [2, 12]
<22> *Streptomyces lividans* [2]
<23> *Desulfovibrio thermophilus* [5, 33]
<24> *archaebacterium* (VC-16 [6]) [6]
<25> *Desulfobulbus propionicus* [7]
<26> *Desulfobacter postgatei* [7]
<27> *Desulfococcus multivorans* [7]
<28> *Desulfosarcina variabilis* [7]
<29> *Chlorobium vibrioforme* (f. sp. thiosulfatophilum [8]) [8]
<30> *Archaeoglobus fulgidus* [9, 20, 27, 29, 33]
<31> *Rhizobium meliloti* [28]
<32> *Chlorobium limicola* [16]
<33> *Desulfovibrio sp.* [17]
<34> *Desulfovibrio africanus* [17]
<35> *Desulfomicrobium baculatus* [33]
<36> *Arabidopsis sp.* (Columbia ecotype [25]) [25]
<37> *Allochromatium vinosum* [21]
<38> *Burkholderia cepacia* (DBO1 [22]) [22]
<39> *Pseudomonas aeruginosa* [22, 28]
<40> *Pseudomonas putida* [22]
<41> *Ralstonia pickettii* [22]
<42> *Rhizobium tropici* [22]
<43> *Mycobacterium sp.* [22]
<44> *Bacillus subtilis* [22]
<45> *Lemna minor* [23, 35]
<46> *Arabidopsis thaliana* [23, 26]
<47> *Enteromorpha intestinalis* [24]
<48> *Tetraselmis sp.* [24]
<49> *Dunaliella salina* [24]
<50> *Thalassiosira weissflogii* [24]
<51> *Thalassiosira oceanica* [24]
<52> *Isochrysis galbana* [24]
<53> *Emiliania huxleyi* [24]
<54> *Heterocapsa triquetra* [24]

3 Reaction and Specificity

Catalyzed reaction

AMP + sulfite + acceptor = adenylyl sulfate + reduced acceptor (<30>, catalysis involves a nucleophilic attack of the N^5-atom of reduced FAD on the sulfur atom of the enzyme [9]; <3,8,10,30>, mechanism [18,27])

Reaction type

oxidation
redox reaction
reduction

Natural substrates and products

S adenylyl sulfate + reduced acceptor <1, 2, 3, 4, 5, 6, 7, 8, 9, 10, 30, 36, 38, 39, 40, 41, 42, 43, 44, 45> (<1, 2, 3, 4, 5, 6, 7, 8, 9>, enzyme reduces adenylyl sulfate that is formed during dissimilatory reduction of sulfate [1]; <10>, enzyme is involved in vectorial electron transport of cells grown on hydrogen plus sulfate as the sole energy source [11]; <21>, the enzyme is preferably required during the most rapid phase of growth [12]; <3, 10, 13, 14, 16, 17, 18, 33, 34>, enzyme is involved in dissimilatory sulfate reduction [17]; <37>, constitutive enzyme participates in oxidation of sulfite via the adenylyl sulfate pathway, that yields ATP through substrate-level phosphorylation [21]; <36,38,39,40,41,42,43,44>, assimilatory sulfate reduction [22,25,28]; <3,8,10,30>, key step in sulfur cycle [27]; <45>, key enzyme of assimilatory sulfate reduction [35]) (Reversibility: ? <1, 2, 3, 4, 5, 6, 7, 8, 9, 10, 27, 30, 36, 38, 39, 40, 41, 42, 43, 44, 45> [1, 11, 12, 17, 21, 22, 25, 27, 28, 35]) [1, 11, 12, 17, 21, 22, 25, 27, 28, 35]
P AMP + sulfite + acceptor

Substrates and products

S ADP + sulfite + ferricyanide <25, 30> (<30>, 50% of the activity with AMP [29]) (Reversibility: ? <25, 30> [7, 29]) [7, 29]
P ? + ferrocyanide
S AMP + sulfite + acceptor <30> (Reversibility: r <30> [9]) [9]
P adenylyl sulfate + reduced acceptor
S AMP + sulfite + cytochrome c <6, 10, 11, 25> (<25>, reaction with horse heart cytochrome c, no activity with cytochrome c from Candida krusei [7]; <30>, no activity with cytochrome c [29]) (Reversibility: ? <6, 10, 11, 25> [1, 7, 13, 15, 19, 29, 32]) [1, 7, 13, 15, 19, 29, 32]
P adenylyl sulfate + ?
S AMP + sulfite + ferricyanide <3, 6, 8, 10, 11, 13, 24, 25, 29, 30, 32> (Reversibility: ? <3, 6, 8, 10, 11, 13, 24, 25, 29, 30, 32> [1, 2, 3, 4, 6, 7, 8, 13, 16, 18, 19, 27, 29, 32, 34]) [1, 2, 3, 4, 6, 7, 8, 13, 16, 18, 19, 27, 29, 32, 34]
P adenylyl sulfate + ferrocyanide
S ATP + sulfite + ferricyanide <24, 25, 30> (<24>, 24% of the activity with AMP [6]; <25>, 12.8% of the activity with AMP [7]; <30>, 24% of the activity with AMP [29]) (Reversibility: ? <24, 25, 30> [6, 7, 29]) [6, 7, 29]
P ? + ferrocyanide

S CMP + sulfite + cytochrome c <11> (<11>, 7% of the activity with AMP [19]) (Reversibility: ? <11> [19]) [19]
P cytidylyl sulfate + ?
S CMP + sulfite + ferricyanide <6, 10, 11, 13, 24, 25> (<13>, 41% of the activity with AMP [3]; <24>, 24% of the activity with AMP [6]; <25>, 7.6% of the activity with AMP [7]; <11>, 8% of the activity with AMP [19]; <8>, no activity with CMP [32]) (Reversibility: ? <6, 10, 11, 13, 24, 25> [3, 6, 7, 15, 19, 32]) [3, 6, 7, 15, 19, 32]
P cytidylyl sulfate + ferrocyanide
S GMP + sulfite + cytochrome c <6, 11> (<11>, 9% of the activity with AMP [19]) (Reversibility: ? <6, 11> [19, 32]) [19, 32]
P guanylyl sulfate + ?
S GMP + sulfite + ferricyanide <8, 10, 11, 13, 21, 24, 25, 29, 30> (<13>, GMP is as active as AMP [3]; <24>, 68% of the activity with AMP [6]; <25>, 29.4% of the activity with AMP [7]; <29>, 54% of the activity with AMP [8]; <21>, 94% of the activity with AMP [12]; <11>, 51% of the activity with AMP [19]; <30>, 68% of the activity with AMP [29]) (Reversibility: ? <8, 10, 11, 13, 21, 24, 25, 29, 30> [3, 6, 7, 8, 12, 15, 19, 29, 32]) [3, 6, 7, 8, 12, 15, 19, 29, 32]
P guanylyl sulfate + ferrocyanide
S IMP + sulfite + cytochrome c <11> (<11>, 62% of the activity with AMP [19]) (Reversibility: ? <11> [19]) [19]
P ?
S IMP + sulfite + ferricyanide <10, 11, 24, 30> (<24>, 41% of the activity with AMP [6]; <11>, 90% of the activity with AMP [19]; <30>, 74% of the activity with AMP [29]) (Reversibility: ? <10, 11, 24, 30> [6, 8, 15, 19, 29]) [6, 15, 19, 29]
P ? + ferrocyanide
S UMP + sulfite + cytochrome c <11> (<11>, 9% of the activity with AMP [19]) (Reversibility: ? <11> [19]) [19]
P uridylyl sulfate + ?
S UMP + sulfite + ferricyanide <6, 10, 11, 13, 21, 25> (<13>, 36% of the activity with AMP [3]; <25>, 7.6% of the activity with AMP [7]; <21>, 33% of the activity with AMP [12]; <11>, 8% of the activity with AMP [19]; <8>, no activity with UMP [32]) (Reversibility: ? <6, 10, 11, 13, 21, 25> [3, 7, 12, 15, 19, 32]) [3, 7, 12, 15, 19, 32]
P uridylyl sulfate + ferrocyanide
S adenylyl sulfate + GSH <12, 45, 47> (Reversibility: ? <12, 45, 47> [24, 31, 35]) [24, 31, 35]
P AMP + sulfite + GSSG
S adenylyl sulfate + dithioerythritol <12, 39, 45> (Reversibility: ? <12, 39, 45> [23, 28, 31]) [23, 28, 31]
P AMP + sulfite + ?
S adenylyl sulfate + dithionite <13, 30> (Reversibility: ? <13, 30> [3, 27]) [3, 27]
P AMP + sulfite + ?
S adenylyl sulfate + dithiothreitol <46> (Reversibility: ? <46> [26]) [26]

P AMP + sulfite + ?

S adenylyl sulfate + ferrocytochrome c_3 <10> (Reversibility: ? <10> [30]) [30]

P AMP + sulfite + ferricytochrome

S adenylyl sulfate + reduced methyl viologen <1, 2, 3, 4, 5, 6, 7, 8, 9, 10, 13, 30> (Reversibility: ? <1, 2, 3, 4, 5, 6, 7, 8, 9, 10, 13, 30> [1, 3, 15, 18, 27]) [1, 3, 15, 18, 27]

P AMP + sulfite + methyl viologen <10> [18]

S adenylylsulfate + thioredoxin <39> (Reversibility: ? <39> [22]) [22]

P AMP + sulfite + ?

S dAMP + sulfite + cytochrome c <11> (<11>, 80% of the activity with AMP [19]) (Reversibility: ? <11> [19]) [19]

P deoxyadenylyl sulfate + ?

S deoxyAMP + sulfite + ferricyanide <6, 11, 24, 25, 30> (<24>, 74% of the activity with AMP [6]; <25>, 82.7% of the activity with AMP [7]; <11>, 128% of maximal activity [19]; <30>, 74% of the activity with AMP [29]) (Reversibility: ? <6, 11, 24, 25, 30> [6, 7, 19, 29, 32]) [6, 7, 19, 29, 32]

P deoxyadenylyl sulfate + ferrocyanide

S guanylyl sulfate + reduced methyl viologen <10> (Reversibility: ? <10> [15]) [15]

P GMP + methyl viologen + sulfite

S Additional information <45> (<45>, adenosine 5'-phosphosulfate sulfotransferase and adenosine 5'-phosphosulfate reductase are identical enzymes [35]) [35]

P ?

Inhibitors

2'-AMP <46> [26]

3'-phosphoadenosine-5'-phosphate <46> [26]

5'-AMP <46> [26]

AMP <8, 10, 25, 29> (<25>, inhibits reaction with cytochrome c above 0.1 mM, inhibits reaction with ferricyanide above 1.6 mM [7]; <29>, above 2 mM [8]; <8>, above 0.2 mM [27]) [7, 8, 27, 30]

NEM <6, 8, 11, 29> (<29>, 5 mM, 49% inhibition [8]; <11>, 10 mM, 45% loss of activity with ferricyanide, complete loss of activity with cytochrome c [19]; <6>, 5 mM complete inhibition [32]; <8>, 10 mM, complete inhibition [32]) [8, 19, 32]

NaN_3 <11> (<11>, 10 mM, no loss of the activity with ferricyanide, 52% loss of the activity with cytochrome c [19]) [19]

PCMB <6, 8, 11, 29> (<29>, 1 mM, 30% inhibition [8]; <11>, 1 mM, 23% loss of the activity with ferricyanide, complete loss of activity with cytochrome c [19]; <6>, 1 mM, complete inhibition [32]; <8>, 0.1 mM, 30% inhibition [32]) [8, 19, 32]

iodoacetamide <11, 29> (<29>, 5 mM, 54% inhibition [8]; <11>, 10 mM 2% loss of the activity with ferricyanide, 19% loss of the activity with cytochrome c [19]) [8, 19]

potassium phosphate <6> (<6>, inhibits reaction with cytochrome c [13]) [13]
sodium arsenite <11> (<11>, 10 mM, no loss of the activity with ferricyanide, 88% loss of the activity with cytochrome c [19]) [19]

Cofactors/prosthetic groups

FAD <3, 6, 8, 10, 11, 13, 14, 24, 29, 30, 32, 35> (<3>, enzyme contains 0.74 mol of FAD per mol of enzyme [2]; <3,6,8,10,11,13,14,24,30,35>, enzyme contains 1 mol of FAD per mol of enzyme [1,3,4,6,8,13,15,19,20,29,33]; <6,8,30>, non-covalently bound [29,32]; <32>, contains 1.4 mol of FAD per mol of enzyme [16]; <10>, enzyme contains FAD [18]; <30>, located nearby Fe-S center I, pointing ist C-8 methyl group to the iron-sulfur center [27]; <10>, electron mediator [30]) [1, 2, 3, 6, 8, 13, 15, 16, 19, 20, 27, 29, 30, 32, 33]
$FADH_2$ <10> (<10>, can donate electrons directly to adenylyl sulfate reductase [30]) [30]
$FMNH_2$ <10> (<10>, can donate electrons directly to adenylyl sulfate reductase [30]) [30]
dihydroriboflavin <10> (<10>, can donate electrons directly to adenylyl sulfate reductase [30]) [30]
flavin <25, 30> (<25>, contains 1 mol of flavin per mol of enzyme [7]; <30>, flavoenzyme [9]; <45,46>, flavin is absent [23]) [7, 9]
heme <11> (<11>, contains 2 heme groups of cytochrome c characterper mol of enzyme [19]) [19]

Activating compounds

menadione <10> (<10>, stimulates oxidation of ferrocytochrome c_3 with adenylyl sulfate and FAD as electron mediator [30]) [30]

Metals, ions

iron <3, 6, 8, 10, 11, 13, 15, 23, 25, 29, 30, 32, 35, 39, 45> (<10>, contains 6-8 gatom of nonheme iron [1]; <6>, contains 8-10 gatom of nonheme iron [1]; <8>, contains 6-11 gatom of nonheme iron [1]; <11>, contains 4 gatom of nonheme iron and 2 gatom of heme iron [1]; <3>, enzyme contains 6.3 mol iron per mol of enzyme [2]; <13>, iron sulfur protein, contains 8 iron atoms per 93000 Da protein, two [4Fe-4S] clusters [3]; <8>, enzyme contains 6-11 non-heme Fe per mol [4]; <25>, contains 8 mol of iron per mol of enzyme [7]; <29>, contains 4-6 mol of non-haem iron per mol of enzyme [8]; <30>, the two electrons required for the adenylyl sulfate reduction are transferred via two [4Fe-4S] clusters from the surface of the protein to FAD [9]; <10>, enzyme contains 12.7 mol of iron per 22000 Da enzyme [15]; <10,32>, contains nonheme iron [16,18]; <11>, contains 4 mol of non-heme iron per mol of enzyme [19]; <3>, contains 7 mol of Fe per mol of enzyme [20]; <45>, 4 paramagnetic iron sites, which form a diamagnetic [4Fe-4S]2+ cluster [23]; <10>, contains 5.6 iron per α,β heterodimer, novel Fe-S structure, possibly with an iron-nuclearity greater than four [27]; <30>, two [4Fe-4S] clusters, one cluster is deeply buried in the protein matrix, Fe-S center I, center II is located at the surface of the protein [27]; <39>, contains a [4Fe-4S] cluster [28]; <30>, contains 8 mol of nonheme iron per mol of enzyme [29]; <6>,

contains 8-10 mol of non-heme iron per mol of enzyme [32]; <8>, contains 8-13 mol of non-heme iron per mol of enzyme [32]) [1, 2, 3, 4, 7, 8, 9, 15, 16, 18, 20, 23, 27, 28, 29, 32, 33]
Additional information <46> (<46>, stimulation by high ionic strength [26]) [26]

Turnover number (min^{-1})
Additional information <3> [2]

Specific activity (U/mg)
0.6 <35> [33]
1.25 <10> [33]
1.65 <30> [33]
2.39 <24> [6]
2.5 <14> [33]
5.75 <10, 32> [15, 16]
5.8 <39> [22]
6 <13> [3, 33]
6.4 <6> [32]
8.93 <25> [7]
13 <3> [20]
13.15 <29> [8]
30-40 <45> [23]
40 <47> [24]
Additional information <3, 11, 21, 30> (<21>, specificity runs though a maximum after 30 h of growth [12]) [2, 12, 19, 29]

K_m-Value (mM)
0.0025 <6> (AMP, <6>, reaction with cytochrome c [32]) [32]
0.0065 <47> (adenylyl sulfate) [24]
0.01 <21> (AMP) [12]
0.01 <6> (GMP, <6>, reaction with cytochrome c [32]) [32]
0.015 <30> (adenylyl sulfate) [27]
0.015 <10> (riboflavin) [30]
0.017 <6> (sulfite, <6>, reaction with cytochrome c [32]) [32]
0.019 <10> (adenylyl sulfate) [15]
0.02 <8> (AMP) [27]
0.021 <25> (cytochrome c) [7]
0.025 <10> (guanylyl sulfate) [15]
0.033 <11> (cytochrome c) [19]
0.041 <8> (AMP, <8>, reaction with ferricyanide [32]) [4, 32]
0.05 <10, 11> (AMP, <11>, reaction with cytochrome c [19]) [19, 27]
0.071 <25> (sulfite, <25>, reaction with cytochrome c [7]) [7]
0.073 <11> (sulfite, <11>, reaction with ferricyanide [19]) [19]
0.089 <25> (ferricyanide) [7]
0.09 <25> (AMP, <25>, reaction with ferricyanide [7]) [7]
0.091 <25> (AMP, <25>, reaction with cytochrome c [7]) [7]
0.091 <32> (ferricyanide) [16]

0.093 <11> (sulfite, <11>, reaction with cytochrome c [19]) [19]
0.1 <6> (AMP, <6>, reaction with ferricyanide [32]) [32]
0.13 <29> (AMP) [8]
0.13 <11> (ferricyanide) [19]
0.13 <10> (sulfite) [27]
0.16 <3> (AMP) [33]
0.16 <13, 21> (ferricyanide) [12, 33]
0.17 <3> (AMP) [27]
0.17 <29> (sulfite) [8]
0.18 <10> (AMP, <10>, reaction with cytochrome c [15]) [15]
0.18 <3> (sulfite) [27]
0.2 <32> (AMP) [16]
0.2 <10> (sulfite) [27]
0.25 <29> (adenylyl sulfate) [8]
0.3 <10> (AMP, <10>, reaction with ferricyanide [15]) [15, 27]
0.31 <14> (AMP) [33]
0.34 <10, 13> (sulfite) [27, 33]
0.4 <3> (AMP) [2]
0.4 <24> (ferricyanide) [6]
0.6 <10> (sulfite) [33]
0.63 <8> (GMP, <8>, reaction with ferricyanide [32]) [32]
0.76 <14> (sulfite) [33]
0.78 <3> (sulfite) [33]
0.8 <10> (IMP, <10>, reaction with ferricyanide [15]) [15]
0.91 <32> (sulfite) [16]
1 <24> (AMP) [6]
1.2 <8> (sulfite) [27]
1.25 <10> (GMP, <10>, reaction with cytochrome c [15]) [15]
1.3 <25, 30> (sulfite, <25>, reaction with ferricyanide [7]) [7, 33]
1.4 <47> (GSH) [24]
1.5 <10> (IMP, <10>, reaction with cytochrome c [15]) [15]
1.5 <8, 11> (sulfite, <8, 11>, reaction with ferricyanide [19, 32]) [4, 19, 32]
1.7 <21> (sulfite) [12]
2.5 <6> (sulfite, <6>, reaction with ferricyanide [32]) [32]
3 <10> (GMP, <10>, reaction with ferricyanide [15]) [15]
3.1 <10> (CMP, <10>, reaction with cytochrome c [15]) [15]
4.2 <29> (ferricyanide) [8]
4.5 <10> (UMP, <10>, reaction with cytochrome c [15]) [15]

pH-Optimum

7 <6> (<6>, ferricyanide assay, phosphate buffer [32]) [32]
7-7.5 <10> (<10>, oxidation of methyl viologen [15]) [15]
7.2 <8> (<8>, reaction with AMP, SO_3^{2-} and ferricyanide [1]) [1, 32]
7.4 <3, 6, 10, 13> (<10>, reaction with AMP, SO_3^{2-} and ferricyanide [1]; <6>, ferricyanide assay, Tris-HCl buffer [32]) [1, 3, 32, 33]

7.5 <10, 14> (<10>, reaction with cytochrome c and UMP or CMP [15]) [15, 33]
7.6 <23> [33]
7.7 <25> (<25>, reaction with ferricyanide [7]) [7]
8 <11, 24, 29, 30> (<11>, reaction with AMP, SO_3^{2-} and ferricyanide [1]; <11>, reaction with ferricyanide [19]) [1, 4, 6, 8, 19, 33]
8.5 <46> (<46>, reaction with adenylyl sulfate and dithiothreitol [26]) [26]
8.7 <32> [16]
8.8 <25> (<25>, reaction with cytochrome c [7]) [7]
9 <10, 11> (<11>, reaction with AMP, SO_3^{2-} and cytochrome c [1]; <10>, reaction with cytochrome c and IMP or GMP [15]; <11>, reaction with cytochrome c [19]) [1, 15, 19]
9.5 <6, 10> (<6>, reaction with AMP, SO_3^{2-} and cytochrome c [1]; <10>, reaction with cytochrome c and AMP [15]; <6>, reaction with cytochrome c as electron acceptor, Tris-HCl buffer [32]) [1, 15, 32]

Temperature optimum (°C)
85 <24> [6]

Temperature range (°C)
70-90 <24> (<24>, 70°C: about 65% of maximal activity, 90°C: about 90% of maximal activity [6]) [6]

4 Enzyme Structure

Molecular weight
90000-124000 <3> (<3>, dynamic light scattering [20]) [20]
160000 <13, 24, 30> (<24,30>, gel filtration [6,29]) [6, 29, 33]
170000 <3, 6, 15, 35> [1, 33]
175000 <25, 23> (<25>, gel filtration [7]) [7, 33]
180000 <8, 11, 29> (<29>, gel filtration [8]) [1, 8]
186000 <10> (<10>, smallest catalytic active subunit, also aggregations of 324000 Da and 569000 Da are detected, gel filtration [27]) [27]
190000 <3> (<3>, gel filtration [2]) [2]
200000 <8> (<8>, estimation from sedimentation coefficient [4]) [4]
210000 <32> (<32>, gel filtration [16]) [16]
220000 <10> [1, 2]
400000 <13> (<13>, gel filtration [3]) [3]
439500 <10> (<10>, equilibrium sedimentation [15]) [15]

Subunits
? <3, 10, 13> (<10>, x * 72000 + x * 20000, SDS-PAGE [15]; <3,13>, x * 70000 + x * 26000, SDS-PAGE [2,3]) [2, 3, 15]
dimer <3, 24> (<3>, α,β, 1 * 75000 + 1 * 20000, SDS-PAGE [20]; <24>, 2 * 80000, SDS-PAGE [6]) [6, 20]

tetramer <10, 30> (<30>, $\alpha_2\beta_2$, 2 * 73300 + 2 * 17100, SDS-PAGE [29]; <10>, $\alpha_2\beta_2$, 2 * 67800 + 2 * 25600, smallest catalytic active unit, SDS-PAGE [27]) [27, 29]

5 Isolation/Preparation/Mutation/Application

Source/tissue

cell suspension culture <12> (<12>, axenic cell suspension culture [31]) [31]
leaf <36> (<36>, 3fold decline in activity during development can be attributed to a reduction of enzyme during aging of individual leaves, the highest activity occurs in the youngest leaves and the lowest in fully expanded leaves [25]) [25]
Additional information <10> (<10>, extracts of sulfite-grown cells have lower activity than extracts of sulfate-grown cells [14]) [14]

Localization

chloroplast <36> (<36>, exclusively located in chloroplast [25]) [25]
chromatophore <21> [12]
cytoplasm <10, 13, 23> [5, 10]
cytoplasmic membrane <10> (<10>, 87% of the activity is located on the cytoplasmic side of the cytoplasmic membrane [11]) [11]
spheroplast <13> [10]
Additional information <46> (<46>, mitochondrial or cytoplasmic localization [26]) [26]

Purification

<3> (G100A [2]) [2, 20, 27]
<6> [13, 32]
<8> [12, 32]
<10> [15, 27]
<11> [19]
<13> [3]
<24> [6]
<25> [7]
<29> [8]
<30> [27, 29]
<32> [16]
<39> [22]

Crystallization

<30> [9]

Cloning

<3> [20]
<12> (expression in Escherichia coli [31]) [31]
<38> [22]
<39> [22, 28]

<45> (overexpression in Escherichia coli [23]) [23]
<46> (overexpression in Escherichia coli [23]) [23, 26]
<47> [24]

Engineering

C129S <45> (<45>, maximal velocity dropps to 50% of that of the wild-type enzyme [23]) [23]
C248S <45> (<45>, mutation completely abolishes enzyme activity [23]) [23]

Application

synthesis <8> (<8>, production of adenosine 5'-phosphosulfate labeled with either ^{14}C or ^{35}S [34]) [34]

6 Stability

Temperature stability

22 <32> (<32>, 2 h, more than 30% loss of activity [16]) [16]
60 <6, 8, 32> (<8>, 1 h, stable [4]; <32>, heating for longer periods results in protein precipitation with simultaneous loss of activity [16]; <6>, 6 min, stable [32]) [4, 16, 32]
65 <25> (<25>, 5 min, complete loss of activity [7]) [7]
70 <32> (<32>, 2 min, 50% loss of activity [16]) [16]
75 <8> (<8>, thermotolerant up to [4]) [4]

General stability information

<32>, 30-50% loss of activity by repeated freezing and thawing [16]
<45>, enzyme is rapidly denatured by even micromolar concentrations of ferricyanide [23]

Storage stability

<6>, 5°C, 5 days, 20% loss of activity [32]
<8>, frozen, protein concentration above 5 mg/ml, stable for several months [32]
<13>, -20°C, much loss of activity after 30 days [3]
<25>, -18°C, 88-90% loss of activity after 1 week, 68% loss of activity after 16 days, in presence of 50% glycerol, 46% loss of activity after 16 days [7]
<25>, 4°C, more than 20% loss of activity after 8 h [7]
<32>, -20°C, stable for several weeks [16]
<32>, 4°C, more than 2 days, more than 30% loss of activity [16]
<45>, 4°C, continous loss of activity [23]
<46>, 4°C, continous loss of activity, precipitation after 2-3 days [23]

References

[1] Hatefi, Y.; Stiggall, D.L.: Metal-containing flavoprotein dehydrogenases. The Enzymes, 3rd Ed. (Boyer, P.D., ed.), **13**, 175-297 (1976)

[2] Odom, J.M.; Jessie, K.; Knodel, E.; Emptage, M.: Immunological cross-reactivities of adenosine-5'-phosphate reductases from sulfate-reducing and sulfide-oxidizing bacteria. Appl. Environ. Microbiol., **57**, 727-733 (1991)
[3] Lampreia, J.; Moura, I.; Teixeira, M.; Peck, H.D.; LeGall, J.; Huynh, B.H.; Moura, J.J.G.: The active centers of adenylylsulfate reductase from Desulfovibrio gigas. Characterization and spectroscopic studies. Eur. J. Biochem., **188**, 653-664 (1990)
[4] Taylor, B.F.: Thermotolerance of adenylsulfate reductase from Thiobacillus denitrificans. FEMS Microbiol. Lett., **59**, 351-354 (1989)
[5] Kremer, D.R.; Veenhuis, M.; Fauque, G.; Peck, H.D.; LeGall, J.; Lampreia, J.; Moura, J.J.G.; Hansen, T.A.: Immunocytochemical localization of APS reductase and bisulfite reductase in three Desulfovibrio species. Arch. Microbiol., **150**, 296-301 (1988)
[6] Speich, N.; Trueper, H.G.: Adenylylsulphate reductase in a dissimilatory sulphate-reducing archaebacterium. J. Gen. Microbiol., **134**, 1419-1425 (1988)
[7] Stille, W.; Trueper, H.G.: Adenylylsulfate reductase in some new sulfate-reducing bacteria. Arch. Microbiol., **137**, 145-150 (1984)
[8] Khanna, S.; Nicholas, D.J.D.: Substrate phosphorylation in Chlorobium vibrioforme f. sp. thiosulfatophilum. J. Gen. Microbiol., **129**, 1365-1370 (1983)
[9] Fritz, G.; Roth, A.; Schiffer, A.; Buchert, T.; Bourenkov, G.; Bartunik, H.D.; Huber, H.; Stetter, K.O.; Kroneck, P.M.H.; Ermler, U.: Structure of adenylylsulfate reductase from the hyperthermophilic Archaeoglobus fulgidus at 1.6-A resolution. Proc. Natl. Acad. Sci. USA, **99**, 1836-1841 (2002)
[10] Odom, J.M.; Peck, H.D.: Localization of dehydrogenases, reductases, and electron transfer components in the sulfate-reducing bacterium Desulfovibrio gigas. J. Bacteriol., **147**, 161-169 (1981)
[11] Badziong, W.; Thauer, R.K.: Vectorial electron transport in Desulfovibrio vulgaris (Marburg) growing on hydrogen plus sulfate as sole energy source. Arch. Microbiol., **125**, 167-174 (1980)
[12] Schwenn, J.D.; Biere, M.: APS-reductase activity in the chromophores of Chromatium vinosum strain D. FEMS Microbiol. Lett., **6**, 19-22 (1979)
[13] Adachi, K.; Suzuki, I.: A study on the reaction mechanism of adenosine 5'-phosphosulfate reductase from Thiobacillus thioparus, an iron-sulfur flavoprotein. Can. J. Microbiol., **55**, 91-98 (1977)
[14] Kobayashi, K.; Morisawa, Y.; Ishituka, T.; Ishimoto, M.: Biochemical studies on sulfate-reducing bacteria. XIV. Enzyme levels of adenylylsulfate reductase, inorganic pyrophosphatase, sulfite reductase, hydrogenase, and adenosine triphosphatase in cells grown on sulfate, sulfite, and thiosulfate. J. Biochem., **78**, 1079-1085 (1975)
[15] Bramlett, R.N.; Peck, H.D.: Some physical and kinetic properties of adenylyl sulfate reductase from Desulfovibrio vulgaris. J. Biol. Chem., **250**, 2979-2986 (1975)
[16] Kirchhoff, J.; Trueper, H.G.: Adenylylsulfate reductase of Chlorobium limicola. Arch. Microbiol., **100**, 115-120 (1974)
[17] Skyring, G.W.; Trudinger, P.A.: A comparison of the electrophoretic properties of the ATP-sulfurylases, APS-reductases, and sulfite reductases from

cultures of dissimilatory sulfate-reducing bacteria. Can. J. Microbiol., **19**, 375-380 (1973)
[18] Peck, H.D.; Bramlett, R.; DerVartanian, D.V.: On the mechanism of adenylyl sulfate reductase for the sulfate-reducing bacterium, Desulfovibrio vulgaris. Z. Naturforsch. B, **27**, 1084-1086 (1972)
[19] Trueper, H.G.; Rogers, L.A.: Purification and properties of adenylyl sulfate reductase from phototrophic sulfur bacterium, Thiocapsa roseopersicina. J. Bacteriol., **108**, 1112-1121 (1971)
[20] Fritz, G.; Buchert, T.; Huber, H.; Stetter, K.O.; Kroneck, P.M.H.: Adenylylsulfate reductases from archaea and bacteria are 1:1 $\alpha\beta$-heterodimeric iron-sulfur flavoenzymes - high similarity of molecular properties emphasizes their central role in sulfur metabolism. FEBS Lett., **473**, 63-66 (2000)
[21] Sanchez, O.; Ferrera, I.; Dahl, C.; Mas, J.: In vivo role of adenosine-5'-phosphosulfate reductase in the purple sulfur bacterium Allochromatium vinosum. Arch. Microbiol., **176**, 301-305 (2001)
[22] Bick, J.A.; Dennis, J.J.; Zylstra, G.J.; Nowack, J.; Leustek, T.: Identification of a new class of 5'-adenylylsulfate (APS) reductases from sulfate-assimilating bacteria. J. Bacteriol., **182**, 135-142 (2000)
[23] Kopriva, S.; Buchert, T.; Fritz, G.; Suter, M.; Weber, M.; Benda, R.; Schaller, J.; Feller, U.; Schurmann, P.; Schunemann, V.; Trautwein, A.X.; Kroneck, P.M.; Brunold, C.: Plant adenosine 5'-phosphosulfate reductase is a novel iron-sulfur protein. J. Biol. Chem., **276**, 42881-42886 (2001)
[24] Gao, Y.; Schofield, O.M.E.; Leustek, T.: Characterization of sulfate assimilation in marine algae focusing on the enzyme 5'-adenylylsulfate reductase. Plant Physiol., **123**, 1087-1096 (2000)
[25] Rotte, C.; Leustek, T.: Differential subcellular localization and expression of ATP sulfurylase and 5'-adenylylsulfate reductase during ontogenesis of Arabidopsis leaves indicates that cytosolic and plastid forms of ATP sulfurylase may have specialized functions. Plant Physiol., **124**, 715-724 (2000)
[26] Setya, A.; Murillo, M.; Leustek, T.: Sulfate reduction in higher plants: molecular evidence for a novel 5'-adenylylsulfate reductase. Proc. Natl. Acad. Sci. USA, **93**, 13383-13388 (1996)
[27] Fritz, G.; Buchert, T.; Kroneck, P.M.H.: The function of the [4Fe-4S] clusters and FAD in bacterial and archaeal adenylylsulfate reductases: evidence for flavin-catalyzed reduction of adenosine 5'-phosphosulfate. J. Biol. Chem., **277**, 26066-26073 (2002)
[28] Kopriva, S.; Buchert, T.; Fritz, G.; Suter, M.; Benda, R.; Schunemann, V.; Koprivova, A.; Schurmann, P.; Trautwein, A.X.; Kroneck, P.M.H.; Brunold, C.: The presence of an iron-sulfur cluster in adenosine 5'-phosphosulfate reductase separates organisms utilizing adenosine 5'-phosphosulfate and phosphoadenosine 5'-phosphosulfate for sulfate assimilation. J. Biol. Chem., **277**, 21786-21791 (2002)
[29] Dahl, C.; Truper, H.G.: Sulfite reductase and APS reductase from Archaeoglobus fulgidus. Methods Enzymol., **331**, 427-441 (2001)
[30] Yagi, T.; Ogata, M.: Catalytic properties of adenylylsulfate reductase from Desulfovibrio vulgaris Miyazaki. Biochimie, **78**, 838-846 (1996)

[31] Prior, A.; Uhrig, J.F.; Heins, L.; Wiesmann, A.; Lillig, C.H.; Stoltze, C.; Soll, J.; Schwenn, J.D.: Structural and kinetic properties of adenylyl sulfate reductase from Catharanthus roseus cell cultures. Biochim. Biophys. Acta, **1430**, 25-38 (1999)
[32] Taylor, B.F.: Adenylylsulfate reductases from thiobacilli. Methods Enzymol., **243**, 393-400 (1994)
[33] Lampreia, J.; Pereira, A.S.; Moura, J.J.G.: Adenylylsulfate reductases from sulfate-reducing bacteria. Methods Enzymol., **243**, 241-260 (1994)
[34] Adams, C.A.; Warnes, G.M.; Nicholas, D.J.D.: Preparation of labeled adensosine 5'-phosphosulfate using APS reductase from Thiobacillus denitrificans. Anal. Biochem., **42**, 207-213 (1971)
[35] Suter, M.; Von Ballmoos, P.; Kopriva, S.; Op den Camp, R.; Schaller, J.; Kuhlemeier, C.; Schurmann, P.; Brunold, C.: Adenosine 5'-phosphosulfate sulfotransferase and adenosine 5'-phosphosulfate reductase are identical enzymes. J. Biol. Chem., **275**, 930-936 (2000)

Hydrogensulfite reductase 1.8.99.3

1 Nomenclature

EC number
1.8.99.3

Systematic name
trithionate:(acceptor) oxidoreductase

Recommended name
hydrogensulfite reductase

Synonyms
bisulfite reductase
bisulfite reductase (P582)
carbon monoxide-binding pigment P582 <11> [23]
desulfofuscidin
desulforubidin
desulfoviridin
dissimilatory sulfite reductase
dissimilatory-type sulfite reductase
reductase, bisulfite

CAS registry number
42612-25-9
85876-01-3
9045-15-2
9059-42-1

2 Source Organism

<1> *Desulfovibrio thermophilus* (DSM 1276 [1]; sulfite reductase [4]) [1, 4]
<2> *Desulfovibrio baculatus* (strain DSM 1743, desulforubidin [2]; DSM 1743 [36]) [2, 36]
<3> *Desulfovibrio gigas* (desulfoviridin [2]) [2, 4, 14, 19, 21, 26]
<4> *Desulfobacterium anilini* (sulfite reductase, no desulfoviridin [3]) [3]
<5> *Desulfovibrio vulgaris* (strain Hildenborough [4,29,32,33,34]; Miyazaki F [7]; 2 enzyme forms DSF-I and DSF-II [12]; strain Miyazaki [18]) [4, 7, 12, 16, 17, 18, 20, 24, 25, 29, 32, 33, 34]
<6> *Thermodesulfobacterium mobile* [35]
<7> *Desulfovibrio sulfodismutans* (strain ThAc01, desulfoviridin [5]) [5]
<8> *Desulfovibrio africanus* (Benghazi, NCIB 8401 [6]) [6]

<9> *Desulfonema limicola* (strain 5ac10 [8]) [8]
<10> *Thermodesulfobacterium commune* [9, 35]
<11> *Desulfotomaculum nigrificans* [10, 21, 22, 23]
<12> *Thiobacillus denitrificans* [11]
<13> *Chromatium vinosum* [13, 15]
<14> *Desulfovibrio desulfuricans* (Norway strain [27]; strain Essex [31,32]) [14, 25, 27, 31, 32, 36]
<15> *Desulfotomaculum ruminis* [14]
<16> *Desulfosarcina variabilis* (desulforibidin [28]) [28]
<17> *Desulfovibrio baarsii* [28]
<18> *Desulfotomaculum acetoxidans* [28]
<19> *Bilophila wadsworthia* (RZATAU, desulfoviridin [30]) [30]

3 Reaction and Specificity

Catalyzed reaction

trithionate + acceptor + 2 H_2O + OH^- = 3 hydrogen sulfite + reduced acceptor

Reaction type

oxidation
redox reaction
reduction

Natural substrates and products

S sulfide + electron acceptor <14> (<14>, although the bacterium is unable to grow with sulfide as electron donor the dissimilatory sulfite reductase catalyzes the oxidation of sulfide in vivo [31]) (Reversibility: ? <14> [31]) [31]
P ? + oxidized electron donor
S sulfite + electron donor <2, 5, 6, 8, 9, 10, 11, 14, 19> (<2,5,6,8,9,10,14,19>, dissimilatory reduction of sulfite or disulfite [6,8,9,16,30,31,32,34,35,36]; <11>, reduction of bisulfite to sulfide by a dissimilatory pathway involving trithionate and thiosulfate as intermediate [10]; <19>, enzyme is involved in energy conservation by reducing sulfite, which is formed during the degradation of taurine as an electron acceptor to sulfide [30]) (Reversibility: ? <2, 5, 6, 8, 9, 10, 11, 14, 19> [6, 8, 9, 10, 16, 30, 31, 32, 34, 35, 36]) [6, 8, 9, 10, 16, 30, 31, 32, 34, 35, 36]
P ? + oxidized electron donor

Substrates and products

S bisulfite + electron donor <5, 11> [17, 20, 23, 24, 25]
P ? + oxidized electron donor
S hydroxylamine + electron donor <5, 10, 11> (<5>, electron acceptor: reduced methyl viologen [20]) (Reversibility: ? <5, 10, 11> [12, 20, 21, 24, 34, 35]) [12, 20, 21, 24, 34, 35]
P NH_3 + H_2O + oxidized electron donor

S nitrite + electron donor <5, 10> (<5>, electron acceptor: reduced methyl viologen [20]) (Reversibility: ? <5, 10> [9, 20, 24, 34, 35]) [9, 20, 24, 34, 35]

P NH_3 + H_2O + oxidized electron donor <10> [9]

S sulfide + electron acceptor <14> (<14>, electron acceptor: cytochrome c_3 [31]) (Reversibility: ? <14> [31]) [31]

P ?

S sulfite + electron donor <1-19> (<2,3,5,8,10,11,13,14>, electron donor: methyl viologen [6,7,9,13,20,21,24,25,26,27,35,36]; <10>, electron donor: benzyl viologen [9,35]; <12>, electron donor: reduced viologen dyes [11]; <5,14>, electron donor: cytochrome c_3 [25,31]) (Reversibility: ? <1-19> [1-36]) [1-36]

P trithionate + thiosulfate + sulfide + polythionate + oxidized electron donor <5, 8, 10, 11, 13> (<8>, forms mainly trithionate besides thiosulfate and sulfide in sulfite reduction coupled to hydrogenase and methyl viologen [6]; <5>, formation of trithionate, thiosulfate and sulfide when coupled with a hydrogenase-methyl viologen system [7]; <10>, with reduced methyl viologen as electron donor, the major product of sulfite reduction is trithionate [9]; <5,11>, trithionate and thiosulfate as intermediate compounds [10,16]; <12,13>, sulfide, thiosulfate and trithionate [11,13,20]; <5,10,11>, enzyme produces mainly trithionate, besides thiosulfate and sulfide [12,21,23,35]; <5>, the major product trithionate is formed at 5 to 10 times more than thiosulfate and sulfide [17]; <3>, trithionate, thiosulfate, tetrathionate and sulfide, variations in assay conditions cause variations in the end product of sulfite reduction [19]) [6, 7, 9, 10, 12, 15, 16, 17, 19, 20, 21, 23, 35]

S trimethylamine N-oxide + electron donor <5> (Reversibility: ? <5> [20]) [20]

P ? + oxidized electron donor

Inhibitors

arsenite <5> (<5>, 10 mM, 68% inhibition [20]) [20]

Activating compounds

cyanide <5> (<5>, 1 mM, slight activation [20]) [20]

Additional information <5> (<5>, active form of enzyme binds and catalyzes substrate reduction, inactive enzyme form exists in resting cells [29]) [29]

Metals, ions

iron <1, 2, 3, 5, 8, 10, 11, 12, 13, 14, 15, 16> (<13>, enzyme contains siroheme [13]; <1>, enzyme contains 4 sirohemes and 8 [4Fe-4S] centers per mol of protein [1]; <2>, contains exchange-coupled siroheme-[4Fe-4S] units [2]; <8>, iron is the only significant chelatable metal contained in the chromophore sirohydrochlorin [6]; <10>, contains siroheme as prosthetic group, 20-21 iron atoms per molecule of protein, native protein contains 4 sirohemes and 4 [4Fe-4S] cluster per molecule [9]; <12>, contains siroheme, 24 Fe per enzyme molecule [11]; <5>, enzyme form DSF-I contains 0.125 mM of Fe per g of protein, enzyme form DSF-II contains 0.116 mM of Fe per g of protein

[12]; <13>, contains 51 Fe atoms per enzyme molecule [13]; <14,15>, enzyme contains sirohem and Fe-S clusters, reaction of reductase with reduced methyl viologen results in reduction of only the siroheme [14]; <11>, tetrahydroporphyrin-based prosthetic group sirohydrochlorin [22]; <16>, contains two fully metalated sirohemes and Fe-S clusters, 15 Fe and 19 S^{2-} [28]; <14>, enzyme contains siroheme and Fe-S centres in the active site [31]; <14>, contains 24 Fe and 18 S^{2-} per 200000 Da enzyme, 5% of the siroheme moiety is metallated [32]; <5>, characterization of the redox thermodynamics of the Fe4S4-siroheme prosthetic center [33]; <5>, two pairs of [Fe4S4] centers and siroheme units [34]; <6,10>, 4 mol of siroheme per mol of enzyme [35]; <6>, 32 iron atoms per molecule [35]; <10>, 20 iron atoms per molecule [35]; <2,14>, contains 4 [4Fe-4S] clusters, each high-spin ferric siroheme is exchange coupled to a $[4Fe-4S]^{2-}$ cluster, a solitary iron center is not associated with either the sirohemes or [4Fe-4S] clusters [36]) [1, 2, 3, 6, 9, 11, 12, 13, 14, 22, 28, 31, 32, 33, 34, 35, 36]

Turnover number (min^{-1})

Additional information <5, 12, 13> (<5>, 18.6 mol SO_3^{2-} * min^{-1} * mol heme-1, 2.28 mol nitrite * min^{-1} * mol heme-1, 1740 mol hydroxylamine * min^{-1} * mol heme-1 [34]) [11, 13, 34]

Specific activity (U/mg)

0.041 <14> (<14>, soluble enzyme [32]) [32]
0.128 <14> (<14>, membranous enzyme [32]) [32]
0.163 <13> [13]
0.2 <10> [35]
0.26 <5> [24]
0.331 <12> [11]
0.41 <14> [27]
0.632 <3> [26]
1.47 <1> [1]
2 <10> [35]
Additional information <5, 8, 13> [6, 7, 15]

K_m-Value (mM)

0.028 <5> (nitrite) [34]
0.06 <5> (sulfite) [34]
3.6 <5> (sulfite) [20]
48 <5> (hydroxylamine) [34]

pH-Optimum

5.5-6 <5> (<5>, reduction of sulfite [20]) [20]
5.8 <1> [1]
5.8-6 <10> [35]
6 <3, 10, 11, 12, 13> [9, 11, 13, 21, 26]
7.5 <5> (<5>, reduction of hydroxylamine [20]) [20]

pH-Range

5-6.5 <5> (<5>, pH 5.0: about 70% of maximal activity, pH 6.5: about 60% of maximal activity, reduction of sulfite [20]) [20]
6-8 <5> (<5>, pH 6.0: about 50% of maximal activity, pH 8.0: about 90% of maximal activity, reduction of hydroxylamine [20]) [20]

Temperature optimum (°C)

65-70 <10> [9, 35]

Temperature range (°C)

35-70 <10> (<10>, 35°C: very low activity below 35°C, 65-70°C: maximal activity at [9,35]) [9, 35]

4 Enzyme Structure

Molecular weight

160000 <12> (<12>, gel filtration [11]) [11]
167000 <10> (<10>, equilibrium sedimentation [9,35]) [9, 35]
175000 <1> (<1>, gel filtration [1]) [1]
180000 <5, 13> (<5>, gel filtration [7]; <13>, disc electrophoresis [15]) [7, 15]
190000 <1> (<1>, equilibrium sedimentation [1]) [1]
200000 <3, 5, 14, 16> (<5,14,16>, gel filtration [12,28,32]; <3>, equilibrium sedimentation [26]) [12, 26, 28, 32]
280000 <13> (<13>, gel filtration [13]) [13]
285000 <19> (<19>, gel filtration [30]) [30]

Subunits

? <14, 19> (<14>, $\alpha_2\beta_2\gamma_n$, 2 * 500000 + 2 * 450000 + x * 11000, SDS-PAGE [32]; <19>, $\alpha_2\beta_2\gamma_n$, 2 * 49000 + 2 * 54000 + x * 11000, SDS-PAGE [30]) [30, 32]
hexamer <16> (<16>, $\alpha_2\beta_2\gamma_2$, 2 * 50000 + 2 * 42500 + 2 * 12000, SDS-PAGE [28]) [28]
tetramer <1, 5, 10, 12, 13> (<13>, $\alpha_2\beta_2$, 2 * 37000 + 2 * 42000, SDS-PAGE [13]; <12>, $\alpha_2\beta_2$, 2 * 38000 + 2 * 43000, SDS-PAGE [11]; <5>, $\alpha_2\beta_2$, 2 * 50000 + 2 * 39000, SDS-PAGE [7]; <5>, x * 42000 + x * 500000, SDS-PAGE [24]; <1>, 4 * 44000-48000, SDS-PAGE [1]; <5>, 2 * 45000 + 2 * 55000, SDS-PAGE [25]; <10>, $\alpha_2\beta_2$, x * 48000, SDS-PAGE [9]; <10>, x * 48000, four similar but not identical subunits, SDS-PAGE [35]) [1, 7, 9, 11, 13, 24, 25]

5 Isolation/Preparation/Mutation/Application

Source/tissue

culture condition:sulfite-grown cell <5> (<5>, activity is highest in sulfite-grown cells [18]) [18]

Localization

cytoplasm <1, 3, 5> [4]
membrane <14> (<14>, two putative transmembrane α-helices are postulated on each β-subunit [31]) [31]
soluble <14> [32]

Purification

<1> [1]
<3> [26]
<5> [7, 17, 24, 25, 34]
<6> [35]
<8> [6]
<10> [9, 35]
<11> [21]
<12> [11]
<13> [13]
<14> [27, 36]
<16> [28]
<19> [30]

6 Stability

Temperature stability

70 <10> (<10>, stable to, rapidly denatured above [9]) [9]

Storage stability

<5>, -20°C, 6 months, stable [34]
<13>, -20°C, 1 month, no loss of activity [15]
<13>, 4°C, 1 month, 60% loss of activity [15]

References

[1] Fauque, G.; Lino, A.R.; Czechowski, M.; Kang, L.: DerVartanian, D.V.; Moura, J.J.G.; LeGall, J.; Moura, I.: Purification and characterization of bisulfite reductase (desulfofuscidin) from Desulfovibrio thermophilus and its complexes with exogenous ligands. Biochim. Biophys. Acta, **1040**, 112-118 (1990)
[2] Moura, I.; LeGall, J.; Lino, A.R.; Peck, H.D.; Fauque, G.; Xavier, A.V.; DerVartanian, D.V.; Moura, J.J.G.; Huynh, B.H.: Characterization of two dissimilatory sulfite reductases (desulforubidin and desulfoviridin) from the sulfate-reducing bacteria. Moessbauer and EPR studies. J. Am. Chem. Soc., **110**, 1075-1082 (1988)
[3] Schnell, S.; Bak, F.; Pfennig, N.: Anaerobic degradation of aniline and dihydroxybenzenes by newly isolated sulfate-reducing bacteria and description of Desulfobacterium anilini. Arch. Microbiol., **152**, 556-563 (1989)

[4] Kremer, D.R.; Veenhuis, M.; Fauque, G.; Peck, H.D.; LeGall, J.; Lampreia, J.; Moura, J.J.G.; Hansen, T.A.: Immunocytochemical localization of APS reductase and bisulfite reductase in three Desulfovibrio species. Arch. Microbiol., **150**, 296-301 (1988)
[5] Bak, F.; Pfennig, N.: Chemolithotrophic growth of Desulfovibrio sulfodismutans sp. nov. by disproportionation of inorganic sulfur compounds. Arch. Microbiol., **147**, 184-189 (1987)
[6] Seki, Y.; Nagai, Y.; Ishimoto, M.: Characterization of a dissimilatory-type sulfite reductase, desulfoviridin, from Desulfovibrio africanus Benghazi. J. Biochem., **98**, 1535-1543 (1985)
[7] Aketagawa, J.; Kojo, K.; Ishimoto, M.: Purification and properties of sulfite reductase from Desulfovibrio vulgaris, Miyazaki F. Agric. Biol. Chem., **49**, 2359-2365 (1985)
[8] Widdel, F.; Kohring, G.W.; Mayer, F.: Studies on dissimilatory sulfate-reducing bacteria that decompose fatty acids. III. Characterization of the filamentous gliding Desulfonema limicola gen. nov. sp. nov., and Desulfonema magnum sp. nov.. Arch. Microbiol., **134**, 286-294 (1983)
[9] Hatchikian, E.C.; Zeikus, J.G.: Characterization of a new type of dissimilatory sulfite reductase present in Thermodesulfobacterium commune. J. Bacteriol., **153**, 1211-1220 (1983)
[10] Akagi, J.M.: Reduction of bisulfite by the trithionate pathway by cell extracts from Desulfotomaculum nigrificans. Biochem. Biophys. Res. Commun., **117**, 530-535 (1983)
[11] Schedel, M.; Trueper, H.G.: Purification of Thiobacillus denitrificans siroheme sulfite reductase and investigation of some molecular and catalytic properties. Biochim. Biophys. Acta, **568**, 454-467 (1979)
[12] Seki, Y.; Kobayashi, K.; Ishimoto, M.: Biochemical studies on sulfate-reducing bacteria. XV. Separation and comparison of two forms of desulfoviridin. J. Biochem., **85**, 705-711 (1979)
[13] Schedel, M.; Vanselow, M.; Trueper, H.G.: Siroheme sulfite reductase isolated from Chromatium vinosum. Purification and investigation of some of ist some of ist molecular and catalytic properties. Arch. Microbiol., **121**, 29-36 (1979)
[14] Liu, C.L.; DerVartanian, D.V.; Peck, H.D.: On the redox properties of three bisulfite reductases from the sulfate-reducing bacteria. Biochem. Biophys. Res. Commun., **91**, 962-970 (1979)
[15] Kobayashi, K.; Katsura, E.; Kondo, T.; Ishimoto, M.: Chromatium sulfite reductase. I. Characterization of thiosulfate-forming activity at the cell extract level. J. Biochem., **84**, 1209-1215 (1978)
[16] Drake, H.L.; Akagi, J.M.: Dissimilatory reduction of bisulfite by Desulfovibrio vulgaris. J. Bacteriol., **136**, 916-923 (1978)
[17] Drake, H.L.; Akagi, J.M.: Product analysis of bisulfite reductase activity isolated from Desulfovibrio vulgaris. J. Bacteriol., **126**, 733-738 (1976)
[18] Kobayashi, K.; Morisawa, Y.; Ishituka, T.; Ishimoto, M.: Biochemical studies on sulfate-reducing bacteria. XIV. Enzyme levels of adenylylsulfate reductase, inorganic pyrophosphatase, sulfite reductase, hydrogenase, and ade-

nosine triphosphatase in cells grown on sulfate, sulfite, and thiosulfate. J. Biochem., **78**, 1079-1085 (1975)
[19] Jones, H.E.; Skyring, G.W.: Effect of enzymic assay conditions on sulfite reduction catalysed by desulfoviridin from Desulfovibrio gigas. Biochim. Biophys. Acta, **377**, 52-60 (1975)
[20] Kobayashi, K.; Seki, Y.; Ishimoto, M.: Biochemical studies on sulfate-reducing bacteria. 8. Sulfite reductase from Desulfovibrio vulgaris - mechanism of trithionate, thiosulfate, and sulfide formation and enzymatic properties. J. Biochem., **75**, 519-529 (1974)
[21] Akagi, J.M.; Chan, M.; Adams, V.: Observation on the bisulfite reductase (P582) isolated from Desulfotomaculum nigrificans. J. Bacteriol., **120**, 240-244 (1974)
[22] Murphy, M.J.; Siegel, L.M.: Siroheme and sirohydrochlorin. The basis for a new type of porphyrin-related prosthetic group common to both assimilatory and dissimilatory sulfite reductases. J. Biol. Chem., **248**, 6911-6919 (1973)
[23] Akagi, J.M.; Adams, V.: Isolation of a bisulfite reductase activity from Desulfotomaculum nigrificans and its identification as the carbon monoxide-binding pigment P_{582}. J. Bacteriol., **116**, 392-396 (1973)
[24] Lee, J.P.; LeGall, J.; Peck, H.D.: Isolation of assimilatory- and dissimilatory-type sulfite reductases from Desulfovibrio vulgaris. J. Bacteriol., **115**, 529-542 (1973)
[25] Kobayashi, K.; Takahashi, E.; Ishimoto, M.: Biochemical studies on sulfate-reducing bacteria. XI. Purification and some properties of sulfite reductase, desulfoviridin. J. Biochem., **72**, 879-887 (1972)
[26] Lee, J.P.; Peck, H.D.: Purification of the enzyme reducing bisulfite to trithionate from Desulfovibrio gigas and its identification as desulfoviridin. Biochem. Biophys. Res. Commun., **45**, 583-589 (1971)
[27] Lee, J.P.; Yi, C.S.; LeGall, J.; Peck, H.D.: Isolation of a new pigment, desulforubidin, from Desulfovibrio desulfuricans (Norway strain) and its role in sulfite reduction. J. Bacteriol., **115**, 453-455 (1973)
[28] Arendsen, A.F.; Verhagen, M.F.J.M.; Wolbert, R.B.G.; Pierik, A.J.; Stams, A.J.M.; Jetten, M.S.M.; Hagen, W.R.: The dissimilatory sulfite reductase from Desulfosarcina variabilis is a desulforubidin containing uncoupled metalated sirohemes and S = 9/2 iron-sulfur clusters. Biochemistry, **32**, 10323-10330 (1993)
[29] Lui, S.M.; Cowan, J.A.: Conformational gating of the dissimilatory sulfite reductase from Desulfovibrio vulgaris (Hildenborough). Synthesis, characterization, and stopped-flow kinetics studies of 1,5-IAEDANS-labeled desulfoviridin. Biochemistry, **33**, 11209-11216 (1994)
[30] Laue, H.; Friedrich, M.; Ruff, J.; Cook, A.M.: Dissimilatory sulfite reductase (desulfoviridin) of the taurine-degrading, non-sulfate-reducing bacterium Bilophila wadsworthia RZATAU contains a fused DsrB-DsrD subunit. J. Bacteriol., **183**, 1727-1733 (2001)
[31] Steuber, J.; Kroneck, P.M.H.: Desulfoviridin, the dissimilatory sulfite reductase from Desulfovibrio desulfuricans (Essex): new structural and func-

Printing: Krips bv, Meppel
Binding: Litges & Dopf, Heppenheim